GUIDANCE AND CONTROL
1989

CONFERENCE COMMITTEE

Bob Lewis, Chairperson . . . . (303) 939-4130
Ball Aerospace Systems Group

Paul Shattuck, Co-Chairperson . . . . (303) 971-6079
Martin Marietta Astronautics Group

Arlo Gravseth, Co-Chairperson . . . . (303) 977-6147
Martin Marietta Astronautics Group

Don Parsons, AAS National . . . . (303) 971-7591
Martin Marietta Astronautics Group

Bob Barsocchi, Registration and Finance . . . . (303) 979-5504
Consultant

COMMITTEE MEMBERS

Rex Agler . . . . Martin Marietta Astronautics Group

Pam Alstott . . . . Martin Marietta Astronautics Group

Robert Culp . . . . University of Colorado (Boulder)

John Durrett . . . . Martin Marietta Astronautics Group

Carl Henrickson . . . . Ball Aerospace Systems Group

Terry Kelly . . . . Ball Aerospace Systems Group

Ken Lebsock . . . . Consultant

James McQuerry . . . . Ball Aerospace Systems Group

Lou Morine . . . . Martin Marietta Astronautics Group

John Sand . . . . Ball Aerospace Systems Group

Kirk Sterling . . . . Honeywell, Inc.

AAS PRESIDENT
    E. Larry Heacock   NOAA/NESS

VICE PRESIDENT - PUBLICATIONS
    Walter Froehlich   International Science Writers

SERIES EDITOR
    Dr. Horace Jacobs   Univelt, Incorporated

EDITORS
    Dr. Robert D. Culp   University of Colorado
    Robert A. Lewis   Ball Aerospace Systems Group

ASSOCIATE EDITOR
    Robert H. Jacobs   Univelt, Incorporated

<u>Front Cover Illustration</u>:

Artist's concept of an Aeroassist Flight Experiment entering the atmosphere to evaluate braking effects after being released from the Space Shuttle. Art by Scott Kahler (Courtesy of Ball Space Systems Division).

<u>Frontispiece</u>:

National Aerospace Plane viewing the Space Station (Courtesy of Department of the Air Force, Wright Research and Development Center).

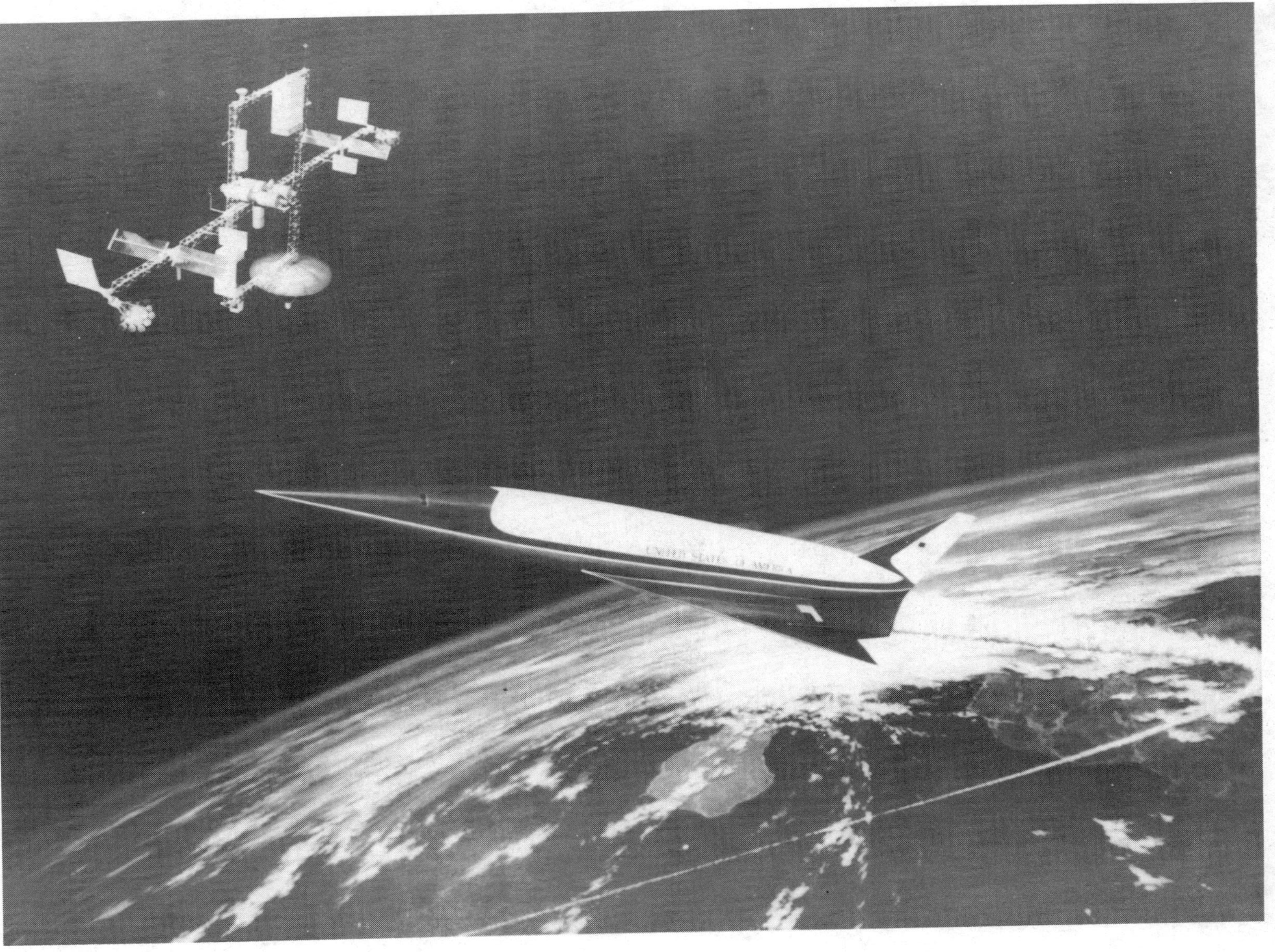

GUIDANCE AND CONTROL 1989

Volume 68
ADVANCES IN THE ASTRONAUTICAL SCIENCES

Edited by
Robert D. Culp
Robert A. Lewis

Proceedings of the Annual Rocky Mountain Guidance and Control Conference held February 4-8, 1989, Keystone, Colorado

Published for the American Astronautical Society by Univelt, Incorporated, P.O. Box 28130, San Diego, California 92128

FOREWORD

HISTORICAL SUMMARY

The Annual Rocky Mountain Guidance and Control Conference began as an informal exchange of ideas and reports of achievements among local guidance and control specialists. Since most area guidance and control experts participate in the American Astronautical society, it was natural to gather under the auspices of the Rocky Mountain Section of the AAS.

In 1977, Bud Gates, Don Parsons, and Bob Culp organized the conference formally and began the annual series of meetings the following winter. In March, 1978, the Annual Rocky Mountain Guidance and Control Conference, sponsored by the Rocky Mountain Section of the American Astronautical Society, met at Keystone, Colorado. It has met there annually ever since. The twelfth Conference was held this year.

The first Conference was chaired by Bud Gates, co-chaired by Bob Culp, with arrangements by Don Parsons. Bob Culp was Section Chairman. The local chairmen were Bob Barsocchi, Carl Henrikson, Lou Morine; session chairmen were Sherm Seltzer, Pete Kurzhals, and Lou Herman. Other members of the original organizing committee were Ed Euler, Joe Spencer, and Tom Spencer.

A tradition from the beginning has been the opening banquet: this provides a chance to meet friends, renew acquaintances, and lay plans for the five-day meeting. It is an elegant feast, still the evening is marked by informality and good cheer. A general interest speaker has become a popular feature. These have been:

1978	Sherm Seltzer, NASA MSFC, told a joke
1979	Sherm Seltzer, Control Dynamics, told another joke
1980	Andrew J. Stofan, NASA Headquarters, "Recent Discoveries through Planetary Exploration"
1981	Jerry Waldvogel, Cornell University, "Mysteries of Animal Navigation"
1982	Robert Crippen, NASA astronaut, "Flying the Space Shuttle"
1983	James E. Oberg, author, "Sleuthing the Soviet Space Program"
1984	W. J. Boyne, Smithsonian Aerospace Museum, "Preservation of American Aerospace Heritage: A Status on the National Aerospace Museum"
1985	J. B. Irwin, NASA astronaut (retired), "In Search of Noah's Ark
1986	Roy Garstang, University of Colorado, "Halley's Comet"
1987	Kathryn Sullivan, NASA astronaut, "Pioneering the Space Frontier"
1988	William E. Kelley and Dan Koblosh, Northrop Aircraft Division, "The Second Best Job in the World, The Filming of *Top Gun*"
1989	Brig. Gen. Robert Stewart, U.S. Army Strategic Defense Command, "Exploration in Space: A Soldier-Astronaut's Perspective"

The Rocky Mountain Section of the Society established a broad-based Conference Committee, the Rocky Mountain Guidance and Control Committee, chaired ex-officio by the next Conference Chairman, to run the Annual Conference. The Conference has been a success from the start. Currently, the meeting attracts over 200 of the nation's top specialists in space guidance and control:

Year	Conference Chairman	Attendance
1978	Robert L. Gates	83
1979	Robert D. Culp	109
1980	Louis L. Morine	130
1981	Carl Henrikson	150
1982	W. Edwin Dorroh, Jr.	180
1983	Zubin Emsley	192
1984	Parker S. Stafford	203
1985	Charles A. Cullian	200
1986	John C. Durrett	186
1987	Terry Kelly	201
1988	Paul Shattuck	244
1989	Robert A. Lewis	201

In addition to providing for an annual exchange of the most recent advances in research and technology of astronautical guidance and control, the Conference brings in a guest speaker of special prominence to present a lecture series in one specific area of current interest. These speakers and their topics have been:

1978	Professor Daniel Debra, Stanford University, "Navigation"
1979	Professor William L. Brogan, University of Nebraska, "Kalman Filters Demystified"
1980	Professor J. David Powell, Stanford University, "Digital Control"
1981	Professor Richard H. Battin, Massachusetts Institute of Technology, "Astrodynamics - A New Look at Old Problems"
1982	Professor Robert E. Skelton, Purdue University, "Interactions of Dynamics and Control"
1983	Professor Arthur E. Bryson, Stanford University, "Attitude Stability and Control of Spacecraft"
1984	Dr. William B. Gevarter, NASA Ames, "Artificial Intelligence and Intelligent Robots"
1985	Dr. Nathaniel B. Nichols, The Aerospace Corporation, "Classical Control Theory"
1986	Dr. W. G. Stephenson, Science Applications International Corp., "Optics in Control Systems"
1987	Professor Daniel Debra, Stanford University, "Guidance and Control: Evolution of Spacecraft Hardware"
1988	Professor Arthur E. Bryson, Stanford University, "Software Application Tools for Modern Controller Development and Analysis"
1989	Professor John L. Junkins, Texas A&M University, "Practical Applications of Modern State Space Analysis in Spacecraft Dynamics, Estimation, and Control"

In recent years, under the leadership of Don Parsons, the Rocky Mountain Section of the AAS has strongly supported students interested in the aerospace program. Every year approximately twenty college students are invited as guests of the Conference. They obtain a valuable, first-hand exposure to the state of the art in space guidance and control. The Conference has been a major sponsor of the University of Colorado's student-run Get-Away-Special Space Shuttle Experiment. Finally, the

Section every year gives $1000 to two scholarships at the University of Colorado: one in Aerospace Engineering Sciences, and one in Electrical and Computer Engineering. These are recognized at the opening banquet the year following the award. Since they have been given, the winners have been:

	AERO	ECE
1981	Jim Chapel	
1982	Eric Seale	
1983	Doug Stoner	John Mallon
1984	Mike Baldwin	Paul Dassow
1985	Bruce Haines	Steve Piche
1986	Beth Swickard	Mike Clark
1987	Tony Cetuk	Fred Ziel
1988	Mike Mundt	Brian Olson

These scholarship winners are honored guests of the AAS at this Conference. The Rocky Mountain Section has made a commitment to continue this support through a $15,000 endowment, which will assure the existence of these scholarships indefinitely.

In 1983, Don Parsons, Joe Spencer, Ed Dorroh, and Bob Culp, at the request of the National Board of the American Astronautical Society, prepared a charter which established the Annual Conference as the AAS Annual Rocky Mountain Guidance and Control Conference. The Conference will be sponsored by the National Organization for the AAS, will meet annually at Keystone, Colorado, and will be run by the Rocky Mountain Section.

A National AAS Guidance and Control Technical Committee, with national representation, was established to provide oversight to the local conference committee. W. Edwin Dorroh, Jr., was the first chairman of the AAS Guidance and Control Technical Committee. At the 1985 Conference, Robert L. Gates took over as the chairman of the National AAS Guidance and Control Committee. This committee meets every year at the Keystone Conference, and also at either the summer Guidance and Control Conference, or the fall AAS Annual Meeting.

Thus, the AAS Rocky Mountain Guidance and Control Conference continues as the premier conference of its type anywhere. As a National Conference sponsored by the AAS, it promises to be the preferred idea exchange for guidance and control experts for years to come.

The Conference Committee extends congratulations to the 1989 Conference Chairman, Bob Lewis, for an excellent conference and a job well done. His report follows.

Robert D. Culp
University of Colorado
Boulder, Colorado

Proceedings of the Rocky Mountain
Guidance and Control Conferences are available as follows:

Guidance and Control 1989, Volume 68
    Advances in the Astronautical Sciences;

Guidance and Control 1988, Volume 66,
    Advances in the Astronautical Sciences;
    Microfiche Supplement: Volume 56, AAS Microfiche Series

Guidance and Control 1987, Volume 63,
    Advances in the Astronautical Sciences;

Guidance and Control 1986, Volume 61,
    Advances in the Astronautical Sciences;
    Microfiche Supplement: Volume 53, AAS Microfiche Series

Guidance and Control 1985, Volume 57,
    Advances in the Astronautical Sciences;
    Microfiche Supplement: Volume 50, AAS Microfiche Series

Guidance and Control 1984, Volume 55,
    Advances in the Astronautical Sciences;
    Microfiche Supplement: Volume 48, AAS Microfiche Series

Guidance and Control 1983, Volume 51,
    Advances in the Astronautical Sciences;
    Microfiche Supplement: Volume 44, AAS Microfiche Series

Guidance and Control 1982, Volume 48,
    Advances in the Astronautical Sciences;
    Microfiche Supplement: Volume 38, AAS Microfiche Series

Guidance and Control 1981, Volume 45
    Advances in the Astronautical Sciences;
    Microfiche Supplement: Volume 36, AAS Microfiche Series

Guidance and Control 1980, Volume 42,
    Advances in the Astronautical Sciences;

Guidance and Control 1979, Volume 39,
    Advances in the Astronautical Sciences;
    Microfiche Supplement: Volume 31, AAS Microfiche Series

Guidance and Control 1978, Volume 29,
    AAS Microfiche Series
    (Includes one paper from earlier conference)

Order from Univelt, Inc., P.O. Box 28130, San Diego, CA 92128

PREFACE

The Twelfth Annual AAS Rocky Mountain Section Guidance and Control Conference was held at Keystone, Colorado.

The conference began Saturday evening, February 4, with our kickoff banquet. The guest speaker, Brigadier General Robert Stewart, presented a most interesting talk entitled "Exploration in Space: A Soldier-Astronaut's Perspective". In this talk, he communicated how it feels to be in space, traveling at 17,000 miles per hour, 100 nautical miles above the Earth's surface. General Stewart's speech was well received and got the conference off to a great start.

Professor John Junkins, of Texas A&M University, presented Sunday's tutorial session. His topic: "Practical Applications of Modern State Space Analysis in Spacecraft Dynamics, Estimation, and Control" was timely in that many challenging spacecraft missions are about to be initiated. A compilation of his viewgraph presentation is included in this proceedings volume. Also, it may be published as a separate book at a later date.

The first technical session, Advances in Guidance, Navigation, and Control, was chaired by Colonel Charles Heimach of the Air Force Space Division and co-chaired by Philip Turner of the Jet Propulsion Laboratory. The seven papers presented provided insights into recent advances in component technology and work being done in spacecraft autonomy involving cost reduction techniques and implementing artificial intellegence.

The ever popular Storyboards were included as the second technical session where conference attendees had the opportunity to review 15 displays. Many of the presentations included hardware demonstrations of relevant guidance and control subsystems. A buffet, sponsored by many of the prominent aerospace companies involved in the conference, was provided during this three-hour session.

Attitude Referenced Pointing Systems were the subject of the third technical session. The session, chaired by Sam Hollander of the Naval Research Laboratory and co-chaired by Bob Laskin of the Jet Propulsion Laboratory, contained six very interesting papers. Several of the papers addressed problems associated with pointing a payload to microradian and submicroradian accuracies while being subjected to base motion disturbances.

The fourth session, held on Tuesday evening, was entitled "Guidance, Navigation, and Control Systems for Specialized Missions". Control issues for missions including the National Aerospace Plane (NASP), Pegasus, Magellan, and Mars Rover were presented and discussed. This session was chaired by Harold Schofield of

NASA/Marshall Space Flight Center and co-chaired by Major Wiley Larson of the U.S. Air Force Academy.

Recent Experiences, the final conference session, was held on Wednesday, February 8. The session was chaired by Brian McGlinchey of the Jet Propulsion Laboratory and co-chaired by Bob Williamson of The Aerospace Corporation. Six excellent papers were presented which covered not only successes and failures of recent space experience but also effects associated with significant launch delays that have resulted from the 1986 space shuttle accident. Topics included Delta 181, Galileo, Eutelsat 5, AMSAT-OSCAR-13, and the Polar BEAR missions and spacecraft.

A meeting of the AAS Guidance and Control Technical Committee was held concurrent with the story board session of the conference. Acting chairman, Bob Culp, solicited many ideas for suggested format and content for next year's conference. The 1990 AAS Guidance and Control Conference will be held February 3-7, 1990, at Keystone, Colorado, and will be chaired by Arlo Gravseth, Martin Marietta Astronautics Group, P.O. Box 179, Denver, Colorado 80201, Telephone: (303) 977-6147. The Co-Chairman will be Bob Lewis, Ball Electro-Optics/Cryogenics Division, P.O. Box 1062, Boulder, Colorado 80306, Telephone: (303) 939-4130.

Reiterating what I stated in the closing remarks of the conference, we readily accept any suggestions on improving the quality of the conference. The Rocky Mountain Guidance and Control Conference was originally formed, and continues to support, the application of guidance and control to the space community. So drop Arlo or me a line and let us know your opinions for, without your support, this conference would never have achieved its current stature. I thank all of you and look forward to seeing you at next year's conference.

Robert A. Lewis
Ball Electro-Optics/Cryogenics Division
Chairman
1989 AAS Guidance and Control Conference

CONTENTS

Page

Section I
ADVANCES IN GUIDANCE, NAVIGATION AND CONTROL

SESSION I

Chairperson:	Colonel Charles Heimach U.S. Air Force Space Division
Co-Chairperson:	Philip Turner Jet Propulsion Laboratory
Local Chairperson:	Arlo Gravseth Martin Marietta Astronautics Group

The following paper numbers were not assigned:

AAS 89-008 to -009

A LOW COST AUTONOMOUS NAVIGATION SYSTEM

Frank Tai and Peter D. Noerdlinger[*]

The Microcosm Autonomous Navigation System (MANS) is capable of achieving 3-sigma positioning accuracies of 100 m to 1.5 km with straightforward modifications to existing Earth sensing hardware. Because it uses hardware that would be on board the spacecraft for attitude determination, the additional cost for achieving fully autonomous navigation will be quite low. The modifications will allow the sensor to measure the size of the Earth and sense the relative positions in the spacecraft sky of the Earth, Moon, and Sun. Because the measurements can be made with a single sensor many of the principal bias terms can be eliminated or greatly reduced. The proposed system is expected to have a significant impact on ground operations costs, mission definition and design, survivability, and the potential development of very low cost fully autonomous spacecraft.

INTRODUCTION

Autonomous navigation is the ability to determine one's position and velocity onboard in real time and independently of external data. Its development received renewed impetus when the need arose for a satellite constellation that could provide global prioritized data-voice service during peacetime and essential communications during crises. For this purpose a large number of satellites with high survivability was required, which would ordinarily entail very costly continuous ground support. To keep down this cost, it became essential to turn to autonomous navigation. Besides dramatically reducing the cost of ground support, autonomous navigation contributes directly to survivability of the constellation by making its integrity independent of jamming or even destruction of the ground stations. Our work was done in this context.

Autonomous satellite navigation has been of interest since the beginning of space flight. Flight applications began as early as 1963. Various complements of sensors have been used including horizon scanners, known landmark trackers, unknown landmark trackers, star sensors, strapdown gyros, mosaic sensors, and space sextants. Reference 1 traces the history of satellite autonomous navigation development.

Recent activities in the development of autonomous satellite navigation have emphasized four methods: 1) direct Earth horizon sensing, 2) inferred Earth horizon sensing through stellar refraction, 3) satellite crosslinks, and 4) use of Navstar or the Global Positioning System (GPS). Fig. 1[1] represents actual and projected position

---

* Microcosm, Inc., 2601 Airport Drive, Suite 230, Torrance, California 90505.

accuracies using these methods. Method 2 requires the use of onboard star sensors. Method 3 requires the instrumentation necessary to make satellite-to-satellite ranging measurements. Method 4 is not fully autonomous as it requires ground-based monitor stations and updates from GPS satellites. Only method 1, direct Earth horizon sensing, remains as a means for providing truly low cost and autonomous satellite navigation.

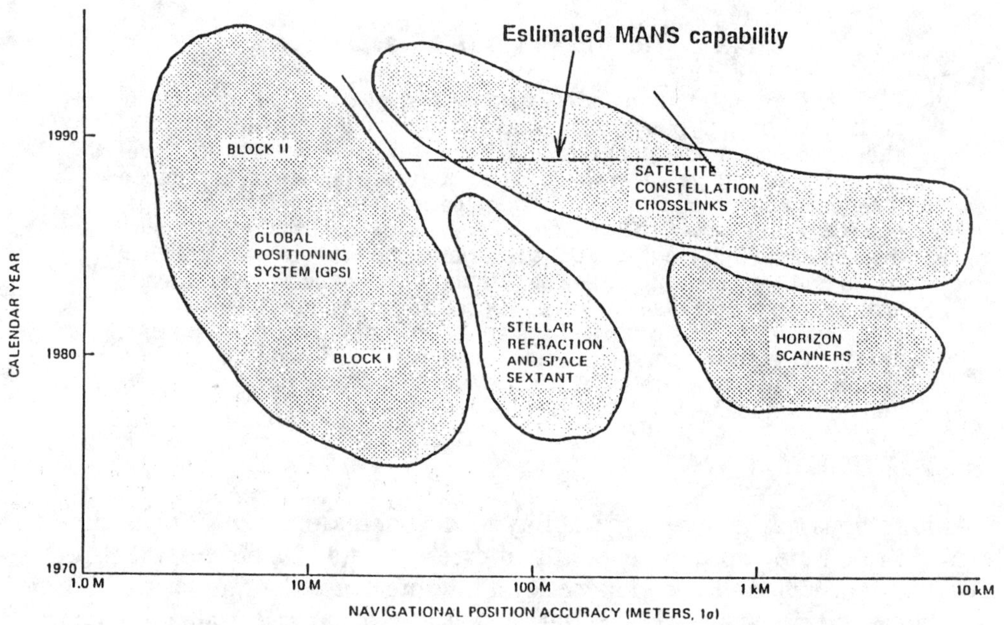

Fig. 1   Overview of Autonomous Satellite Navigation Capabilities (Chory[1])

NAVIGATION CONCEPT

The concept behind the proposed low cost autonomous navigation system is to determine the satellite's orbit based on the distance to the Earth, as found from the apparent Earth size, and the directions to the Earth, Sun and Moon. There are six orbital elements. The three in-plane elements (semi-major axis, eccentricity, and mean anomaly) are all determined by successive observations of the Earth radius (which fixes the altitude at any instant). The orbital inclination, line of nodes, and argument of the perigee are found from fixes on the Sun and Moon.

In the section on the Autonomous Navigation Simulator it is shown that the angle of the Sun or Moon to the orbital pole can be found from the time variation of the Earth/Sun or Earth/Moon angle respectively, as seen from the spacecraft. Fig. 2 shows circles about the directions S1 and S2 to the Sun at different times and about the position of the Moon at some favorable time. The three circles intersect in a point, indicating redundant determination of the orbit pole on the celestial sphere. Because the Moon is dimmer and determination of its true center in visible light requires an additional algorithm, it would be simpler if one could use only the Sun and Earth. If, however, only one Sun observation is available, the position of the orbit pole can lie anywhere on the corresponding circle. Thus, use of the Sun and Moon, or two Sun observations at different times, is required to resolve this uncertainty. The argument of the perigee is determined by phase information in the variation of the angles, completing the determination of the orbit.

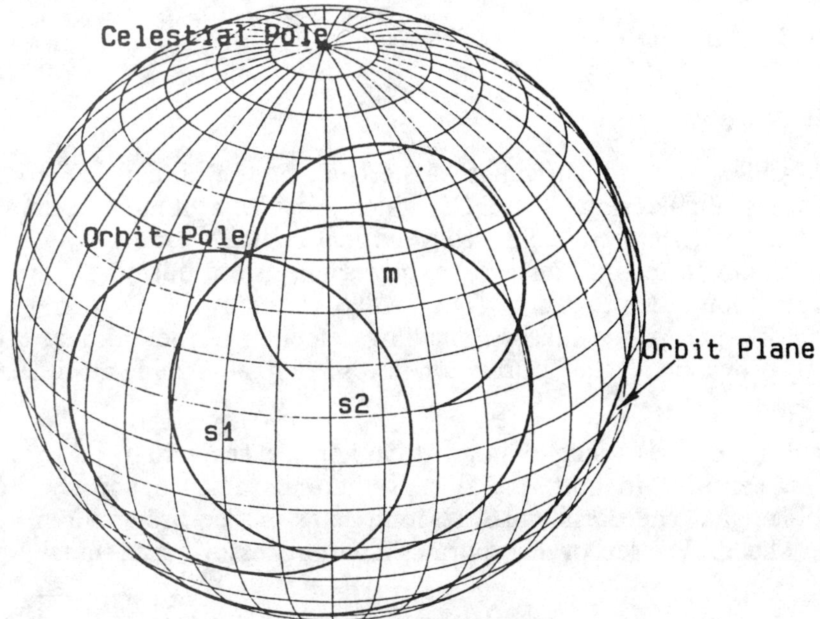

**Fig. 2   Determining the Orbit Pole from its Angle Relative to
the Sun and Moon Vectors**

MANS LOW COST, HIGH ACCURACY CONCEPT

The Microcosm Autonomous Navigation System (MANS) approach is closest in concept to those using star sensors, but it relies on dual scan Earth horizon sensors modified to provide accurate sensing of the Sun and Moon. Such hardware can be integral with that used for attitude determination, thereby reducing cost. Software cost is minimal as well, because there is no need for star maps or landmark-recognition algorithms and because the MANS separates orbit and attitude determination. Significant advantages are gained when navigation sensing is performed by a *single* sensor: the total number of sensors and therefore costs are reduced, the major source of systematic biases is eliminated thus improving overall accuracy, data interface requirements are minimized, and processing complexity is reduced.

A Kalman filter implements overall orbit determination from the incoming data. It calculates *corrections* to an extrapolated orbit as new data become available. Because it incorporates a continuum of measurements, the Kalman filter can in principle use data from a *moving* celestial body at two different *times* in place of *simultaneous* data from two different bodies. Great improvements over the usual results for horizon scanners are achieved because of the very accurate sensing of the angular radius of the Earth, because of the inclusion of Sun and Moon data, because redundant measurements facilitate better bias removal, and because the Kalman filter enables the incorporation of hundreds of measurements in a consistent fashion. This work was done for a low earth orbit three-axis stabilized satellite, but it works well at higher altitudes, for eccentric orbits, and for spinning spacecraft.

SYSTEM DESCRIPTION

The MANS Sensor

The MANS sensor is a derivative of the Dual Cone Scanner[3] (DCS) manufactured by Barnes Engineering. The DCS, which provides attitude determination for three-axis stabilized spacecraft, is a conventional conical scanner which has been modified by adding a second Earth cone, increasing its attitude sensitivity and enabling determination of Earth radius. The MANS scan sensor is further enhanced by the addition of a pair of fan sensors, enabling detection of the Sun and Moon, and consequently providing all the information necessary to determine attitude and position.

The sensor employs a single, motor-driven optical scanning head with multiple fields-of-view sensitive to both the Earth's thermal radiance and the Sun's and Moon's visible light. The Earth and Sun/Moon fields-of-view are provided within the single optical head. The sensor incorporates microprocessor-based signal processing.

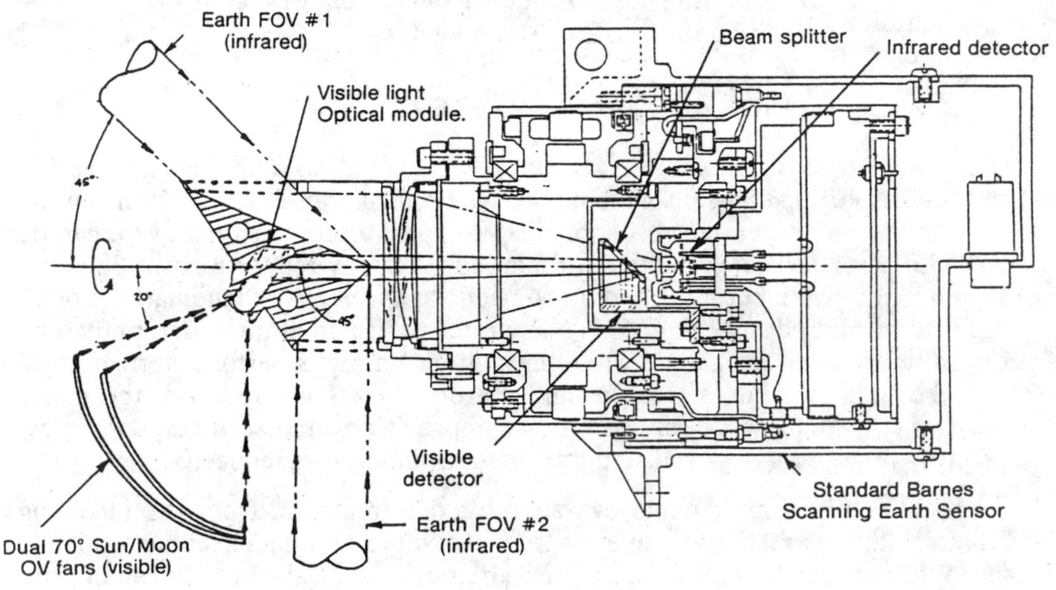

Fig. 3   Mechanical and Optical Configuration of the MANS Sensor

The sensor power and weight requirements are similar to those of the unit from which it is derived. The total power estimate is 11 watts. The weight is estimated at 2.75 pounds for the sensor head and 6.75 pounds for the electronics, including augmentation for navigation capability. Fig. 3 shows the mechanical and optical configuration of the MANS sensor.

Sensor Heritage

The MANS is based on flight proven components from previously flown sensors. It utilizes scan motors, bearings and lubrication systems from USAF/Lockheed Agena (and related) programs. It also employs pyroelectric infrared detectors from the Combined Earth Sensor for the USAF/Rockwell GPS/Navstar program.

The Dual Cone Scanner (DCS) represents the starting point for the sensor design. The following straightforward modification was made to the basic optical/mechanical design to achieve the enhancements in features and performance described herein. A silicon photodiode detector, used in the DCS to detect Sun presence near the IR field-of-view, will determine the attitude of the Sun or Moon using a pair of fan sensors. The visible light detector is given a clear field-of-view through the scanning mirror and a direct view of the Sun or Moon. The 5mm field-of-view of the silicon detector is split into two halves and directed into different directions away from the IR scan plane. The fans are set to a minimum separation of 5° at the sensor equator and are tilted 15 degrees with respect to the scanner spin axis. Each half is expanded with a negative lens into a 70 degree long fan-shaped field of view. When it rotates, it sweeps out an entire hemisphere minus a 20 degree cone about the spin axis. The sensor detects the presence of the Sun or Moon and computes its two-axis attitude based on the timing of this data. The detector has multiple light intensity thresholds to distinguish between the Sun and Moon, and for discrimination of the Earth. Fig. 4 illustrates the IR and sweeping fan sensor fields of view on the sky as seen from the spacecraft, and Fig. 5 a photograph of a prototype of the sensor.

Sensor Configuration and Mounting

Two sensing configurations have been evaluated: a single MANS providing moderate performance at lowest cost, and a dual MANS configuration providing enhanced performance at slightly greater cost. When two sensors are mounted at appropriate angles, the sky coverage is excellent resulting in superior navigation accuracies.

Mounting of the sensor(s) on the satellite was optimized to achieve maximum viewing of the Sun, Moon and Earth. Fig. 6 shows the mounting orientation for the dual MANS approach. Both sensors are mounted with the rotation axes in the orbit plane, 80° from nadir . The single MANS configuration simply eliminates one of the two, and so represents a failure mode configuration as well. The infrared horizon sensing cone angles were selected to provide good Earth altitude and attitude sensitivity with a wide coverage range about the nominal 400 nmi altitude and nadir pointing attitude. The Sun fan limit and orientation were selected to provide the widest coverage of the Sun and Moon while considering Earth and satellite blockage.

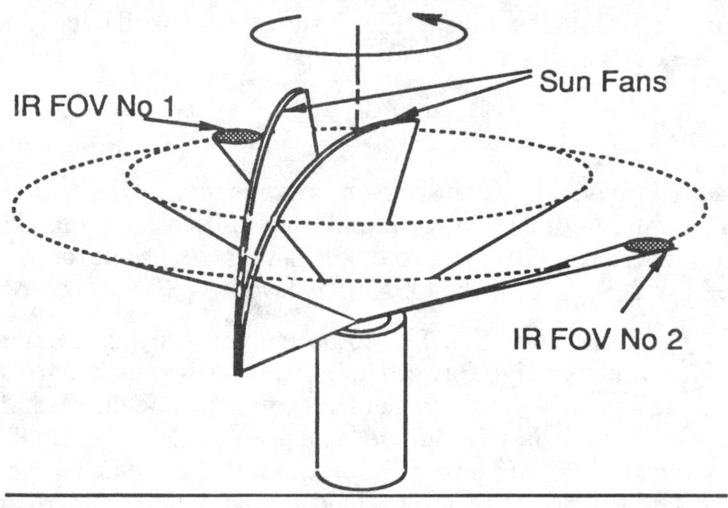

Fig. 4   MANS Earth (IR) and Sun/Moon Fan (Visible Light) Fields of View

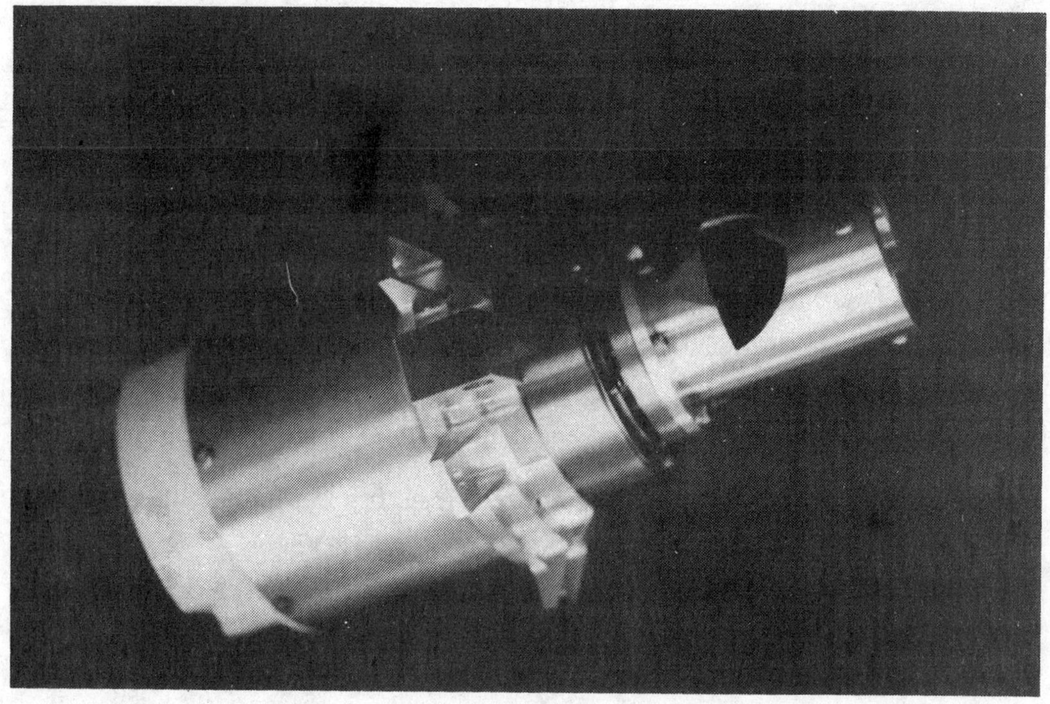

Fig. 5   Photograph of a Prototype MANS Sensor

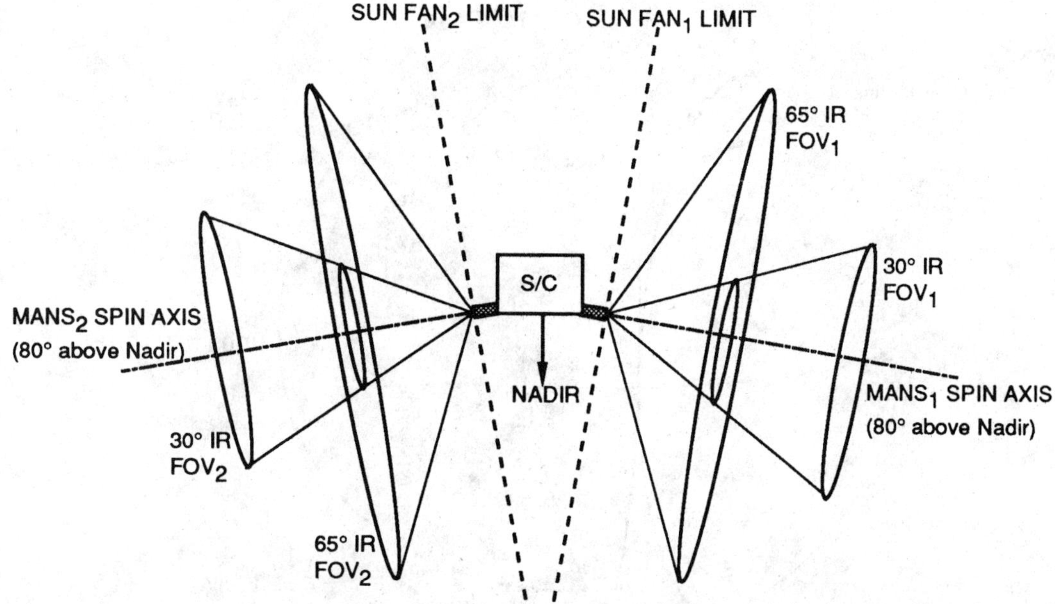

Fig. 6 Sensor Mounting Configuration

SENSOR PERFORMANCE

Sensor Coverage

The Earth sensors are designed to provide four horizon crossings or two Earth widths per sensor per scan for all attitudes within the design limits. In greater extremes of pitch, one sensor will generally provide roll, pitch, and Earth radius information.

The Sun and Moon fans will lose coverage for the following reasons:

1. Eclipse of the object by the Earth

2. Blockage of field-of-view by the satellite (unimportant for our mounting angle)

3. Object is outside fan limits; $0 < El < 70$ deg

4. The Moon signal is washed out when the Earth is in the field-of-view.

The region of the sky within which the Moon is visible is shown for a typical orbit in Fig. 7, with the Earth horizon (solid circle) tangent to the left fan. The fan spins rapidly about the sensor axis and can detect the Moon only if none of the four conditions listed above occurs. In the position shown, it is on the verge of losing the Moon from condition 4. The Sun coverage region looks almost the same but includes the region within the wedge defined by the fans. Note that the "outer radius" or limit of the fans is the sensor equator. The "inner radius" is $20°$, the complement of the $70°$ in condition 3. Fig. 8 shows the coverage of the Sun or Moon ignoring condition 4.

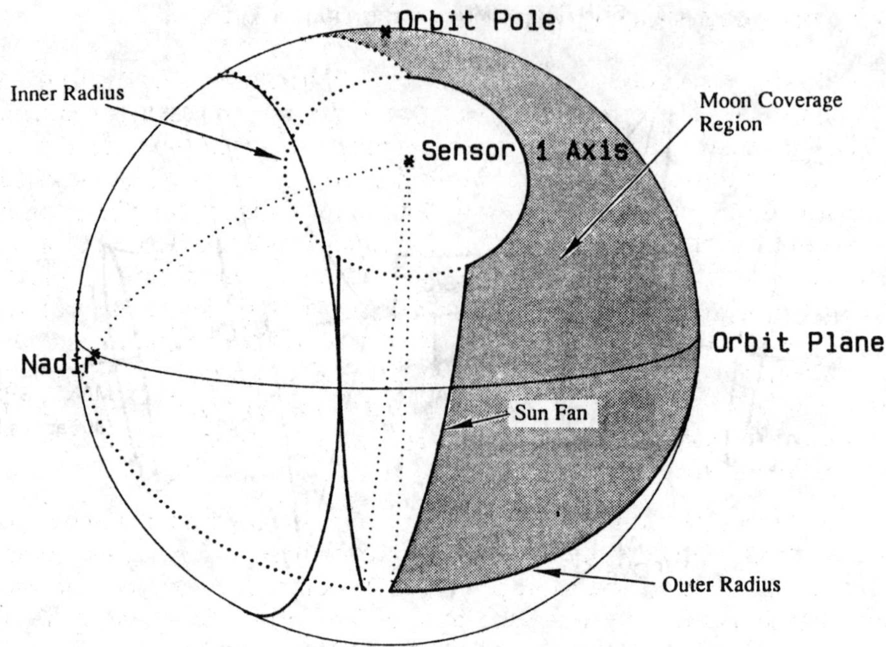

Fig. 7   Moon Coverage Region Defined by Fan and by Earth Interference

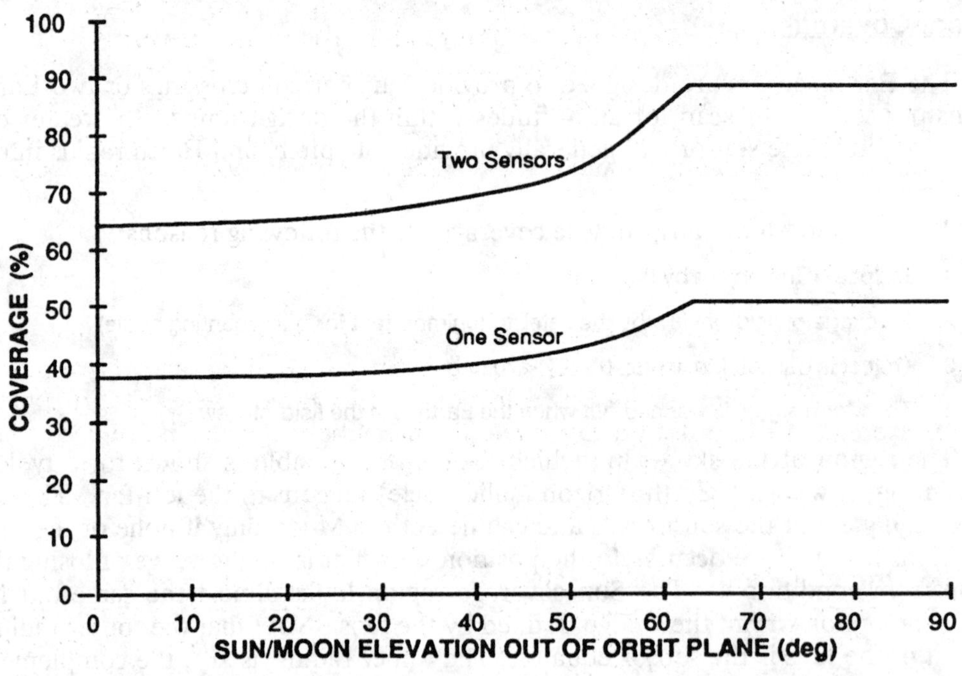

Fig. 8   Sun/Moon Coverage for One and Two MANS Sensors

ERRORS AND DATA SMOOTHING -- GENERAL

The following discussions of errors and smoothing applies equally to the Earth, Sun, or Moon data. The errors consist of two kinds, systematic error and random noise. The systematic part can be further broken down into errors that can be calibrated out and those that cannot. Mounting errors lead to constant biases that can be calibrated out, but changes due to thermal or deformation effects often cannot. We assume that the constant part of the biases can be calibrated so accurately that the other errors dominate. In the error budgets to be presented "bias" terms refer to internal sensor biases and not to mounting biases. There are two kinds of noise, both of which tend to average out over repeated measurements. Digitization introduces errors generally independent of the signal. Thermal noise contaminates the signals and so introduces error in determining when the signal crosses the triggering threshold or "locator". The effect is largest for the Moon, intermediate for the Earth, and negligible for the Sun.

Individual measurements will contain considerable random error and also arrive at too high a rate for processing by present microprocessors. Therefore, some number N of measurements will be averaged together before they are fed into the main processing algorithms to determine the attitude and orbit. This averaging process must be more sophisticated than a simple average, because attitude changes and orbital motion alter the measurements during the observation period. The algorithm can also provide some error checking against noise outside the 3 sigma level at low computational cost. In this report we assume that suitable algorithms will be developed so that error propagation through the averaging process can be handled as if an ordinary average were being taken. The random errors then decrease as $N^{-1/2}$, while the systematic errors are fixed, no matter how large N. We chose a 25 sec smoothing time but have also analyzed 50 sec and 12.5 sec times, with similar results. In the ensuing discussion, azimuth, elevation, meridian, etc., here are referred to sensor coordinates, such that elevation 90° is along the boresight. A 240 rpm scan rate is assumed.

EARTH SENSING

Table 1 breaks down Earth sensor error for a single horizon crossing. The error for the Earth width is 1.41 times larger as it comes from the difference of two pulse times, while the azimuth error for one Earth arc is 0.707 times as large, as it comes from the average of two pulse times. While the azimuth of the Earth center can be found from one arc, there will generally be four such arcs from the two sensors, each with two conical scans. The error is then halved again. All errors are in degrees and all are 3 sigma.

The errors for Earth radius (implicitly--for altitude) and for elevation are obtained by dividing the errors in pulse width by the dual scanner gains. In this report, we define gain to be the change in sensed pulse width or phase per unit change in satellite elevation or azimuth or Earth radius. These are obtained as follows: Equation (11-39) of Wertz[4] relates the Earth radius *rho*, the elevation *eta*, the cone angle *gamma*, and a single scan Earth width (arc along the scan cone) *omega* by:

$$\cos(rho) = \cos(gamma) * \cos(eta) + \sin(gamma) * \sin(eta) * \cos(omega_i/2)$$

When this is applied to two Earth crossings, it is possible to solve for *rho* and *eta*. The gains are found from the changes in *rho* or *eta* with small changes in the *omega*$_i$ values. Table 2 gives the dual scanner gains for our mounting configuration. They are used to convert nominal measurement errors within a scan into accuracies in Earth centroid position.

Table 1
EARTH SENSOR ERROR BUDGET

Bias Errors per horizon crossing

Earth Radiance Variation	0.012 deg
Phase Reference Pickup nonlinearity	0.005 deg
Aging and Temperature Effects	0.008 deg
Internal Alignment Uncertainty	0.006 deg

RSS bias error	0.016 deg

Noise Error for N crossings $\qquad$ (0.104 deg)/$N^{1/2}$

RSS of bias and noise for 25 sec (100 crossings): $\qquad$ 0.019 deg

Table 2
DUAL SCAN GAINS
(30 °/65° IR HALF CONES, 80° FROM NADIR, 400 NMI ALT)

	Elevation	Radius
Single MANS	1.00	1.20
Dual MANS	5.8	7.1

Assuming four arcs for azimuth and dividing the stated error of 0.027 deg in pulse width by the gains yields the errors for centroid position and Earth radius shown in Table 3.

Table 3
MANS EARTH ERRORS FOR 25 SEC AVERAGES

Sensor	Centroid Elevation	Centroid Azimuth	Radius
Single MANS	0.027	0.014	0.023
Dual MANS	0.0047	0.007	0.0039

SUN AND MOON POSITIONS AND ERRORS

Sun and Moon Positions

Pulses are detected from the two fans at times t_1 and t_2, relative to a fiducial mark on the sensor. The range of each time is from 0 up to the rotation period P. All timings are converted to degrees of sensor rotation. Let A_1 and A_2 be the angular measures of t_1 and t_2 in degrees, $A_i = 360\, t_i/P$, where $i = 1$ or 2. The azimuth is found from

$$AZ = (A_1 + A_2)/2 \tag{1}$$

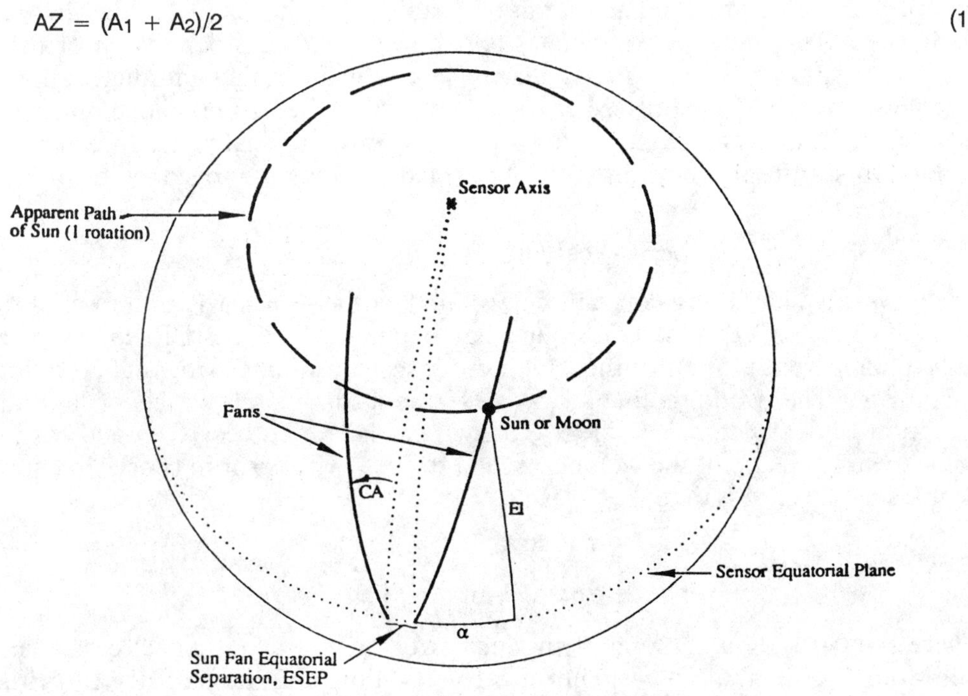

**Fig. 9   The Geometry for Finding Sun or Moon Elevation
in Sensor Coordinates**

To find the elevation, we need alpha, the angle from the Sun or Moon to the meridional plane through the intersection of a fan plane with the equator. From Fig. 9, one sees that

$$\text{alpha} = (ESEP/2) - (A_2 - A_1)/2 \tag{2}$$

where ESEP is the angular distance between the intersections of the fans with the sensor equatorial plane. In the present case, ESEP = MSEP, the minimum separation of the fans, since the outer radius is 90 deg. Then elevation is found from the spherical triangle in Fig. 9:

$$\sin(\text{alpha}) = \tan(El)\,\tan(CA) \tag{3}$$

where CA is the 15° cant angle of the fans from the sensor axis. Differentiating this we get

$$d(El) = \cot(CA)\, \cos^2(El)\, \cos(alpha)\, d(alpha), \tag{4}$$

which will be needed for the error analysis.

Errors in the Sun and Moon Positions

The biases for either the Sun or Moon pulse timings are the same as in Table 1 except that the Earth Radiance variation is absent, so that the RSS bias error is only 0.011 deg. The noise for the Sun is 0.01 deg per pulse, independent of the Sun's elevation as its signal is so strong that this error is all timing and digitization error. The noise for the Moon pulses depends on elevation, because the locator error is not negligible. The locator timing error is inversely proportional to the rate at which the Moon crosses the fan (the rate of change of the orthogonal distance from Moon to fan), namely

$$HD = \cos(alpha)\, \cos(CA)\, \cos(El) \tag{5}$$

where we have ignored a constant determined by the sensor rotation rate, as the underlying error is expressed in equivalent degrees. The $\cos(El)$ comes from the reduced apparent motion of the Moon in sensor coordinates when its elevation (El) is nonzero. The Moon then appears to execute a small circle (circle of "latitude") at elevation El as the sensor rotates, as shown in Fig. 9. The $\cos(CA)$ and $\cos$ (alpha) come from the angle of the Moon's path to the slit. The error in t_1 or t_2 in equivalent degrees in one scan is then:

$$\text{Pulse noise error} = \begin{cases} 0.01 \text{ deg} & \text{for Sun} \\ (0.04 \text{ deg}/HD) + 0.01 \text{ deg for Moon} \end{cases} \tag{6}$$

where the 0.04 deg is the underlying locator error in a hypothetical case where the full Moon crosses the slit normally at zero elevation, and the 0.01 deg is digitization and timing error. Combining the bias and noise errors and averaging over N measurements, we then have:

Table 4
RSS TIMING ERROR, N PULSES

Sun	$[(0.011 \text{ deg})^2 + (0.01 \text{ deg})^2/N]^{1/2}$
Moon	$[(0.011 \text{ deg})^2 + (\{0.04 \text{ deg}/HD\} + 0.01 \text{ deg})^2/N]^{1/2}$

The azimuth is found from two pulses per rotation so N is set to double the number of rotations in one smoothing period to get the azimuth error. The error in alpha is the same as that in the azimuth. The elevation error is found from the error in alpha through Eq. (4). Figures 10 and 11 show the derived position errors in sensor coordinates.

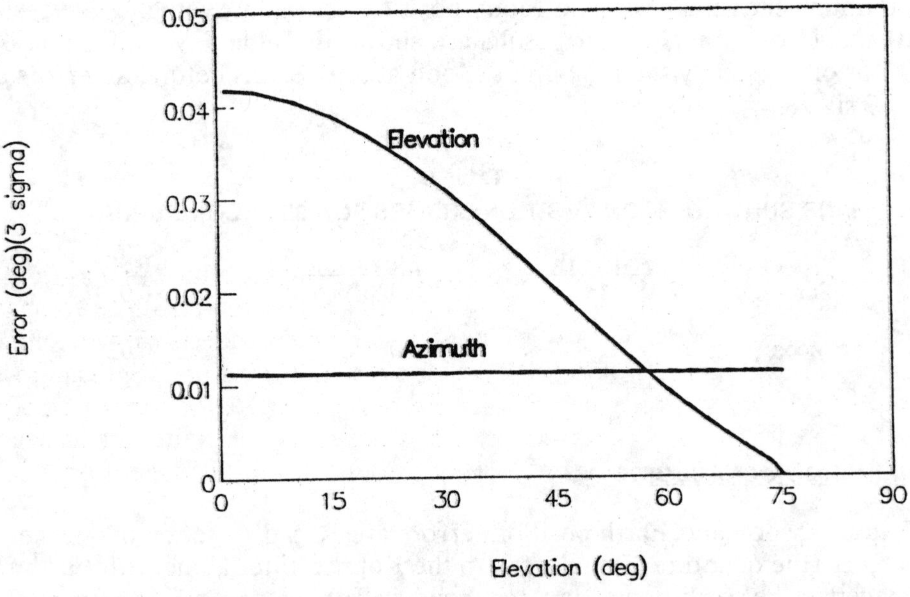

We obtain the mean Sun and Moon position errors by averaging the results in Figs. 10 and 11 over angle. The results are shown in Table 5, valid for one or two sensors - as one can only see the Sun or Moon at a time. As before, all errors are in degrees, 3 sigma.

Table 5
MANS SUN AND MOON POSITION ERRORS FOR 25 SECOND UPDATES

	Azimuth	Elevation	RSS
Sun	0.011	0.020	0.023
Moon	0.012	0.022	0.025

Overall Interobject Accuracies

The Sun, Moon and Earth position errors are RSS'd to get errors in the interobject angles (the quantities actually fed to the Kalman filter along with Earth radius for propagating the satellite position), as shown in Table 6.

Table 6
FINAL INTEROBJECT ANGLE ERRORS FOR 25 SECOND UPDATES

Interobject Angle	Error Sources	Single MANS Error	Dual MANS Error
Earth Radius	Earth Radius	0.023	0.0039
Earth/Sun Angle	Earth Attitude Sun Attitude	0.038	0.024
Earth/Moon Angle	Earth Attitude Moon Attitude	0.039	0.026

AUTONOMOUS NAVIGATION SIMULATIONS

The MANS simulation program follows a spacecraft in Earth orbit, modeling how it tracks its orbital motion using Earth, Sun, and Moon sensors. The program contains two main parts: an orbit propagator, and a sensor package/Kalman filter component that propagates the estimated state, simulates measurement of the appropriate angles and computes corrections to the estimated state parameters.

The spacecraft orbital dynamics were modeled allowing for the J2 term and certain nongravitational forces. The Kalman filter modeled the J2 term, but not the nongravitational forces. This was done to simulate the condition in which one would model known forces, leaving small unmodeled effects for the Kalman filter to compensate. The program simulates the actual coverage of the Earth, Moon, and Sun, using analytic ephemerides of the Sun and Moon. Simulator inputs include the mounting angle bias values and the sensor noise levels. The Moon sensor will function poorly or not at all when the Moon is near "new" phase, i.e., not bright enough. The program allows selection of the cutoff Moon/Sun angle within which the Moon data will be unavailable.

The program was designed primarily for LEO orbits but worked very well at higher altitudes. At present, the geocentric parallax of the Moon is ignored, so the Moon-Sun (M/S) angle varies only on the time scale of a month, and is useless for navigation. An on-board Kalman filter should include the correction for the geocentric parallax of the Moon but should probably not use the variation in the M/S angle over the orbit for navigation unless the altitude is of order one Earth radius or more, because the informational content is low.

Navigation is accomplished by fitting the estimated orbit to the simulated data. Two salient problems prevent us from using classical methods based on a few observations in the current instance. One is that our sensors are less accurate than the desired precision in the orbital parameters, so that many observations must be averaged or filtered. Second, there are forces on the craft that cause the elements to change with time. Solar and Lunar gravity perturbations, higher harmonics of the Earth gravity field, residual atmospheric drag, solar radiation pressure, and Earth IR and albedo radiation forces are potentially significant and can be modeled to various levels of accuracy. The hardest forces to model are radiation forces and atmospheric drag. Radiation forces are sensitive to attitude, and absorption and reflection coefficients. Atmospheric drag in LEO is sensitive to the variability of the exosphere.

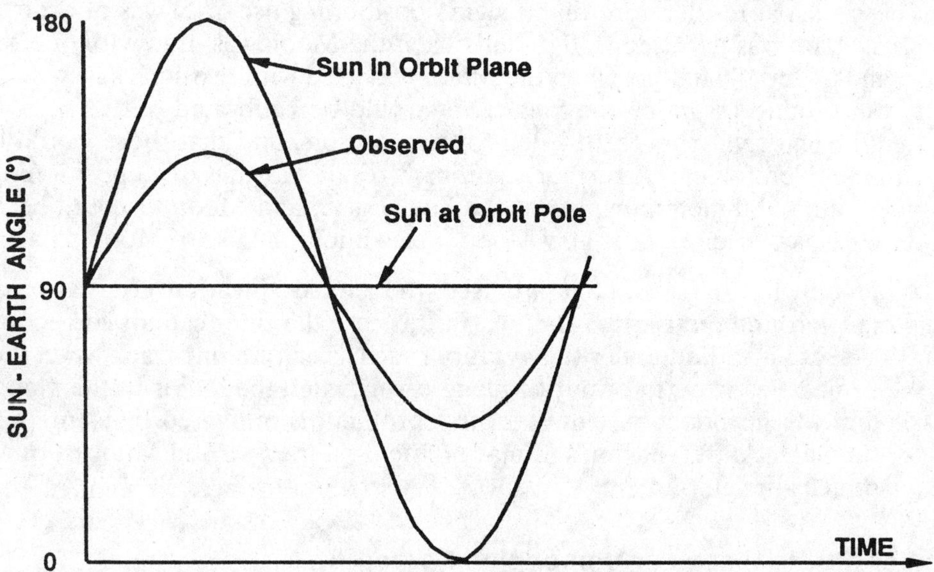

Fig. 12   Determination of Angle from Sun to Orbit Plane and In-Track Position from Variation of Sun/Earth Angle

Fig. 12 shows how the angle of the Sun to the orbit plane can be determined from the amplitude of the variation of the Sun-Earth angle. The phase also gives in-track position. Because of errors in the data and unmodeled forces, the orbital equations cannot be fitted exactly to the data. Numerous techniques have been devised to obtain approximate fits. Most least squares methods have to be applied to fixed batches of data and are not readily adaptable to cases such as autonomous navigation where new data arrive continually or sporadically. We desire a method that continues to make some use of the orbital history but that incorporates new data as they arrive. We selected the Kalman filter (Ref. 4, pp. 459 - 469) for our modeling.

The instantaneous state of the craft was described by a State Vector XTRUE = (X, Y, Z, V_x, V_y, V_z). The initial errors in each component of the estimated state vector X were chosen randomly from a rectangular distribution with mean XTRUE and with halfwidth 4×10^{-5} of the true value. The vector y_{obs} of observations consisted of a true part y plus an error component y_{err}, yielding $y_{obs}(i) = y(i) + y_{err}(i)$. The three components to y_{obs} and y were Earth radius, Earth-Sun angle, and Earth-Moon angle. A Gaussian random number generator produced the input error values. The estimated state vector and error covariance matrix were propagated according to Ref. 4. The process noise figure was adjusted according to the unmodeled forces and biases. (Too large a value produces noisy tracking, but too small a value in the presence of unmodeled forces or bias leads to locked in constant or periodic navigation error, or to divergence.)

NAVIGATION PERFORMANCE

The principal results were that 3-sigma positioning accuracy was of order 1 km when the Moon was not used, 130 meters when the Moon was used with one MANS sensor, and under 100 meters when the Moon was used with two such sensors. These results were found assuming the bias terms would be calibrated out to better than 0.02 deg for one sensor and 0.014 deg for two sensors, and that there would be no major unmodeled forces. A further significant result was that once positioning was obtained within 100 meters by use of the Sun, Earth, and Moon, it could be maintained for 10 days or more near new Moon (i.e., without the use of Moon data).

When only the Earth and Sun are used, the data are invariant to rotation of the orbital plane around the Earth-Sun line. In this case the filter cannot correct orbital errors which are invariant under this rotation, such as error along track when the Sun is at the orbit pole, or periodic out-of-plane errors when the Sun is in the plane. As remarked in the introduction, however, this problem is mitigated by using position data for the Sun at different times in place of data on the Sun and Moon at one time to resolve such ambiguities.

Results with No Biases or Unmodeled Forces

We shall first discuss the overall orbital accuracy with no sensor biases or unmodeled forces. These are the results that would be anticipated after bias calibration and with unmodeled forces well below the level of the noise. There are four main subdivisions, according to whether there are 1 or 2 sensors and according to whether the Moon is visible. Recall that near new Moon autonomous navigation will have to rely on Sun and Earth measurements only. Within these four categories variations in navigation accuracy occur because of variations in the angle of the Sun and Moon out of the orbit plane and the relative phase of the Sun and Moon about orbit normal and differing initial errors.

Fig. 13 shows the resulting accuracies for the four cases described above. On the ordinate we show the 3 sigma navigational error in meters. For the abscissa, to represent the accuracy of the sensor package as a whole, we used the RMS of the sensor errors for all three sensed angles.

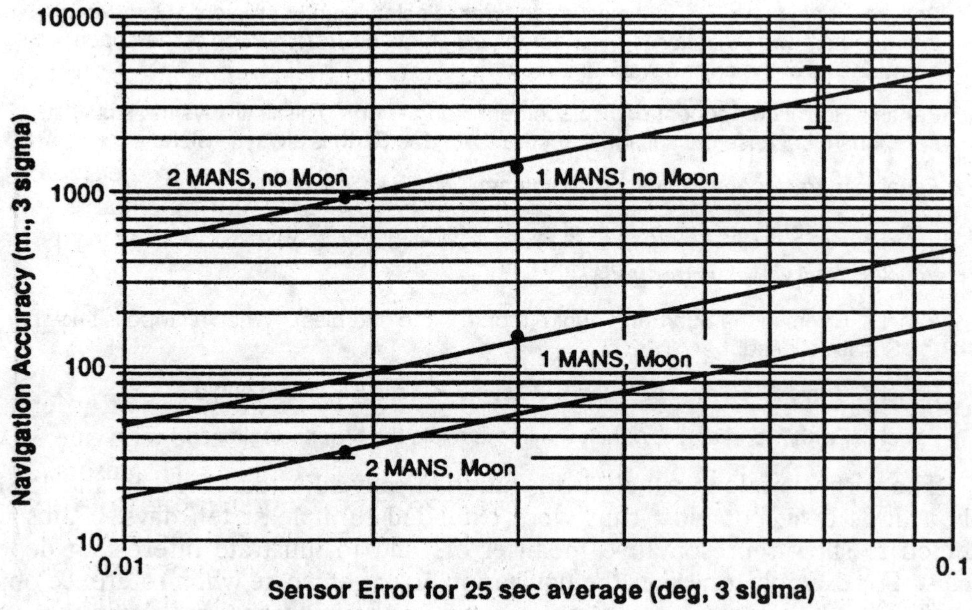

Fig. 13   Navigation Accuracy with Different Sensor Coverage

The individual curves are averaged over several runs with different angles of the Sun (and/or Moon) to the orbital plane, different accuracies of the initial position and velocity estimates, and different values of process noise. The variation from run to run is indicated by the vertical bar at the upper right, but in the best case (2 MANS, Moon visible) the variations were less than half as big. Linear dependence of navigation error on sensor error is expected and was verified by a number of runs at half, double, and four times the nominal sensor error values. The two "no-Moon" cases coincide within the width of the curves. This means that when only the Earth and Sun are visible, the additional sky coverage provided by the second sensor does not help, per se; only the improved angular resolution helps.

The most striking result of Fig. 13 is the increased accuracy when the Moon is visible. To further study this difference, extensive runs were made starting at full Moon and going through new Moon. These runs showed that good results were maintained through new Moon, when two MANS were used. The loss of accuracy during that period was negligible, due to the excellent Sun coverage. In summary, for best results the initial errors must be reduced by starting with Moon data, and in that case good Sun coverage will provide excellent navigation accuracy. Of course, any other means to reduce the initial errors would presumably work as well. It is useful to compare to single-fix accuracies. From Fig. 13 the mean 3 sigma accuracy varies from less than 100 meters (two sensors using Earth, Sun, and Moon) to 1.4 km (one sensor, no Moon).

We present next results from typical runs in sequences of frames of eight panels. Numbering these panels first down the left column, then down the right, they represent:

Panel 1: *Plain curve* = Total navigation error (displacement) in meters. *Curve interspersed with the letter "s"* = theoretical 3 sigma error from trace of the space components of the error covariance matrix.

Panel 2: Radial component of displacement along $\mathbf{r}^0$ (m). This is always the smallest component because the sensor is accurate and the Earth is always visible.

Panel 3: In-track component of displacement, along $\mathbf{s}^0$ (m).

Panel 4: Out-of-plane component of displacement, along $\mathbf{w}^0$ (m).

Panel 5: Total velocity error (m/s).

Panel 6: Errors in the Sun/Earth angle (Monte Carlo and bias). When a stripe is shown, the Sun is eclipsed.

Panel 7: *Curve 1* = True Earth width. *Curve 2* = True Earth/Sun angle.

Panel 8: Earth/Moon angle. When the line is thick, the Moon is eclipsed.

The vertical scale in panels 1-5 is automatically adjusted. Figs. 14 and 15 show only a day's data, but most runs were extended to at least ten days. Runs were selected to show representative mean errors and to illustrate interesting dependencies. For example, one sees the navigation error get worse when reference objects are eclipsed. In longer runs one can even notice the in-plane error deteriorate most when an in-plane object is eclipsed or moves out-of-plane.

Results Using Earth and Sun Data

Fig. 14 shows the beginning of a typical run with one sensor and no Moon available. A small amount of process noise was added and is evident in the in-track error. The Sun is only 28 deg from the orbital plane in this case, which helps control in-track error, but it happens to be eclipsed 63% of the time. These eclipses have, of course, the *same* periodicity as the orbit itself, which also accounts for the persistence of in-track error. With the Sun so close to the orbit plane, out-of-plane error is controlled surprisingly well. The tendency of eclipses to have the same period as the orbit hinders correction of the out-of-plane error. When a second sensor is used, the eclipses are shorter mainly due to better sky coverage and the Sun is 43 deg below the orbit plane. Without the process noise, the in-track error is quite persistent in sign.

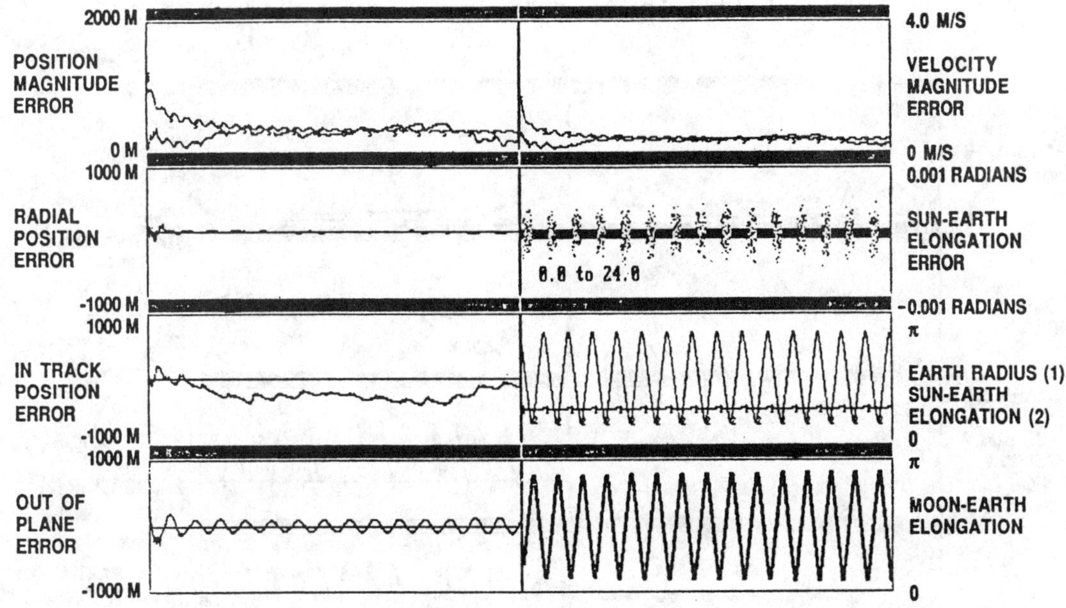

Fig. 14   Outputs for 1 Sensor and no Moon.

The 80 deg inclination orbits precess slowly, and the Sun moves only a degree a day. This means that it may take a long time for the changing Sun angle to resolve the ambiguity of rotation about the Earth/Sun line. Thus it is best to have at least occasional Moon data for this reason and because unmodeled forces could harm the process.

Results Using Earth, Sun, and Moon Data

The Moon offers two major advantages: i) it serves as a third reference point, and unless nearly collinear with the Sun, this resolves the rotation ambiguity; ii) it moves 12 degrees a day in the sky, so that periodicities in its visibility are not normally commensurate with orbital periodicities. This tends to reduce errors that are quite persistent in cases without the Moon. Fig. 15 shows a run with two sensors and the Moon visible some of the time. The vertical scales of panels 1-5 in this figures are half of those in Fig. 14. The presence of even sparse Moon data is quite significant, bringing the navigational error down by a factor 10. When using only 1 sensor, many runs gave 3 sigma navigational errors of order 100 meters or less but the variation from run to run was large.

Results with Biases

Biases can significantly degrade navigation performance and ordinarily will set the real performance limit. The residual bias errors, and therefore the navigation accuracy, will be determined by the level of in-flight calibration that is applied. This decision will be based on the accuracy versus the cost to achieve specific mission objectives.

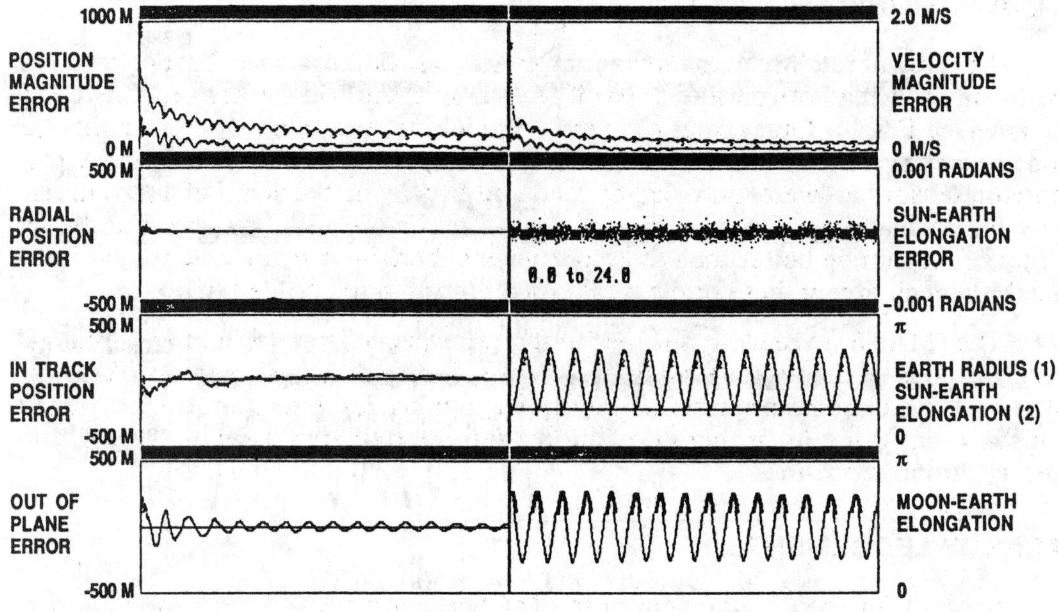

Fig. 15   Outputs for 2 Sensors with Moon

For the present study we assumed on-orbit bias determination to approximately 0.01 deg. The biases were assumed to be 0.01 deg in Earth radius, 0.01 deg in Sun/Earth angle, and 0.0075 for the Moon/Earth angle. In these runs, the errors tended not to have much randomness nor dependence on the initial data. Thus it is meaningless to quote a 1 sigma or 3 sigma error and we give instead the navigation error, which is the RMS displacement over a typical orbit. The navigation error for these cases typically stabilized at 1 km, mostly out-of-plane.

Results with Unmodeled Forces

To test the ability of the Kalman filter to track in the presence of unmodeled forces, several runs were done with the force of solar radiation pressure present. An acceleration away from the Sun was added with a fixed value in sunlight and zero in eclipse. The force was only in the true orbit propagator, not in the Kalman filter propagator. A large force level was chosen to produce a perturbation sufficient to drive the filter to divergence when no process noise was included. Again, the actual navigation error rather than a 3-sigma level will be given. For an unmodeled force of 10^{-6} of the Earth's surface gravity force, the navigation error stabilized at 250 meters one sensor was used and the Moon was present. With two sensors the error was reduced to 150 meters--a result quite consistent with the linear dependence of navigation error on sensor error shown in Fig. 13. These are very encouraging results, considering the opportunity to model perturbative forces in detail.

CONCLUSIONS

A survey of satellite autonomous navigation capabilities shows that previous low cost systems using horizon sensors provide 1 sigma accuracies on the order of 1.5 to 25 km (see Fig. 1). Other types of systems provide better accuracy, but are either far more expensive or are not truly autonomous. In contrast, the MANS uses low cost horizon sensor hardware, only slightly modified to provide the additional data necessary for navigation . The MANS accuracy, as plotted in Fig. 1, is shown to be an order of magnitude better than those of previous horizon scanner systems and it exceeds them all except those of the most expensive and sophisticated systems.

The MANS exceeded our performance expectations. Requiring minimal modifications to the attitude control hardware already flown on most satellites, this system shows extreme promise for widespread application based on its high benefit for low cost. It merits further development for other missions such as transfer, and geosynchronous missions.

ACKNOWLEDGEMENTS

Support for the development of the MANS concept has been provided in part by the Rome Air Development Center (RADC) of the U.S. Air Force under contract No. F30602-87-C-0056, and in part by Microcosm, Inc. Support for the definition of the hardware modifications necessary to create the MANS sensor has been provided by the Barnes Engineering Division of the EDO Corporation. We wish specifically to thank Lt. Steve Kabelis of RADC for his interest, support, and long term assistance in bringing about this study. We thank Gerry Falbel, Robert Barnes, and Robert Savoca of Barnes Engineering for their personal involvement in the hardware issues, and Dr. James Wertz of Microcosm, Inc. for development of the MANS concept.

REFERENCES

1. Chory, M.A., Hoffman, D.P., LeMay, J.L., "Satellite Autonomous Navigation - Status and History," Proceedings IEEE PLAN (Position, Location, And Navigation) Symposium, November, 1986

2. Chory, M.A., et al, "Autonomous Navigation -- Where We Are In 1984", Paper No. AIAA-84-1142-CP

3. Tai, F., and Barnes, R., "The Dual Cone Scanner: An Enhanced Performance, Low Cost Earth Sensor,," AAS-89-013, Proceedings of the Annual Rocky Mountain Guidance and Control Conference, February 7, 1989.

4. Wertz, J.R., and Mullikin, T.L., "Reducing the Cost and Risk of Orbit Transfer," Paper No. AIAA-87-0172 presented to the AIAA 25th Aerospace Sciences Meeting, Reno, Nevada, Jan. 12-15, 1987.

5. Wertz, J.R., ed. *Spacecraft Attitude Determination and Control*, D. Reidel Publishing Company, Dordrecht, Holland, 1978, 1980, 1984.

CONCEPT DESIGN AND PERFORMANCE TEST OF A MAGNETICALLY SUSPENDED SINGLE-GIMBAL CONTROL MOMENT GYRO

Katsumi Kito[*], Hiroshi Kanki[†], and Shinya Ishii[†]

Space vehicles are growing large in size and weight and attitude control systems require higher torque and momentum capability. As a primary attitude actuator for future platform type space vehicles, concept design model of a single-gimbal control moment gyro was built and tested. Based on the consideration of potential requirements for long life, low energy consumption, and vibration reduction, this concept design model adopts electromagnetic bearings which suspend two wheels. This single-gimbal control moment gyro has nominal angular momentum capacity of 100 Nms. The concept design model was tested and showed satisfactory results.

[*] Space Systems Engineering Department, Nagoya Aircraft Works, Mitsubishi Heavy Industries, Ltd., 1-1-chome, Daiko-cho, Minato-ku, Nagoya 461, Japan.

[†] Takasago Research and Development Center, Mitsubishi Heavy Industries, Ltd., Takasago, Japan.

INTRODUCTION

Single-gimbal control moment gyros are known as attitude actuators of high torque and momentum capacity. Since space vehicles are growing large in size and weight, and are requiring more maneuver capability, such actuators become useful. From the standpoint of attitude control systems, concept design of a single-gimbal control moment gyro (gyro actuator) was performed and a concept design model was built. This model adopts electromagnetic bearings which suspend two wheels. These are based on the consideration for long life, low energy consumption and vibration reduction.
The single gimbal system is selected because the mechanism is simple and the reliability is relatively high, although its control method is rather complicated.
Nominal angular momentum of the concept design model is designed to be 100 Nms, however it can be easily extended by adjusting wheel speed. Nominal wheel speed is designed from 7000 to 10000 rpm.

In this paper, electromechanical system concepts and preliminary performance test results are presented.

ELECTROMECHANICAL SYSTEM CONCEPT

System Concept

Recently, the application of electromagnetic bearings for rotating machinery have been increased in many fields. And the reliability of the system increased step by step.

Fig.1 Mechanical Assembly of Model CMG

For control moment gyros (CMG'S), ball bearings have been applied as main bearings. However, ball bearings have several problems such as lubrication and bearing life in assuring reliability for long life operation in space. As an alternative design, we tried to apply electromagnetic bearing system for the CMG. Since the output torque of CMG is transmitted to the gimbal through the bearing, electromagnetic bearing loads of CMG are much severer compared with those of reaction wheel.

Bearing load = Torque / Bearing Span

High stiffness as well as load capacity are required for electromagnetic bearings of CMG. This is because small inclination of spin axis results in large output error.

As a first step, we performed the concept design of the CMG specified in Table 1. The CMG is comprised of a mechanical assembly and three electronic units. Electronic units are comprised of spin drive electronics, magnetic bearing control electronics and gimbal drive electronics. The photograph of the mechanical assembly is shown in Fig.1. Fig.2 shows the block diagram of the CMG.

Mechanical Assembly

Mechanical assembly of the concept design model of CMG was made to realize high reliability. The engineering points are covered by high potential of rotor dynamics and many development experiences of high speed turbo machinery.

Rotor design was performed to establish compactness and effective application of electromagnetic bearing.

Figure 3 shows the natural frequencies of rotor-wheel system. Critical speeds are about 5000 rpm and 12000 rpm for rigid modes. These rigid modes can be easily shitted by tuning electromagnetic bearing control parameters.

Casing of the mechanical assembly acts as a gimbal frame. The casing configuration is almost sphere for getting sufficient stiffness and compactness.

Table 1

PRIMARY CHARACTERISTICS OF THE CONCEPT DESIGN CMG

Type	Single Gimbal Type
Angular Momentum	100 Nms
Output Torgue	50 Nm
Rotor Speed	7000 to 10000 rpm
Bearing Type	Electromagnetic Bearing

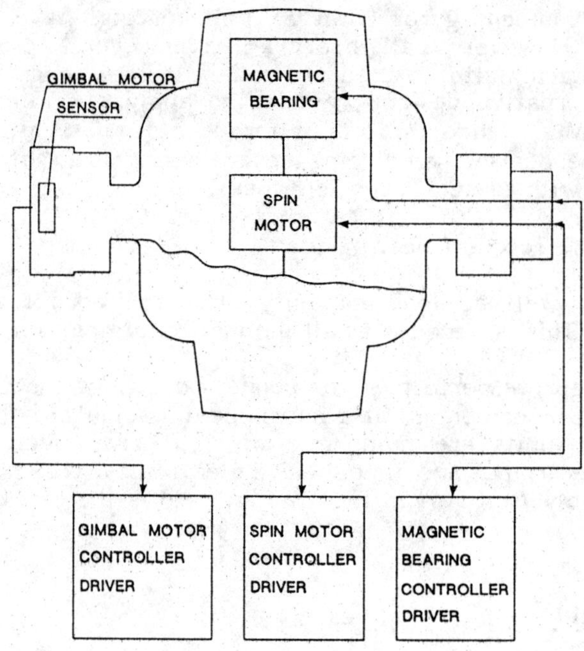

Fig.2 Block Diagram of Model CMG

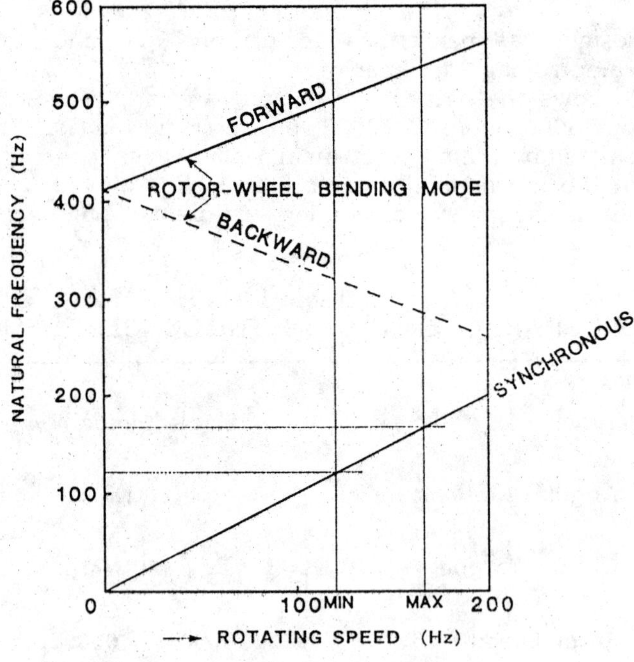

Fig.3 Rotor-wheel Bending Mode Natural Frequency

<u>Spin Control</u>

AC induction motor is applied for driving the wheels. The spin axis rotating speed is controlled by a variable frequency driver.

Magnetic Bearing System

We applied elctromagnetic bearing system.
The location of the radial bearing is decided by the compromise between space saving and reduction of bearing load.

The control system of the electromagnetic bearing is independent for individual axis, and a block diagram of one axis is shown in Figure 4. Gap sensors are applied for the rotor position measurement.
The typical frequency characteristics from the compensator to the sensor of the bearing-rotor system is shown in Figure 5.
The stability of the system is assured for maximum rotating speed and for higher bending modes.

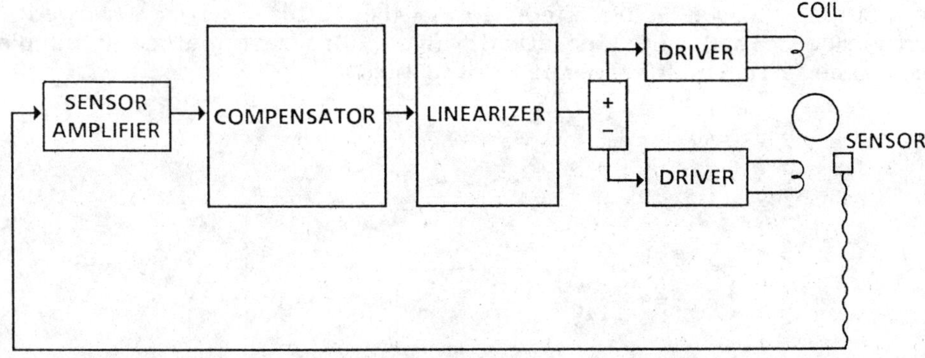

Fig.4 Block Diagram of Magnetic Bearing Control

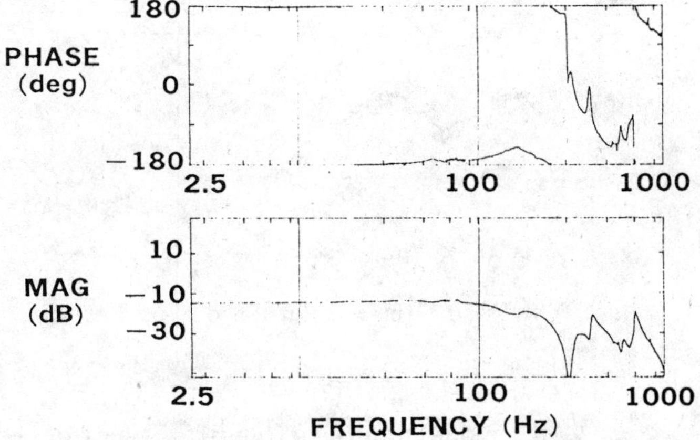

Fig.5 Frequency Characteristics of Radial Magnetic Bearing System

<u>Gimbal Control</u>

For the gimbal control of the concept design model of CMG, AC brushless servomotor is applied for precise control.

The velocity feedback control is applied. Relatively large output motor is applied for wide range of performance test.

PERFORMANCE TEST

Output torque characteristics of the concept design model were evaluated under the normal operating condition.

Since this scale of CMG is expected to be used in relatively small space vehicle, normal torque level is at most an order of 1 Nm if excessive maneuver is not required. Simple test unit using an air bearing table is prepared. This test unit is designed to estimate output torque from the table motion.

Test was performed setting the CMG on the table and driving the gimbal angular velocity sinusoidally. Typical test result for output torque of 1.0 Nm is shown in Fig.6.

Output torque calculated from the table motion showed good correspondence with the theoretical value. This test method is simple but found to be sufficient for the unit level testing.

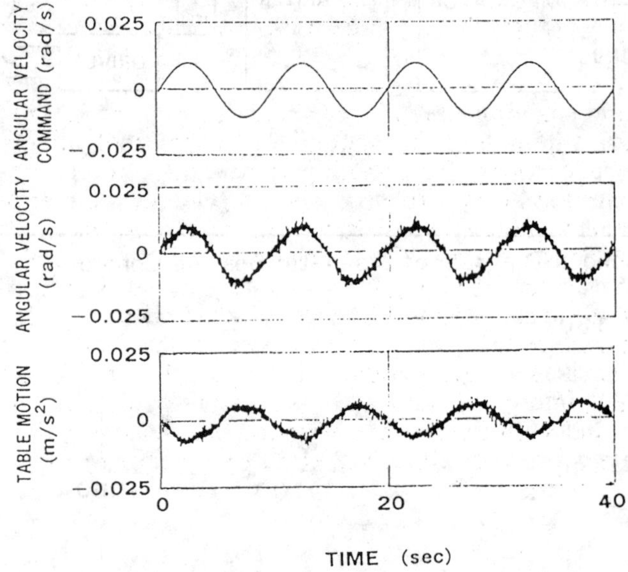

Fig.6 Typical Test Result for Output Torque of 1.0 Nm

CONCLUSIONS

A concept design CMG model applying electromagnetic bearings was successfully built and tested. This model can be extended up to 300 Nms angular momentum system withuot large change of the design principle. Further efforts for proper refinement of the design are planned for the next phase.

AUTONOMOUS STAR REFERENCED
ATTITUDE DETERMINATION

R. W. H. Van Bezooijen[*]

A star pattern recognition algorithm is described enabling a state-of-the-art star tracker to quickly and reliably determine its attitude about all three axes without requiring any a-priori attitude knowledge. This device, referred to as a "Full-sky Autonomous Star Tracker (FAST)," can serve as a key sensor in future autonomous attitude control and navigation systems. Having a field of view of 11.5 by 11.5 degrees and an accuracy of 8 arcsec (1 sigma), the ASTROS II star tracker developed by JPL for use with the Mariner Mark II planetary spacecraft is ideally suited to serve as a FAST. Its large field of view allows the number of guide stars in the all-sky data base of the tracker to be limited to a manageable number, while its high accuracy ensures that the pattern formed by the observed guide stars is unique.

In addition to enabling a FAST, the recognition algorithm can also be used for automating the acquisition of celestial targets by astronomy telescopes, for updating the attitude of gyro based attitude control systems autonomously (planned for Mariner Mark II), and for automating ground based attitude reconstruction. Monte Carlo simulations and a much faster, highly accurate quasi-analytical method are used for predicting the success rate of the algorithm as a function of the many parameters required to define the sky, the star catalog from which the guide stars are extracted, the tracker, and the control parameters of the algorithm. It is shown that an ASTROS II based FAST, integrated with an all-sky database of some 4,100 guide stars, and 840K bytes of memory, can determine its attitude in approximately 1 second with a success rate very close to 100%.

INTRODUCTION

The attitude of spacecraft can be determined with high precision using star sensing devices such as star trackers and star mappers. Fig. 1 illustrates the attitude determination problem. Due to attitude errors, the coordinate system associated with the star sensing device (X_o, Y_o, Z_o) deviates from its expected inertial orientation (X_i, Y_i, Z_i). By storing the position and brightness of a-priori selected stars, referred to as "guide stars," and using a star pattern recognition algorithm it is possible to determine which of the observed stars within the field(s) of view of the star sensing device correspond to which of the guide

[*] Jet Propulsion Laboratory, California Institute of Technology, 4800 Oak Grove Drive, Pasadena, California 91109.

stars. If at least two of the observed stars are guide stars, and these stars are correctly identified, the attitude about all three axes can be determined. The area(s) in which guide stars are selected, called the "uncertainty area(s)," needs to be sufficiently large to ensure that it includes the field(s) of view of the star sensing device.

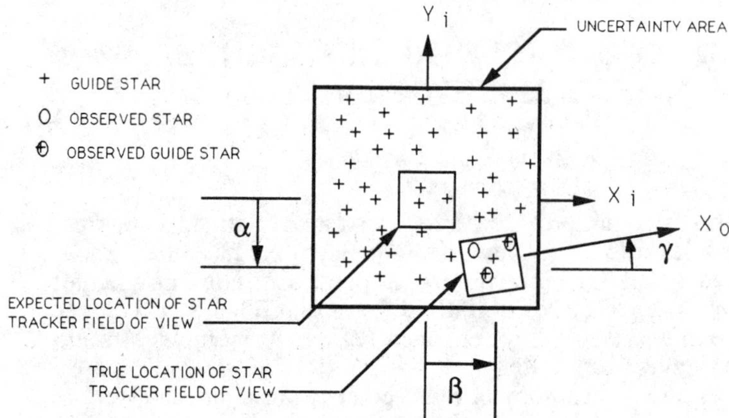

Fig. 1 Attitude Determination Problem

This paper describes a robust star pattern recognition algorithm that is very fast due to the fact that it is basically non iterative. Although not restricted to, the algorithm is intended for use with star trackers (STRs) equipped with two-dimensional solid state array detectors (e.g., CCDs) capable of accurately measuring the position and brightness of multiple stars within the field of view (FOV). The algorithm has a wide range of applicability. It can be used for automating the acquisition of celestial targets by astronomy telescopes, for autonomous updating of gyro-based attitude control systems, and for ground based attitude reconstruction. Most interestingly, the recognition algorithm, enables the construction of a Full-sky Autonomous Star Tracker, capable of accurately determining its attitude about all three axes without requiring any a-priori attitude knowledge (i.e., the uncertainty area in Fig. 1 measures 4π steradian).

Having a FOV of 11.5 by 11.5 degrees and a spatial accuracy of 8 arcsec (1 σ, goal), the redundant Charge-Coupled Device ASTROS II star tracker[1], developed by JPL for use with the Mariner Mark II planetary spacecraft[2], is ideally suited to serve as a Full-sky Autonomous Star Tracker (FAST). Its large FOV allows the number of guide stars in the database of the tracker to be limited to a manageable number (4,100), while its high accuracy ensures that the pattern formed by the observed guide stars is unique. Furthermore, this tracker is integrated with a radiation hard microcomputer capable of performing the star identification in approximately 1 second.

In addition to the aforementioned general recognition algorithm, a special algorithm will be described applicable to cases where the attitude error is of the same order as the radius of the STR FOV. The special algorithm, developed for use with the Space Infrared Telescope Facility[3] (SIRTF), allows an important reduction of the memory space required. SIRTF will carry a 1 m class cryogenically cooled infrared telescope nearly three orders of magnitude more sensitive than the current generation of infrared telescopes. The primary attitude data used for pointing SIRTF will be provided by a CCD STR with a 15 arcmin diameter FOV, integrated in the detector bay of the telescope as is shown in Fig. 2 (the Fine

Guidance Sensor is the STR). Following each slew to a new target, the observed stars are identified, permitting determination of the attitude (angles α, β, and γ in Fig. 1), after which the spacecraft is rotated to the desired attitude where the stars used for pointing control are acquired. Because of its small FOV, use of guide stars up to magnitude 15 is required. These stars can be extracted from the catalog prepared for the Hubble Space Telescope[4].

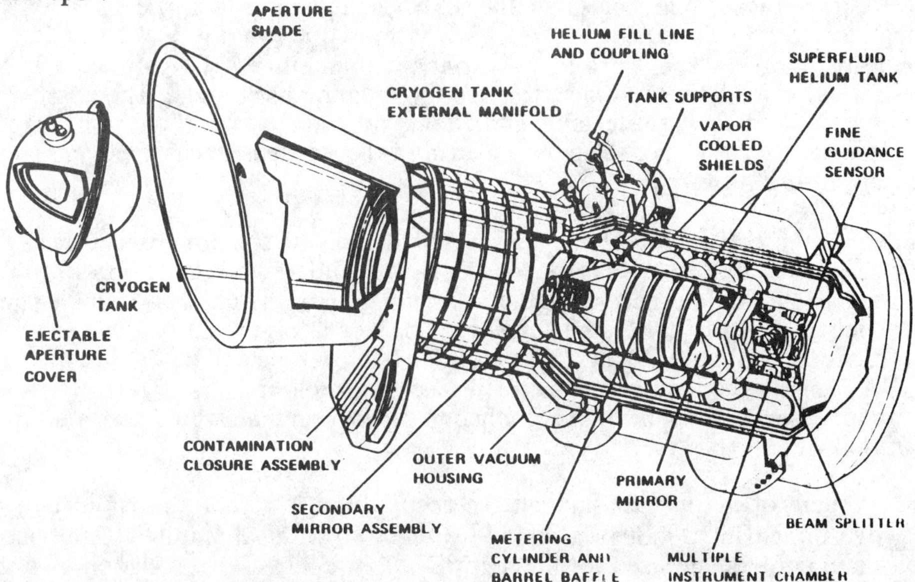

Fig. 2 Telescope of The Space Infra Red Telescope Facility (SIRTF) with the Star Tracker (Fine Guidance Sensor).

Monte Carlo simulations were used for establishing the performance of both recognition algorithms, employing parameters applicable to SIRTF. Since all failed cases can be diagnosed, the simulations are very important for developing and perfecting recognition algorithms. However, in order to obtain sufficient statistical confidence it is necessary to perform many evaluations (typically at least 1000 per case), requiring very long compute times. This motivated the development of a quasi-analytical method, described in detail in Ref. 5, for obtaining the recognition success probability. The quasi-analytical method, which is typically two to three orders of magnitude faster, yields results that closely match the values obtained with the simulations.

Being insensitive to the size of the uncertainty area, the quasi-analytical method is used to show that a wide angle STR like JPL's ASTROS II can be developed into a Full-sky Autonomous Star Tracker, capable of determining its attitude with a success rate very close to 100%. Applied to Mariner Mark II, it is planned to use ASTROS II in a semi-autonomous mode for updating the spacecraft attitude. Upon receiving its expected attitude plus attitude error from the attitude control processor, the tracker will establish the uncertainty area, extract guide stars in this area from its all-sky catalog (<3000 stars), perform the identification and compute its attitude. The paper is finalized by presenting an estimate for the computational requirements of an ASTROS II based FAST.

PREVIOUS RELATED RESULTS

In the past, primitive versions of automated star identification have been used onboard spacecraft. The identification process was simple, involving only a few guide stars per attitude determination. Planetary spacecraft typically used Canopus as a single guide star. Being the brightest star in the area searched, identification of Canopus is fairly simple. The Astronomical Netherlands Satellite[6] (ANS), employing an image dissector star tracker integrated in the instrument bay of the UV telescope, used guide star pairs for acquiring celestial targets. The Infra Red Astronomical Satellite[7,8] (IRAS), had a V-slit type star sensor in the focal plane of the telescope for updating its attitude. The number of guide stars used per update was selectable and could vary from 1 to 5. For the foregoing three examples, one component of the pointing error of the star sensor was kept very small by using a sun sensor.

Galileo[9] will use a V-slit type star sensor mounted on the spinning part of the dual-spin spacecraft. Although use of an onboard all-sky guide star catalog was initially planned[10], memory limitations necessitated a vast reduction of the number of guide stars. It is now planned to typically use 6 guide stars for a particular orientation of the spin axis. A star identification method utilizing CCD trackers has been described by Junkins and Strikwerda[11]. Their method is iterative and is based on matching pairs of observed stars with pairs of guide stars. Its application is limited to cases where the attitude error is small relative to the FOV of the tracker.

The problem of automated star pattern recognition is rather similar to that of recognizing two-dimensional industrial parts[12] where known local features of the parts, such as holes and corners, are used for the identification process. The location of these features relative to each other is fixed as is true for stars in a given group. The signature of each feature, such as the diameter of a hole, is analogous to the brightness of stars. Partial obscuration of a part due to another part is analogous to the unobservability of guide stars that may be caused by such factors as brightness errors, nonoptimal selection of the brightness threshold of the star sensor, or the use of non-existent guide stars caused by star catalog reliability problems.

STAR PATTERN UNIQUENESS

Successful star identification requires that the pattern of "valid observed guide stars" is unique. Here uniqueness is defined as a condition where no other group of observed stars of a size equal to the number of valid observed guide stars matches a pattern of guide stars in the uncertainty area. The members of the largest group of observed guide stars meeting all criteria imposed by the recognition algorithm are defined as the valid observed guide stars. The uniqueness probability is maximized if the algorithm exploits the full information contents of a pattern. This condition is satisfied if the algorithm checks all independent variables defining a pattern.

The number of independent variables required for defining a pattern, referred to as the "pattern dimension," is shown in table 1 as a function of the number of stars in the pattern. When there are more than two stars, the geometrical contribution consists of the distance between two of the stars, a sign for defining a coordinate system relative to these two stars, and two coordinates for defining the position of each of the remaining stars. If the color of the stars were used, which is impractical in most case, the dimension would be expanded. From table 1 it may be seen that the pattern dimension, and hence the uniqueness probability, increases rapidly with the size of a pattern.

Table 1

DIMENSION OF STAR PATTERN

Number of Stars	Contribution to Dimension			Dimension
	Sign	Coordinates	Brightness	
1	0	o	1	1
2	0	1	2	4
3	1	3	3	8
4	1	5	4	11
k	1	2k-3*	k	3k-1**

* for $k > 1$, ** for $k > 2$

RECOGNITION ALGORITHMS

<u>Match Criteria</u>

The objective of the recognition algorithm is to find the largest group of observed stars that matches a group of guide stars. The "basic" recognition algorithm assumes that a group of observed stars matches a group of guide stars if all of the following criteria are met:

1. The measured angular distance of each pair of observed stars matches the predicted angular distance of the corresponding pair of guide stars to within the distance tolerance (*tdis*).
2. The measured magnitude of each of the observed stars matches the predicted magnitude of the corresponding guide stars to within the magnitude tolerance (*tmag*).
3. The geometry of the group of observed stars is not the mirror image of that of the corresponding group of guide stars.

In the above, the angular distance of a star pair is defined as the angle between the two stars of the pair. It may be seen that the basic algorithm uses the full information content of the pattern.

If the attitude error is small, as is true for SIRTF, an additional "rotational" criterion is imposed. This criterion states that the rotation angle of the tracker about the boresight (angle γ in Fig. 1) needed to align the pattern of observed stars with the pattern of guide stars needs to be less than the FOV rotation tolerance (*trot*).

<u>Special Algorithm</u>

If the attitude error is small (requiring an uncertainty area of typically 10 times the tracker FOV), as is the case for SIRTF, the special algorithm can be applied. This algorithm finds the largest group of observed stars that can be matched with a group of guide stars by a rotation about the axis perpendicular to the expected location of the STR boresight (axis Z_i). The recognition process starts by identifying all guide star pairs having a mutual distance less than the greatest dimension of the tracker FOV (diagonal or diameter), which could be done during the slew to the target. Next, it is determined which pairs of guide stars match each pair of observed stars. A star pair match occurs if all of the following conditions are satisfied:

1. The angular distance of the guide star pair differs less than *tdis* from that of the observed star pair.
2. The rotation about the boresight of the STR needed to align the guide star pair with the observed star pair is less than *trot*.
3. Each guide star of the pair has a predicted instrument magnitude that differs less than *tmag* from that of the corresponding observed star.

Following the detection of a match, a match matrix linking the observed stars (lines) with the guide stars (columns) is updated.

This process is illustrated using Fig. 3 which shows a situation where four (stars 1, 2, 4, and 7) of the seven observed stars are guide stars (stars 14, 26, 51, and 3 respectively). The matching process for one of the pairs of observed stars (*b*) is illustrated in Fig. 4. Five guide star pairs match the distance of the observed star pair, two of which also meet the FOV rotation constraint. However, only the guide stars of one of the pairs (*b'*) match the brightness of the observed stars. Because pair *b* matches pair *b'*, the match matrix elements 1,14 and 4,51 are incremented (see Fig. 5). Following a successful matching of observed star pairs *a* through *f* with guide star pairs *a'* through *f*, four matrix elements will have a value of 3. The value of each element of the match matrix is referred to as the "confirmation value" of the star match, because it indicates how often the observed star was found to be linked to the guide star.

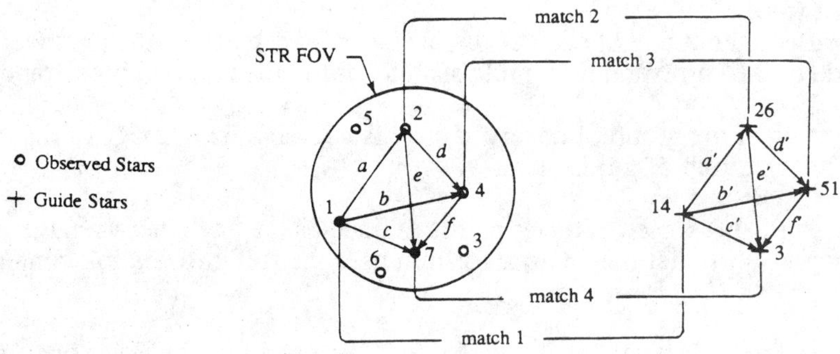

Fig. 3 Star Match Example (Four of the Observed Stars are Guide Stars)

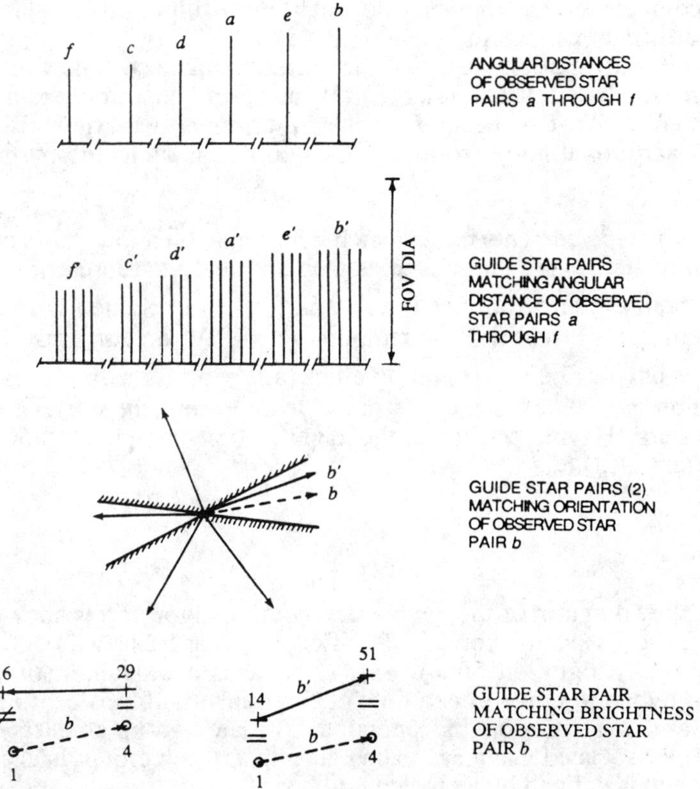

Fig. 4 Illustration of Process Used for Generation of the Match Matrix and the Base Match Matrices.

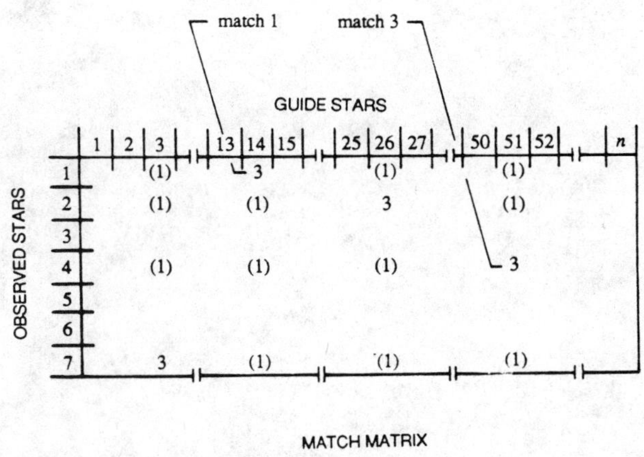

Fig. 5 The Match Matrix Used for the Special Recognition Algorithm.

Because of coincidence matches of guide star pairs with observed star pairs there will be more non zero matrix elements than shown in Fig. 5. In addition, coincidence matches can cause the confirmation values of the correct matches of observed stars with guide stars to be greater than 3. If neither a FOV rotation constraint nor a brightness constraint was imposed, the match check of the pairs *a* through *f* with the pairs *a'* through *f'* would result in 12 additional non zero matrix elements as is indicated by the numbers within parentheses.

In the following step, elements in the match matrix having a low value compared to the largest confirmation values are eliminated as they are likely to represent spurious star matches. Next, the pointing error (angles α and β in Fig. 1) associated with each of the remaining matrix elements is established, assuming a zero FOV rotation error. The largest group of star matches having "equal" α and β values (allowing for differences consistent with the FOV rotation tolerance) is then assumed to represent the correct matches of observed with guide stars. Having performed the identification, computation of the attitude is trivial (use least squares fit).

General Algorithm

A high level flow diagram of the general recognition algorithm is shown in Fig. 6. Defining recognition success as the correct identification of at least two of the observed stars, it follows that the recognition fails if only one star is observable, something that could happen due to dynamic range constraints of the tracker. If two or more stars are observed a list of "base" match groups is generated. A match group consists of a kernel match and a number of associated matches. Each match in a match group links an observed star with a guide star and is defined by its match number.

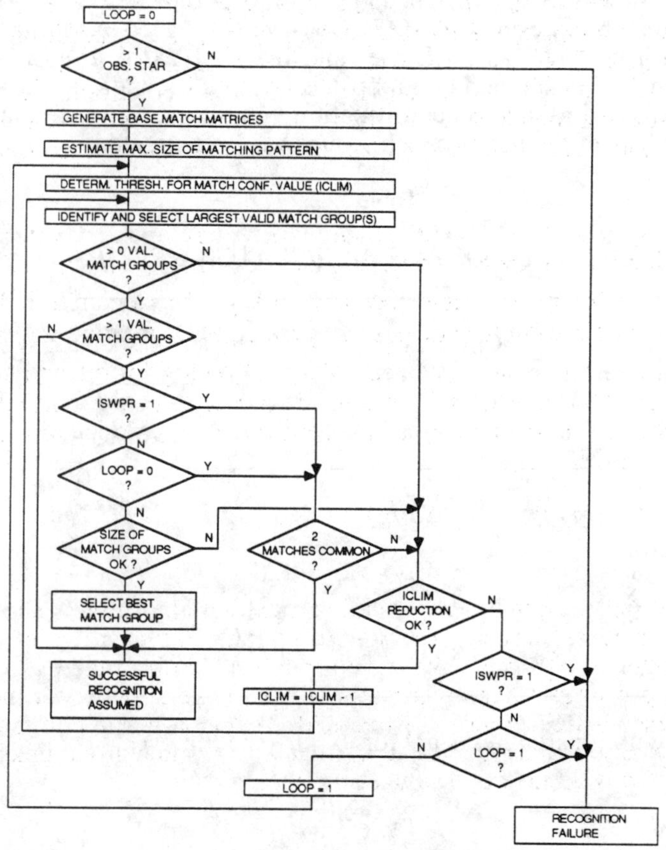

Fig. 6 High Level Flow Diagram for the General Recognition Algorithm.

As for the special method, all guide star pairs need to be identified first. For FAST all guide star pairs are permanently stored in memory in sequence of increasing angular distance, which makes the search for guide stars matching the angular distance of each observed star pair as efficient as possible. Next, applying the same criteria as used for the special algorithm, the guide star pairs matching each pair of observed stars are identified. Obviously, step 2 in that process is omitted in the case of FAST. It is most convenient to use the vector distance between stars (stars on celestial sphere with unit radius) as a measure of the angular distance, rather than the angle itself.

Using the example of Fig. 3 again, it will be assumed that of the 21 observed star pairs, the six a through f will be evaluated first, and in that sequence. In the absence of coincidence matches, the first four lines of the list of base match groups would be as shown in table 2. Following the match of guide star pair a' with observed star pair a, the match of observed star 1 with guide star 14 is designated match 1, while that of observed star 2 with guide star 26 is designated match 2. Table 3 shows how each star match (i) is linked to an observed star, $OMAT(i)$, and a guide star, $GMAT(i)$. In addition, match group 1 is created with match 1 as its kernel $(KMAT(1) = 1)$, and match 2 as the first match associated with this kernel $(MAT(1,1) = 2)$. The confirmation value of the match group is set to 1 $(NASS(1) = 1)$. Similarly, the second match group is created with match 2 as its kernel, match 1 as its first associate match, and a confirmation value of 1. After matching b' with b, the third match is defined $(OMAT(3) = 4$ & $GMAT(3) = 51)$, the third match group is created with match 3 as its kernel and match 1 as its first associate and a confirmation value of 1. Further, match group 1 is updated by adding a second associate match $(MAT(1,2) = 3)$ and raising its confirmation value to 2. After the last match of f with f', the first four lines of the list of match groups has been created.

Table 2

EXAMPLE OF LIST OF BASE MATCH GROUPS

Match Group	Kernel Match	Confirmation Value	Matches Associated with Kernel		
			First	Second	Third
ig	$KMAT(ig)$	$NASS(ig)$	$MAT(ig,1)$	$MAT(ig,2)$	$MAT(ig,3)$
1	1	3 (2)	2 (2)	3 (4)	4 (-)
2	2	3 (3)	1 (1)	3 (3)	4 (4)
3	3	3 (2)	1 (2)	2 (4)	4 (-)
4	4	3 (2)	1 (1)	2 (2)	3 (3)

Note: The numbers within parentheses are for $b = b'$ (see Fig. 3)

Table 3

EXAMPLE OF MATCHED OBSERVED AND GUIDE STARS

Match Number	Observed Star	Guide Star
1	1	14
2	2	26
3	4	51
4	7	3

Due to coincidence matches, there will usually be many more match groups, while the first four match groups may have spurious associated matches and thus greater confirmation values. All information of the base match groups is contained in the vectors $KMAT(ig)$, $NAS(ig)$, $OMAT(i)$, $GMAT(i)$, and the matrix $MAT(ig,j)$, where ig refers to the match group. This set of matrices is referred to as the base match matrices. If pair b' does not match pair b, the base match matrices are different as is indicated by the numbers within parentheses. The original base match matrices are retained and any further processing is performed on a copy of them.

Next, the maximum size of a match group is estimated by noting that in order to have n matches, there need to be at least n match groups with a confirmation value of at least n-1 each. Having found the maximum size (e.g., n), a threshold (ICLIM) is defined that is initially set to n-1-ICMAXR, where ICMAXR is a control parameter that needs to be at least 1. Using the threshold, all match groups having a confirmation value equal or less than the threshold are eliminated as they are suspected to be spurious. Furthermore, the remaining match groups are adjusted to account for any associated matches that were kernels of eliminated match groups.

In the following step, the remaining match groups have to pass a geometry test, where it is checked if the angular distance of all observed star pairs matches that of the corresponding pair of guide stars. For match group 1 in the example, which has been associated with matches 2, 3, and 4, it implies that the distances associated with match 2 & match 3, match & match 4, and match 3 & match 4 are checked. The first of these three tests is executed by checking if match $MAT(1,3)$, = match 3, is an associate of the kernel of match group $MAT(1,2)$, = match 2, which is indeed the case. This process is continued until it has been verified that all 6 distances associated with match group 1 do match.

The geometry test is finalized by checking if the pattern of guide stars is not the mirror image of the pattern of observed stars. This is done by verifying that the signs of the vector product of $\underline{a}$ and $\underline{b}$ and that of $\underline{a}'$ and $\underline{b}'$ are the same. If all distances match, the end result of the geometry test is shown in table 4, case a. If distance b differs from b' by more than $tdis$, then the end result would be as case c of table 4. Because b was found not to match b', match group 2 split into two match groups, 2a with matches 3 & 4, and 2b with matches 1 & 4. The same happened to match group 4.

As the possibility exists that the distances do not match because of a "high sigma case," and it is noted that the recognition success probability drops with a reduction of the number of stars in the pattern, a "relaxed distance tolerance" $tdis1$ $(tdis1 > tdis)$, can be specified. If the diference between b and b' is less than this tolerance, the match groups shown as case b in table 4 would result. Otherwise, the groups shown as case c would be generated.

Next, redundant match groups are eliminated, implying that one group is left in case a of table 4, tw in case b, and two in case c. This is followed by extraction of the largest remaining match group(s), being just one in cases a and b, but two in case c. Usually, there is only one match group left at this point, and the matches of this group are assumed to be correct . If there are more than one match group left, and these groups have at least two matches in common (see case c), then successful recognition is also assumed.

If there is no remaining match group, or multiple groups were found without common matches, then it is assumed that the threshold ICLIM was set too conservatively and the process is repeated after decrementing the threshold. If this yields no solution after having lowered the threshold to its minimum value, and the control switch ISWPR was set to 1, the algorithm reports that it cannot perform the recognition. However, if ISWPR had been set to any other value, a second recognition attempt would be performed upon having set the LOOP parameter to 1 (initialized at 0).

Table 4

LIST OF MATCH GROUPS PASSING GEOMETRICAL CHECKS

Case	Match Group	Kernel Match	Confirmation Value	Matches Associated with Kernel		
				First	Second	Third
	ig	$KMATT(ig)$	$NASS1T(ig)$	$MATT(ig,1)$	$MATT(ig,2)$	$MATT(ig,3)$
a	1	1	3	2	3	4
	2	2	3	1	3	4
	3	3	3	1	2	4
	4	4	3	1	2	3
b	1	1	2	2	4	-
	2	2	3	1	3	4
	3	3	2	2	4	-
	4	4	3	1	2	3
c	1	1	2	2	4	-
	2a	2	2	1	4	-
	2b	2	2	3	4	-
	3	3	2	2	4	-
	4a	4	2	1	2	-
	4b	4	2	2	3	-

If during this second attempt the situation is encountered where there is more than one remaining match groups, the match group with the highest quality is selected and the algorithm assumes the recognition to be successful. The sum of the differences between the angular distances of the observed star pairs and the corresponding guide star pairs is used as the quality index. The group with the lowest index is assumed to contain the correct matches.

Having identified the stars, determination of the attitude is trivial (least squares solution). By retaining the base match matrices, the recognition algorithm is very fast, even if iterations are required.

SUCCESS RATE OF RECOGNITION ALGORITHMS

<u>Test Cases and Evaluation Methods</u>

The general star pattern algorithm is evaluated for two extreme cases, SIRTF with its 15 arcmin diameter STR FOV, and a FAST with a 100 square degree circular FOV. SIRTF was also used for testing the special algorithm. For SIRTF, both algorithms were evaluated using Monte Carlo simulations. These simulations were also utilized for improving the general algorithm. A quasi-analytical method[5] was developed and used for predicting the recognition success rate of the general algorithm as applied to SIRTF and FAST. The latter method, which asssumes that the algorithm uses the full information contents of a pattern, is analytical except for the fact that it employs two interpolation tables. Because it incorporates conservative approximations and also cannot account for some refinements in the algorithm, the quasi analytical method yields conservative results.

The sky model used for both test methods is based on the Bahcall and Soneira galaxy model[13] with extensions based on information from Allen[14]. Fig. 7 shows the assumed integrated star density as a function of the apparent visual magnitude for sky areas of mean, maximum and minimum density. When performing the tests, the log of the density is assumed to be a linear function of magnitude with a slope equal to the value of the faintest guide star used.

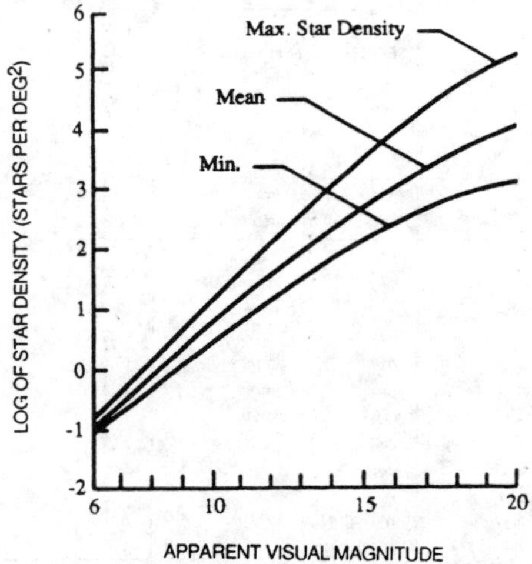

Fig. 7 Integrated Star Density vs Apparent Visual Magnitude for Sky Regions
with Maximum, Mean, and Minimum Star Densities.

The parameters used for the Monte Carlo simulations of the SIRTF case are shown in table 5, while table 6 lists the parameters employed for the quasi-analytical evaluation method of both the SIRTF and FAST cases.

Table 5

INPUT PARAMETERS FOR SIMULATING THE SIRTF REFERENCE CASE

PARAMETER				
Group	Description	Symbol	Unit	Value
Sky	10 log of integrated star density at 0 magnitude	A	per square deg	-3.14
	Slope of 10 log star density vs magnitude	B	per magnitude	0.38
Star	Catalog completeness	CC	fraction	0.95
Catalog	Catalog reliability	CR	fraction	0.95
	Magnitude limit		magnitude	15
STR	Radius of field of view	$rfov$	arcmin	7.5
	Limit to number of observed stars	$nosl$	per FOV	20
	Dynamic range	mdr	magnitude	5.0
	Sensitivity error	$EMMAG$	magnitude	0.0
Catalog,	Star position error (1 σ)	$epos$	arcsec	1.00
STR and	Star magnitude error (1 σ)	$EMAG$	magnitude	0.45
Sky				
	Magnitude margin for close disturbing stars	dmd	magnitude	1.75
	Avoidance distance for close disturbing stars	rd	arcsec	9.0
	Clumping factor	fc		3.0
Attitude	Pointing error component (1 σ)	$epoint$	arcmin	5.34
Error	FOV rotation error (1 σ)	$erot$	arcmin	5.34
Control	Tolerance on star magnitude	$tmag$	magnitude	1.16
Parameters	Expected relative brightness threshold	$THMAG$	magnitude	0.67
	Absolute true STR brightness threshold		magnitude	17.0
	Guide star density	ngs	per FOV	9.07
	Tolerance on FOV rotation	$trot$	arcmin	17.6
	Tolerance on angle between guide stars	$tdis$	arcsec	3.78
	Relaxed tolerance on angle between guide stars	$tdis1$	arcsec	3.78
	Radius of uniqueness area	ru	arcmin	23.75
	Limit to number of stars used for recognition	$nbosl$		7
	Match confirmation threshold reduction			
	Initial	$ICMAXR$		1
	Maximum	$ICMAXM$		5
	Choose best match in case of confusion?			No
	Minimum number of stars in best pattern			3
Program	Magnitude limit for bright stars	mbl	magnitude	14.8
Parameters	Number of evaluations	$NEVA$		1000

Space Infra Red Telescope Facility Case

For each evaluation, the Monte Carlo method randomly generates stars in a field around the target (assuming a uniform distribution) that is large enough to ensure that the STR FOV will land inside of it. Using a fixed magnitude limit (17 in table 5), the distribution of the stars generated for each evaluation is Poissonean. Next, magnitudes are assigned to each of the stars, consistent with the star density function used (parameters A and B of table 5), the desired number of guide stars are selected within the uncertainty area (brightest stars, defined by ngs), and the STR FOV is placed randomly on the field assuming normal distributions for each component of the pointing error and the rotation error about the boresight. This is followed by the assignment of position and magnitude errors to the guide stars within the STR FOV, assuming normal distributions.

Table 6

PARAMETER VALUES FOR EVALUATING THE RECOGNITION SUCCESS RATE
FOR SIRTF AND THE FULL-SKY AUTONOMOUS STAR TRACKER USING THE
QUASI-ANALYTICAL METHOD

| PARAMETER | | | | CASE | |
Group	Description	Symbol	Unit	SIRTF Case	Autonomous STR
Sky	Slope of 10 log of star density vs mag.	B		0.38	0.45
Star Catalog	Catalog completeness	CC	fraction	0.95	0.95
	Catalog reliability	CR	fraction	0.95	0.95
STR	Limit to number of stars per FOV	$nosl$		20.0	20.0
	Dynamic range	mdr	magnitude	5.0	5.0
	Sensitivity error	$EMMAG$	magnitude	0.0	0.0
Catalog, STR and Sky	Star position error (1 σ)	$epos$	FOV dia	0.0011	0.00025
	Star magnitude error (1 σ)	$EMAG$	magnitude	0.45	0.30
	Magnitude margin of close disturbing stars	dmd	magnitude	1.75	1.75
	Avoidance of close disturbing stars	rd	FOV dia	0.01	0.01
	Clumping factor	fc		3.0	3.0
Attitude Error	Pointing error (1 σ)	$epoint$	FOV dia	0.356	n/a
	FOV rotation error (1 sss)	$erot$	arcmin	5.34	n/a
Control Parameters	Tolerance on star magnitude	$tmag$	magnitude	1.16	0.77
	Expected prob. of meeting mag. tol.	$PRMAG$	fraction	0.99	0.99
	Expected relative brightness threshold	$THMAG$	magnitude	0.67	0.42
	Expected prob. of exceeding threshold	$PRTH$	fraction	0.99	0.99
	Guide star density	ngs	per FOV	9.07	10.0
	Expected prob. of 3 prevalid guide stars	$PVGS$	fraction	0.99	0.995
	Tolerance on FOV rotation	$trot$	arcmin	17.6	n/a
	Predicted prob. of meeting rotation tol.	$PROT$	fraction	0.999	n/a
	Relative FOV rotation constraint used?			no	yes
	Tolerance on angle between g.s.'s	$tdis$	FOV dia	0.0042	0.00091
	Probability of meeting angle tolerance	$PDIS$	fraction	0.99	0.99
	Uniqueness area	CA	FOV area	10.0	413.0
	Probability FOV in uniqueness area	PIN	fraction	0.99	1.0
	Limit to no. of stars used	$nbosl$	per FOV	7.0	8.0
Auxiliary Output	True prob. of meeting magnitude tol.	$PRMAGT$	fraction	0.99	0.99
	True brightness threshold (rel.)	$THMAGT$	magnitude	0.67	0.42
	True prob. of exceeding bright. threshold	$PRTHT$	fraction	0.99	0.99
	Density of prevalid guide stars	$nvgs$	per FOV	8.43	9.31
	Density of threshold exceeders	nte	per FOV	17.7	16.2
	Probability of 1 star due to dyn. range	$DYNF$	fraction	0.0123	0.0056
	Fraction of useful stars	pvg	fraction	0.84	0.88
Computed Option	Limit to no. threshold exceeders	lte		20.0	20.0
	Method, L (long) or S (short)	L, S		L	L
	Predicted recognition success potential	RSP	fraction	0.966	0.982

The number of guide stars may be less than desired due to the magnitude limit of the catalog, while, in addition, not all of the brighter stars are selected and a number of spurious guide stars are generated consistent with the defined catalog completeness and reliability. Next, the stars within the STR FOV satisfying the dynamic range constraint and being brighter than the threshold (affected by the STR sensitivity error) are designated as being the observed stars. If their number exceeds the limit to the number used for the recognition process *(nbosl)*, then the brightest *nbosl* stars are selected for use of the recognition process.

An observed guide star will be disqualified if there is a disturbing star within a distance *rd* of it, where *rd* depends on the point spread function, which will be a function of the pixel size of the detector. A star is deemed disturbing if it is within the avoidance distance *rd*, and not at least *dmd* magnitude fainter than the observed guide star. In order to account for the fact that the stars are in reality not uniformly distributed but tend to be grouped in binary and multiple systems, the mean star density within an area with radius *rd* around each observed guide star can be biased using the clumping factor *fc*. An fc factor of three triples the star density and thus the probability of the presence of a disturbing star.

In addition to the control parameters mentioned, the size of the uniqueness area (ru), and the mode of the algorithm (choose best match in case of ambiguity ?, ISWPR in Fig. 6) need to be defined.

For the parameters of table 5, which evaluated the performance of the algorithm at a sky area with mean density, it turned out that the algorithm successfully identified the stars in 968 of the 1000 cases. Line number 3 of table 7, which pertains to this case, gives an assessment of the failed cases. In only one of the 1000 cases, the algorithm assumed a successful recognition, while in reality it was a failure, which demonstrates the robustness of the algorithm. Two of the three unrecognized failures would have resulted in correct attitudes. They were deemed failures because one of the stars in a large pattern was incorrectly identified. Thirteen failures were caused by the dynamic range limiting the number of observable stars to just one. Of the 16 remaining failures, the failure causes include lack of a sufficient number of valid observed guide stars (not enough guide stars and guide stars with out of tolerance magnitude or position errors), and lack of uniqueness of the pattern of observed guide stars.

Table 7

EFFECT OF ALGORITHM REFINEMENTS ON STAR PATTERN RECOGNITION
PERFORMANCE (SIRTF CASE)

Limit to Number of Obs. Stars Used	Select Best Match ?	Star Magnit. Tolerance	Relaxed Distance Tolerance	Successes (out of 1000)	Failures			
					Unrecognized		Dynamic Range Problem	Other
					Okay *	True		
3	No	1.16	3.78	921	1	1	13	64
5	No	1.16	3.78	966	3	2	13	16
7	No	1.16	3.78	968	2	1	13	16
10	No	1.16	3.78	973	3	2	13	9
3	Yes	1.52	4.67	950	2	3	13	32
5	Yes	1.52	4.67	967	4	1	13	15
7	Yes	1.52	4.67	975	3	1	13	8
10	Yes	1.52	4.67	977	4	1	13	5

* Attitude computed correctly.

By comparing line 7 with line 3 of table 7, it may be seen that the number of successful cases (including okay failures) is raised from 970 to 978 by using a wider magnitude tolerance, a relaxed distance tolerance, and the selection of the best match in case of ambiguity.

Using the same parameters, it was found that the special algorithm resulted in a successful recognition in 975 out of the 1000 case, which is quite good considering the significant reduction in software that its use permits.

Table 7 also shows how the performance is effected by the limit to the number of observed stars used for the recognition process. A low value of this limit allows a software reduction and the use of a simpler tracker at the cost of a reduction in performance.

The quasi-analytical method, implemented as an interactive computer program, allows very fast computation of the success rate of the basic general recognition algorithm. As may be seen from table 6, it allows specification of the same sky, catalog, STR, attitude error, and control parameters as the Monte Carlo simulations. Furthermore, it permits the control parameters to be specified either directly or in terms of a probability associated with it, as is shown in table 6.   If the second option is selected, which usually makes more sense, the program computes the value of the control parameter.

The advantage of the quasi-analytical method is illustrated in table 8, where it is compared with the Monte Carlo simulations for three sky areas. Whereas the Monte Carlo simulations (5000 evaluations), require a compute time ranging from 250 to 510 minutes, the quasi-analytical method, though tending to be slightly conservative, yields practically the same results in a mere 0.6 to 1.8 minutes. The success rate is different for the different sky areas due to the fact that the star catalog is insufficiently deep at the galactic poles, limiting the guide star density to 6.7 stars per STR FOV. In addition, the performance is also effected by the A and B parameters characterizing the sky, which are -3.96 and 0.5 for the sky area with maximum density (case 2), while for the area with minimum density (case 3) the values are -2.37 and 0.30.

Table 8

COMPARISON OF QUASI-ANALYTICAL METHOD WITH MONTE CARLO
SIMULATIONS

Case	Sky Region In Terms of Star Density	Guide Star Density (per FOV)	Predicted Recognition Success Probability		Required Computation Time*	
			Quasi-Anal Method	Monte Carlo Simulations	Quasi-Anal Method (min)	Monte Carlo Simulations (min)
1	Mean	9.1	0.966	0.970	1.1	450
2	Maximum	9.1	0.978	0.978	1.8	510
3	Minimum	6.7	0.929	0.934	0.6	250

* For a computer with a MC68000 microprocessor operating at 12 Mhz and 5000 evaluations.

Fig. 8 compares the recognition success probability predictions as obtained with the quasi-analytical method with those generated by the Monte Carlo simulations for three sky areas. The success rate is shown as a function of the limit to the number of observed stars used for the recognition. The filled in circles refer to the aforementioned simulation cases, while the open circles were generated using 1000 evaluations per case. Again, the quasi-analytical method is shown to be very accurate, and correctly predicts a reduction in performance with an increase of the limit to the number of observed stars used. This is due to the fact that as this limit is increased, the fraction of guide stars among the observed stars is reduced, causing a reduction of the uniqueness probability of the pattern of observed guide stars.

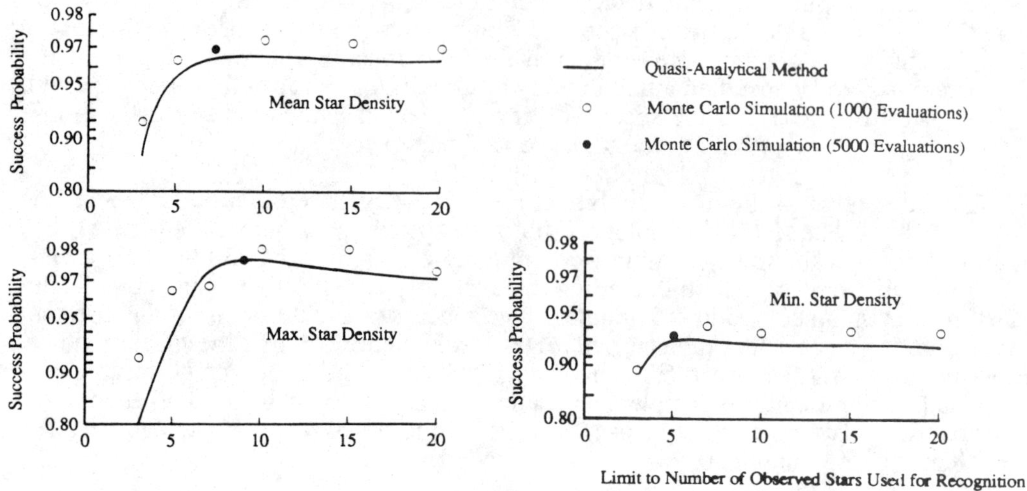

Fig. 8   Star Pattern Recognition Success Probability as Predicted Using the Quasi-Analytical Methjod and as Obtained Through Monte Carlo Simulations.

Finally, it should be mentioned that the recognition success probability for cases like SIRTF can be increased to practically 100% by performing two or three independent recognition attempts in adjacent parts of the sky at the cost of some slew time.

Full-sky Autonomous Star Tracker Case

The FAST that is evaluated using the quasi-analytical method has a circular FOV with an area of 100 square degrees, an accuracy of 10 arcsec (1 σ) relative to a detector attatched coordinate system, a dynamic range of 5 magnitudes, uses a guide star density of 10 stars per FOV, and needs to be sufficiently sensative to observe stars with a magnitude of 7.5. Table 6 shows the parameters used for FAST. Since the FOV is 100 square degrees, it follows that the size of the uniqueness area is 413. Since the guide stars are fairly bright, it is appropriate to assume a magnitude error of 0.3 (1 σ). it turned out to be optimum to limit the number of observed stars used for recognition to the brightest 8. Hence, the FAST needs to adjust its threshold to a value where 8 stars are observed, unless this is prevented by dynamic range problems.

A FAST with these parameters will have a recognition success probability of at least 98.2%. If the dynamic range is increased to 10 magnitudes, the success rate goes up to 99.3%. Hence, of the 1.8% failure rate, 1.1% can be attributed to dynamic range problems. Since successful recognition requires at least 3 guide stars within the FOV, while a significant fraction of cases with three guide stars will fail, it is important to note that the probability of having less than 3 guide stars is 0.28%, while the probability of

48

having 3 guide stars is 0.76%. This is true because the quasi-analytical method assumes a Poisson distribution for the number of guide stars within the FOV. Hence, the percentage of failures due to a lack of guide stars is expected to be in the range from 0.28 to 1.04%. Therefore, virtually all failures can be attributed either to dynamic range problems or to the absence of a sufficient number of guide stars within the FOV.

It is expected that the dynamic range can effectively be increased either by fast readout of the detector and integration in solid state memory or by not using the brightest star(s) in the field. If guide stars are properly selected, it is likely that the distribution of the number of guide stars within the FOV is much more favorable than the Poisson distribution assumed. Hence, by implementing provisions that effectively increase the dynamic range, and by careful selection of guide stars, the failure rate for FAST can be reduced to a small fraction of one percent.   A reduction of the accuracy of the tracker from 10 to 40 arcsec would increase the failure rate by 0.35%.

Because ASTROS II does not seem to have a dynamic range problem, has an equivalent FOV, and is expected to be more accurate (8 arcsec vs 10), a Full-sky Autonomous Star Tracker based on ASTROS II is expected to exhibit a success rate very close to 100%.

COMPUTATIONAL REQUIREMENTS FOR FAST

The memory required for FAST will be computed assuming use of ASTROS II. The memory space is required for the recognition algorithm, which consists of approximately 500 lines of Fortran code occupying less than 32K bytes, and array space required for the matrices associated with the algorithm. In order to compute the amount of array space required, it is necessary to know the number of guide stars, the upper limit of the match confirmation value, and the number of guide star pairs. In order to limit the number of guide star pairs, it is best to only use that part of the ASTROS II FOV that lies within the circle inscribed within the square FOV. This implies that the useful part of the FOV measures some 100 square degrees. It also means that for a guide star density of 10 stars per FOV, 4125 guide stars will be required.

Each guide star has a number of guide star neighbors within a distance equal to the diameter of the STR FOV that averages 4 times the guide star density. Hence, the total number of guide star pairs is approximately equal to the product of 4 times the guide star density and the number of guidestars divided by two.   This implies that the number of guide star pairs with angular distance less than the FOV diameter is approximately 82,500. Using 2 byte integers, it takes a total of 6 bytes per pair to specify the angular distance and to identify the guide stars associated with each pair, resulting in a total of 495,000 bytes.

Since the integrated probability versus the angular distance of the guide star pairs and the observed star pairs are quadratic functions, it follows that on the average, the number of guide star pairs matching a pair of observed star pairs is equal to 4/3 times twice the distance tolerance times the total number of guide star pairs. Hence, assuming a distance tolerance equal to 8/10 times 0.00091, it follows that on the average, each observed star pair matches the angular distance of 160 guide star pairs.

For the magnitude tolerance used (0.77) and the slope of the log of the integrated star density (0.45), the probability of a guide star matching the brightness of an observed star can be found to be less than 0.55. Hence, the probability for each of the two orientations to match is equal to 0.3. This implies that the number of star matches resulting from a guide star pair matching the angular distance of an observed star pair is equal to 2 x 0.3 x 2= 1.2.

Since the number of observed stars used for the recognition process is limited to 8, the maximum number of observed star pairs is equal to 28. Therefore, the upper limit for the number of star matches is estimated at 1.2 x 28 x 160 = 5376. From simulations it was found that spurious matches can increase the confirmation value from 7 to 11. This implies that the matrix $MAT(ig,j)$ and its copy measure 5376 x 11 x 2 = 118,272 bytes each. Since these matrices are very sparse, it is possible to replace each by a number of smaller matrices reducing the memory space by a factor of 2.8.

This brings the array space for the guide star pairs and the MAT matrices to 579,480 bytes. Since these matrices represent roughly 70% of the space needed, it follows that the total amount of array space is approximately 810K bytes, bringing the total required to 840K bytes.

Assuming fixed point arithmatic and the use of a MC68000 class microprocessor operating at 12 Mhz with no wait states, an analysis showed that generation of the base match matrices would take approximately 0.55 s. The additional time required for performing the recognition is estimated at 0.1 s, and will certainly be less than 4.1 s. In summary, the memory space required for a FAST based on ASTROS II is estimated at 840K bytes, while the compute time is expected to be 1 s, assuming the ASTROS II microprocessor to be as fast as the MC68000 at 12 Mhz. An increase of the FOV would lead to a roughly equal decrease of the amount of memory required.

CONCLUSIONS

Using the general star pattern recognition algorithm described in this paper, JPL's redundant, microcomputer equipped ASTROS II CCD star tracker could be developed into a Full-sky Autonomous Star Tracker (FAST), capable of determining its attitude about all three axes without requiring any a-priori attitude knowledge. This would allow ASTROS II tobe integrated as a key sensor in future autonomous attitude control and navigation systems.

Having an 11.5 by 11.5 field of view and an accuracy of 8 arcsec (1 sigma, goal), an ASTROS II based FAST requires the use of approximately 4,100 guide stars, necessitating a memory of 840K bytes. With space qualified 256 K bit ROM and RAM chips becoming avalable in the near future, this amount of memory can be provided by less than 50 memory chips (including EDAC). The SA 3300 microprocessor of ASTROS II is expected to be capable of performing the attitude determination in approximately one second.

Although a quasi-analytical performance evaluation method predicts a success rate of 98.2% for the ASTROS II based FAST, it is expected that dynamic range limitations, which were assumed at 5 magnitudes, can be solved raising the success rate to 99.3%. Most of the remaining 0.7% failure rate is caused by the fact that the quasi-analytical method assumes a Poisson distribution for the number of guide stars within the tracker FOV, which is highly pessimistic. Hence, it is expected that an ASTROS II based FAST will have a success rate that differs from 100% by only a small fraction of 1 percent.

Using both Monte Carlo simulations and a quasi-analytical method, it was shown that the general recognition algorithm and a less software intensive special algorithm, can be used to reliably automate the acquisition of celestial targets by astronomy telescopes (e.g., SIRTF for which the method was tested), a process currently often performed by a human operator.

In addition to enabling the construction of a FAST and automating the acquisition of celestial targets by astronomy telescopes, the recognition algorithm can also be used for autonomous updating of gyro based attitude control systems (as is planned for Mariner Mark II), and for automating ground based attitude reconstruction.

JPL is currently applying neural network technology to the problem of autonomous star pattern recognition with some encouraging early results.

ACKNOWLEDGEMENT

The author greatfully acknowledges the NASA Ames Research Center and the Jet Propulsion Laboratory, California Institute of Technology for their partial support of the research described in this paper under NASA grants NCC-2-29 and NCA2-IR745-407. He also would like to thank Professor J.D. Powell of Stanford University, Dr. K.R. Lorell of Lockheed Palo Alto Research Laboratory, and Dr. E.W. Dennison of the Jet Propulsion Laboratory for their help and advice on the star pattern recognition research.

REFERENCES

1.   E.W. Dennison, R.H. Stanton, and K. Shimada, "The Development of a Charge-Coupled Device Tracker for Spacecraft," Paper AAS 87-007, 10th Annual AAS Guidance and Control Conference, Keystone, Colorado, Feb 1987.

2.   R.F. Draper, "The Mariner Mark II Program," Paper AIAA 88-0067, AIAA 26th Aerospace Sciences Meeting, Reno, Nevada, January 1988.

3.   P. Eisenhardt and G.G. Fazio, "The SIRTIF Concept and its Realization," NASA Ames Research Center, Moffett Field, CA, November 17, 1988.

4.   B.M. Lasker, H. Jenker, J.L. Russell, "The Guide Star Catalog," Proceedings of IAU Colloquium No. 133, Paris, France, June 1987.

5.   R.W.H. van Bezooijen, "Success Potential of Automated Star Pattern Recognition," Paper AIAA-86-0254, AIAA 24th Aerospace Sciences Meeting, Reno, Nevada, Jan 6-9, 1986.

6.   W.J. Christis, "The Optical Sensors of the Netherlands Astronomical Satellite (ANS), the Star Sensor," Philips Tech. Rev., Vol 34, pp 218-224, 1974.

7.   W.I. McLaughlin and W.H. de Leeuw, "Infrared Astronomical Satellite," Spaceflight, Vol 20, pp 187-191, May 1978.

8.   R.W.H. van Bezooijen, "IRAS Attitude Control Subsystem," ESA SP-128, pp 187-191, Nov. 1977.

9.   R. Rhoads Stephenson, "The Galileo Attitude and Articulation Control System: A Radiation-Hard, High-Precision, State-of-the-Art Control System," Tenth IFAC Symposium on Automated Control in Space, Toulouse, France, June 1985.

10.   E.C. Wong and W.G. Breckenridge, "Inertial Attitude Determination for a Dual-spin Planetary Spacecraft," AIAA J. of Guidance, Control, and Dynamics, pp 491-498, 1983.

11.   J.T. Junkins and T.E. Strikwerda, "Autonomous Star Sensing and Attitude Estimation," Paper AAS 79-013, Annual AAS Guidance and Control Conference, Keystone, Colorado, 1979.

12.   R.C. Bolles and R.A. Cain, "Recognizing and Locating Partially Visible Objects: The Local -feature-focus-Method," Technical Note 262, SRI International, Menlo Park, California, 1982.

13.   J.N. Bahcall and R.M. Soneira, "The Universe at Faint Magnitudes, I. Models for the Galaxy and the Predicted Star Counts," The Astrophysical Journal Supplement Series, Vol 44, pp 73-110, 1980.

14.   C.W. Allen, Astronomical Quantities, 3rd ed., The Atholone Press, University of London, London, 1976.

ADVANCES IN SPACECRAFT AUTONOMY USING ARTIFICIAL INTELLIGENCE TECHNIQUES

Lorraine M. Fesq and Amy Stephan[*]

Increasing requirements for autonomy in space missions have promoted much work in the applications of Artificial Intelligence (AI) to spacecraft. Initial studies of expert systems using the traditional rule-based approach rapidly produced systems, but they were found to be shallow in scope and unwieldly when expanded to handle broader-based problems. To solve this, more recent studies have turned to model-based systems as they utilize a much deeper knowledge base. By applying predetermined causal relationships between the hardware components of a satellite system, these systems have demonstrated more power and flexibility than rule-based systems. A model-based AI system for doing Satellite Attitude Control System (ACS) fault-management is examined in detail. Most recently, innovative research in satellite AI systems have used more sophisticated techniques, including deriving causal relationships and autonomous control system synthesis. These techniques have demonstrated great potential benefits of autonomous satellite operations using AI. This paper compares and contrasts Rule-based versus Model-based reasoning systems, gives an expose of TRW's ACS fault management system, and gives insight into the role of AI in future spacecraft.

INTRODUCTION

Increasing demands for enhanced spacecraft autonomy issue a challenge to develop more advanced, intelligent systems. Through innovations in on-board computing hardware and software, the ground-dependent satellites of twenty-five years ago have evolved into spacecraft that can control event sequences, change operation modes and identify and correct simple faults. Despite their increased robustness and processing power, today's spacecraft are unprepared for tomorrow's missions (Fig. 1). Increased demands for spacecraft survivability, autonomy and complexity will require satellites to process more information, more quickly and with less ground support. The autonomous satellite must be able to survive failures without ground contact, while more complex spacecraft will produce large amounts of data difficult to process even with ground support. To meet these demands, significant strides must be made in both hardware and software. Spacecraft that can continue their missions without ground support will require software able to respond quickly to unexpected data. New techniques in software development, including artificial intelligence, will be called upon to provide some of these capabilities.

---

[*]   TRW Space and Technology Group, One Space Park, Redondo Beach, California 90278.

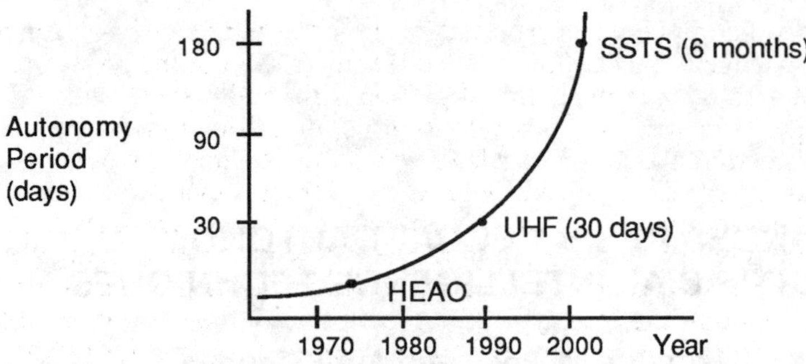

Figure 1. Projected Spacecraft Autonomy Requirements

Artificial intelligence (AI) is a widely misunderstood term. It does not offer any magical solution, nor does it necessarily provide any more "intelligence" than conventional programming. What artificial intelligence offers is a different approach to software problems (Fig. 2). Conventional programs organize knowledge procedurally. Based on values of certain variables, a procedural program will grind down a known path to a pre-determined solution. AI programs can manipulate data independent of its type or value; it does so by manipulating data symbolically. Artificial intelligence systems contain information about the relationships between these symbols, ranging from how a component is connected to another element in a circuit to the fact that an illness causes a certain symptom. AI programs use their knowledge about relationships between data items to generate and examine possible solutions to a problem. Often AI programs will be able to produce more than one solution to a given problem. In general, they are more flexible and better able to handle unanticipated situations than their conventional counterparts.

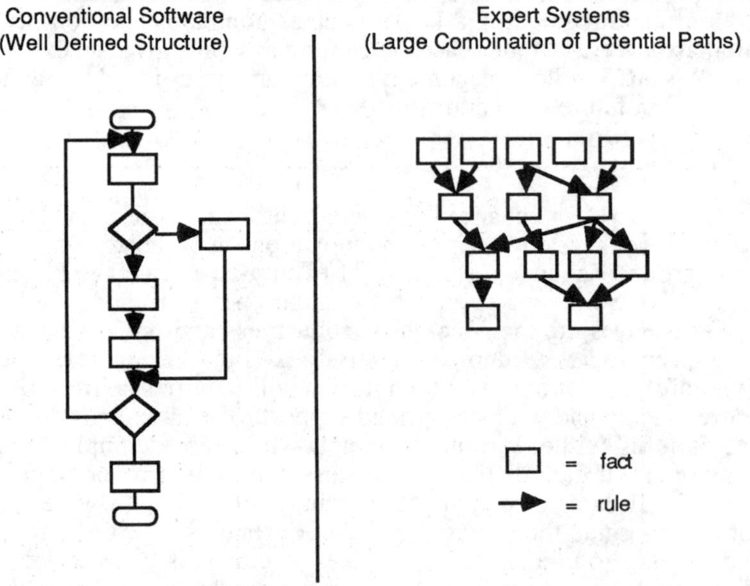

Figure 2.   Conventional vs AI Software Techniques

The main distinction between artificial intelligence and conventional programming lies in the way knowledge is encoded, not in the language used or the type of problem. To enhance the ease of symbol manipulation, AI programs normally are written in LISP, a list-processing language. However, it is possible to write an artificial intelligence program in FORTRAN or a procedural program in LISP. It also is possible to solve most problems using either programming technique, although some problems that would be quite difficult using conventional programming methods are solved easily using artificial intelligence and vice versa. Artificial intelligence programming, then, does not replace but can complement conventional software. For example, procedural programs are better suited for the attitude control and determination functions typically performed by today's on-board computers. Artificial intelligence programs can better handle tasks such as monitoring the health of a system, diagnosing faults and constructing solutions. Both capabilities are needed for long-range, complex missions now being planned. This paper describes some artificial intelligence techniques now being developed and how they can be applied to enhance autonomous fault management capabilities of future spacecraft.

AI TECHNIQUES APPLIED TO SPACECRAFT FAULT MANAGEMENT

Artificial intelligence embodies many techniques and terms, some quite loosely defined. AI programs range from simple operator's advisors acting on a limited body of knowledge to sophisticated adaptive or learning systems, able to derive causal relationships about a system and dynamically alter their internal models of that system. Research is underway to develop artificial intelligence applications for performing fault management functions of fault detection, isolation, and correction for diagnosing spacecraft failures. Each of the systems under development displays strengths and weaknesses, and many will have a place in the future of spacecraft control. AI techniques to perform fault management tasks initially will be inserted in ground stations to serve as operator's assistants. These systems eventually will migrate on board the spacecraft to perform fault management tasks autonomously. The remainder of this paper examines several AI techniques, their capabilities and potential applications, summarized in Table 1.

Table 1.
OVERVIEW OF AI FAULT MANAGEMENT TECHNIQUES

AI TECHNIQUE	DESCRIPTION	APPLICATION
Rule-based Reasoning	• Based on experiential knowledge • Encoded in if-then rules	• Operator's assistants • Training tool for new operators
Set Covering	• Based on surface knowledge • Uses inverted fault-symptom table	• Operator's assistants • Training tool for new operators
Fault Modeling	• Contains model of system • Models failure modes of system components	• Fault Detection, Diagnosis • Autonomous systems with well-defined component failures
Constraint Suspension	• Models behavior of system components based on constraints. • Removes constraints to isolate failure	• Fault Detection, Isolation • Autonomous systems with built-in redundancies and ill-defined component failures

Current Implementations - Operator's Assistants

Operations support teams rely on their experience with a system and their ability to quickly understand large amounts of data when responding to system failures. As spacecraft complexity grows, so does the amount of data the operator must process. Operator's assistants now under development will help ground crew members sort through the telemetry, quickly recognize and isolate faults and recover from failures. These systems train and advise operators, but have no direct control over the spacecraft.

Because these systems must recognize and react immediately to the data present, and because operators generally respond using rules-of-thumb rather than numerical algorithms, it would be difficult to build such a system using conventional methods. However, a number of artificial intelligence techniques, including rule-based reasoning and set covering, are well-suited to this application. Systems built using these techniques are based directly on an expert's knowledge of a problem domain and offer speed and flexibility, but may be unable to handle unanticipated failures and are therefore unsuitable for fully autonomous applications.

Rule-based Reasoning. Perhaps the best known and most widely used application of AI is the rule-based expert system. These systems are built around the knowledge of a so-called expert on the problem domain. Generally the programmer, referred to as a knowledge engineer, asks the expert to describe a set of system problems and his response to these conditions. The programmer then develops a series of if-then rules which link the system's conditions to the expert's response to these conditions. This type of expert system most closely resembles conventional programs, in that its reasoning ability is hard-coded into the rules; it does not make novel inferences. Rule-based expert systems however, differ from conventional code in that they are data-driven, not procedural. As soon as data indicating a particular condition appear, all rules handling that condition attempt to execute. Rather than waiting, as in conventional code, for the program to arrive at a given if-then or case statement and test for the appropriate condition, a rule-based system contains a set of opportunistic rules, ready to execute when the appropriate conditions are met. The heart of a rule-based system is its inference engine, which manages the rules and determines which has priority when two or more are ready to execute simultaneously.

Applications ranging from configuring computer networks to advising doctors in a medical diagnosis employ rule-based expert systems. As a simple example, consider a rule-based expert system which acts as an assistant to an auto mechanic. The knowledge engineer would discuss with the expert, a master mechanic, how to diagnose various auto malfunctions. For example if the car won't start, the mechanic might suggest that either the battery is low or the starter is broken. He may further state that if the starter is broken, the car will make a clicking sound when the ignition key is turned, while a car whose battery is weak may produce certain gasping noises. Weak headlights also indicates a battery failure. The programmer abstracts this information into facts containing information about the state of the car and rules that act upon these facts. For example, this rule would check for a failed starter:

```
IF      [ car-starts                    FALSE]
AND  [ noise-when-key-turned    CLICKING]
THEN
        [ Suggest                 FAILED-STARTER]
```

Another rule could account for a failed battery:

```
IF      [car-starts                          FALSE]
AND  [OR  [ noise-when-key-turned   GASPING]
               [ headlight-strength              WEAK]]
THEN
            [ Suggest                      WEAK-BATTERY]
```

A third rule could ask the operator for further information:

```
IF       [ car-starts                                    FALSE]
AND    [ noise-when-key-turned                  UNKNOWN]
THEN   [ tell-operator    "it may be the starter or the battery"]
            [ask-operator "enter noise when ignition key turned"]
```

This idea of if-then statements is not unfamiliar to conventional programmers, and in fact this simple mechanic's advisor easily could be built using conventional languages. Several points however, place these type of rule-based systems in the realm of artificial intelligence. First, this system responds dynamically to the data in the system, looking for certain conditions rather than following a pre-determined algorithm. Second, the data has gone through a level of abstraction and is processed symbolically, not quantitatively. Last, the system is flexible; new rules and conditions can be added at any time and the system can request more information or suggest possibilities when unsure of the solution.

Rule-based systems run quickly and are straightforward to build, but have many limitations. Because they are based on an expert's knowledge of a system and not the structure or behavior of the system itself, they are said to contain shallow or experiential knowledge about how the system responds to certain conditions rather than deeper knowledge about how the system actually functions. Rule-based systems are limited to those cases explicitly described by the expert. For example, the auto mechanic system does not handle a car that is out of gas. Rule-based programs do not degrade gracefully, nor can they respond to an unanticipated condition based on a knowledge of the system itself. They also are unable to check the validity of the responses they provide. Rule-based systems tend to be very domain specific; they seldom can be generalized for use on a similar problem.

This inability to respond to unanticipated conditions make rule-based systems poor candidates for autonomous control or on-board applications. These systems do, however, have their place as operators assistants for ground-based centers and training tools for mission operations specialists. Rule-based expert systems offer an easy way to organize and retain human knowledge about a fairly straightforward problem domain. The responses of an experienced mission operations specialist to a number of conditions can be encoded in rule-based system. New operators could use this system as a training aid, while experienced analysts could use it to verify their responses to anomalies.

Set-Covering Technique. A second technique for building expert systems is set covering. Unlike simple rule-based systems which encode an expert 's response to various system conditions in a series of if-then rules, set covering requires the designer to develop a table linking all known system failures to the symptoms by which they manifest themselves. The designer then inverts this table, linking recognizable symptoms with possible faults (Fig. 3). A set of rules will consider system anomalies and seek an explanation, based on the fault-symptom links, which can best account for all symptoms present. Set covering may employ information about which faults never cause certain symptoms and which symptoms always result from a given fault, as well as how likely a given fault will cause a particular symptom associated with it. These likelihoods, often encoded using Bayesian probability, allow the system to better find the optimal explanation for a given anomaly, even when no explanation can account for all symptoms present. Expert systems using set covering are suited especially well for diagnostic applications. They are fairly robust and can be adapted to consider multiple failures with minimal work.

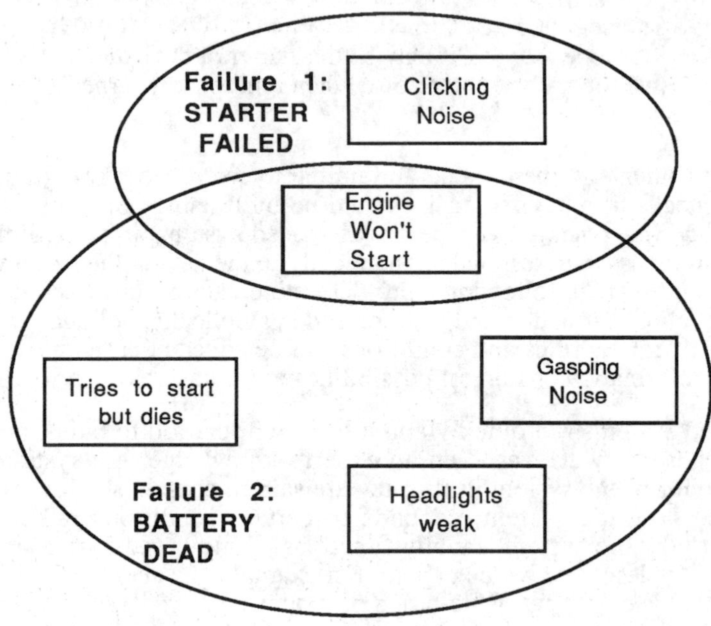

FIGURE 3.   Set Covering Technique

Consider the auto mechanic expert system developed using set covering methods. First the designer would list all the symptoms caused by various mechanical breakdowns:

STARTER FAILED   =>   engine won't turn over
                      clicking sound when key turns
BATTERY DEAD   =>   engine won't turn over
                      car tries to start when ignition turned
                      lights weak or dead

This information could then be inverted into a symptom-failure data base:

Engine won't turn over   =>   starter failed,
                                                         battery dead
Clicking sound when key turned   =>   starter failed
Car tries to start, but dies   =>   battery dead
Lights weak or dead   =>   battery dead

Suppose one enters into this system the fact that the car won't start. The system would produce two possible explanations for this failure, a failed starter or a dead battery, each equally likely. If the system were to decide which failure occurred or which were more plausible, further sophistication would be necessary. Suppose, for example, that given a non-starting car and no other information, a dead battery were known to be three times more likely to cause this failure than a bad starter. Further, one could include the information that a car with a bad starter will never try to start when the ignition is turned and a car with a dead battery always will have weak headlights. This information could be added to our data base as follows:

Engine won't turn over   =>   starter failed [1]
                                                    battery weak[3]
Car tries to start, but dies   =>   NOT [starter failed]
                                                  battery dead [5]
Clicking sound when key turned   =>   starter failed[6]
Lights weak or dead   =>   ESSENTIAL [battery weak]

This additional information allows the system to determine what failures must be ruled out based on the presence or absence of certain key symptoms. The system also can provide an ordered list of likely solutions, even when all symptoms indicating a given failure have not been recognized.

As one can see from this simple example, both the organization of the fault-symptom database and the rules which manipulate it must be fairly sophisticated if the set-covering system is to demonstrate any reasoning ability. Knowledge acquisition for set-covering systems presents additional challenges, as experts may be unable to list all the symptoms a given failure exhibits and may have trouble assigning meaningful probabilities to the fault-system pairs. Although more difficult to develop, expert systems using set covering are more dynamic and powerful than simple rule-based systems. They also run quickly and are able to produce an ordered set of possible explanations. Both set covering and rule-based expert systems, however, lack the ability to self-check their answers, a capacity which more sophisticated systems demonstrate. For this reason, set-covering expert systems, like rule-based systems, are unsuitable for autonomous applications, but quite useful in the role of operator's assistants.

Systems Under Development - On-board Autonomous Control

Expert systems which are to operate autonomously must be able to react to situations unforeseen by the designer, continue to analyze data in the event of a failure and validate their responses. Set covering and rule-based techniques cannot by themselves provide these capabilities, as they only recognize and react to system conditions defined when they are built. Model-based expert systems, however, contain knowledge about the structure and behavior of the system itself. These systems are said to reason from first principles; they contain knowledge about how the system operates, which enables them to predict its behavior under certain conditions and compare the predicted behavior to actual data. From the discrepancies between the predicted and actual system data, a model-based program can reason about where and why the anomaly occurred.

All model-based expert systems contain descriptions of each component's behavior and a description of the system's internal structure - how the components are related. The program uses this information to predict the system's behavior and compares these predictions to the system's actual behavior. When a discrepancy occurs, the expert system will attempt to locate its cause and after doing so, will continue to predict the system's behavior with the failed component. Model-based expert systems primarily use two approaches when deciding what has caused a discrepancy between the actual and the predicted data. The first, the fault modelling method, includes pre-selected models for failed components used both to identify the failure and to model future behavior of the failed component. Systems which use the second approach, the constraint suspension method, model the system's behavior as a set of constraints and identify failures by determining which constraints must be removed to make the model consistent with observed behavior.

Fault Modelling Systems Fault modelling techniques are applicable when the ways in which each system component can fail are readily enumerated and the effects of these failures on the component's behavior are well understood. Fault modelling systems require that all possible failures for each component be identified and modeled. This is not as difficult a task as it may seem. In general, the fault modes of individual components are easier to extract than the fault modes of a complex mechanism. Because the model has knowledge of how each component affects the system as a whole, a model of a failed component will be propagated throughout the system. A model-based system reduces the large number of ways a system can fail to a manageable number of ways in which each component type can fail. Model-based systems are therefore more likely to be complete than their rule-based cousins which must exhaustively consider all cases of failure for the entire system. Rule-based reasoning, however, can serve as part of a larger model-based system. A fault modelling system identifies faults by modelling various failures until one produces outputs that match actual system data. If-then rules may examine the actual data and suggest possible failures to model or eliminate a set of failures from consideration. These rules, called heuristics, may be supplemented by a directed search of possible failures, thus combining the power of a model-based system to handle unusual or unanticipated conditions with the speed of a rule-based system when considering routine failures.

Model-based systems not only model a system's behavior, they also must compare the predicted values with actual data, generate hypothesized failures, model and test the accuracy of these hypothesized failures, determine which of several promising hypotheses is correct and decide what corrective action should be taken. These functions are best understood through an example. At TRW's Space and Technology Group, a model-based expert system has been developed to diagnose and correct failures in an attitude control system. This system, called Expert Systems on Spacecraft (ESS), analyzes real-time telemetry data provided by a vehicle dynamics simulator and conventional on-board control system software (Fig. 4). The control system program handles the number-crunching and algorithmic processing normally done by an on-board computer. This program transmits processed data to the expert system which performs fault management functions using abstracted, higher level information.

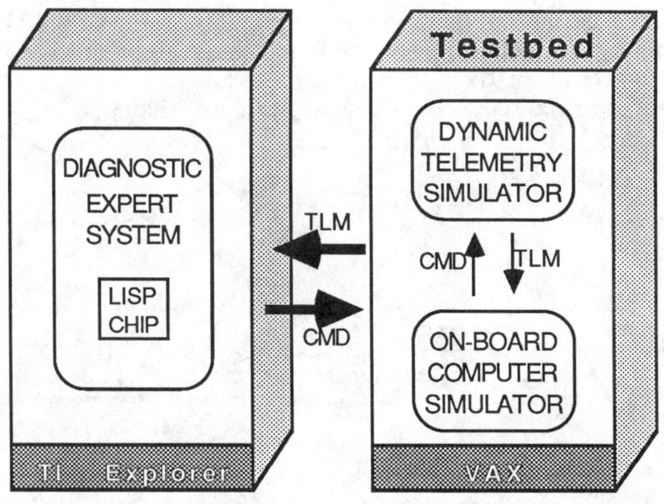

Figure 4. ES - Test Bed Configuration

ESS, shown in Fig. 5, is composed of five functional blocks: the Model, the Comparator, the Heuristic Mechanism, the Problem Solver, and the Correction Mechanism. The Model of the control system components and their interactions produces predicted telemetry. The Comparator module performs the role of fault detection and compares actual telemetry to the predictions once per second. Thresholds of tolerance are set around the predicted telemetry, allowing dynamic limit checking of the data. When the Comparator detects a condition which is out of tolerance, the system enters fault isolation mode. The Problem Solver initiates a directed search for the cause of the anomaly, manipulating the model to determine the effects of possible failures on the system. Solutions that seem promising, i.e. whose outputs correspond to actual telemetry for the time modeled, are saved and compared to new telemetry as it is generated. The Heuristics Mechanism examines trends in data miscompares and suggests likely explanations. These suggestions guide the Problem Solver in determining which hypotheses should be examined. The Problem Solver considers suggested explanations more plausible than those generated by the directed search and continues to examine them even after several mismatches. The Problem Solver can manipulate the Model temporally, and examine the effects of the same component failure occurring at different times. It continues to model possible failures until a hypothesis agrees with the actual telemetry data for a pre-specified length of time. The system accepts this hypothesis as an accurate model of the current state of the spacecraft and passes it to the Correction Mechanism as a diagnosis. The Correction Mechanism determines and executes recovery actions based on the type of failure, the current spacecraft configuration and past history of failures.

This system is able to specify what component failed, how it failed and at what time. It can check diagnoses against its understanding of the system's behavior and is able to provide a new, completely specified model, even after a component failure. It can handle multiple and unexpected faults, making it much more flexible than traditional, rule-based systems. Like rule-based systems, however, the fault modelling system is limited to a pre-determined set of failures. The accuracy of the model used - and all models necessarily contain inaccuracies - also limits the performance of these systems. Data discrepancies apparent in long-term trending caused by a failure slow to manifest itself can be confused with model inconsistencies. Despite their limitations, fault modelling systems are robust and accurate enough to be used for autonomous control applications when the system's components can be modeled easily and their faults are well-known. Systems that are more difficult to model, or for which failure models can not be developed, can be monitored using another technique known as constraint suspension.

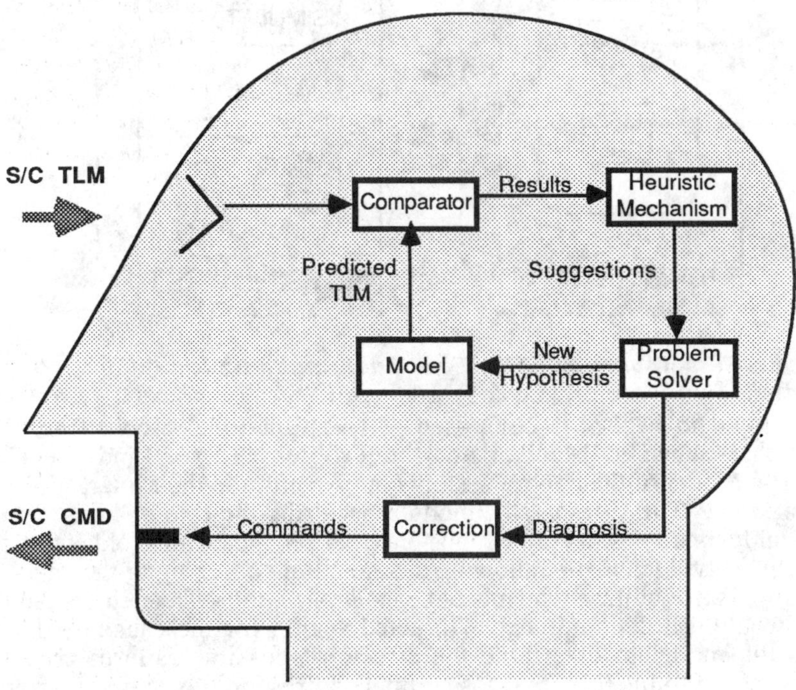

Figure 5.   ESS Fault Modelling Technique

<u>**Constraint Suspension Systems.**</u> Constraint suspension systems model the behavior of each component as a set of constraints. The system models both individual components and larger structures of these components by placing constraints on relationships among the values at the input and output nodes of a device (Fig. 6). The constraints allow the program to both derive a component's output from a given input and determine if a given input or set of inputs could have produced a known output. Constraint suspension systems model the functional rather than the physical relationship between system components. Actual system data - both inputs and outputs - are fed into the model and allowed to propagate through the model in both directions. The program uses this actual data to calculate values at all device nodes, based on the device constraints. If the model's prediction based on the system input differs from the expected value based on the observed output, the program will recognize an inconsistency and attempt to identify the source of the problem. The system can be designed with a hierarchical structure, allowing the program to identify failure candidates at a higher level and only explore the substructure of these candidates. Because data propagate in both directions, a single failure often will cause more than one inconsistency and the component at fault may appear to be accurate. The system must determine what failure or set of failures can account for all inconsistencies in the model.

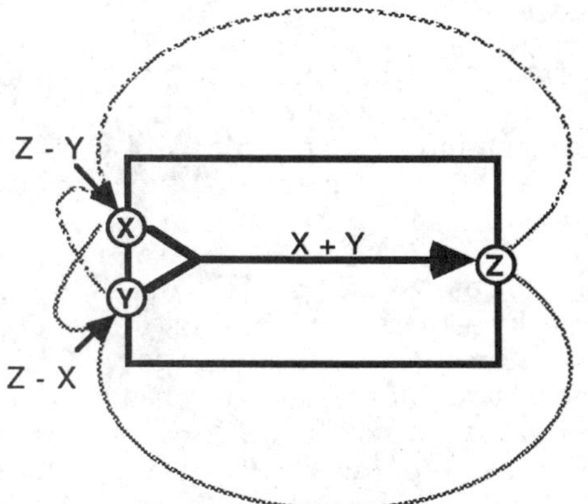

Figure 6. Constraint-based Modelling

Constraint suspension attempts to diagnose failures by determining not where the system has failed, but where the model has failed to match the actual system behavior. Because the program will model normal healthy components correctly, a component which the program fails to model accurately must be anomalous. A failed component no longer obeys the model's constraints, and the program therefore cannot predict its behavior. Constraint suspension involves lifting the constraints on a component or set of components, eliminating all assumptions about their behavior. The program will propagate observed data through this reduced system in both directions. If this does not result in an inconsistency, the suspect component or components are indeed the source of the problem. If an inconsistency occurs, the component being examined cannot alone account for the anomaly. The program then will replace the constraints on the current suspect and using built-in knowledge about the system, determine what component to test next.

Constraint suspension systems reason about whether a component failed without knowing how it failed and without specifically modelling the effect of the failure. This flexibility has its price: these systems can provide only general diagnoses. They cannot explicitly determine how something failed, only that it failed. Constraint suspension systems are useful, however, when the failure modes of each component are indeterminate or when the way in which a component failed is inconsequential. They are appropriate for systems with built-in redundancies where fault diagnosis is not as critical as fault isolation; e.g., spacecraft subsystems.

Fault modelling and Constraint Suspension techniques meet many of the requirements placed on an autonomous fault management system. They contain the underlying knowledge of how a monitored system operates. The accuracy of this knowledge, however, limits the capabilities of these systems. Discrepancies between the model and the actual which occur from model inaccuracies may be incorrectly diagnosed as a failure. Truly autonomous systems of tomorrow must learn from their environment and adapt to new unforeseen situations.

Current Research - Adaptive Autonomous Control

The on-board software systems of tomorrow must be able to not only handle changing environments, but also learn from and adapt to the new environment. Research is underway to explore techniques to achieve these capabilities. Two projects of particular interest could be used to enhance autonomous fault isolation and correction functions, and are described in the following paragraphs.

Fault Isolation by Deriving Causal Relationships. Work by [Doyle,1988] has produced a system which autonomously forms hypotheses explaining the internal mechanisms of devices. The system, Justified Assertion of Causal Knowledge (JACK), is given external observations about a device and a vocabulary of primitive mechanisms. The system contains information about physical and causal principles. Based on a set of constraints drawn from these principles, it forms hypotheses about the internal configuration of the device. The JACK system develops causal models of the internals of a "black box" which are consistent with the externally observable behavior of the device.

The causal modelling system uses mechanisms such as thermal expansion, fluid flow, gravity, springs, and valves, as the primitive causal relationships from which the model of a device is constructed. These mechanisms are used to map causes to effects. The JACK system hypothesizes paths of mechanisms through the black box of the device. These causal paths map the primitive causes of a device or its inputs, to its final effects or outputs.

A specific example of this system is the formation of hypotheses to describe how a tire gauge works. Here is a mechanism that behaves in a way that is somewhat difficult for most people to explain. The internal slide moves in response to air pressure, but only moves out a certain distance and does not extend all the way out of the cylinder. In addition, the slide can easily be pushed back into the cylinder when the gauge is removed from the air pressure.

A spring can explain the motion out of the cylinder, but if it is attached to the slide, the slide would immediately retract when removed from the air pressure. A ratchet could explain the motion out of the cylinder and the fact that the slide remains extended when removed from the tire; however, this can not explain how the slide can be pushed easily back into the cylinder when the gauge is off the tire.

A description of the observable behavior of a tire gauge was given to the JACK system. This description consists of a timeline in which changes in the observable quantities of the device are recorded, such as the changing position of the slide. The JACK system produced, among others, the following two hypotheses. The first explanation consists of a spring which is attached to a piston inside the cylinder rather than to the slide. The piston pushes, but is not attached to the slide. This allows the slide to move out a certain distance and remain out after the gauge is removed from the tire. It also accounts for the ability to push the slide back into the cylinder. This is a how most tire gauges actually work. The second explanation is an illustration of JACK's ability to form novel hypotheses. It involves two channels of unequal length. Air emerges from the shorter channel first and pushes the slide. The second channel diverts air back in the opposite direction. When air emerges from the second channel, an equilibrium is reached, and the slide stops. When the gauge is removed from the tire, the slide can be pushed back in since the force of air in the channels has been removed. Both of these hypotheses fully explain the observable behavior of the tire gauge device. The JACK system could use information such as the sizes of components, costs, weights, etc. to further distinguish between hypotheses.

The significance of this work, as related to autonomous functions of future satellites, is the capability to generate models about the internal workings of a box based solely on observable behavior. Observable behavior is exactly what an on-board system has to work with -- telemetry points containing information about the satellite's behavior or functionality. This capability could be used to enhance fault diagnosis functions by allowing an on-board system to form hypotheses about changes to the system which are not already captured by existing fault models. This is similar to how satellite anomalies are currently diagnosed; satellite experts hypothesize about various alternate models that could explain the anomalous observed behavior.

Fault Correction - Autonomous Control System Redesign. A research project investigating fault correction techniques is being performed by [Nakasuka, Tanabe, 1988] in Japan. This work suggests an innovative concept for spacecraft control system redesign. The concept involves autonomously redesigning the entire control system (as opposed to minor system reconfigurations) in the event of a component failure. The system, called the Autonomous Controller Design System (ACDS), creates a search space of various control system designs based on available resources. ACDS generates and traverses through this space to determine a new control system that will meet the specified control requirements and constraints.

The search space is pruned by using a knowledge base containing expertise in the field of control system design. The expertise is obtained by autonomous self learning. The computer generates control problems (control objects, requirements, and constraints) and through an automatic trial and error design process, it analyzes the relation between the design parameters and their effects. The system abstracts useful information and stores it in a knowledge base. This knowledge base directs the redesign process in the event of failures.

The ACDS learns by experimentation: it capitalizes on the computer's ability to abstract information from large amounts of data. This information includes relationships between operators (e.g., lead compensators) and their effects on the behavior of a control system. Another example of learned information would be the sequence in which requirements should be met to produce the minimum order control system. The sequence selection is based on the order of the compensator and the number of trials to achieve desired performance, both of which are minimized. For example, stability requirements require controller compensator selection, and should be satisfied before requirements such as gain margin which only requires parameter adjustment. The system would choose this sequence after testing several possibilities and determining that it required the least number of trials to achieve a given performance level.

Once the ACDS collects this information, it stores it in a knowledge base. In the event of a failure, the ACDS produces a search space containing nodes which represent intermediate design stages. Operators such as "Add compensator" and "Tune Parameter" are applied to the nodes. These operators are applied in an order which is guided by the information stored in the knowledge base, allowing only useful nodes to be extended. This minimizes the size of the search space and optimizes the design process. The ACDS incrementally designs a completely new controller which satisfies the given requirements and constraints.

This research demonstrates an innovative approach to handling failures autonomously on board a spacecraft. A system such as ACDS could be trained on the ground to learn particular parameter/effect relationships. The more trial runs, or examples, given to the system, the stronger the relationships that will be produced. After the training period, the ACDS could be placed on board to perform fault correction functions. It would use the learned knowledge to handle control system failures by autonomously redesigning the control system based on available resources, requirements, and constraints.

CONCLUSION

We have examined a number of AI techniques and assessed their strengths, weaknesses, and potential fault management applications in spacecraft projects. Rule-based systems based on experiential knowledge and systems using techniques such as set covering are fairly straightforward to develop and handle predetermined failure scenarios extremely well. These systems are appropriate in the role of the operators assistant and can function interactively with a user. The difficulty with building these systems is the knowledge acquisition: the system is only as good as your expert and his ability to verbalize his knowledge.

Model-based systems are much more powerful, and use the underlying causal relationships of a system to determine the health of that system. Model-based systems are better able to meet the demands of an on-board diagnostic system. They respond to situations that were not anticipated, they can be modified to reflect new failure configurations, and they can check the correctness of their conclusions by modelling them and comparing to the actual state of the system. These systems contain many of the capabilities necessary to perform on-board autonomous fault management functions, but have a significant limitation. They are only as good as the model of the system, and are not yet able to overcome their inherent inaccuracies.

Recent research has developed techniques that can enhance and supplement model-based fault diagnosis systems. This includes advances in automated modelling techniques, both in how a system is currently configured as well as determining the best reconfiguration of a failed system. These techniques will produce systems able to dynamically adjust the model to account for inaccuracies as well as a changing environment.

ACKNOWLEDGEMENTS

The research described in this paper was performed at TRW, Space and Technology Group, Redondo Beach, CA, under Independent Research and Development funding. We wish to thank Frank Tai for his much needed insights into control systems, and Richard Doyle for his invaluable contributions both to this paper and to the field of Artificial Intelligence.

BIBLIOGRAPHY

1.*Autonomous Spacecraft Design and Validation Methodology Handbook*, NASA JPL, prepared for AF Space Headquarters, 1984

2. Davis, R., Hamscher, W., "Model-based Reasoning: Troubleshooting," *Proceedings of AAAI-88*, St. Paul, Minnesota, 1988

3. Davis, R., "Diagnosis Based on Structure and Behavior," *AI Journal*, Dec., 1984

4. Doyle, R., "Hypothesizing Device Mechanisms: Opening Up the Black Box," TR-1047, Artificial Intelligence Laboratory, Massachusetts Institute of Technology, 1988

5. Doyle, R., et al. "Enhancing Aerospace Systems Autonomy Through Predictive Monitoring," 27th Aerospace Sciences Meeting, AIAA, Reno, NV, Jan. 1989

6. Dvorak, D., Kuipers B., "Model-Based Monitoring of Dynamic Systems," submitted to 11th IJCAI, Detroit, 1989

7. Nakasuka, S., Tanabe T., "Autonomous Controller Redesign - A New Approach to Spacecraft Control System Reconfiguration," IFAC Workshop in Spacecraft Autonomy, Pasadena, CA., 1988

8. Newmark, J., *Statistics and Probability in Modern Life,* 4th ed., Saunders College Publishing, New York, 1988

DESIGN OF THE TOPEX EARTH POINTING SAFE HOLD MODE

C. J. Dennehy[*], R. J. Williams[†], B. B. Lee[‡] and R. V. Welch[**]

This paper presents a technical description of the design and analysis of the Earth Pointing Safe Hold Mode (EPSHM) for The Ocean Topography Experiment (TOPEX) satellite. The EPSHM serves as the primary hardwired analog backup controller for safe-haven satellite operations in the event of on-orbit anomalies. The EPSHM being designed by Fairchild Space Company for the TOPEX satellite is a modified version of the baseline Multimission Modular Satellite (MMS) EPSHM used on the LANDSAT-4 mission. Modifications to the MMS baseline EPSHM used on LANDSAT-4 were required to satisfy the unique TOPEX operational requirements. One such TOPEX operational requirement is that the EPSHM provide a spacecraft yaw-axis slewing capability to maintain adequate illumination of the solar array. The EPSHM architecture, constituent hardware components, performance requirements and predicted on-orbit performance will be described in this paper.

INTRODUCTION

The TOPEX/POSEIDON remote sensing mission is a joint scientific program sponsored by the National Aeronautics and Space Administration (NASA) and the Centre National d'Etudes Spatiales (CNES)[††]. The TOPEX Project is being managed by the Jet Propulsion Laboratory (JPL) for the NASA Office of Space Science and Applications. The POSEIDON Project is being managed by the Toulouse Space Laboratory of CNES. The Phase C/D hardware implementation program for the TOPEX satellite was initiated at Fairchild Space Company in June 1987. Fairchild serves as the prime TOPEX contractor responsible for the design, development, integration, test and launch of the TOPEX satellite under contract to JPL.

This paper describes the Earth Pointing Safe Hold Mode (EPSHM) being designed and developed for the The TOPEX Attitude Determination and Control Subsystem (ADCS).

---

[*]   Staff Engineer, ACS Department, Fairchild Space Company, Sherman Technology Center, 20301 Century Blvd., Germantown, Maryland 20874-1181.

[†]   Principal Engineer, ACS Department.

[‡]   Senior Engineer, ACS Department.

[**]   Senior Director, ACS Department.

[††]   The French National Space Studies Center.

The EPSHM serves as the primary hardwired analog backup controller for safe-haven satellite operations in the event of on-orbit anomalies. A top-level mission overview is provided in the first section followed by a description of the TOPEX satellite. Subsequent sections will describe the EPSHM functional and performance requirements, the EPSHM hardware components, the EPSHM controller design methodology, analysis and predicted on-orbit performance.

TOPEX MISSION SCIENCE OVERVIEW

A renewed emphasis is being placed on developing space-based systems for monitoring the world's oceans for both scientific study and for practical benefits in such areas as weather and climate prediction, coastal storm warning and maritime safety.

The TOPEX satellite is being designed to make highly accurate measurements of the sea surface elevations over entire ocean basins for several years. The joint U.S./French TOPEX/POSEIDON mission will provide the international oceanographic research community with a space-based capability to determine global ocean circulation patterns via high-resolution radar altimeter measurements of sea-surface topography. The primary TOPEX/POSEIDON mission science requirement is to make geocentric measurements of the global ocean sea level with an accuracy of ± 14.0 cm and a precision of ± 2.4 cm along a fixed ground track that is repeated every 10 days for a period of three years, with a possible mission extension to five years.

The satellite altimetry data generated during the TOPEX/POSEIDON mission will contribute to the World Climate Research Program's World Ocean Circulation Experiment (WOCE)[1]. This mission will play a key role by providing oceanographers with the data needed to permit a fundamental breakthrough in their comprehension of the functioning of the global oceanic system and its interaction with the Earth's atmosphere. When the TOPEX mission data is integrated with surface and sub-surface measurement data, researchers will gain an improved knowledge of ocean dynamics, leading to a more complete understanding of the relationship between ocean circulation patterns and the global climate. This improved knowledge of interannual oceanic circulation dynamics will result in improved predictions of weather patterns a season or more in advance.

TOPEX MISSION ORBIT

Mission plans call for TOPEX to be launched using an ARIANE-42P Expendable Launch Vehicle in April of 1992. Expected to have an initial on-orbit weight of approximately 2400 Kg (5280 Lbm), the three-axis stabilized satellite will operate in a 1335.5 Km (721 Nm) circular, non-Sun-Synchronous Earth orbit with an inclination of 64.6-65.6 Degrees. Table 1 defines the TOPEX baseline operational orbit characteristics.

Due to orbital geometry constraints, providing the correct solar array orientation in order to generate adequate power for mission operations is a more complicated matter in the TOPEX mission than it was on previous MMS missions. Since the TOPEX operational orbit is not Sun-Synchronous the Sun inclination angle relative to the spacecraft orbital plane, β', varies over the range of $\pm$ 87 Degrees during the mission as shown in Figure 1. Furthermore, for reasons of reliability and cost-effectiveness the baseline TOPEX design incorporates a single non-canted four-panel solar array wing with a single-axis bi-directional Solar Array Drive Assembly (SADA) similar to those used on other low-Earth orbiting.

Analysis has shown that for β' greater than $\pm 10°$ a yaw axis rotation of the vehicle is required to achieve adequate solar array pointing to the sun. Note, from Figure 1, that β' is greater than $\pm 10°$ for approximately 90% of the TOPEX mission. Thus, a two-axis solar array pointing control capability is achieved by this yaw axis spacecraft maneuvering in tandem with the pitch axis rotation provided by the SADA. Figure 2 depicts the spacecraft yawing requirements for perfect solar array sun pointing.

Table 1: TOPEX Operational Orbit Characteristics

Orbit Parameter	Value
Altitude	1335.5 Km
Inclination	65.1°
Eccentricity	<0.001°
Phasing	Ground track over the Calibration Sites
Nodal Period	112.4 min
Ascending Node Crossings	12.7 per day
Ascending Node Longitude Increments	28.35°
Repeat Cycle (for ground track coverage)	9.91 days (127 revs)
Ground Cross Track Repeatability at Equator	±1.0 Km

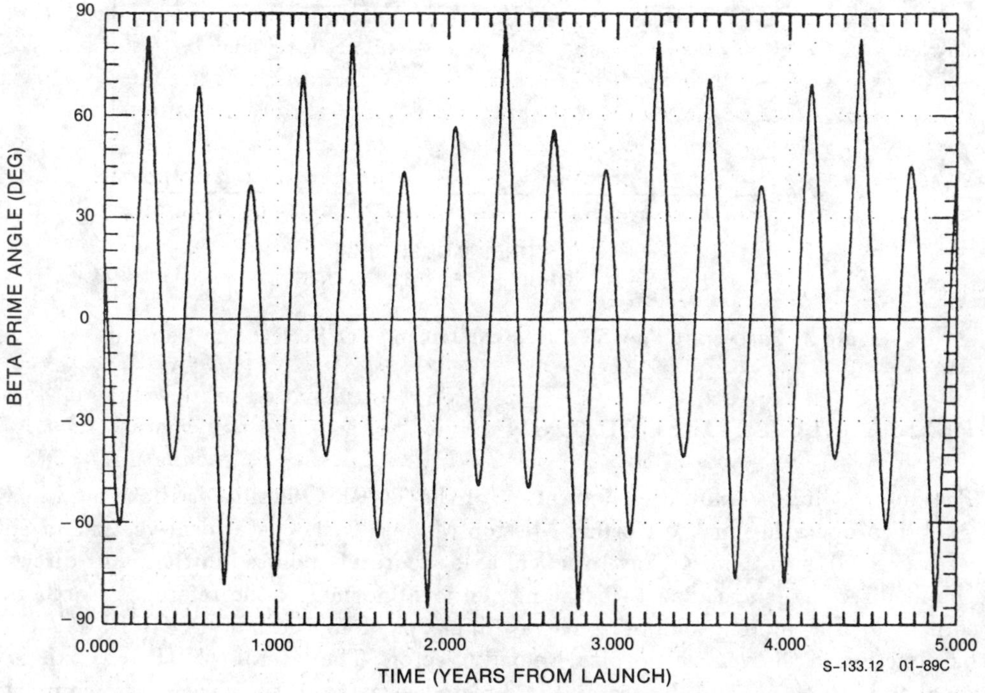

Figure 1. Sun Angle From Orbit Plane (5–Year Mission)

71

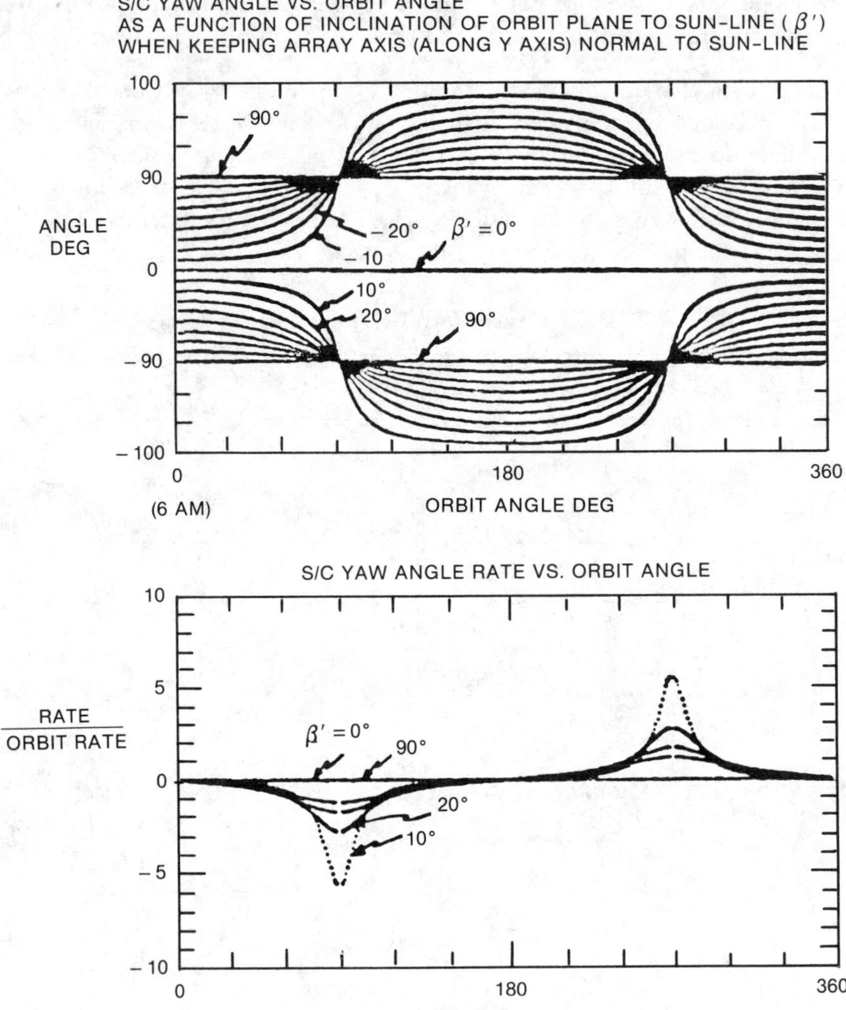

Figure 2. Spacecraft Yaw Slewing Requirements for Perfect Sun Pointing

TOPEX SATELLITE DESCRIPTION

The on-orbit fully-deployed configuration of the TOPEX satellite is illustrated in Figure 3. When operating in the Normal Mission Mode TOPEX is a local-vertical oriented spacecraft (S/C) with the +Z (Yaw) vertical axis accurately pointed in the nadir direction. The nadir direction is specified to be along the local normal of the reference Earth geoid which is defined as an ellipsoid. At zero yaw angle, the longitudinal +X (Roll) axis points in the direction of the vehicle's orbital velocity vector. The lateral +Y (Pitch) axis is orthogonal to both the X and Z axes and is directed away from the solar array, completing the spacecraft's right-handed coordinate frame.

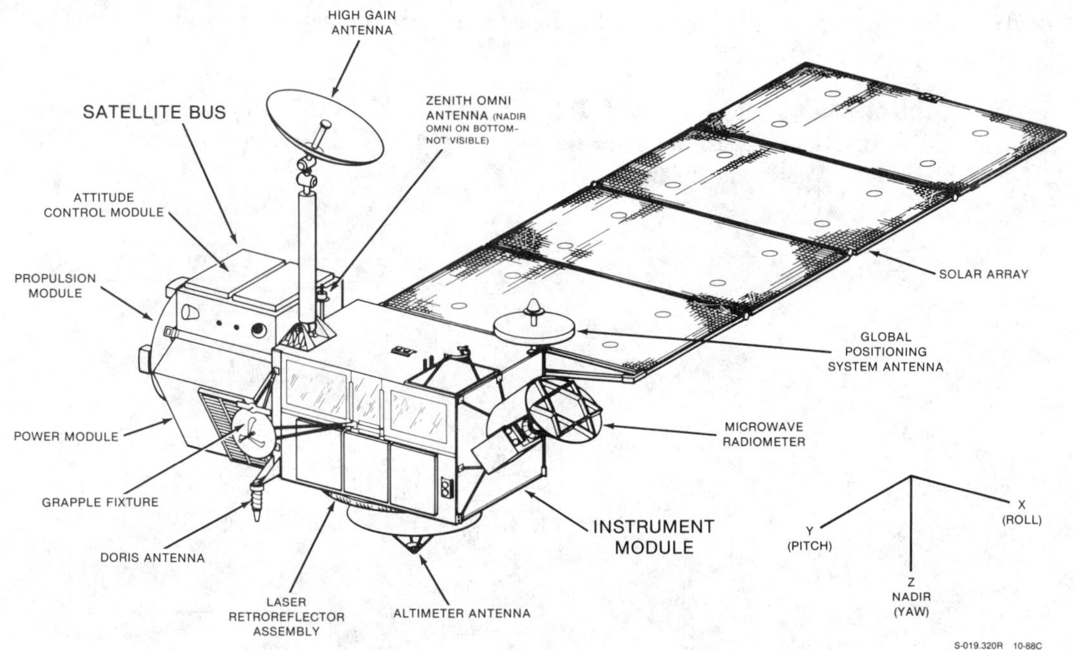

Figure 3. Deployed TOPEX Satellite

As shown in Figure 3, the TOPEX satellite consists of two basic elements:

- the Instrument Module (IM) which houses the six mission sensors and associated electronics and provides the mounting provisions for both the deployable solar array and the deployable High Gain Antenna (HGA) used for communications with the Tracking and Data Relay Satellite (TDRS)

- the TOPEX MMS bus which accomodates the essential subsystems required to support mission payload operations.

Multimission Modular Spacecraft Bus

The MMS bus being designed and developed by Fairchild for the TOPEX mission is derived from similar designs used on previous space missions. This MMS concept was initially conceived as a cost-effective approach to providing common bus functions to unique missions with design versatility built-in[2,3]. To date six spacecraft have incorporated or are incorporating elements of this MMS design: the Solar Maximum Mission (SMM), LANDSAT-4 and LANDSAT-5, Gamma Ray Observatory (GRO), Upper Atmospheric Research Satellite (UARS) and the Explorer Platform (EP)[4]. In addition, an augmented MMS ADCS conceptual design has been analyzed and evaluated for possible use on the Space Infrared Telescope Facility (SIRTF)[5] mission.

The MMS bus design provides four modularized subsystems: the Modular Attitude Control Subsystem (MACS) module, the Modular Power Subsystem (MPS) module, the Command and Data Handling (C&DH) module and the Propulsion Module (PM). All are mounted to a triangular Module Support Structure (MSS) with the MACS, MPS, and

C&DH module components being housed in nearly identical box structures as illustrated in Figure 4.

Figure 5 illustrates the layout of the TOPEX ADCS hardware components within the MACS module. Insofar as possible, to preserve the MMS reliability heritage, the TOPEX ADCS utilizes flight proven hardware.

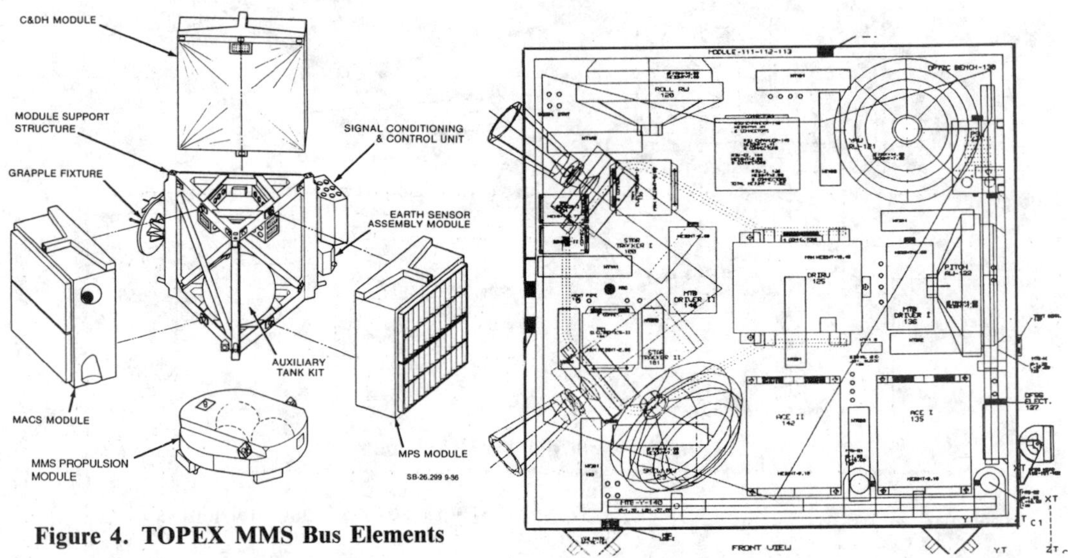

Figure 4. TOPEX MMS Bus Elements

Figure 5. TOPEX MACS Module Component Layout

TOPEX ADCS DESCRIPTION

The TOPEX satellite is an Earth-oriented, three-axis body-stabilized vehicle. The zero-momentum TOPEX ADCS is derived from the flight-proven SMM and LANDSAT-4[6] designs. The TOPEX ADCS contains all the equipment required for the functions of attitude stabilization, attitude determination, attitude acquisition, attitude control and hydrazine thruster firing control during all phases of the mission[7]. The TOPEX ADCS consists of the MACS module supported by the digital On-Board Computer (OBC), the Earth Sensor Assembly Module (ESAM), which is mounted externally on the Earth-viewing face of the satellite and externally mounted Coarse Sun Sensors (CSS).

The functional block diagram of the TOPEX ADCS is given in Figure 6. The TOPEX ADCS is designed so that the controlling sensor and actuator hardware sets can be command selected as appropriate for the particular mission phase. Table 2 summarizes the ADCS modes of operation. The Computer Controlled sub-modes provide the required ADCS functions for all the primary pointing and orbit maintenance operations. The Safe Hold Modes (SHM) serve as hardwired analog backup controllers for safe-haven satellite operations in the event of on-orbit anomalies. In the case of TOPEX, the required Safe Hold Mode consists of two sub-modes: Earth-Pointing SHM and Sun-Pointing SHM.

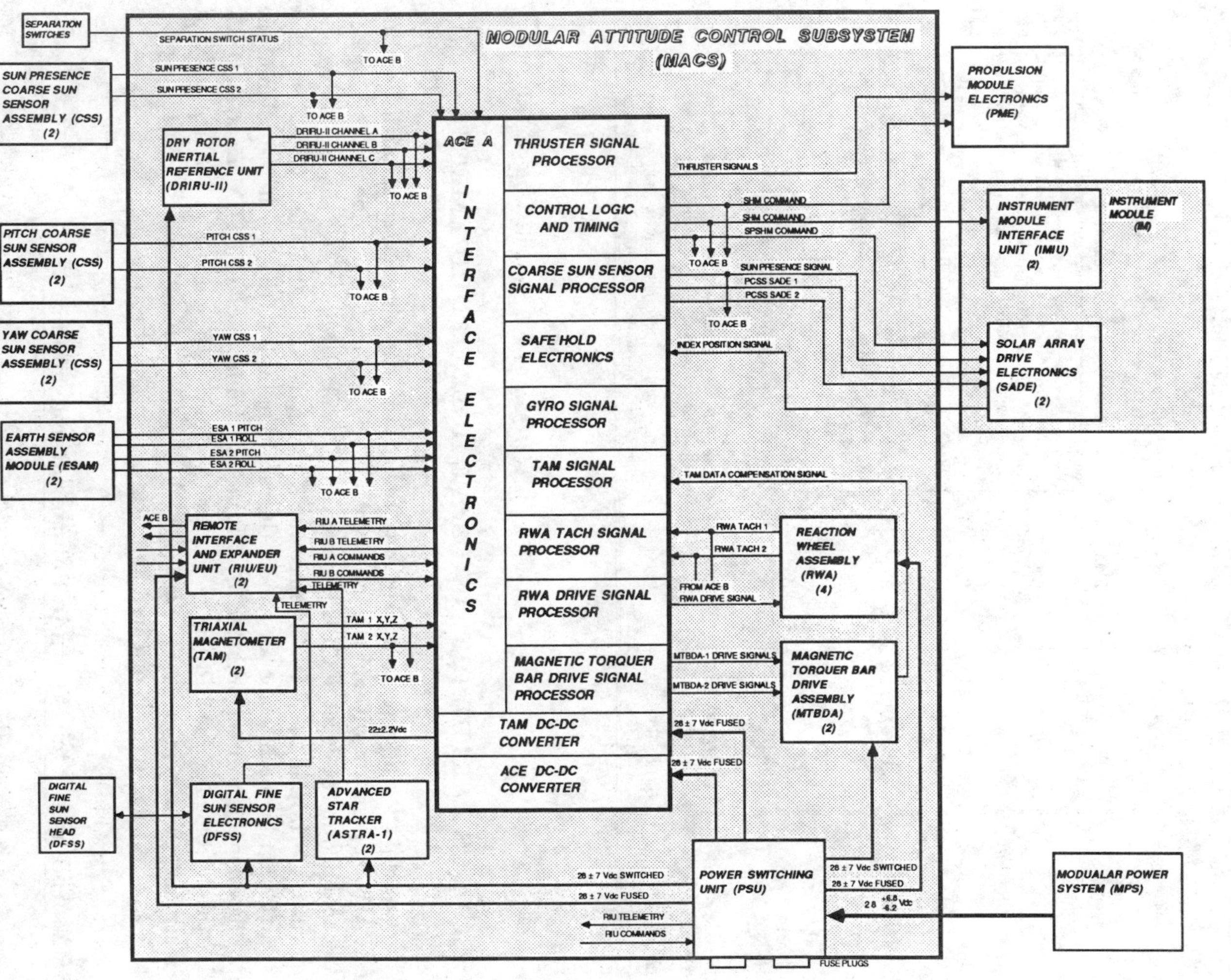

Figure 6. TOPEX ADCS Functional Block Diagram

Table 2. TOPEX Required Operational Modes

Operational Mode	Primary Sensors	Primary Actuators	Description
Computer Controlled Modes			
Earth Acquisition	DRIRU-II,ESA	RWA	Uses ESA pitch and roll outputs to acquire the Earth and DRIRU-II rate outputs to rate damp the roll, pitch and yaw axes.
Stellar Acquisition	DRIRU-II,TAM, DFSS,ASTRA	RWA	Follows Earth or Fine Sun Acquisition. Uses DFSS and TAM data to compute initial attitude and then uses DRIRU-II data to refine the inertial attitude information.
Orbit Adjust	DRIRU-II	Thrusters	Uses DRIRU-II data and thrusters to hold inertial attitude during orbit adjust burns. Low level thrusters used for roll control, high level thrusters modulated on/off for pitch/yaw control.
Calibration	DRIRU-II,ASTRA	RWA	Uses ASTRA data and special maneuvers to calibrate the DRIRU-II, then uses DRIRU-II and special maneuvers to calibrate the ASTRA and/or other special optical sensors including the P/L sensors and instruments.
Normal Mission	DRIRU-II,ASTRA, DFSS	RWA	Uses ASTRA and DRIRU-II data for precision Earth pointing. Requires high accuracy ephemeris data to achieve desired pointing accuracy.
Safe Hold Modes			
Earth Pointing (Gyrocompass)	ESA,DRIRU-II	RWA	Uses pitch and roll data from ESA for position control and the analog roll rate from the DRIRU-II for yaw control. Rate information is obtained from the DRIRU-II.
Earth Pointing (Yaw-Slew)	ESA,DRIRU-II, YCSSA	RWA	Uses pitch and roll data from the MSS mounted ESA for position control and the YCSSA for yaw position control. Rate information is obtained from the DRIRIU-II
Sun Pointing	DRIRU-II,PCSSA, YCSSA	RWA	Uses the PCSSA and YCSSA for pitch and yaw position inputs. Rate information is obtained from the DRIRU-II.

TOPEX ADCS EARTH POINTING SAFE HOLD REQUIREMENTS

The capability for an autonomously initiated, power-safe/thermal-safe Earth Pointing mode of ADCS operation without the use of the OBC is required to provide a viable backup control capability in the event of OBC anomalies[8]. The EPSHM provides this capability and is the primary SHM for the TOPEX mission. This mode maintains the TOPEX satellite in an Earth pointing orientation for a smooth transition back to the Normal Mission Mode (NMM) by allowing a direct transfer of attitude control from the EPSHM back to OBC-based attitude control. The EPSHM is required to control the pitch and roll axes so that the spacecraft yaw axis is oriented towards the Earth while simultaneously controlling the yaw axis so that solar array is oriented approximately normal to the Sunline.

EPSHM Roll and Pitch Attitude Pointing Requirements

During on-orbit EPSHM operations, the spacecraft pitch and roll axes must be controlled such that the spacecraft Z-axis is maintained towards nadir to within $\pm3°$ [9]. No significant modification to the existing LANDSAT-4 EPSHM design was required to satisfy this typical performance requirement.

76

<u>**EPSHM Yaw Attitude Pointing Requirements**</u>

The spacecraft yaw axis must be controlled during EPSHM on-orbit operations such that the solar array normal is, on average during a given orbit, pointed along the Sunline to within $\pm15°$ [9]. A standard orbital gyrocompassing technique (where the roll axis rate signal is used as a measure of yaw position) provides the necessary yaw attitude pointing control capability for those portions of the mission when β' is less than $\pm10°$. When β' is greater than $\pm10°$ a yaw axis attitude control mode is implemented, where the yaw CSSA output signal is used, in place of the roll rate signal, to accomplish the desired yaw attitude slewing maneuver. This is *the most significant change* to the baseline LANDSAT-4 EPSHM design for the TOPEX mission.

TOPEX EPSHM DESCRIPTION

The EPSHM maintains the spacecraft in an Earth Pointing orientation using the scanning infrared MMS mounted ESA's. The EPSHM is preselected by ground commanding the SHM Sensor Select registers in the Attitude Control Electronics (ACE) to process the ESA roll and pitch errors signals for position control. The control loop block diagram using the roll and pitch error signals for position control is shown in Figure 7. Yaw position control is also preselected by ground command and consists of one of the two following sub-modes:

Gyrocompass EPSHM Sub-Mode

The *Gyrocompass EPSHM Sub-Mode* is selected by configuring the ACE SHM Sensor Select registers to use the Dry Rotor Inertial Reference Unit-II (DRIRU-II) analog roll angular rate signal to develop a yaw position error signal when the sun inclination angle is within $\pm10°$ of the orbit plane. In this sub-mode the positive roll axis of the spacecraft is maintained along the spacecraft's velocity vector when the sun is between $0.0°$ and $+10.0°$ and the negative roll axis of the spacecraft is maintained along the velocity vector when the sun is between $0.0°$ and $-10.0°$. This maneuvering is performed to maintain the array axis normal to the Sunline and to prevent any shadowing of the spacecraft onto the array. In order to maintain the proper sense of the roll rate error to yaw position error when the above conditions are being employed, two additional commands are required when selecting this submode:

- FLY FORWARD - This command is used when the sun inclination angle β' is between $0.0°$ and $10.0°$ above the orbit plane. This command will cause a positive roll rate error signal to represent a negative yaw position error signal.

- FLY BACKWARD - This command is used when the sun inclination angle β' is between $0.0°$ and $10.0°$ below the orbit plane. This command will cause a positve roll rate error signal to represent a positive yaw position error signal.

The Gyrocompass Sub-Mode control loop block diagram illustrating the roll rate being used for yaw position control and the switch for Fly Forward and Fly Backward is shown in Figure 8.

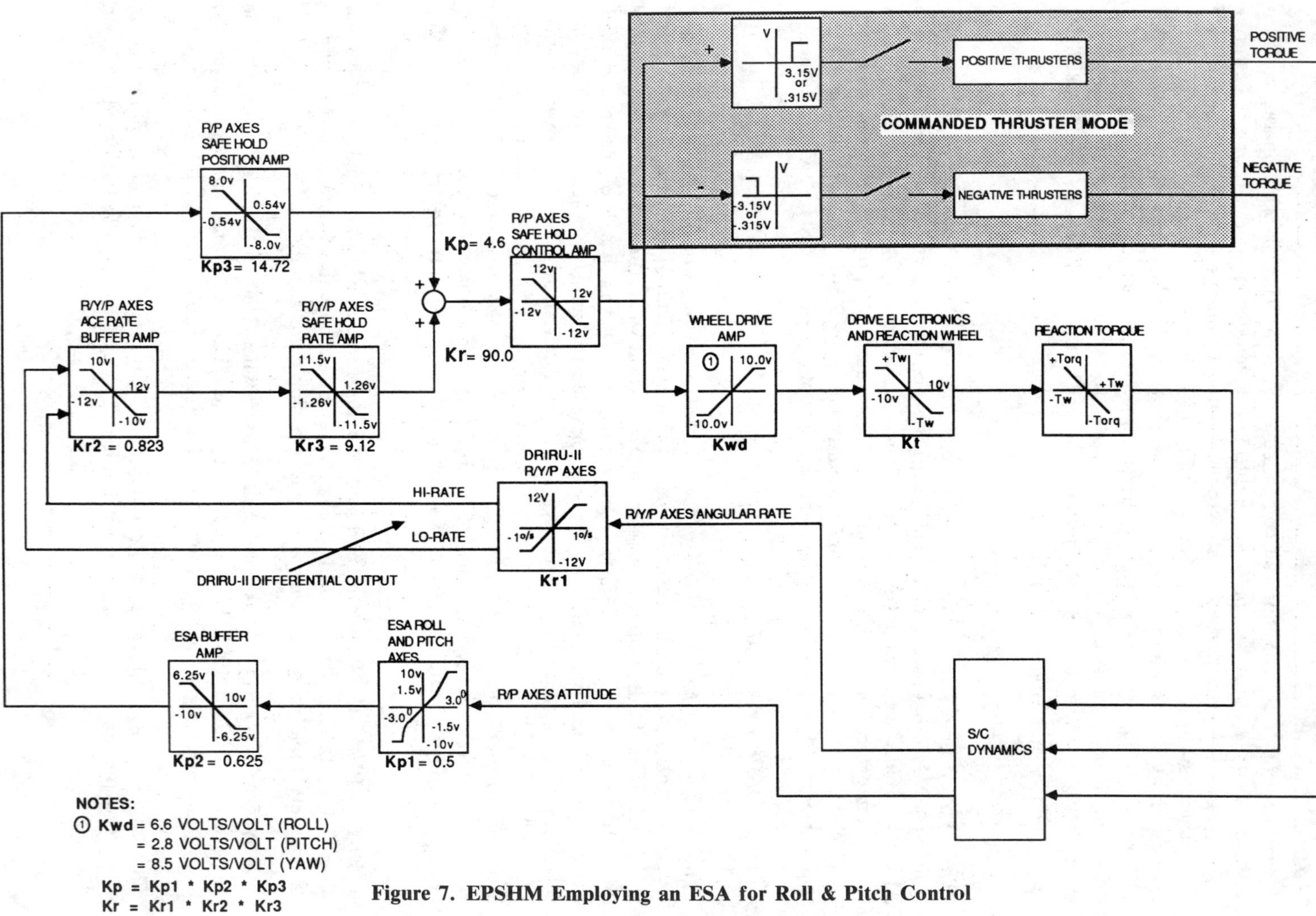

Figure 7. EPSHM Employing an ESA for Roll & Pitch Control

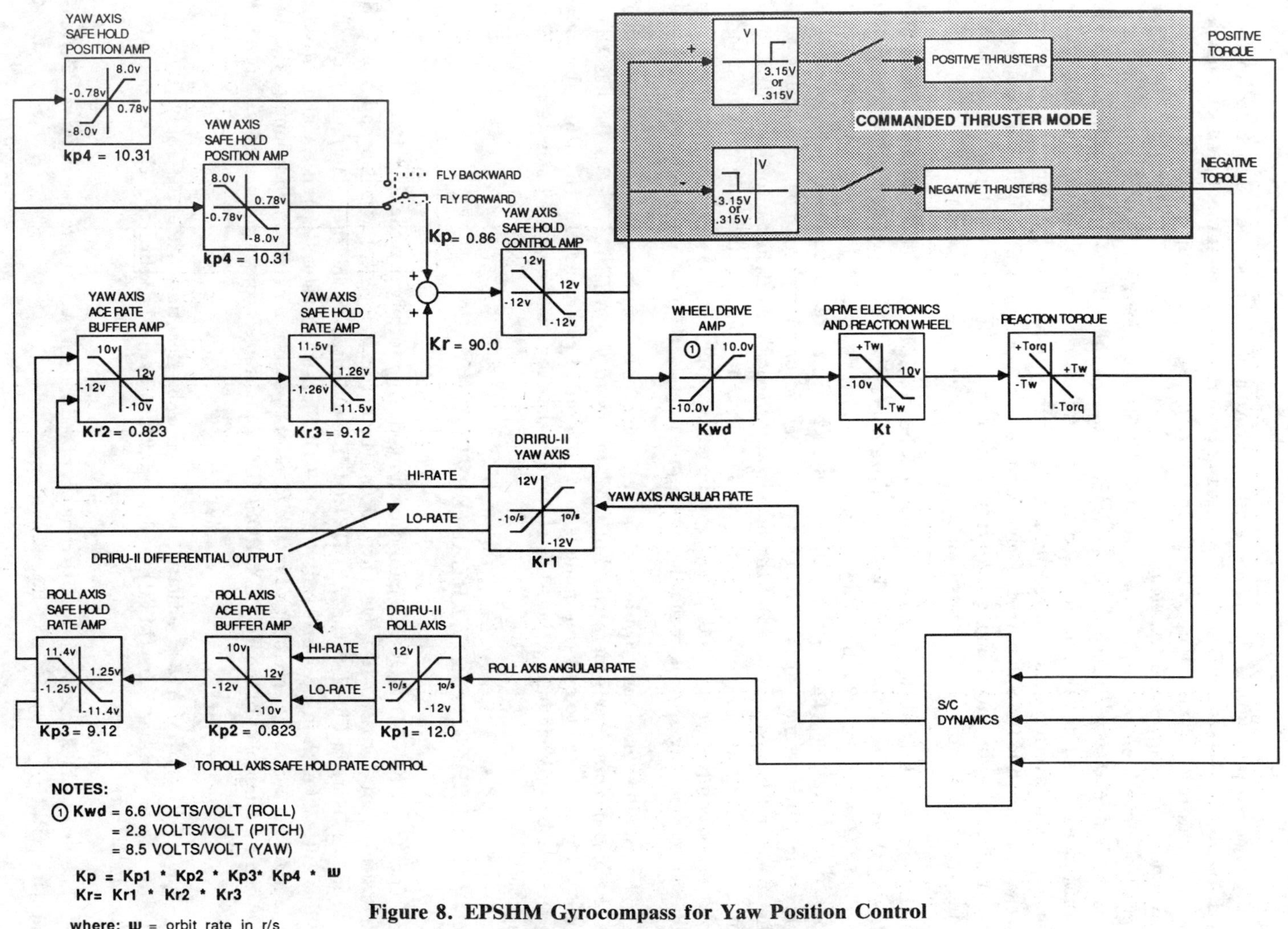

NOTES:
1. **Kwd** = 6.6 VOLTS/VOLT (ROLL)
   = 2.8 VOLTS/VOLT (PITCH)
   = 8.5 VOLTS/VOLT (YAW)

$Kp = Kp1 * Kp2 * Kp3 * Kp4 * \omega$
$Kr = Kr1 * Kr2 * Kr3$

where: ω = orbit rate in r/s

Figure 8. EPSHM Gyrocompass for Yaw Position Control

The *Yaw-Slew EPSHM Sub-Mode* is an outgrowth of the baseline MMS EPSHM used on LANDSAT-4 to satisfy the TOPEX mission requirement of maintaining adequate solar array power during EPSHM operations. In order to meet this requirement, body-mounted yaw-axis Coarse Sun Sensor Assemblies (YCSSA) are used to develop a yaw position error signal for direct control of the yaw-axis. The YCSSA is selected when the sun β' angle is greater than $\pm 10.0°$ with respect to the orbit plane. In this sub-mode the sunline vector is maintained in the roll-yaw vehicle plane by slewing the spacecraft about the yaw axis. This yaw-slew maneuvering acts as a second gimble for maintaining the solar array axis normal to the sunline while the Solar Array Drive Assembly (SADA) provides pitch-axis pointing of the array to the sun, maximizing the array output power. The control loop block diagram using the YCSSA for yaw-axis position control is shown in Figure 9.

In addition, this mode requires that a bias be applied in the control loop to continue slewing the spacecraft about yaw during eclipse operation. As shown in the block diagram of Figure 9, a bias of +3.8 deg/min or -3.8 deg/min is selected by ground command prior to eclipse operation. This bias is "Or'ed" with the Sun Presence sensor signal such that when the Sun Presence signal falls below a threshold of 150 microamps, (indicating that the spacecraft has entered eclipse) the bias is automatically switched into the control loop until the spacecraft has reentered sunlight. This bias has been included to prevent a large solar array pointing error buildup during eclipse. By applying this slew bias the worst case solar array pointing error is reduced from approximately 160° to 20° at the end of eclipse.

The satellite will enter SHM by ground command, OBC command or automatically due to the loss of computer generated "I'm OK" pulses to the ACE[11]. Once SHM has been entered the only way to exit this mode is through ground command. An EPSHM sub-mode is selected by preconfiguring the ACE storage registers with the desired selected position registers.

TOPEX EPSHM HARDWARE OVERVIEW

The TOPEX SHM ADCS sensing functions are performed by the Dry Rotor Inertial Reference Unit (DRIRU-II), the Three-Axis Magnetometer (TAM), ESA's and the CSS assemblies. Note in Figure 5 that the DRIRU-II is mounted on the thermally/mechanically stable optical bench in close proximity to the ADCS optical Master Reference Cube (MRC). The MACS control axes are defined by the faces of this optical MRC. In addition pre-launch mechanical alignment of the TOPEX/POSEIDON mission sensors is performed with respect to the MRC.

The TOPEX SHM ADCS actuation devices are the Reaction Wheel Assembly (RWA's), the Magnetic Torquer Bars (MTB), and the 0.2-Lbf hydrazine thrusters (backup only) of the MMS PM.

The SHM ADCS electronics suite consists of redundant ACE units and a Power Switching Unit (PSU).

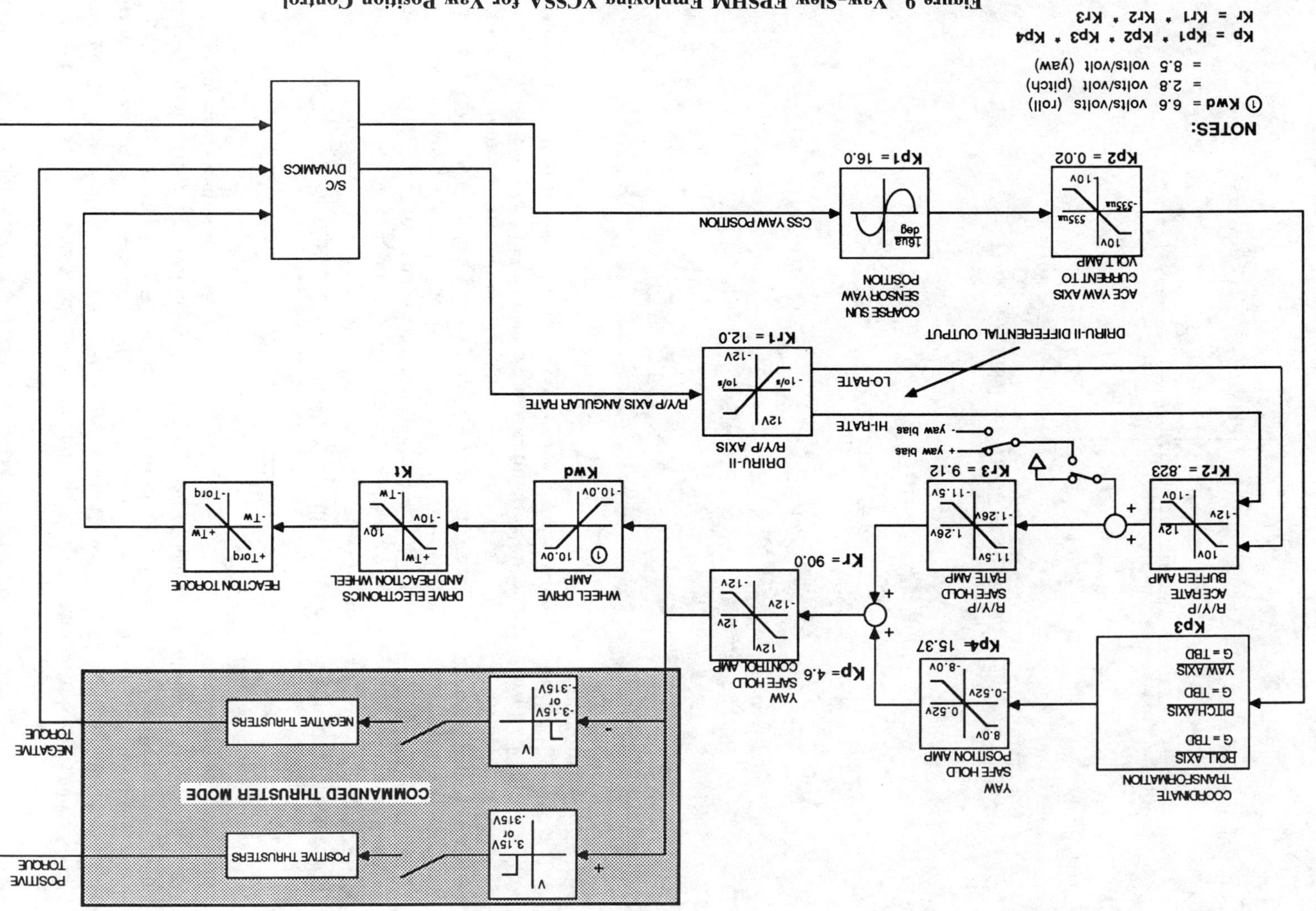

Figure 9. Yaw-Slew EPSHM Employing YCSSA for Yaw Position Control

81

TOPEX ADCS EPSHM HARDWARE DESCRIPTION

All of the hardware, except for the Pitch and Yaw CSS Assemblies, ESA'S (the ESA's are contained in the ESAM) and the 0.2-Lbf hydrazine thrusters used by the TOPEX satellite during the EPSHM are housed in the MACS Module. As shown in the functional block diagram in Figure 6, the MACS Module contains all of the sensors and actuators which are required in EPSHM. The following is a description of the hardware used in EPSHM and their performance requirements.

Pitch and Yaw Coarse Sun Sensor Assemblies (CSSA)

The CSSA detectors used on the TOPEX satellite consist of a single photo-detector, mounting fixture and two terminal leads. The CSS provides an analog output signal which indicates the position of the sun with respect to the detector. The location and mounting arrangement of the Pitch and Yaw CSS's are shown in Figure 10. Each detector has approximately a hemispherical Field-of-View (FOV). The eighteen individual CSS detectors, which together make up the primary and backup Pitch and Yaw CSSA's, are arranged to give 4π steradian coverage about the pitch and yaw axis of the TOPEX satellite. The detector outputs are combined to give a transfer function output as shown in Figure 11. This output is then used in the ACE for the yaw-slewing operation of EPSHM.

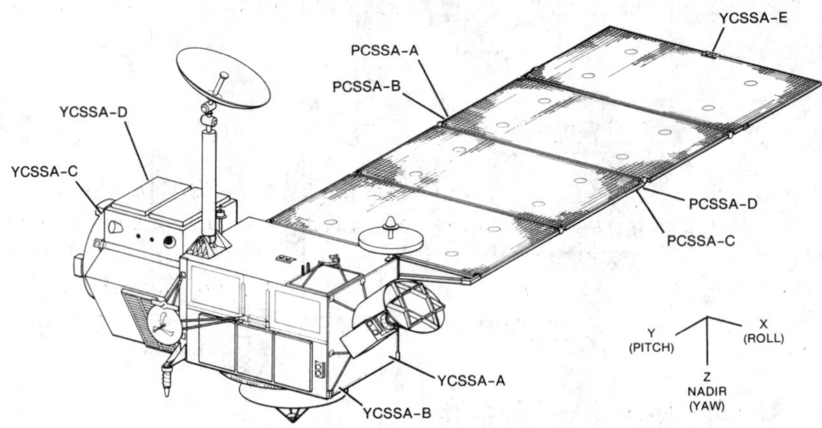

Figure 10. Pitch and Yaw CSSA Locations

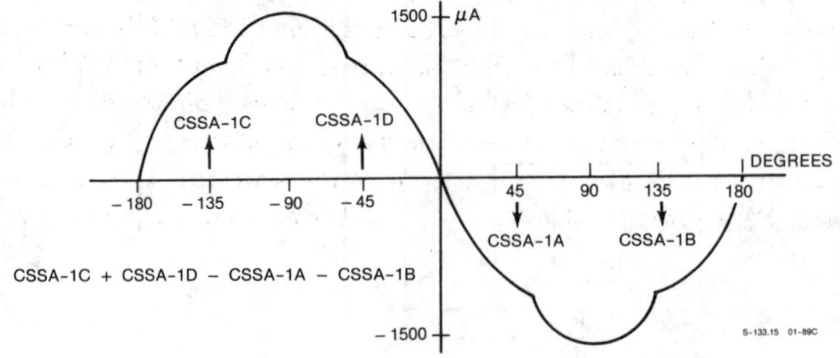

Figure 11. Representative Pitch & Yaw CSSA Output Transfer Function

Earth Sensor Assembly (ESA)

The TOPEX satellite employs redundant ESA's which provide spacecraft roll and pitch attitude error signals to the ACE to be used in EPSHM. The ESA's are contained in the ESAM located on the MMS bus, mounted on the +Z (nadir) face of the MSS between the C&DH and the MPS Modules as shown in Figure 4. One ESA is oriented with its scanner boresight axis in the X-Z plane tilted down 30° from the -X axis. The other ESA is oriented with its scanner boresight in the spacecraft Y-Z plane, tilted down 30° from the +Y axis.

Dry Rotor Inertial Reference Unit-II (DRIRU-II)

The NASA Standard DRIRU-II is a flight proven inertial sensor package[10], which supplies the analog rate data required for attitude control in the EPSHM. The DRIRU-II is mounted on the MACS optical bench which provides a thermally stable reference for the three totally independent gyro channels contained in the DRIRU-II. Each channel supplies two orthogonal axes of information, derived from a dedicated gyro with two degress of freedom. The three gyros are aligned with the MACS Module rectangular coordinate axes and supply differential analog rate information relative to the MACS coordinate reference frame during EPSHM operations. A full-scale range of ±2.0 deg/sec with a linear range of ±1.0 deg/sec is provided by the DRIRU-II for EPSHM rate operation. The DRIRU-II also provides $\delta\theta$ pulses for use in OBC and Inertial Hold SHM which is not described in this paper.

Three Axis Magnetometers(TAM)

The MACS Module contains redundant TAM's whose axes are parallel to within 1.0° to the MACS Module rectangular coordinate reference frame. The TAM outputs are analog signals proportional to the three axis components of the Earth magnetic field vector. This analog output is used in the ACE momentum unloading loop during SHM. Each TAM provides a voltage output for the sensed magnetic field intensity with a $\pm$ 1.0 gauss linear range.

Reaction Wheel Assemblies(RWA's)

The primary actuators for attitude control during EPSHM are the three orthogonal RWA's aligned with the MACS roll, pitch and yaw axes. A fourth RWA aligned equiangular to the other three (at 54.7°) can also be used during EPSHM by selecting it through ground command, this wheel is used as a backup in the event of a failure of one of the primary wheels. Each RWA consists of a motor, inertia wheel, thermistor, redundant analog tachometers and drive electronics enclosed in a sealed container. The purpose of the RWA is to develop reaction torque in response to commands from the ACE and to store angular momentum. Each RWA provides 0.1035 Ft-Lbf of torque and can store ±20 Ft-Lbf-Sec of momentum.

Magnetic Torquer Bar Assembly(MTBA)

The MACS Module contains a MTBA which consists of redundant Magnetic Torquer Bar Driver Assemblies (MTBDA), Roll and Pitch Magnetic Torquer Bars (MTB) each with redundant windings and redundant Yaw MTB's. The four MTB's are employed in an orthogonal axes configuration aligned with the MACS Module reference axes and develop the required magnetic dipole moment of 140,000 pole-cm about each of the spacecraft's three orthogonal axes. These magnetic dipoles interact with the earth's magnetic field to create a torque on the spacecraft. The MTBA accepts magnetic torquer dipole commands from the momentum unload control loop in the ACE and outputs an analog signal to the ACE, proportional to the MTB dipole moment, used to compensate the TAM's for MTB induced fields measured by the TAM's. A significant feature of the MTBA's is that each winding of the Roll and Pitch MTB's can produce the full 140,000 pole-cm dipole, so that even after a failed winding (or driver for that winding) the full dipole can be provided. On the Yaw axis, with loss of a winding, the capability is reduced to 90,000 pole-cm.

Attitude Control Electronics(ACE)

The MACS Module contains redundant ACE's whose primary purpose is to provide the interface between the ADCS sensors and actuators and the OBC. In addition to this primary function, the ACE contains analog Safe Hold Electronics (SHE) for processing analog sensor signals, computing analog error signals and generating wheel drive control signals for automatic spacecraft rate stabilization and attitude control independent of the OBC. Attitude control signals are provided by the ACE in SHM to orient the spacecraft to an Earth pointing orientation (EPSHM) using MMS mounted ESA's or to orient the solar array cell side positive along the sun vector in response to CSS inputs (Sun Pointing SHM), to an inertial reference (Inertial Hold SHM) or rate damped reference (Rate Damped SHM) furnished by the DRIRU-II. This paper will only discuss the EPSHM operation.

During SHM operation the ACE provides the capability for automatic generation of magnetic torquer drive commands for wheel unloading on each axis. A simple analog cross-product control law is implemented in the SHE which generates wheel drive commands in response to the sensed Earth magnetic field and wheel tachometer output.

The ACE contains the electronics to enter SHM from the OBC mode by commands from either RIU (originating at the OBC or ground) or automatically by the ACE Computer Status Monitor (CSM) electronics which monitors the computer "I'm OK" pulses[11].

EPSHM CONTROLLER DESIGN AND ANALYSIS

Controller Design

In designing the TOPEX Safe Hold controllers the guiding philosophy was to use simple electronic hardware. These controllers consist of a position path and a rate path with no integral control and are intentionally designed with a low bandwidth to avoid destabilizing controls/structures interactions. In addition, a single low-pass filter is used in this control system to attenuate the effects of high-frequency structural flexible body vibration modes.

The linear model for the design of the EPSHM control system is given by the block diagram shown in Figure 12. The closed-loop command-following response for this loop is governed by the transfer function:

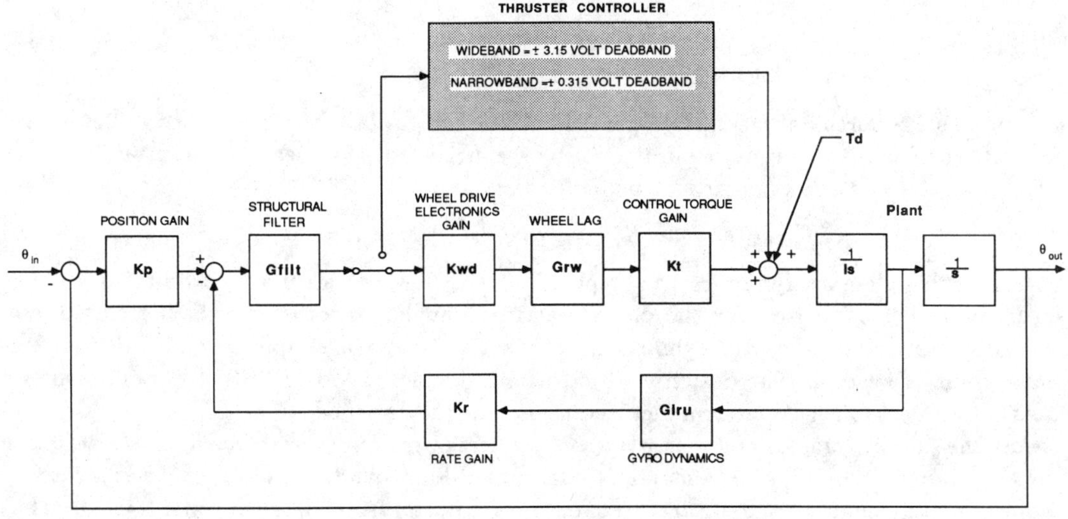

Figure 12. SHM Control Loop Block Diagram

$$\frac{\theta_{S/C}(s)}{\theta_{cmd}(s)} = \frac{\frac{K_p K_{wd} K_t}{I}}{s^2 + \frac{K_r K_{wd} K_t}{I}s + \frac{K_p K_{wd} K_t}{I}} \tag{1}$$

and the closed-loop disturbance response is likewise governed by the transfer function:

$$\frac{\theta_{S/C}(s)}{T_{dist}(s)} = \frac{\frac{1}{I}}{s^2 + \frac{K_r K_{wd} K_t}{I}s + \frac{K_p K_{wd} K_t}{I}} \tag{2}$$

$$
\begin{aligned}
where: \quad I &= \quad Spacecraft\ inertia\ (ft - lb - sec^2) \\
K_{wd} &= \quad Wheel\ drive\ amplifier\ gain\ (Volt/Volt) \\
K_t &= \quad Wheel\ torque\ constant\ (ft - lb/Volt) \\
K_p &= \quad Position\ gain \\
K_r &= \quad rate\ gain
\end{aligned}
$$

The closed-loop command-following transfer function shown in Equation (1) has the standard second-order form:

$$\frac{\omega_n^2}{s^2 + 2\zeta\omega_n s + \omega_n^2} \tag{3}$$

Therefore the following expressions can easily be written for the undamped natural frequency, ω_n, and damping ratio, ς, in terms of the vehicle inertia and the various loop gains:

$$\omega_n^2 = \frac{K_p K_{wd} K_t}{I} \tag{4}$$

$$\varsigma = \frac{K_r K_{wd} K_t}{2\omega_n I} \tag{5}$$

In this Proportional-Derivative (PD) EPSHM control law, the on-axis commanded reaction wheel control torque is simply a linear function of the measured attitude position error and the attitude angular rate error for that axis:

$$T_{CMD}(t) = K_{POSITION}\,\theta(t) + K_{RATE}\,\dot{\theta}(t) \tag{6}$$

Assuming single-axis dynamics this type of control law will result in a damped second-order equation of motion for the on-axis attitude angle. In other words the closed-loop EPSHM control system will exhibit the characteristics of a simple second-order spring-mass-damper system. The design goal is to *tune* the closed-loop EPSHM system response so that the established performance requirements are satisfied. *The design process is to select the parameters ω_n and ς to obtain the desired second-order response characteristics.* In this design, the EPSHM controller -3dB closed-loop bandwidth, ω_c, is assumed to be approximately equal to the natural frequency, ω_n. Recall from classical control theory that this assumption is true only for damping ratios in the 0.5-0.7 range. Typically, the choice of ω_c is driven by the performance requirements, such as response time, and ς is chosen to minimize overshoot. In the case of EPSHM there are no rapid maneuvering and associated settling time requirements. The controller problem is one of maintaining the desired attitude in the face of low frequency environmental disturbances.

ANALYSIS

In general, for an Earth-oriented spacecraft such as the TOPEX satellite being considered in this paper, the vehicle roll and yaw axes are dynamically coupled due to the orbital angular rate, but are independent of any coupling with the vehicle pitch axis dynamics. In the following analysis it will be assumed that this dynamic coupling between the roll and yaw axes is negligible and classical single-axis linear analysis will be used to select the EPSHM control loop gains. Note, this assumption is not true for the yaw-slew operation in EPSHM, where all three axes are dynamically coupled.

The concept of *Orbital Gyrocompassing* is based on the assumption that under perfect control, the roll rate is zero, allowing the output from the roll gyro to function as a yaw attitude sensor. The DRIRU-II measures the spacecraft body rate along the roll axis and for small attitude angle errors, this signal is

$$\omega_x = \dot{\phi} + \Omega\psi \tag{7}$$

$$\begin{aligned}
where: \quad \omega_x \; &= \; Spacecraft\ roll\ rate\\
\phi,\ \psi \; &= \; Spacecraft\ roll\ and\ yaw\ Euler\ angles,\ respectively\\
\Omega \; &= \; Spacecraft\ orbit\ rate
\end{aligned}$$

This signal contains two components: the time rate of change of roll angle, and a term which is proportional to yaw angle. Obviously, when gyrocompassing is used to develop a yaw position signal, the roll/yaw equations are coupled and cannot be separated. Note that this violates one of the basic assumptions used for the linear analysis of the EPSHM gyrocompass sub-mode. Therefore three-axis time domain simulations were used to examine the stability of the EPSHM gyrocompass control sub-mode and to aid in the selection of loop gains.

Figure 12 shows the block diagram used for this analysis. For the second-order Proportional/Derivative (PD) controller, the rate and position gains can be expressed as functions of the desired system closed-loop bandwidth and damping ratio, where:

$$K_p \;=\; Position \;\; gain \;=\; \frac{\omega_c^2 I}{K_{wd} K_t} \tag{8}$$

$$K_r \;=\; Rate \;\; gain \;=\; \frac{2\varsigma\omega_c I}{K_{wd} K_t} \tag{9}$$

$$where: \quad \omega_c \;=\; Desired \;\; bandwidth \;\; (rad/sec)$$
$$I \;=\; Spacecraft \;\; inertia \;\; (slug - ft^2)$$
$$K_{wd} \;=\; Wheel \;\; drive \;\; amplifier \;\; gain \;\; (Volt/Volt)$$
$$K_t \;=\; Wheel \;\; torque \;\; constant \;\; (ft - lbf/Volt)$$
$$\varsigma \;=\; System \;\; damping \;\; ratio$$

The selection of the EPSHM controller system gains was based on the derived requirement to provide a closed-loop system damping ratio of 0.5 and a closed-loop system bandwidth of 0.008 Hz. The wheel drive amplifier gain (K_{wd}) is unique for each axis; the gain is selectable, and must be less than $10\frac{Volts}{Volt}$. The wheel drive amplifier gain stage is in the control loop after the switch for thruster control. Increasing this amplifier gain, increases the thruster position and rate deadbands. In other words, this gain determines the thruster phase-plane switching lines. In order for the thruster deadbands to be identical for all three axes, the ratio of on-axis inertia to wheel drive amplifier gain is selected to be the same for all axes. The value for the wheel drive amplifier gain is chosen to provide satisfactory limit cycle performance. Since the spacecraft inertia is largest in the yaw axis, the wheel drive amplifier gain in that axis was initially selected at $8.5\frac{Volts}{Volt}$ (allowing room for inertia growth). The resulting initial control loop position gain (K_p) was selected to be $4.6\frac{Volts}{deg}$ and the rate loop gain (K_r) was selected to be $90.0\frac{Volts}{deg/sec}$ for all axes. These values were then used to select the wheel drive amplifier gains, K_{wd}, for the roll and pitch axes as $2.8\frac{Volts}{Volt}$ and $6.6\frac{Volts}{Volt}$ respectively.

Gyrocompass EPSHM Gain Selection.

As described above, linear, single-axis analysis selected the control loop position gain at $4.6\frac{Volts}{deg}$, and the rate loop gain at $90.0\frac{Volts}{deg/sec}$; these gains preserved system bandwidth and damping for all safe hold modes in all three axes. Gyrocompass simulations for this set of gains were performed and showed that there was a stable limit cycle in the yaw axis. This stable limit cycle was caused by saturating the yaw control loop amplifiers. This saturation effectively lowers the gain in that particular path, causing stable oscillatory behavior. This resulted in an unacceptable design, and the yaw axis gains were modified as described below.

This orbital gyrocompassing technique (and therefore the yaw position loop) is unique to the EPSHM mode. All other position loops are common to the other analog control modes. Changes to any control path other than the yaw position loop will impact other safe hold control modes. Since limit cycles are caused by saturation, the yaw axis position amplifier gain was reduced to $10.31\,\frac{Volts}{Volt}$, resulting in the following gain distribution:

$$K_{P_{roll,pitch}} = \underbrace{0.5\frac{Volts}{deg}}_{ES} \times \underbrace{0.625\frac{Volts}{Volt}}_{ACE\ Buf\ Amp} \times \underbrace{14.72\frac{Volts}{Volt}}_{SH\ Posn\ Amp} = 4.6\ \frac{Volts}{deg} \qquad (10)$$

$$K_{P_{yaw}} = \underbrace{12.0\frac{Volts}{\frac{deg}{sec}}}_{IRU} \times \underbrace{0.823\frac{Volts}{Volt}}_{Rate\ Buffer\ Amp} \times \underbrace{9.12\frac{Volts}{Volt}}_{SH\ Rate\ Amp} \times \underbrace{10.31\frac{Volts}{Volt}}_{SH\ Posn\ Amp} \times \underbrace{\Omega\frac{rad}{sec}}_{Orbit\ Rate} \qquad (11)$$

$$K_{P_{yaw}} = 0.86\ \frac{Volts}{deg}$$

$$K_r = \underbrace{12.0\frac{Volts}{\frac{deg}{sec}}}_{IRU} \times \underbrace{0.823\frac{Volts}{Volt}}_{Rate\ Buffer\ Amp} \times \underbrace{9.12\frac{Volts}{Volt}}_{SH\ Rate\ Amp} = 90.0\ \frac{Volts}{\frac{deg}{sec}} \qquad (12)$$

Yaw-Slew EPSHM Gain Selection.

When β' exceeds 10 degrees, EPSHM yaw axis control is switched from the gyrocompass sub-mode to the yaw-slew sub-mode using coarse sun sensors. For design and analysis this sub-mode can be considered as a single axis, therefore gains for the yaw-slew sub-mode were chosen by linear analysis. The gain distribution for this mode is:

$$K_{P_{roll,pitch}} = \underbrace{0.5\frac{Volts}{deg}}_{ES} \times \underbrace{0.625\frac{Volts}{Volt}}_{ACE\ Buf\ Amp} \times \underbrace{14.72\frac{Volts}{Volt}}_{SH\ Posn\ Amp} = 4.6\ \frac{Volts}{deg} \qquad (13)$$

$$K_{P_{yaw}} = \underbrace{16.0\frac{\mu amps}{deg}}_{CSS} \times \underbrace{0.0187\frac{Volts}{\mu amps}}_{ACE\ Buffer\ Amp} \times \underbrace{15.374\frac{Volts}{Volt}}_{SH\ Posn\ Amp} = 4.6\ \frac{Volts}{Volt} \qquad (14)$$

$$K_r \;=\; \underbrace{12.0\frac{Volts}{\frac{deg}{sec}}}_{IRU} \times\; \underbrace{0.823\frac{Volts}{Volt}}_{Rate\ Buffer\ Amp} \;\times\; \underbrace{9.12\frac{Volts}{Volt}}_{SH\ Rate\ Amp} \;=\; 90.0\;\frac{Volts}{\frac{deg}{sec}} \qquad (15)$$

The controller bandwidth (0.008 Hz) and damping ratio (0.5) are the same in all axes.

EPSHM Control Loop Analysis.

There are three paths associated with stability margins; these paths are through the (1) position loop, (2) the rate loop, or (3) the path which is common to both rate and position (through the wheel drive amplifier). For that reason, three transfer functions were examined to determine gain and phase margins for the system (see Figure 12):

$$Position\ \ Loop\ \ Open \;=\; \frac{G_1 K_p}{s(Is \;+\; G_1 K_r G_{iru})} \qquad (16)$$

$$Rate\ \ Loop\ \ Open \;=\; \frac{G_1 K_r G_{iru} s}{Is^2 \;+\; G_1 K_p} \qquad (17)$$

$$Control\ \ Loop\ \ Open \;=\; \frac{G_1}{Is^2}(K_r G_{iru} s \;+\; K_p) \qquad (18)$$

$$where: \qquad G_1 \;=\; G_{filter} K_{wd} G_{rw} K_t$$

Gain and phase margins are shown in Table 3. The transfer function linear analysis was performed using the INteractive Controls Analysis (INCA) software package[12]. The INCA generated Bode plots for these single-axis systems are shown in Figures 13-15.

Table 3. Linear Analysis Results – Gain and Phase Margins

Transfer Function	Gain Margin (db) Upper/Lower	Phase Margin (degrees)
Position Loop Open Loop	27.7 / none	50.9
Rate Loop Open Loop	44.9 / -27.7	83.4
Control Loop Open Loop	44.5 / none	48.7

EPSHM CONTROLLER PERFORMANCE

The EPSHM controller was implemented in a three-axis non-linear time-domain computer simulation to evaluate pointing performance capabilities. This simulation models the ADCS sensor/actuator hardware and the TOPEX spacecraft rigid-body three-axis dynamics and kinematics. The controller algorithm in the simulation includes the individual amplifier saturation limits. Both the Gyrocompass and Yaw-Slew sub-modes use the earth sensors for roll and pitch attitude information and as in all the safe hold control modes, the on-axis DRIRU-II analog rate signals are used by the controller for rate control.

The commercially available MODEL (Multi Optimal Differential Equation Language) computer software package was used to develop the FORTRAN code for this time-domain

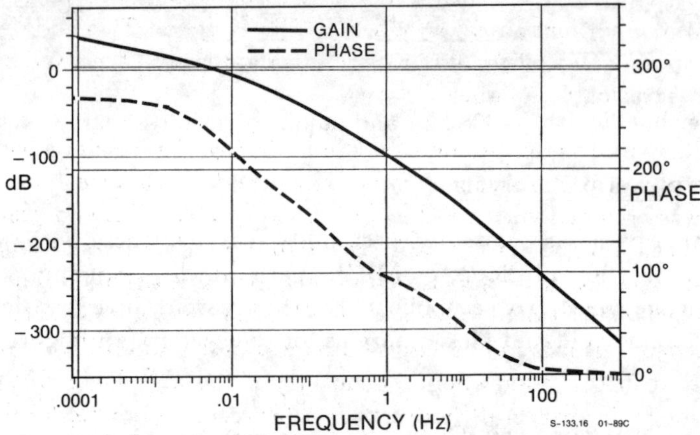

Figure 13. Bode Plot–Position Loop Open Loop

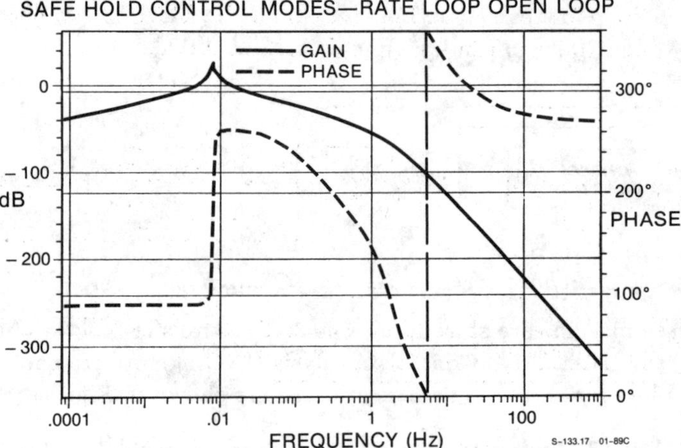

Figure 14. Bode Plot–Rate Loop Open Loop

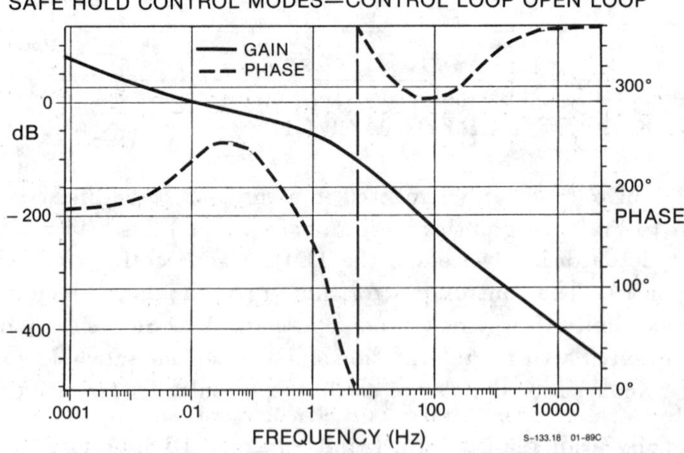

Figure 15. Bode Plot–Control Loop Open Loop

90

simulation of the TOPEX EPSH controller[13]. The MODEL software translates the user-defined differential equations into *optimal* FORTRAN code. This code generates the numerical, time-domain solution for the system equations using the fixed-step, fourth order Runge-Kutta integration technique.

Several fundamental assumptions were made during the development of this EPSHM simulation. The spacecraft is assumed to be a rigid body with a constant inertia tensor. A circular Earth was assumed with a radius of 6378 kilometers. Gravity gradient disturbance torques are modeled but the aerodynamic and solar radiation pressure disturbance torques are not included. The Earth's magnetic field is modeled as a simple tilted dipole. The RWA model does not include frictional losses and wheel torque is limited to 0.1035 Ft-Lbf. In this simulation all sensor outputs are assumed to be noiseless. All EPSHM controller amplifier limits are modeled in this simulation. Shadowing of the individual CSS eyes (due to the interference of the main spacecraft body or the solar array) is not modeled. The integration step size was selected to be 0.5 seconds and the DRIRU-II dynamics, ESA dynamics, and RWA dynamics are not included. Figures 16 and 17 show the logic flow for the simulation code. The control loop for the roll and pitch axes for the EPSH controller is shown in Figure 18. The EPSHM controller parameters are defined in Table 4.

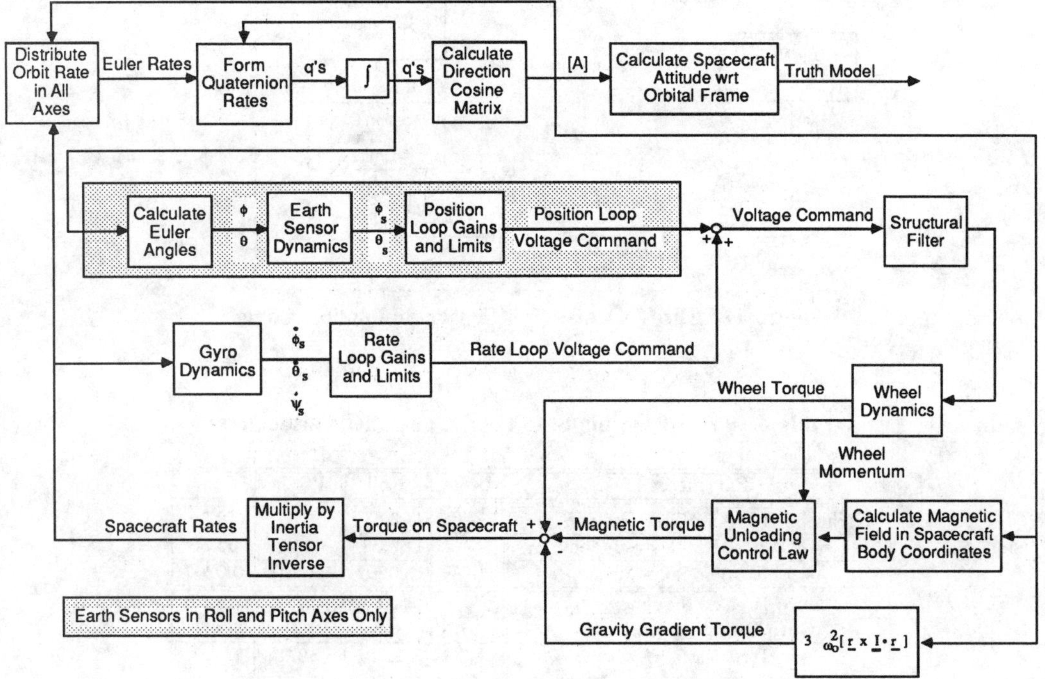

Figure 16. EPSH Simulation Logic

During actual spacecraft operation, ground command will switch the EPSH mode controller from Gyrocompass to the Coarse Sun Sensor mode (or visa versa). The change from one sub-mode to the other depends on the angle between the sun and the orbit plane (β'). When the magnitude of β' is less than 10 degrees, the spacecraft is controlled with the

EPSH/gyrocompass mode; for β' above 10 degrees, the spacecraft must track the sun, and ground command switches control to the EPSH/CSS mode. For simulation purposes, the orbit is fixed; changes to β' are implemented by changing the initial conditions of two orbit parameters: the right ascension of the ascending node, and the angle to the sun in the ecliptic plane.

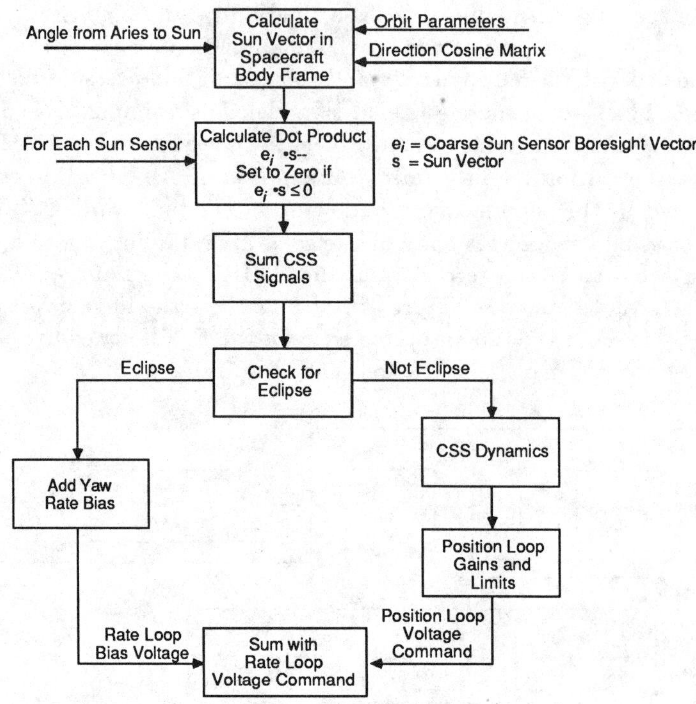

Figure 17. EPSH Coarse Sun Sensor Simulation Logic

Table 4. Earth Pointing Safe Hold Controller Parameters

Parameter	Units	Roll	Pitch	Yaw
Wheel Drive Amplifier Gain, K_{wd}	$\frac{Volts}{Volt}$	6.6	2.8	8.5
Reaction Wheel Break Frequency, ω_w	$\frac{rad}{sec}$	62.38	62.38	62.38
Wheel Torque Constant, K_t	$\frac{ft-lb}{Volt}$	0.0207	0.0207	0.0207
Gyro Break Frequency, ω_g	$\frac{rad}{sec}$	12.57	12.57	12.57
Gyro Damping Ratio, ς_g	nd	0.6	0.6	0.6
Gyro Gain, K_{iru}	$\frac{volts}{deg}$	12.0	12.0	12.0
Rate Buffer Amp. Gain, K_{rbe}	$\frac{Volts}{Volt}$	0.823	0.823	0.823
Earth Sensor Break Frequency, ω_e	$\frac{rad}{sec}$	3.16	3.16	3.16
Earth Sensor Damping Ratio, ς_e	nd	0.7	0.7	
Earth Sensor Gain, $K_e s$	$\frac{Volts}{deg}$	0.5	0.5	
Position Buffer Amp. Gain, K_{pbe}	$\frac{Volts}{Volt}$	0.625	0.625	
Coarse Sun Sensor Gain, K_{css}	$\frac{\mu amps}{deg}$			16.0
CSS Buffer Amp. Gain, K_{p2}	$\frac{Volts}{\mu amps}$			0.0187
Structural Filter Break Frequency, ω_f	$\frac{rad}{sec}$	1.25	12.5	1.25

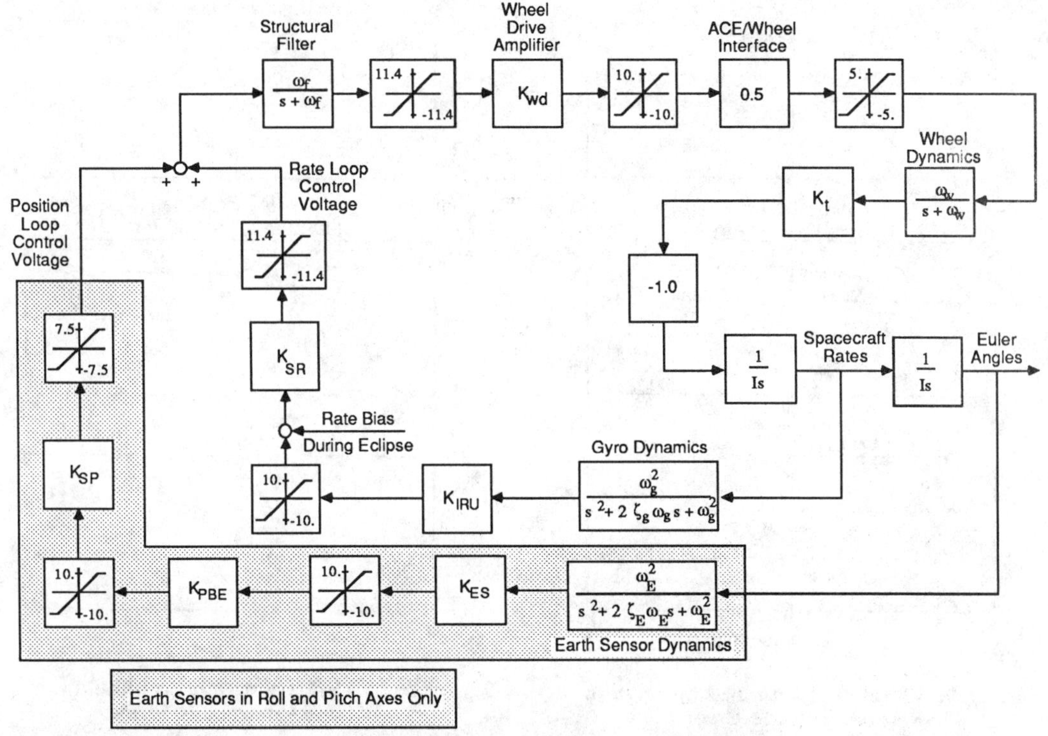

Figure 18. EPSH Simulation Block Diagram

EPSHM SIMULATION RESULTS

Gyrocompass EPSHM Simulation Results

EPSHM controller gain margins, as shown in Table 5, were determined by non-linear time-domain simulation. Figures 19-21 show the EPSHM gyrocompass controller response to an initial condition attitude error of 5 degrees in all axes. The EPSHM gyrocompass controller brings the roll attitude error to within 0.5 degrees in less than 500 seconds; the control of the attitude error in yaw is slower, but reaches steady state in less than 700 seconds. Yaw attitude error lags since the roll axis rate is used for the yaw position signal.

Table 5. EPSH/Gyrocompass Mode – Gain Margins

Transfer Function	Gain Margin (db)
Roll Axis Position Loop	-40.0
Pitch Axis Position Loop	infinite
Yaw Axis Position Loop	12.7
Roll Axis Rate Loop	-34.0
Pitch Axis Rate Loop	-34.0
Yaw Axis Rate Loop	-12.0
Roll Axis Control Loop	-26.0
Pitch Axis Control Loop	-34.0
Yaw Axis Control Loop	-26.0

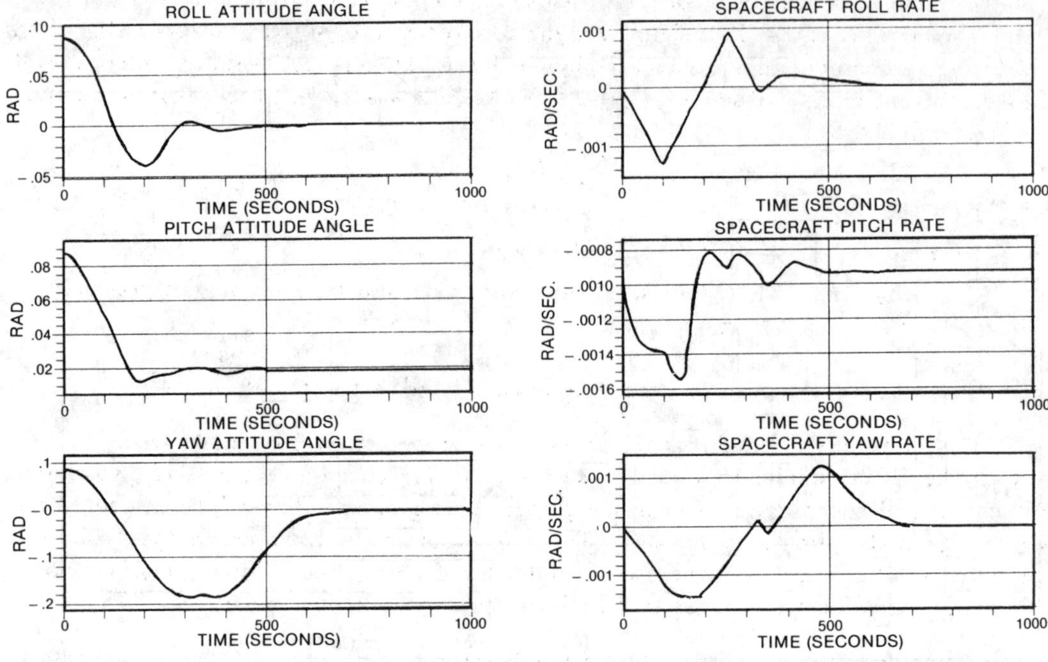

**Figure 19. EPSH/Gyrocompass Simulation
Results–Attitude Error**

**Figure 20. EPSH/Gyrocompass Simulation
Results–Spacecraft Rates**

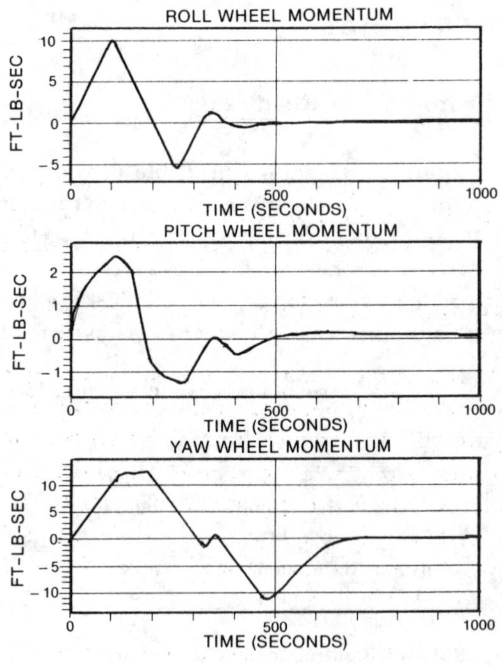

**Figure 21. EPSH/Gyrocompass Simulation
Results–Wheel Momentum**

In the pitch axis, there is a constant (i.e. a hangoff) attitude error. The pitch axis gyro senses inertial body rates. This signal also includes orbit rate, which is always in the pitch axis during gyrocompassing. Unless the pitch axis rate signal is modified (i.e. null out orbit rate), a constant pitch attitude error results. In steady state, the attitude error in the pitch axis is determined by

$$\theta = \frac{K_r}{K_p}\Omega \tag{19}$$

$$where: \quad \theta = Pitch\ attitude\ error$$
$$K_p,\ K_r = Pitch\ axis\ rate\ and\ position\ loop\ gains,\ respectively$$
$$\Omega = Orbit\ rate$$

For the present rate and position gains, the pitch attitude error is 1.045 degrees.

Since β' is usually greater than 10 degrees, the spacecraft will normally track the sun while in the EPSH mode; this causes orbit rate to (1) change sign, and (2) distribute between the roll and pitch axes. For the gyrocompass mode, nulling orbit rate requires a bias voltage in the pitch axis; for the coarse sun sensor mode, however, compensation for orbit rate requires major ACE hardware modification. Since the predicted attitude error was small, no modification to the ACE SHE hardware was recommended.

Additionally, there is a steady-state attitude error in the yaw axis. This is due to the ratio of the gains for the gyrocompass mode. In steady state, the rate and position control loop voltages are equal, but of opposite sign. In the yaw axis, the signals are

$$K_p\psi + K_r\phi = 0.0 \tag{20}$$

Solving for yaw angle,

$$\psi = \frac{K_r}{K_p}\phi \tag{21}$$

In the yaw axis, the ratio of rate to position gain is 78 to 1; this causes the roll attitude angle to be magnified in the yaw axis. Roll attitude error is extremely small, therefore the steady state error in yaw is still well below 1 degree as shown in Figure 19.

Yaw-Slew EPSHM Simulation Results

The Yaw-Slew EPSHM tracks a commanded yaw attitude angle which is generated by the yaw CSS's. These CSS's are body mounted in the spacecraft x-y plane. The sensors are arranged as shown in Figure 10. The signals from these sensors are combined to indicate the commanded yaw attitude angle. During eclipse, this signal disappears, and without compensation, the controller will drive the spacecraft yaw rate to zero, causing unacceptable attitude errors at sunrise. A bias voltage, which is equivalent to a constant command rate of 3.8 $\frac{deg}{min}$, is therefore applied during eclipse. For β' angles above 57 degrees, the spacecraft is in full sun and no eclipse occurs.

Simulation results for a β' of 10 degrees are shown in Figures 22-25 (this is an orbit which experiences eclipse). The peak yaw attitude error is 40 degrees for this β' angle, which reduces to 20 degrees for β' of 20 and 30 degrees. Wheel momentum is largest in

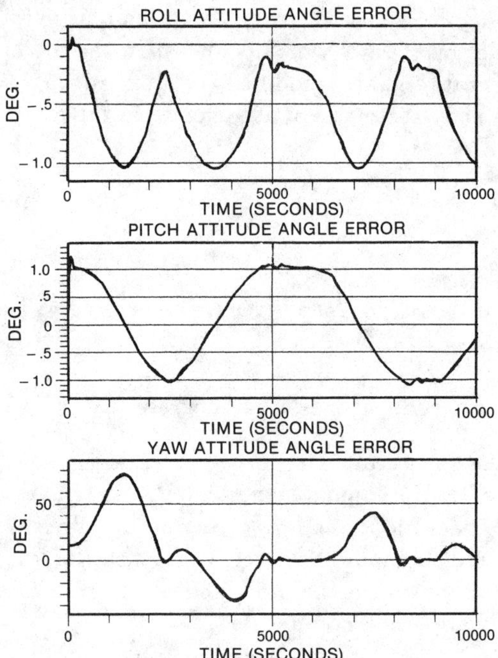

Figure 22. EPSH/CSS Simulation Results–$\beta' = 10$ degrees Spacecraft Attitude Error

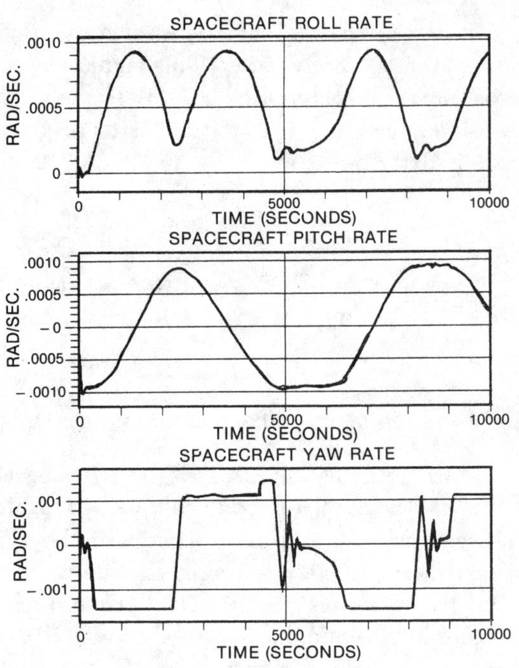

Figure 23. EPSH/CSS Simulation Results–$\beta' = 10$ degrees Spacecraft Rates

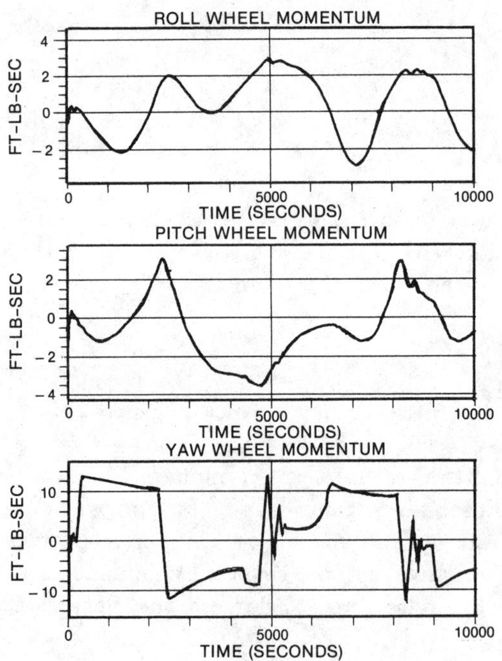

Figure 24. EPSH/CSS Simulation Results–$\beta' = 10$ degrees Wheel Momentum

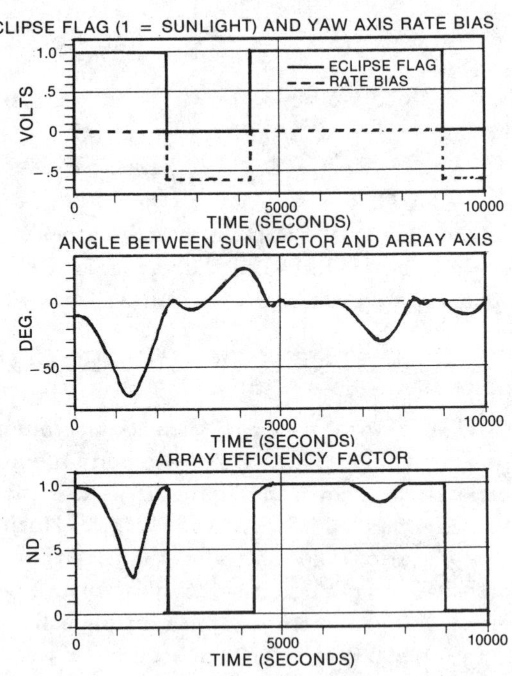

Figure 25. EPSH/CSS Simulation Results–$\beta' = 10$ degrees Sun Pointing Parameters

the yaw axis, but remains within 15 Ft-Lbf-Sec. Figures 26-29 show simulation results for a β' angle of 80 degrees. Yaw attitude error stays below one degree, and yaw axis wheel momentum is less than 4 Ft-Lbf-Sec.

The Yaw-Slew EPSHM sub-mode is used to point the spacecraft to keep the solar array axis normal to the sunline. Simultaneously, the SADA controller orients the array towards the Sun about the pitch axis. Any measure of EPSHM performance must indicate how well this task is performed. As part of the simulation code, the sun line vector, $\hat{s}$, is resolved into the spacecraft body frame. The angle between the array axis and the sun vector can be calculated by:

$$\phi = arcsin(s_y) \tag{22}$$

$$where: \quad \phi = Angle\ between\ sun\ and\ array\ axis$$
$$s_y = y-axis\ component\ of\ the\ sun\ vector$$

Defining the array efficiency factor, η, as:

$$\eta = \cos\phi \tag{23}$$

we can then calculate the power generation factor (P_{gf}), as:

$$P_{gf} = 1.0 - \frac{\int_{t_1}^{t_2} \eta\ dt}{(t_2 - t_1)} \tag{24}$$

$$where: \quad (t_2 - t_1) = Time\ spacecraft\ is\ in\ sunlight$$

When η is zero, the array axis is perpendicular to the sun vector, and the power generation factor is one. When η is one, the array axis and sun vector are parallel; no power can be generated, and the power generation factor is zero. These calculations assume that the array is perfectly pointed in pitch and the albedo error at sunrise (which could cause attitude perturbations of up to 10 degrees) has not been considered. This factor gives us a true measure of performance for the spacecraft control system. Since the power generation factor is a cosine function, large attitude errors do not cause a significant decrease in array efficiency as shown in Table 6. More than 95% of the maximum power can still be generated which for safe hold control is acceptable performance.

Table 6. EPSH/CSS Simulation Results

β' (deg) (degrees)	Peak Yaw Attitude Error (degrees)	Solar Array Pointing Error (deg)	Power Generation Factor (degrees)
10	40	30/20	0.976
20	20	15/10	0.994
30	20	15/10	0.998
40	30	20/20	0.996
50	30	20/20	0.997
57	<1	0/0	1.0
80	<1	0/0	1.0

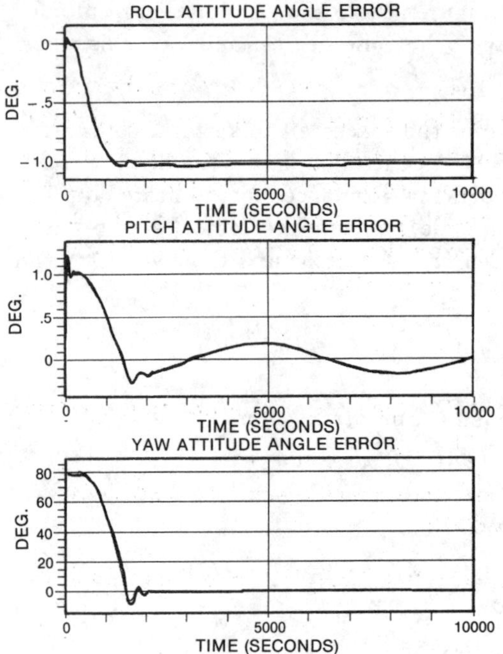

Figure 26. EPSH/CSS Simulation Results–
$\beta' = 80$ **degrees Spacecraft Attitude Error**

S–133.05

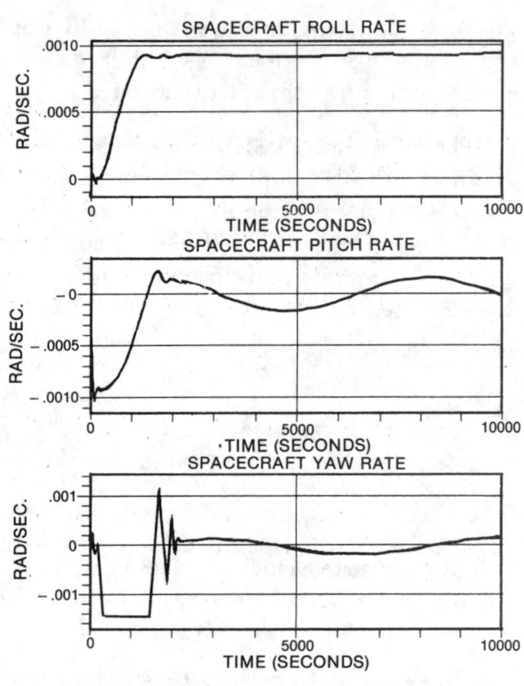

Figure 27. EPSH/CSS Simulation Results–
$\beta' = 80$ **degrees Spacecraft Rates**

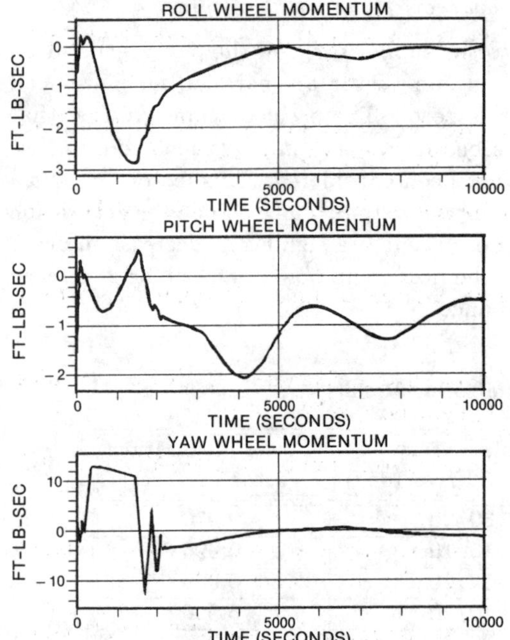

Figure 28. EPSH/CSS Simulation Results–
$\beta' = 80$ **degrees Wheel Momentum**

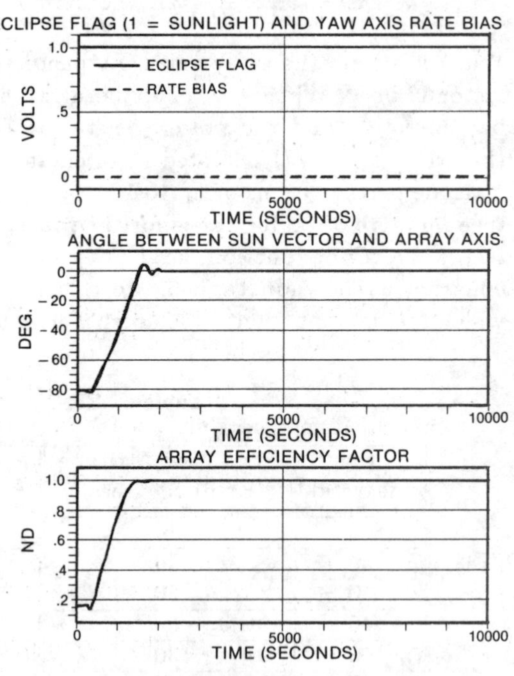

Figure 29. EPSH/CSS Simulation Results–
$\beta' = 80$ **degrees Sun Pointing Parameters**

98

CONCLUSIONS

In this paper the TOPEX ADCS has been described in general and the design and performance of the ADCS in the Earth Pointing Safe Hold Mode has been reviewed in detail. The ADCS for the TOPEX satellite has been designed to satisfy the broad range of mission requirements. Performance of the EPSHM controller meets the specified pointing control requirements while providing sufficient rigid body stability margins. A unique design feature of the TOPEX ADCS, sinusoidal yaw attitude slewing, was highlighted. In particular it was shown how this capability was implemented in the EPSHM.

REFERENCES

1. Stewart, R.A., and M. Lefebvre, "TOPEX/POSEIDON: A Contribution to the World Climate Research Program," AAS Paper 86-306

2. Multimission Modular Spacecraft External Interface Specification and User's Guide, NASA/GSFC Document S-700-11, February 1978

3. Falkenhayn, E., "Multimission Modular Spacecraft (MMS)," AIAA Paper 88-3513, June 1988

4. Simpson, R.F., "Explorer Platform," AIAA Paper 88-0066, January 1988

5. Class, B.F., F.H. Bauer, K. Strohbehn, and R.V. Welch, "Space Infrared Telescope Facility/Multimission Modular Spacecraft Attitude Control System Conceptual Design," AAS Paper 86-031, February 1986

6. Das, A., "The On-Orbit Attitude Determination and Control System for the LANDSAT-D Spacecraft," AIAA Paper 82-310

7. Dennehy, C.J., R.V. Welch and T. Kia, "Attitude Determination and Control Subsystem for the TOPEX Satellite," AIAA Paper 88-4129, August 1988

8. TOPEX Satellite System Performance Requirements, NASA/JPL Document D-2226, June 1987

9. TOPEX Satellite System Design Specification, Fairchild Space Company Document 968-PF1000C, September 1988

10. Irvine, R.B. and Ritter, J.W.,"DRIRU-II : The NASA Standard High Performance Inertial Reference Unit," AAS Paper 79-021, February 1979

11. Welch, R.V.,"Autonomous Failure Detection and Correction on LANDSAT-4," AIAA Paper 83-2265, August 1983

12. Bauer, F.H. and, and Downing, J.P., "Interactive Controls Analysis (INCA) Volumes 1-4," Program No. GSC-12998, COSMIC, University of Georgia, 1985

13. "MODEL (Multi Optimal Differential Equation Language) Users Guide", NASA Goddard Space Flight Center, Benjamin Zimmerman, January 1982.

ACKNOWLEDGEMENTS

The work described in this paper was carried out by the Fairchild Space Company under Contract No. 957849 with the Jet Propulsion Laboratory, California Institute of Technology.

The authors greatfully acknowledge the contributions of the many individuals at both the Jet Propulsion Laboratory and the Fairchild Space Company who participated in the design of the TOPEX ADCS. Their assistance in preparing and reviewing this paper is appreciated. Special thanks is extended to Teofilo A. Almaguer, Jr., the TOPEX Satellite System Program Manager at JPL and Dr. Tooraj Kia, the TOPEX ADCS Work Unit Manager at JPL. The authors would also like to express their gratitude to William A. Johnston, Jr., the TOPEX Program Manager at Fairchild for his support during the preparation of this paper.

DESIGN, FABRICATION AND TEST OF A PROTOTYPE DOUBLE GIMBAL CONTROL MOMENT GYROSCOPE FOR THE NASA SPACE STATION

Joseph Blondin[*], Eric Hahn[*], John Kolvek[*],
Lewis Cook[†], Paul Golley[†], and Henning Krome[†]

Recognizing the need to develop future technologies in support of the Space Station, NASA's Advanced Development Program (ADP) placed as its goal the design and fabrication of a prototype 4750 Newton-meter-second (3500 ft-lb-sec) Control Moment Gyroscope (CMG). The CMG uses the principle of momentum exchange to impart control torques for counteracting vehicle disturbances. This paper addresses the selection of the double gimbal CMG over the single gimbal and describes the major subassemblies of the prototype design. Particular attention is given to the choice of the materials, fabrication and design details dictated by the man-rated mission requirement.

Physical characteristics and the results of functional testing are presented to demonstrate the level of system performance obtained. Comparisons are made of the measured system responses against design goals and predictions generated by computer simulation.

INTRODUCTION AND HISTORY

NASA initiated the ADP in parallel with the Phase B definition and preliminary design efforts for the Space Station. This program focused on technologies applicable to the initial Space Station with the goal of accelerating these technologies to meet the proposed operational schedule for the station. Other objectives of the ADP were enhancement of the performance of the Space Station, reduction of life cycle costs during the operating phase and reduction of risks encountered during the development phase.

In connection with the above, the Attitude Control and Stabilization (ACS) team of the ADP proposed a number of technical activities; one of these being the design, fabrication and test of a prototype control moment gyro (CMG). Using a CMG for control of the Space Station Freedom seemed obvious since a CMG has a replenishible momentum capability which is achieved through appropriate gravity gradient desaturation maneuvers and requires no consumables. NASA's experience with this type of control was amply demonstrated on the Skylab program in the mid-seventies. Skylab used three double gimbal CMGs (DGCMGs) for attitude control and employed gravity gradient maneuvers and a thruster system for momentum bias desaturation.

---

* Allied-Signal Aerospace Company, Guidance Systems Division, Teterboro, New Jersey 07608.

† NASA Marshall Space Flight Center, Huntsville, Alabama 35816.

The first decision to be faced by the ACS team was whether to develop a DGCMG or a single gimbal CMG (SGCMG) for the Space Station prototype.

The advantages of DGCMGs include:

o   Much simpler control laws without elaborate singularity avoidance

o   No impact of unit failure on control laws

o   No impact of failure on spherical momentum envelope shape

o   Growth capability by adding individual DGCMGs without impacting control laws

o   Simpler vehicle mounting geometry

SGCMGs have an advantage in that they can provide greater torque capability for the same angular momentum. Since the Space Station Freedom has no rapid maneuvering requirements necessitating high torques, this did not prove to be an important consideration for this application.

System level trade studies involving weight, size, power and reliability produced no advantage to either type since the SGCMGs require oversizing to produce the same angular momentum envelope as DGCMGs. The flexibility of the DGCMG to support a large variation in vehicle inertia, especially during station build-up, ultimately proved to be the main consideration for the selection. Since the ACS control laws are not affected by the number of units employed, the initial manifest need not contain the full complement of units. As the station assembly configuration changes, additional CMGs could be added at any time to support the ACS requirements.

The prototype CMG design parameters were extrapolated from the Skylab CMG experience and improvements were made in a number of areas based primarily on the momentum storage capability and the long life required for the Space Station Freedom application. The Marshall Space Flight Center (MSFC) was selected to lead the CMG development effort since that center had been responsible for the Skylab CMG development and had the technical expertise and testing capability for continued development in this area. Subsequently, MSFC selected the Guidance Systems Division (GSD) of the Allied-Signal Aerospace Company (formerly The Bendix Corporation) to design, develop and . fabricate a prototype CMG under NASA contract NAS8-36628. The final concept, shown in Figure 1, is a double gimballed system with unlimited freedom for the outer gimbal and a 50 percent increase in angular momentum over the Skylab CMG. Mechanical features include an on-orbit servicing capability, power and signal transfer through rotary transformers and fiber optics respectively and an active oil lubrication system for the spin bearings. The three year development has resulted in a prototype CMG which will undergo verification and life testing at MSFC.

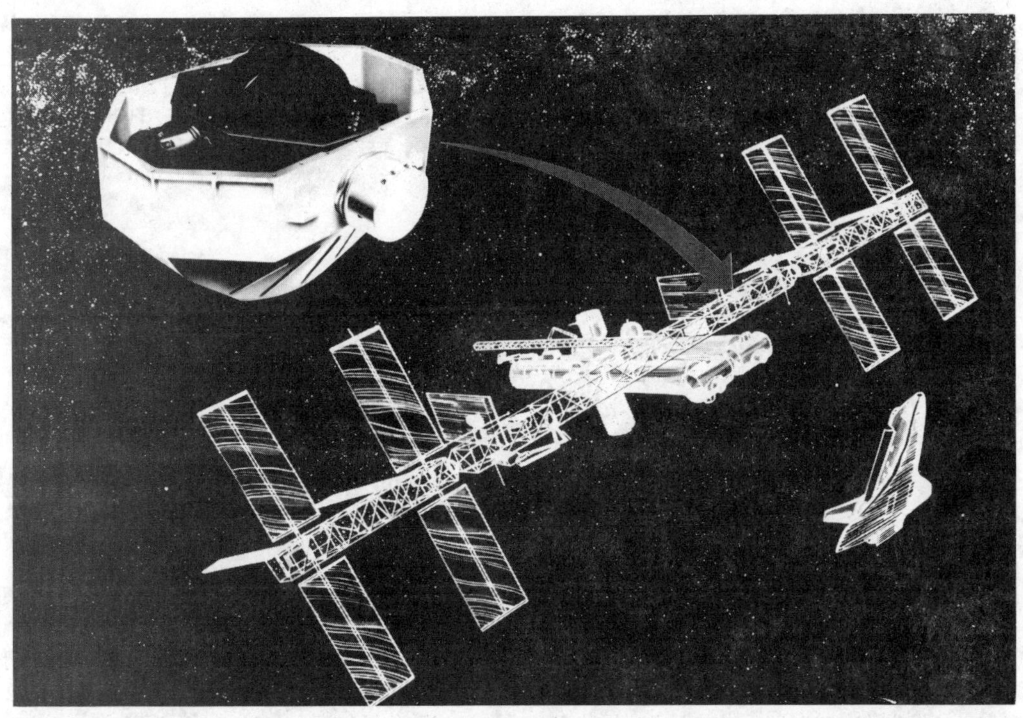

Figure 1 Space Station Prototype DGCMG

<u>System Requirements</u>

The CMG was designed to meet the following requirements:

o Angular momentum of 4750 N-m-s (3500 ft-lb-s) at a speed of 6600 rpm

o Peak output torque applied shall be equal to or greater than 274 N-m (200 ft-lbs)

o Unlimited angular freedom for the outer gimbal

o Angular freedom of ± 1.57 rad (90 deg) for the inner gimbal

o Rotor design to have a safety factor of 4 on yield stress

o High reliability and a ten year operational life for overall CMG design

103

CMG Configuration Trade Studies

A trade study was performed to optimize the selection of rotor material. The study compared a wide range of candidate materials for the following mechanical properties:

o   Material strength

o   Fracture toughness

o   Stress corrosion resistance (MSFC-SPEC-522B)

o   Producibility

In addition to the above parameters, the following constraints were also considered:

o   Rotor diameter of 63 cm (25 in.)

o   Rotor speed less than 9000 rpm

o   No maraging steel

o   Use of materials with published data base.

To make comparisons between various materials, three candidate systems were established and analyzed. Each of the designs was analyzed for weight, inertia, momentum and yield stress. In addition, rotor stiffness and resonant frequencies were examined for each case. The conclusion of this study led to the selection of Custom 455 stainless steel as the optimum rotor material.

A second trade study was performed to determine the feasibility of replacing CMG rotor bearings during the mission. The long duration mission proposed for the Space Station program requires that the CMG rotor spin bearings must perform consistently for a minimum of ten years. Even with a theoretical reliability of 0.999, the possibility of bearing deterioration or failure exists. Typically, degradation in bearing performance is characterized primarily by an increase in average friction torque with a corresponding increase in motor power consumption. In the case of serious bearing degradation, a bearing replacement would prevent loss of the CMG and its corresponding impact on the mission. As a result of the trade study, a bearing configuration has been incorporated into the CMG rotor bearing design which could support replacement of the bearings on-orbit if deterioration is detected.

<u>Repairability and ORU Design Concept</u>

The prototype CMG was designed using the orbital replaceable unit (ORU) concept to make repairs and component replacement in space as convenient as possible. For CMG removal, the mounting pads were designed for captive bolts which prevents bolt loss after removal. Sufficient clearance exists between electrical connectors to allow insertion or removal by an astronaut wearing a space suit. The CMG is provided with handles to allow relative ease of on-orbit handling.

Electronic assemblies can be replaced without disturbing the mounting of the CMG. However, for safety considerations, it is recommended to remove power from the CMG prior to this replacement. Captive bolts and quick disconnect connectors will facillitate the replacement of the ORUs.

Spin bearing replacement was designed to be achieved after placement of the CMG into a Space Station "shirt-sleeve" work area. The actual replacement procedure requires some mechanical acumen and extensive training to become familiar with the unit, tooling, and assembly sequence.

DESIGN DESCRIPTION

The design of the CMG is an evolution based on many CMG and momentum exchange devices built and flown since the Skylab program. Most of the major components of the system have a successful heritage and design base in keeping with the man-rated mission requirements of the Space Station Freedom. The system shown in Figure 2 consists of a rotor mounted within two sets of orthogonal gimbals, permitting the spin axis of the rotor to be oriented in any desired direction. All the drive and support electronics are mounted on the mounting ring and gimbal structures to minimize signal transfer across the gimbal pivots.

A block diagram of the system is presented in Figure 3. The Outer Gimbal Electronics Assembly (OGEA) accepts the external electrical interface in the form of power and a MIL-STD-1553 serial communications link. In the OGEA a microprocessor channels the communication of command and telemetry signals and thus requires no analog signal processing of any type. Power is transferred across the outer gimbal pivot using redundant rotary power transformers. Communication between the OGEA and Inner Gimbal Electronics Assembly (IGEA) is performed via a Fiber Optic Rotary Joint (FORJ). In this manner no contacting signal transfer is employed where life limitations might be of concern.

The Rotor Electronics Assembly (REA) controls and monitors the speed of the rotor and is mounted on the inner gimbal. Signal transfer is accomplished from the IGEA to the REA by a limited motion twist capsule that is protected by gimbal stops that limit the inner gimbal motion to 90 degrees. The major electrical functions are cross-strapped to improve overall system reliability and minimize the orbital replacement operation.

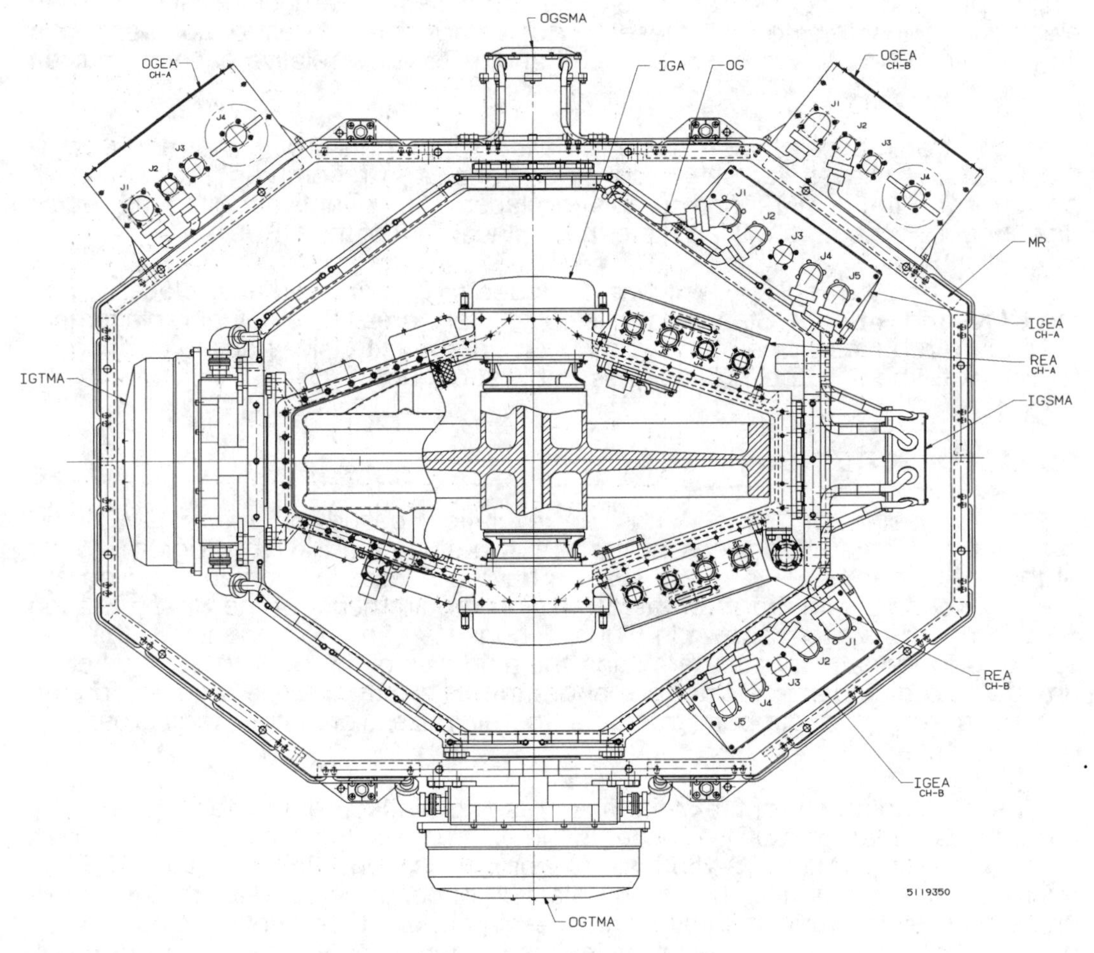

Figure 2 Layout of Prototype DGCMG System

106

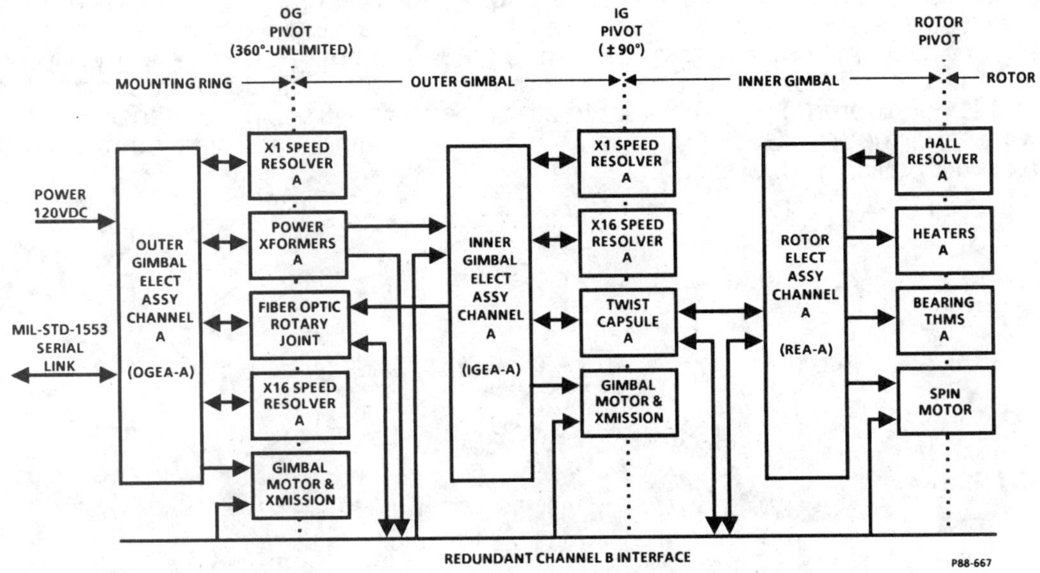

Figure 3 System Block Diagram

The major interface characteristics are described below:

o Size:

   - Length: 1.21 m (47.8 in)
   - Width: 1.16 m (45.9 in)
   - Height: 1.15 m (45.5 in) - With Outer Cover In Place

o Weight: 279 kg (615 lb)

o Mounting: 4 Point C.G. (Gimbal Axes Lie In The Mounting Plane)
o Power: (120 volts DC)

   - Quiescent: 95 watts
   - Spin-Up (Peak):   240 watts

The following sections will describe in further detail the major components of the system and what requirements influenced the design or fabrication activities.

Rotor Design And Safety Analysis

A cross-section of the rotor installed in the Inner Gimbal Assembly (IGA) is presented in Figure 4. The rotor is a single-web wheel forged from Custom 455 stainless steel. It is supported at each end by a single angular contact ball bearing. Outside diameter of the rim is 0.635 meters (25.0 in.) and the overall shaft length is 35 cm. (13.85 in.). Custom 455 is a precipitation hardenable steel and is considered highly resistant to stress corrosion cracking. Extensive testing was performed by GSD to properly qualify this material and verify the physical and mechanical properties.

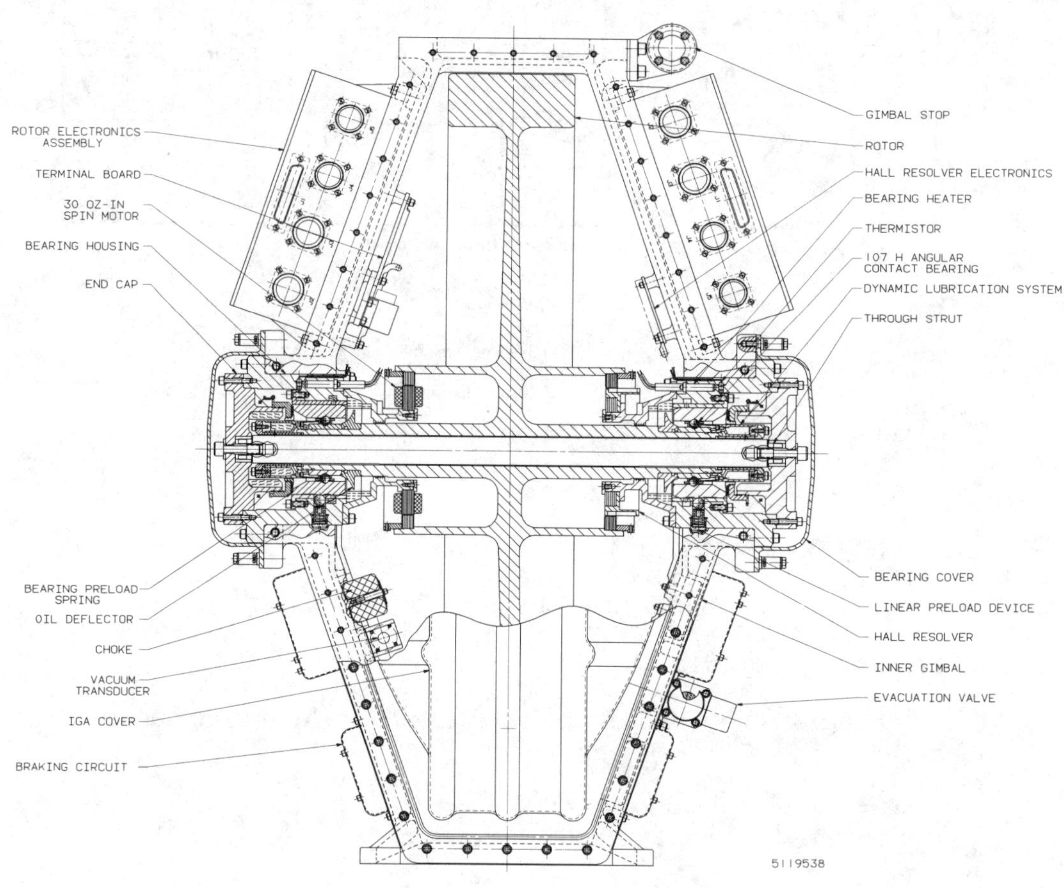

Figure 4 Inner Gimbal Assembly

A finite element model of the rotor was created and analyzed for stresses and deflections using NASTRAN. The centrifugal loading of the rotor at 6600 rpm produces the maximum steady operational forces. Gyroscopic stresses on the rotor are much less of a concern due to the low level of output torque that is required of the system. A modified Goodman diagram presented in Figure 5 shows that the combined centrifugal (steady) and gyroscopic torquing (cyclic) stresses are well within the infinite life region of the graph.

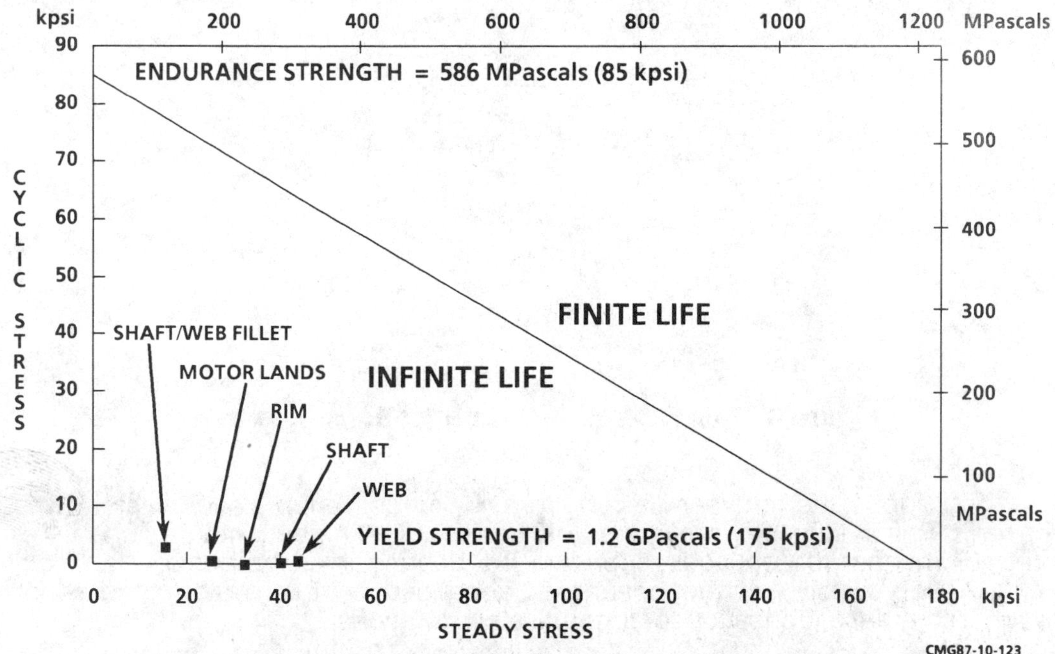

Figure 5 Modified Goodman Diagram Of Rotor Stresses

A design requirement was placed on the rotor to provide a safety factor 4 on yield stress at 105% of the nominal wheel speed (6930 rpm). An analysis of the rotor stress results in a peak value in the web of 296 MPascals (43 kpsi). In comparing this to the 1.2 GPascals (175 kpsi) yield strength of the material produces a safety factor greater than required. During component test, the rotor was subjected to an overspeed of 1.33 times the nominal speed (8800 rpm).

Spin Bearings And Lubrication System

The spin bearings used in the design are angular contact type 107H size ball bearings with special retainers. This configuration has been used previously and dates back to the Skylab CMG. The bearing retainers have been modified to provide proper distribution of the lubricant to the ball and race contact zone. The material for the races and balls is VIM-VAR 52100 chrome steel and the retainer is fabricated from a cotton-based phenolic impregnated with bearing lubricant.

109

When assembled, as illustrated in Figure 6, the bearings are preloaded by a constant force Belleville spring that ensures the bearings remain preloaded under all conditions. Low-level heaters are provided in the housing for low temperature operation if required.

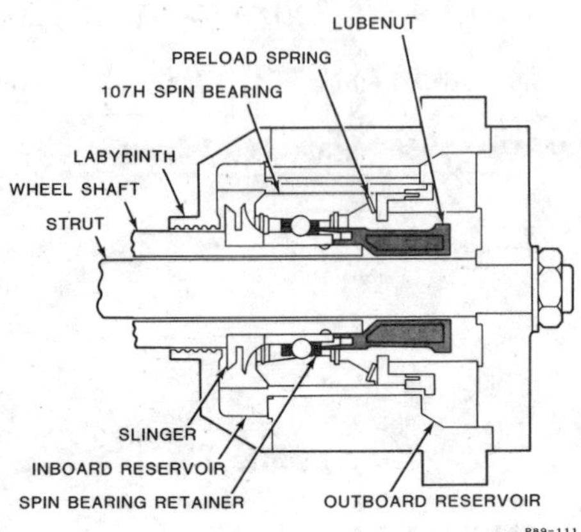

Figure 6 Spin Bearing And Lubrication System

To support the ten-year life requirement, an active lubrication system was chosen. This system will provide a flow of new KG-80 lubricant in a "one-time-through" manner to continually lubricate the bearing over the design life of the CMG. The reliability of the bearing is enhanced by this system and is far superior to grease lubrication for long mission durations.

Torque Motor And Transmission

To develop the required torque of 272 N-m (200 ft-lbs), a torque motor and geared transmission are utilized. The motor is a brushless DC type design capable of developing 12 N-m (9 ft-lbs) of torque. An ironless stator is employed which produces no hysteresis or eddy current losses and thus eliminates magnetic cogging and drag torques for better system performance.

The gear train employed in the transmission is shown in Figure 7 and consists of a two-stage, parallel-path spur gear arrangement. Windup of one gear train path with respect to the other provides a preload that effectively eliminates backlash in the transmission. A gear ratio of 27.76 to 1 allows the motor to achieve the required torque level. This type of configuration has been employed on previous designs including a unit that has accumulated six years of special life testing under severe duty cycle operations.

The torque motor and transmission components are housed in the Torquer Module Assembly (TMA) along with a multi-speed resolver used for rate feedback. Shown in cross-section in Figure 8, this assembly is identical for both gimbal pivots.

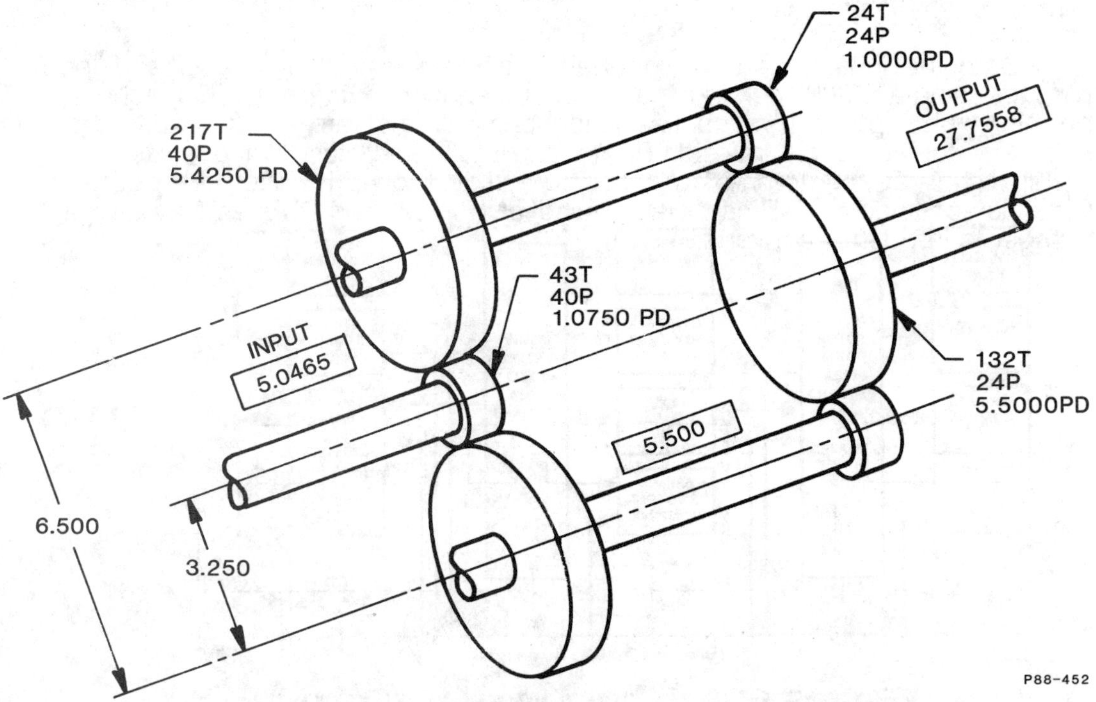

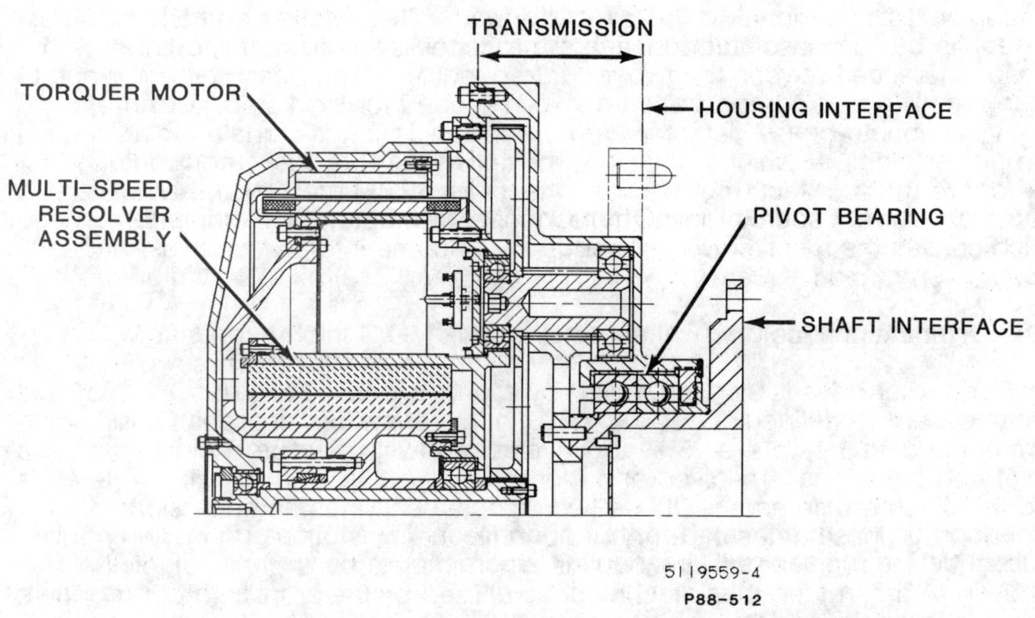

Figure 7 Parallel Path Gear Train Transmission

Figure 8 Torquer Module Assembly Cross-Section

Gimbal Drive Electronics

Both the inner and outer gimbals are rate controlled in a closed loop manner using a phase-locked-loop technique shown in Figure 9. This technique permits high input command resolution and precise rate control without the necessity for precision low-signal-level analog electronics. In operation, the 16-bit digital rate command is applied to a digital low-pass filter, the output of which is accepted by a Binary Rate Multiplier (BRM). The BRM acts as a digital number to frequency converter.

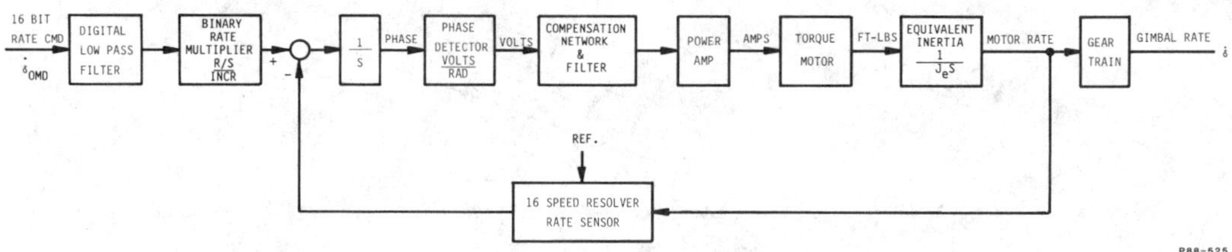

Figure 9 Gimbal Control Loop Block Diagram

A 16-speed resolver acts as a rate sensor and produces an output whose frequency is proportional to the gimbal speed. The resolver and BRM outputs are applied to a phase detector which produces an output proportional to the phase difference between the two input frequencies. The phase detector output is then applied to a compensation network needed for loop stability and a notch filter to attenuate phase detector carrier harmonics. This signal is then applied to a power amplifier which contains a multiplier unit to achieve commutation for the torque motor. A current feedback technique is utilized by the amplifiers to produce a current-source drive. In response, the motor accelerates to a speed which causes the resolver output frequency to come into exact correspondence with the command frequency.

Rotor Drive Electronics

The rotor drive is also controlled by a phase-locked-loop design. Nominal wheel speed is defined as 6,600 rpm, but the system is capable of being commanded to operate at 5% above and below this value. A frequency is generated by a Hall resolver and compared to the commanded reference frequency. The difference in these two signals generates an error signal that is applied to a phase detector, amplified and frequency shaped. It is then applied to the PWM current amplifier which is commutated by the output of the Hall resolver. A current feedback technique is utilized by the amplifiers to produce a current source-drive to the spin motor.

MATHEMATICAL MODELING

For design and analysis purposes, the behavior of the CMG can be characterized by a 6-mass model. This model represents both the inner and outer gimbal loops which are coupled as a function of the inner gimbal angle. In general, the loops are designed as high-gain wide-bandwidth rate loops to enhance small signal performance and damp the gear train resonance. A digital prefilter is used to provide the overall bandwidth characteristics as viewed by the vehicle control loops. When the proper stabilization networks are employed, the inner gimbal loop produces the frequency response characteristics shown in Figure 10.

From this linear model, a nonlinear representation was developed that permits a more accurate determination of CMG performance. A simulation was generated using the Boeing Computer Services EASY-5 Analysis Program and incorporates nonlinear effects such as: torque motor saturation, gear train compliance, electronic limits and pivot friction. The model can be exercised for any type of input command (i.e. sinusoidal, step, impulse or impulse train).

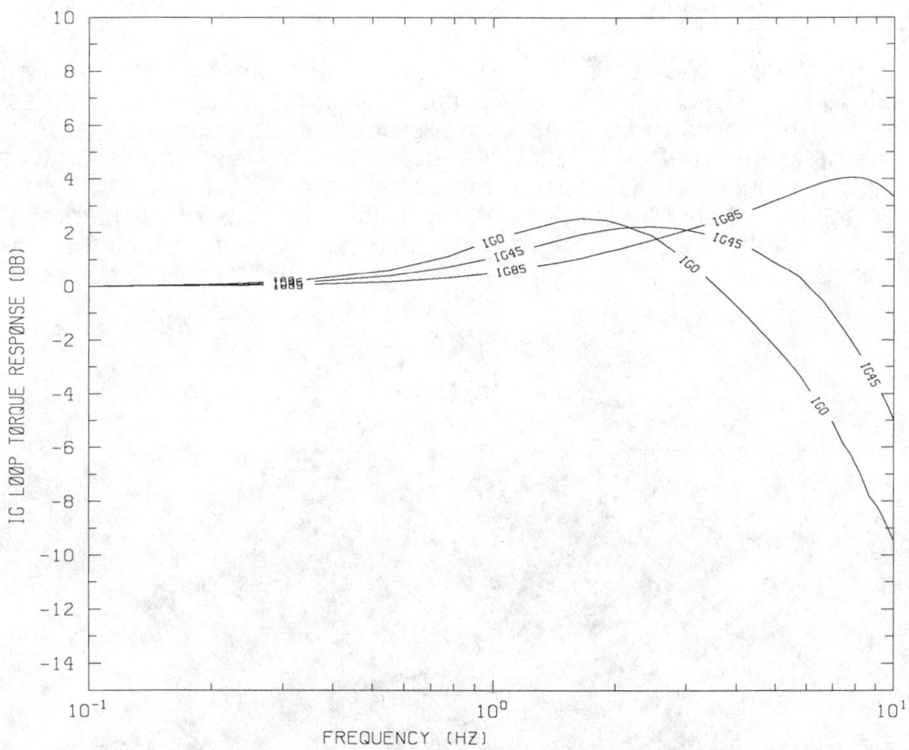

Figure 10 Math Model Prediction Of IG Loop Response

TEST RESULTS

<u>Support Equipment</u>

The CMG system is supported during testing by an automated computer-controlled test station. Interface to each of the two channels of the CMG is via a single cable which supplies power and provides a dual redundant serial data link. The station uses an IBM PC/AT computer equipped with a 30 Mbyte hard disk drive. All operator interface and monitoring of the CMG is provided by a MIL-STD-1553 serial communications bus which plugs directly into the computer. Power requirements to the system are provided by a 120 VDC supply that is controlled and monitored by the test station computer.

The computer displays the command status and health of the system on a CRT monitor. Response data from the unit is processed and various flags, alarms and shutdowns are automatically implemented by the station. Hard copy of the display may be obtained on command or at regular intervals. Test data can be stored or transferred to floppy disk for post processing.

<u>Force And Moment Table</u>

To measure output performance, a force and moment table was designed and built for the CMG system. This table, illustrated in Figure 11 uses four piezoelectric three-axis force sensors whose outputs are summed according to their mounting geometry to produce the three forces and moments that completely describe the system mechanical output. Using a digital signal analyzer, these signals can be displayed in either the time or frequency domain to characterize the output performance levels as described in the following paragraphs.

Figure 11 DGCMG On Force And Moment Table

<u>Frequency And Step Response</u>

Figure 12 shows the results obtained for the frequency response test of the inner gimbal loop. This is typical of results obtained for both loops and demonstrates the dependency on the inner gimbal angle. In general, the agreement is excellent when compared with the predicted results given earlier, although the test data has slightly more peaking. This effect was also noticed in the step response behavior of the system in both the overshoot and settling characteristics observed. This difference has been attributed to the gear train stiffness being lower than anticipated.

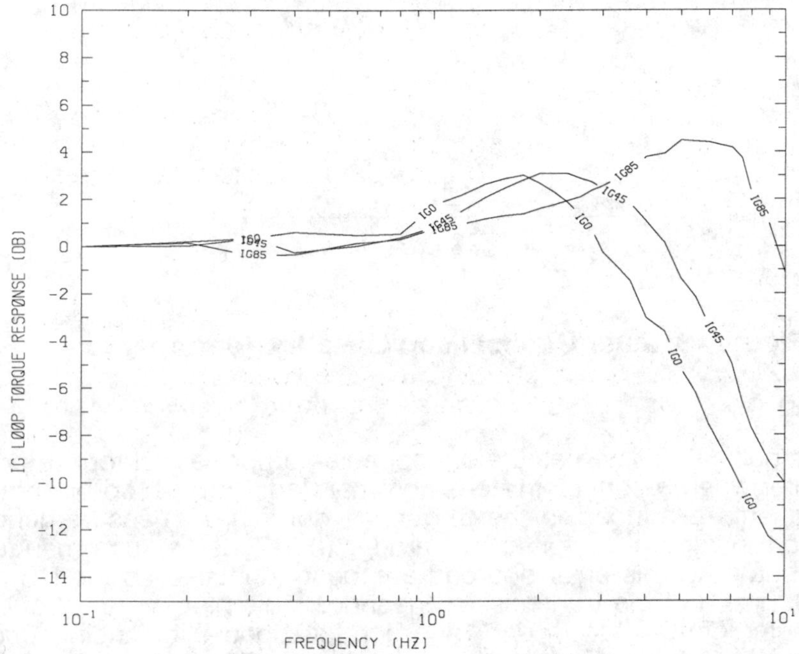

Figure 12 Measured IG Loop Frequency Response

<u>Gimbal Rate Linearity</u>

Gimbal loop scale factor and linearity were measured for various commanded rates. The test consisted of commanding a constant rate for a known period of time and computing the actual rate from the change in gimbal angle. Figure 13 presents a plot of the difference between the measured rate and linear fit of the data. The results are typical for both loops and yield a linearity below the design goal of 0.1%. Due to the manner in which the test was performed, these results represent errors in the test method employed rather than the system accuracy. This is consistent with the expected performance for the phase-locked-loop implementation as well as the commanded zero rate drift which is below the threshold of what could be measured.

115

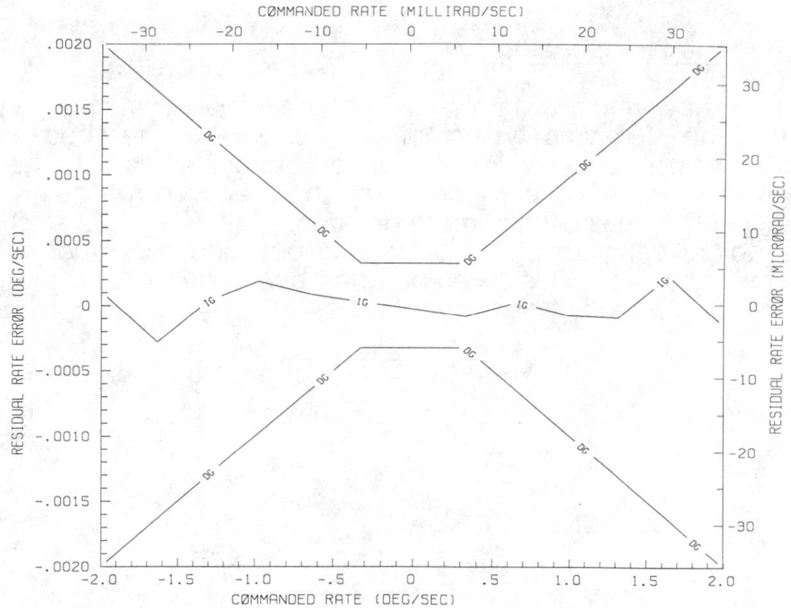

Figure 13 Inner Gimbal Loop Linearity Performance

Torque Noise

Torque noise is defined as the undesirable component of torque produced by the actuator when a constant rate is commanded. Expressed in terms of the RMS components produced in the frequency domain, a measurement of this noise is given in Figure 14 for a 5.7 mrad/s (0.327 deg/s) commanded rate. Major contributors to this error source have been identified as the DC offset in the drive voltages and the transmission gearing. This plot is typical of both the inner and outer gimbal loops and incorporates special balancing circuits to minimize the effect of the offset of the drive voltages. Total noise when viewed in the time domain produces a value of approximately 4% RMS for the case given.

Induced Vibration

Another performance parameter important to the operation of the Space Station is the induced vibration of the CMGs. Concerns exist for potential coupling to the inertial sensors, located on the same pallet, as well as the effect produced on the station micro-gravity environment. The dominant contributor to this performance is the balance of the rotor wheel as it rotates at the nominal speed of 6600 rpm. A full characterization of the system would consist of the three forces and moments that would be seen at the mounting interface for a host of gimbal positions. Typical values for these parameters have been measured and result in forces that range from 0.9 to 5.8 N (0.2 to 1.3 lbs) and moments that range from 0.68 to 2.9 N-m (0.5 to 2.1 ft-lbs).

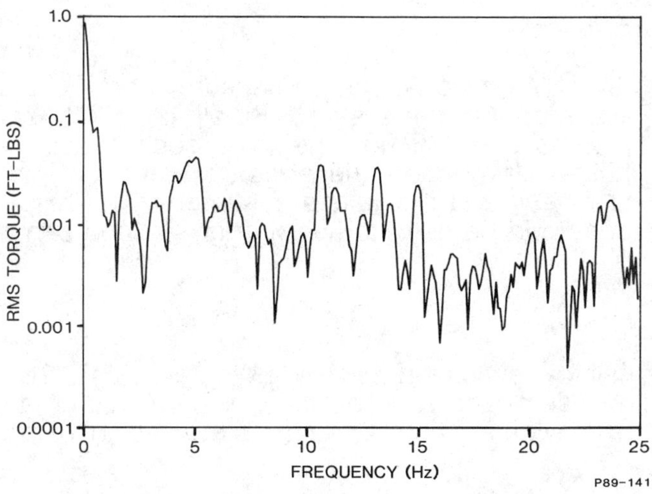

Figure 14 Torque Noise Performance

<u>Gimbal Angle Readout</u>

A readout of the gimbal positions is provided as a system output to provide necessary information to the momentum management and vehicle control laws. Measurements of the readout accuracies of both gimbals were below 1.7 mrad (0.1 deg) for all cases measured.

<u>Wheel Speed Control</u>

During all operations of the CMG, the wheel speed was monitored and the performance of the loop has exceeded the design goal of 0.1 %. Actual speed never exceeds 2 rpm from the commanded value even during maximum rate conditions.

PROPOSED FUTURE TESTS

The results obtained to date represent the current state of the system characterization. Future work is planned to improve the response and evaluate other performance parameters.

<u>Cross-compensation</u>

Cross-coupling, between the inner and outer gimbal servo loops, causes variations in the frequency response as a function of the inner gimbal angle. Although these bandwidth variations appear to be acceptable for the accuracy requirements of Space Station, a proposed improvement is to use a variable cross-feed compensation. The expected results would produce a frequency response characteristic nearly independent of inner gimbal angle.

117

<u>Small Signal Characterization</u>

Results for relatively large signal performance have previously been shown to agree with the linear model of the system. Small signal rate commands will be used to determine the effect of non-linearities such as dead-zone, if any exists, and gimbal friction. If required, these test results could then be used to modify both the model itself and the values assumed. The result would be a high-fidelity model that could be used to assist analyses and simulations of the ACS.

<u>Life Tests</u>

Current plans call for verification and life testing at MSFC. The support test equipment has been designed to simulate the duty cycle commands expected for Freedom and thus provide a means to address the design life performance.

CONCLUSIONS

The development of the prototype DGCMG has provided NASA with a working design that meets or exceeds the goals of the Advanced Development Program. In addition to providing a safe design for man-rated missions, this device can contribute to the Attitude Control System definition and address concerns at a relatively early stage in the Space Station development. Major accomplishments of the program include the following:

o Successful demonstration of compliance to all the system
   requirements and design constraints imposed

o Concurrence of analytical models and simulation results to the
   measured performance

o Development of the necessary test station and the measurement
   equipment needed to characterize output performance

ACKNOWLEDGEMENTS

The work presented in this paper was performed under contract NAS8-36628 with the Marshall Space Flight Center of the National Aeronautics and Space Administration. The authors wish to express their appreciation to the entire CMG program staff for the dedicated efforts put forth in the design and fabrication of the prototype unit.

DISCRETE-TIME CONTROL OF A SPACECRAFT WITH RETARGETABLE FLEXIBLE ANTENNAS

Leonard Meirovitch[*] and Martin E. B. France[†]

This paper is concerned with the control of a spacecraft consisting of a rigid platform and retargetable flexible antennas. The mission consists of a minimum-time maneuver of the antenna(s) to coincide with predetermined line(s) of sight, while stabilizing the platform in an inertial space and suppressing the elastic vibration of the antenna(s). The system is modeled by a set of linearized, time-varying equations of motion. A discrete-time approach permits consideration of the time-varying nature of the system in designing the digital control law. Several control techniques were investigated and results from numerical examples involving a spacecraft with a single flexible antenna are presented.

INTRODUCTION

As the complexity of future space systems and missions increases, e.g., NASA's Space Station, the problem of maneuver and control of flexible spacecraft takes on added importance. While the orientation of the line of sight of simpler flexible spacecraft involves slewing the entire vehicle (Refs. 1-3), the new generation of spacecraft may have several substructures requiring independent, simultaneous retargeting (Ref. 4). The spacecraft considered in Ref. 4 consists of a rigid platform stabilized in an inertial space and several appendages hinged to the platform so as to allow rotation about two orthogonal axes relative to the platform. Such a system can be described by a set of hybrid, linear differential equations with

---

[*] University Distinguished Professor, Department of Engineering Science and Mechanics, Virginia Polytechnic Institute & State University, Blacksburg, Virginia 24061.

[†] Captain, U.S. Air Force; Graduate Student, Department of Engineering Science and Mechanics, Virginia Polytechnic Institute & State University, Blacksburg, Virginia 24061.

time-varying coefficients. This paper is concerned with the
control of a system such as in Ref. 4 during independent
maneuvering of flexible antennas.

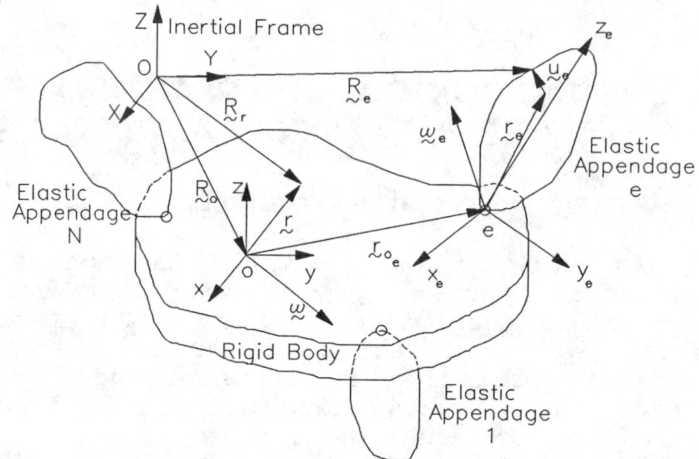

Fig. 1   Rigid Platform with Flexible Appendages

Figure 1 shows a spacecraft of the type described above
consisting of a rigid main body and several flexible appendages.
The equations of motion are derived in Ref. 4 and a summary
of the equations is presented below. The control problem is
treated in Refs. 4 and 5 under the assumption that the time-
varying terms in the coefficients are sufficiently small that
they can be ignored for the purpose of control design. This
is based on the premise that the appendages are relatively
small and that the maneuvers are "slow."

This paper extends the work of Refs. 4 and 5 by including
the time-varying terms in the control design. A discrete-time
approach permitting control design with a step-varying plant,
as well as better computer simulation of the response, is used.

As in Refs. 4 and 5, the maneuver of each antenna is
carried out open-loop using a bang-bang control law, while
simultaneously performing platform stabilization and appendage
vibration suppression. To this end, a variety of control
techniques are considered. A discrete-time linear quadratic
regulator method is first used, followed by linear decentralized
control and another decentralized control using a combination
of linear and nonlinear control. Each approach is demonstrated
by means of a numerical example involving a spacecraft with
a single flexible antenna undergoing a 45° reorientation
relative to the platform, while the platform is stabilized
relative to an inertial space. The results for the various
cases are then compared and discussed.

EQUATIONS OF MOTION

The equations of motion for the system shown in Fig 1 were derived in Ref. 4 using a new formulation of Lagrange's equations for flexible bodies in terms of quasi-coordinates (Ref. 6). Following is a summary of these equations. Referring to Fig 1, we identify a set of inertial axes XYZ, a set of body axes xyz attached to the rigid platform and another set of axes $x_e y_e z_e$ embedded in the typical flexible appendage e. The position vector of a point in the rigid body is given by $\underline{R}_r = \underline{R}_o + \underline{r}$ and that in the appendage can be written as $\underline{R}_e = \underline{R}_o + \underline{r}_{oe} + \underline{r}_e + \underline{u}_e$, $(e = 1, 2, ..., N)$, where $\underline{R}_o$ is the radius vector from O to o, $\underline{r}$ is the position vector of a point in the rigid body relative to xyz, $\underline{r}_{oe}$ is the radius vector from o to e, $\underline{r}_e$ is the position vector of a nominal point in the undeformed appendage relative to $x_e y_e z_e$ and $\underline{u}_e$ is the elastic displacement of that point. Vector $\underline{R}_o$ is expressed in terms of components along XYZ, $\underline{r}$ and $\underline{r}_{oe}$ in terms of components along xyz and $\underline{r}_e$ and $\underline{u}_e$ in terms of components along $x_e y_e z_e$. The velocity vector of o can be written in terms of components along xyz in the form $\underline{V}_o = C\dot{\underline{R}}_o$, where C is the matrix of direction cosines between xyz and XYZ and $\dot{\underline{R}}_o$ is the velocity vector of o in terms of components along XYZ. The angular velocity vector of axes xyz in terms of components along xyz is given by $\underline{\omega} = D\dot{\underline{\theta}}$, where $\dot{\underline{\theta}}$ is a vector of angular velocities $\dot{\theta}_i$ and D is a matrix depending on the angular displacements θ_i $(i = 1, 2, 3)$. In view of the above, the velocity vector of a point on the rigid body in terms of components along xyz is

$$\underline{V}_r = \underline{V}_o + \underline{\omega} \times \underline{r} \tag{1}$$

and that of a point in appendage e in terms of components along $x_e y_e z_e$ is

$$\underline{V}_e = E_e(\underline{V}_o + \underline{\omega} \times \underline{r}_{oe}) + (E_e\underline{\omega} + \underline{\omega}_e) \times (\underline{r}_e + \underline{u}_e) + \underline{v}_e, \quad e = 1, 2, ..., N \tag{2}$$

where $\underline{\omega}_e$ is the angular velocity of axes $x_e y_e z_e$, E_e is a matrix of direction cosines between axes $x_e y_e z_e$ and xyz and $\underline{v}_e$ is the elastic velocity of the point in the appendage relative to $x_e y_e z_e$, $\underline{v}_e = \dot{\underline{u}}_e$. For the proposed maneuver, the angular velocity vectors $\underline{\omega}_e$ of $x_e y_e z_e$ relative to xyz are given, so that the rotational motions of the appendages relative to the platform do not add degrees of freedom to the system.

The equations describing the rigid-body translations and rotations of the system are ordinary differential equations and those for the elastic motions of the appendages are partial differential equations, so that the equations of motion are hybrid. Because control design for systems described by hybrid equations is not feasible, we must discretize the system in space. This is accomplished by expressing the elastic displacements as linear combinations of space-dependent admissible functions multiplied by time-dependent generalized coordinates of the form

$$\underline{u}_e(\underline{r}_e,t) = \Phi_e(\underline{r}_e)\underline{q}_e(t), \quad e = 1,2,\dots,N \tag{3}$$

where Φ_e is a matrix of admissible functions and $\underline{q}_e$ is a vector of generalized coordinates.

The equations of motion are given in detail in Ref. 4 and will not be repeated here in full. Assuming small motions of the stabilized rigid platform, a linearized version of these equations has the state form

$$\underline{\dot{x}}(t) = A(t)\underline{x}(t) + B(t)\underline{f}(t) + D(t)\underline{d}(t) \tag{4}$$

where $\underline{x}(t) = [\underline{R}_o^T \underline{\theta}^T \underline{q}_1^T \underline{q}_2^T \cdots \underline{q}_N^T \underline{V}_o^T \underline{\omega}^T \underline{\dot{q}}_1^T \underline{\dot{q}}_2^T \cdots \underline{\dot{q}}_N^T]^T$ is a state vector, in which $\underline{\theta}$ is a symbolic vector of angular displacements of the platform and

$$A(t) = \left[\begin{array}{c|c} 0 & I \\ \hline -M^{-1}(t)K(t) & -M^{-1}(t)G(t) \end{array}\right] \tag{5a}$$

$$B(t) = \left[\begin{array}{c} 0 \\ \hline M^{-1}(t)B^*(t) \end{array}\right] \qquad D(t) = \left[\begin{array}{c} 0 \\ \hline M^{-1}(t) \end{array}\right] \tag{5b,c}$$

are coefficient matrices. Moreover, $\underline{d}(t)$ is a vector of disturbances caused by the maneuver and $\underline{f}(t) = \left[\underline{F}_o^{*T} \underline{M}_o^{*T} \underline{f}_{11}^T \underline{f}_{12}^T \cdots \underline{f}_{1n_1}^T \underline{f}_{21}^T \underline{f}_{22}^T \cdots \underline{f}_{2n_2}^T \underline{f}_{31}^T \cdots \underline{f}_{Nn_N}^T\right]^T$ is the control force vector, in which $\underline{F}_o^*$ and $\underline{M}_o^*$ are actuator force and torque vectors acting on the rigid platform and $\underline{f}_{ei}$ are the actuator forces acting on appendage e at point i, where appendage e has n_e actuators. Explicit expressions for the matrices $M(t)$, $K(t)$, $G(t)$ and $B(t)$ are given in the APPENDIX.

TIME-VARYING LQR WITH DISTURBANCE ACCOMMODATION

We propose to control the system using a combination of open- and closed-loop controllers, taking into account the time-varying nature of the spacecraft and the fact that for an open-loop maneuver the inertial disturbances are known. Hence, the control is separated into open and closed-loop components as follows:

$$\underline{f}(t) = \underline{f}_o(t) + \underline{f}_c(t) \tag{6}$$

Then, the open-loop control is chosen to satisfy

$$B^*(t)\underline{f}_o(t) + D(t)\underline{d}(t) = \underline{0} \tag{7}$$

so that the open-loop control is given by

$$\underline{f}_o(t) = -[B^*(t)]^\dagger D(t)\underline{d}(t) \tag{8}$$

where

$$[B^*(t)]^\dagger = [(B^*(t))^T B^*(t)]^{-1}[B^*(t)]^T \tag{9}$$

is the pseudo-inverse of $B^*(t)$.

For the closed-loop control, we consider a discrete-time linear-quadratic optimal control. The discrete-time equivalent of the system state equations is (Ref. 7)

$$\underline{x}(k+1) = \mathcal{A}(k)\underline{x}(k) + \mathcal{B}(k)\underline{u}(k), \quad k = 0,1,2,\dots \tag{10}$$

where k denotes the time $t = t_k = kT$, in which T is the sampling period, and

$$\mathcal{A}(k) = e^{A(k)T} = I + \frac{A(k)T}{1!} + \frac{A^2(k)T^2}{2!} + \cdots + \frac{A^i(k)T^i}{i!} + \cdots \tag{11}$$

$$\mathcal{B}(k) = \left[IT + \frac{A(k)T^2}{2!} + \frac{A^2(k)T^3}{3!} + \cdots + \frac{A^i(k)T^{i+1}}{(i+1)!} + \cdots \right] B(k) \tag{12}$$

are discrete-time coefficient matrices. Equations (11) and (12) are obtained by regarding $A(t)$ and $B(t)$, Eqs. (5a,b), as being constant over the steps $kT < t < kT + T$, $(k = 0,1,2,\dots)$. The discrete-time optimal feedback control can be written as

$$\underline{u}(k) = \mathcal{K}(k)\underline{x}(k), \quad k = 0,1,2,\dots \tag{13}$$

where $\mathcal{K}(k)$ is the discrete-time control gain matrix, obtained by minimizing the discrete-time quadratic cost function

$$J = \underline{x}^T(N)P(N)\underline{x}(N) + \sum_{k=0}^{N-1}[\underline{x}^T(k)Q(k)\underline{x}(k) + \underline{u}^T(k)R(k)\underline{u}(k)] \quad (14)$$

The weighting matrices $P(N), Q(0), Q(1), \ldots, Q(N-1)$ and $R(0), R(1), \ldots, R(N-1)$ are symmetric. Moreover, $Q(k)$ and $R(k)$ are positive definite and $P(N)$ is positive semidefinite, and for our purposes they are assumed to be constant and diagonal for all N control steps with $P(N) = Q(N-1)$. Also, it is assumed throughout this paper that all system states are available for feedback, and that there is no estimation error involved.

The solution to this state-feedback problem is based on the work of either Kalman or Bellman and can be found in a variety of texts . Following the procedure based on Bellman's Principle of Optimality described in Ref. 7, the optimal feedback gains can be calculated backwards in step. Beginning with $i = 1$ and the known $P(N)$, we can write

$$\mathcal{K}(N-i) = -[\mathcal{B}^T(N-i)P(N+1-i)\mathcal{B}(N-i) + R(N-i)]^{-1} \quad (15)$$
$$\times \mathcal{B}^T(N-i)P(N+1-i)\mathcal{A}(N-i)$$

$$P(N-i) = [\mathcal{A}(N-i) + B(N-i)\mathcal{K}(N-i)]^T P(N+1-i)[\mathcal{A}(N-i) \quad (16)$$
$$+ \mathcal{B}(N-i)\mathcal{K}(N-i)] + \mathcal{K}^T(N-i)R(N-i)\mathcal{K}(N-i) + Q(N-i)$$

and continue until $\mathcal{K}(0)$ has been calculated. For our system, the final step will occur several time steps after the slewing maneuver is completed, at which time the system can be regarded as time-invariant. Therefore, the gains calculated soon after the completion of the maneuver will be treated as constant, leading to steady-state regulation.

In implementing such a control law, the LQR gains are calculated prior to maneuver execution based on the known system parameters and slewing profile and stored in the computer for use during and after the maneuver. The disturbance accommodating open-loop control is calculated during the maneuver and updated with each time step.

In evaluating the utility of such a control law versus the time-invariant approximation, we must weigh the improvements in the time response of the spacecraft against the added computational complexity involved in a time-varying plant.

DECENTRALIZED CONTROL

Implementing a decentralized control law has many significant advantages for the type of system under investigation. Especially in the case of a spacecraft with several antennas, using many simple independent controllers could reduce significantly the computational requirements relative to a global control system. The computations can also be carried out completely "on-line," as opposed to the significant amount of premaneuver computations required by global controls. This computational savings could allow for faster, more accurate sampling in the case of a discrete-time control system.

In this paper, we investigate two different types of decentralized control, the first in which individual points on the antennas are controlled with collocated sensors and actuators (Collocated Control) and the second in which individual modes of the independently modeled substructures are controlled (Substructure Decentralized Control). A method of substructure control based on the Component Mode Synthesis method has been proposed by Young (Ref. 8). In the case of Substructure Decentralized Control, each appendage has an individual controller whose actuator forces are based solely on that appendage sensor output. In both cases, the rigid-body motions of the entire system are controlled independently via actuators and sensors located on the main platform. In the case of Collocated Control, we assume that both displacements and velocities relative to the appendage body axes are available. Likewise, in the case of Substructure Decentralized Control, all states associated with the flexible motion of the antennas are available. Though designed using a decentralized approach, the response in both cases is calculated by applying the decentralized control law to the full spacecraft.

Collocated Control

In this case, we first consider a linear control law for the controllers at all points. Control gains for each controller are designed and implemented using a discrete-time approach considering the time-varying nature of the plant. Each of the six controllers responsible for the rigid body motion of the spacecraft are designed separately, each with a closed-loop frequency chosen well above the fundamental frequency of the flexible appendage and damping equal to 70% of the critical damping. The controllers are designed on the basis of the discrete-time version of the state equations. The continuous-time equations for the rigid-body motion are

$$\frac{d}{dt}\begin{bmatrix} x_i(t) \\ \dot{x}_i(t) \end{bmatrix} = \mathbf{A}_i \begin{bmatrix} x_i(t) \\ \dot{x}_i(t) \end{bmatrix} + \mathbf{B}_i(t)u_i(t), \quad i = 1, 2, \ldots, 6 \tag{17}$$

where

$$A_i = \begin{bmatrix} 0 & 1 \\ 0 & 0 \end{bmatrix}, \quad B_i(t) = \begin{bmatrix} 0 \\ m_{ii}^{-1}(t) \end{bmatrix}, \quad i = 1,2,\ldots,6 \qquad (18a,b)$$

in which m_{ii} is mass or moment of inertia corresponding to the rigid body mode i. Using full state feedback for each controller, the closed-loop difference equation is

$$\underline{x}(k+1) = [\mathcal{A}(k) + \mathcal{B}(k)\mathcal{K}(k)]\underline{x}(k), \quad k = 0,1,2,\ldots \qquad (19)$$

where $\mathcal{K}(k)$ is chosen so as to yield satisfactory closed-loop response.

For the point controllers located on the appendage(s), the control law is based on the concept of Uniform Damping Control (Ref. 9). This technique is attractive because the control design is independent of the system stiffness. When n discrete actuators are used on an antenna, the uniform damping control forces at points at points P_r are given by

$$\underline{F}_r(t) = -\alpha^2 M_r \underline{U}_r(t) - 2\alpha M_r \underline{\dot{U}}_r(t), \quad r = 1,2,\ldots,n \qquad (20)$$

where $\underline{U}_r(t) = \underline{u}(P_r,t)$ and $\underline{\dot{U}}_r(t) = \underline{\dot{u}}(P_r,t)$ are measurements of displacement and velocity, respectively, at points P_r and $M_r = \int_{D_r} \rho(P)\,dD_r$ is the mass associated with the region on which $\underline{F}_r(t)$ acts; α is the desired exponential decay rate of the closed-loop response.

Next, we implement a nonlinear (bang-off-bang) control law for the point controllers on the appendage. A combination of velocity and position feedback is used in conjunction with a deadzone. A dual elliptical deadzone is used to reduce control interaction between subsystems and slow limit-cycle behavior, thus conserving fuel. The two deadzones are such that the "off" deadzone is smaller and lies completely within the "on" deadzone (Ref. 10). An example of this proposed switching curve is shown in the phase plane of Fig. 2.

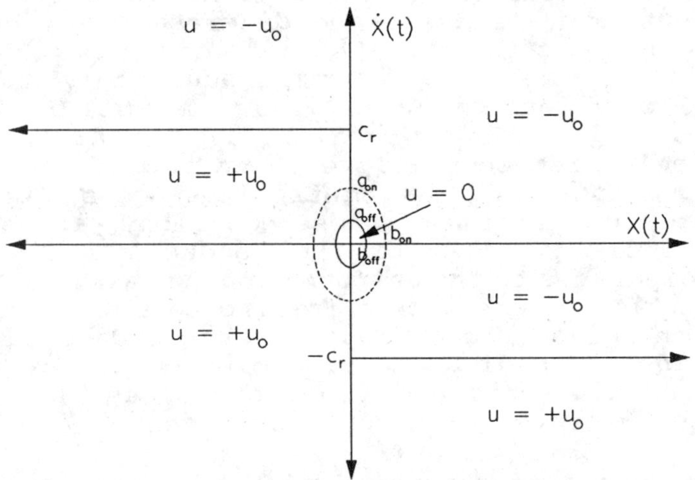

Fig. 2   Antenna Collocated Control Switching Curve

Substructure Decentralized Control

We propose to control each flexible substructure by means of actuator forces depending on state measurements associated with a given substructure alone. For example, in the case of an antenna modeled as a flexible beam hinged to a rigid platform, we wish to use a controller that bases the feedback forces on displacement and velocity information at discrete points on the antenna to control a finite number of antenna cantilever modes.

As in the case of Collocated Control, we first consider linear control for both the rigid-body and the flexible motion. The control law used for rigid-body motion is unchanged from the Collocated Control cases. The equations of motion for substructure e can be expressed as the infinite set of decoupled second-order equations

$$\ddot{q}_i(t) + \omega_i^2 q_i(t) = Q_i(t), \quad i = 1, 2, \dots \tag{21}$$

or as the set of first-order state equations

$$\frac{d}{dt}\begin{bmatrix} q_i(t) \\ \dot{q}_i(t) \end{bmatrix} = \begin{bmatrix} 0 & 1 \\ -\omega_i^2 & 0 \end{bmatrix}\begin{bmatrix} q_i(t) \\ \dot{q}_i(t) \end{bmatrix} + \begin{bmatrix} 0 \\ Q_i(t) \end{bmatrix}, \quad i = 1, 2, \dots \tag{22}$$

where ω_i^2 is the natural frequency and $Q_i(t)$ the generalized control force, each associated with the ith mode.

Of course, it is neither necessary nor possible to control all the substructure modes using discrete controllers. The first step, then, is to truncate these equations, controlling enough modes to simulate the motion adequately and minimize control spillover into the uncontrolled modes. Using a number of actuators larger than or equal to the number of controlled modes also helps reduce spillover. We choose feedback gains for the equations of the controlled modes so that the discrete-time formulation of these equations yields a damping factor and settling time for the fundamental mode of each appendage comparable to the uniform damping case. Lower factors for the higher modes are picked leading to a settling time for each equation comparable to that of the fundamental mode for that appendage. The discrete-time closed-loop system matrix can then be constructed and the eigenvalues checked at key points in the maneuver to assure convergence.

The r generalized control forces that result can then be transformed to actual control forces for each antenna using the transformation

$$\underline{f}_e(k) = \left[\Phi_e^T\left(\underline{r}_{e_1}\right) \quad \Phi_e^T\left(\underline{r}_{e_2}\right) \quad \cdots \quad \Phi_e^T\left(\underline{r}_{e_r}\right)\right]^T \underline{Q}(k) \tag{23}$$

where $\Phi_e^T\left(\underline{r}_{e_i}\right)$ is the substructure participation matrix, i.e., a matrix of admisssible functions evaluated at the location of actuator e_i, and $\underline{Q}(t) = [Q_1 Q_2 \cdots Q_r]^T$ is the vector of generalized controls. The coefficient matrix for the closed-loop system can then be expressed as

$$\mathcal{A}_{cl}(k) = \mathcal{A}(k) + \mathcal{B}(k)\mathcal{K} \tag{24}$$

where

$$\mathcal{K} = \left[-\frac{\mathcal{K}_{11\,6\times6}}{\mathbf{0}_{6\times6}} + \frac{\mathbf{0}_{6\times10}}{\mathcal{K}_{22\,6\times10}} \;\middle|\; \frac{\mathcal{K}_{13\,6\times6}}{\mathbf{0}_{6\times6}} + \frac{\mathbf{0}_{6\times10}}{\mathcal{K}_{24\,6\times10}} \right] \tag{25}$$

in which $\mathcal{K}_{11}$ and $\mathcal{K}_{13}$ are diagonal, the first with elements $c_{ii}(t)$ $(i = 1, 2, \ldots, 6)$ corresponding to discrete-time position feedback gains for each the ith rigid-body mode and the second with elements $b_{ii}(t)$ $(i = 1, 2, \ldots, 6)$ corresponding to velocity feedback gains. Moreover, $\mathcal{K}_{22}$ and $\mathcal{K}_{24}$ are the result of transforming diagonal matrices (with each diagonal element corresponding to a feedback term from the substructure's modal equation similar in form to the rigid body equations) to the configuration space using the substructure matrix of admissible functions; they are given in detail in the APPENDIX. Of course, these gains are constant since they are based on the independent, time-invariant substructure, and nonzero for the controlled modes only.

The nonlinear case for Substructure Decentralized Control is similar to the Collocated Control case for controlling the rigid-body motion. The flexible motion of each controller is controlled by bang-off-bang controllers associated with substructure modes. Because the natural frequencies of a substructure are only indirectly related to the eigenvalues of the overall system, basing the switching curves solely on substructure eigenvalues would have little value. We add nonlinear damping to each mode (i.e., velocity feedback) using an actuator force similar to that used in Ref. 12, i.e., using the control law

$$f_r = \begin{cases} -k_r, & \dot{q}_r \geq d_r \\ 0, & |\dot{q}_r| < d_r \\ k_r, & \dot{q}_r \leq -d_r \end{cases} \qquad , \qquad r = 1,2,3 \tag{26}$$

However, if the fundamental frequency of the system with stabilized rigid body motion is so low that settling time requirements can be met only through added nonlinear stiffness (i.e., position feedback), then a more appropriate control law is

$$f_r = \begin{cases} -k_r, & q_r \geq c_r \ \cup \ [q_r \geq -c_r \ \cap \ \dot{q}_r \geq d_r] \\ 0, & |q_r| < c_r \ \cap \ |\dot{q}_r| < d_r \\ k_r, & q_r \leq -c_r \ \cup \ [q_r \leq c_r \ \cap \ \dot{q}_r \leq -d_r] \end{cases} \qquad , \qquad r = 1,2,3 \tag{27}$$

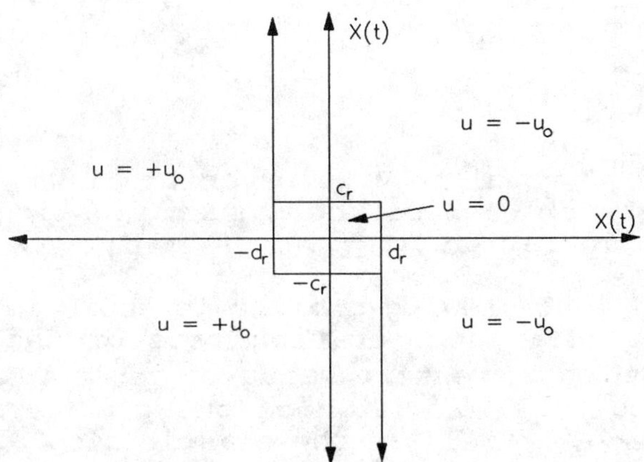

Fig. 3   Antenna Substructure Control Switching Curve

In this case, position feedback is used when large displacements occur, while velocity feedback is employed for small amplitude motions. Figure 3 shows this criterion on the phase plane.

NUMERICAL EXAMPLE

The mathematical model consists of a rigid platform with one flexible antenna, represented by a uniform beam, hinged to the platform at one end and free at the other (Fig. 4). The antenna is slewed through a 45° angle about the x_e axis. The beam is discretized in space using five admissible functions for each displacement component. The admissible functions for the y_e displacement component are taken in the form the of cantilever modes

$$\phi_{y_j} = -(\cos\beta_j z - \cosh\beta_j z) + C_j(\sin\beta_j z - \sinh\beta_j z), \quad j = 1,2,3,4,5$$

This is consistent with the fact that the flexible antenna represents a slewing cantilever beam, where the slewing angle is a given function of time. The admissible functions for the x_e displacement component have the same form. The platform has one actuator corresponding to each rigid body degree-of-freedom and the antenna has three sets of collocated actuators and sensors at $z_i = l_e/3, 2l_e/3, l_e$ for both the y_e and x_e directions. The following data was used in the computer simulation:

$$m_r = 10.0\,kg \qquad m_e = 1.0\,kg$$

$$l_e = 5.0\,m \qquad r_{o_e} = 2.0\,\hat{k}\,m \qquad EI_e = 122.28\,N \cdot m^2$$

$$\mathbf{I}_e = \begin{bmatrix} 8.33 & 0 & 0 \\ 0 & 8.33 & 0 \\ 0 & 0 & 0 \end{bmatrix} kg \cdot m^2 \qquad \mathbf{I}_r = \begin{bmatrix} 20.0 & 0 & 0 \\ 0 & 20.0 & 0 \\ 0 & 0 & 20.0 \end{bmatrix} kg \cdot m^2$$

Structural damping equal to 0.2% of critical was added to the simulation model for each of the system elastic modes in the premaneuver configuration.

The time-optimal slewing profile is shown in Fig. 5. For all cases the angular acceleration rate of the antenna was $\pm\pi/8$ rad/s, so that the entire maneuver is completed in about 2.8 s. The uncontrolled motion of the platform relative to the inertial space, as well as the elastic displacement of the antenna tip is shown in Fig. 6.

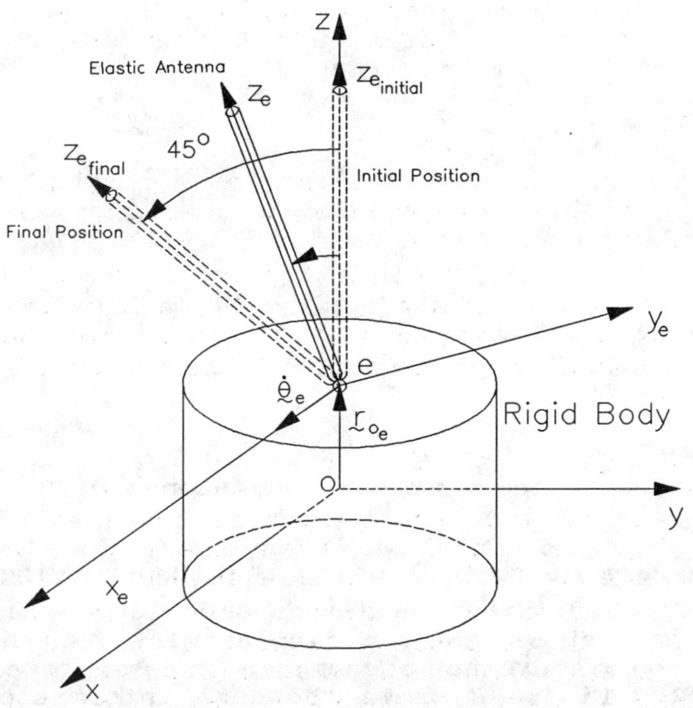

Fig. 4   Numerical Example System

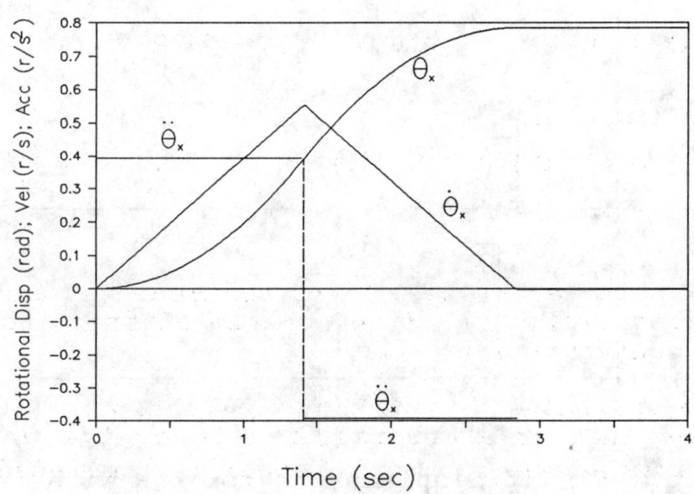

Fig. 5   Antenna Slew Profile

131

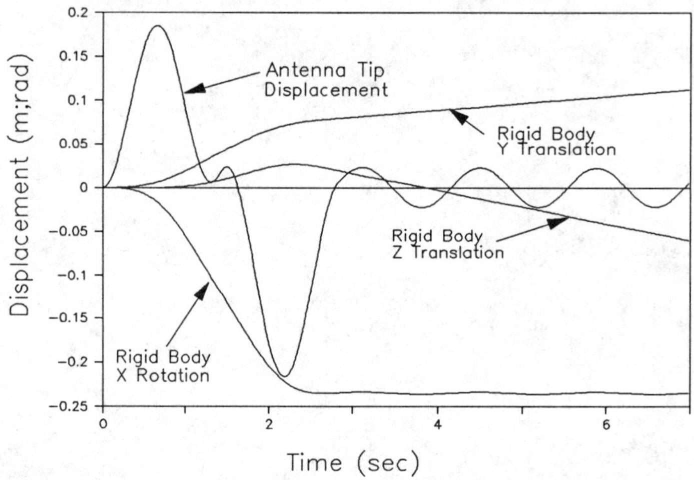

Fig. 6   Uncontrolled Platform and Antenna Displacements

In the case of control using the step-varying LQR, the weighting matrices $P = Q = 100 \times [I]_{24 \times 24}$ and $R = .001 \times [I]_{12 \times 12}$ were used.   Figure 7 shows the tip displacement for the case in which open-loop disturbance accommodation is employed together with the LQR.   Figure 8 shows the same information without disturbance accommodation.

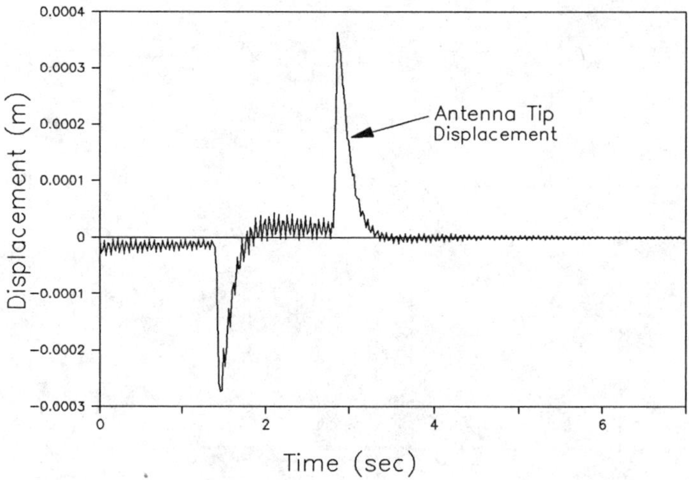

Fig. 7   Tip Displacement for LQR with
Disturbance Accommodation

132

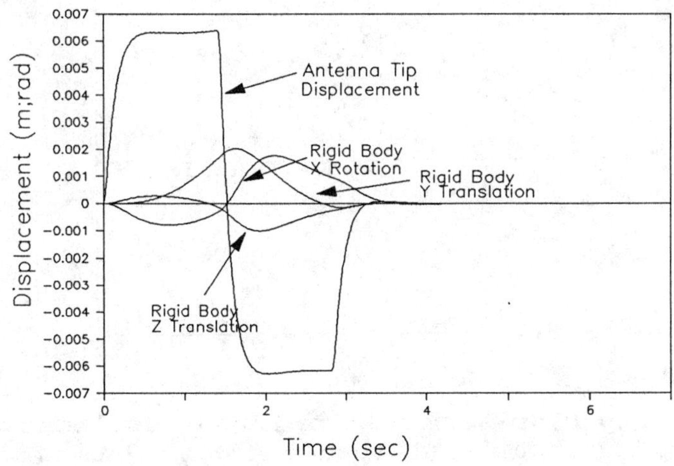

Fig. 8   Tip Displacement for LQR without
Disturbance Accommodation

Linear Collocated Control using uniform damping was implemented using a decay rate $\alpha = 5.0$.   The rigid-body motion was controlled using a linear control law with $\omega_n = 10\,\mathrm{rad/s}$ and $\zeta = 0.70$.   Figures 9 and 10   show displacements with and without the open-loop disturbance accommodating control, respectively.

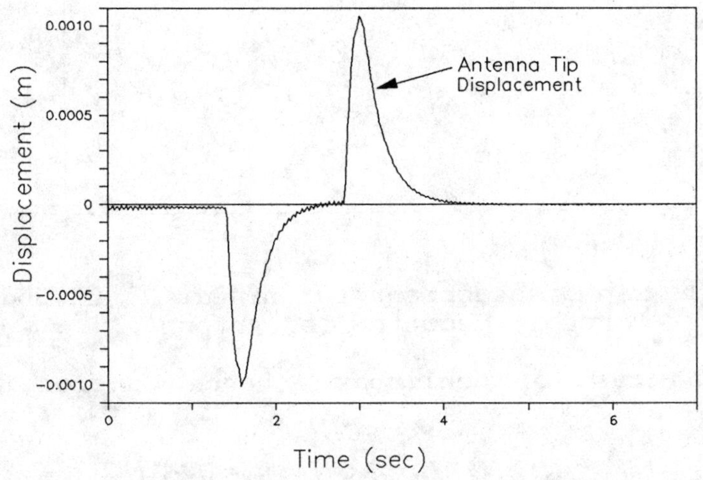

Fig. 9   Tip Displacement for Linear Collocated
Control with Disturbance Accommodation

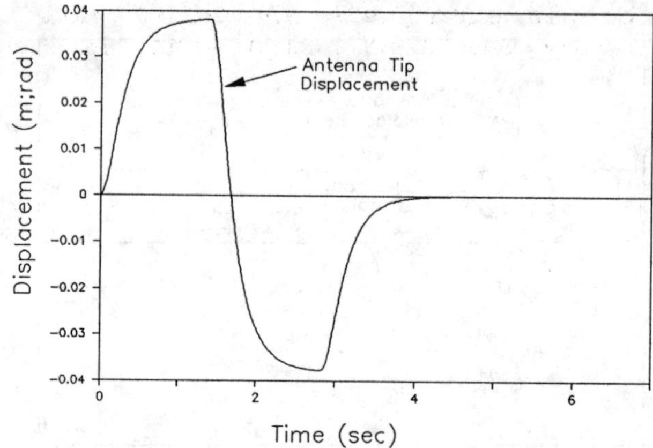

Fig. 10 Tip Displacement for Linear Collocated
Control without Disturbance Accommodation

When the rigid-body modes only are controlled linearly,
the resulting uncontrolled tip displacement of the antenna in
the y_e direction is quite large, as can be seen in Figure 11.

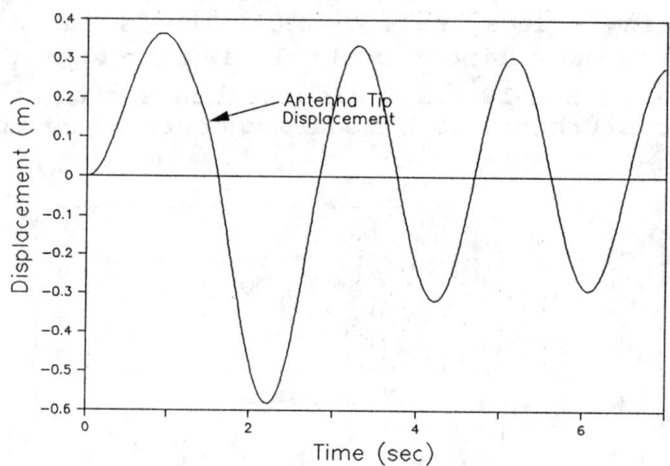

Fig. 11 Tip Displacement for Linear Rigid Body
Control, Uncontrolled Antenna

In the case of nonlinear antenna control using the
parameters

$$f_1 = 0.18N \quad f_2 = 0.5N \quad f_3 = 0.22N$$

$$c_r = .05\,m/s, \quad a_{off_r} = .001\,m/s \quad a_{on_r} = .01\,m/s$$

$$b_{off_r} = .003\,m/s \quad b_{on_r} = .03\,m/s, \quad r = 1,2,3$$

134

the resulting tip displacement is as shown in Figure 12. Figures
13, 14 and 15 display the time history of the substructure
actuator forces.

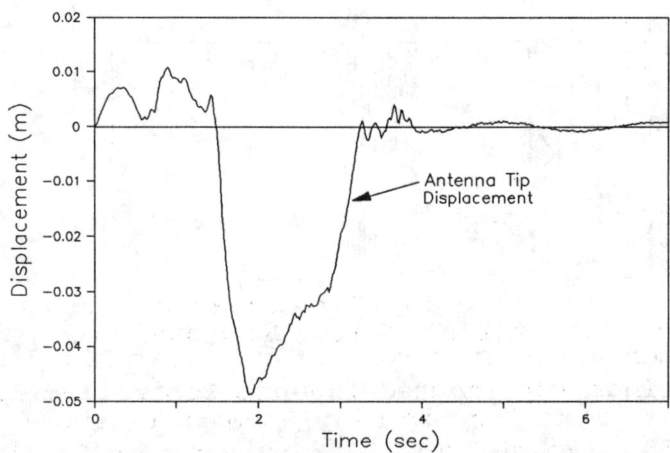

Fig. 12   Tip Displacement for Linear Rigid Body Control,
Nonlinear Collocated Antenna Control

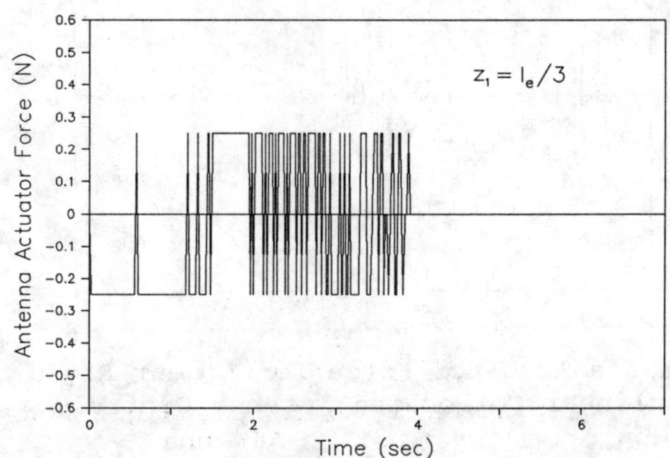

Fig. 13   Antenna Actuator Force for Linear Rigid Body
Control, Nonlinear Collocated Antenna Control, $z_1 = l_e/3$

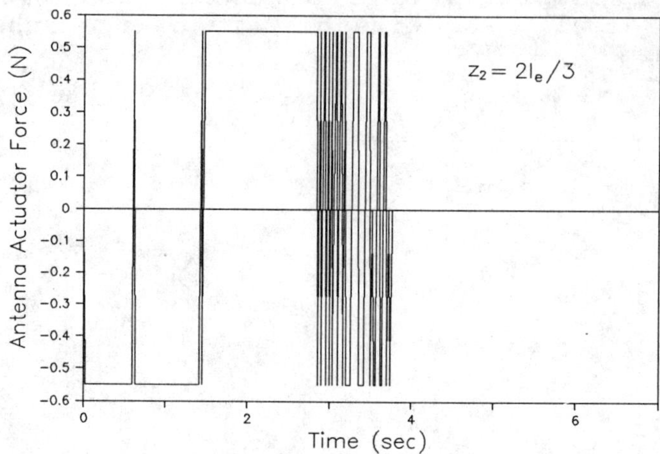

Fig. 14   Antenna Actuator Force for Linear Rigid Body
Control, Nonlinear Collocated Antenna Control, $z_2 = 2l_e/3$

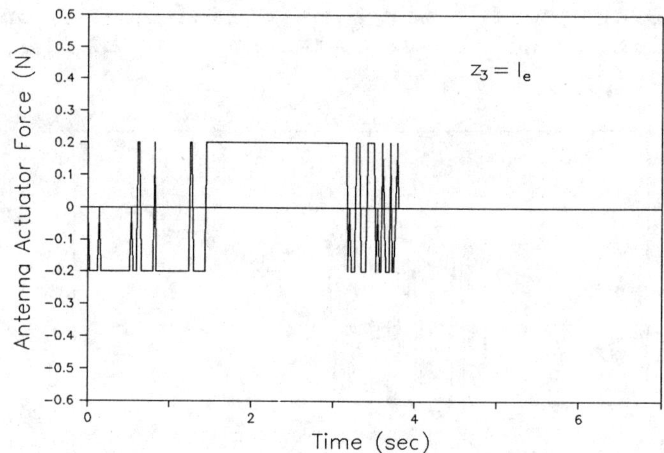

Fig. 15   Antenna Actuator Force for Linear Rigid Body
Control, Nonlinear Collocated Antenna Control, $z_3 = l_e$

In the case of substructure decentralized linear control,
the discrete-time gains for the control laws corresponding to
each of the substructure's three controlled modes were designed
so as to yield an exponential decay rate of $\alpha = 5.0$. The antenna
tip displacement with and without disturbance accommodation
is shown in Figures 16 and 17, respectively.

136

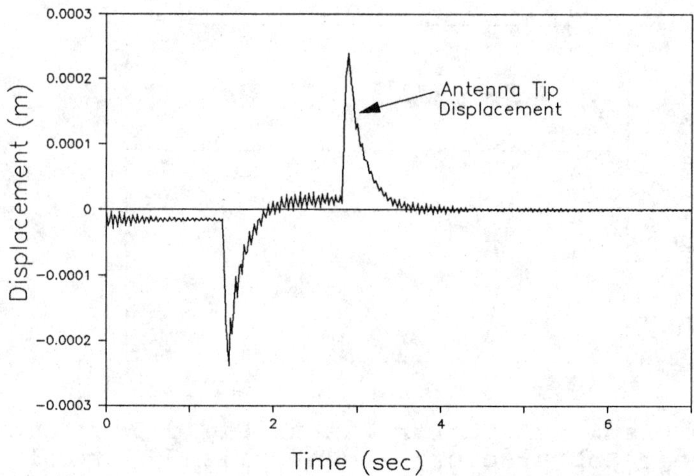

Fig. 16 Tip Displacement for Linear Substructure Decentralized Control with Disturbance Accommodation

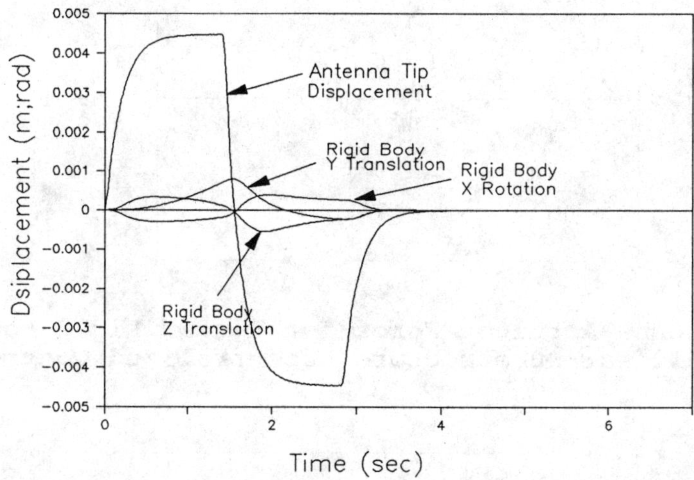

Fig. 17 Tip Displacement for Linear Substructure Decentralized Control without Disturbance Accommodation

The substructure decentralized linear control was next replaced with a nonlinear control with the following switching parameters:

$$c_1 = c_2 = c_3 = .01\,m/s\,, \quad F_{q_1} = 0.15\,N\,, \quad F_{q_2} = F_{q_3} = 0.05\,N$$

Nonlinear control, as described earlier, was used for the first substructure mode with $d_1 = 0.01\,m$. The resulting displacements are shown in Figure 18 and the actuator force histories are shown in Figures 19, 20 and 21.

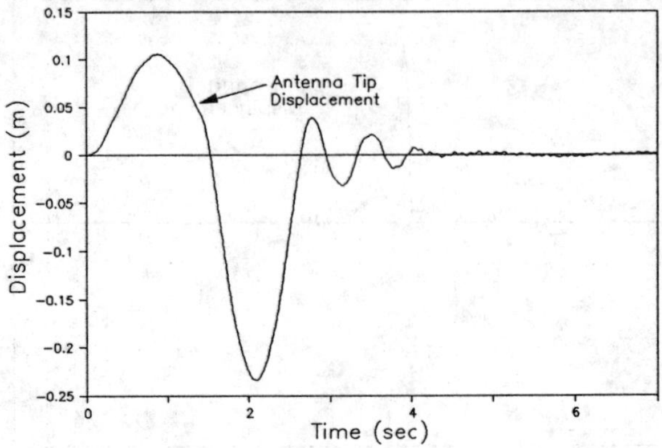

Fig. 18 Tip Displacement for Linear Rigid Body Control, Nonlinear Substructure Decentralized Control

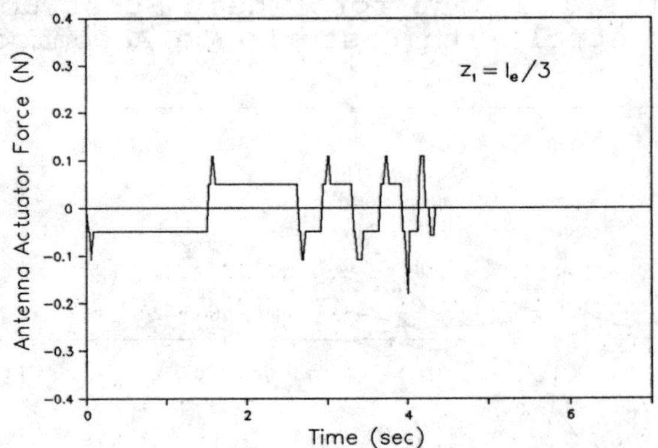

Fig. 19 Antenna Actuator Force for Linear Rigid Body Control, Nonlinear Substructure Decentralized Control
$$z_1 = l_e/3$$

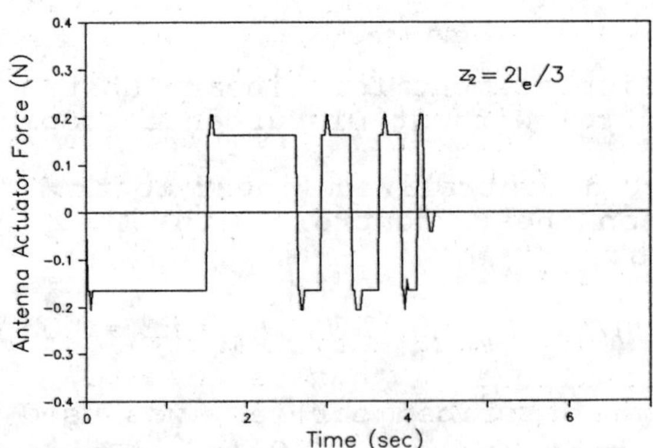

Fig. 20 Antenna Actuator Force for Linear Rigid Body Control, Nonlinear Substructure Decentralized Control
$$z_2 = 2l_e/3$$

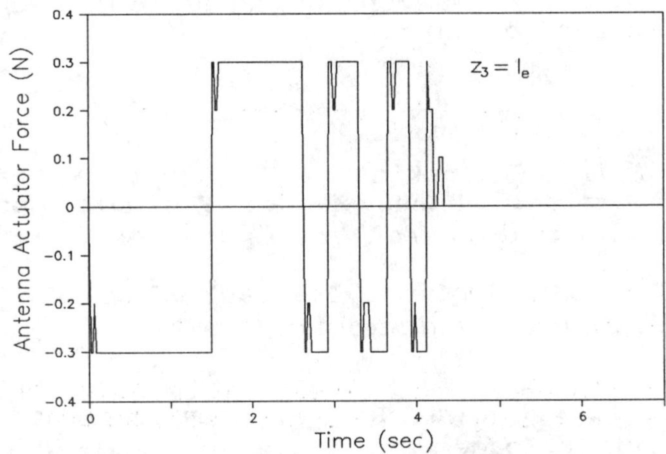

Fig. 21   Antenna Actuator Force for Linear Rigid Body
Control, Nonlinear Substructure Decentralized Control
$$z_3 = l_e$$

CONCLUSIONS

Control of a spacecraft with retargetable flexible antennas
using discrete-time techniques is considered.   First, a
discrete-time LQR approach is used to control the time-varying
system model, whereby the global step-varying feedback gains
are precalculated and stored for use during a known spacecraft
maneuver.   Next, two decentralized approachs were used -
Collocated Control and Substructure Decentralized Control.
Both approaches offer the advantage of low computational effort,
while still accounting for the time-varying nature of the
system.   Whereas Collocated Control suppresses the elastic
motion at specific points, Substructure Decentralized Control
suppresses the elastic appendage modes, taken as cantilever
modes. In each of the decentralized cases, linear and nonlinear
(on-off) schemes were presented for controlling the elastic
motion of the antennas. Note that on-off controls translate
into quantized actual controls in the Substructure Decentralized Control case.   In each of the linear cases,
disturbance-accommodation control based on the known antenna
maneuver profile was also included.

A numerical example using a system with one antenna was
presented and the results of each case compared. In each of
the linear cases, the spacecraft was controlled quite well,
with both decentralized methods offering performance comparable
to the more computationally intensive step-varying LQR. While,
the linear controls performed significantly better than the
nonlinear, as might be expected, both nonlinear control methods

139

performed adequately, reducing tip motion for the 5 m antenna
to under 4 mm within one second of maneuver completion, with
minimal limit-cycle behavior.

REFERENCES

1.   Turner, J.D. and Junkins, J.L., "Optimal Large-Angle
Single-Axis Rotational Maneuvers of Flexible Spacecraft,"
Journal of Guidance and Control, Vol 3, No 6, 1980, pp. 578-585.

2.   Breakwell, J.A., "Optimal Feedback Slewing of Flexible
Spacecraft," *Journal of Guidance and Control*, Vol 4, No 5, 1981,
pp. 472-479.

3. Meirovitch, L. and Quinn, R.D., "Maneuvering and Vibration
Control of Flexible Spacecraft," *The Journal of the Astronautical Sciences*, Vol 35, No 3, 1987, pp. 301-328.

4.   Meirovitch, L. and Kwak, M.K., "Dynamics and Control of
a Spacecraft with Retargeting Flexible Antennas,"
*Proceedings of the 29th AIAA/ASME/ASCE/AHS/ASC Structures,
Structural Dynamics and Materials Conference*, Williamsburg, VA,
April 18-20, 1988, pp. 1584-1592.

5. Meirovitch L. and Kwak, M.K., "Control of Spacecraft with
Multitargeted Flexible Antennas," *AIAA/AAS Astrodynamics
Conference*, Minneapolis, MN, August 15-17, 1988.

6.   Meirovitch, L., "Equations of Motion for Flexible Bodies
in Terms of Quasi-Coordinates," *IUTAM/IFAC Symposium on
Dynamics of Controlled Mechanical Systems*, Zurich, Switzerland, May 30 - June 3, 1988.

7.   Hostetter, G.H., *Digital Control System Design*, Holt, Rinehart and Winston, Inc., New York, NY, 1988.

8.   Young, K.D., "A Distributed Finite Element Modeling and
Control Approach for Large Flexible Structures," *Proceedings
of the AIAA Guidance, Navigation and Control Conference*, August
15-17, 1988, Minneapolis, MN, pp. 253-263.

9.   Silverberg, L., "Uniform Damping Control of Spacecraft,"
*Dynamics and Control of Large Structures: Proceedings of the
Fifth VPI&SU/AIAA Symposium*, Blacksburg, VA, June 12-14,
1985, pp. 145-162.

10.   Meirovitch, L., *Dynamics and Control of Structures*, Wiley
Interscience, 1989 (to appear).

11.   Kirk, D.E., *Optimal Control Theory*, Prentice-Hall Inc., Englewood Cliffs, NJ, 1970.

12. Meirovitch, L., Baruh, H., Montgomery, R.C., and Williams, J.P., "Nonlinear Natural Control of an Experimental Beam," *Journal of Guidance, Control, and Dynamics*, Vol 7, No 4, July-August 1984, pp. 437-442.

13. Meirovitch, L., *Computational Methods in Structural Dynamics*, Sijthoff & Noordhoff, The Netherlands, 1980.

APPENDIX

From Ref. 4, the matrices $M(t)$, $K(t)$ and $G(t)$ have the form

$$
M(t) = \begin{bmatrix}
m_t I & \tilde{S}_t^T & E_1^T \overline{\Phi}_1 & \cdots & E_N^T \tilde{\Phi}_N \\
\tilde{S}_t & I_t & E_1^T + \tilde{r}_{o_1} E_1^T \overline{\Phi}_1 & \cdots & E_N^T \tilde{\Phi}_N + \tilde{r}_{o_N} E_N^T \overline{\Phi}_N \\
\overline{\Phi}_1^T E_1 & \tilde{\Phi}_1^T E_1 + \overline{\Phi}_1^T E_1 \tilde{r}_{o_1}^T & M_1 & \cdots & 0 \\
\cdot & \cdot & \cdot & \cdots & \cdot \\
\cdot & \cdot & \cdot & \cdots & \cdot \\
\overline{\Phi}_N^T E_N & \tilde{\Phi}_N^T E_N + \overline{\Phi}_N^T E_N \tilde{r}_{o_N}^T & 0 & \cdots & M_N
\end{bmatrix}
$$

$$
K(t) = \begin{bmatrix}
0 & 0 & E_1^T(\dot{\tilde{\omega}}_1 + \tilde{\omega}_1^2)\overline{\Phi}_1 & \cdots & E_N^T(\dot{\tilde{\omega}}_N + \tilde{\omega}_N^2)\overline{\Phi}_N \\
0 & 0 & \tilde{r}_{o_1} E_1^T(\tilde{\omega}_1^2 + \dot{\tilde{\omega}}_1)\overline{\Phi}_1 + E_1[\tilde{\omega}_1 J_1(\underline{\psi}_1) + J_1(\underline{\dot{\omega}}_1)] & \cdots & \tilde{r}_{o_N} E_N^T(\tilde{\omega}_N^2 + \dot{\tilde{\omega}}_N)\overline{\Phi}_N + E_N[\tilde{\omega}_N J_N(\underline{\psi}_N) + J_N(\underline{\dot{\omega}}_N)] \\
0 & 0 & K_1 + \overline{H}_1(\underline{\omega}_1) + \tilde{H}_1(\underline{\dot{\omega}}_1) & \cdots & 0 \\
\cdot & \cdot & \cdot & \cdots & \cdot \\
\cdot & \cdot & \cdot & \cdots & \cdot \\
0 & 0 & 0 & \cdots & K_N + \overline{H}_N(\underline{\omega}_N) + \tilde{H}_N(\underline{\dot{\omega}}_N)
\end{bmatrix}
$$

$$
G(t) = \begin{bmatrix}
0 & 2\sum_{e=1}^{N} E_e^T[\tilde{S}_e \underline{\widetilde{\omega}}_e] & 2E_1^T \tilde{\omega}_1 \overline{\Phi}_1 & \cdots & 2E_N^T \tilde{\omega}_N \overline{\Phi}_N \\
0 & G_{22} & G_{23}^1 & \cdots & G_{23}^N \\
0 & \left[\tilde{\Phi}_1^T \tilde{\omega}_1^T - J_1^T(\underline{\omega}_1)\right]E_1 & 2\tilde{H}_1(\underline{\omega}_1) & \cdots & 0 \\
\cdot & \cdot & \cdot & \cdots & \cdot \\
\cdot & \cdot & \cdot & \cdots & \cdot \\
0 & \left[\tilde{\Phi}_N^T \tilde{\omega}_N^T - J_N^T(\underline{\omega}_N)\right]E_N & 0 & \cdots & 2\tilde{H}_N(\underline{\omega}_N)
\end{bmatrix}
$$

where

$$m_t = m_r + \sum_{e=1}^{N} m_e \qquad \underline{S}_e = \int_{m_e} \underline{r}_e \, dm_e \qquad \bar{S}_t = \sum_{e=1}^{N} \left(m_e \tilde{r}_{o_e} + E_e^T \bar{S}_e E_e \right)$$

$$I_r = \int_{m_r} \tilde{r} \tilde{r}^T \, dm_r \qquad I_e = \int_{m_e} \tilde{r}_e \tilde{r}_e^T \, dm_e$$

$$I_t = I_r + \sum_{e=1}^{N} \left(m_e \tilde{r}_{o_e} \tilde{r}_{o_e}^T + E_e^T I_e E_e + \tilde{r}_{o_e}^T E_e^T \bar{S}_e E_e + E_e^T \bar{S}_e E_e \tilde{r}_{o_e} \right)$$

$$M_e = \int_{m_e} \Phi_e^T \Phi_e \, dm_e \qquad \bar{\Phi}_e = \int_{m_e} \Phi_e \, dm_e \qquad \tilde{\Phi}_e = \int_{m_e} \tilde{r}_e \Phi_e \, dm_e$$

$$\bar{H}_e(\underline{\omega}_e) = \int_{m_e} \Phi_e^T \tilde{\omega}_e \Phi_e \, dm_e \qquad \bar{H}_e(\underline{\omega}_e) = \int_{m_e} \Phi_e^T \tilde{\omega}_e^2 \Phi_e \, dm_e$$

$$J_e(\underline{\omega}_e) = \int_{m_e} (\tilde{r}_e \tilde{\omega}_e + \widetilde{[\tilde{r}_e \tilde{\omega}_e]}) \Phi_e \, dm_e$$

$$K_e = [\Phi_e, \Phi_e]$$

in which the symbol [,] represents an energy inner product (Ref. 13).

$$G_{22} = \sum_{e=1}^{N} \left(E_e^T (2\tilde{\omega}_e I_e - \mathrm{tr}\, I_e \tilde{\omega}_e) E_e + 2\tilde{r}_{o_e} E_e^T [\widetilde{\bar{S}_e \tilde{\omega}_e}] E_e \right)$$

$$G_{23}^e = 2\tilde{r}_{o_e} E_e^T \tilde{\omega}_e \bar{\Phi}_e + E_e^T \tilde{\omega}_e \tilde{\Phi}_e + E_e^T J_e(\underline{\omega}_e)$$

The matrix $B^*(t)$ is given by

$$B^*(t) = \begin{bmatrix} I & b^1 & b^2 & \cdot & \cdot & \cdot & b^N \\ 0 & c^1 & 0 & \cdot & \cdot & \cdot & 0 \\ 0 & 0 & c^2 & \cdot & \cdot & \cdot & 0 \\ \cdot & \cdot & \cdot & \cdot & \cdot & \cdot & \cdot \\ \cdot & \cdot & \cdot & \cdot & \cdot & \cdot & \cdot \\ 0 & 0 & 0 & \cdot & \cdot & \cdot & c^N \end{bmatrix}$$

where

$$b^e = \begin{bmatrix} E_e^T & E_e^T & \cdots & E_e^T \\ \tilde{r}_{o_e} E_e^T + E_e^T \tilde{r}_{e_1} & \tilde{r}_{o_e} E_e^T + E_e^T \tilde{r}_{e_2} & \cdots & \tilde{r}_{o_e} E_e^T + E_e^T \tilde{r}_{e_{n_e}} \end{bmatrix}$$

$$c^e = \begin{bmatrix} \Phi_e^T(\underline{r}_{e_1}) & \Phi_e^T(\underline{r}_{e_2}) & \cdots & \Phi_e^T(\underline{r}_{e_{n_e}}) \end{bmatrix}$$

The inertial disturbance vector is defined as

$$\underline{d}(t) = \begin{bmatrix} \sum_{e=1}^{N} E_e^T (\tilde{S}_e \dot{\underline{\omega}}_e + \tilde{\omega}_e \tilde{S}_e \underline{\omega}_e) \\[2ex] \sum_{e=1}^{N} \left[\tilde{r}_{o_e} E_e^T (\tilde{S}_e \dot{\underline{\omega}}_e + \tilde{\omega}_e \tilde{S}_e \underline{\omega}_e) - E_e^T (I_e \dot{\underline{\omega}}_e + \tilde{\omega}_e I_e \underline{\omega}_e) \right] \\[2ex] -\tilde{\Phi}_1 \dot{\underline{\omega}}_1 + \int_{m_1} \Phi_1^T \tilde{\omega}_1 \tilde{r}_1 \underline{\omega}_1 \, dm_1 \\[1ex] \vdots \\[1ex] -\tilde{\Phi}_N \dot{\underline{\omega}}_N + \int_{m_N} \Phi_N^T \tilde{\omega}_N \tilde{r}_N \underline{\omega}_N \, dm_N \end{bmatrix}$$

Moreover,

$$\mathcal{K}_{22} = \left[\Phi_e^T\left(\underline{r}_{e_1}\right) \quad \Phi_e^T\left(\underline{r}_{e_2}\right) \quad \cdots \quad \Phi_e^T\left(\underline{r}_{e_{n_e}}\right) \right]^T [k_{22}]$$

$$\mathcal{K}_{24} = \left[\Phi_e^T\left(\underline{r}_{e_1}\right) \quad \Phi_e^T\left(\underline{r}_{e_2}\right) \quad \cdots \quad \Phi_e^T\left(\underline{r}_{e_{n_e}}\right) \right]^T [k_{24}]$$

where, in the case of one antenna,

$$k_{22} = \begin{bmatrix} c_{x_{11}} & 0 & \cdot & \cdot & \cdot & \cdot & \cdot & 0 \\ 0 & c_{x_{22}} & 0 & \cdot & \cdot & \cdot & \cdot & 0 \\ \cdot & 0 & \cdot & 0 & \cdot & \cdot & \cdot & 0 \\ \cdot & \cdot & 0 & c_{x_{rr}} & 0 & \cdot & \cdot & 0 \\ \cdot & \cdot & \cdot & 0 & c_{y_{11}} & 0 & \cdot & 0 \\ \cdot & \cdot & \cdot & \cdot & 0 & c_{y_{22}} & 0 & 0 \\ \cdot & \cdot & \cdot & \cdot & \cdot & 0 & \cdot & 0 \\ 0 & 0 & 0 & 0 & 0 & 0 & 0 & c_{y_{ss}} \end{bmatrix}$$

$$k_{24} = \begin{bmatrix} b_{x_{11}} & 0 & \cdot & \cdot & \cdot & \cdot & \cdot & 0 \\ 0 & b_{x_{22}} & 0 & \cdot & \cdot & \cdot & \cdot & 0 \\ \cdot & \cdot & 0 & \cdot & 0 & \cdot & \cdot & 0 \\ \cdot & \cdot & \cdot & 0 & b_{x_{rr}} & 0 & \cdot & 0 \\ \cdot & \cdot & \cdot & \cdot & 0 & b_{y_{11}} & 0 & \cdot & 0 \\ \cdot & \cdot & \cdot & \cdot & \cdot & 0 & b_{y_{22}} & 0 & 0 \\ \cdot & \cdot & \cdot & \cdot & \cdot & \cdot & 0 & \cdot & 0 \\ 0 & 0 & 0 & 0 & 0 & 0 & 0 & b_{y_{ss}} \end{bmatrix}$$

Note that $c_{x_{ii}} = b_{x_{ii}} = 0$ if the ith mode in the y_e direction is controlled, and the same can be said for modes in the x_e direction.

Section II
GUIDANCE AND CONTROL STORYBOARD DISPLAYS

SESSION II

Chairperson:	Pam Stewart Ball Aerospace Systems Group
Co-Chairperson:	Scott Dahl Martin Marietta Astronautics Group

The following papers were not available for publication:

AAS 89-010   A Laser Test Bed for Control Experiments, H. Gross, D. Banko, and J. Duricy, Dept. of Astronautics, U.S. Air Force Academy

AAS 89-011   The Multi-Hex Prototype Experiment, D. Hyland, Harris Corporation

AAS 89-012   Large Dynamic Range, Submicroradian Precision Pointing and Tracking, A. Goodrich, B. Biesterfeld, Ball Aerospace Systems Group

AAS 89-016   Experiment Teleoperations, an Approach to Remote Operations of Space Instruments, T. Sparn, J. Faber, Laboratory for Atmospheric and Space Physics (LASP), University of Colorado

AAS 89-018   Advanced Ring Laser Gyro IMU Systems, Dr. Meyers, Honeywell

AAS 89-019   Acquisition, Tracking and Pointing for Optical Communication Systems, D. Southwood, of Rutkowski, K. Taylor, TRW Space and Technology Group

AAS 89-020   A Test Facility for Control Systems Validation of Large Flexible Structures, D. Eldred, Jet Propulsion Laboratory

AAS 89-022   Star Identification Using Neural Network, A. San Martin, P. Alvelda, C. Bell, J. Barhen, Jet Propulsion Laboratory

AAS 89-023   SATCOM: A Expert System Application to Improve Satellite Autonomy, John Barry, Rockwell International

The following paper numbers were not assigned:

AAS 89-026 to -029

THE DUAL CONE SCANNER: AN ENHANCED PERFORMANCE, LOW COST EARTH SENSOR

Frank Tai[*] and Robert Barnes[†]

The single cone Conical Earth Sensor (CES) has become a standard attitude sensor for Earth orbiting satellites because of its low cost and wide altitude and attitude range. By utilizing two conical scans, the Dual Cone Scanner (DCS) offers additional advantages in redundancy, in reduction of ground support and in reduction of systematic errors, all for the weight, power, and cost of a single sensor. Unlike the CES, the DCS provides both satellite attitude and altitude determination data without the need for ephemeris information, it provides redundant data allowing automatic rejection of Sun contaminated measurements, and it operates over extended attitude and altitude ranges. In addition, because it employs only a single optical head and electronics, it retains the distinct advantages of low power and weight while eliminating inter-sensor misalignment errors. This paper describes the DCS hardware, its features, and its applications.

OVERVIEW

The Dual Cone Scanner (DCS)[‡] is a straightforward modification to an existing Conical Earth Sensor (CES) design which offers substantial improvements in performance. By substituting a mission-specific optical module which splits the scanner field-of-view into two halves, 180 degrees out of phase and at two distinct selectable cone angles, four Earth horizon crossings are obtained per sensor revolution. Thus a single DCS provides the data output equivalent of two concentric single cone sensors while maintaining the low power, weight, and cost of a single CES.

By utilizing two conical scans, a Dual Cone Scanner makes an independent determination of the altitude of the spacecraft in real time as well as pitch and roll. Unlike a CES, a single DCS provides fully automatic Sun rejection and continues to provide high accuracy, altitude independent attitude information even when the Sun is directly on the sensed horizon. Redundancy is enhanced and the need for ephemeris data is eliminated. The cone angles can be selected to assure that usable Earth crossings are maintained during extreme attitude maneuvers or over very large altitude ranges. The use of a four horizon crossing algorithm enhances the accuracy

---

[*]   Microcosm, Inc., 2601 Airport Drive, Suite 230, Torrance, California 90505.

[†]   Barnes Engineering Division, EDO Corp., P.O. Box 867, 88 Long Hill Cross Roads, Shelton, Connecticut 06484-0867.

[‡]   U.S. Patent Number 4, 792,684.

of attitude determination and the single unit design eliminates intersensor bias errors.

In order to introduce the Dual Cone Scanner, we begin with a review of the single scan Conical Earth Sensor hardware, features, and operating principles. This is followed by a description of the DCS hardware, function, and operational features. The performance benefits provided by a DCS on various spacecraft programs are then highlighted by presenting four mission examples:

<u>Low Earth Orbit Coverage</u> -- application demonstrating the extremely wide acquisition capability from any initial attitude

<u>Low Earth Orbit Precision</u> -- optimization of attitude determination accuracy for a LEO spacecraft

<u>LEO to GEO Transfer Orbit</u> -- attitude sensing over a substantial change in Earth apparent diameter

<u>Geosynchronous Earth Orbit Precision</u> -- stabilization with very high precision nadir determination

CONICAL EARTH SENSORS

To understand the operation of the Dual Cone Scanner (DCS), it is helpful to review the operation of single field-of-view Conical Earth Sensors (CES).

CES History

The CES is a proven spacecraft attitude determination instrument which has been in continuous use since the earliest days of the space program when conical scanners stabilized the X-15 rocket research aircraft and the experimental launches leading to the first manned flights. Project Mercury astronauts relied upon CES hardware to stabilize their space capsules on orbit and during reentry. Since then six hundred CES units have been produced and flown on missions ranging from military reconnaissance to deep space science.

CES Sensor Construction

Typically, a CES consists of an optical head and a separate signal processing electronics assembly. The heart of the system is an infrared detector sensitive to the thermal radiation of the Earth in the 14 to 16 micron wavelength region where the normal variations in infrared radiance are lowest. A motor rotates an optical module which contains either a mirror or a refractive optical element which diverts the field-of-view of the detector at a specific half angle such that the locus of scanned points describes a cone as the sensor spins.

CES Operating Principles

As a CES rotates, it scans the field-of-view of its infrared detector across the Earth from horizon to horizon, creating an electrical Earth pulse signal with the leading and trailing edges (horizon crossings) defining the geometric intersection of the scan cone with the Earth. The length of this Earth pulse is often referred to as the Earth width. Assuming that a single Conical Earth Sensor is mounted with its spin axis pointing along the orbital path, it is evident that the angular phase of the Earth pulse with regard to a fixed sensor/spacecraft reference is a direct measure of spacecraft roll. A simple photoelectric pickup can provide the index point.

Pitch can also be determined if spacecraft altitude, which influences the apparent angular size of Earth, is known. Changes in the pitch of the spacecraft will result in discernibly longer or shorter Earth pulses so long as the sensor's field-of-view intersects the Earth disk. The sensitivity of this pitch attitude measurement is greatest when the scan nears the Earth disk's edge and is at a minimum when the scan crosses the Earth's diameter. Thus, if altitude is known, the length of the Earth pulse (Earth width) at any given time is a measure of spacecraft pitch. A CES is generally insensitive to yaw motions of the satellite.

Formulas for conversion of CES Earth pulses to spacecraft attitudes are given in the Appendix 1. Several refinements, such as the effect of Earth oblateness are possible, but are not addressed here for the sake of brevity.

Use of Multiple CES

Generally, two CES units are mounted on a spacecraft to increase the amount of attitude data available and to extend the range over which the spacecraft can maneuver without having the sensor's scan cones leave the Earth. Two CES units, usually placed at different azimuth angles on the spacecraft, provide sufficient data to allow the determination of spacecraft pitch and roll without the need for an independent altitude reading. Two CES units also provide a measure of redundancy, although, in the event of a failure, the remaining sensor must be provided with altitude information from another source. Additional CES can be mounted to provide full redundancy or to extend even further the range over which the spacecraft can maneuver without having the scan cones leave the Earth's disk. When more than one CES is employed, great care must be taken in mounting them to assure that their reference axes are precisely aligned. Residual inter-sensor installation errors or changes in alignment on orbit contribute significantly to the overall attitude determination bias errors. The choice of sensor scan angles and the placement of the sensors on the satellite is highly mission-specific; a large variety of multi-sensor configurations are possible.

CES Sun-On-Horizon Errors

A common problem with all CES applications is that the Sun will also stimulate the infrared detector. Such signals can easily be distinguished when the Sun is encountered well away from the Earth on the basis of either the inherently high radiance of the Sun or by observing that the signal pulse is far too short to be a true Earth pulse. However, when the Sun is within a few degrees of the Earth's horizon, or partially masked by the edge of the Earth (Sunrise/Sunset conditions) there will be an intermixing of the real Earth edge signal with that from the Sun which can cause substantial errors. An ancillary silicon visible light detector sharing the optical path can identify any such contaminated signals so that they can be disregarded, but this results in a loss of sensor function for extended periods when the Sun is near the horizon. The Moon can introduce similar problems.

CES Performance

Table 1 gives specifications for a typical Conical Earth Sensor used in a 660 km polar circular orbit with a 1 second integration time.

Table 1
SINGLE CONICAL EARTH SENSOR SPECIFICATIONS

Parameter	Single CES
Attitude Accuracy (3 sigma):	
Earth Radiance Errors:	±0.05°
Alignment Errors:	±0.01°
Noise Equivalent Angle (jitter):	±0.02°
Sun-On-Horizon Errors:	±1.5°
Maximum Maneuvering Range:	
Pitch:	±180°
Roll:	-30° to +60°
Scan Cone Angle:	45°
Instantaneous field-of-View, circular	2.5°
Spectral Band:	14- to 16-μm
Weight:	8.5 pounds
Power:	10 Watts

DUAL CONE SCANNER

The previous section describes the operation of the single optical path Conical Earth Sensor (CES). This section will introduce the Dual Cone Scanner (DCS) and highlight its features and performance advantages.

CES to DCS -- Straightforward Optical Module Modifications

The Dual Cone Scanner can easily be implemented due to the modular optical design of current space-qualified scanning sensors produced for commercial programs such as LANDSAT and military satellites such as the SDIO Relay Mirror Experiment (RME). In fact, the standard CES supplied to the RME program is an appropriate starting point for a discussion of the DCS. The RME sensor's rotating optical module consists of a single mirror which simply diverts the optical scan path to create the 60 degree cone angle which is optimum for that mission. To convert the sensor to a Dual Cone Scanner this rotating optical module is replaced by one with a "V" shaped scanning mirror with reflective facets set so that the optical aperture is divided in half. In this manner, two independent coaxial scan cones are created. The scanned fields-of-view are 180 degrees apart in sensor rotation angle so that one scans open space while the other is scanning the Earth. The facets on the mirror are cut so that the two scan cones are at different angles chosen to provide the optimum sensor performance for a specific mission.

The following three figures show the mechanical layout of a DCS, a drawing of its scan cones and scan paths, and a photograph of an engineering model.

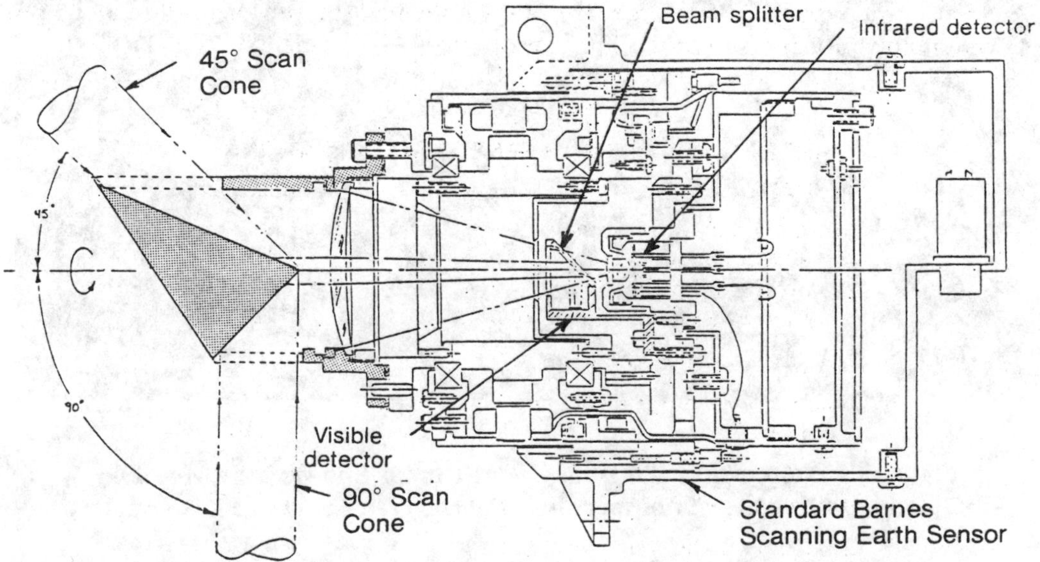

Figure 1   Layout of Dual Cone Scanner with Mirror Angles Selected to Produce One Scan Cone at 45 and One at 90 (a planer scan)

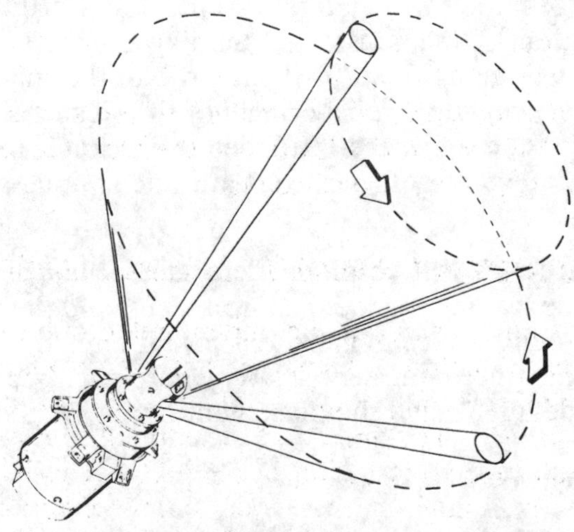

Figure 2   Isometric Drawing of a DCS with Scan Cones Set at 20 Degrees and 45 Degrees.

**Figure 3   Photograph of a Dual Cone Scanner Showing the Modified
Optical Module with Two Apertures 180 Degrees Apart in Phase Angle**

DCS Operating Principles

At nominal flight attitudes, when both fields of view intersect the Earth disk, the Dual Cone Scanner generates two laterally separated Earth scans (four horizon crossings) per revolution of the sensor. Assuming once again that the sensor is mounted with its spin axis along the orbital path, roll of the spacecraft will be sensed directly as identical changes in the phase angle of the signals from both these scans relative to a spacecraft fixed reference. Roll determination is independent of orbital altitude and nearly independent of spacecraft attitude so long as its nadir reference remains on the Earth.

Pitch of the spacecraft will result in detectable changes in the Earth pulse lengths measured by the two independent conical scans, and the pitch angle is, to first order, derived from the difference in these Earth width lengths. Since the sensitivity of this pitch attitude measurement depends on where each scan crosses the disk of the Earth, the two cone angles and the placement of the DCS on the spacecraft are chosen for the mission. This is in order to assure that the two Earth chords will be in an area of high sensitivity over the planned range of altitude and attitude changes.

Formulas for conversion of DCS Earth widths to spacecraft attitude are given in the Appendix 1.

Sensor Mounting Considerations

In optimizing a Conical Earth Sensor for a particular mission, one has control over its cone angle and mounting orientation on the spacecraft. The Dual Cone Scanner offers an additional controllable variable for the optimization of particular missions. In addition to the choice of a spacecraft azimuth location for the sensor and the elevation of its spin axis, the mission analyst can select two cone angles instead of the single angle of the CES. In the subsequent sections of this paper, four different missions are used as examples to demonstrate the effectiveness of the DCS in meeting diverse mission requirements. Considerable flexibility in the selection of DCS scan cone angles allows the optimization of specific performance parameters and the elimination of sensing singularities over any defined maneuvering and altitude range.

Four Point Local Vertical Location Algorithm

One approach to understanding the operation of the DCS is to treat the four horizon crossings as independent data points. Since the cone angles are fixed, and the sensor is rigidly mounted to the spacecraft, the relative geometry of these points can be used to establish the Earth's local vertical directly. Any three of the four points establish a circle and thus the direction to the center of the Earth disk can be defined without any knowledge of spacecraft altitude. Since four data points are available, this remains true even when one horizon crossing is contaminated by the Sun. Another benefit of this data set is that when all four points are usable, attitude determination can be done with even higher precision using a four point algorithm.

Direct Measurement of Altitude in Real Time

A single Dual Cone Scanner offers another important output, a real time measurement of the altitude of the spacecraft. This information can also be calculated directly from any three of the four horizon crossing points, and thus is available at any time, even during Sunrise/Sunset conditions. Spacecraft altitude is of particular importance to lightsats, in autonomous navigation applications or anywhere ground-based updates are not available. Since the altitude measurement is made directly in real time, it provides particular advantages on missions involving elliptical orbits or high impulse maneuvers where altitude is changing rapidly.

DCS Avoidance of Sun-On-Horizon Problems

As was described in the section on the operation of the Conical Earth Sensor, significant errors can occur when the Sun appears at or near the point on the Earth's horizon where the sensor is making measurements. Taking advantage of the fact that the Dual Cone Scanner produces four horizon crossing points and only three are needed to determine spacecraft attitude (and instantaneous orbit altitude), the DCS can disregard the contaminated crossing and continue to produce full accuracy pitch and roll data even during Sunrise and Sunset conditions.

The contaminated horizon crossing is unambiguously identified by the use of the silicon photodiode detector described earlier. This visible light detector observes space through a germanium beam splitter installed in the convergent optical beam just ahead of the pyroelectric infrared detector (see Fig. 1). The silicon detector subtends a circular field-of-view slightly larger than the 2.5 degree infrared field-of-view. Thus it will produce a "guard band" surrounding the infrared field-of-view and will alert the system to the presence of the Sun (or Moon) when it might contaminate the infrared signal from the Earth's horizon. Due to the high intensity of the Sun's signal and the sensitivity of the silicon detector, the Sun will be identified even when it is nearly obscured by the edge of the Earth disk.

Ability to Provide Data During Extreme Maneuvers

The cone angles of a Dual Cone Scanner can be selected to allow a spacecraft to operate at very large pitch or roll angles while maintaining Earth attitude reference measurement capability. In missions where it becomes necessary to exceed the range over which both DCS scans intersect the Earth, the DCS can rely on the data from only one of its two scan cones and continue to provide pitch and roll information in the same way that a Conical Earth Sensor operates. When operating with only one scan cone on the Earth, the DCS must be provided with spacecraft altitude or data from a second DCS must be available. In these circumstances, the DCS will operate essentially the same as a CES with a comparable cone angle setting and the remaining horizon crossing points will be vulnerable to Sun signal contamination. Nevertheless the ability to operate over a wider flight attitude envelope is of paramount importance in certain missions.

Redundancy Advantages

An important feature of the Dual Cone Scanner is that one sensor can provide all the attitude (and altitude) determination data required for spacecraft operation. Therefore, on a spacecraft where two DCS units are flown for enhanced mission reliability, only one of the two DCS units need be powered. The other can be held in standby reserve, thus avoiding wearout conditions as well as conserving spacecraft power. The reliability predictions computed in accordance with MIL-HDBK 217D for a five year mission with the redundant unit in passive standby show a 98.2% probability of mission success compared to a 96.6% probability of success when both scanners are operating in parallel. To achieve the same level of redundancy using CES, it would be necessary to use at least three units since one alone would not provide satellite attitude without an external altitude input.

Inter-Sensor Misalignment Errors Avoided

The two scan cones of the Dual Cone Scanner are inherently co-aligned in all axes since they share the same rotating assembly and photoelectric phase angle reference pickup. The relative angles on the rotating mirror are machined and polished in a single, solid block of aluminum and can be checked for precise alignment. Therefore, a single alignment procedure, aided by a factory-set optical reference cube machined into the DCS casting, is all that is necessary to assure that the DCS attitude determination sensor is co-aligned with the spacecraft payload. Since both pitch and roll attitude determination data are generated in the same unit, there need be no concern about bias terms in these readings due to residual intersensor installation errors or subsequent changes in the structure of the spacecraft due to thermal warping or zero-gravity recovery.

Comparative Performance of CES and DCS

Table 2 gives a direct comparison of the performance of a Dual Cone Scanner (45°/90° half cone angles, tilted up 55° from nadir) with a pair of Conical Earth Sensors (45° half cone angle, each tilted up 75° from nadir back to back) for 1 second integration time. Both configurations were designed to maximize geometrical gain for roll sensing while still allowing considerable roll maneuvering range. The performance numbers are for a 660 km (350 n-mi) circular polar mission (all other sensor parameters are essential equal).

Table 2

TWO CES VERSUS SINGLE DCS PERFORMANCE

Parameter	Two CES	Single DCS
Attitude Accuracy (3 sigma):		
Earth Radiance Errors:	±0.035°	±0.035°
Alignment Errors:	±0.014°	±0.010°
Noise Equivalent Angle (jitter):	±0.011°	±0.021°
Altitude Accuracy (3 sigma):	±620 m*	±680 m
Maximum Maneuvering Range:		
Pitch:	±180°	±180°
Roll:	±60°	-30° to +100°
Total Power:	20 Watts	10 Watts
Total Weight:	17.0 lbs	8.7 lbs
Total Cost[†]	2.0	1.1

*Both CES must be operating
[†]Cost in units of that of a single CES

Fig. 4 illustrates the configuration and mounting of two of the DCS covered above, optimized for a Low Earth Orbit application.

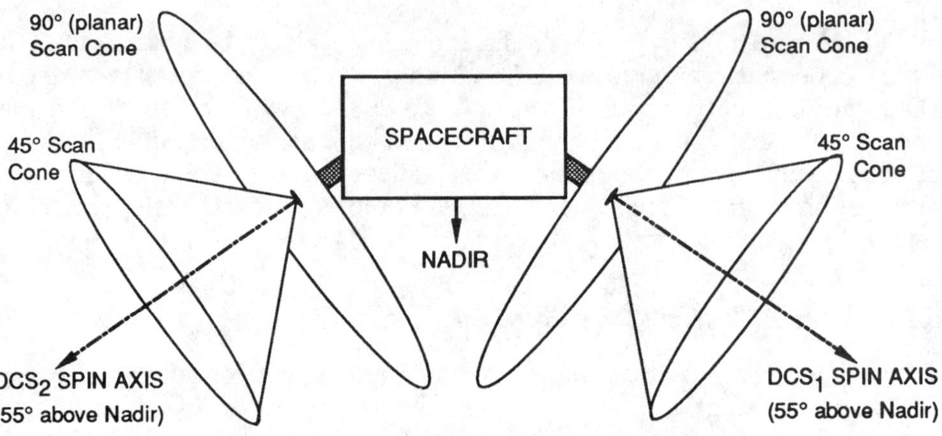

Figure 4   Low Earth Orbit Dual Cone Scanner Application; 45° and 90° Scan Cone Half-Angles and Spin Axes Oriented 55° Above Nadir in Spacecraft Pitch/Yaw Plane (spacecraft velocity vector is out of paper)

OPERATIONAL ADVANTAGES OF DUAL CONE SCANNER

The preceding sections described the construction and operation of the Dual Cone Scanner (DCS). This section will show the operational performance of the DCS in four distinct mission cases.

The cone angles and sensor mounting angle of the DCS can be optimized to maximize Earth coverage over the sky, to minimize attitude and altitude errors, or some combination of these. Sensor configurations are presented in this paper for a number of representative satellite missions objectives. They assume the satellite will employ two identical DCS which are mounted back to back at the same elevation angles relative to nadir. The following four reference missions were selected to demonstrate the capabilities of the DCS:

Reference mission one corresponds to the attitude acquisition phase of LEO missions following satellite deployment from the shuttle or ELV. This DCS configuration (i.e. its scan cone angles and mounting orientation on the satellite) is designed to maximize coverage of the Earth from LEO.

Reference mission two applies to the on-orbit phase of LEO Earth observation missions which require accurate satellite nadir pointing. This DCS configuration is designed to minimize attitude errors from LEO near nadir.

Reference mission three corresponds to the orbit transfer phase of GEO missions which boost the satellite from shuttle or ELV altitude to GEO altitude. This DCS configuration is designed to maximum the range of apparent Earth size, or altitude, for which attitude determination is provided.

Reference mission four applies to the on-orbit phase of GEO communication missions which require accurate satellite nadir pointing. This DCS configuration is designed to minimize attitude errors from GEO near nadir.

Table 3 summarizes these reference missions, the DCS configurations designed to best meet the requirements, and the resultant performance capabilities. It shows that pairs of DCS can accommodate a wide range of applications by merely adjusting scan angles and mounting orientations on the satellite.

Table 3
MISSION APPLICATIONS FOR DCS PAIRS

Mission	Scan Angles	Mounting Angle	Sky Coverage	Attitude Accuracy
LEO acquisition	$45°/90°$	$55°$	100% (97% alt-indep)	$0.02°$ at nadir $0.06°$ typical elsewhere
LEO pointing	$45°/90°$	$55°$	100% (97% alt-indep)	$0.02°$ at nadir $0.06°$ typical elsewhere
Orbit transfer	$60°/72°$	$66°$	100% at LEO	0.05° at LEO 0.02° at GEO
GEO pointing	$60°/72°$	$66°$	62.5% at GEO	0.02°

While designing the DCS for the four different missions, configurations were found to converge on two distinct solutions. The $45°/90°$ DCS was found to best meet the requirements of the LEO acquisition and the LEO pointing missions and the $60°/72°$ DCS best met the requirements of the Orbit Transfer and the GEO missions. The specific performance characteristics of these DCS configurations and their ability to meet the requirements of the reference missions will be covered in the remainder of this paper.

Spacecraft-Centered Celestial Sphere Diagrams

Throughout the remainder of this paper, we will present a diagrammatic representation of Dual Cone Scanner performance through the use of a presentation technique employing the spacecraft-centered celestial sphere. The spacecraft occupies a position at the center of a unit radius, celestial sphere. Any point on the sphere represents a direction in space. Therefore, the perimeter of the Earth can be drawn as a circle or disk on the sphere and it blocks a portion of the celestial sphere as seen by the spacecraft. For a low altitude mission, that Earth disk will occupy nearly half the "sky;" at higher altitudes, proportionally smaller disks represent the Earth. Equally, the position of the Sun, Moon, and other celestial objects can be applied in their proper position and size on the sphere. The scan pattern of a CES or DCS can be represented by a circular path on the celestial sphere. If oriented correctly, each scan will cut across the Earth disk producing two horizon crossing points (corresponding to the end points of the electrical Earth pulse signal generated by the sensor). Figs. 5, 6, 10, 11, and 12 employ this celestial sphere presentation technique.

<u>**Low Earth Orbit Dual Cone Scanner**</u>

The DCS specified for both the LEO precision and coverage missions has half-cone angles of 45° and 90°, and they are mounted back-to-back with their spin axes tilted up 55° from nadir. Fig. 4 illustrates these cone angles and the orientations of the DCS on the satellite. Fig. 5 shows the satellite-centered celestial sphere with the DCS scans and the 130° diameter disk of the Earth as viewed by the satellite from an altitude of 660 km (350 n-mi). In this application, the sensors are mounted in the pitch-yaw plane of the satellite (tilted 55° from nadir towards the orbit normal and anti-normal). They could equally well be rotated about the spacecraft yaw (i.e. to lie in the roll-yaw plane) to meet specific accuracy and coverage requirements or to eliminate spacecraft interference, so long as their 55° elevations above nadir are maintained.

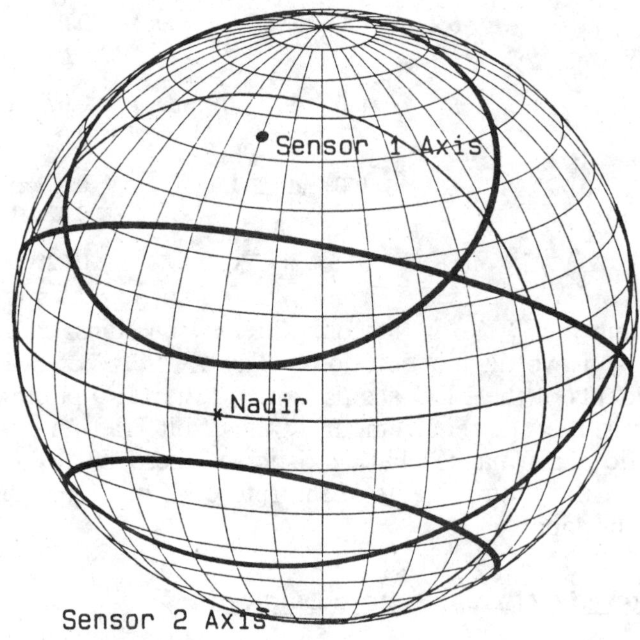

**Figure 5   Scan Cones for LEO Dual Cone Scanner and Earth Disk
as Viewed from LEO in Satellite Sky**

To maximize Earth coverage in LEO missions, the angular separation between the four IR cones (two from each DCS) were designed to intersect the Earth when it is located over as much of the spacecraft sky as possible. This corresponds to maximizing the range of satellite orientations for which the DCS will determine the satellite's attitude. For attitudes in which two or more scans intersect the Earth, altitude-independent attitude data is provided with full Sun rejection capability. Over the range in which only one scan intersects the Earth, satellite altitude is required for attitude determination. Fig. 6 shows the coverage properties of the dual $45^o/90^o$ DCS combination mounted on a satellite in a 660 km (350 n-mi) altitude orbit. The darker shaded areas show the portions of the sky in which two or more scans will intersect the Earth and therefore altitude-independent attitude is provided (97% of the total sky). The lightly shaded regions show those portions of the sky where one scan intersects the Earth and therefore altitude is required (3% of the total sky). There are no portions of the sky where no scans intersect the Earth and therefore where attitude is not provided. Thus, this sensor combination will provide attitude determination and acquisition capability for any satellite orientation, and altitude is required in only a small region.

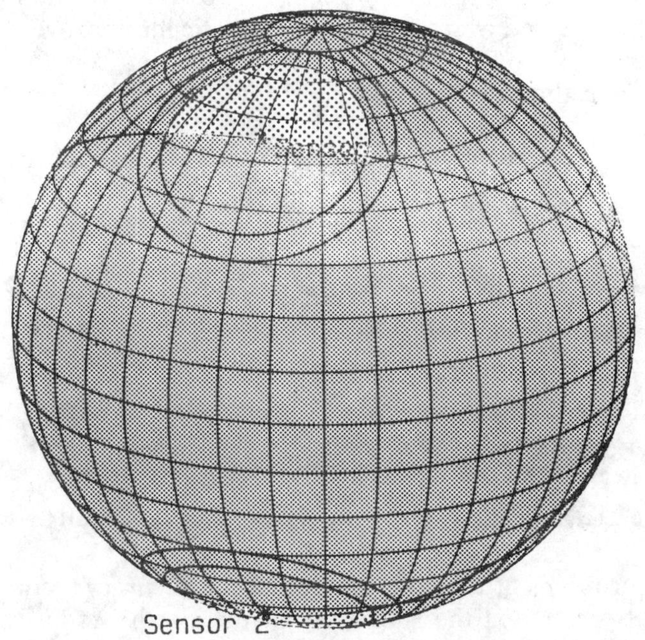

**Figure 6   Earth Coverage Portions of Satellite Sky
for LEO Dual Cone Scanner**

The attitude accuracies of the DCS are computed from the Earth width measurement accuracy of the scanner and the sensitivity of the attitude algorithm to measurement variations. Assuming that Earth oblateness will be modelled, the scanner's measurement accuracy is determined from the error sources given in Table 4.

The sensitivity of the attitude outputs to measurement errors is determined from the geometry of the scans on the Earth for the specific optical design and mounting of the DCS. This sensitivity is the change in pitch or roll attitude for unit changes in Earth width measurements. The output quantity of interest is a single parameter attitude error estimate[*] or the RMS of the pitch and roll errors. Figs. 7 and 8 show the accuracy characteristics of the LEO DCS pair for 1 second time

---

[*]   Wertz, J.R., ed. Spacecraft Attitude Determination and Control, D. Reidel Publishing Company, Dordrecht, Holland, 1978, 1980, 1984, pg 381.

averages. To demonstrate the applicability of the DCS as a wide coverage sensor, Fig. 7 shows its attitude accuracies to be 0.02^{o} (3-sigma) or less for pure pitch or roll deviations of up to 35° about nadir. When the roll angle exceeds 35°, the error increases sharply which corresponds to one of the DCS 90° scans no longer intersecting the Earth. However even while using only the remaining three scans, altitude-independent attitude is determined to better than 0.035^{o}. These DCS accuracy numbers are contrasted with the 0.06^{o} attitude accuracy typically quoted for a pair of CES and labelled as "baseline accuracy for two CES" in the plot. To demonstrate the applicability of the DCS for precision pointing, Fig. 8 shows its improvement in accuracy gained with increased averaging times. Significant improvement is noted for several seconds converging on a residual value which comes from the bias portions of the sensor errors given above.

Table 4

CONICAL SCANNER ERROR SOURCES

Error Source	3 Sigma Error (per pulse)
Noise Equivalent Angle (NEA) per horizon crossing	0.060^{o}
Earth radiance variations after 1st order corrections	0.035^{o}
Phase Reference Pickup (PRP) nonlinearity (affects pitch only)	0.014^{o}
Component aging and temperature effects (affects pitch only)	0.008^{o}
Alignment uncertainty	0.014^{o}

Similar to determining attitude accuracy, altitude accuracy is computed from the DCS measurement errors and the sensitivity of the Earth's radius or satellite's altitude to these measurements. Again, for the specific geometry of the scans on the Earth, this sensitivity is simply the change in determined altitude for unit changes in Earth width measurements. Fig. 9 shows the altitude determination accuracy (1 second average) of the LEO DCS pair for roll excursions about nadir. Altitude determination accuracy is almost entirely independent of satellite pitch attitude for moderate excursions about nadir since the scan geometry changes little for pitch changes. Thus, satellite altitude is determined to accuracies within 500 m by the $45^{o}/90^{o}$ DCS for moderate attitude variations about the nominal nadir attitude. Similar to the improvement of attitude error with averaging time shown in Fig. 8, the altitude determination accuracy will improve with increased averaging time.

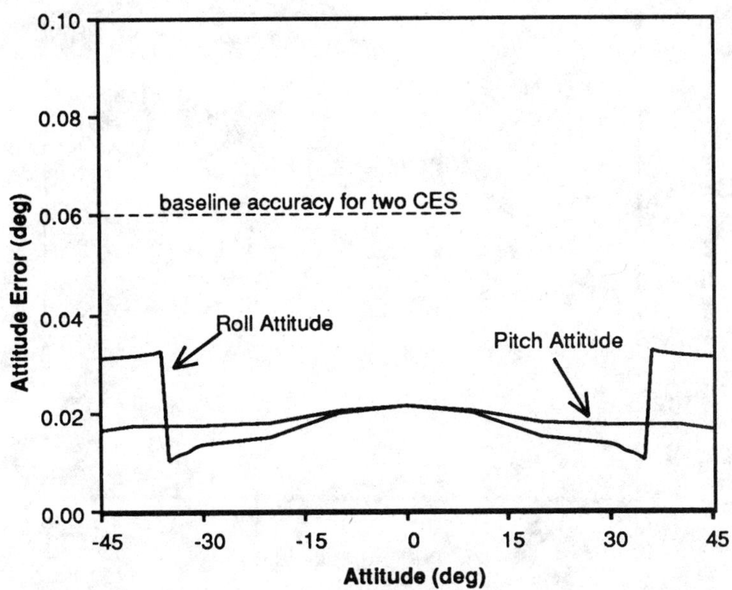

Figure 7   Attitude Accuracy of LEO Dual Cone Scanner as a Function of Pure Satellite Pitch and Roll Attitudes

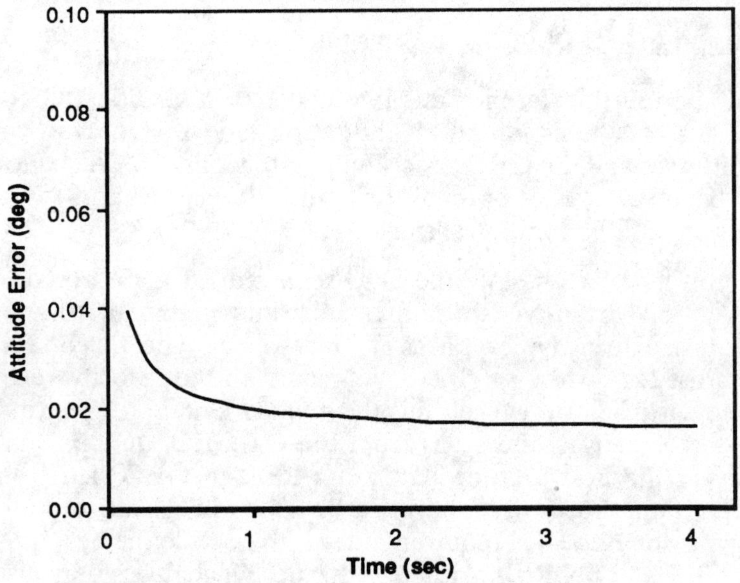

Figure 8   Attitude Accuracy of LEO Dual Cone Scanner as a Function of Integration Time

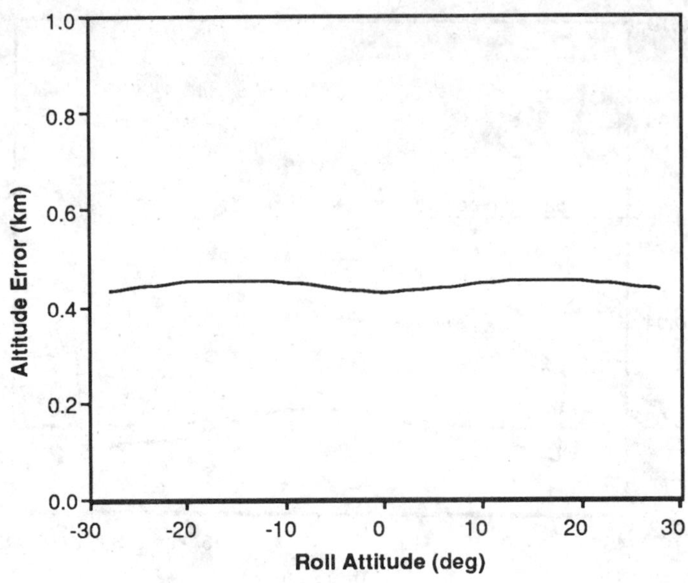

**Figure 9   Altitude Resolution of LEO Dual Cone Scanner
for Satellite Roll Attitudes**

Orbit Transfer and GEO Dual Cone Scanner

The third and fourth reference missions chosen for specific DCS designs were the Orbit transfer (OT) mission from Shuttle altitude to Geosynchronous Earth Orbit (GEO) altitude and the GEO precision pointing mission. A single sensor configuration has been selected as optimal for both GEO applications: high precision attitude sensing at nadir <u>and</u> orbit transfer from LEO.

For the orbit transfer mission, the DCS requirement is to provide the best attitude data over this extensive altitude range while in the orientation for delta velocity operations. During typical burns at apogee or perigee, this orientation is one in which the thrust axis of the satellite is aligned with the velocity vector. For the purposes of our Earth sensor, we call this the nadir orientation. Therefore, to meet the orbit transfer objectives, the scans from the OT/GEO DCS pair are placed to cover and provide the best attitude data from LEO to GEO (i.e. from the small Earth disk to the large Earth disk shown in Fig. 10) while nadir oriented. For the GEO pointing mission, the DCS requirement was to provide the best attitude data at that altitude while nadir pointed. Although this design goal was only a subset of the goals for the orbit transfer mission, it was best met by using the same DCS configuration.

The DCS specified for these missions has half-cone angles of 60° and 72°. They are mounted back to back with their spin axes tilted up 66° from nadir. Fig. 10 shows the satellite-centered celestial sphere, the paths of the scan cones from both DCS, and the disks of the Earth as seen from LEO and GEO. The scan paths are 120° and 144° in diameter and are centered on sensor axes, each of which are oriented 66° away from nadir. The concentric circles centered on nadir represent the Earth as viewed from altitudes of 195 km (105 n-mi) and 35,700 km (19,300 n-mi). As with the LEO DCS application, their spin axes are assumed to be in the pitch-yaw plane of the satellite, or tilted 66° from nadir towards the orbit normal and the anti-normal.

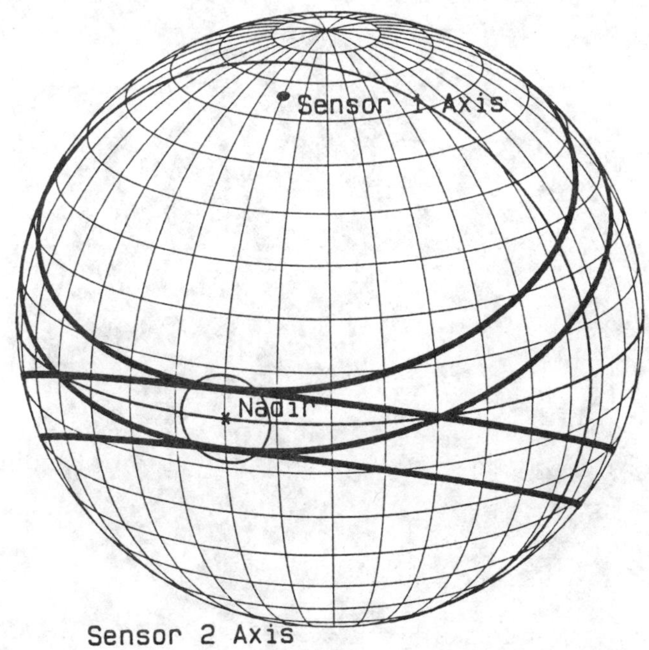

Figure 10   Scan Cones for Orbit Transfer/GEO Dual Cone Scanner and Earth Disks as Viewed from LEO and GEO in Satellite Sky

Figs. 11 and 12 show the coverage properties of the 60°/72° DCS in LEO and GEO, respectively. Fig. 11 shows that near full sky coverage is provided at LEO, mostly without need for altitude information. The darker shaded portion of the globe shows that for 73% of all satellite orientations, two or more scans intersect the Earth and therefore altitude-independent attitude is provided. The lightly shaded bands show that over an additional 11% of satellite orientations attitude data will be available from a single Earth scan, and altitude information will be required. The unshaded regions show that only over the remaining 16% of the sky will no scans intersect the Earth. For the orbit transfer mission, these regions of no attitude information are of little concern for they are far from the nominal nadir satellite orientation maintained during typical delta-velocity maneuvers.

Fig. 12 shows the Earth coverage provided by the 60°/72° DCS pair while in GEO. In GEO, Earth coverage is much reduced from that at LEO because of the reduced apparent size of the Earth. For this DCS complement, Earth coverage is near full sky over much of the lower altitudes (It achieves true full sky coverage at altitudes above the 195 km case shown in the previous figure before coverage begins dropping as the apparent size of the Earth shrinks appreciably). In GEO, the shaded regions show that attitude information is provided for 63% of the satellite's orientations. Of this, 15% of the orientations require altitude information to determine attitude. Like the coverage at LEO, the satellite's attitude range over which data are provided encompasses the attitude which is maintained for orbit transfer operations (either plus or minus delta-velocity) and for nadir pointing.

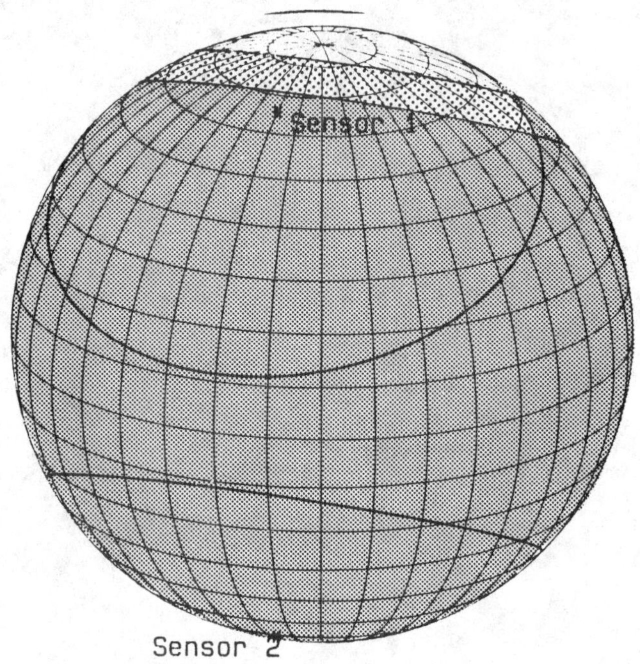

**Figure 11   Earth Coverage Portions of Satellite Sky While in LEO for
Orbit Transfer/GEO Dual Cone Scanner**

Fig. 13 shows the attitude accuracy of the $60^\circ/72^\circ$ DCS pair from LEO to GEO (1 second time average). To meet the requirements of a realistic orbit transfer operation, the DCS was designed to provide optimal accuracy at the starting and ending altitudes in which thrusting will be performed. At the starting shuttle altitude of 195 km (105 n-mi), attitude accuracy is maintained at 0.05° (3-sigma) and at the ending altitude of 35,700 km (19,300 n-mi) it is maintained at 0.02°. The rise in attitude errors in the intermediate altitudes occur when the Earth width measurements lie near a diameter of the Earth where it is less sensitive to changes in satellite roll. Nevertheless, attitude accuracy is maintained at less than 0.075°. In the worst case conditions, since pitch error remains relatively constant, the majority of the error is in roll which is the axis of no concern during delta-velocity maneuvers.

Fig. 14 shows the altitude determination resolution of the $60^\circ/72^\circ$ DCS pair during the LEO to GEO transfer (1 second average). The log-log plot shows the altitude error to monotonically increase with satellite altitude as one might expect. This is because the error in the sensor can be thought of as an angular portion of its scan over which the horizon is uncertain. This horizon uncertainty, although constant from the sensor's point of view, resolves into increased Earth radius uncertainty (i.e. increased altitude uncertainty) as the sensor's perceived size of the Earth decreases. Similar to the improvement of attitude error with averaging time shown in Fig. 8, the altitude determination accuracy will also improve with time.

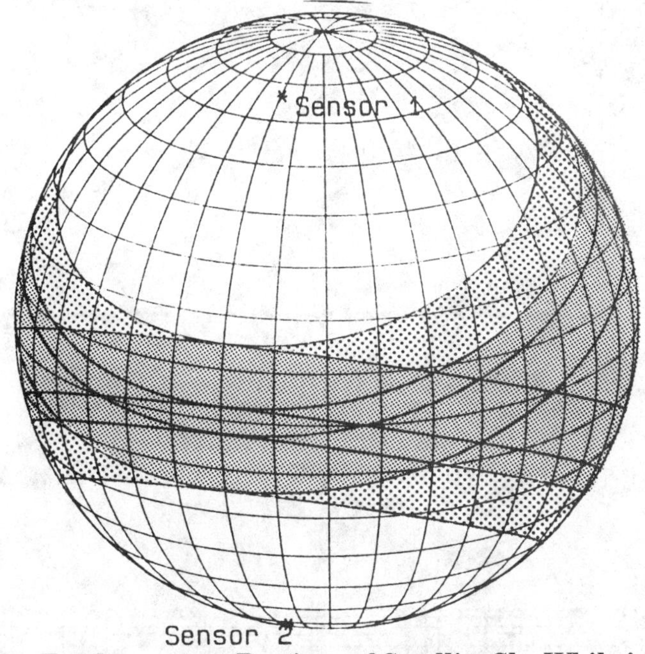

Figure 12   Earth coverage Portions of Satellite Sky While in GEO for Orbit Transfer/GEO Dual Cone Scanner

SUMMARY AND CONCLUSIONS

The Conical Earth Sensor (CES) is a well established satellite attitude sensor which is often used in multiples because of the additional accuracy and sensing range that is provided. The Dual Cone Scanner (DCS) is a technically straightforward means for combining the function of two CES into a single instrument. In addition to the accuracy and sensing range enhancements, the DCS also provides attitude data which does not depend on satellite ephemeris information, it provides an altitude determination capability, and it allows complete rejection of Sun contaminated measurements while eliminating a majority of typical inter-sensor bias errors.

The Dual Cone Scanner has been shown to meet the attitude sensing requirements for a wide range of Earth orbiting satellite missions. Its independent scan angles are easily adjusted by making simple modifications to its scan optics. These scan cone angles plus the mounting orientation of the DCS on the satellite can be optimized to meet particular mission objectives. To demonstrate the versatility and power of this sensor by tuning these parameters, configurations were designed and performance was analyzed for four sets of mission objectives.

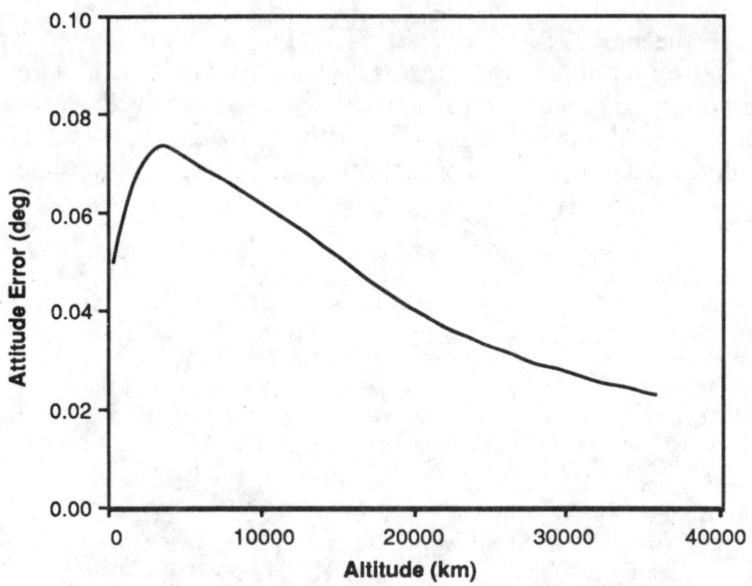

Figure 13   Attitude Resolution of Orbit Transfer/GEO Dual Cone Scanner
from Shuttle to GEO Altitude

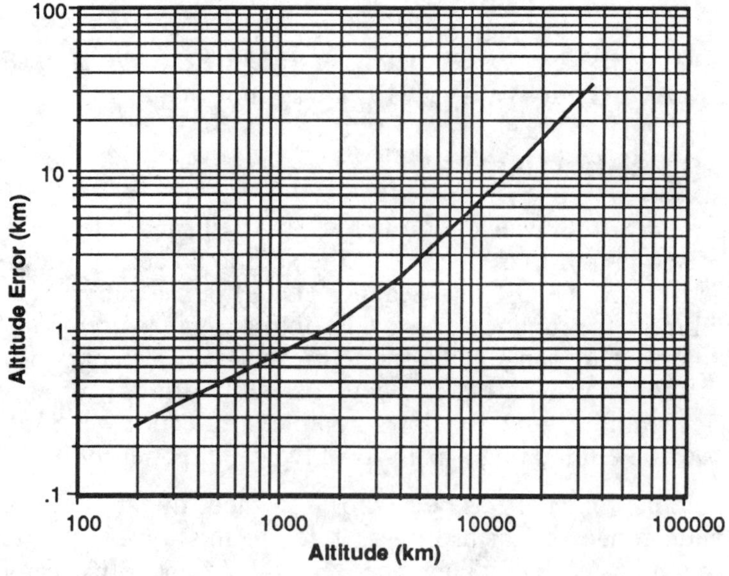

Figure 14   Altitude Resolution of Orbit Transfer/GEO Dual Cone Scanner
from Shuttle to GEO Altitude

Mission one requires the satellite to acquire or determine its attitude after being deployed from the shuttle or ELV in LEO, conditions under which its initial orientation could be totally unpredictable. Two $45^\circ/90^\circ$ DCS, each tilted 55° up from nadir, were found to fully meet the goals of this mission by providing full-sky attitude coverage.

Mission two calls for precision pointing by a LEO satellite. The $45^\circ/90^\circ$ DCS combination used above was found to provide optimal determination accuracies of 0.02° for attitude and better than 500 meters for altitude. (Both are 3 sigma and are for 1 second time average.)

Mission three entails transfer of a satellite from initial orbits as low as the shuttle's to final orbits as high as GEO. A $60^\circ/72^\circ$ DCS pair, each tilted 66° up from nadir, was found to meet these requirements by providing accurate and wide coverage satellite attitude and altitude determination from 195 km to 35,700 km. While in LEO, they determine attitude from almost any satellite orientation to nominal accuracies of 0.05° for attitude and better than 300 meters for altitude (3 sigma and 1 second time average). In GEO, they determine attitude for 63% of satellite orientations to nominal accuracies of 0.02° for attitude and 35 km for altitude (3 sigma and 1 second time average).

Mission four requires precision pointing of a satellite in GEO. The $60^\circ/72^\circ$ DCS combination used above was found to provide optimal determination accuracies of 0.02° for attitude and 35 km for altitude (3 sigma and 1 second time average).

The performance evaluations given in this paper show the effectiveness of the DCS, not only as an enhanced conical sensor, but as a mission tunable sensor accomplished by making minor mechanical modifications. The DCS should prove to be an effective solution for many emerging satellite programs which require improved performance, but are highly constrained in power, weight and cost. Thus, the DCS should fulfill a majority of the upcoming space program needs for Earth sensors, and higher precision attitude sensors.

ACKNOWLEDGEMENT

We wish to acknowledge the EDO Corporation for funding the development of the Dual Cone Scanner. The Barnes Engineering and Microcosm staff were instrumental in providing the analysis and background information that was used. Of particular significance were the inputs of James Wertz and Steve Collins who provided the mathematical appendix and sensor determination errors, and Gerald Falbel who provided the instrument errors. Special recognition should go to Bob Savoca who conceived the design of the Dual Cone Scanner.

APPENDIX 1

FORMULAS FOR COMPUTING PITCH AND ROLL FROM CES OR DCS DATA

We first present formulas for the position of the Earth in sensor coordinates for both CES and DCS sensors followed by formulas for the position of the Earth in spacecraft coordinates and the position of the spacecraft in Earth coordinates. We have assumed the sensor axis is in the nominal pitch/yaw plane. All formulas assume a spherical Earth, i.e., that oblateness, varying triggering height, and other non-spherical corrections have been previously applied. With this assumption, the formulas are exact and applicable to all altitudes and attitudes.

<u>Known or Measured Parameters</u>

Parameter	Range (deg)	Definitions
γ	0 to 180	Sensor scan cone radius (known parameter)
M	-90 to +90	Mounting angle from sensor axis to nominal nadir (known parameter)
ρ	0 to 90	Earth angular radius (assumed known in CES, solved for in DCS)
W	0 to 180	Measured Earth pulse width divided by 2
ϕ	-180 to +180	Measured phase of center of Earth pulse relative to internal mark representing nominal nadir direction (Azimuth of Earth in sensor coordinates)

<u>Computed Parameters</u>

Parameter	Range (deg)	Definitions
η	0 to 180	Angle from sensor axis to center of Earth (also referred to as nadir angle)
R_e	-90 to +90	Roll coordinate of Earth in spacecraft coordinates
R_{spc}	--	Roll coordinate of spacecraft in orbit coordinates
P_e	-180 to +180	Pitch coordinate of Earth in spacecraft coordinates
P_{spc}	--	Pitch coordinate of spacecraft in orbit coordinates

<u>Computing nadir angle (h) from CES data</u>

$$\eta = \eta_a \pm \eta_b \qquad\qquad \text{(2 valid solutions)} \qquad (1)$$

where

$$\tan \eta_a = \tan \gamma \cos W \qquad\qquad [0° < \eta_a < 180°] \qquad (2)$$

$$\cos \eta_b = \cos \rho \, / \, \cos K \qquad\qquad [0° < \eta_b < 180°] \qquad (3)$$

$$\sin K = \sin \gamma \sin W \qquad\qquad [K \text{ in same quadrant as } W] \qquad (4)$$

<u>Computing nadir angle (h) from DCS data</u>

$$\eta = \eta_1 + H_1 \qquad\qquad (5)$$

$$\cos \rho = \cos C_1 \cos \eta_1 \qquad\qquad [0° < \rho < 90°] \qquad (6)$$

where

$$\tan \eta_1 = \left(\frac{\cos C_1}{\cos C_2} - \cos (H_2 - H_1) \right) \Big/ \sin (H_2 - H_1) \qquad [-90° < \eta_1 < +90°] \qquad (7)$$

$$\tan H_{1or2} = \cos W_{1or2} \, \tan \gamma_{1or2} \qquad\qquad [0° < H < 180°] \qquad (8)$$

$$\sin C_{1or2} = \sin \gamma_{1or2} \, \sin W_{1or2} \qquad\qquad [C \text{ in same quadrant as } W] \qquad (9)$$

<u>Computing Earth position in spacecraft coordinates from nadir angle (h) and azimuth (f)</u>

$$\sin R_e = \sin M \cos \eta - \cos M \sin \cos \phi \qquad\qquad [-90° < R_e < +90°] \qquad (10)$$

$$\cos P_e = \frac{\cos \eta - \sin M \sin R}{\cos M \cos R} \qquad\qquad [0° < P_e < 180°] \qquad (11)$$

$$\text{if } \phi < 0, \text{ then } P_e = - P_e$$

<u>Transforming to spacecraft position in orbit coordinates</u>

Spacecraft roll and pitch in orbit coordinates are not well defined by the Earth sensor because the sensor does not measure yaw and non-zero yaw will result in a mixing of apparent roll and pitch. Exact algorithms for the coordinate transformations if yaw is known are given in Ref. (4). If roll, pitch, and yaw are all approximately 0, then:

$$R_{spc} \approx -R_e \qquad\qquad (12)$$

$$P_{spc} \approx -P_e \qquad\qquad (13)$$

ON DEVELOPING THE LOCAL RESEARCH ENVIRONMENT OF THE 1990S: THE SPACE STATION ERA

Robert Chase and Fred Ziel[*]

We have performed a requirements analysis for the Space Station's Polar Platform data system. Based upon this analysis, we have designed, developed and now evaluate a cluster, layered cluster, and layered-modular implementation of one specific module within the Eos Data and Information System (EosDIS), an Active Data Base for satellite remote sensing research. We find that a distributed system based on a layered-modular architecture and employing current generation workstation technologies has the requisite attributes ascribed by the remote sensing research community. Although, based on benchmark testing, probabilistic analysis, failure analysis and user-survey technique analysis, we find that this architecture does present some operational shortcomings that will not be alleviated with new hardware or software developments. Consequently, we have also evaluated the potential of a fully-modular layered architectural design for meeting the needs of Eos researchers, concluding that it would be well suited to the evolving requirements of this multi-disciplinary research community. Finally, from the perspective of our research and development experience, we conclude that a fully-modular layered architecture could be implemented (although not cost effectively in the near-term) with current generation technology.

INTRODUCTION

The local computational and analysis environment, including hardware, software, and data and information, comprises an Active Data Base [1, 2]. It is the focal point for a researcher's everyday data processing and analysis and preparation of reports and papers. Consequently, the planning and design of this class of computer system must be undertaken with some care since the success or failure of the tool itself can play a major role in determining the scientific productivity of a given researcher or

---

[*] Colorado Center for Astrodynamics Research, University of Colorado, Campus Box 429, Boulder, Colorado 80309-0429.

group of researchers and ultimately, the success of the Space Station and associated Polar Platforms.

With this in mind, we must ask if there are robust system designs which will easily accommodate hardware and research focus changes that lie in the future of any given research group. Can a system be configured which readily supports evolution, growth and change? Will it be dynamically responsive to changing research problems, new technologies and data processing and analysis requirements? We believe the answer to these questions is a qualified "yes." To provide the basis for this conclusion, we examine four research-oriented Active Data Base implementations and explore design changes which we believe yield a system more capable of meeting the demands and rigors of the contemporary remote sensing research environment.

We have examined the advantages and disadvantages of each configuration from an operational point of view [3]; as well, we have distilled the characteristics of each configuration into an "abstract implementation." These abstractions have been normalized to present a similar number of facilities except as required by innate differences among the configurations. Our goal is to demonstrate the characteristics of each architecture rather than those of any particular implementation. As a compromise between the requirements of the various configurations and the requisite analysis software, each abstraction includes three terminals, two mass stores, two tape drives, and one printer. The remaining components and the device interconnections are determined by the specific architecture being modeled.

We have performed a reliability analysis on each abstract implementation to demonstrate the availability of system resources utilized in normal research activities (as a function of system component reliability.) The research actions modeled by the reliability analysis are computing usage, tape drive access and printer access. The analysis was performed using a state enumeration method [4, 5] with the computer reliability varying between 0.90 and 1.00 while the reliability of the other devices remained fixed at 0.99. The analysis required a connected terminal, computer, and mass store for computing access, while tape drive and printer usage required access to a tape drive or printer in addition to computer access. The selection of peripherals described above allows, for the configurations studied, the tape access availability to represent the availability of all doubly available resources while the printer access availability represents the availability of singly available resources.

A CLUSTER CONFIGURATION

Fig. 1 schematically depicts a cluster configuration which we have tested. The abbreviations used on this and following figures are listed in the appendix. The nucleus of the cluster was a Digital Equipment Corporation VAX 11/750 which included an FPS AP 120B array processor and a Gould-DeAnza IP8500 image processor. The cluster contained 16 terminal ports with hard-wired dumb terminals, a high speed line printer, two 1/2"-9 track tape drives, 16MBytes of main memory, and 1.25 GBytes of hard disk.

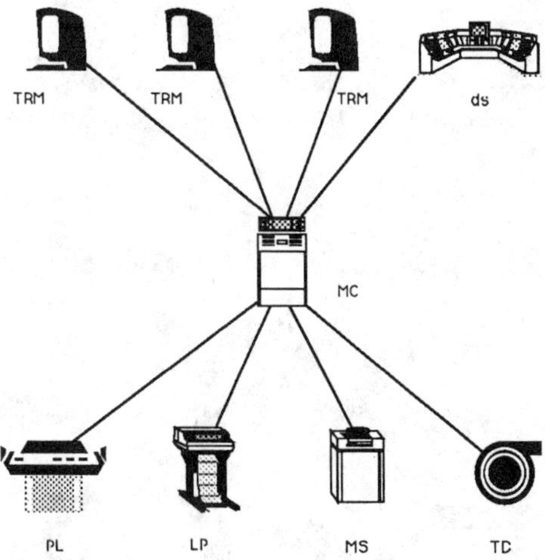

Figure 1: A cluster configuration for an Active Data Base

Although this cluster Active Data Base implementation employs contemporary system design, we noted significant difficulty in accommodating hardware modifications and software implementation requests. The cluster architecture does not readily and gracefully support evolution and expansion. As well, this system remains a virtual "one-person" machine due to display station limitations. We also find that CPU power is inadequate to meet user needs when the terminals are being fully utilized.

In addition to the operational difficulties we encountered with this configuration, the availability analysis indicates further limitations. The abstract implementation (Fig. 2) is a straightforward transcription of the cluster configuration (Fig. 1), modified to meet the normalization specifications described above. The reliability analysis (Fig. 3) indicates that the parameters are all linearly dependent on computer availability. This result is due both to the non-replication of computing resources as well as the computer's role as the connecting point for the entire system.

Also of interest in this and the other configurations is the difference in availability between the printer (singly available) and the tape drive (replicated). In this configuration, the effect of the redundant tape drive is noticeable but, again, bounded by the reliability of the computer. (The difference between the tape drive availability and computer reliability is approximately 2×10^{-4} which is unnoticeable on the graph.) The system does, however, benefit from a minimal number of components and interfaces so that with high computer reliability (the usual case), resource availability is reasonable providing, of course, that the number of users is minimized.

LAYERED CLUSTER CONFIGURATION

We reconfigured the cluster ADB into a layered cluster architecture (Fig. 4) by functionally integrating a PACX terminal server to the system. No other major changes in either hardware or software were undertaken since this configuration can support little additional peripheral equipment.

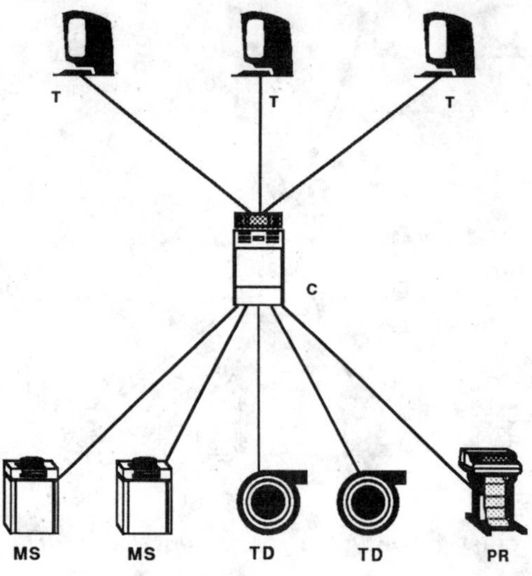

Figure 2: Abstract implementation for the cluster configuration

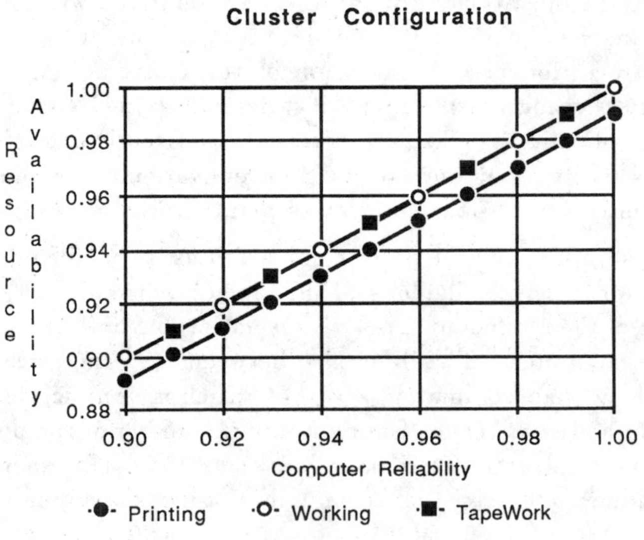

Figure 3: Reliability analysis for the cluster configuration

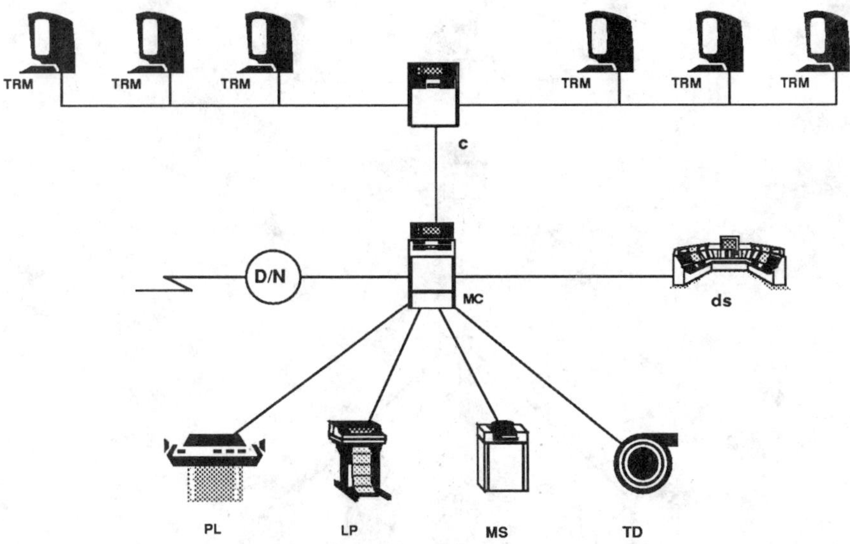

Figure 4: Layered cluster configuration for an Active Data Base

The functional switch from the cluster configuration to a layered cluster architecture theoretically should have increased user access, yielding a quasi-multiuser system (within the limitations of a single display station). On the other hand, "improved" access was effectively limited to the PACX terminal server since the number of time-sharing users seeking access to the machine now can easily fully saturate the CPU and available memory.

Consequently, while the layered-cluster architecture provides a feasible, although highly sub-optimal, multiuser environment, it also has some significant problems. When the central minicomputer experiences a failure, users not only lose their computational capability but also access to their files. For users on hardwire connections, access to other computer facilities is terminated. Additionally, this configuration is at best limited to peripheral upgrades even though it is CPU-limited and does not support straightforward, inexpensive growth. To further upgrade the computational capacity of the system would require its complete replacement.

The abstract implementation (Fig. 5) is essentially that of the cluster (Fig. 2) with the addition of the terminal server. The reliability analysis (Fig. 6) is, as would be expected, quite similar to that for the cluster except that all parameters are lower by 1%. This is due to the introduction of the terminal server which represents a single point failure and therefore factors directly into all the values. In an actual implementation, the terminal server would serve additional terminals, so the likelihood that all the terminals would be nonfunctional is reduced. This would, to some extent, counteract the decreased availability caused by the insertion of the terminal server. The abstraction therefore slightly exaggerates the decrease in availability since the number of terminals is unchanged.

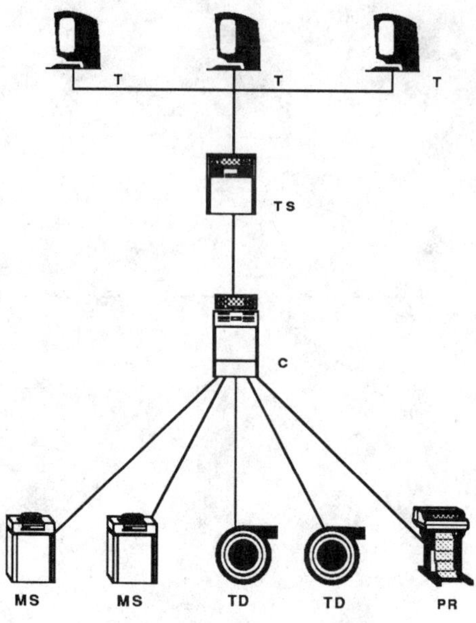

Figure 5: Abstract implementation for the layered cluster configuration

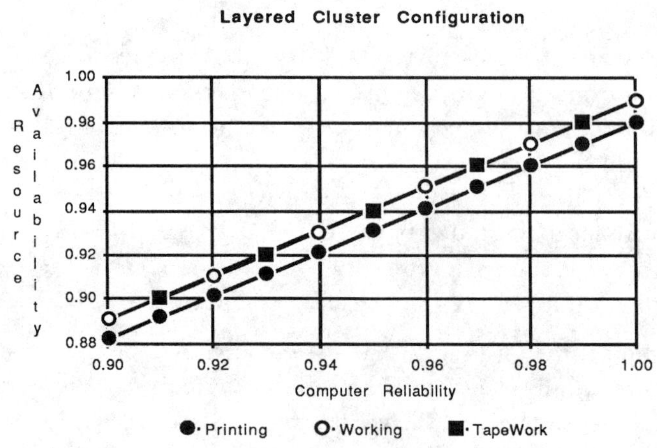

Figure 6: Reliability analysis for the layered cluster configuration

LAYERED MODULAR ARCHITECTURES

Based upon the recommendations of the Eos Data Panel [2], we have pursued the design, implementation, and testing of two layered, modular architectural concepts. In general, the Active Data Base module consists of five layers or strata: user, man-machine interface, compute, utility, and service. The user strata contains the entire user population for the Active Data Base. The man-machine interface (MMI) strata consists of an advantageous blend of interactive input facilities such as workstations, personal computers, and terminals. The compute strata supports all computing facilities such as microcomputers, minicomputers, and supercomputers. The utility strata provides system-wide facilities for on-line data storage and retrieval, such as mass storage, tape drives, printers and plotters, and archive. The service strata provides facilities for on-demand and off-line data storage and retrieval such as deep archives, documentation, manual ingest, and internetwork access. The various modules of the system are linked by a high-speed local area network. Partitioning system functions (based on user requirements derived from [2] and [6]) into modules was based, to the extent practical and feasible, on minimizing known difficulties associated with distributed computing systems (e.g., [1, 7]).

Layered Modular Cluster

Fig. 7 depicts a layered modular cluster which we have implemented and tested. This configuration supports five layers, including user, man-machine interface, compute, utility, and service. It is largely based upon a network of Sun Microsystems, Inc. workstations. Using Sun's Network File System (NFS) protocol, each of the MMI interactive input devices on the local area network can function as a complete image processing display station and has the ability to access the entire system peripheral pool.

This system has two peripheral servers. A Sun 3/280S acts as the primary system server. It controls two gigabytes of magnetic disk storage, a 1/2-inch tape drive, a color thermal transfer printer, a laser printer and two 12-inch Write Once Read Many-times (WORM) optical disk drives. A Sun 4/260, to a lesser extent, acts as a server as well. It supports 1/4-inch cartridge tape access and an additional 500 MBytes of disk storage. Users may log in to either of these machines and use them as compute engines or they may use their own workstation's computational power, relying on the prime system server to provide access to disk space, the printers, tape drive, and other system-supported utility devices and services.

This configuration addresses many of the problems presented by the earlier cluster and layered cluster configurations. If a user's workstation or terminal is down, he may log in from another device and have access to the same files and peripherals. Adding peripherals is essentially the same as on the pure cluster except that all users on the local area network now have full access to that peripheral. If more computational power is needed, a more powerful workstation may be transparently added to the network without impairing any other researcher's access. This new workstation would

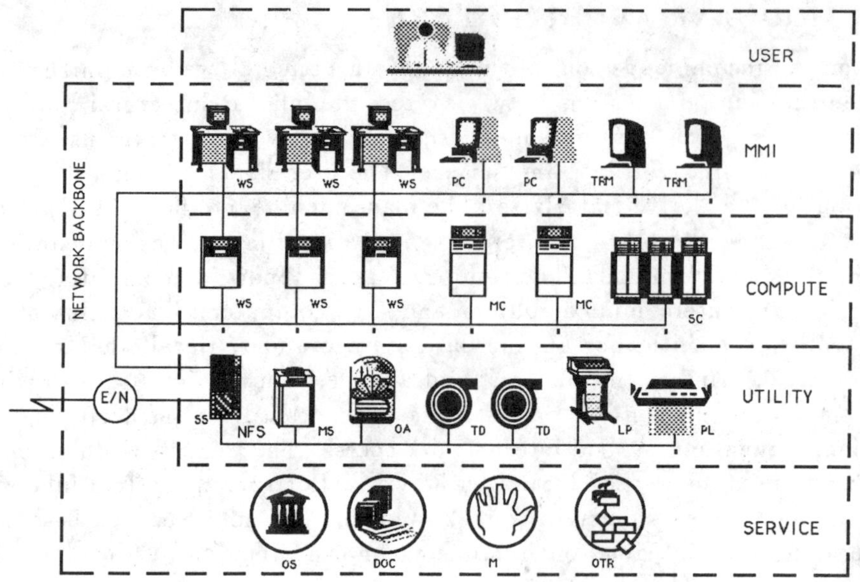

Figure 7: A layered modular configuration of an Active Data Base

have access to all of the existing peripherals and services, and other users may also access this added computational resource from their own MMI input device.

The key advantages of this system are essentially unlimited local computational power, transparent and graceful evolution, a pool of system peripherals that may be accessed by all of the computers on the network, and remote access from virtually any location via Internet. Taken together, it seems quite clear that a layered modular architecture promotes significantly more flexibility than other tested architectural configurations. As well, since all of our testing and evaluation has occurred during the installation phase of the project, we have been able to simulate utility-strata device failures with ease. While the "loss" of one of these system-unique devices presents inconveniences, it is far from debilitating. Instead, a straightforward reconfiguration using replicate devices located on other campus local area networks is adequate to meet user needs until the device in question is restored to service.

However, since this is not a fully modular, layered configuration, it shares some vulnerabilities with the cluster configuration and introduces a few new ones. A failure of the single system server is not catastrophic in terms of computational power but it does result in a loss of peripheral access for the entire network. Since some of the workstations are diskless nodes, a server failure effectively takes these particular workstations off line, forcing users to log in via the secondary server. Additionally, most disk requests pass over the network, which is noticeably slower than access to a local disk and results in increased network traffic.

The abstract implementation (Fig. 8) reflects the concepts of the tested implementation (Fig. 7) with the removal of devices to meet the normalization criteria.

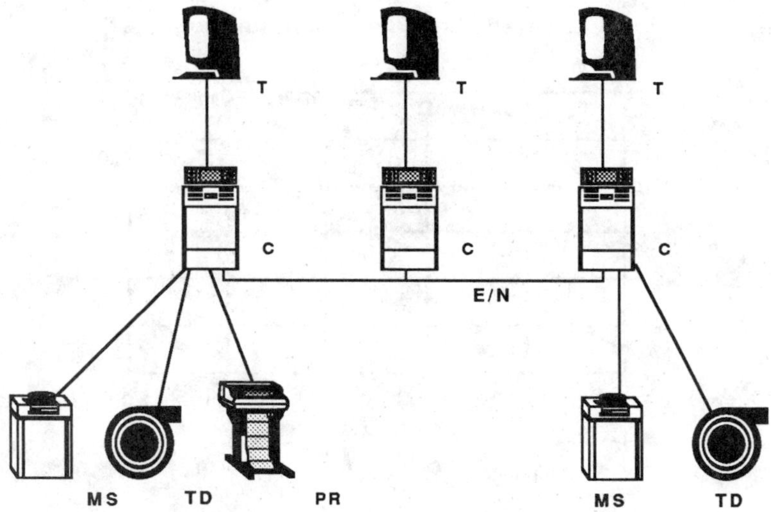

Figure 8: Abstract implementation for the layered modular configuration

The reliability analysis (Fig. 9) for this implementation reflects both the advantages and limitations of this configuration. The computer access availability is uniformly high (>98%) and only slightly dependent on the reliability of the individual computers. This is due to the lack of single point failures with respect to the compute strata: the computers, the mass stores, and the paths in between are replicated.

The difference between the behavior of the tape drive access and the printer access in his configuration is most striking. As with computer access, the tape drives benefit from a lack of single point failures. The printer however, is essentially unchanged from the cluster configuration since access requires that the controlling computer and printer be functional.

Fully-Modular Layered Architecture

A fully-modular layered architecture presents another step forward in distributed system configurations. As depicted in Fig. 10, each device on the local area network serves a specific purpose. While in the other configurations, the failure of a single CPU or network interface could result in the loss of access to all of the peripherals on the network (if not the entire system), in this case a failure results in the loss of only the affected device. This configuration is also more flexible. For example, disk space could be added to the network by adding another drive either locally or within the utility strata. In either case, there is, at most, minimal negative impact on any researcher's ongoing activities.

Additionally, modular layered architectures support the placement of significant computational resources under the direct control of each user. This has a strong, positive psychological (hence productivity) impact on users. With an appropriate selection of MMI devices (e.g., workstation with local disk and inexpensive printer),

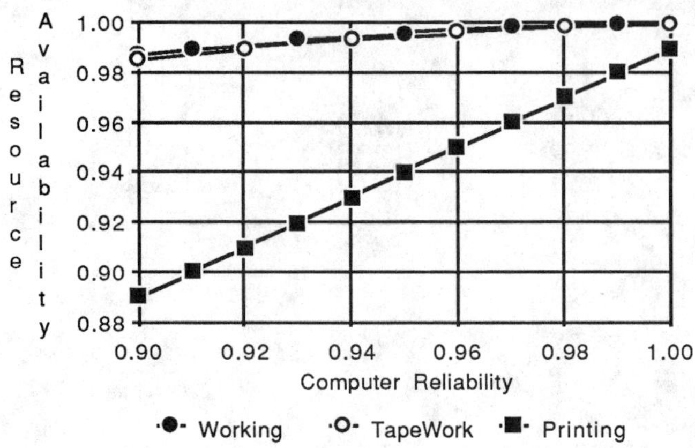

Figure 9: Reliability analysis for the layered modular configuration

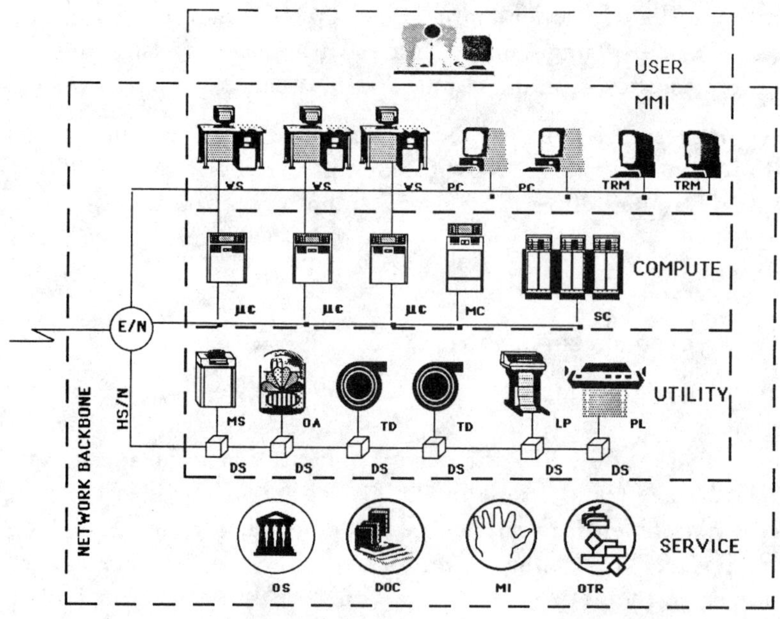

Figure 10: A fully-modular layered configuration for an Active Data Base

members of the same multidisciplinary research team, located at geographically distributed institutions, can participate in joint research activities with very minimal negative impact.

The primary drawbacks to this architecture are: (1) the required network bandwidth; and (2) the needed active intelligence for each device integrated onto the local area network (cf. [8]). Since all user and peripheral interaction would be carried on the network, the traffic loading will be higher, and the transactions which previously occurred over high speed dedicated interfaces must now be handled by the network. The advent of broad bandwidth fiber optical cable may eventually prove fiscally reasonable; this would effectively minimize or obviate perceived difficulties with local area network loading.

In terms of the second perceived limitation, in order to connect with the local area network, each device must have a network interface as well as native intelligence to generate and process network requests using reasonably sophisticated protocols. To achieve this, each device would require roughly the capabilities of a current-generation, low-end workstation to act as an individual or customized device server (depicted as DS in Fig. 10). In the near-term, this limits the cost-effectiveness of this architecture since the capital expense of replicating current-generation system servers to act only as interface controllers for each peripheral would be prohibitively expensive. However, network interfaces for stand-alone utility devices are already beginning to appear on some laser printers and optical drives. As these interfaces gain acceptance, development of more cost effective network interface controllers should soon follow.

The reliability analysis (Fig. 11) for the abstract implementation (Fig. 12) supports, for the most part, the claimed benefits of the fully-modular layered architecture (note the change in ordinate scale). The availabilities are nearly independent of computer reliability and are uniformly as good or better than the availabilities for the other configurations. This is understandable since the computers are no longer acting as the primary connection mechanism for the system. The system is, however, quite dependent on network and device server reliability. This is shown in Fig. 13 by varying the device server reliability rather than the computer reliability. The system also includes more interfaces which has a noticeable effect on devices which aren't replicated, in this case, the printer.

CONCLUSION

Based on our experience with various implementations, we find that while careful planning and certain compromise are required, the basic functionality and characteristics of layered-modular architectures are not beyond the reach of current technology. As well, systems based on a layered-modular architecture were found to provide significantly better support for evolution, growth and change; these are the prime factors affecting research-oriented Active Data Bases [2]. As can be anticipated, the most difficult and important part of the design is the decomposition of system functions

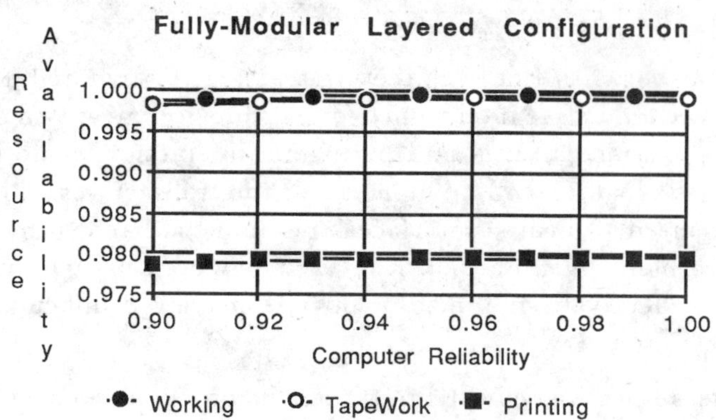

Figure 11: Reliability analysis for the fully-modular layered configuration

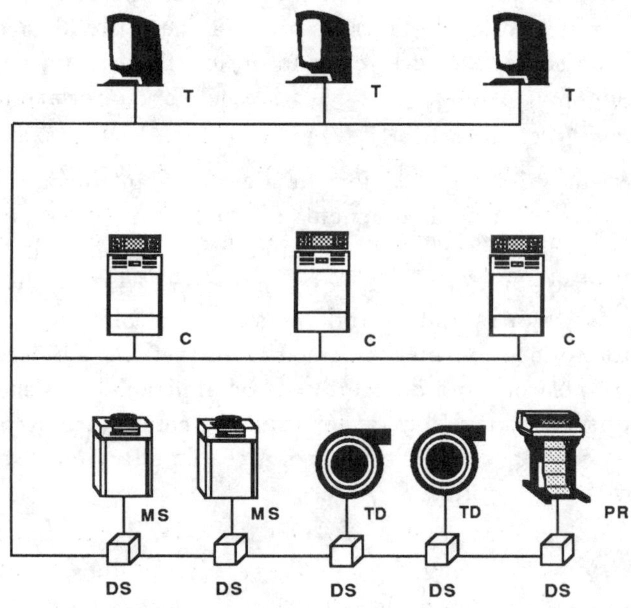

Figure 12: Abstract implementation for the fully-modular layered configuration

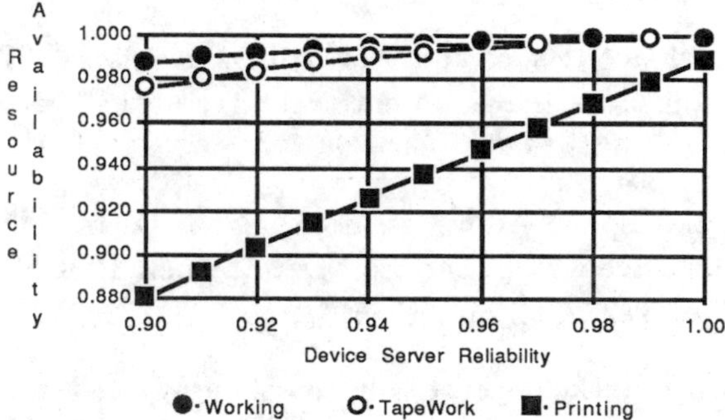

Figure 13: Reliability analysis for the fully-modular layered configuration versus device server reliability.

into appropriate modules and the specification of the interfaces between these modules. From the apparent lack of debilitating problems in the layered modular cluster (as opposed to deviances from the fully-modular layered architectural concept), it appears that the decomposition which we selected was appropriate. However, as we suggest, the interface standards will need to be further specified.

The reliability analysis, while for the most part enforcing common sense, is useful in pinpointing the exact nature of system deficiencies. For example, the analysis indicates that a hybrid system utilizing fully-modular layered concepts for devices replicated in the system and layered modular concepts for singly available devices would provide better cost performance without loss of availability. We are continuing this analysis and plan to describe our methods and findings in further detail in a forthcoming paper.

Finally, this analysis concentrates primarily on the interfacing and hardware concerns for a current-generation Active Data Base. Further study is necessary to fully define optimal software configurations for layered modular architectures.

Acknowledgments

Funding for this research, and development of the layered modular cluster, was provided by the National Aeronautics and Space Administration under Grant NAGW-1191. We sincerely wish to thank the Eos Data Panel for their initial consideration of layered modular architectures and Kelly Luetkemeyer (University of Colorado) and Rob Hall (Image Understanding Systems, Inc.) for helpful discussion and comments during the implementation phase of the Active Data Base. We also thank Mike Neely (CTA, Inc.) for providing the original icons and Patrick Allen (University of Colorado) for preparing the schematics.

References

[1] National Academy of Science. Issues and Recommendations associated with distributed computation and data management systems for the space sciences. National Academy Press, 111 pp., 1986.

[2] National Aeronautics and Space Administration. Earth Observing System Data and Information System: Report of the Eos Data Panel; R. Chase, Chairman, *NASA Tech. Memo. TM-87777*, 62 pp., 1986.

[3] Chase, R., F. Ziel, and K. Luetkemeyer. Evolution of an Active Data Base: Toward Developing the Local Research Environment of the 1990's, submitted, *J. Geosci. Remote Sensing*, 1989.

[4] Henley, J. Graph theory in modern engineering; computer aided design, control, optimization, reliability analysis. Academic Press, New York, 303 pp., 1973.

[5] Colbourn, C. J. The combinatorics of network analysis. Oxford University Press, Oxford, 160 pp., 1987.

[6] Chase, R., Towards a complete Eos Data and Information System. *IEEE J. Geosci. Remote Sensing*, in press. 1989

[7] National Academy of Science. Selected issues in space science data management and computation. National Academy Press, 66 pp., 1988.

[8] National Academy of Science. Data management and computation, volume 1: issues and recommendations. National Academy Press, 167 pp., 1982.

KEY TO FIGURES

USER = user strata	MMI = man/machine interface strata
COMPUTE = computational strata	UTILITY = utility strata
SERVICES = on-demand services	E/N = Ethernet local area network
D/N = DECNET communications	NFS = network file system
HS/N = high-speed local area network	

DOC = documentation	OTR = other services	ds = display station
PC = personal computer	DS = device server	PL = plotter
LP = line printer	SC = super-computer	MC = mini-computer
SS = system server	MI = manual ingest	TD = tape drive
MS = mass store	TRM = dumb terminal	OA = optical archive
WS = workstation	OS = off-line storage	uC = micro-computer

OPTIMIZED ORBITS FOR THE CONSTELLATION OF SPACE STATION POLAR PLATFORMS

Robert Chase and Michael Mundt[*]

The orbital configuration of the Eos polar-platform constellation is examined to determine if baseline orbital parameters are appropriately specified to meet the mission task of measuring global oceanic mesoscale variability. Using the Nyquist sampling theorem, spatial and temporal measurement requirements are first established for the constellation. Based upon these requirements, the optimum solution indicates that a total of five spacecraft equipped with nadir-pointing altimeters, or two multibeam-equipped platforms, would be required to meet the mesoscale mapping task. Additional analyses were performed on both one- and two-platform configurations as well as the complete four-platform Eos constellation. The results of these calculations show that the preferred scenario would be to equip all four Eos polar platforms with multibeam altimeters. However, we have determined that the inclusion of just one multibeam altimeter would very nearly meet statistically significant sampling criteria for mapping global oceanic mesoscale variability without having significant impact on mission goals, objectives or synergistic sampling strategies.

INTRODUCTION

In their report [1], the Earth Observing System (hereafter Eos) Science and Mission Requirements Working Group recommended that measurement of the oceanic mesoscale become one of the prime geophysical goals of the Eos mission. The technical basis for pursuing the task of global mesoscale mapping is provided in the Eos Altimetric System Panel Report [2]. The Panel recognized that the measurement of mesoscale variability has been demonstrated under certain restrictive (i.e., collinear trace) orbital cases [e.g., 3,4]. They then argued that "multiple platforms deployed at different phases can be used to provide dense spatial resolution topographic coverage needed to adequately map oceanic mesoscale."

[*]   Colorado Center for Astrodynamics Research, University of Colorado, Campus Box 429, Boulder, Colorado 80309-0429.

As a consequence, and given the synergistic sampling strategies mandated for the Eos mission [5] together with co-deployment of passive optical instruments on the same platforms as the radar altimeters [6], we have performed a detailed analysis to determine: 1) what the optimal configuration of a polar platform constellation would need to be to support the task of mesoscale mapping, 2) what the performance of this configuration would be with respect to statistically significant sampling criteria, and 3) what effects a multibeam altimeter might have, vis a vis the nadir single-beam altimeter recommended by the Altimetric System Panel.

PRELIMINARY CONSIDERATIONS

Before proceeding with calculations to optimize the orbits of the Eos constellation of polar platforms for oceanic mesoscale mapping, we have established sampling criteria based on a somewhat arbitrary definition of "mesoscale," examined the nominal orbits for the platforms in terms of repeat periods, tidal aliasing and ground trace crossing angles, considered how best to computationally handle descending ground traces, and defined a criterion for determining relative performance of an optimized orbital configuration [7].

Sampling Requirements

For our purposes, warm-core rings provide a basis for defining sampling criteria in this analysis because they are smaller than cold-core rings [8] and therefore will be more difficult to measure with statistical certainty. In order to adequately sample a warm-core ring, the Nyquist criterion must be invoked. Using this theorem, the spatial sampling required to resolve a feature with a characteristic length scale of 100 km is no greater than 50 km (i.e., 100 km/2). The sampling frequency needed to resolve a feature with a frequency of 0.043 d^{-1} should be no less than 0.086 d^{-1} (i.e., $2*0.043$ d^{-1}). This yields directly the temporal sampling constraint that oceanic mesoscale must be measured at least once every 11.6 days if its natural period is 23.2 days.

About the Satellites

The spacecraft considered for mapping the mesoscale are the four Eos polar platforms currently planned for deployment in the mid-1990s. The two NASA-supplied afternoon platforms (i.e., those platforms whose nodal crossing times are after local noon) currently are configured so that a relative constraint exists between them. The constraint is such that one platform lags the other by one-quarter of a revolution. At present, these two afternoon platforms, and a morning platform to be supplied by the European Space Agency, have nominal baseline orbital configurations of:

semi-major axis (a)   =7202 km
inclination angle (i)   =98.7°
eccentricity (e)   =0.00114
argument of perigee (ω)=90.0°
orbits per repeat   =227
repeat period   =16 days
spatial repeat precision =± 1 km

Descending Ground Traces

It seems reasonable that, when considering a satellite's or a constellation of platform's ground-track spacing and repeat times, it is advantageous to include the descending orbital traces. These ground tracks could, theoretically, double a satellite's sampling effectiveness. The problem is, however, that the descending traces aid both the spatial and temporal resolution, but only in an average sense [7]. Because of the variable nature of sampling improvements provided by the descending ground tracks, it is difficult to integrate descending orbital traces in a simple and straightforward manner into an overall scheme for orbit optimization. As a result, the descending ground traces are omitted when ranking the relative performance of a given platform configuration.

Sun-synchronous Orbits and Tidal Aliasing

Nominally, the Eos polar platforms will be in orbits which are sun-synchronous, meaning that they always cross the equator at the same local time. Because of this, sun-synchronous orbits alias tidal signals into very low (even zero) frequencies. This, in turn, causes errors in mesoscale height measurements because it makes the differentiation of tidal signals from those due to other geophysical phenomena difficult, at best, to achieve. We assume that, by the time the platforms are deployed, the astronomical tides will have been accurately measured globally so that tidal aliasing need not be a further consideration in this analysis [cf. 9].

Ground Track Crossing Angles

The acute crossing angle between ascending and descending ground tracks of sun-synchronous orbits becomes an important factor when trying to resolve two components of geostrophic velocity at the ocean surface. In order to determine the component velocities with as small an error as possible, the acute angle between ground tracks should be as close to 90 degrees as possible [10]. For a typical sun-synchronous crossing angle of 25 degrees, the error in measuring the meridional component will be small. However, the error in measuring the zonal component may be quite large. Based on the criteria from Parke et al. [10], the zonal component error may be too large. Conversely, in this case the actual zonal component of velocity would be much larger than the meridional component [11] and the relative accuracy level of each velocity component would then be approximately equal.

Measuring Performance

Two criteria are used to measure the performance of a system of satellites. They are the minimum resolvable size of a feature and the maximum resolvable velocity of that feature. We use a single, simple dimensionless index (referred to as the performance index, I_p) to relate how the actual spatial and temporal resolution of a given satellite configuration ranks in terms of the necessary sampling requirements for statistical significance.

OPTIMUM SOLUTION TO THE MESOSCALE SAMPLING PROBLEM

The next step in analyzing the performance of a constellation of polar platforms for mapping oceanic mesoscale is to determine the number of satellites, equipped with nadir altimeters, needed to accomplish this task in the most efficient manner possible. We refer to this case as the Optimum Solution.

The simplest method of meeting the defined sampling requirements is to configure the required N satellites into identical orbits. In this way, all of the repeat times will be identical and the Earth's equatorial circumference (and all other latitudes, for that matter) will be divided into equal spacings. The only other constraint is that, since the polar platforms are in sun-synchronous orbits, the Optimum Solution will be found from among the suite of sun-synchronous low Earth orbits. The problem then, is to solve for N, the number of satellites required. For N satellites in identical orbits, we have [cf. 12]:

$$D_s = (2 * \pi * R_e) / (T_s * G * N), \text{ where} \tag{1}$$
$$D_s = \text{sampling distance at equator (equatorial spacing)},$$
$$R_e = \text{mean radius of Earth},$$
$$T_s = \text{sampling time (equal to repeat time of orbit), and}$$
$$G = \text{number of revolutions per day}.$$

For these N spacecraft, all in identical orbits, T_s will be the same for all of the satellites, and D_s will be scaled by a factor of $1/N$. Therefore, if the desired sampling time and distance are specified, N can be solved for directly. In this case, the required sampling interval is the reciprocal of the Nyquist frequency. It will be designated T_c (i.e., $T_c = 11.6$ days). The required sampling distance, similarly, is the Nyquist sampling distance, designated D_c ($D_c = 50.0$ km). Finally, the characteristic velocity, V_c, can be introduced, where V_c is defined as D_c/T_c and is equal to 5.0 cm s^{-1}. The number of satellites required, N, will itself be a function of altitude and inclination angle, because G is a function of both. However, because of the desire for a sun-synchronous orbit, T_s must be an integer number of days less than T_c (i.e., $T_s = 11$) and only one degree of freedom remains. As a result, N can be expressed as a function of altitude only.

For all the orbits within the Eos altitude envelope [viz., 5], the resulting values for N are given in Table 1. Since for obviously practical purposes it is impossible

Altitude (Km)	Inclination Angle (Degrees)	Number of Satellites Needed	Days to Repeat	Orbits per Repeat
477.95	97.32	4.77	11	168
505.35	97.42	4.80	11	167
533.03	97.53	4.83	11	166
589.23	97.75	4.89	11	164
617.75	97.86	4.92	11	163
646.58	97.97	4.95	11	162
675.70	98.09	4.98	11	161
705.12	98.21	5.01	11	160
734.85	98.33	5.04	11	159
764.89	98.46	5.07	11	158
795.26	98.59	5.11	11	157
825.94	98.72	5.14	11	156

Table 1: Optimum Solution for nadir-beam constellation

to deploy a fraction of a satellite, only orbits for which N is already an integer, or very close to one, can be selected. Thus, either the 675.7 km orbit or the 705.1 km orbit are viable choices. Either selection yields a 5-satellite constellation capable of sampling oceanic mesoscale with statistical certainty.

In order to characterize the number of satellites, equipped with multibeam altimeters, required to map the oceanic mesoscale, a similar procedure can be invoked. However, in this case, the equatorial, inter-satellite spacing, which was equivalent to the sampling distance for the nadir-looking case, is no longer appropriate. With the introduction of a seven-beam, pushbroom altimeter, with beams spaced at 25 km intervals to either side of the satellite ground track (yielding a 150 km swath width) [2], the inter-satellite distance can now be much greater. Assuming a swath width of 150 km, an additional 50 km can also be added to this in order to position the two closest beams of adjacent satellites 50 km apart. This 50 km resolution implies that features 100 km and larger can be resolved. Consequently, a 200 km, equatorial, inter-satellite, ground-track spacing is required. Allowing this to be D_s, and using T_s equal to 11 days, one can solve for the number of satellites required at a particular altitude. This is shown in Table 2. Approximately 1.28 satellites are required at an altitude of 826 km, which is close to the nominal Eos altitude of 824 km. Since only an integer number of satellites can be deployed, this result indicates that two satelites are needed to optimally map the oceanic mesoscale with statistical certainty when equipped with multibeam altimeters.

The resulting improvement is four-fold over the nadir-looking configuration. This is because the multibeam altimeter, for the purpose of achieving 100 km resolution,

Altitude (Km)	Inclination Angle (Degrees)	Number of Satellites Needed	Days to Repeat	Orbits per Repeat
477.95	97.32	1.19	11	168
505.35	97.42	1.20	11	167
533.03	97.53	1.21	11	166
589.23	97.75	1.22	11	164
617.75	97.86	1.23	11	163
646.58	97.97	1.24	11	162
675.70	98.09	1.24	11	161
705.12	98.21	1.25	11	160
734.85	98.33	1.26	11	159
764.89	98.46	1.27	11	158
795.26	98.59	1.28	11	157
825.94	98.72	1.28	11	156

Table 2: Optimum Solution for multiple-beam constellation

essentially has four beams, spaced equally apart at 50 km intervals. This predicted four-fold improvement is evidenced by the fact that about 5.14 satellites are required at 826 km for the nadir-looking case, whereas now only 1.28 are needed.

OPTIMIZING ORBITS FOR A SATELLITE CONSTELLATION

Based on our analysis, five spacecraft equipped with nadir altimeters are needed to meet the specified sampling criteria for mapping oceanic mesoscale with statistical certainty. Consequently, the maximum allotment of four polar platforms will not be adequate to achieve this monitoring task [cf. 2]. Therefore, in terms of selecting a configuration for a sub-optimal constellation, we will in essence need to seek simultaneous maxima in both spatial resolution and resolvable velocity.

Configuring the four available platforms in such a manner that the impact on sampling distance will equal the impact on resolvable velocity, we obtain the following:

$$D_s'/D_s = V/V', \tag{2}$$

where D_s' and V' are the spatial resolution and resolvable velocity, respectively. Then, in terms of quantitatively measuring the performance of this constellation, a dimensionless parameter, the performance index (I_p), can be defined by recalling that maxima in spatial resolution and resolvable velocity are the criteria for judging orbit optimization. Therefore, for our purposes, let:

$$I_p = (D_s'/D_c) * (V_c/V'). \tag{3}$$

For the Optimum Solution, $I_p = 0.948$; the value is less than unity because the optimum solution requires a smaller repeat period than the original criteria mandated.

Now, to find the necessary repeat time for placing the available Eos platforms into the best possible sub-optimal configuration for mapping oceanic mesoscale, it is first necessary to define the governing equations. For the ideal configuration of five spacecraft, the sampling time and distance are T_s and D_s. Let N denote the number of satellites found in the optimum configuration (i.e., N = 5). For the N' platform case, where N' is an integer number of spacecraft (in this case four) other than the ideal configuration, let the sampling time and distance again be denoted by T_s' and D_s', respectively. These six variables are related by the following equation:

$$T_s * D_s * N = T_s' * D_s' * N'. \tag{4}$$

The respective resolvable velocities, V and V', are given by

$$V = D_s/T_s, \text{ and} \tag{5}$$
$$V' = D_s'/T_s'. \tag{6}$$

Using Eqs. 2, 4, 5, and 6, we have four equations in the four unknowns, D_s', T_s', V and V'. These can be readily solved, with the following results:

$$D_s' = D_s * K^{1/3}, \tag{7}$$
$$T_s' = T_s * K^{2/3}, \text{ and} \tag{8}$$
$$V' = V/K^{1/3}, \text{ where} \tag{9}$$
$$K = N/N' \tag{10}$$

and is hereafter referred to as the constellation constant.

Phasing of the Platforms

Nominally, the two afternoon platforms have a one-quarter revolution separation between them [13]. Because of this phasing, the orbits of the two platforms will overfly the same ground tracks during each repeat cycle. This occurs when

$$X * D = I, I = 0,1,2,3,..., \tag{11}$$

where

$$X = \text{the fractional orbital lag,}$$
D = the number of days to repeat for either satellite (since they are the same),

and I is an integer. For the Eos case, X=0.25 and D=16. Thus, I=4. I is significant because it indicates how many subintervals are skipped between the passing of the first platform and the passing of the second.

To position the two NASA-supplied afternoon platforms optimally for mapping oceanic mesoscale with nadir-pointing altimeters, the ground traces of the first platform should pass 1/2 subinterval away from the ground traces of the second platform. This will provide even global ground-track spacing. To do this requires re-phasing the platforms by increasing or decreasing the nominal 25.4-minute lag time between the two platforms by approximately 3.2 minutes.

Similarly, in order to position four polar platforms optimally, each platform's orbital traces should pass 1/4 subinterval away from the adjacent platform's traces. This requires increasing or decreasing the lag time between the two afternoon spacecraft so that the second platform either samples 4.25 subintervals or 3.75 subintervals away from the ground tracks of the first. For the baseline one-quarter of a revolution separation, the lag time is 25.4 minutes; increasing or decreasing this by 1.6 minutes would achieve the desired spacing.

OPTIMIZING ONE POLAR PLATFORM

Since only 1.28 satellites equipped with multibeam altimeters are required to map the mesoscale at an altitude close to that planned for the polar platforms, it is certainly worthwhile to consider the performance of only one satellite. With the introduction of the swath width instead of the transect sampling, the equations that define resolvable velocity and size must include the size of the swath. The resulting equations are:

$$V = (D_s\text{-}150)/T_s, \tag{12}$$
$$V' = (D_s'\text{-}150)/T_s', \tag{13}$$
$$V/V' = (D_s'\text{-}150)/(D_s\text{-}150), \tag{14}$$

and Eq. 4 listed previously.

Note the change in the definitions of resolvable velocities in Eqs. 12 and 13. This arises from the different usage of D_s and D_s' when applied to multibeam altimeter configurations. Because the largest inter-sampling distance will occur between the outer fringes of adjacent swaths, and also because the swath is 150 km wide, the maximum sampling distance can be written in terms of the equatorial, inter-satellite distance, D_s', and the swath width, 150 km, as $(D_s'\text{-}150)$. Identical reasoning produces the term $(D_s\text{-}150)$. It is also important to realize that Eq. 4 is only valid for a constant altitude. Thus, D_s', V', and T_s' will correspond to the altitude implicit in T_s, D_s, and V. Given these four equations with four unknowns, V, V', T_s', and D_s', all four variables can be solved for explicitly. A convenient method is to obtain an equation in D_s':

$$(D_s'\text{-}150)^2/(D_s\text{-}150) = (V * T_s * N * D_s)/(N' * D_s'). \tag{15}$$

Note that Eq. 15 is a cubic equation in D_s'. In general, this equation may have three real roots, so that a proper initial guess for D_s' must be made. This can be done by noting that D_s' will be slightly larger than D_s, which is 200 km for this case. A guess greater than 200 km is justified since D_s' must grow larger than D_s for any N' less than N (in order to preserve the assumptions made in Eq. 14). Thus, a guess of $D_s' = 210$ km will converge to the proper solution. This results in:

$$D_s' = 204.53 \text{ km},$$
$$\text{sampling distance} = 54.53 \text{ km},$$
$$V' = 3.96 \text{ km d}^{-1}, \text{ and}$$
$$T_s' = 13.77 \text{ days}.$$

Subtracting the 150 km swath width from D_s', the resulting value is 54.54 km, which means that features 109.06 km or larger can be adequately resolved. Also note that the value obtained for D_s' indeed checks, since $(D_s'-150)/(D_s-150) = V/V'$. Using the previous definition of performance index given in Eq. 3, and modifying D_s' and D_s so that the 150 km swath width is first subtracted from the nominal values, $I_p = 1.19$, which is the minimum possible value of I_p for a one-spacecraft configuration.

Now, consider the performance of one satellite in the nominal Eos orbit, which is a 16-day, 227-orbit repeat at an altitude of 824 km. The following parameters result:

$$D_s' = 176.54 \text{ km},$$
$$\text{sampling distance} = 50.00 \text{ km},$$
$$T_s' = 16 \text{ days},$$
$$V' = 3.13 \text{ km d}^{-1}, \text{ and}$$
$$I_p = 1.38.$$

It should be emphasized that the sampling distance is chosen to be 50.00 km because the original criteria only specified that level of accuracy; choosing a value larger than the minimum sampling distance (26.54 km in this case) results in an increase in resolvable velocity. The value of the performance index is identical to that of a four-satellite, 16-day case using nadir altimeters [7]. This is because the multibeam altimeter is effectively acting as a four-satellite system. Essentially, one multibeam altimeter could efficiently replace four conventional, nadir-looking altimeters on the Eos platforms.

OPTIMIZING TWO POLAR PLATFORMS

For the two-satellite case (N' = 2) with nadir-looking altimeters, the constellation constant (K) equals 2.50. Substituting this value for the constellation constant into Eqs. 7, 8, and 9, the following values for D_s', T_s', and V' result:

$$D_s' = 67.86 \text{ km},$$
$$T_s' = 20.26 \text{ days, and}$$
$$V' = 3.88 \text{ cm s}^{-1} \text{ (or 3.35 km d}^{-1}).$$

Since the resulting sampling distance and resolvable velocity are significantly different than those desired, we can consider changing either platform altitude or repeat period (or both, for that matter). It seems somewhat impractical to significantly change the baseline altitude of the polar platforms to the optimal value of 705 km (even though this is within the Eos altitude envelope that extends from 475 km to 850 km) principally because many of the instruments are being planned and designed for an 824 km deployment. Furthermore, extending the repeat time is neither feasible nor desirable since many of the physical processes to be studied with the radar altimeters, as well as other platform instruments, have characteristic time scales that are less than this. Consequently, the only viable option remaining is to consider two of the platforms in their nominal orbital configurations.

The current configuration of the two afternoon platforms, with one lagging the other by one-quarter of a revolution, results in overlapping orbital traces. If however, 3.2 minutes was added to (or subtracted from) the current 25.4 minute time difference between the two satellites, the two platforms would generate individual ground tracks, all evenly distributed around the Earth. This would result in an equatorial sampling distance of 88.28 km, meaning that features larger than 177 km could be resolved. With a 16-day repeat time, the maximum resolvable velocity is 5.52 km d^{-1}, or 6.39 cm s^{-1}. The resulting performance index ,I_p, is equal to 1.38. Although the equatorial spatial resolution does not meet the spatial sampling criterion, it is important to note that as latitude increases, the sampling distance will decrease by the cosine of the latitude angle.

Using the procedure outlined for optimizing one multibeam platform, the same steps can be applied to a two-platform constellation, but now with N'=2. However, as N' increases, the number of orbits to repeat for each satellite, defined as N_R, must decrease so that the swaths don't overlap (non-overlapping swaths are essential for symmetrical coverage of the Earth without redundancy). This, in turn, will necessitate a reduction in the repeat time, T_s, because larger values of T_s result in finer spatial sampling due to the inverse relation shown in Eq. 4. Moreover, the required repeat time will roughly decrease in a manner inversely proportional to N'. By extrapolating from the results obtained for the optimum one-satellite case, this implies that the required repeat time for each multibeam satellite in a two-platform configuration will be approximately 7 days. Given that the Eos platforms are currently configured in 16-day repeats, it seems exceedingly unlikely that there is a possibility of altering these orbits substantially. A new approach can be considered which takes advantage of the overlapping satellite swaths in order to optimize mesoscale mapping.

Oversampling Considerations

Consider three adjacent ground traces of a single satellite in the vicinity of the equator (three traces that are adjacent after the satellite has completed one full cycle of its repeat orbit, not three traces that are successive in time). This is illustrated in Fig. 1. If a time tag is given to each orbit in terms of the nearest day of occurrence,

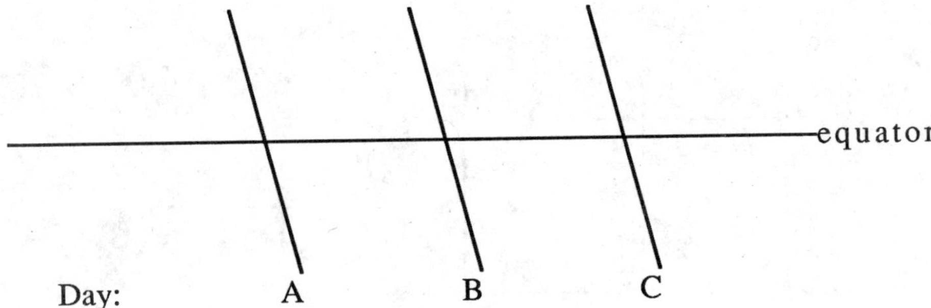

Figure 1: Three adjacent equatorial ground traces for a single satellite

there will be a pattern laid down for entire satellite ground track. In general, two adjacent orbits will differ by δ, where $\delta = |\text{Day A - Day B}|$, or $\delta = |\text{Day B - Day C}|$. The two different values of δ are related by:

$$\delta_1 + \delta_2 = T, \tag{16}$$

where T is the number of nodal days for the satellite to repeat. One value of δ will always be larger than T/2, and the other will always be smaller than T/2. By definition, let δ_1 be less than T/2, and let δ_2 be greater than T/2. By drawing the ground traces that are laid down within a fundamental interval, it is relatively easy to find the desired value of δ (a fundamental interval is the interval bounded by successive orbits of a single satellite; the fundamental interval is divided into T subintervals in T days).

As well, there is another method that can be followed in order to find the two values of δ. This is given by:

$$\delta = (kT + 1)/s, \tag{17}$$

where s is the skip number of the exact repeat orbit [10] , T is the repeat period in nodal days, and k is an integer. To determine δ, start with k=0 and increment k until δ becomes an integer. Note that this will introduce some error, because the number of days between orbits is not an integer. This arises due to the fact that the fractional remainder of N_R divided by T consists of a numerator and denominator that are prime numbers relative to each other. In other words, their least common multiple is their product [10]. In essence, there exists an inherent error in δ that vanishes only when δ=T (which cannot occur). Instead, the actual number of days can be found from:

$$\gamma = (mN_R + 1)/N_D, \tag{18}$$

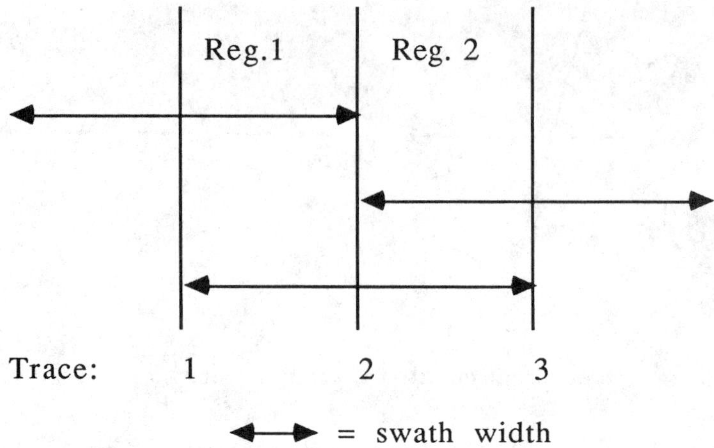

Figure 2: Region 1 is sampled by traces 1 and 2 in T days. Region 2 is sampled by traces 2 and 3 in T days.

where N_R is the number of orbits to repeat, N_D is the number of days to repeat, m is again an integer (though different from k), and γ is the spacing between two adjacent ground traces in terms of the number of orbits. To convert γ to days, simply multiply by N_D/N_R. This results in:

$$\gamma^* = m + 1/N_R,\tag{19}$$

where γ^* is now the adjacent ground track spacing in terms of days. Comparing this with the approximation to the nearest day given by Eq. 17, and noting that both δ and m are integers, it is evident that m and δ must be equivalent as long as the proper δ is used. Thus, the error in using the day approximation is simply the difference of these two values, which is $1/N_R$ days. This amounts to less than 0.01 days for any orbit with $N_R > 100$. We consider this accuracy to be more than sufficient and use the approximation to the nearest day for all subsequent calculations.

Now, with the information provided by δ, we can find the reduced, or effective, repeat time of a satellite constellation for which there is overlapping sampling. This may occur for any multiple-satellite multibeam configuration, and will be critical in determining the relationship that governs the reduction in repeat time. To see how this reduced repeat time arises, we provide Fig. 2. For an arbitrary N'-satellite configuration, the ground-trace pattern is intermeshed as shown in the figure. Along with the number of satellites, N', the other important parameter will be the distance to either side of the swath that any one satellite can cover in terms of the distance between adjacent ascending nodes. It is convenient to introduce η:

$$\eta = \text{INT}(W_S/L)/2,\tag{20}$$

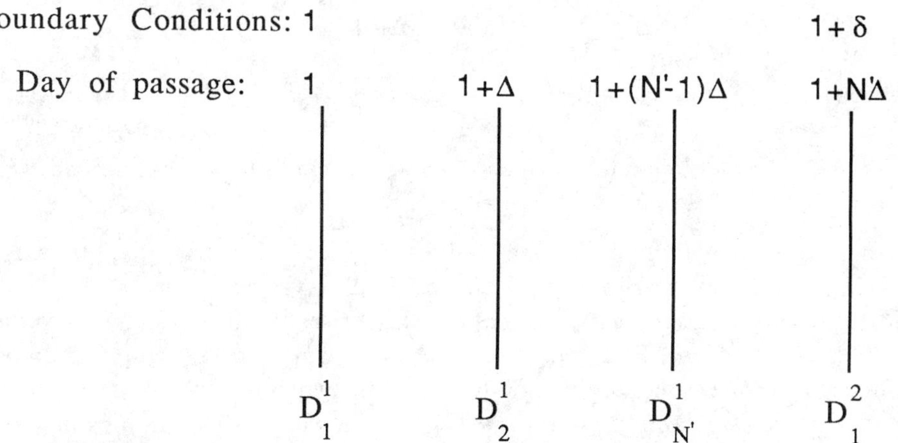

Figure 3: Illustration of boundary conditions of satellite ground traces

where W_S is the multibeam swath width and L is the equatorial spacing between adjacent ascending nodes of nearest-neighbor satellites. Given these two variable parameters, a system of equations can be derived that will illustrate all options available for reducing the effective repeat time. From Fig. 3 (where the subscript denotes the satellite and the superscript denotes the trace number of a given satellite), equating the two values at the second ground track of satellite #1, Δ can be found in terms of δ, where Δ is the optimal temporal spacing, between different satellites, of nearest-neighbor ground traces. It is important to note, however, that the values of $(1+\Delta)$, $(1+2\Delta)$, etc., are by necessity modulated with respect to the repeat period of the satellites, T. This allows several different relationships between δ and Δ. To determine the number of equations needed in order to fully represent the entire spectrum of different satellite configurations available without redundancy, it is simply necessary to realize that there is a symmetry, or folding, in the values of Δ about T/2. The number of equations necessary for a given value of N' can be determined from:

$$XT + \delta_{max} \geq N'T/2, \text{ where } \delta_{max} = T. \tag{21}$$

By substitution,

$$(X+1)/N' \geq 1/2, \text{ or } X \geq N'/2 - 1, \tag{22}$$

where X must be an integer. Thus, it is only necessary, given N', to solve for X, where $(X+1)$ equations are necessary to describe the system entirely with no redundancy. The $(X+1)$ equations can be represented in vector form:

$$[A]\Delta = 1/N'\{[B]T + [A]\delta\}, \tag{23}$$

where $[A]$ is an $((X+1) \times 1)$ vector that has elements $[1, 1, ..., 1]$, and $[B]$ is another $((X+1) \times 1)$ vector whose elements are $[0, 1, 2, ..., X]$. This set of equations entirely describe the range of possible configurations than an N'-satellite system can assume with uniform, global time reduction.

The final step is to account for different values of η for an arbitrary value of N'. Any point on the Earth will be sampled 2η times per T days, regardless of the value of N'. However, there is a limit of η in terms of N'; this arises because, after η becomes larger than $N'/2$, swaths from the same satellite are covering mutual sampling space, and it is impossible to configure a previously-defined orbit with respect to itself. Thus, the analysis presented here is only valid for $\eta \leq N'/2$. However, this will be sufficient for the swath width of 150 km. Now, for a region of space that is sampled 2η times within T days, the minimum effective repeat time is given by:

$$T^*_{min} = T/(2\eta). \tag{24}$$

Once η is found, then the desirable T^*_{min} can be found from Eq. 24 above. From an analysis of Δ, and noting that Δ is the number of days between the passing of adjacent ground tracks, the following relations for T^* can be derived:

$$T^* = T - (2\eta\text{-}1)\Delta, \text{ or} \tag{25}$$
$$T^* = \Delta, \tag{26}$$

whichever is larger . Note that they are equal when:

$$T - (2\eta\text{-}1)\Delta = \Delta, \text{ or} \tag{27}$$
$$\Delta = T/2\eta, \tag{28}$$

which corresponds to T^*_{min}. Thus, given an arbitrary N'-satellite system, it is desirable to use a value for Δ, found from the array of Eq. 23, that matches as closely as possible the value for T^*_{min} for the particular value of η being considered.

For two multibeam polar platforms which are phased such that the ascending nodes of one satellite bisect the equatorial spacings between the ascending nodes of the other, the distance between adjacent ascending nodes will be 88.28 km. Thus, using Eq. 20, $\eta=0$. However, the swath width is only 13.28 km narrower than one that would result in a value of $\eta=1$. Furthermore, this 13.28 km is much smaller than the desired resolvable mesoscale size of 100 km. Therefore, the Earth will be sampled two times per T days. As a result, the analysis is completed assuming a value of $\eta=1$ for the two-satellite configuration. Since $N'=2$, the application of Eq. 22 indicates that only one equation is required to completely specify the choices of satellite configurations. Utilizing the vector relation of Eq. 23:

$$\Delta = \delta/2. \tag{29}$$

Furthermore, Eq. 28 suggests that the minimum repeat time is T/2, which can be confirmed by letting $\delta \to$ T. For the polar platforms, the two values of δ are $\delta_1 = 5$ and $\delta_2 = 11$. Note that the sum of the two values is indeed equal to 16, which is the repeat time of the two platforms in days. Substituting each value into Eq. 29 yields two solutions:

$$\Delta = 2.5 \text{ days, and}$$
$$\Delta = 5.5 \text{ days.}$$

From the general guidelines established above, it is known that, for $\eta = 1$, the optimal value for Δ will be T/2. This is a guide in choosing the proper value of Δ. In this case, $\Delta = 5.5$ days, giving the smallest corresponding value for T*. Thus, the effective repeat time, T*, is given by:

$$T^* = 16\text{-}5.5 = 10.5 \text{ days.}$$

Now, in order to judge the performance of this configuration with a reduced effective repeat time, it is necessary to invoke the performance index, I_p, given by Eq. 3. There are some modifications that need to be made to the variables in this equation, however. D_s' can now be chosen to be anywhere between 25 and 50 km; this arises because the sampling density of the multibeam swath is 25 km, but the original criteria mandated only 50 km sampling. V_c will still be defined to be D_c/T_c, but V' will now be defined as $V' = D_s'/T^*$, where, as stated above, D_s' can be chosen anywhere in the range of 25 to 50 km, and T* is the effective repeat time. If the definitions of V_c and V' above are now substituted into Eq. 3, the following results:

$$I_p = T^*/T_c. \tag{30}$$

Consequently, for any effective sampling time less than the originally specified sampling time of $T_c = 11.6$ days, the criteria are met and the satellite system can sample the desired temporal and spatial scales with statistical certainty.

Applying this to the two-satellite configuration currently being considered, the following values result:

$$I_p = 0.905,$$
$$D_s' = 47.57 \text{ km, and}$$
$$V' = 4.53 \text{ km d}^{-1}.$$

Thus, this system is completely capable of mapping the mesoscale within the criteria for sampling that were originally specified. It should be noted that the system will have a larger (and thus less desirable) performance index than that for a two-satellite system that has no overlapping swaths (e.g., the 7-day repeat configuration discussed earlier). This is because the geometry of the satellite orbits themselves prevent the maximum possible reduction in time.

OPTIMIZING FOUR POLAR PLATFORMS

For the case of four platforms with nadir altimeters, the constellation constant, K, is equal to 1.25. The following values for D_s', T_s', and V' derive from Equations 7, 8 and 9:

$$D_s' = 53.86 \text{ km,}$$
$$T_s' = 12.76 \text{ days, and}$$
$$V' = 4.89 \text{ cm s}^{-1} \text{ (or 4.22 km d}^{-1}).$$

Using the nominal orbits (i.e. 16 days, 227 orbits) for the polar platforms, we assume that it is possible to configure the constellation so that no platform overflies the same ground tracks as any other platform. Further, we assume that the platforms can have their orbits altered so that all ground traces are evenly spaced. With this arrangement, there would be four 227-orbit, 16-day repeat periods producing evenly spaced traces around the equator.

The resulting equatorial spacing is 44.14 km, and the repeat period is 16 days for all four platforms. These map directly into a sampling distance of $D_s' = 44.14$ km and a sampling time of $T_s = 16$ days. The maximum resolvable velocity of any feature is given by D_s'/T_s'. For this case, the maximum resolvable velocity, V', is 2.76 km d^{-1}, or 3.19 cm s^{-1}. The minimum resolvable size of a feature is $2*D_s$, which equals 88.28 km. However, the original criteria only required a minimum resolvable size of 100 km. If mapping features smaller than 100 km is not desired, a larger resolvable velocity can be obtained. Performing this trade-off, the maximum resolvable velocity is 50 km/T_s, which results in V' equal to 3.13 km d^{-1}, or 3.62 cm s^{-1}. In terms of the performance of this four-satellite system, all spatial requirements are met. However, the maximum resolvable velocity of a mesoscale feature is only 72.3% of the required velocity (this is with the second choice defined above). The resulting performance index is 1.38.

With the procedure already outlined for two multibeam-equipped platforms, it is straightforward to extend the analysis to N'=4. Using Eq. 20, the resulting maximum value for η is $\eta=3/2$. Furthermore, application of Eq. 22 indicates that two equations will be required to fully represent the different possible configuration. They can be written as follows:

$$\Delta = \delta/4, \tag{31}$$
$$\text{and } \Delta = (T+\delta)/4. \tag{32}$$

For each equation, δ can take on the values $\delta_1=5$ and $\delta_2=11$. Thus, this gives four possible states of the system. Substitution into Eqs. 31 and 32 gives the four values:

$$\Delta=1.25,\ 2.75,\ 5.25,\ \text{and}\ 6.75\ \text{days}.$$

For $\eta=3/2$, Eqs. 24 and 28 indicate that the desired value is $T^*=\Delta=T/3=5.33$ days. Thus, the best choice is $\Delta=5.25$ days. Using this, the following values result:

$$T^*=16\text{-}2(5.25)=5.5\ \text{days},$$
$$I_p=0.474,$$
$$D_s'=34.43\ \text{km, and}$$
$$V'=6.26\ \text{km d}^{-1}.$$

Here, D_s' and V' have been configured, based on the value of I_p given above, to satisfy the relation given in Eq. 2. As expected, this system is extremely overqualified in its ability to map mesoscale features with the prespecified space and time scales. In fact, the value for I_p above indicates that there is approximately a two-fold overall increase in performance beyond what is required.

CONCLUSIONS

Comparing the performance levels achieved by the various satellite configurations equipped with nadir-looking altimeters reveals a marked increase in performance for the multibeam-equipped satellites. For the nadir-looking cases, the best performance achieved by satellites in the nominal Eos orbits is for the $N'=4$ scenario. The resulting performance index is $I_p=1.38$. However, for the multibeam case, an equivalent level of performance is already achieved for the $N'=1$ case, for which the performance index has a value of $I_p=1.38$ for $T=16$ days. Again, one multibeam-equipped satellite essentially performs as well as a system of satellites equipped with conventional altimeters. This can easily be seen by recognizing that, ignoring the second, fourth, and sixth beams of the altimeter (since they only serve to provide additional and unnecessary spatial resolution for a system with no swath overlapping, the multibeam instrument is acting effectively as four conventional altimeters that have equatorial beam spacings of 50 km. This explains how such performance can be achieved from a single multibeam satellite. For the $N'=2$ and $N'=4$ multibeam constellations, there is no direct comparison to any configuration of nadir-looking altimeters considered. It is sufficient to note that both constellations achieve a level of performance beyond the original criteria, and they are far superior to the performance of any constellation comprised of conventional, nadir-pointing altimeters for comparable values of N'.

Certainly, the optimum scenario would be the deployment of four Eos platforms equipped with multibeam altimeters in order to provide the most complete mesoscale mapping possible. However, it has been shown that the addition of even one multibeam altimeter aboard a polar platform would provide substantial coverage, equalling that of four conventional altimeters. Furthermore, the implementation of two multibeam altimeters would provide the minimum number necessary in order to meet the originally specified sampling criteria. Because of the promising performance expected from all of these different configurations, all would be viable alternatives, with preferences tending toward maximizing the number of deployed altimeters.

Acknowledgments

Funding for this research was provided by the National Aeronautics and Space Administration under Grant NAGW-1191. We wish to thank George Born and George Rosborough (both of the Colorado Center for Astrodynamics Research at the University of Colorado) for helpful discussions.

References

[1] NASA "Earth Observing System, Science and Mission Requirements Working Group Report Volume I," D. Butler, Chairman, NASA Tech. Memo.86129, 1984, 58 pp.

[2] NASA "Altimetric system panel report: Earth Observing System Volume IIh," R. Chase, Chairman, NASA Tech. Memo. 86129, 1987, 61 pp.

[3] COLTON M.T. and CHASE, R. "The interaction of the Antarctic Circumpolar Current with bottom topography: An investigation using satellite altimetry," J. Geophys. Res., 88, 1983, pp. 1825-1843.

[4] CHENEY, R.E., MARSH, J.G. and BECKLEY, B.D. "Global mesoscale variability from collinear tracks of SEASAT altimeter data," J. Geophys. Res., 88(C7), 1983, pp. 4343-4354.

[5] NASA "From pattern to process: The strategy of the Earth Observing System, Eos Science Steering Committee Report, Volume II," D. Butler, Chairman, NASA Tech. Memo. 86129, 1987, 140 pp.

[6] NASA "The Earth Observing System Announcement of Opportunity, Part 1," A.O. No. OSSA-1-88, 1988, 42 pp.

[7] CHASE, R. and MUNDT, M. "On Optimizing the Orbits of a Constellation of Altimetric Satellites for Measuring Global Oceanic Mesoscale Variability," submitted J. Astronaut. Sci., 1988.

[8] RICHARDSON, P.L. "Gulf Stream rings," in Robinson, A.R. (editor), Eddies in Marine Science, Springer-Verlag, Berlin, 1983, pp. 31-39.

[9] NASA "Satellite altimetric measurements of the ocean, report of the TOPEX Science Working Group," Jet Propulsion Tech. Rep. 400-111, 1981, 78 pp.

[10] PARKE, M.E., STEWART, R.H., FARLESS, D.L. and CARTWRIGHT, D.E. "On the choice of orbits for an altimetric satellite to study ocean circulation and tides," J. Geophys. Res., 92(C11), 1987, pp. 11,693-11,707

[11] BROWN, O.B., CORNILLON, P.C., EMMERSON, S.R. and CARLE, H.M. "Gulf Stream warm rings: a statistical study of their behavior," Deep-Sea Res., 33(11/12), 1986, pp. 1459-1473.

[12] CUTTING, E., BORN, G.H. and FRAUTNICK, J.C. "Orbit analysis for SEASAT-A," J. Astronaut. Sci., 26(4), 1978, pp. 315-342.

[13] NASA "The Earth Observing System Announcement of Opportunity, Part 3," A.O. No. OSSA-1-88, 1988, 40 pp.

REMOTE ATTITUDE MEASUREMENT SENSOR ADVANCES SMALL-MOTION DETECTION CAPABILITY

Mary L. Sullivan[*] and Hugh W. Davis[†]

As space structure size becomes larger to accommodate multiple payloads, increased weight may exceed most launch-vehicle load capacity. To meet weight constraints, lighter, more flexible structures are being built. Alignment and attitude determination have become a real problem in controlling these flexible structures.

Through a study supported by NASA Goddard Space Flight Center (GSFC) and further development effort, Ball Aerospace Systems Group (BASG) developed a position sensor for monitoring arc-second motions. This sensor is a proof-of-concept and demonstration model that verifies the high accuracies and update rate.

BASG's Remote Attitude Measurement Sensor (RAMS) is a non-contacting, optical-position sensor designed to measure the small motion of vibrating structures or surfaces by monitoring many locations in the structure.

This paper examines the current RAMS prototype -- including sensor design, processing electronics, and optical configuration options. Many applications suitable for remote sensing are surveyed, and RAMS enhancements planned to meet these growing needs are outlined.

REMOTE ATTITUDE MEASUREMENT SENSOR (RAMS)

There are many space applications which require or benefit from the use of high-accuracy, high-bandwidth sensors such as RAMS. Space structures, especially large, flexible structures, require a means of measuring both translation and rotation of numerous points on the structure for systems identification and for active control of the surface figure or alignment of the structure. Deployable structures also benefit from sensors that monitor deployment, control its sequences and rates, and verify proper alignment and latching when deployment is completed.

When payload instruments are mounted on flexible platforms, pointing knowledge can no longer be derived from the spacecraft attitude control system (ACS). Payload pointing-accuracy is degraded by on-board disturbances that excite resonance modes above ACS cutoff (approximately 0.1 Hz) and below the region

---

[*]   Associate Design Engineer, Ball Aerospace Systems Group, P.O. Box 1062, Boulder, Colorado 80306.

[†]   Principal System Engineer, Ball Aerospace Systems Group.

where intrinsic structural damping become effective (approximately 100 Hz). Isolation of payloads from these base-motion disturbances requires precise, high-bandwidth position data such as that provided by RAMS.

Other equally important uses of RAMS include alignment calibration, coalignment of payload instruments, and transfer of alignment/attitude information between inertial reference units and payload mounting surfaces. In addition, a variety of space vehicles planned for support of space station operations, such as the Orbital Transfer Vehicle (OTV), Orbital Maneuvering Vehicle (OMV), and polar and co-orbiting platforms, will require proximity knowledge for docking, interface, and avoidance.

Primary attributes sought in any large space structure (LSS) sensor include:

o High accuracy

o Multiple monitoring locations (targets)

o High update rates

o Low cost

o Unobtrusive (non-contacting)

o Low complexity

o Reliability.

Although performance characteristics vary with each specific application, accuracy in the submillimeter range and update rates of at least 250 Hz are generally required to support systems identification and active control of LSS with disturbances of 25 to 50 Hz. The greater the number of targets (structural locations) monitored, the better is characterization of its dynamic behavior. Low complexity leads to higher reliability and maintainability. Sensors having an unobtrusive nature can be used to study photovoltaic solar arrays and other film-like structures without the sensor mass influencing the dynamics of the structure. It is desirable that the LSS sensor have broad application and be versatile to accommodate configuration changes easily.

RAMS was developed using space-proven technology directly applicable to LSS. Proprietary algorithms developed as part of our star tracker programs permit subpixel interpolation and focus compensation. RAMS heritage is based on the BASG-built retroreflector field tracker (RFT), the first solid-state sensor flown in space. The RFT measured dynamic structural behavior as part of the Solar Array Flight Experiment (SAFE).

RAMS Demonstration Description

This working display of the prototype RAMS, Fig. 1, has been assembled to demonstrate RAMS capabilities to measure the dynamic behavior of a vibrating structure. This demonstration has been intentionally simplified to show low-frequency vibrations with a few targets. However, the RAMS design is capable of monitoring as many as 100 targets and measuring arc-second position changes.

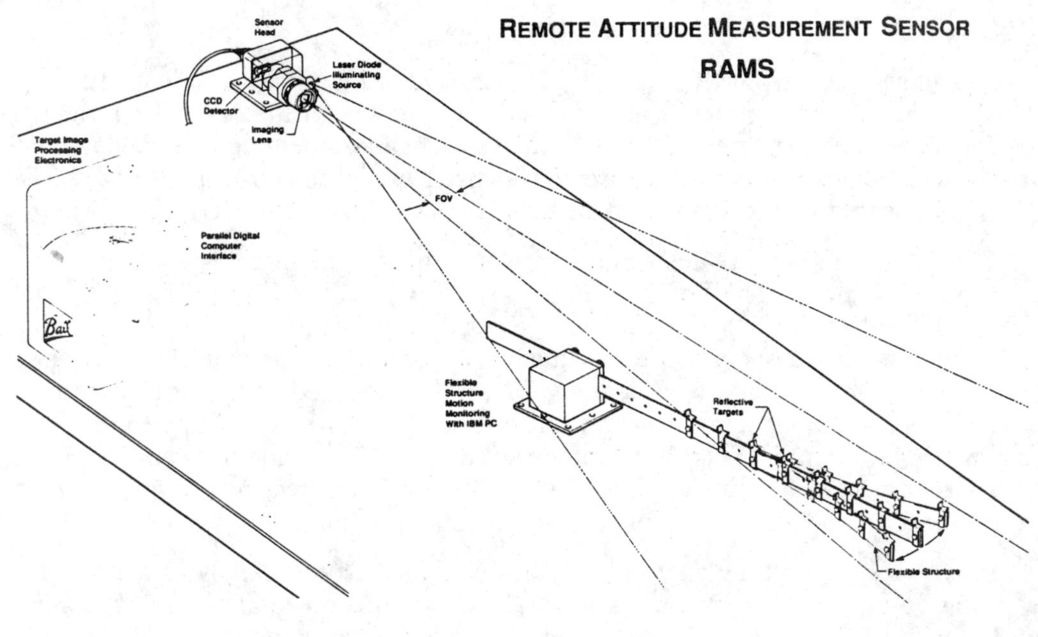

**Figure 1   RAMS demonstrates the ability to measure the dynamic behavior
of a vibrating structure**

This demonstration uses a 3-mW laser diode to illuminate retroreflective tape targets distributed along the length of a model beam. The beam is designed to exhibit a first-mode resonance frequency of approximately 1 to 2 Hz. A Bodine variable-speed, dc-gearmotor is used to excite the structure through a push-rod assembly.

The sensor head includes a 0.106m-focal-length lens and a 2048-pixel linear (single row of pixels) charge-coupled device (CCD). The interpolation algorithms, implemented in the processing electronics, are used to define the image shape, precisely locate its center, and compensate for changes in focus.

The retroreflective tape used for targets has an effective retroreflective (observation) angle of approximately 3 deg. This defines the allowable separation between the illumination source and the centerline of the sensor optics. Intensity of the reflected target image is fairly constant at any angle up to 30 deg off the normal and decreases rapidly beyond 30 deg.

A computer interface has been added to the processing electronics to provide near real-time motion monitoring of the vibrating structure. The data is transferred to the computer to a 16-bit parallel interface and graphically displayed using Turbo Pascal software. The display shows target location displacements from null when the structure is excited. This displacement information provides vital structural analysis knowledge in a development effort.

Sensor Concept

A light source -- typically a laser diode -- is used to illuminate a reflective target, such as a mirror or retroflector, attached to the structure of interest. The reflected image is focused by a cylindrical lens in the sensor head onto a linear CCD detector. Displacement of the target will cause the focused image to move along the CCD, giving an indication of the angular change.

The cylindrical lens creates a line image and ensures the target is focused on the narrow CCD detector. This configuration provides a sensitive plane, Fig. 2, parallel to the length of the CCD, and a nonsensitive plane perpendicular to it. Thus, one CCD and lens assembly provides a single-axis displacement sensor. If two axis measurements are needed, two linear CCDs can be mounted orthogonally, with individual lens assemblies inside a single sensor head, giving the capability to measure two axis displacements. This same configuration can also provide data about a third axis, rotations about the line-of-sight (LOS), provided that two retroreflective mirrors are used with adequate spacing to ensure proper resolution.

Each pixel subtends an angle equivalent to 25 arc seconds. Applying a 2 percent interpolation factor, the sensor's resolution is 0.5 arc second rotation and 0.0003 in. translation at an 11-ft range. When the target image is focused on the CCD, its energy is spread over several pixels, as shown, Fig. 2. The interpolation algorithms identify the radiometric center of this distributed image.

It is an important design condition that each target must have a clear LOS to the sensor head. The placement of targets on the structure must consider the geometrical view of each target by the sensor, so that no two targets overlap or obscure other targets from view. This condition is usually achievable with little sacrifice in configuration.

Data Processing

RAMS is capable of measuring the position of 100 targets simultaneously and updating each target at 250 Hz. This update rate is achieved with off-the-shelf electronics parts.

A "target present" condition is met when the image signal on the CCD exceeds a preset threshold level. Due to focusing and target size, the image shape is quite critical to interpolation accuracy; so the optical geometry is foremost in proper detection. The interpolation algorithms are implemented in an analog pipeline processor, Fig. 3, rather than a software-controlled microprocessor, to achieve higher update rates and to accommodate a large dynamic range.

From the display controller, direct position-readout is available so that no integration of the output data is necessary. This data may be analyzed by a computer interface similar to that currently implemented or the data may be fed directly into a feedback network for structural control.

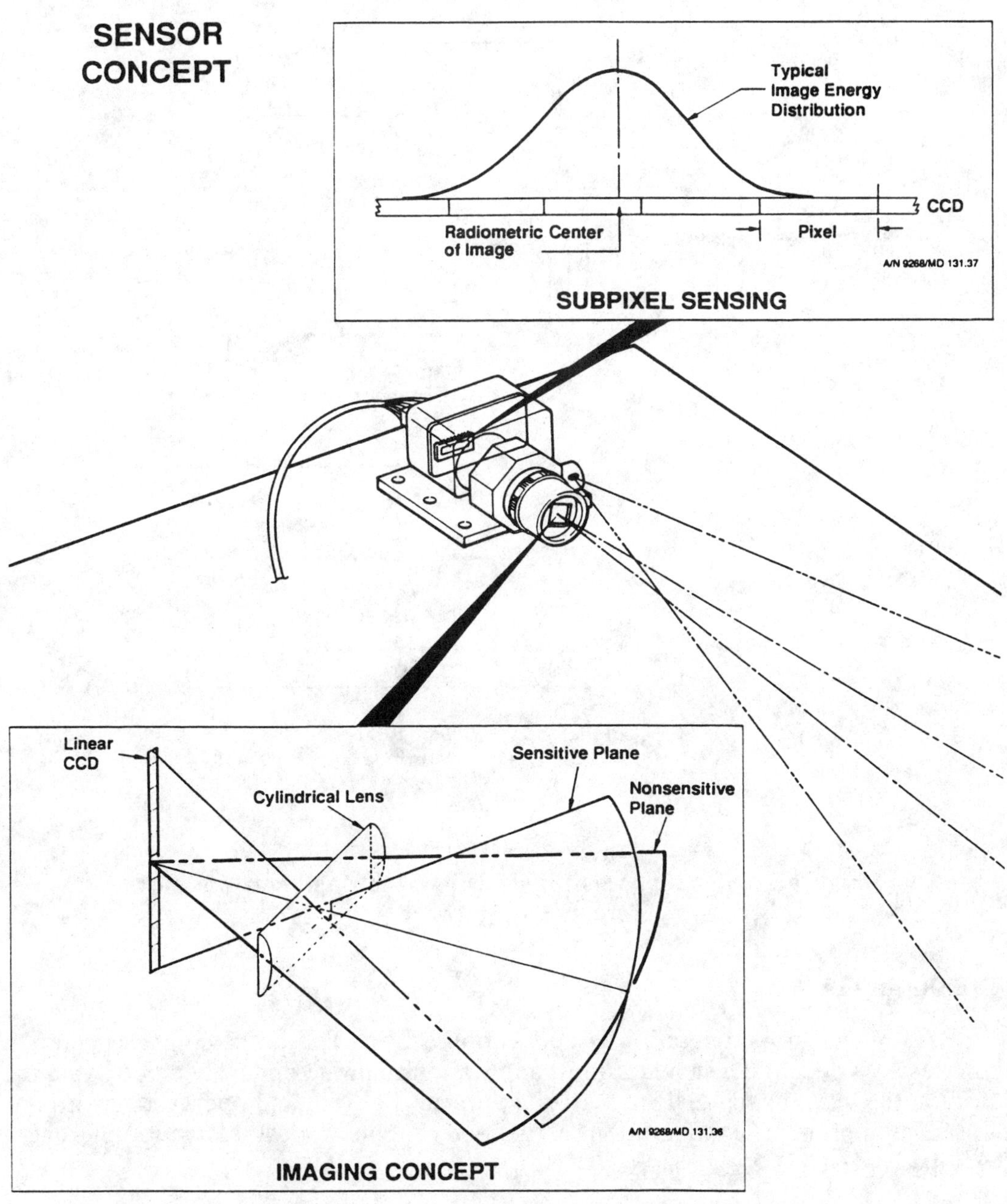

Figure 2   RAMS sensor images target motion onto a CCD, proper image focus is critical for high accuracy

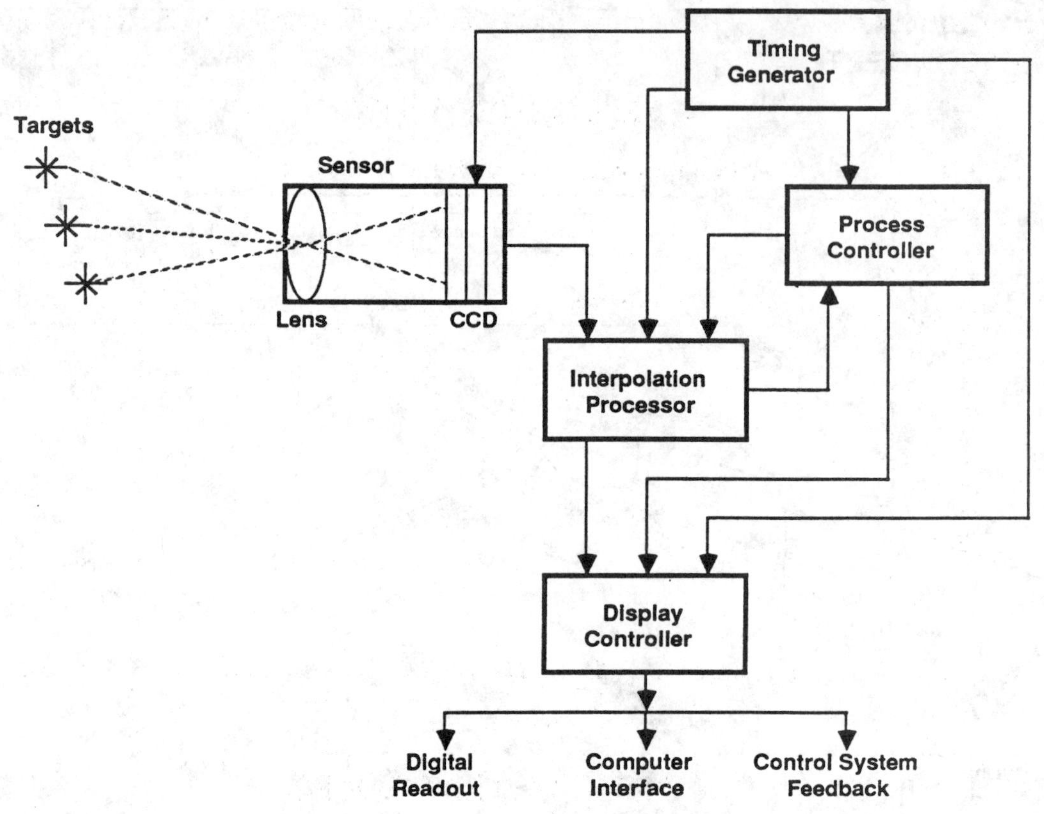

**Figure 3   Signal processing is achieved through hardware
rather than software**

Configurations

RAMS optical configuration is application-driven. The processing electronics and interfaces remain the same for most applications, but the optical geometry varies based on the viewing angle, light source, accuracies desired, and operating environment. The optimum configuration is defined by an optical software package developed at BASG.

The graphs shown, Fig. 4, have been included to accommodate specific questions about accuracy and field of view (FOV). Using a displacement equivalent of 1 arc second for system accuracy, the translation sensitivity can be determined at ranges from zero to 40 m. Using an effective focal length (EFL) of 106 mm, the displacement sensitivity is degraded less than 0.2 mm (1 arc second) at a 40 m range.

In any optical configuration, there is a tradeoff between FOV and EFL. The second graph depicts these optical tradeoffs for the RAMS sensor geometry. FOV can be increased by reducing EFL; however, reduced EFL results in degraded accuracy, so these parameters must be evaluated closely to achieve the optimum design.

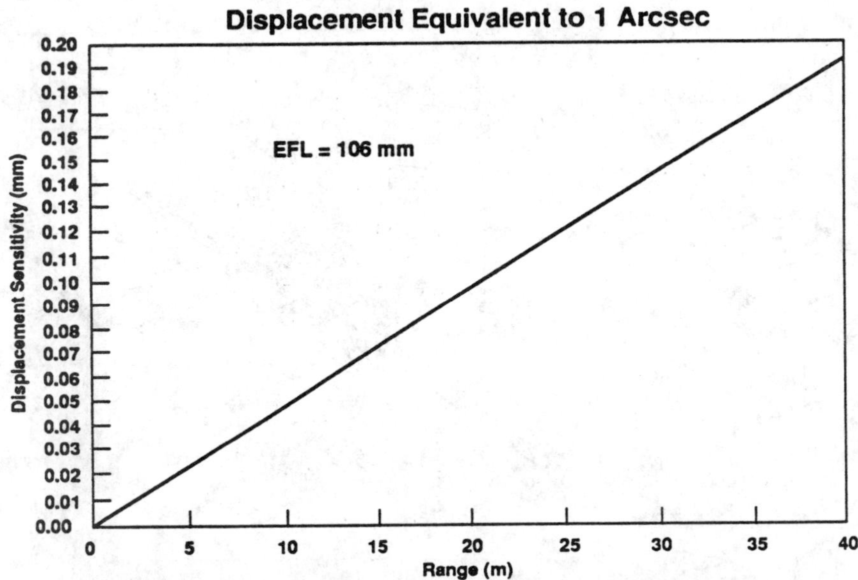

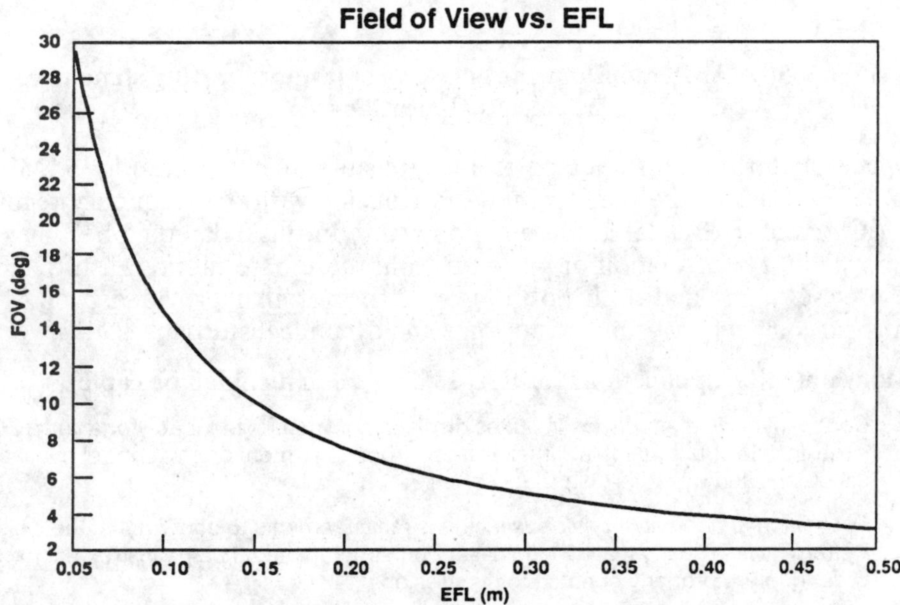

Figure 4   Optical configuration tradeoffs optimize design

211

<u>**RAMS Applications**</u>

A typical application for RAMS is depicted in Fig. 5, which shows a flexible boom supporting an antenna that must be accurately pointed. RAMS provides feedback data on the position of specific targets attached to the boom, and permits determination of boom shape and active control of the antenna orientation.

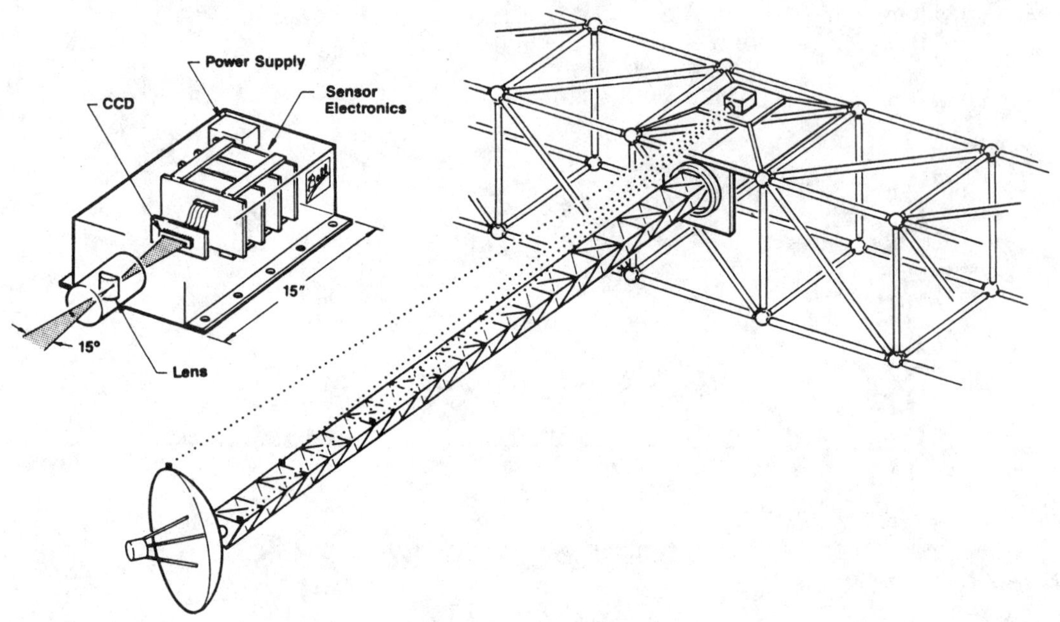

Figure 5   RAMS monitors the behavior of large, vibrating structures at several locations

Space platforms are subjected to many disturbance factors, Fig. 6. Note that natural frequencies of the platform tend to coincide with the frequency regime between ACS cutoff (0.01 Hz) and the region where intrinsic damping becomes effective (above 100 Hz). Isolation of payloads from these base-motion disturbances requires precise, high-bandwidth position data, such as that provided by RAMS, to monitor and control quasi-static distortions and dynamic disturbances.

Many potential applications for RAMS have been included in Table 1.

o *Control of flexible structures* includes displacement sensor feedback for a variety of large flexible structures (platforms, trusses, booms) to monitor and control quasi-static distortions and dynamic disturbances.

o *Calibration/alignment of payloads* includes measurements to confirm that the payload instrument is properly mated with its support structure and that the instrument axes (or boresight) are properly oriented for its mission.

o *Coalignment of multiple payloads* includes concurrent alignment measurements for two or more instruments to verify that all are pointed at the same target.

o   *Altitude transfer* implies measuring the relative attitude differences between an instrument payload and an inertially referenced fixture, and calculating the inertial attitude of the instrument based on these differences.

o   *Surface figure measurement* includes measuring the position of numerous points on the surface of an antenna or other structure to verify the maintenance of proper shape or figure.

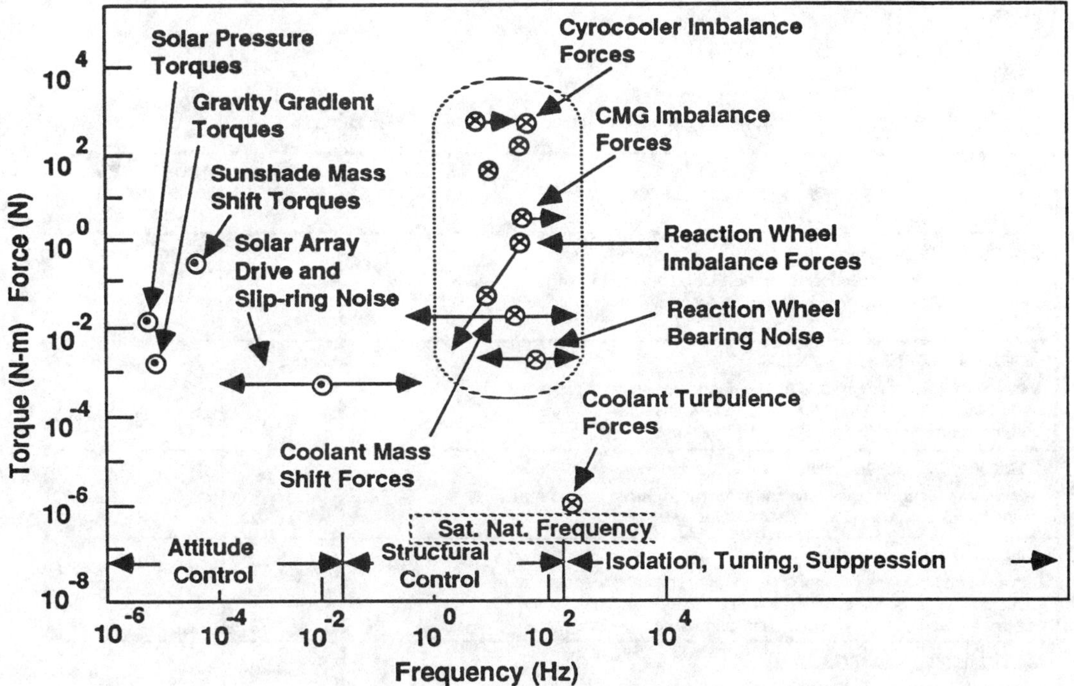

Figure 6   Spacecraft disturbances cover a broad spectrum and make isolation difficult

Planned Improvements

The RAMS design will be improved and expanded during 1989 to include some capabilities needed to accommodate additional applications. This continuing improvement program will be accomplished using both independent research and development (IR&D) and contract efforts to meet the growing needs of the aerospace community.

RAMS improvements are planned in each of these general areas:

o   Analog processing circuitry. Minor design changes will result in reduced sensitivity to temperature changes and background illumination changes and will improve the update rate significantly beyond 250 Hz for all targets.

o   Two-axis prototype sensor. Fabrication of a full-scale, two-axis prototype sensor will permit extended characterization testing and demonstration in the user's laboratory.

o   Opto-mechanical changes. Enhancements to certain optical and mechanical design features will result in improved focusing capability and a more rugged design to accommodate transportability and use in a high-vibration environment.

In addition, BASG intends to continue its review of industry and government needs for large, space-structure sensors, and to meet as many of these needs as possible. We will also closely monitor new developments in electro-optical components, especially detectors and laser diodes, to take advantage of these in future RAMS improvements.

Table 1

RAMS POTENTIAL APPLICATIONS ARE BROAD

SPACE APPLICATIONS FOR RAMS	TYPICAL RANGES (m)	ACCURACY AT MAXIMUM RANGE (mm)
Control of Flexible Structures: Displacement sensor feedback for monitoring and controlling quasi-static distortions and dynamic disturbances of large flexible structures	5 to 75	0.36
Calibration/Alignment of Payloads: Measurements to confirm that the payload instrument is properly mated with its support structure and that the instrument's axes (or boresight) are properly oriented for its mission	1 to 10	0.10
Coalignment of Multiple Payloads: Concurrent alignment measurements for two or more instruments to verify that all are pointed at the same target	1 to 15	0.14
Attitude Transfer: Implies measuring the relative attitude differences between an instrument payload and an inertially referenced fixture, and calculating the inertial attitude of the instrument based on these differences	5 to 100	0.48 (1 arc sec)
Surface Figure Measurement: Measuring the position of numerous points on the surface of an antenna or other structure to verify the maintenance of proper shape or figure	1.5 to 20	0.10

A COMPARISON OF HARDWARE AND SOFTWARE SIMULATION OF AN AGILE POINTING SYSTEM

Wade Scherer[*], Peter Van Atta[*], John Miller[*], Gerard Steiner[*] and Ray Briggs[†]

A critical factor in the feasibility of a multiple target laser ranging system is the ability to very accurately, and very quickly repoint the laser beam at the succession of targets. A near term solution includes the use of a light weight telescope, with an on-gimbal laser oscillator and off-gimbal power amplifier, and an agile gimbal system. The total operation of this device will be validated through a hardware demonstration/test of a scaled, prototype unit. As part of the initial definition and design of the unit, software simulation/analysis tools (e.g. TUTSIM, MATRIX$_x$) have been used to assess the combined performance of the selected control laws and the physical design demonstration of the hardware. The SDI sponsored R2P2 facility provided a means to extend the simulation/analysis procedure to a hardware environment. The tests provided a positive comparison between the two types of simulations identifying some aspects of the system that had not adequately been included in the software simulation. The final results showed excellent agreement between the software and hardware representations and provide added confidence in the planned Hardware-in-the-loop test of the gimbals and control system.

Introduction

The Agile Pointing Subsystem

Various space based defense systems may use tracking LADARs to provide their fire control calculation with autonomous data on the target location. The requirements for an Agile Pointing System (APS) are driven by the update rates for multiple target information at long range.   Azimuth and elevation angles are provided by the telescope position, while range is

---

[*]   Martin Marietta Astronautics, P.O. Box 179, Denver, Colorado 80201.

[†]   Aerojet ElectroSystems, 1100 W. Hollyvale Avenue, Azusa, California 91702.

obtained from laser pulse transit time. The defense spacecraft is then autonomous in that it is not dependent on binocular data from two widely spaced sensor platforms as required by passive systems.

If the APS is to perform this mission adequately, it must be capable of passively tracking target plumes, and slewing between multiple targets in rapid succession. In addition, it must be capable of settling on a particular target with a jitter less than the laser transmit beamwidth. Furthermore, the APS must be capable of maintaining a high degree of accuracy on targets that possess changing angular acceleration with respect to the APS inertial system.

The process of acquisition, tracking and ranging on the targets is shown in figure 1. The sensor passively acquires the target with a mid wave infrared (MWIR) scanning focal plane. The APS servos are then commanded to center the target within the telescope field of view. The target is tracked using a small staring array (located adjacent to the scanning linear array) with the central four pixels configured as a quad cell. The ranging is then performed by pulsing the laser at 40 PPS and detecting the active signature. At least two returns are required to establish an acceptable range for the target. This sequence of events takes place within one second to provide range data at the rate of one target per second.

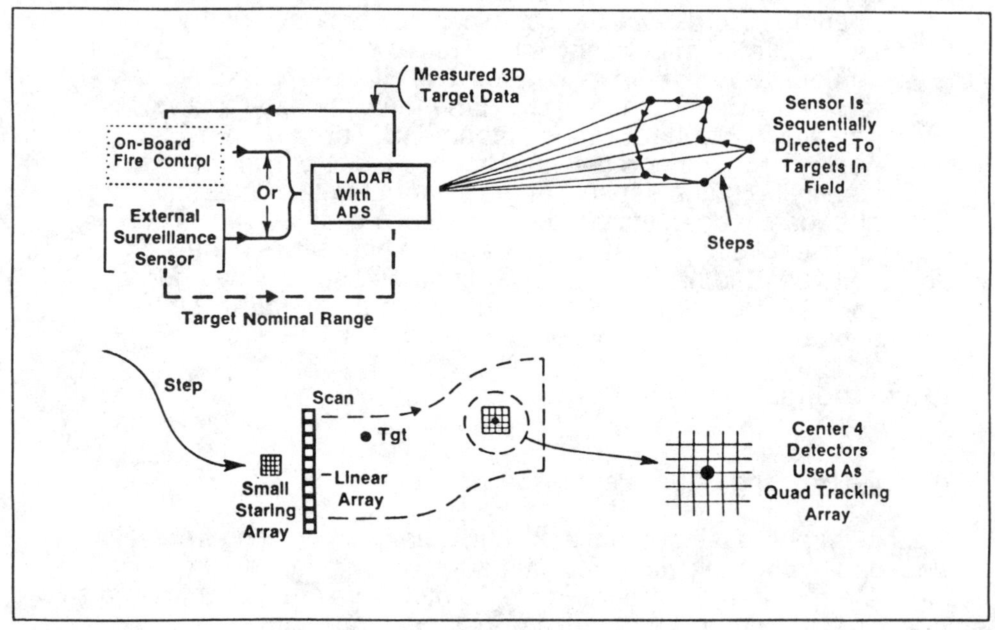

Figure 1 Target Acquisition Sequence

The LADAR uses a medium energy source to illuminate the target and actively determine its range. The critical aspects such as aperture and laser energy are driven primarily by the minimum target signatures.

The APS uses a gimballed telescope concept to minimize large optical elements and ease the pointing control requirements (figure 2). The telescope optical axis is slaved to the spin axis of a two-degree-of-freedom tuned rotor gyro. The pointing of the telescope is controlled by torquing the gyro about the appropriate axis. The telescope inner gimbals are driven by direct drive brushless torquers over a small angular freedom of less than ±0.1 radians. The inner gimbals are mounted on coarse outer gimbals which provide the large field of regard coverage. The gimbals are mounted in a (coarse AZ): (coarse EL): (fine EI): (fine AZ) configuration. The gyro is mounted in the inner gimbal and as this gimbal follows the gyro (because it is torqued), the gyro itself requires only a very small gimbal freedom.

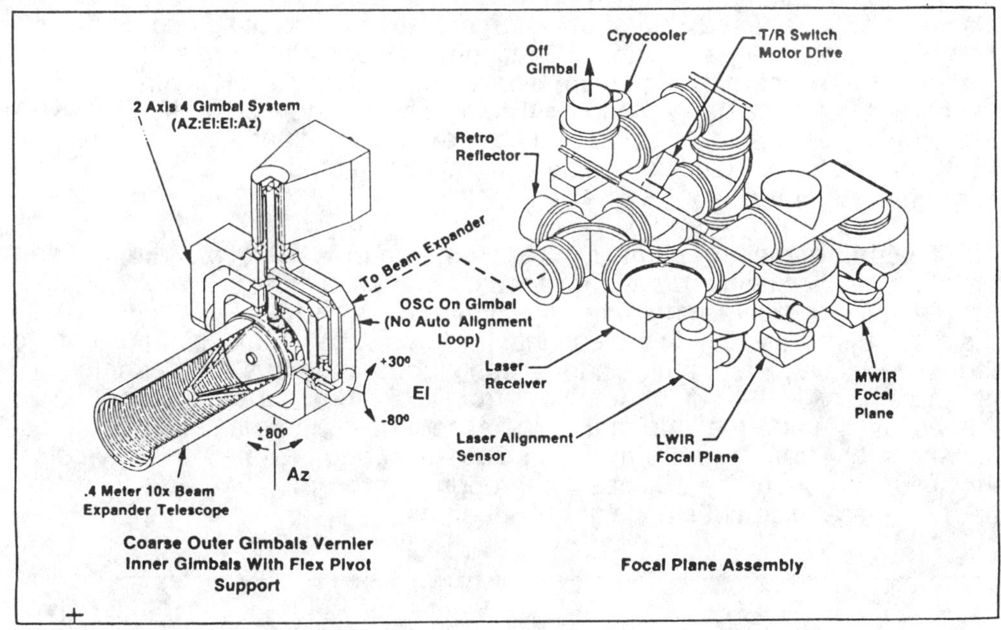

Figure 2   Gimballed Telescope Concept

The purpose of the inner vernier gimbals is to provide precision control and stabilization of the line of sight. The step from one target to the next is also achieved using these vernier gimbals. After the step is achieved, the coarse outer gimbals are used to follow up the vernier gimbals so that they do not exceed their limits. The small gimbal freedom of the vernier gimbal makes it possible to use flex pivot supports, which, with the brushless torque motors, essentially eliminate gimbal friction.

Software Control Simulation Tools

As part of the intial definition and design of the system, software simulation/analysis tools were used to assess the combined performance of the selected control laws and the physical hardware. TUTSIM was used as a quick check tool during the initial design and MATRIXx as a final check using all the known characteristics of the design and the R2P2 plant prior to the hardware experiment.

TUTSIM

The initial simulations of the APS control system were done using TUTSIM, a program for simulation of continuous dynamic systems originally written at the Twenty University of Technology in the Netherlands. The version used was run on an IBM PC. The program allows the user to construct the system by creating and specifying the interconnection of blocks in a diagram format. The straightforward use of the program allows the rapid construction and simulation of the system for intitial design checks.

MATRIXx

The APS experiment was simulated using a beta test release of MATRIXx Workstation V1.2 installed on an Apollo Domain Series 3000 workstation. In this version, the user can construct a large system by creating and connecting blocks in a block diagram format. Continuous (s-domain) blocks can be connected with discrete (z-domain) blocks to form a multi-rate hybrid system. Switching logic, time delays, and nonlinear effects such a saturation and quantization can also be included. Blocks are user definable and groups of blocks called "Super-blocks" can be defined and nested by the user to yield a hierarchial structure.

To perform the time-domain simulation, the MATRIXx software internally regroups the blocks in subsystems according to sampling interval and first sample time as shown schematically in figure 3 (Note that a continuous block has a sampling interval of zero seconds) These subsystems communicate to each other via a global, multi-channel continuous "data bus" For example, if the input to a block is the output of a block from a different subsystem, the connection is made (i.e. information is passed) using the data bus. Each discrete subsystem input from the bus passes through a sampler. Each discrete subsystem output to the bus passes through a zero-order hold (ZOH), which places a piece-wise constant signal on the bus.

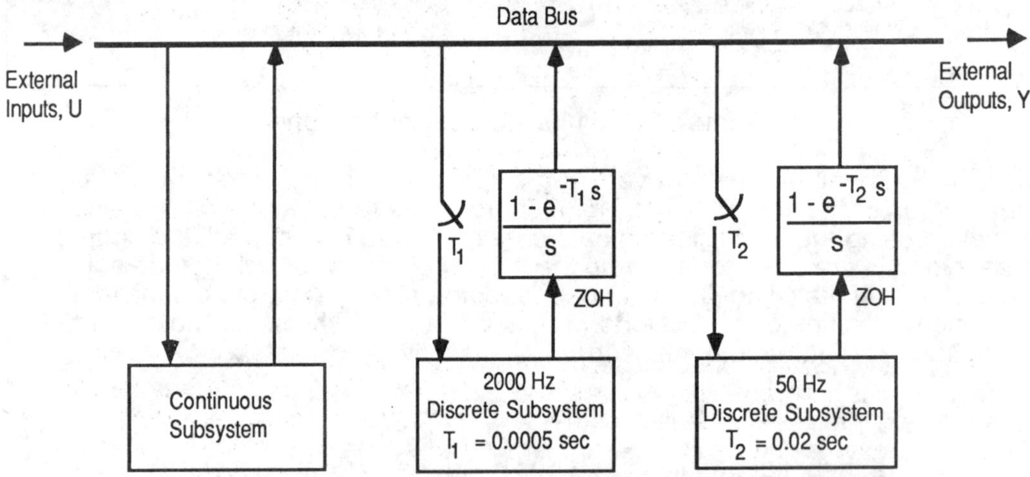

Figure 3 - MATRIX$_X$ Subsytems and Data Bus

Continuous blocks are converted into state equations and state variables and output variables are propagated through time by numerically integrating the continuous subsystem state equations. Six integration algorithms are available.

For the results presented in this paper the variable Kutta-Merson algorithm was used. In this algorithm the integration step is optimized to provide the largest step while remaining within the user-specified error tolerances.

<u>R2P2 Hardware/Software Simulation Facility</u>

The Rapid Retargeting and Precision Pointing (R2P2) facility at Martin Marietta's Waterton, Co. location was constructed as a national facility for simulating subsystems required for ultra accurate pointing and tracking in space. The design approach was to replicate the acquisition, pointing, tracking, and vibration isolation elements of SDI systems by direct analogy, (figure 4).   It features a large payload structure or central truss which is an adjustable flexible body to simulate the forward steerable portion of a spacecraft. The spacecraft bus or aft body simulates the maneuverable support body. A pointer/isolator (gimbalflex) controls movement between the two bodies and serves to isolate optical elements from spacecraft vibrations. The optical system provides simulated targets, precision track and scoring. The computer fire control system manages surveillance, acquisition, track, and retargeting.

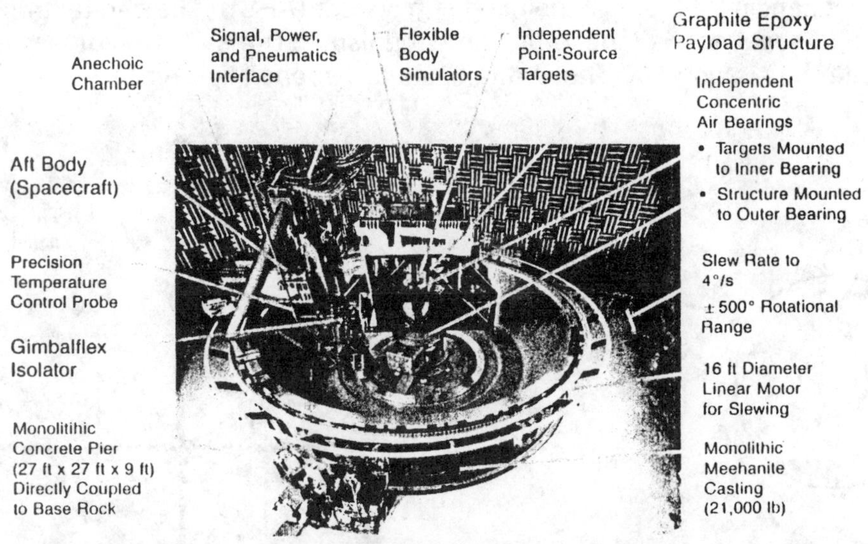

Figure 4 R2P2 Simulation Facility

The facility is built on a sandstone formation which is embedded in silt resulting in extremely low seismic background noise. Structural characteristics of a space system are simulated by the flexible body simulators. Target position, velocity, and acceleration are simulated by target simulators which reside on the innermost of the concentric air bearings. The aft body is driven by the linear motor and the graphite epoxy structure is driven by the outer bearing and the gimbalflex. Gimbalflex also isolates the aft body noise from the sensitive payload. All operations are performed within a temperature-controlled, anechoic environment on the quiet pier base, to ensure reliable collection of data uncontaminated by outside error sources. Refer to AAS 88-019 from the 1988 G&C Conference.

Purpose of the experiment

The purpose of evaluating the APS pointing performance in the dynamic enviroment simulation afforded by R2P2 was to test and evaluate the architecture and individual elements of the controller design by providing early data on the pointing performance and modifying the design to overcome problems revealed by actually controlling hardware. The experiment was to perform a planar (single axis) simulation of the APS track, slew, and settle maneuvers on a stationary target starting from an offset position.

Experiment Description

Functional Description

Mechanical

The hardware simulation used the carriage/aft body to simulate the outer gimbal mechanism of the agile pointing subsystem (APS). The central truss was used to simulate the inner gimbal mechanism.   The seismic pier was used to simulate the spacecraft. These analogies are seen in figure 5.

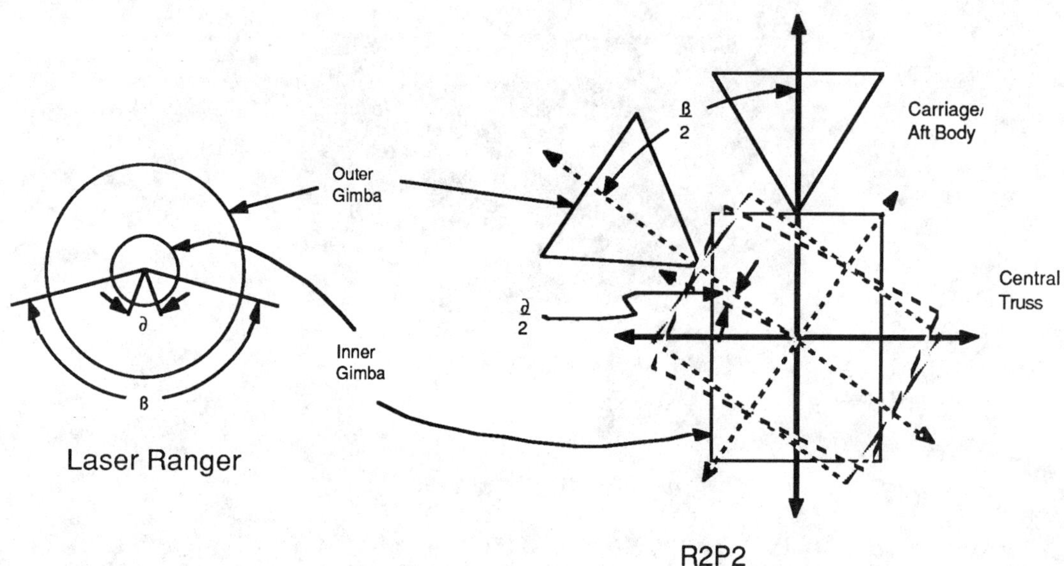

Figure 5. Agile Pointing Subsystem Concept - R2P2 Analogy

During the pointing/ranging phase of the mission the spacecraft will be relatively motionless. This characteristic allowed us to use the R2P2 Seismic Pier to model the spacecraft.

The outer gimbal mechanism provides the APS with a wide field of regard (WFOR) platform for the gimballed APS. The WFOR required for this platform was modelled on R2P2 by using the carriage/aft body. If the carriage and aft body are coaligned and locked in this position (lock journal air bearing), they have the same point of rotation as the central truss, corresponding with the inner and outer gimbals of the APS (see figure 6). The carriage/aft body does not have the necessary rates and accelerations to model a worst case APS retargeting scenario, however it was able to perform a subset of the maneuvers described for the APS inner and outer gimbal mechanisms.

The inner gimbal mechanism provides the APS with a fine pointing ability. The range of motion relative to the outer gimbal required of this platform is reproducible by the R2P2 gimbalflex. The gimbalflex modelled the flex pivots used to isolate the APS inner gimbal mechanism from the APS outer gimbal mechanism. The central truss modelled the laser ranger tracker supported by the inner gimbal flex pivots. The central truss motor was used to drive the simulated inner gimbal. The central truss/central truss motor system provided accelerations and rates scaled to the acceleration and rate of the actual APS. These rate and acceleration limitations led to scaling the slew angle and expected slew time to a 1.5 degree slew to target in about .4 seconds.

A tuned rotor gyro is used on the inner gimbal as a stable platform to provide a pseudo target for the inner gimbal. The inner gimbal is slaved to the stable platform by a position control loop. The gyroscopes used on the R2P2 Simulator are not capable of the high rate maneuvers required of the APS. The time constants associated with these gyros are also too large. Subsequently this gyro was simulated in software at 2kHz by an ideal integrator.

Optical

We are primarily concerned with evaluating the ability to point the APS to the desired accuracy. Therefore, we only modelled the APS subsystem on R2P2. The magnification sensors simulated the APS staring array. The optics on the APS are coaligned with the inner gimbal platform (simulated by the R2P2 central truss), as are the coarse tracker and magnification sensor.

Figure 6 presents a view of the optical layout on the current R2P2 and the sensors necessary for simulation of the agile pointing subsystem. The coarse trackers and magnification sensors provided the inertial angle and the target angle for the agile pointing subsystem, respectively.

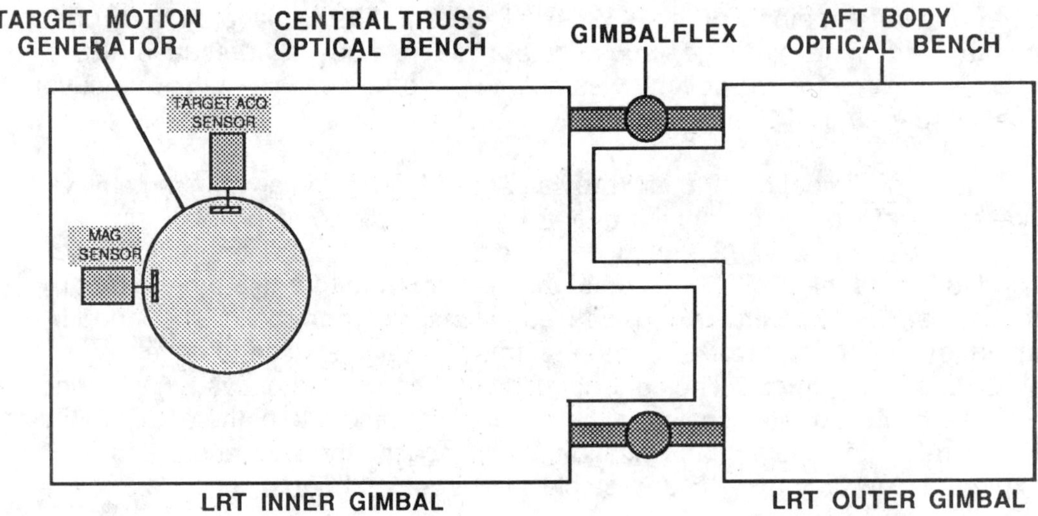

Figure 6. R2P2 - Laser Ranger Agile Pointing Subsystem Schematic

R2P2 currently does not possess optical array sensors. The APS receives target information from the Fire Control Sensor (FCS), which is mounted on a separate gimbal. The target data provided by the FCS (scanning array) for the APS was provided by the R2P2 Coarse Tracker.

The APS uses a passive staring, focal plane array to obtain target information directly. The R2P2 Magnification Sensor was used to simulate the APS FPA. The field of view of the APS FPA is approximately the same as the field of view of the magnification sensor. The mag sensor has a smaller quantization level than the APS FPA. Only point source targets are observable in this configuration.

Alignment of the coarse tracker was checked relative to the magnification sensor and adjusted to reduce the error seen between these two sensors. This provided an ideal handover situation between the settle and track modes for this simulation. The alignment error present during testing was approximately 25 microradians. Errors in the actual system may be as large as 100 microradians.

Controls

A system of controllers is used to control the APS throughout its mission, from slewing to a target to achieving final pointing accuracy. This system consists of several controllers whose general forms and purposes is discussed below. A block diagram of the control system and the R2P2 plant is shown in Figure 7.

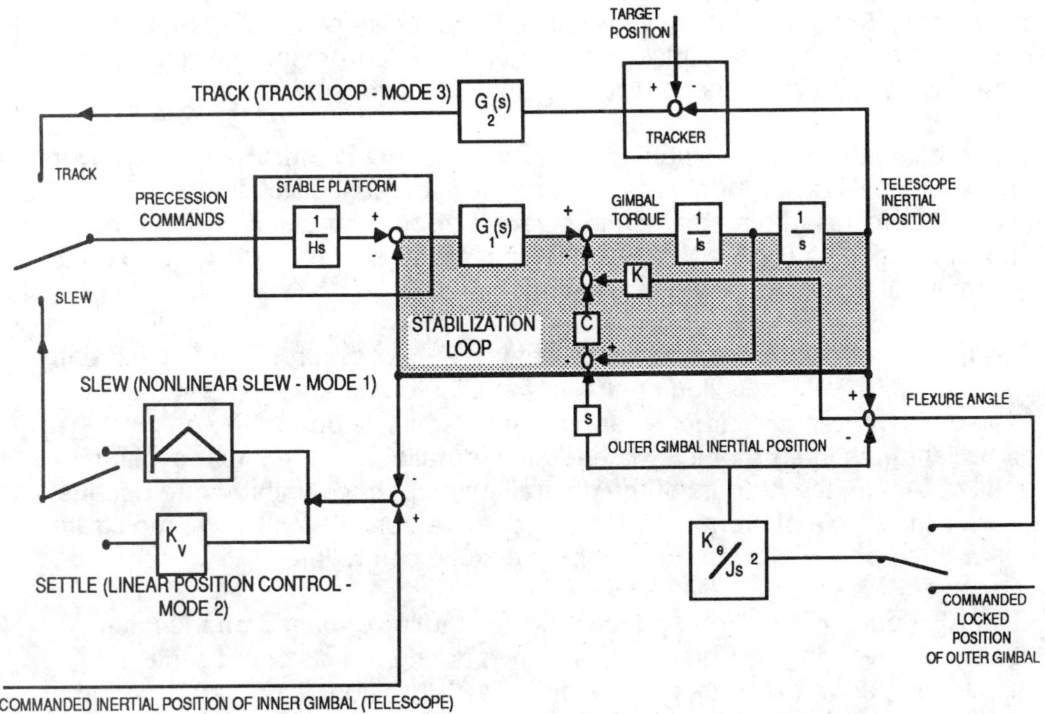

Figure 7. R2P2 APS Control System Block Diagram

The stable platform drives the stabilization loop. Inputs to the stable platform in the form of precession commands have the units of angular rate. Output is in units of angular position. The stabilization loop is a relatively high bandwidth controller which slaves the inner gimbal position to the position of the stable platform (tuned rotor gyro) mounted on the APS. This controller has the form of a lead compensator. When driven in a closed loop configuration on the R2P2 plant, the inner gimbal follows the position error presented to the stabilization loop by the stable platform.

The APS inner gimbal (coaligned with optics) will initially be driven open loop (bang-bang control) to the target.   Slew termination is executed when the inner gimbal (central truss) rate passes through zero and system control is handed over to settle mode.

The purpose of the settle mode is to reduce the position error present at the end of the slew mode to a level suitable for handover to the track mode. The settle mode has the form of a linear position controller in which the position error of the system is simply multiplied by a constant. This product, now in the form of a rate (the constant multiplier has units of 1 per second) is fed directly to the stable platform as a precession command. After the settle loop has been allowed to settle the system for a specified length of time, inertial angle and rate are checked to insure handover to the track loop with a reasonable accuracy (based on the APS FPA field of view, since the track loop is closed on this sensor). This handover also includes the changeover from inertial space error (coarse tracker) to target space error (magnification sensor).

After the settle mode has settled the system, system control is handed over to the track mode. Track mode uses the magnification sensor to simulate the APS passive IR detector which obtains direct target error information. The track controller's task is to reduce the pointing error to the accuracy required to obtain range information on the target of interest.

The track controller is a combination of two types of controllers. On first entry to track mode, a linear position controller (similar to settle, except with a lower gain) is used. When the target error reaches a small value (on the order of 10 microradians), an integrator is switched-on in parallel with the linear position controller. Again, the outputs of the controllers must have units of rate because they drive the stable platform. The track controller runs at a 50 Hz sample rate and is a lower bandwidth controller than the settle controller.

The APS outer gimbal loop will be a position follower based on the inner gimbal's flex pivot angle. This loop is inactive during the slew and settle portions, and part of the track portion, of the APS mission. During this period of inactivity, the outer gimbal is effectively locked up to provide a stiff connection to the relatively large inertia that the inner gimbal must torque against. After the inner gimbal gets into track mode and some angular error criterion is satisfied, the outer gimbal is released and commanded to reduce the angle in the flexure between the two gimbals. During the R2P2 experiments, the outer gimbal (carriage/aft body) was locked-up by airing down the carriage air bearings.

All of the control loops listed above were implemented at 2 kHz on the R2P2 PRIME Computer System, with the exception of the track and outer gimbal controllers. This was done to coincide as nearly as possible with the chosen operating frequencies for the APS

<u>Experimental Limitations</u>

There were several experimental limitations identified at the beginning of the APS experiment which were dictated by equipment , schedule, and budget limits. None of the limits were considered to reduce the value of the validation experiment. This simulation did not use the actual gimbals and gyroscope. This may have affected the fidelity of the pointing performance achieved in testing. Simulating the spacecraft body with an infinite mass object (the seismic pier) was an idealization of the system; however, with the APS outer gimbal locked to provide a "stiff" connection, the space platform does provide an effectively "infinite" reaction mass for the APS inner gimbal. The use of point source targets and differential sensors was not a direct analogy to the LRT's focal plane array LWIR sensors, but did provide adequate sensor inputs for pointing performance assessment. Inner and outer gimbal rates necessary for the APS could not be fully achieved on R2P2. However, the rates achievable are adequate for pointing performance testing, and results can be scaled in time. Only a static target case was simulated in this experiment.

Simulations

<u>Hardware Experiment Results</u>

The experiment results achieved the goals desired for the control system and each subsystem. The subsystems were tested individually and as an integrated system. Test results were compared with results obtained from the MATRIXx simulation. Subsystem tests were verified against MATRIXx simulation results.

The overall system goals in terms of a timeline were to be able to slew 1.5 degrees in 1.5 seconds, including settling on a target to a specified accuracy. Figures 8-13 present various results allowing comparison of MATRIXx and hardware experiment results. Figures 8 and 9 present inner gimbal rate profiles during the slew mode. Figures 10 and 11 present position profiles during the settle mode. Figures 12 and 13 present position profiles during the track mode.

The control loop validation was completed in two iterations.

The first iteration was driven by a stabilization loop (G1 controller) design based on a theoretical R2P2 plant model. System testing of this stabilization loop design appeared satisfactory until the settle loop was closed. Using the gains initially designed for this loop led to instability in the system. Lower gains eliminated the instability; however, degraded performance was still evident (the settle mode itself required 1.6 seconds to settle the system to allow handover to the track mode). These results led to a closer examination of the entire system. Upon closer examination of the R2P2 plant model, a discrepancy was found between the theoretical model and an experimentally determined model. The theoretical model did not include a structural mode present at 43 Hz.

The structural mode was included in the new plant transfer function definition used on MATRIXx. This model was used to retune the stabilization loop so that the structural mode was not excited (as it was by the previous stab loop controller).

The second iteration of testing described in several of the tests is based on the experimental model of the R2P2 plant and the retuned stabilization loop associated with this plant. This retuning consisted of gain adjustment and shifting of pole/zero locations. An additional iteration was done on a few of the subsystems to fine-tune the controllers for optimal performance. Retesting based on the enhancements and corrections described above, showed that the system slewed to the target and settled on it to the specified accuracy in less than 1.4 seconds, thereby satisfying the time and pointing requirements.

<u>MATRIXx Model Description</u>

The R2P2 team performed a computer simulation of the APS/R2P2 experiment using the software simulation tool MATRIXx.

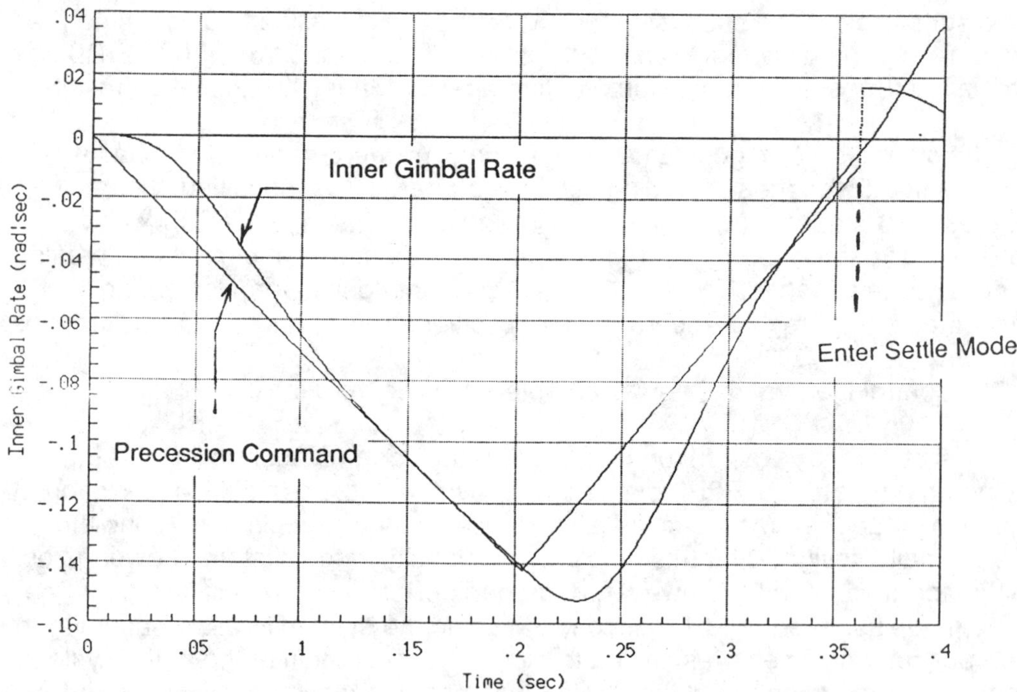

Figure 8. MATRIXx Simulation: Slew Mode Inner Gimbal Angular Rate and Stable Platform Precession Command

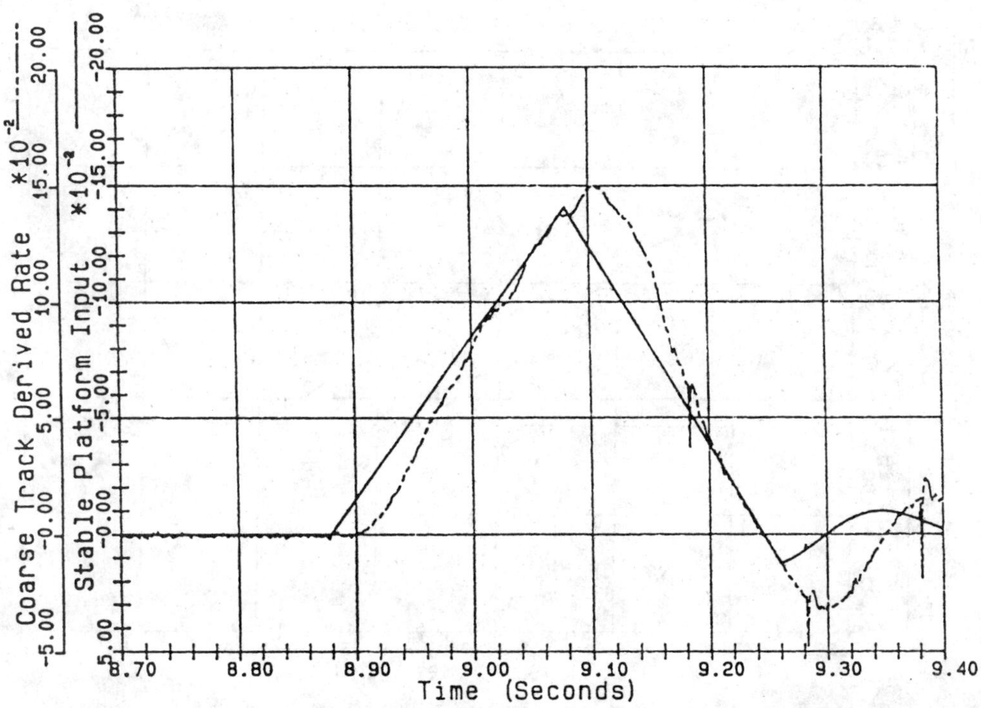

Figure 9. Hardware Simulation: Slew Mode Inner Gimbal Angular and Commanded Rates

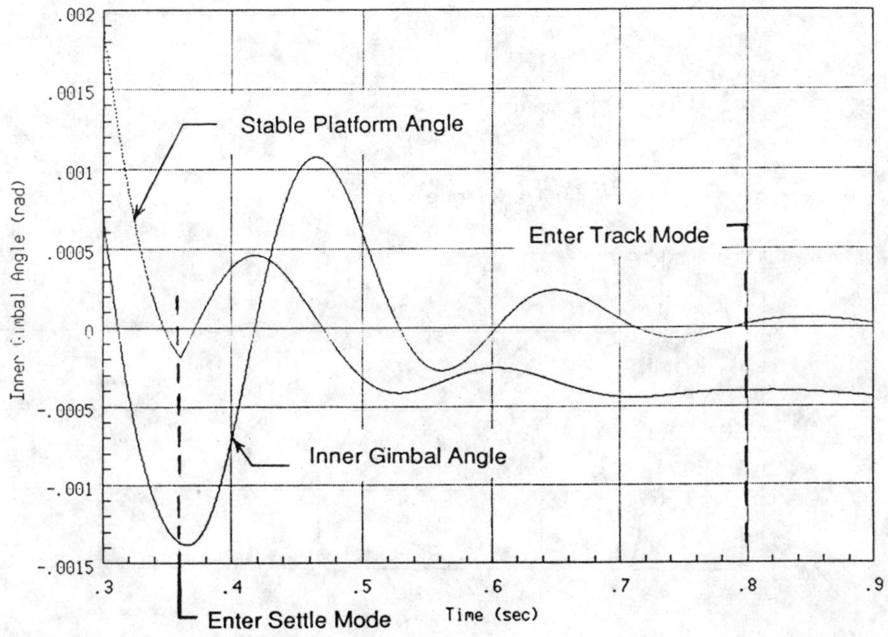

Figure 10. MATRIXx Simulation: Settle Mode Inner Gimbal Angle and Stable Platform Angle

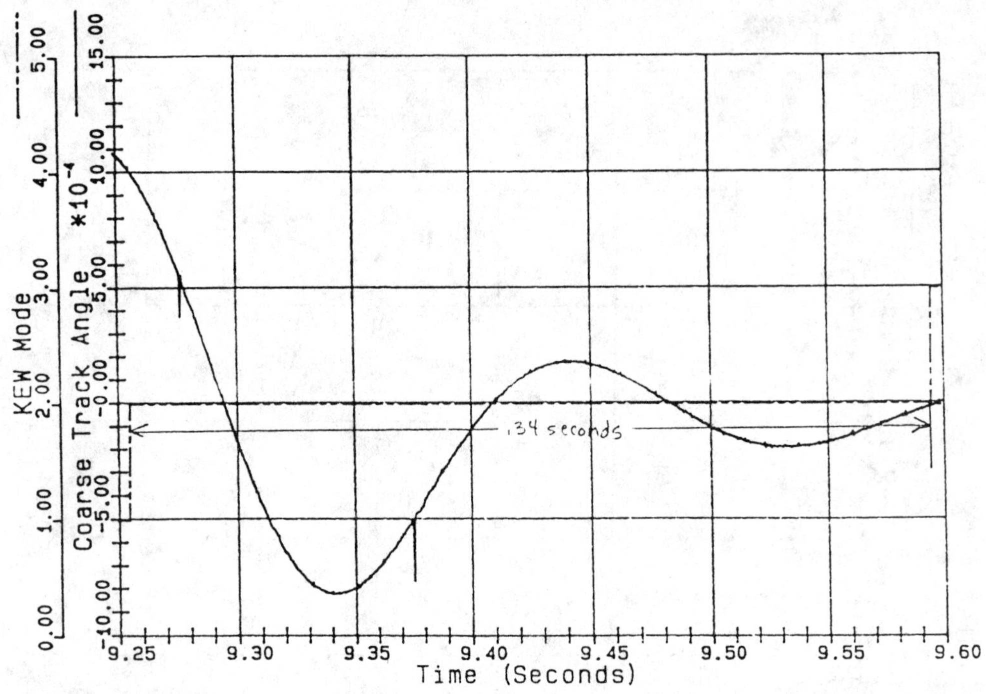

Figure 11. Hardware Simulation: Settle Mode Inner Gimbal Angle and Mode Flag

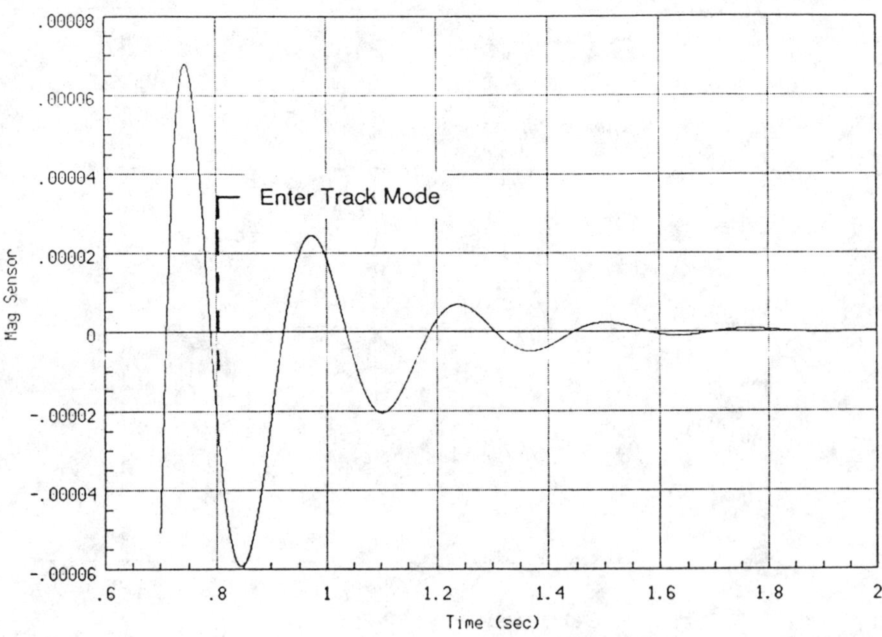

Figure 12. MATRIXx Simulation: Track Mode Inner Gimbal Angle

228

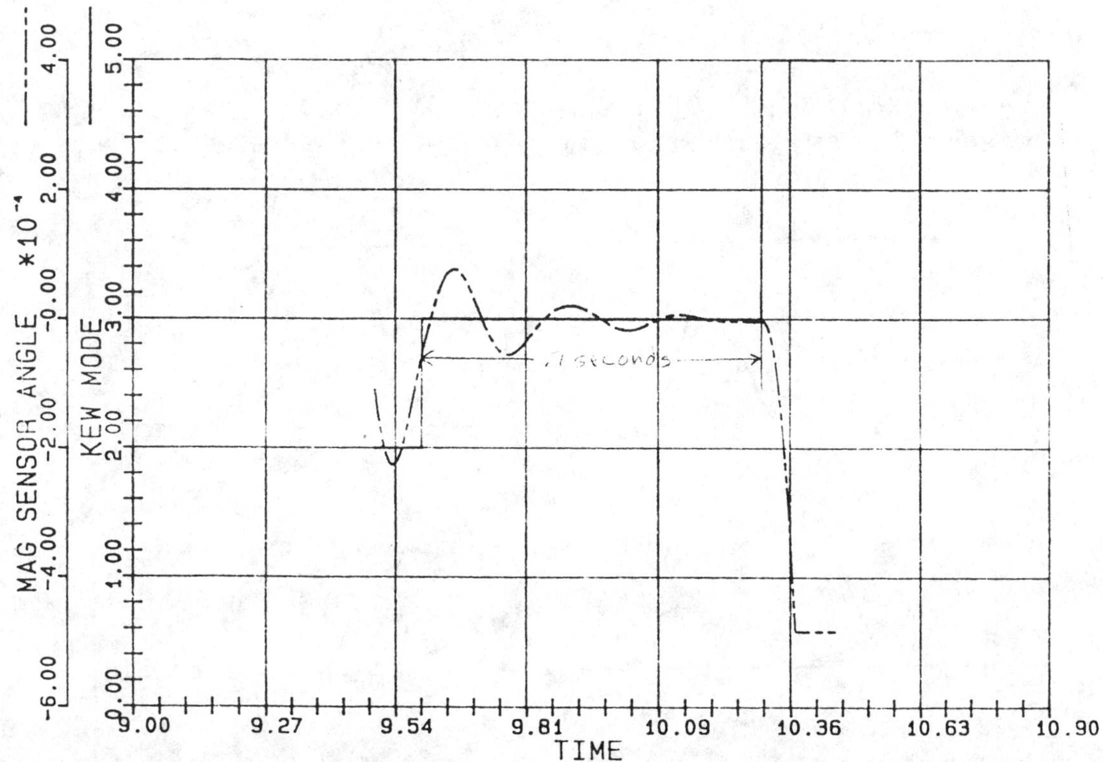

Figure 13. Hardware Simulation: Track Mode Inner Gimbal Angle and Mode Flag

An experimentally obtained transfer function for the R2P2 plant was used in the MATRIXx model.

All digital controllers were simulated in discrete blocks using the same coefficients and sampling rates that were used in the actual experiment. As in the actual hardware, a one-time-frame computational delay was included in the 50 Hz digital processing and in the 2 kHz digital processing.

<u>TUTSIM Model Description</u>

Aerojet ElectroSystems performed a computer simulation of the APS/R2P2 experiment using the software simulation tool TUTSIM. The experimentally obtained transfer function for the R2P2 plant was provided to the Aerojet team and used in TUTSIM.

All digital controllers were simulated in discrete blocks using the same coefficients and sampling rates that were used in the actual experiment. As in the actual hardware, a one-time-frame computational delay was included in the 50 Hz digital processing and in the 2 kHz digital processing.

<u>MATRIXx / TUTSIM Result Correlation</u>

In general, the TUTSIM simulation results agree with the MATRIXx results. A compraison of Figures 9 and 14 shows the same precicted rate response of the system.

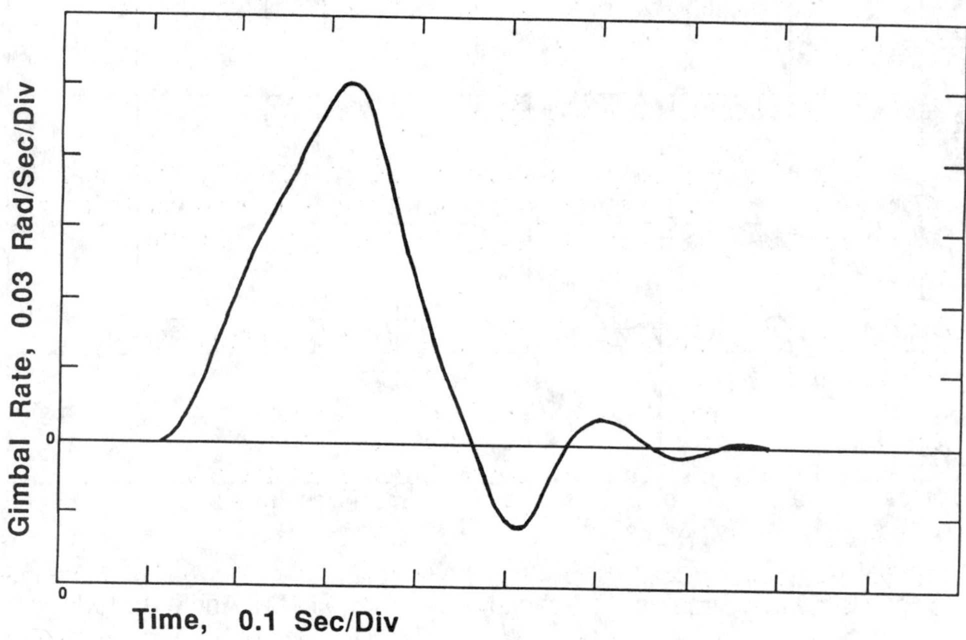

Figure 14. TUTSIM Simulated Rate Profile

On the inner gimbal position profile the TUTSIM slew rise time is approximately 0.31 seconds as is the MATRIXx rise time. The peak time for both simulations is approximately 0.375 seconds and the overshoot is 1.5 to 2.0 milliradians.

Track error on the TUTSIM simulation remains below 10 microradians after approximately 0.89 seconds. It remains below 2 microradians after approximately 1.02 seconds. On the MATRIXx simulation the track error remains below 10 microradians after 1.06 seconds and remains below 2 microradians after 1.50 seconds.   Another difference is that after 0.90 seconds the TUTSIM track error exhibits the effect of the integrator in the track loop whereas the effect is not as pronounced in the MATRIXx simulation.

Conclusion

In summary the system achieved better than satisfactory results based on the ultimate goals described for the experiment. The desired goals were a 1.5 slew to target, with settling on the target to sufficient accuracy to satisfy target ranging requirements. The experiment performed the 1.5 degree slew to the target and settled on it to the specified accuracy in less than 1.4 seconds, thereby satisfying the time and pointing requirements.

<u>Validation of control loop design and logic</u>

Validation of the control system design was accomplished by implementing the general form of the APS control system on R2P2. This includes the general form of each controller used and all of the switching logic used in the control system.

Thus the following controls were implemented in the following manner.

Slew	Tuned bang-bang acceleration profile
Settle	Linear Position Command (gain block)
Track	Hybrid - Linear Position Command/Integrator
Stabilization	Lead Compensator

Switching between the controllers was accomplished in the 50 Hz executive.

As tests were run on each of the subsystems, the test results were compared with results from the MATRIXx simulation (which had been verified by comparison with TUTSIM results provided by Aerojet).

Tests showed that the original controller parameter values were not adequate to control the system (degraded performance occured in the settle loop). However retuning the stabilization loop lead compensator based on an experimentally derived R2P2 plant model improved the overall system performance and made it possible to achieve the desired response time and pointing performance for the system. Tuning was also required on the gain of the settle loop after the stabilization loop was required.

The slew controller required tuning of the deceleration level to reduce the position error present when handover (zero rate at end of slew) to settle mode occured. The tuning of these loops to achieve the desired performance levels did not include changing the form of these controllers (i.e., the stabilization, slew and settle loops maintained the forms of a lead compensator, time optimal slew (bang-bang acceleration profile) and linear position controller respectively).

The stabilization loop is sensitive to the structural modes of the body it is controlling. Thus retuning of this loop was necessary to accomodate the 43 Hz structural mode present on the R2P2 central truss. A similar mode (47 Hz) is expected on the APS and must be given careful consideration when this controller is designed; as the 72 Hz mode expected for the APS must also be considered.

The settle loop is sensitive to the position error presented to it by the slew mode, and its performance may be degraded (ringing occurs) if the combination of the loop gain and the initial position error presented to this loop are too large.

The position error presented by the slew to the settle loop can be reduced by tuning the acceleration and deceleration levels of the slew routine. Torquer saturation can have detrimental effects here and must be considered. Also the R2P2 experiment examined this tuning only for a 1.5 degree slew maneuver. Other magnitude slew maneuvers should be examined to determine how the acceleration profile is affected by these different angles. It may be necessary to set different acceleration/deceleration ratios for different slew angles.

The track loop uses a linear position control loop similar to that used by the settle loop. The residual error at the settle/track handover is not large enough to cause the ringing witnessed in the settle loop. Switching-in the integrator degraded this loop's performance to a small degree. The experiment was run with the integrator active when the final pointing and time goals were met. The R2P2 experiment did not examine a moving target case and the effects of a moving target should be considered with this integrator.

Overall, the design of these loops was adequate to achieve the time and pointing goals established for this experiment.

<u>Lessons learned from controlling hardware</u>

The use of actual hardware often presents discrepancies between the model and the actual hardware. This experiment turned up two areas that were affected by the manner in which the hardware performed.

The primary point of interest here was the effect of the stabilization loop on the 43 Hz structural mode. The stabilization loop had to be retuned to accomodate this mode. The first definition of this loop actually amplified the modal response and the loop bandwidth had to be reduced to eliminate this effect.

In addition, the slew mode had to be tuned to reduce the position error presented to the settle mode. Simulation results showed tuning acceleration and deceleration to 0.7 and -0.72 rads/sec/sec, respectively, provided sufficient reduction in this error. On the simulator, these levels were tuned to 0.7 and -0.85, respectively. This was a considerable difference from the TUTSIM results.

These results should not be interpreted to invalidate the value of software simulation results. Meaningful studies can be made in this manner on the general characteristics of a system. However, these studies do not replace the value of an actual hardware experiment, as the actual characteristics of the hardware can have profound effects on the performance of an individual controller or controller suite.

Acquisition of early data on APS pointing performance

This experiment showed the ability of the current APS control system design to point to the desired pointing accuracy within a stressing time requirement. The experiment provided substantial insight beyond software analysis alone and provided valuable information to prepare for a prototype hardware-in-the-loop experiment.

ZERO-LOCK^{TM} LASER GYRO[*]

Manny Fernandez, Bob Ebner and Neal Dahlen[†]

Litton's ZLG™ (Zero-Lock™ Laser Gyro) is a "non-dithered" RLG (Ring Laser Gyro). It is a multioscillator type RLG that effectively circumvents the frequency-locking phenomenon by removing it (within the optical system) from the gyro's angular-rate input operation-region. Conventional RLGs use mechanical "dithering" or "rate bias-ing" for circumvention.

The ZLG has the following advantages for space satellite Attitude Control System (ACS) and Pointing and Tracking (P&T) applications:

- A truly solid-state gyro with no moving mechanical parts
- No mechanical noise
- A low-noise, high-bandwidth sensor
- Quantization to within nanoradian resolution
- An extremely accurate scale factor
- A sensor relatively immune to performance transients under high-dynamic environments
- A high slew-rate capability
- Excellent long-term drift stability
- Excellent reliability/life
- Low power
- Typically no requirement for temperature control
- Negligible acceleration sensitivity
- Rapid reaction

Low noise, high bandwidth, low quantization and immunity to tran-sients under high dynamic environments are features of the ZLG that make it especially applicable to space satellite ACS and P&T applica-tions. The paper presents test data that demonstrates the superior performance of the ZLG for these applications.

---

[*] Zero-Lock and ZLG are Litton trademarks.

[†] Litton Guidance and Control Systems, Woodland Hills, California 91367.

INTRODUCTION

Litton's ZLG™ (Zero-Lock™ Laser Gyro) is a "non-dithered" RLG (Ring Laser Gyro). It is a multioscillator type RLG that effectively circumvents the frequency-locking phenomenon by removing it (within the optical system) from the gyro's angular-rate input operation-region. Conventional RLGs use mechanical "dithering" or "rate biasing" for circumvention.

The ZLG has the following advantages for space satellite Attitude Control System (ACS) and Pointing and Tracking (P&T) applications:

- A truly solid-state gyro with no moving mechanical parts
- No mechanical noise
- A low-noise, high-bandwidth sensor
- Quantization to within nanoradian resolution
- An extremely accurate scale factor
- A sensor relatively immune to performance transients under high-dynamic environments
- A high slew-rate capability

The first five advantages relate to conventional RLGs and the last two to spinning-wheel gyros. Scale-factor improvement stems directly from circumvention of the non-linear frequency-locking phenomenon. Spinning-wheel gyros are prone to performance transients under high-dynamic environments (severe case angular motion) which significantly affect performance. Other advantages of the ZLG common to the RLG are as follows (relative to spinning-wheel gyros):

- Excellent long-term drift stability
- Excellent reliability/life
- Low power
- Typically no requirement for temperature control
- Negligible acceleration-sensitivity
- Rapid-reaction

Low-noise, high-bandwidth and immunity to upsets under high-dynamic environments are features of the ZLG that make it especially applicable to space satellite ACS and P&T applications. In order to successfully apply the ZLG to these applications, the performance characteristics of the ZLG need to be understood as discussed below.

DRIFT PARAMETERS

A space satellite ACS or P&T gyro can be characterized by the following non-acceleration sensitive bias drift parameters.

- NEA (Noise Equivalent Angle), in arcseconds
- Random drift, in degrees per hour
- Drift stability, in degrees per hour
- Trend, in degrees per hour/hour
- Turn-on to turn-on repeatability, in degrees per hour

The NEA parameter may incorporate the overall effects of the random drift and drift stability parameters. Typically, trend and turn-on repeatability parameters are handled via periodic recalibrations and realignments, respectively. The frequency of these periodic recalibrations/realignments can be an important mission limiting parameter. In this regard, the ZLG provides less need for periodic recalibrations and realignments due to its excellent drift and scale-factor stabilities.

The following definitions apply.

NEA

For a constant rate input to the gyro, obtain data of gyro angle output versus time. Derive a best-fit sloping straight line to the data. The variance of the residuals is NEA. NEA is a measure of how accurate the gyro can point for the length of the run selected. Note that data sampling rate, length of run, and bandwidth of gyro readout-electronics have a significant impact on the resulting NEA. Therefore, care has to be taken in selecting these parameters if meaningful results are to be obtained.

Random Drift

For a constant rate input to the gyro, obtain data of gyro rate output versus time. Derive a best-fit (flat) straight line to the data (mean). The variance of the residuals is random drift. Random drift sets a limit to measuring the gyro drift and, thus, how well the gyro can be calibrated. Note that here again data sampling rate, length of run and bandwidth of the gyro readout electronics are important parameters to be selected.

Drift Stability

If we take a gyro rate output run with a time representative of the application and obtain the mean and variance, the variance is what can be termed "drift stability" during a representative run. Environmental considerations (such as temperature variations, magnetic variations, and acceleration and vibration inputs) should be taken into account if meaningful data are to be obtained.

Trend

If we fit a best-fit (sloping) straight line to the drift stability data we can extract "trend." Trend effects may or may not be significant depending on the length of the run. Note that drift stability here is measured by two parameters: trend and variance.

Turn-on to Turn-on Repeatability

As the name implies, this is the gyro drift change from turn-on to turn-on, allowing for drift to stabilize after each turn-on and measuring the variance of many representative nonoperating/operating (turn-on) periods.

It should be noted in the above characterization of drift parameters that we imply that the gyro has stabilized before we take data, since typically the gyro drift changes during its stabilization period. In contrast to spinning-wheel gyros that are sometimes temperature controlled at a set temperature, the RLG or ZLG are typically operated with short reaction times (stabilization periods) with thermal modeling compensation.

For the ZLG, the ARW (Angle Random Walk) defines the minimum drift measurement error. The ARW is caused by the quantum limit defined by the spontaneous emissions in the plasma.

There is a need to understand the ARW nature of the ZLG. ARW is the direct result of white noise at the angular rate level being measured by the ZLG at an angular output level. ARW contributes to NEA and to "derived" angular rate measurements. The units of ARW are degrees per root-hour.

Since the ZLG does not use mechanical "dither," the ARW is limited by the quantum limit rather than the uncompensated mechanical "dither" contribution. In a dithered RLG the gyro goes through the frequency lock-in region twice during each period. This produces a higher ARW. The ZLG, on the other hand, does not require mechanical dither or rate-biasing and therefore its output is not corrupted by these effects. Other things being equal, the ZLG will have a lower ARW that is set by the quantum limit rather than by these residual effects caused by extraneous mechanical motion.

Furthermore, in a conventional RLG, which uses mechanical "dither" or "rate biasing," the output must be compensated for mechanical motion due to dither or rate-biasing in order to extract the desired inertial angular rate. This compensation can never be perfect; therefore, the compensated output will have errors that manifest themselves as measurement errors and/or noise.

SCALE FACTOR PARAMETERS

A space ACS or P&T gyro can be characterized by the following scale factor parameters:

- Scale Factor Stability
- Scale Factor Linearity
- Scale Factor Asymmetry
- Scale Factor Repeatability
- Scale Factor Resolution

Since accurate scale factors are difficult to recalibrate in space, the ZLG's extremely accurate scale factor provides a significant mission benefit. The following definitions apply.

Scale Factor Stability

For a range of discrete plus and minus angular rate inputs, the variance of the deviations from the average at each discrete input is defined as scale factor stability.

Scale Factor Linearity

For a range of discrete plus and minus angular rate inputs, the variance of the deviations from a plus or minus average scale factor is defined as scale factor linearity.

Scale Factor Asymmetry

For a range of discrete plus and minus angular rate inputs, the difference between the plus and minus average scale factor is defined as scale factor asymmetry.

Scale Factor Repeatability

Scale factor repeatability is the turn-on to turn-on repeatability of the absolute scale factor. It can be quantified as the variance of the absolute scale factor over many representative non-operating/operating (turn-on) periods.

Scale Factor Resolution

Scale factor resolution is defined as the angular output resolution (i.e. the least discernable output angle change). In a conventional RLG the resolution is corrupted by quantization.

In an RLG low values of quantization (arcseconds per count output) require large pathlengths and, hence, high weight. In the ZLG it is possible to provide resolution enhancement in the unit's electronics to reduce the quantization by several orders of magnitude. Therefore, for a specified quantization, the ZLG has a weight advantage over the conventional RLG. Since quantization manifests itself as angular rate measurement error and/or angle measurement error, the ZLG with resolution enhancement provides not only a weight advantage but a performance advantage in reduced measurement errors and low noise due to quantization effects.

ZLG™ TEST DATA

Table 1 summarizes ZLG™ test data in the areas of noise, drift and scale factor.

Table 1
ZLG™ TEST DATA (25 CM PATH LENGTH)

Noise

• *NEA, arcseconds, peak-to-peak, 0.37 sec*	*0.03*
• *Angle random walk, deg/root-hr*	*0.0007-0.0008*

Drift

• *Drift stability (16 hr) (includes ARW and Trend)*	
deg/hr, peak-to-peak	*0.0015-0.0018*

Scale Factor

• *Scale factor stability, PPB*	
• *Scale factor linearity, PPB*	*Below 10.0 PPB*
• *Scale factor asymmetry, PPB*	*measurement error*
• *Scale factor resolution (in arcseconds)*	
(without resolution enhancement)	*1.5*
(with resolution enhancement)	*0.001*

Figure 1 shows the effectiveness of resolution enhancement in reducing noise. Figures 2 and 3 show angular rate threshold and angular resolution data. For the angular resolution data of Figure 3, the gyro is measuring a component of earth's rate and a quasi-sinusoid input at approximately 44 Hz with an amplitude of approximately 2 arcseconds, peak-to-peak. The sampling rate for these data is 1600 times per second.

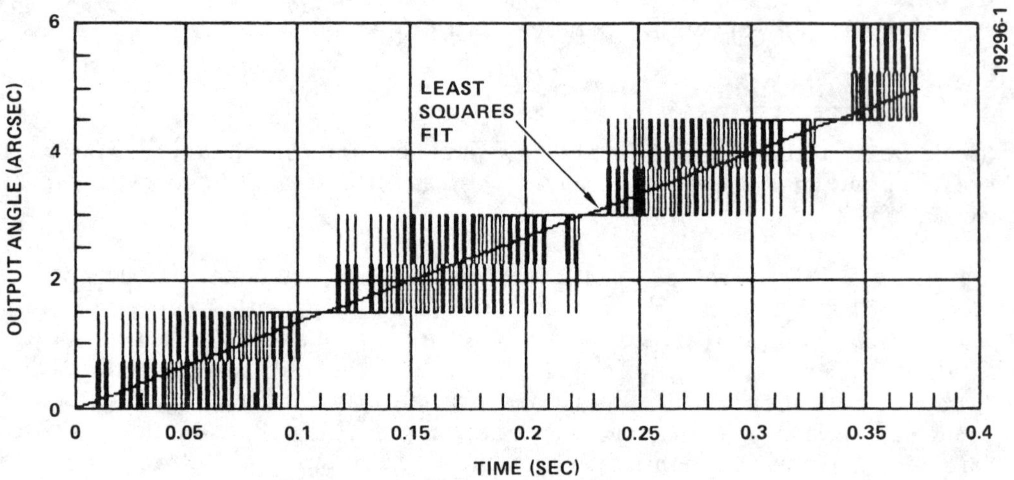

a. GYRO OUTPUT ANGLE WITHOUT RESOLUTION ENHANCEMENT

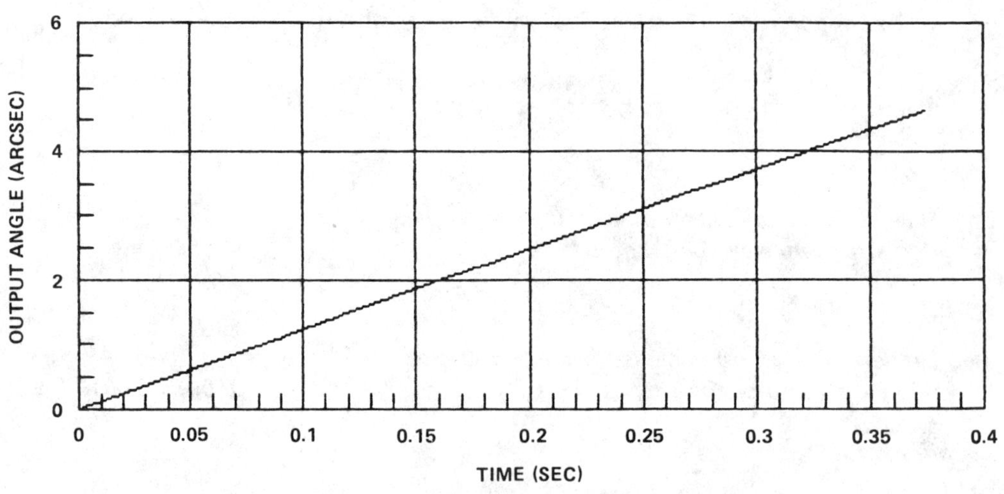

b. GYRO OUTPUT ANGLE WITH RESOLUTION ENHANCEMENT

Fig. 1 Effectiveness of Resolution Enhancement

INPUT (DEG/HR)	DATA (DEG/HR)	DIFFERENCE (DEG/HR)
• 0.043439388	0.044340035	0.0009 (σ = 0.0033 DEG/HR)
• 0.0018954	0.001600	0.0002954 (σ = 0.006 DEG/HR)

NOTE: ARW CONTRIBUTES TO σs

Fig. 2   Angular Rate Threshold

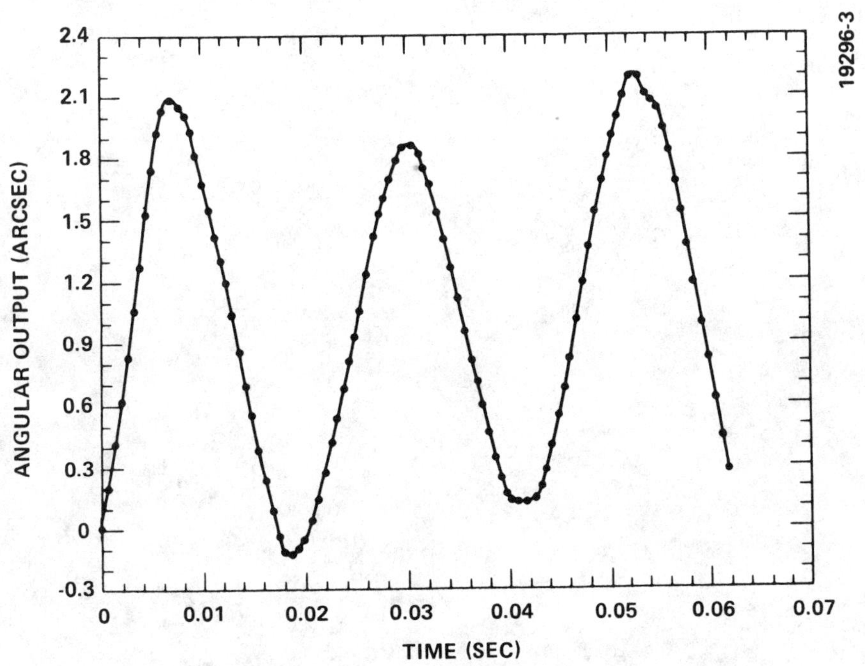

Fig. 3   Angular Resolution

CONCLUSION

The test data presented demonstrates that the ZLG provides superior performance for space satellite ACS and P&T applications. Key parameters are resolution to
the nanoradian level (0.001 arcseconds or 5 nanoradians) and low noise in the submicroradian level (0.15 microradians or 0.03 arcseconds, peak-to-peak, 0.37 sec). In
addition, the ZLG provides an extremely accurate scale factor and excellent long-term
drift stability.

EXPERT SYSTEM APPLICATION
TO IMPROVE SATELLITE AUTONOMY

John M. Barry[*], Linas Raslavicius[†] and Thomas Gathmann[‡]

This paper describes an expert system architecture which Rockwell's Satellite & Space Electronics Division (S&SED) is developing to improve satellite autonomy. This paper reviews the significant progress Rockwell made since earlier publications reported initial system concepts. This expert system will first assist satellite controllers to dynamically schedule the allocation of on board satellite resources and mission activities, analyze anomalies, and provide convenient access to critical engineering and operational data. Rockwell research is aimed at embedding some form of this architecture into future satellites to increase autonomy. This innovative architecture represents an advanced satellite control concept which is dependent not only on embedded satellite sensor readings, but also on data interpretation. This architecture is implemented as a dynamic prototype in the Planning & Analysis System (PAS) to portray the functions, processes and human interface necessary to reduce the workload of the ground control segment. PAS is being designed to take advantage of the latest advancements in Artificial Intelligence and hypertext technology to assist in the command, control and maintenance of orbiting satellites. The PAS prototype demonstrated at the Conference is available to all AF and NASA personnel on a 5 1/4 diskette through Rockwell. The Rockwell PAS differs from the current approaches to satellite control because it takes a system viewpoint that recommends action based on the interpretation of various knowledge sources rather than display of raw data on individual subsystems. These sources include knowledge about the design and operation of the satellite. This viewpoint encompasses the entire satellite and its interaction and coordination with ground control. This global perspective differs from other approaches which only report and control individual subsystem sequences. The PAS is being designed to monitor and diagnose within and across subsystems so that on-board coordination of resources on both a subsystem and system level can be accomplished.

INTRODUCTION

Satellite control is accomplished through hardware designs that allow automatic sequencing and safing or responses to uploaded software commands. Most satellite

---

[*]   Research & Technology Program Development Manager, Satellite & Space Electronics Division, Rockwell International Corp., 2600 Westminster Blvd., P.O. Box 3644, Seal Beach, California 90740-7644.

[†]   Research & Technology Supervisor, Software & Simulations Systems Department, Rockwell International Corp., 2600 Westminster Blvd., P.O. Box 3644, Seal Beach, California 90740-7644.

[‡]   Member of the Technical Staff, Software & Simulations Systems Department, Rockwell International Corp., 2600 Westminster Blvd., P.O. Box 3644, Seal Beach, California 90740-7644.

control is done by sending software commands via communications from ground control centers to the satellite. These centers monitor and control satellite "health and status" and uplink new commands to control the satellite subsystems and mission. The present method of managing these resources is through interpretation of digital satellite data, consulting orbit operation handbooks, manually creating new schedules, then uplinking software commands to the satellite to execute these schedules. Controlling the satellite through uplinking commands is often referred to as the "software screwdriver." This method currently dominates satellite operations. The application of Artificial Intelligence techniques represents an opportunity to enhance spacecraft operations and reduce costs.

Controlling satellites involves a variety of data sources. Analyzing this data requires extensive knowledge of the system and subsystem operations, operational constraints, and satellite design and configuration. This process requires highly trained experts anywhere from several hours to several weeks to accomplish. The process is done through "brute force" - that is, examining cryptic mnemonic data "off line" to interpret the "health and status" of the satellite. Interpreting, diagnosing, and controlling satellite resources and activities is presently a labor intensive and time consuming ground operations task. Diagnosing a malfunction, recommending corrective action and developing a software fix requires extensive knowledge of the system and subsystem operations, operational constraints, and satellite design and configuration. This process requires highly trained experts anywhere from several hours to several weeks to accomplish. The fixes are formulated either as the result of practical operator experience or heuristics - that is "rules of thumb." Orbital operations must become more productive in the future to reduce life cycle costs and decrease dependence on ground control. This reduction is required to increase autonomy and survivability of future systems. The design of future satellites require that many current ground control functions be transferred from ground to on board systems.

Even using the software screwdriver, operating and maintaining a satellite in orbit is a large, expensive, and complex task which requires many people, diverse skills, and coordination of various contractor and government organizations. Air Force studies indicate that an average of 8 controllers are required to operate and maintain 1 satellite. However, this figure is just the "tip of the iceberg". Backing up these controllers are government and contractor "back room support" personnel such as orbital analysts, computer operators, programmers, systems engineers and so forth. This support easily expands into 200-300 people per satellite system.

Future costs are prohibitively high when the current satellite operations practices are scaled to the expected number of satellites for future space operations. We can no longer afford to control future spacecraft missions in the manner that we support present Air Force and NASA programs. Studies indicate that the satellite operations costs will rise dramatically if we continue these present methods. These increasing cost trends clearly indicates a need to increase satellite autonomy. This need is the primary reason that the 1988 Space Division and Space Command Satellite Architecture Study stressed that increasing satellite autonomy is one of the top three design priorities. Ignoring current cost trends will severely limit our ability to afford the acquisition, deployment and control of future space programs. Therefore, reducing satellite command and control costs is a way to make more money available to develop future space programs.

Cost is not the only motivation for changing the present method of controlling satellites. A loss of communication from the control centers due to war, terrorism or natural disaster would leave the satellite in a position where its mission might be degraded. The design of future satellites require that the many of the health and status functions be transferred from ground to on-board systems to increase autonomy, survivability, adaptability and reduce costs and response time. Less time required to control the satellite means more time is available to analyze and respond to critical mission data.

TRADITIONAL APPROACHES TO SATELLITE CONTROL

Approaches to improve satellite control traditionally concentrate on automating computational and data reduction tasks, and developing better displays. However, these efforts alone will not solve the satellite control problems. The solution is not trivial because significant engineering judgment and reasoning are required to operate the satellite and resolve anomalies. Satellite control is complex because of the limited amount of on-board resources available such as electrical power or propellant. This situation is further complicated by multimission satellites which must share these resources among a variety of systems. Sharing resources requires consideration of multiple constraints and the sequencing of operations and availability of resources.

The management and planning of missions is presently accomplished by manually or automatically translating, sorting, and analyzing large amounts of digital data and displaying trends. Current satellite consoles display contains only limited, cryptic alphanumeric data that the operator must decipher. Some satellite operations centers then transmit this data to other computers for off-line analysis to display trend and graphical data. However, trend analysis is insufficient to accurately predict and correct satellite anomalies. Such analysis cannot process and interpret multidimensional information efficiently to recommend potential solutions to predict or correct anomalies.

INNOVATIVE ROCKWELL APPROACH

Rockwell Satellite & Space Electronics Division is developing an integrated approach to assist the current generation of ground satellite control and design for the future implementation of such systems on-board satellites. The Planning and Analysis System (PAS) is an architecture designed to capitalize on existing satellite data to streamline current satellite operations. The PAS provides a multidimensional data and analysis view to increase the effectiveness and reduce the response time of satellite controllers and orbital analysts. The PAS was developed as a framework to interpret data, diagnosis and predict malfunctions and dynamically schedule the allocation of on-board satellite resources and activities. The PAS is a continuation of prior Rockwell on-board satellite intelligence research concepts.

The PAS views include the controller and "intelligent agents," such as the satellite planner, subsystem specialist and mission analyst. The primary function of the planner is to generate a plan for fulfilling the objectives of a satellite or a group of mission related payloads. The subsystem specialist is responsible for the operational availability of its associated subsystem. The analyst assesses the degree of success or failure of an executed mission. The controller coordinates the generation of an agenda for executing selected missions of a satellite or group of mission related payloads. The controller is being prototyped to substantiate the concept of increasing on-board satellite autonomy. This concept also provides insights to simplify the task of the present satellite operations ground controller and other personnel who support ground control. This simplification will transform vast amounts of cryptic satellite data to create more intelligible information for satellite operators and analysts.

Rockwell developed a framework for the PAS which can integrate AI and conventional software technologies. This framework is being designed to interpret, diagnose, and recommend strategies to fix anomalies, predict failures, and execute mission objectives. Our current architecture involves a involves examining the electronic blackboard concept to allow each of the subsystems to post its current activities and flag out of tolerance conditions to the controller. The controller can then access any of the satellite engineering, heuristic, and case based data to recommend feasible strategies to manage the satellite resources and activities. The engineering data, such as found in the specification and testing data include physical parameters such as temperature and voltage. Heuristics are currently used by ground satellite operations specialists to analyze the satellite's condition and take action. The case based database is an accumulation of knowledge based on engineering data and the relationships to observable cause and effect occurrences monitored during satellite operation. These relationships help predict and solve potential anomalies. Rockwell will transition the ground expert system into future satellite designs once they have been proven and tested in the ground control environment. The initial implementation will be as an assistant to the ground controller. The eventual application will be in the design of new satellite systems to increase on-board autonomy.

The Rockwell approach is based on examining current design and operations of several satellite systems which it is currently designing, producing or operating. These systems include various navigation and surveillance satellites. Our approach starts with a functional examination of the objectives needed to operate and maintain a satellite in the most cost effective manner. The Rockwell concept concentrates on presenting knowledge or formulating advice instead of displaying only raw information to an on-board controller. This knowledge is the result of known constraints, an operational model of the satellite systems, and the judgment developed by experts. Today this knowledge is formulated by the previously mentioned "back room support" personnel. The Rockwell approach is to transform digital data into English phrases to simplify operator understanding. This approach will result in more responsive satellite control and analysis which eventually can be embedded within the spacecraft systems to increase autonomy.

The innovative Rockwell approach described in this paper and demonstrated on a personal computer in this conference covers several facets. These facets include the PAS dynamic prototype, Satellite PAS Concept, and the Enhanced User Interface.

DYNAMIC PROTOTYPING

Rockwell developed a low risk, high confidence approach to the PAS design through dynamic prototyping. Dynamic prototyping is a technique in which one models the visual interface and operation, but not complete functionality of the desired product. This technique is faster for developing functional concepts because it does not produce the pseudo code and functional flows characteristic of rapid prototyping. The controller's perspective was obtained through dialogue and feedback from current Air Force satellite operations personnel. This perspective emphasized the simplification of the user interface and reduction in the number of operational personnel. We used this information and dynamic prototyping to encapsulate the satellite planning and analysis functions. The result was deemed by representatives of the Air Force to accurately reflect the visual cues and data views a satellite controller would like to see. Designing this perspective allows us to simplify ground control mechanisms and functions and understand the processes required to design more autonomy into satellites. The prototype is designed so that it can be readily changed to reflect enhancements to the controller's perspective and true operation of the system without extensive re-coding.

THE SATELLITE PAS CONCEPT

The Rockwell Satellite PAS is an expert system concept to interpret, diagnosis and maintain satellite health and status. This system is being designed to advise on the control, coordination, and management of various subsystem specialists. The primary function of the PAS is to coordinate the generation of an agenda for executing selective satellite missions or mission related payloads. Subsystem specialists control and manage their respective subsystems such as the propulsion, power, attitude control, or communication subsystem. The coordination is achieved through an agenda or common area that either the controller or the subsystem specialists can access. Requests or status of actions are posted on the agenda. This information is used by the controller in creating an initial schedule and in coordinating its execution. Past research projects concentrated on defining the Planning and Analysis System concept to control and simulate the activities performed by the subsystem specialists. Current research is investigating methods to determine the division of the type of knowledge and communications protocols between the controller and the subsystem specialists.

PAS DEVELOPMENT

The PAS is being developed in parallel with the dynamic prototype user interface using CLIPS, a C-based expert system building tool developed by NASA. The design of the system requires us to capitalize on the advancements being made in blackboard architectures, constraint based and model based reasoning and hypertext techniques. The PAS is being designed to accomplish fault diagnosis, mission planning, and resource allocation.

FAULT DIAGNOSIS

A prelude to either mission planning or resource scheduling is the identification, isolation and diagnosis of a fault. This task involves access to and interpretation of data across subsystems. This process is an integral part of anomaly resolution. The PAS dynamic prototype implements a scripted scenario of diagnosing a fault to study the dynamics required to integrate this feature into a rapid prototype.

MISSION PLANNING

The PAS dynamic prototype uses a scripted mission scenario and anomaly to visualize these concepts. Initially the expert system developed will be tested on a ground control workstation. Rockwell is investigating the use their Mission Operation Support Center to construct and test a satellite controller on its Global Positioning System. The ultimate goal is to develop the technology to design this expert system to operate on-board future satellites.

A mission consists of a goal or objective, a start time, a duration, and a priority. The present Rockwell expert system is modeled after a surveillance satellite. A typical mission might be to view a ground location at a prespecified time. A mission is made up of multiple tasks that must be completed in order to satisfy a mission. For a viewing mission, typical tasks that must be scheduled would include operating a sensor, preparing a sensor for operation, shutting down a sensor, and downloading data to a remote tracking station when recorders are full following a mission. General station keeping tasks must also be scheduled such as orbit adjustments, uploading current reference updates when over a remote tracking station, or momentum dumps.

Given multiple, conflicting missions, the expert system will try to schedule as many missions as possible. Currently, priority is the only constraint used to determine which missions will be scheduled first and which missions cannot be scheduled at all. The satellite moves on a path over the earth called a ground track and can move or slew itself several degrees in the plus or minus direction in order to view a location. Therefore, it would be possible to view two locations when on the same ground track by slewing the sensor. It could also move to another ground track to view a location, but this will require resources such as propulsion in order to make the move. In the future, the expert system will incorporate constraint based reasoning to determine how to satisfy as many missions as possible by traveling on a ground track where multiple locations could be viewed at once while minimizing the amount of resources used.

RESOURCE MANAGEMENT

Included in scheduling of tasks and subtasks is scheduling of the on-board resources which enable the task to be completed. Currently, the system will determine if a task can or cannot be scheduled based on available resources. In the future, it will be able to reason about when would be the best time to perform a task based on the resources the task will use. For example, is it better to perform a task on the current ground track, rather than move to another ground track because less propulsion will be used? Currently, three resources are managed: power, propellant, and recording tape. Power is a resource that stays at a fixed level and is reduced or increased when a task is performed or completed or an anomaly occurs. All tasks use power and a minimum amount of power is always used for nominal operations such as station keeping. Propellant starts at a given level and is used as tasks are performed but it is never replenished. Recording tape is used during a sensor operation and is completely used when all recorders are completely full of data. Recorders are replenished when all data has been dumped to the ground when over a remote tracking station.

ENHANCED USER INTERFACE

The prototype of a user interface concentrates on displaying relevant knowledge - i.e. "digested information," meaningful to the operator. This interface can replace digital data from several operator terminals with a single screen partitioned into several windows which display English language phrases and graphical icons of system components. Therefore, the operator is presented with the phrase "sensor 1 slewed 5 degrees" instead of the normal digital data that must be interpreted. This process is more than a simple transformation. It actually involves parsing and interpreting inferred information from the inputs of several systems aboard the satellite.

The PAS interface displays several types of data during its execution: agenda information, satellite controller actions, subsystem health and activity status, and task schedule timelines. The user interface was developed to demonstrate understanding of an on-board design approach, portray a potential ground station controller's workstation, and provide user control of the expert system scripted simulation.

The English phrases transformed from digital data are depicted in several windows. These windows represent several views of the satellite data which correspond to the system, planner, and specialist. The main screen is divided up into a SYSTEMS COMMENTS window, an AGENDA window, Subsystem Icons and a MENU. The AGENDA window displays requests from either a subsystem or the controller. The AGENDA window also shows current ACTIVITIES of the various subsystems. The AGENDA window also shows the current STATUS of the various subsystems of the satellite. The agenda represents the Controller's primary product and is the interface to other intelligent processes. Based on a selected set of missions, the Controller coordinates subsystems and missions. The windows can be activated by a macro key on the terminal which toggles between the two activities. Both windows can be scrolled back to display a time tagged audit trail of activities that have occurred on the satellite. The display has icons on the right side of the screen which activate other macros to allow the operator to "EXAMINE" the schedule, a "HELP" key, and others. In addition, the satellite subsystems are displayed in icons across the bottom of the screen. The individual subsystem icons are highlighted whenever activity is occurring which affects that subsystem. This activation assists the operator in visualizing subsystem status.

The enhanced user interface of the Planning and Analysis System controller replaces numerous satellite control consoles and the digital display mnemonics of current systems. The experience gained from this design will be used to define the data flows for the eventual on-board PAS.

A real user interface to visualize the controller's perspective was previously coded in Turbo C to be used in the PAS evolution. This scheduler, plus the user interface, the blackboard architectures and the various knowledge bases, will provide the platform for PAS architecture.

SUMMARY

The Rockwell Satellite Planning and Analysis System is using Artificial Intelligence technology to develop a concept to reduce future space operational costs and increase effectiveness in controlling satellites. The initial objectives of the Rockwell project are to schedule on-board satellite resources and activities. In the process, Rockwell is developing techniques which can simplify operations and improve the productivity of ground controllers. The Rockwell approach is based on examining system operations and gaining feedback from satellite operators. This feedback was used to construct the prototype demonstrated in this conference on a personal computer.

ACKNOWLEDGEMENTS

The author wishes to acknowledge the Rockwell team responsible for developing the Satellite Controller concept and building the prototype. This team is in the Software and Simulation Systems Department of the Satellite and Space Electronics Division of Rockwell. Linas Raslavicius, the Project Coordinator, and Tom Gathmann, the knowledge engineer, are the original architects of the PAS Concept. Other members involved in the design and evolution of the concept and prototype include Sumalee Johnson, JoAnne Pitts, Jeff Sease, Lisa Nolette and Billie Shannon. Members of the team also provided valuable editorial comments which are reflected in this paper. The author also thanks the many people from the Rockwell Artificial Intelligence Center, Rome Air Development Center, AF Space Command, AF Space Division, NASA, DARPA and the Aerospace Corporation for their encouragement and comments throughout this project. NASA is to be commended for having the foresight and resources to develop CLIPS, the expert system building tool used in this project.

BIBLIOGRAPHY

1. Bratko, Ivan. *Prolog Programming for Artificial Intelligence.* Workingham, England: Addison-Wesley, 1986.

2. Barry, John M. "Applications of Artificial Intelligence Concepts for Employing Logistics Resources in the SDI Environment," AIAA/SOLE First Space Logistics Symposium, March, 1987, Huntsville, Ala.

3. Barry, John M. "Applications of Expert Systems to Command and Control of Satellites," Aerospace Applications of Artificial Intelligence Conference, Oct 1988, Dayton, Oh.

4. Barry, John M., Sary, Charisse. "Increasing Autonomy through

5. Satellite Expert System Scheduling" AIAA/SOLE Second Space Logistics Symposium, Sept 1988, Costa Mesa, Ca.

6. Barry, John M., Thomas Gathmann. "Expert System for On-Board Scheduling and Control." NASA Artificial Intelligence Conference, Nov 1988, Huntsville, Ala.

7. Harmon, P. and D. King. *Artificial Intelligence in Business.* New York: John Wiley, 1985.

8. Harmon, P. and D. King. *Expert Systems: Tools and Applications.* New York: John Wiley, 1987.

9. Hayes-Roth, F., D.B. Lenat, an D.A. Waterman (Eds.) *Building Expert Systems.* Reading, Ma: Addison-Wesley, 1983.

10. Sary, C., Gathmann, T. Kass, J. *Satellite Controller Concept.* FY88 IR&D Final Report, IL No. 792-110-88-083, Sept. 1988.

11. Gathmann, T. Johnson, S., Pitts, J. *Artificial Intelligence (AI); Applications in Satellite Autonomy.* FY88 IR&D Final Report, IL No. 792-110-88-086

Copies of the Rockwell Planning & Analysis demonstration diskette may be obtained for Air Force and NASA employees by writing the author at the following address:

John M. Barry, Satellite & Space Electronics Division, Rockwell International - M/S SK-09, P.O. Box 3644, Seal Beach, CA 90740-7644.

AUTHOR'S BIOGRAPHIES

John Barry is currently the Program Development Manager for Research & Technology at Rockwell's Satellite & Space Electronics Division in Seal Beach, California. Previously he was a member of the technical staff and project manager at Rockwell in the Software and Simulations Systems Department. Prior to joining Rockwell he was an Air Force Officer for 20 years serving the USAF Space Division and Rome Air Development Center. He has a B.S. Degree in Math and Physics, from LeMoyne College, an M.S. Systems Management from the University of Southern California, an M.S. in Logistics Management from the Air Force Institute of Technology and an M.S. in Computer Science from Northrop University. Mr. Barry is an adjunct professor in the graduate program at Northrop University in the Computer Science Department.

Linas Raslavicius is the Supervisor of the S&SED Research & Technology Unit in the Software & Simulation Systems Department and an consultant to the current project. He has a B.S. and M.S. in Electrical Engineering from the University of Illinois. He has over 10 years experience in designing software systems for satellites.

Thomas Gathmann is a Member of the Technical Staff and the current Technical Lead for the Satellite Controller in the Software and Simulations Systems Department at the Satellite & Space Electronics Division. He has B.S. in Aerospace Engineering from Cal Poly, Pomona, and an M.S. in Aerospace Engineering from the University of Southern California. He has over 7 years experience in computer science with 5 years experience in Artificial Intelligence and expert systems at Rockwell. He is a primary instructor of the Rockwell 200 hour Expert Systems course developed by Teknowledge.

Section III
ATTITUDE REFERENCED POINTING SYSTEMS

Chairperson:	Sam Hollander U.S. Naval Research Laboratory
Co-Chairperson:	Bob Laskin Jet Propulsion Laboratory
Local Chairperson:	Larry Germann Ball Aerospace Systems Group

The following papers were not available for publication:

AAS 89-031   Relay Mirror Experiment (RME) Advances Ground-Based Laser Technology, J. Sneary, A. Goodrich, R. Rose, Ball Aerospace Systems Group.

AAS 89-034   Line-of-Sight Pointing and Stabilization Using SAVI Magnetic Suspension Technology, B. Hamilton, P. Wolke, Honeywell Systems Division, D. Founds, Air Force Weapons Laboratory

AAS 89-037   The Spacelab Instrument-Pointing System (IPS) -- Performance Enhancement and Future Applications, A. Woelker, Dornier System

The following paper numbers were not assigned:

AAS 89-038 to -039

MIRROR LINE OF SIGHT ON A MOVING BASE[*]

J. J. Rodden[†]

An acquisition, pointing and tracking technology demonstration is planned for flight in a dedicated Space Shuttle Mission using the NASA Spacelab facility. Crucial for this demonstration is the control of a line of sight through a variety of movable optical elements that isolate the resultant beam from base motion and control direction with sensor detector signals. This paper addresses control laws to maintain a line of sight to an inertially referenced direction through steering of two movable mirrors separated by an optical beam amplifying telescope.

INTRODUCTION

The STARLAB experiment is planned for flight in a dedicated Shuttle mission using the NASA Spacelab. The Air Force Space Division is managing the technical development. The system is being built by Lockheed Missiles and Space Company, Inc. (LMSC) and the Kaman Aerospace Corporation. The NASA Marshal Space Flight Center is responsible for the Spacelab and Shuttle integration.

---

[*] Work performed for U.S. Air Force Space Division, Contract F04701-86-C-0025.

[†] Manager and Chief System Engineer, Starlab Program, Astronautics Division, Lockheed Missiles & Space Company, Inc., P.O. Box 3504, Sunnyvale, California 94088-3504.

The key STARLAB experiment involves a booster engagement. The booster is acquired by an acquisition video camera during the first stage. The plume is tracked in either the visible or infra-red depending on lighting conditions. During staging, the booster is tracked using sensor signals and a predicted trajectory. The upper stage hardbodies are acquired and tracked using an active tracker the laser designator is pointed at an instrumented target board to score the beam positioning accuracy. During the engagement, plume data is collected and the return radiation is also used by a wavefront control experiment to improve the optical quality of the system. Test objects are also to be deployed from the Shuttle to aid in sensor calibration and system checkout.

The experiment hardware is shown in figure 1. The equipment is partitioned between the aft payload bay and the interior of the Spacelab module. An optical schematic is shown in figure 2. The experiment features a fixed telescope and a two axis gimbal mounted optical flat for large angle steering. A visible acquisition camera, an ultraviolet imager, and an infrared imaging camera are located in the aft bay of the Shuttle. The beam from an aft bay located illuminator laser is steered by the gimballed mirror to provide illumination for the active tracking. The visible return from the both actively illuminated and passive luminous objects of interest are collected through the telescope and relayed by fast steering mirrors and beam walk alignment mirrors to the optical bench inside the Spacelab module. On the module optical bench are the wavefront control experiment equipment, the visible tracking sensors, the designator laser, and the self-scoring sensors.

Problem   Statement

The rotation isolation problem is that of processing inertial measurement data on the optical system to generate commands distributed between the pointing flat gimbal and a fast servo mirror so as to maintain the exit line of sight (LOS) transmitted from the detector null through the optical elements to a controlled inertial direction. The commands are to be distributed such that the wider range but limited bandpass signals command the two axis gimballed pointing flat gimbal while the lower amplitude higher bandpass mirror positionings are done by the fast servo mirror. The difficulty in the problem is due to the complexity of the transformations of light ray vectors reflecting from the mirror surfaces.

As part of the line of sight stabilization control, the relative displacement of the telescope and pointing mirror are measured in inertial space, differenced, and applied as feed forward to the fast steering mirror. The solution starts with a representation of the line of sight vector transformations through the various optical elements. Control capability is either the gimballed flat orientation or the fast steering mirror which completely corrects for system rotations within their ranges of operation. The approach taken is to command the low frequency gimbal rate servo to accommodate the entire correction, the actual inertial response of the servo is compared to the command to generate a feedback error signal which is applied through the high frequency FSM to complete the correction. This approach provides a self dividing frequency separation for the correction signals.

The inertial isolation only is considered in this paper as the use of optical sensor detected signals to close the pointing/tracking loops through feedback control is not addressed. The inertial rate to which the line of sight is controlled includes any effective inertial rotations of the test objects with respect to the optical system calculated from the estimated trajectory motion.

System Coordinates

The system coordinates are shown in figure 3. The inertial set is used to reference the Orbiter and space test object trajectories, typically oriented North, and through the Equinox. The Shuttle and optical bench coordinates are assumed coincident with x forward, y the pitch axis, and z the yaw axis. The space object vector, T, is a unit vector from the orbiter to the object. The exit optical line of sight, V_2, is to be maintained coincident to T.

Mirror Transformations

A mirror reflection vector can be expressed as a dyadic or matrix operating on the source beam vector, V_1, as shown in figure 4. The reflection vector, V_2, is transformed by a dyadic which is a function of the surface normal vector, N_m. The dot or inner vector product is denoted as $A*B = A^T B$ where A^T is a transposed vector.

$$V_2 = [V_1 - (V_1*N_m)N_m] - (V_1*N_m)N_m$$

reflection vector	component parallel to mirror surface	component normal to mirror

$$V_2 = V_1 - 2[V_1*N_m] N_m$$

Since $[V_1*N_m] N_m = N_m [N_m*V_1] = [N_m N_m^T] V_1$ then

$$V_2 = [I - 2 N_m N_m^T] V_1$$

where I = 3 X 3 identity,
$( )^T$ = vector transpose
$N_m N_m^T$ = outer product or dyadic

Note that the transformation is equal to its inverse since the relation is unchanged with the input and reflection rays interchanged.

Mirror Rate Relations

The following rate relationships for mirrors are derived as a necessary preliminary to the subsequent development of the mathematical system description.

Define $V = V_2 - V_1$ with magnitude $|V|$.
The mirror normal is $N = V / |V| = V (V^T V)^{-1/2}$
The derivative of the normal is :

$$N' = \omega_m \times N$$

where ω_m is the mirror rate vector in the plane of the mirror. The time derivative is denoted as $( )'$.

$$d/dt (N) = (d/dt\ V) (V^T V)^{-1/2} + V(d/dt(V^T V)^{-1/2})$$

$$\text{or} \quad N' = V' (V^T V)^{-1/2} + V [-1/2(V^T V)^{-3/2} (2 V^T V')]$$
$$= V'/|V| - (V V^T) V'/ |V|^3$$
$$= [I - N N^T] V'/|V|$$
$$= -[N X]^2 V'/|V|$$

The angular rate of the mirror is found from the cross product of the normal vector times its rate :

$$N \times N' = N \times [\omega_m \times N] = -[N \times]^2 \omega_m$$
$$= [I - N N^T] \omega_m$$
$$= \omega_m \text{ since } N [N^T \omega_m] = 0 \text{ with } \omega_m \text{ perpendicular to } N.$$

The cross product of a vector, N, with components N_1, N_2, and N_3, can be expressed in matrix form.

$$[N \times] = \begin{vmatrix} 0 & N_3 & -N_2 \\ -N_3 & 0 & N_1 \\ N_2 & -N_1 & 0 \end{vmatrix}$$

Note that $[N \times]^3 = -[N \times]$ and then $\omega_m = [N \times] V'/|V|$

This result is used to derive the rate commands to the pointing flat to maintain the reflected beam V_1 parallel to the telescope boresight while tracking the test object vector V_2 as it moves due to optical bench rotation rate, ω_b.

$$d/dt(V) = V' = V_2' - V_1' = [\omega_b \times] V_2 - 0$$

With this substitution the expression for the angular rate of the mirror in the form of a triple cross product can be expanded by the vector product rule,

$$[A \times (B \times C)] = B[(A*C) - C(A*B)] \text{ with } A = N, B = V_2, \text{ and } C = \omega_b.$$

$$\omega_m = [N \times] [\omega_b \times] V_2/|V|$$
$$= -(1/|V|) [N \times] [V_2 \times] \omega_b$$
$$= -(1/|V|) \{ N^T V_2 I - V_2 N^T \} \omega_b \text{ , defining a new}$$
matrix, E.
$$= E \omega_b$$

This last relation gives the mirror angular rate vector required to maintain the output vector on a rotating base as a direct transform of the bench angular rate. This inertial rate is in the same coordinates as the optical bench.

System Geometric Transformations

The abbreviated system geometry is shown in figure 5. Incident light rays reflect off the movable pointing flat through the telescope, across folding optics and a fast servo mirror, through the beam walk mirrors and are focused on a focal plane detector. The fixed folding optics transformations and beam walk transfer functions are taken to be unity for this analysis.

The estimated incident object of interest unit vector in inertial coordinates is T and the transformation from inertial to bench coordinates is R. The exit line of sight from the optical bench is V_2 which is to be maintained coincident to T by the motion isolation control system.

The unit vector T or V_2 is related to the light ray at the detector by the product of 3 transformations. Those transformations are that of the pointing mirror flat, the telescope and the fast steering mirror. When the output ray vector, V_2, is properly aligned the the estimated test object vector, T, the unit detector vector, D, has no x or y components.

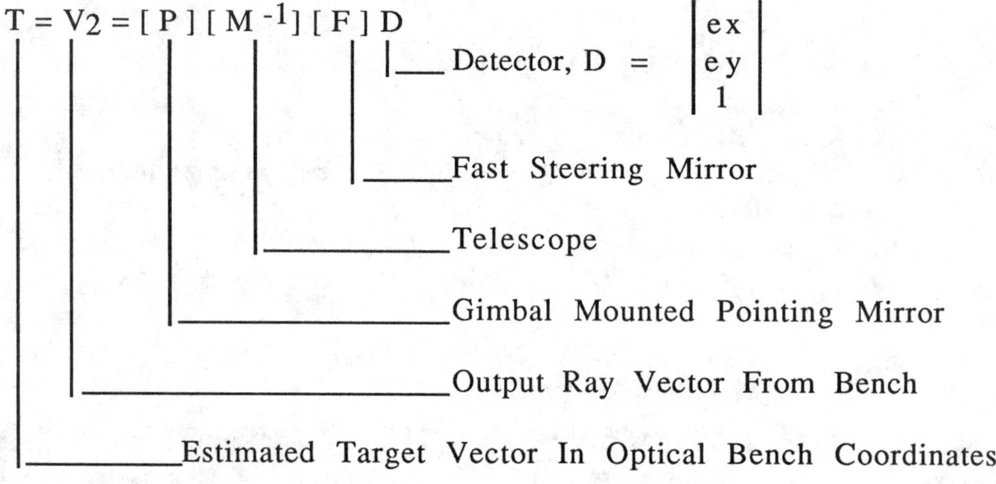

Telescope   Transformation

The telescope boresight is taken to be a unit vector, B, as shown in figure 5. The input ray is V_1 and the output is the vector S.   The input ray vector, V_1, which is the one to be magnified has a component along the boresight and one normal to it.

Boresight in-line component $= (B * V_1) B = [B B^T] V_1$

Boresight normal component $= V_1 - [B B^T] V_1 = [I - B B^T] V_1$

The output ray is magnified by the telescope for rays at small angles from the boresight. The output ray's boresight is unchanged to first order. The boresight normal component is the magnification, m, times that of the input normal component. The net output ray is the vector sum of the components.

$$S = [\, B\, B^T\,]\, V_1 + m[\, I - B\, B^T\,]\, V_1 = m[\, I - (1 - 1/m\,)\, B\, B^T\,]\, V_1$$
$$= M\, V_1$$

or $\quad V_1 = [M^{-1}]\, S$

The telescope magnification matrix and its inverse are :

$$M = \begin{vmatrix} 1 & 0 & 0 \\ 0 & m & 0 \\ 0 & 0 & m \end{vmatrix} \quad \text{and} \quad [M^{-1}] = \begin{vmatrix} 1 & 0 & 0 \\ 0 & 1/m & 0 \\ 0 & 0 & 1/m \end{vmatrix}$$

Transformation For Pointing Mirror

The gimballed pointing mirror normal is nominally a unit vector tilted 45 degrees between the negative z and the positive x axes. The normal vector in mirror coordinates with an x axis normal and a y axis along the elevation axis is transformed to bench coordinates by an elevation rotation, -α, around the y axis followed by an azimuth angle rotation, -β, about the z axis. The following expressions use C( ) and S( ) for cosine and sine of their arguments. Nominally the elevation angle, α, is 45 derees and the azimuth, β, is zero for a z axis input to be reflected into the telescope along the x axis. The mirror normal vector, N_m, in bench coordinates is as follows.

$$N_m = [\, -\beta\,]\, z\ axis * [\, -\alpha\,]\, y\ axis * \begin{vmatrix} 1 \\ 0 \\ 0 \end{vmatrix}$$

where [ ] axis denotes a single axis rotation about the specified axis.

$$N_m = \begin{vmatrix} C(\beta) & -S(\beta) & 0 \\ S(\beta) & C(\beta) & 0 \\ 0 & 0 & 1 \end{vmatrix} \begin{vmatrix} C(\alpha) & 0 & S(\alpha) \\ 0 & 1 & 0 \\ -S(\alpha) & 0 & C(\alpha) \end{vmatrix} \begin{vmatrix} 1 \\ 0 \\ 0 \end{vmatrix}$$

$$N_m = \begin{vmatrix} C(\alpha)\, C(\beta) \\ C(\alpha)\, S(\beta) \\ -S(\alpha) \end{vmatrix} = 1/\sqrt{2} \begin{vmatrix} 1 - (\alpha - \pi/4) \\ -\beta \\ 1 + (\alpha - \pi/4) \end{vmatrix}$$

For small angles ($\alpha - \pi/4$) elevation and β azimuth

The reflection transformation for the pointing mirror is :
$$P = [I - 2N_m N_m^T]$$

$$= \begin{vmatrix} 1-2C^2(\alpha)\,C^2(\beta) & -2C^2(\alpha)\,C(\beta)\,S(\beta) & 2C(\alpha)\,S(\alpha)\,C(\beta) \\ -2C^2(\alpha)\,S(\beta)\,C(\beta) & -2C^2(\alpha)\,S^2(\beta)+1 & 2C(\alpha)\,S(\alpha)\,S(\beta) \\ 2C(\alpha)\,S(\alpha)\,C(\beta) & 2C(\alpha)\,S(\alpha)\,S(\beta) & 1-2S^2(\alpha) \end{vmatrix}$$

Note that the first column of the P matrix is the vector V_2 for a nominal V_1 vector aligned along the x axis parallel to the telescope boresight.

Fast Servo Transformation

The normal of the fast servo mirror is inclined 45 degrees from the x and z axes. This mirror is deflected along two orthogonal axes ; θ_1, about the y axis and θ_2 about a transverse axis in the plane of the mirror half way between the negative x and negative z axes. For small angles, the mirror normal vector, N_f, is :

$$N_f = 1/\sqrt{2} \begin{vmatrix} -(1-\theta_1) \\ \sqrt{2}\,\theta_2 \\ +(1+\theta_1) \end{vmatrix}$$

The reflection vector for an input, D, from the detector along the z axis is :

$$S = [\,I - 2N_f N_f^T\,]\,D = F\,D = \begin{vmatrix} -2\theta_1 & \sqrt{2}\,\theta_2 & 1 \\ \sqrt{2}\,\theta_2 & 1 & -\sqrt{2}\,\theta_2 \\ 1 & -\sqrt{2}\,\theta_2 & -2\,\theta_1 \end{vmatrix} \begin{vmatrix} 0 \\ 0 \\ 1 \end{vmatrix}$$

$$S = \begin{vmatrix} 1 \\ -\sqrt{2}\,\theta_2 \\ -2\,\theta_1 \end{vmatrix} \quad \text{Note } V_1 = [M^{-1}]\,S = \begin{vmatrix} 1 \\ -\sqrt{2}/m\;\theta_2 \\ -2/m\;\theta_1 \end{vmatrix}$$

Gimbal Rate Commands

The gimbal rates are about elevation and azimuth axes. Motion about these axes over specifies the required mirror rates which has no component about the mirror normal. To accommodate this condition, a three axis set of Euler rates is defined, including one about the normal. In this way an exact but non-orthogonal transformation can be formed between the gimbal Euler rates and the required inertial mirror rates.

The Euler axes vectors for elevation, α, azimuth, β, and twist, γ, in optical bench coordinates are as follows:

$$\text{Elevation,}\quad g\alpha = \begin{vmatrix} -S(\beta) \\ C(\beta) \\ 0 \end{vmatrix} \qquad \text{Azimuth,}\quad g\beta = \begin{vmatrix} 0 \\ 0 \\ 0 \end{vmatrix} \qquad \text{Twist,}\quad g\gamma = \begin{vmatrix} C(\alpha)\,C(\beta) \\ C(\alpha)\,S(\beta) \\ -S(\alpha) \end{vmatrix}$$

Note that $g\gamma = N_m$, the pointing mirror normal.

The mirror inertial rates are expressed as a function of the gimbal rates about their respective axes.

$$\omega_m = [\,\{g\alpha\}\ \{g\beta\}\ \{g\gamma\}\,] \begin{vmatrix} \alpha' \\ \beta' \\ \gamma' \end{vmatrix} = \begin{vmatrix} -S(\beta) & 0 & C(\alpha)\,C(\beta) \\ C(\beta) & 0 & C(\alpha)\,S(\beta) \\ 0 & 1 & -S(\alpha) \end{vmatrix} \begin{vmatrix} \alpha' \\ \beta' \\ \gamma' \end{vmatrix}$$

$$\omega_m = B \begin{vmatrix} \alpha' \\ \beta' \\ \gamma' \end{vmatrix}$$

which defines the matrix B with α', β', and γ' as the Euler rates in elevation, azimuth, and twist in optical bench coordinates.

The solution for the Euler rates in terms of the measured bench angular rates is:

$$\begin{vmatrix} \alpha' \\ \beta' \\ \gamma' \end{vmatrix} = [\,B^{-1}\,]\,E\,\omega_b$$

where the matrix E is defined earlier in the relation between the bench rates and mirror.

The nominal output vector, V_2, when the reflection vector, V_1, is perfectly aligned to the telescope is ;

$$V_2 = \begin{vmatrix} 1 - 2C^2(\alpha)\,C^2(\beta) \\ -2C^2(\alpha)\,S(\beta)\,C(\beta) \\ 2C(\alpha)\,S(\alpha)\,C(\beta) \end{vmatrix} \qquad \text{for}\quad V_1 = \begin{vmatrix} 1 \\ 0 \\ 0 \end{vmatrix}$$

Note the magnitude $|V| = 2N_m * V_1 = -2N_m * V_2 = 2\,C(\alpha)\,C(\beta)$

The preceeding relationships and the expressions for the vector components are combined into the Euler rate equation. With some rather tedious algebraic/trigonometric manipulation, the relations for the Euler elevation and azimuth rates are found as a transformation of the bench inertial rates (Aubrun 1986, Rusk 1987).

$$\begin{vmatrix} \alpha' \\ \\ \beta' \end{vmatrix} = \begin{vmatrix} S(\beta) & -C(\beta)+1/(2C(\beta)) & -S(\alpha)\,S(\beta)/(2C(\alpha)\,C(\beta)) \\ \\ -S(\alpha)\,C(\beta)/C(\alpha) & -S(\alpha)\,S(\beta)/C(\alpha) & -1+1/(2C^2(\alpha)) \end{vmatrix} \begin{vmatrix} \omega_{bx} \\ \omega_{by} \\ \omega_{bz} \end{vmatrix}$$

$$= F_3'\,\omega_b \quad \text{defining the matrix } F_3'.$$

Fast Steering Mirror Rates

The direction changes in the ray, V_1, between the telescope and the pointing mirror to maintain the angularly rotating output vector, V_2, are derived directly from the reflection relation for a fixed position of the pointing mirror.

$$V_1 = [\,I - 2N_m\,N_m{}^T\,]\,V_2$$

Differentiating and substituting the rate of V_2 gives:

$$V_1' = [\,I - 2N_m\,N_m{}^T\,]\,V_2' = -[\,I - 2N_m\,N_m{}^T\,]\,[\,V_2\,X\,]\,\omega_b$$

The rate of V_2 is due the rotation rate of the optical bench modified to include trajectory line of sight effects.

$$V_2' = [\,\omega_b\,X\,]\,V_2 = -[\,V_2\,X\,]\,\omega_b$$

Using this expression with those for the normal and nominal output vector gives the following for the two V_1 rates normal to V_1.

$$\begin{vmatrix} V_{1y}' \\ \\ V_{1z}' \end{vmatrix} = \begin{vmatrix} C(\alpha)\,S(\alpha)\,C(\beta) & C(\alpha)\,S(\alpha)\,S(\beta) & C^2(\alpha)-1/2 \\ \\ C^2(\alpha)\,S(\beta)\,C(\beta) & C^2(\alpha)\,S^2(\beta)-1/2 & -C(\alpha)\,S(\alpha)\,S(\beta) \end{vmatrix} \begin{vmatrix} \omega_{bx} \\ \omega_{by} \\ \omega_{bz} \end{vmatrix}$$

$$= F_5'\,\omega_b$$

The V_1 ray vector is changed by moving the fast servo mirror and is transformed through the telescope.

$$V_1 = [M^{-1}] \; F \; D = - \begin{vmatrix} 1 \\ -\sqrt{2}/m \;\; \theta_2 \\ -2/m \;\; \theta_1 \end{vmatrix} \quad \text{for the zero error, } D = \begin{vmatrix} 0 \\ 0 \\ 1 \end{vmatrix}$$

Differentiating this expression and equating to the V_1 rate components gives the FSM angle rates to track the line of sight, V_2, with a fixed pointing flat.

$$F5. \quad \begin{vmatrix} \theta_1' \\ \\ \theta_2' \end{vmatrix} = \begin{vmatrix} 0 & -m/2 \\ \\ -m/\sqrt{2} & 0 \end{vmatrix} \begin{vmatrix} V_{1y}' \\ \\ V_{1z}' \end{vmatrix} = F_5 \; \omega_b \quad \text{defining the matrix}$$

The F_5 matrix is the F_5' matrix with rows weighted by the telescope magnification terms.

Inertial Stabilization of The Pointing Mirror

The pointing mirror is instrumented for inertial stabilization with gyros and inertial angle sensors blended to give two axes of wide bandpass inertial rate feedback. All angular commands to the pointing gimbal servos are necessarily inertial rates coordinatized to gimbal axes. The stabilization process assures that the gimbals rotate to maintain an inertially fixed mirror orientation in the presence of optical bench motion. The external servo commands provide only those additional rates to maintain test object tracking. The inertial rate components of the optical bench along the gimbal axes can be subtracted from the commands derived earlier since those rates are sustained by the gimbal servo with its inertial feedback.

The optical bench components along the gimbal axes are as follows.

$$\begin{vmatrix} \omega_{g\alpha} \\ \\ \omega_{g\beta} \end{vmatrix} = \begin{vmatrix} -S(\beta) & C(\beta) & 0 \\ \\ S(\alpha)\,C(\beta)/C(\alpha) & S(\alpha)\,S(\beta)/C(\alpha) & 1 \end{vmatrix} \begin{vmatrix} \omega_{bx} \\ \omega_{by} \\ \omega_{bz} \end{vmatrix}$$

The total inertial rate to the mirror is :

$$
\begin{vmatrix} \alpha_c' \\ \\ \beta_c' \end{vmatrix} = \begin{vmatrix} \alpha' \\ \\ \beta' \end{vmatrix} + \begin{vmatrix} \omega_{g\alpha} \\ \\ \omega_{g\beta} \end{vmatrix}
$$

Total Rate   Relative   Inertial
Rate   Correction

The commands to be provided externally to the inertially stabilized mirror servo is then:

$$
\begin{vmatrix} \alpha_c' \\ \\ \beta_c' \end{vmatrix} = \begin{vmatrix} 0 & 1/(2C(\beta)) & -S(\alpha)\,S(\beta)/(2C(\alpha)\,C(\beta)) \\ \\ 0 & 0 & 1/(2C^2(\alpha)) \end{vmatrix} \begin{vmatrix} \omega_{bx} \\ \omega_{by} \\ \omega_{bz} \end{vmatrix}
$$

$$
= F_3\,\omega_b \quad \text{defining the matrix } F_3.
$$

Note that no external command is required for pure rotations about the optical bench x axis since the necessary rotation of the pointing flat is provided by the gimbal inertial servo in this case.

Pointing Mirror Error Compensation

The two mirror system for space object pointing has a redundancy of tracking capability since either mirror can provide the necessary line of sight control. The fast servo mirror has a relatively wide bandpass but limited angular range while the pointing mirror servo has a limited bandpass but considerable angular range. The approach taken is to command the slower gimbal servo to respond to the full tracking rates then apply the measured servo error to the fast steering mirror. In this way the difference in frequency response is automatically adjusted and distributed since higher frequency signals beyond the pointing servo capability are detected and relegated to the FSM. Similarly the the higher amplitude lower frequency harmonics inputs are absorbed by the gimbal servo thus screening them from the limited range of the FSM.

The pointing error relation is :

$$
\begin{vmatrix} \delta\,\alpha_p' \\ \delta\,\beta_p' \end{vmatrix} = \begin{vmatrix} \alpha_c' \\ \beta_c' \end{vmatrix} - \begin{vmatrix} \alpha_p' \\ \beta_p' \end{vmatrix}
$$

Pointing Error   Commanded   Sensed Pointing
                            Rate   Mirror Rate

The rate of change of the V_1 vector due to the pointing error rates is found by differentiating the basic reflection relation.

$$
V_1 = [\, I - 2\, N_m\, N_m^T \,]\, V_2
$$

$$
d/dt\,(\,V_1) = -2\,[\, d/dt(N_m)\, N_m^T + N_m\, d/dt(\,N_m^T)\,]\, V_2
$$

Substituting the experssions for V_2 and N_m as functions of the Euler angle pointing error rates give the following.

$$
d/dt = \begin{vmatrix} V_{1y} \\ V_{1z} \end{vmatrix} = F_2' \begin{vmatrix} \delta\,\alpha_p' \\ \delta\,\beta_p' \end{vmatrix} = \begin{vmatrix} 0 & -C^2(\alpha) \\ C(\beta)\ C(\alpha)\ S(\alpha)\ S(\beta) \end{vmatrix} \begin{vmatrix} \delta\,\alpha_p' \\ \delta\,\beta_p' \end{vmatrix}
$$

This derivative of these two components of the V_1 vector have been derived as a relation of the bench inertial rates previously. The Euler angle rates of the pointing mirror to maintain the detector vector, D, at null while holding the inertially stabilized line of sight are determined from the sensed rates. These pointing mirror Euler rates have also been derived earlier.

$$
d/dt \begin{vmatrix} V_{1y} \\ V_{1z} \end{vmatrix} = F_5'\,\omega_b \quad \text{and} \quad d/dt \begin{vmatrix} \alpha \\ \beta \end{vmatrix} = F_3'\,\omega_b
$$

Note that the product $F_2'\,F_3' = F_5'$ The signal command from the bench rate sensors to deflect the V_1 vector by the FSM is the same matrix as for deflecting the V_1 vector using the pointing mirror alone.

The FSM rates can be found to keep the detector at null in the presence of pointing mirror rates being slightly different from that ideally commanded using the relation between the FSM angles and the V_1 vector deflections.

$$\frac{d}{dt}\begin{vmatrix} \theta_1 \\ \theta_2 \end{vmatrix} = \begin{vmatrix} 0 & -m/2 \\ -m/\sqrt{2} & 0 \end{vmatrix} F2' \begin{vmatrix} \delta\alpha_p' \\ \delta\beta_p' \end{vmatrix} = F2 \begin{vmatrix} \delta\alpha_p' \\ \delta\beta_p' \end{vmatrix}$$

System Block Diagram

The block diagram of figure 6 shows the division of the signal path to the gimbal servo and the FSM. The sensed bench motion rate ω_b, modified by the estimated target inertial rate, ω_t, is multiplied by the matrix F3 to form the gimbal command rate. This command passes through the controller compensation and mixed with other control terms. These other terms include FSM off loading or re-centering feedback loops and signals generated from position detecting sensors responding to the target return radiation. The pointing mirror control system torque the gimbal producing rates that are measured with inertial instruments mounted to the pointing mirror. The matrix F4 resolves the gimbal inertial rates to gimbal Euler angle rates. These rates are used in the gimbal servo feedback and also subtracted from the input gimbal command to form a delta rate signal representing the residual rate error to which the gimbal servo was unable to respond. This delta rate error is multiplied by the F2 matrix to form the rate commands for the FSM.

Summary

A design for commanding a cascade of two mirrors through an optically amplifying telescope has been presented. A feedback methodology has been incorporated so that the relative angular ranges and response characterization of the two mirror drives can be used to advantage without overlap of the tracking commands.

Acknowledgements

Appreciation is given for many of the mathematical relations developed by Dr. J. N. Aubrun used in this paper. Mr. E. D. Scott greatly helped the interpretation and presentation of some of the equations regarding mirror rates. Mr. S. J. Rusk has generated an alternate derivation for the isolation equations and a hardware/software design   implementing this system for the STARLAB application.

References

Aubrun, J. N., Baseline Pointing and Tracking Control System, LMSC F155391,   October 7, 1986.

Rusk, S. J., Schmidt, T., Development of Pointing Control Equations For Starlab Experiment, LMSC F185834, July 23, 1987.

Van Allen. Robert L., Dillow, James D., Gurski, Gary F., Directed Energy   Weapons Tracking and Pointing Space Experiments, AAS 87-031,   presented at the 10th Annual AAS Guidance and Control Conference,   January 31-Febuary 4, 1987, Keystone, CO.

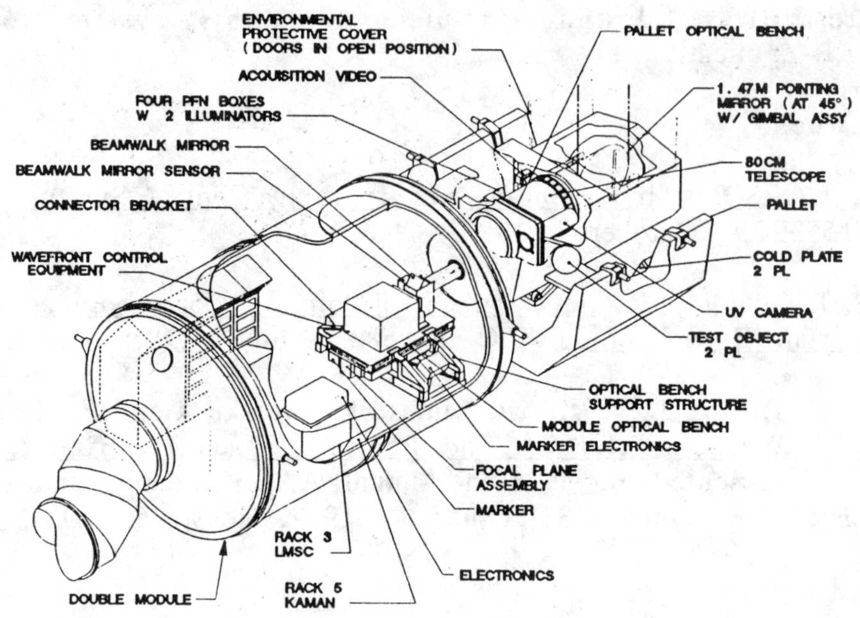

FIGURE 1   POINTING CONTROL SYSTEM EQUIPMENT

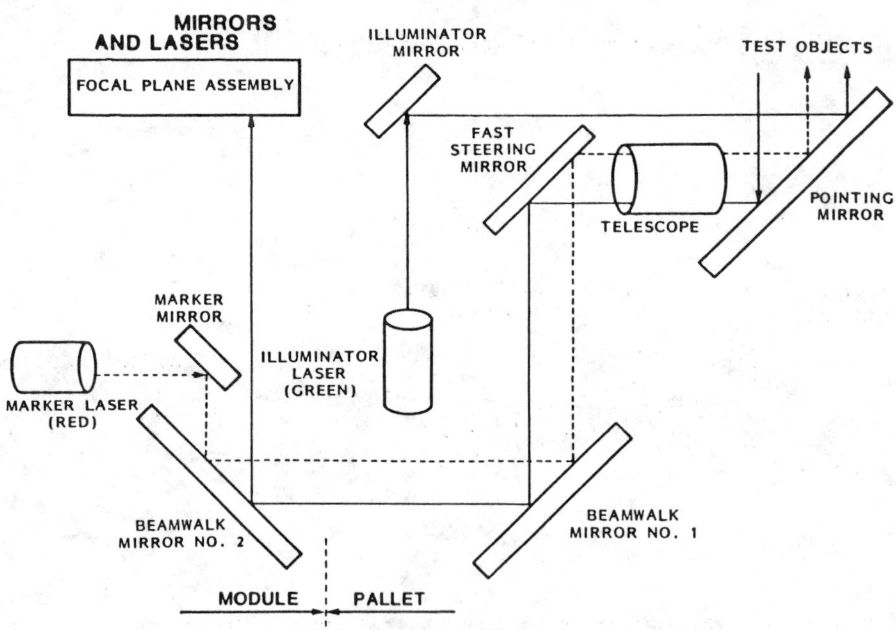

FIGURE 2   OPTICAL SCHEMATIC

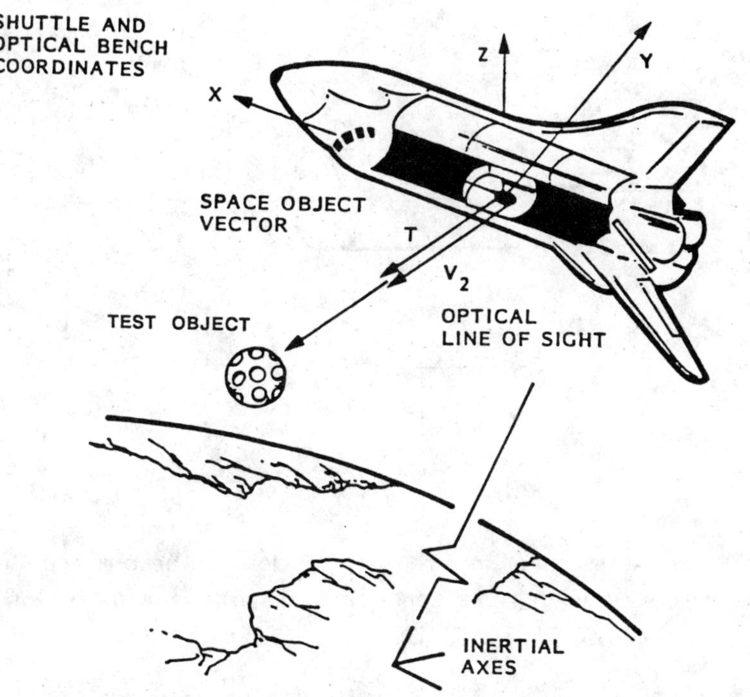

FIGURE 3   SYSTEM COORDINATES

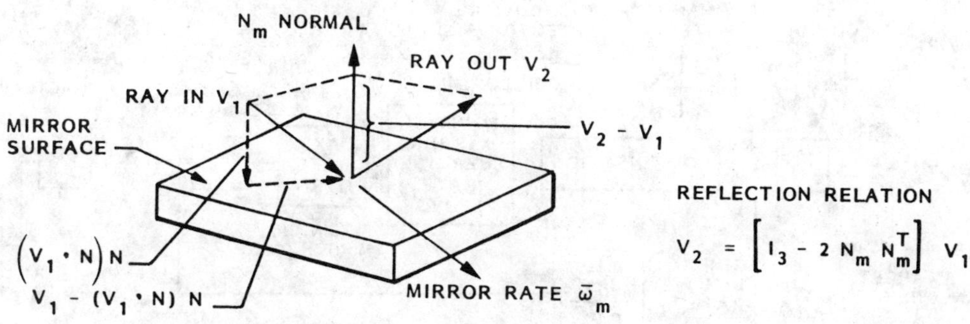

FIGURE 4   MIRROR REFLECTION RELATION

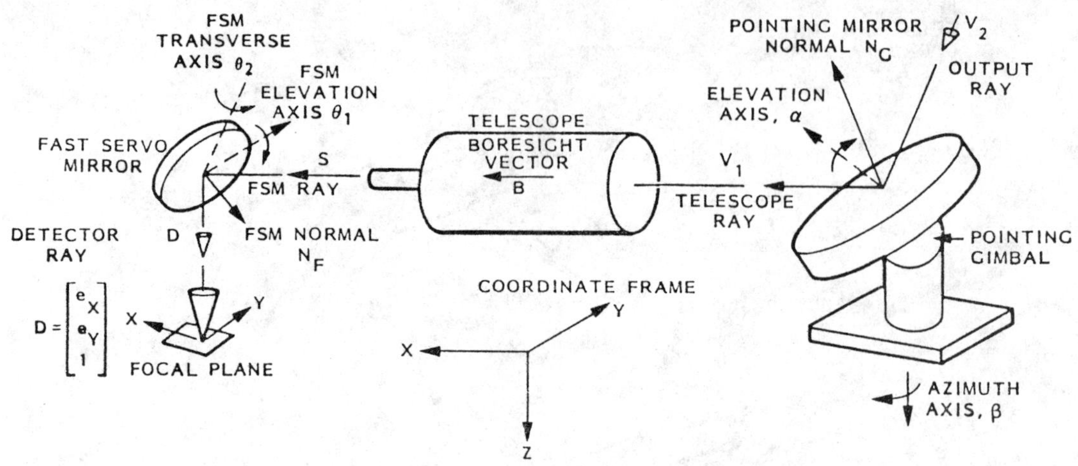

THE DETECTOR RAY, D, PASSES THROUGH 3 TRANSFORMATIONS TO BECOME THE OUTPUT RAY,

V_2 OUTPUT RAY = [POINTING GIMBAL] [TELESCOPE] [FAST SERVO MIRROR] DETECTOR RAY

$$V_2 = \left[I - 2\bar{N}_G \bar{N}_G^T \right] M \left[I - (1-1/M)\bar{B} \bar{B}^T \right] \left[I - 2\bar{N}_F \bar{N}_F^T \right] \bar{D}$$

FIGURE 5   OPTICAL COORDINATES AND TRANSFORMATIONS

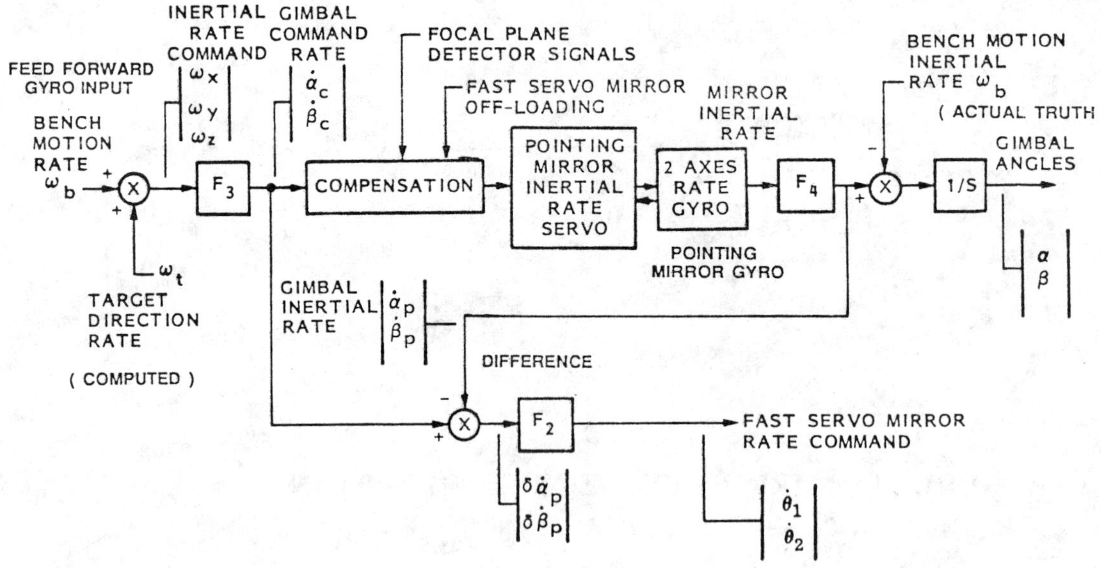

FIGURE 6   BASE MOTION ISOLATION BLOCK DIAGRAM

SPACE-STABILIZED BEAM POINTING FOR A BIFOCAL SATELLITE EXPERIMENT

David C. Redding[*], T. T. Chien[*], Edward H. Kopf[†] and Edward C. Wong[†]

The pointing and tracking control concept for an experimental ground-based laser beam relay satellite is described. The satellite receives a low-power laser beam from a ground-based laser, processes it, and reflects it to an instrumented target. The primary control objective is to achieve extremely accurate and stable pointing at the target, including moving targets. This is achieved using an optical/inertial stabilization technique. The paper describes the satellite optical layout and pointing control mechanization.

INTRODUCTION

One basic approach to ballistic missile defense from space is the Ground-Based Laser (GBL) approach. In a GBL system, lasers based on the ground feed high-energy laser beams to a constellation of relay satellites, which are essentially large mirrors in space. These direct the beam towards the ballistic missile targets.

The relay satellites are typically of two types. Monocle Relay Mirrors use a single, usually segmented, reflector. They are most effective at high altitude, as an intermediate reflection point. They reflect the beam to the Mission Mirrors, which in turn direct the beam to the target. The Mission Mirrors are capable of sensing and correcting beam wavefront error and beam jitter. They must find and acquire the target, illuminate the target, and then rapidly switch to a new target.

In this paper we consider one possible design for an experimental Mission Mirror satellite. In particular, we consider its target tracking and beam pointing control system. The satellite is intended as a testbed and demonstration vehicle for some of the component technologies required for an operational Mission Mirror. It is a so-called "bifocal" Mission

---

* Charles Stark Draper Laboratory, Inc., 555 Technology Square, Cambridge, Massachusetts 02139.

† Jet Propulsion Laboratory, California Institute of Technology, 4800 Oak Grove Drive, Pasadena, California 91109.

Mirror satellite, using beam compression and re-expansion for control of large-aperture beams using small fast-steering and deformable optics. Because it is designed with a single optical axis, we call it the "Single-Axis Bifocal" (SAB) satellite.

We describe the mechanical and optical system conceptual design in the subsequent section. A simplified first-order analysis of the optics demonstrates basic pointing requirements. The third section presents the pointing control system concept. A typical encounter is described. The jitter and tracking control loops are discussed. Some simulation results are presented.

SATELLITE DESCRIPTION

Mechanical Design Concept

As sketched On Fig. 1, the SAB satellite consists of two turrets and a center section. The optics, consisting of beam compressor and beam expander afocal telescopes, a deformable mirror (DM), a fast steering mirror (FSM), and the tracker optics, are all contained in the two turrets. The center section houses spacecraft attitude control actuators, computers, telemetry and other electronics. The Uplink Turret houses the beam compressor and deformable mirror, while the Downlink Turret contains the FSM, beam expander and tracker.

A single large-angle gimbal axis separates the Uplink Turret from the center section, and the center section from the Downlink Turret. It is actuated by 2 large annular single-axis bearings and 2 small-angle Magnetic Bearing Assemblies (MBAs), as sketched. The gimbal axis is nominally the same as the line-of-sight between the deformable and fast-steering mirrors: the beam is passed from the UT to the DT along this optical axis. The MBAs act to isolate the two turrets from vibration generated on the center section, as discussed below.

Successful beam relay requires that the UT be pointed directly at the laser source, the DT be pointed directly at the target, and the optical axis be perpendicular to both lines of sight. The 4 DOF provided by the turret gimbal and 3 spacecraft attitude DOF are fully constrained by these pointing requirements (except for center section attitude) within the range of the small-angle steering mirrors.

This mechanization has the advantage of simplicity: it uses a minimal number of moving parts. The potential disadvantage is that there is a singularity in the motion: a gimbal pole which occurs when the angle from the source beam and target beam is 0° or

180°. This occurs when the beam is passed directly through the spacecraft or returned directly to the target. If the singularity is approached, greater attitude motion is required to preserve track attitude. Analysis shows that this is not necessary for single target tracking.

The small-angle steering mirrors in the SAB beam train provide additional DOF to alleviate this problem over small angles. The singularity could be completely avoided by adding a second or third large-angle gimbal axis (with corresponding mirrors and optical axes), at the cost of more large moving parts and a more complex optical system. For purposes of the SAB experiments, this is not necessary.

An important feature of the SAB mechanical design is the provision of shielding for the optics and electronics. The spacecraft is fully enclosed during non-operational periods to provide protection from radiation, meteoroids, organic contamination, and neutral oxygen erosion. Doors will be opened to expose the primary mirrors and tracker during encounters.

The overall size of the SAB spacecraft will depend on the launch vehicle chosen.

Optical Design Concept

The SAB optical schematic is provided in Fig. 2. The uplink beam compressor is a 2-mirror, off-axis telescope. It is capable of a limited angular range without incurring distortion beyond the range of the wavefront control (WFC) system. The downlink beam expander is a 3-mirror, off-axis design that is capable of a larger range of output angles. This is required for rapid retargeting using the FSM. The main advantage of the 3-mirror design is that beam walk due to beam steering is incurred on the small field mirror rather than the downlink primary mirror.

For packaging reasons, both telescopes are required to be very fast. The telescope design challenge is to maximize the input and output apertures while meeting constraints of keeping distortion due to steering and misalignment within the dynamic range of the wavefront control system and keeping the overall dimensions within launch capabilities.

There are 5 mirrors in the reduced-aperture section of the SAB optics: the secondary mirrors, the DM, the FSM and the field mirror. Depending on the power of the uplink laser, these may or may not be actively cooled.

Beam sensors are provided on the uplink and downlink PMs. The uplink beam sensors are used to provide beam centroiding information to the ground-based laser for cooperative tracking. The downlink sensors are used to control beam walk.

Output wavefront sensing is accomplished using holographic optical elements on the downlink primary mirror, which is the last optic in the main beam train. These feed a large-angle wavefront sensor which senses the local wavefront tilts of the outgoing beam. The tilts are processed to obtain the beam wavefront figure and mean direction. An input wavefront sensor is also provided for uplink beam distortion measurements.

The wavefront map is used to drive the deformable mirror, which corrects the outgoing wavefront errors. This system will correct errors in the arriving wavefront and errors generated internally.

The fast steering mirror is driven by the pointing control system. It corrects residual uplink jitter and beam jitter from optical system vibration, as sensed by the tracker, as described later. The uplink secondary mirror is also steerable, though over a smaller range, and is used for alignment purposes.

The dual line-of-sight (DLOS) tracker is designed to make maximum use of common-mode optics to provide passive rejection of subsystem motion. If the tracker optical train or any element moves in rotation or displacement, the focal-plane images of the target, the downlink beam, and the inertial reference beam are displaced by the same amount. This does not directly disturb the signals of interest, which are the relative displacements rather than the absolute locations of these images.

The key element in the tracker optical train is the annular aperture-sharing mirror. Target return and the reference beam are admitted to the DLOS tracker from backside reflection. The power beam (or power beam beacon) is intercepted by the front side, retro-reflected and admitted to the tracker through the central opening. The aperture-sharing mirror is an all-reflective element to avoid problems with partially-transmissive optics in the power beam train. It can be made of cooled ULE glass if necessary to minimize thermal distortion. The power beam can be kept out of the tracker by the relay beam rejector.

The tracker field-of-view (FOV) is steered by action of the tracker fast-steering mirror(s) (TFSM), which has angular range equal to the power beam FSM, and bandwidth adequate to negate expected beam jitter. Though one mirror is shown, it is likely that a second, small FSM in the telescope will also be required.

The tracker has several focal planes to support the tracking task and for use in booster imaging experiments. Most encounters will be run with imaging in the visible band, but

UV and IR data-collecting focal planes are also included. To support acquisition with large angular uncertainty and also extremely fine pointing, two or three focal planes with different magnifications are required. The progression from coarse to fine track is handled automatically by the PCS. In addition to the imaging focal planes, and co-boresighted with them, are 2 quad cells. These high-bandwidth, non-imaging sensors are tuned to the wavelengths of the power beam (beacon) and to the reference beam. They are used by the PCS to drive the beam stabilization controllers, as discussed below.

The heart of the PCS is an inertially-stabilized reference laser beam, which is generated by the optical reference unit (ORU) and coupled into the tracker common optics via a retro-reflector. The ORU beam provides the reference used by the tracker to measure inertial beam jitter and stabilize the tracker FOV, as described below. The ORU consists of an isolated stabilized gyro platform upon which a laser source is mounted. The platform is capable of small angle motion with respect to the DL turret to the limit of the beam steering capability.

Approximate Pointing Analysis

To demonstrate basic feasibility of the SAB optical and control system concepts we present a summary first-order analysis of the optical train. These equations were derived based on approximate designs of the powered optics, following the method of Ref. 1. They trace the central ray through the beam train, assuming paraxial optics and neglecting the effects of translation of the optics. The beam train is assumed to be nearly aligned, with all motions being small angle except turret and spacecraft attitude motion.

There are three coordinate systems assumed as illustrated in Fig. 2: inertial (I); uplink telescope (UT); and downlink telescope (DT). The DT frame is fixed to the downlink telescope (including the tracker). The UT frame is rotated wrt this frame by the large gimbal angle θ_G. In addition to the large gimbal angle, the uplink turret (with all its optics moving in common) is rotated from the downlink by the small MBA angles, represented as the vector $\vec{\theta}_{MBA}$.

Beam errors are represented as vector differentials, with $d\hat{r}_0$ being the input beam error and $d\hat{r}_3$ that of the crosslink beam (off of the third mirror) connecting the 2 turrets, in UT frame. The output of the uplink turret in UT frame is nominally aligned with the optical axis $\hat{x}_{UT}$, with deflection $d\hat{r}_3$:

$$\hat{r}_3 = \hat{x}_{UT} + d\hat{r}_3 \tag{1}$$

$$d\hat{r}_3 = \begin{bmatrix} 0 & 0 & 0 \\ -7.4 & 0 & 0 \\ 0 & 0 & 9.9 \end{bmatrix} d\hat{r}_0 + \begin{bmatrix} 0 & 0 & 0 \\ 0 & 0 & -19 \\ -19 & 8.9 & 0 \end{bmatrix} \vec{\theta}_{MBA} \tag{2}$$

This expression shows the effect of the roughly 8:1 telescope magnification factor.

The downlink turret output - the downlink beam - is nominally along the DT y-axis. Its deflection is $d\hat{r}_7$, a function of the uplink input $d\hat{r}_3$, the FSM control angle vector $\vec{\theta}_4$, and the large-angle transformation through the gimbal:

$$\hat{r}_7 = \hat{y}_{DT} + d\hat{r}_7 \tag{3}$$

$$d\hat{r}_7 = \begin{bmatrix} 0 & 0 & -0.1 \\ 0 & 0 & 0 \\ -0.07 & 0.07 & 0 \end{bmatrix} \vec{\theta}_4 + \begin{bmatrix} 0 & -0.05 & 0 \\ 0 & 0 & 0 \\ 0 & 0 & 0.07 \end{bmatrix} {}^{DL}_{}\mathbf{T}^{UL} d\hat{r}_3 \tag{4}$$

where

$$ {}^{DL}_{}\mathbf{T}^{UL} = \begin{bmatrix} 1 & 0 & 0 \\ 0 & \cos\theta_G & -\sin\theta_G \\ 0 & \sin\theta_G & \cos\theta_G \end{bmatrix}. \tag{5}$$

There are 3 beams of interest in the tracker. These are projected on tracker focal planes (FPs) and quad cells, which , as they are co-boresighted, are treated as a single FP. The projection process removes the y-component of the reflected beams. The downlink signal is $\vec{A}$, which is a function of the downlink beam deflection $d\hat{r}_7$ and of the TFSM angle vector $\vec{\theta}_{10}$. In addition, tracker errors $\vec{\theta}_T$ effect the spot. In DT frame,

$$\vec{A} = \begin{bmatrix} 0 & 0 & -2 \\ 1 & -1 & 0 \end{bmatrix} \vec{\theta}_{10} + \begin{bmatrix} -1 & 0 & 0 \\ 0 & 0 & -1 \end{bmatrix} d\hat{r}_7 + \frac{\partial \vec{A}}{\partial \vec{\theta}_T} \vec{\theta}_T \tag{6}$$

It is actually more convenient to express the TFSM angles in local mirror coordinates. Choosing $\hat{y}_{TFSM}$ to be the mirror normal, $\hat{z}_{TFSM}$ is the same as for the DT frame, and $\hat{x}_{TFSM} = \hat{y}_{TFSM} \times \hat{z}_{TFSM}$. In these coordinates, the $\hat{y}$ angle is indeterminate, so that we need carry only the $\hat{x}$ and $\hat{z}$ terms of the mirror angle vector, as

$$\vec{\theta}_{TFSM} = \begin{bmatrix} \hat{x}_{TFSM} & \hat{z}_{TFSM} \end{bmatrix} \vec{\theta}_{10} \tag{7}$$

Substituting into Eq. 6,

$$\vec{A} = \begin{bmatrix} 0 & -2 \\ 1.4 & 0 \end{bmatrix} \vec{\theta}_{TFSM} + \begin{bmatrix} -1 & 0 & 0 \\ 0 & 0 & -1 \end{bmatrix} d\hat{r}_7 + \frac{\partial \vec{A}}{\partial \vec{\theta}_T} \vec{\theta}_T \tag{8}$$

Similarly, the target spot is $\vec{B}$, a function of the target beam $d\hat{r}_{11}$ and of the TFSM angle vector $\vec{\theta}_{10}$ and tracker errors $\vec{\theta}_T$:

$$\vec{B} = \begin{bmatrix} 0 & \text{-2} \\ 1.4 & 0 \end{bmatrix} \vec{\theta}_{\text{TFSM}} + \begin{bmatrix} 1 & 0 & 0 \\ 0 & 0 & 1 \end{bmatrix} \vec{dr}_{11} + \frac{\partial \vec{A}}{\partial \vec{\theta}_T} \vec{\theta}_T \qquad (9)$$

The reference spot $\vec{C}$ is a function of the ORU angle vector and the tracker optics:

$$\vec{C} = \begin{bmatrix} 0 & \text{-2} \\ 1.4 & 0 \end{bmatrix} \vec{\theta}_{\text{TFSM}} + \begin{bmatrix} 0 & 0 & 1 \\ \text{-1} & 0 & 0 \end{bmatrix} \vec{\theta}_{\text{ORU}} + \frac{\partial \vec{A}}{\partial \vec{\theta}_T} \vec{\theta}_T \qquad (10)$$

The overall pointing objective is to put the downlink and target beams in the proper "look-ahead" alignment. The look-ahead vector $\vec{\Delta}_{\text{LA}}$ is the offset required to compensate for the finite speed of light and the high relative velocity of the spacecraft and target. It can be expressed as a function of the spacecraft velocity and speed of light c (to first order in v_{sc}/c) as

$$\vec{\Delta}_{\text{LA}} = -\frac{2}{c} \mathbf{P}_{\hat{r}_7} \vec{v}_{\text{SC}} \qquad (11)$$

where $\mathbf{P}_{\hat{r}_7}$ is the projection matrix (in DT frame)

$$\mathbf{P}_{\hat{r}_7} = \begin{bmatrix} 1 & 0 & 0 \\ 0 & 0 & 0 \\ 0 & 0 & 1 \end{bmatrix} \qquad (12)$$

For the SAB experiment, the look-ahead angle is of the order of 50 μrad. Look-ahead computation is discussed at length in Ref. 2.

The pointing objective is met when the the downlink beam leads the target direction as

$$\vec{dr}_7 + \vec{dr}_{11} = \vec{\Delta}_{\text{LA}} \qquad (13)$$

Substituting for $\vec{dr}_7$ and $\vec{dr}_{11}$ in terms of the tracker signals $\vec{A}$ and $\vec{B}$ (from Eqs. 6 and 7), the pointing objective is expressed in terms of these focal-plane signals as

$$\vec{B} - \vec{A} = \begin{bmatrix} 1 & 0 & 0 \\ 0 & 0 & 1 \end{bmatrix} \vec{\Delta}_{\text{LA}} \qquad (14)$$

This expression says that the correct pointing alignment occurs when the downlink and tracker spots are separated by the look-ahead vector. The fact that none of the tracker elements ($\vec{\theta}_{10}$ and $\vec{\theta}_T$) appear in this expression shows the insensitivity to these elements gained by the common-mode DLOS tracker design

POINTING CONTROL SYSTEM CONCEPT

The primary objective of the SAB experiment is to demonstrate extremely accurate and stable pointing of the relayed laser beam on an instrumented target board. This must be accomplished in spite of significant spacecraft vibrational disturbances and the usual sensor and actuator realities of limited accuracy, dynamic range and bandwidth. This is the task of the pointing control system described in this section.

<u>**Encounter Scenario**</u>

A typical encounter begins with the SAB spacecraft moving into view of the ground-based laser site. At this point the spacecraft is in acquisition attitude, with the uplink telescope pointed at the known GBL site, and the DL telescope pointed at the expected target site, within the acquisition angular range of a few mrad.

The GBL site illuminates the spacecraft with a low-power beacon beam, which is acquired in the DLOS tracker coarse focal plane. The PCS then enters closed-loop acquisition mode, driving the beacon spot into the center of the coarse FP using the FSM, where it is picked up on both the fine-track FP and the co-aligned power-beam quad cell. Then the power-beam loop takes over, driving the power beam to null at high bandwidth using the quad cell signal and the FSM.

Meanwhile the tracker reference-beam loop is working to slave the tracker LOS to the ORU reference beam, which it does by driving the reference beam spot to null in the reference-beam quad cell using the tracker FSM. The bandwidth of this loop is greater than that of the power-beam loop by a factor of 6-10. With the power-beam and reference-beam loops closed, the effect is to slave the power beam and tracker FOV both to the stabilized reference beam at high bandwidth. This space-stabilizes both the downlink beam and the tracker FOV. The tracker FOV and power beam can then be steered together by precessing the gyros in the ORU stable platform at low bandwidth.

Having acquired the uplink beam, the PCS moves into target acquisition mode, precessing the ORU platform to drive the tracker FOV towards the target image. The target, which may be passive or active (with a beacon for cooperative illumination), is acquired initially on the coarse FP and then driven to the fine-track FP.

With the target acquired on the fine-track FP, the PCS transitions to closed-loop tracking mode. The track loop sensing is from the fine-track FP, where both the power beam (held at null by the power-beam loop) and the target image are visible. The track loop drives the target image into the required look-ahead offset configuration by precessing the ORU platform gyros. The scoring beam then comes on and the target is illuminated. Received target return can then be used to sharpen the pointing in an active-track mode.

Throughout this sequence the spacecraft turret and attitude control systems act in a relatively low-bandwidth follow-up mode, moving to null the offsets of the various steering mirrors and ORU platform as they track the uplink beam and target. The result is that the spacecraft changes attitude and turret angle to preserve the overall pointing geometry described above, and the mirror and ORU angles remain small.

A more detailed description of the PCS loops follows.

Jitter Control Strategy

The SAB pointing control system utilizes a multiple-stage jitter control strategy. The first stage of jitter control is the passive jitter rejection capability of the common-mode design of the DLOS tracker. The next stage is the optical jitter control system sketched above and developed in a bit more detail below. The optical controller, which directly senses and compensates downlink beam jitter, can be used against both spacecraft-induced and residual uplink jitter, but is of limited capability. To improve performance, two other stages are used to reduce the contribution of spacecraft vibrations.

The next stage uses the turret MBAs to isolate the 2 telescope turrets from the main on-board sources of vibration, which are clustered on the center section. These include the attitude-control CMGs, the turret bearings, tape recorders, and other essential equipment. The MBAs will be an integral part of the turret and spacecraft attitude controllers, being driven in follow-up mode to the steering mirrors and ORU during target tracking. They will function as pointing control devices while providing substantial attenuation of spacecraft-induced vibrations.

This approach is illustrated in the conceptual transfer function from center-section translational vibration to beam jitter of Fig. 3. Overlapping the bandwidths of the MBA rotational control, MBA translational control, and optical jitter control systems should provide several orders of magnitude of attenuation of center section vibration.

Turret structural modes will be excited by spacecraft slew or retargeting maneuvers. The final stage of jitter control is provided by a structural design that will allow - to the maximum extent possible - for rapid passive damping of those high-frequency vibrational transients, especially those that are beyond the optical jitter control system bandwidth.

There are several vibration sources that cannot be put on the center section. These include the active optics - the FSMs and the deformable mirror - and mirror coolant flow if liquid-cooled optics are used. Jitter from these sources, plus uplink beam jitter, can only be counteracted by the optical jitter control system.

<u>**Optical Jitter Sensing and Control**</u>

The ideal beam tracking system would directly sense target motion. This can be done if the target is a bright point source - if it is equipped with a beacon. There are significant problems doing this with a passive target, however. The target is likely to be a dim object in close proximity to a very bright rocket plume. To adequately image the scene, large sensor arrays with slow integration times are necessary. To determine the target's precise location in the scene, significant image processing may be required. The combination of long sensor scan times and image processing delays effectively precludes the possibility of directly using the target image to drive the optical jitter control system.

An alternative reference source is provided by the ORU reference beam: an inertially stabilized laser source. When coupled into the tracker common optics using a retroreflector (Fig. 2), the reference beam ties together all elements of the tracker, as expressed in Eq. 8. At jitter frequencies the ORU angle essentially goes to zero, with the result that $\vec{C}$ directly shows the tracker jitter.

The reference-beam control loop operates at high bandwidth to cancel this jitter. As sketched in Fig. 4, it senses the jitter using the reference beam quad cell and drives the TFSM to null the reference-beam spot:

$$\vec{C} \Rightarrow \begin{bmatrix} 0 \\ 0 \end{bmatrix} \tag{15}$$

Closing the reference-beam loop forces the TFSM to subtract out the effects of tracker error and ORU offset. Substituting Eq. 15 in Eq. 10,

$$\vec{\theta}_{TFSM} \Rightarrow - \begin{bmatrix} 0 & 0.7 \\ -0.5 & 0 \end{bmatrix} \left[\begin{bmatrix} 0 & 0 & 1 \\ -1 & 0 & 0 \end{bmatrix} \vec{\theta}_{ORU} + \frac{\partial \vec{A}}{\partial \vec{\theta}_T} \vec{\theta}_T \right] \tag{16}$$

The effect of closing the reference-beam loop on the target image is (substituting Eq. 6 into Eq. 9):

$$\vec{B} \Rightarrow \begin{bmatrix} 1 & 0 & 0 \\ 0 & 0 & 1 \end{bmatrix} d\hat{r}_{11} + \begin{bmatrix} 0 & 0 & -1 \\ 1 & 0 & 0 \end{bmatrix} \vec{\theta}_{ORU} \tag{17}$$

This shows that the contributors to jitter of the target image are restricted to actual target jitter and jitter of the reference spot. Closing the reference-beam loop eliminates the contribution of tracker jitter. As the reference-beam quad cell is co-boresighted with the tracker FPs and the power-beam quad cell, by eliminating jitter in the reference quad cell it is eliminated in the other sensors as well. This has the considerable benefit of reducing image smear effects in all FPs, allowing for more accurate imaging.

It is important to realize that the FOV is not stabilized about the tracker long axis (y) by this technique. This can lead to pointing command error, as the correct tracking alignment requires the target image to be offset from the tracker null point by the look-ahead angle, or speed-of-light correction factor. An angle error (jitter) about y results in an error in the commanded look-ahead of roughly the magnitude of the angle error times the look-ahead angle of about 50 μrad. As the angle error can be expected to be quite small, this effect should not be significant.

With the reference-beam loop closed, the tracker FOV null is slaved to the inertially stabilized reference beam. The next step is to slave the power beam to the tracker null, thereby also slaving it to the reference beam, albeit at a lower bandwidth. This is done by the power-beam loop sketched in Fig. 4. The power-beam loop nulls $\vec{A}$ by action of the power-beam FSM $\vec{\theta}_4$ via the transformation (in mirror coordinates)

$$\frac{\partial \vec{\theta}_4}{\partial \vec{A}} = \begin{bmatrix} 0 & 10.4 \\ 9.6 & 0 \end{bmatrix} \tag{18}$$

If required, the FSM could be augmented in the high frequency ranges by using the deformable mirror in a steering mode as well as for wavefront correction. With the loop closed,

$$\vec{A} \Rightarrow \begin{bmatrix} 0 \\ 0 \end{bmatrix} \tag{19}$$

Substituting Eqs. 16 and 18 into Eq. 8, the downlink beam is seen to be slaved to the ORU:

$$d\hat{r}_7 \Rightarrow \begin{bmatrix} 0 & 0 & -1 \\ 0 & 0 & 0 \\ 1 & 0 & 0 \end{bmatrix} \vec{\theta}_{ORU} \tag{20}$$

This shows that the downlink beam points where the reference beam points. The only uncompensated contributor to downlink jitter is thus jitter in the ORU.

Errors from several sources will ultimately limit the performance of this jitter control scheme. Primary among these is jitter of the ORU reference beam itself. Errors in the wavefront control system could cause the sampled part of the downlink beam to deviate from the average wavefront. Misalignment of the quad cells from each other could couple y-axis tracker jitter into the downlink. Sensor and actuator resolution, dynamic range and bandwidth will be limiting factors.

Target Tracking

The track loop (Fig. 5) precesses the ORU gyros to move the $\vec{B}$ spot into the look-ahead configuration of Eq. 14, as sensed on the fine-track FP. Because the jitter loops are closed, stabilizing the tracker field-of-view, it can be effective at the relatively low bandwidth of the imaging focal planes: the target image, which moves smoothly in inertial space, moves smoothly also on the focal plane.

With the $\vec{A}$ spot held at null by the power-beam loop, the control objective for the track loop is to drive the $\vec{B}$ spot to $\vec{\Delta}_{LA}$, as

$$\vec{B} \Rightarrow \begin{bmatrix} 1 & 0 \\ 0 & 0 \\ 0 & 1 \end{bmatrix} \vec{\Delta}_{LA} \qquad (21)$$

Substituting Eq. 21 into Eq.17, the required ORU angles are

$$\vec{\theta}_{ORU} \Rightarrow \begin{bmatrix} 0 & 0 & -1 \\ 0 & 0 & 0 \\ 1 & 0 & 0 \end{bmatrix} \left[d\hat{r}_{11} - \vec{\Delta}_{LA} \right] \qquad (22)$$

Substituting Eq. 22 into Eq. 20, the original pointing objective (Eq. 13) is recovered:

$$d\hat{r}_7 \Rightarrow -d\hat{r}_{11} + \vec{\Delta}_{LA} \qquad (23)$$

This demonstrates the basic feasibility of the pointing control concept.

The main source of error for the track loop is in the calculation of the look-ahead $\vec{\Delta}_{LA}$ (Ref. 2). Accurate estimation of $\vec{\Delta}_{LA}$ requires good knowledge of internal alignments, spacecraft attitude and ephemeris, target range and rate, etc. Knowledge of target acceleration characteristics is very helpful.

Once the target is illuminated, the reflected light can be used to improve estimates. A purely active track mode is also possible if the illuminating beam is scanned across the target surface.

Follow-Up Control

The optical actuators used in jitter and tracking control -the FSMs and the ORU - are all small-angle devices. To keep them within their ranges, the spacecraft acts to null their deflections. Spacecraft attitude and turret angle control are commanded by the transformed actuator angles at relatively low bandwidth. In addition, feed-forward commands based on estimated encounter geometry are used.

AN EXAMPLE ENCOUNTER

Figures 6-15 show the simulated time history of several key parameters during an example encounter over Hawaii. The uplink source is located on Maui, with the target placed on Kauai. The simulation conditions are highly simplified - no spacecraft flexible dynamics or noise. The intent is to illustrate the range of parameter motion rather than actual performance. The performance shown is not representative of expected or actual requirements.

Figure 6 shows the line-of-sight angles from the source and target sites to the SAB spacecraft as it flies between them during a 100 sec encounter. Figure 7 shows attitude motion required to maintain pointing during the encounter. Most of the motion is in the pitch axis, with the spacecraft doing an end-over-end tumble. Corresponding body rates are shown on Fig. 8. Figures 9 and 10 show gimbal angle and rate through the encounter.

Figures 11-15 illustrate the first 0.25 sec of the encounter. Figure 11 shows the commanded look-ahead aimpoint as it moves on the focal plane during this time. In this encounter, the SAB pointing system is being driven directly by the target spot, a mode that is available with a cooperative, beacon-equipped target. Figure 12 shows the TFSM angles as they react to the initial target pointing error, starting at 0.1 sec. Motion of the target spot on the focal plane as the PCS drives it to the look-ahead aimpoint is shown on Fig. 13. The power-beam loop is activated at 0.15 sec and immediately drives the power-beam FSM angles as shown on Fig. 14. The power-beam spot trajectory is shown on Fig. 15. Note that the initial motion is away from null, as the TFSM drives the target spot to the look-ahead aimpoint. Once the power-beam loop is activated, the spot rapidly is driven back to null.

The TFSM and FSM servos used in the simulation are simple proportional-integral controllers, with rate limiters. Follow-up control of the spacecraft attitude and gimbal angle occurs at lower bandwidth: there is no appreciable motion in this timescale.

CONCLUSION

The Single-Axis Bifocal satellite experiment is capable of demonstrating high-performance laser-beam relay from a ground source to space and then to an instrumented target board. It offers a versatile testbed for the technologies required for an operational Ground-Based Laser Mission Mirror system, including target acquisition and tracking and beam wavefront control.

ACKNOWLEDGEMENT

This research was performed at the Jet Propulsion Laboratory, California Institute of Technology, under support from the Strategic Defence Initiative Directed Energy Office. The authors thank W. Goss, R. Levin, J. Negro, A. Meinel, M. Miller, R. Korechoff, R. Schlundt and W. Breckenridge for their substantial contributions to this work.

REFERENCES

1. Redding, D.C. and Breckenridge, W.G., "Modeling Reflective Optics for Control Analysis," to appear.

2. Breckenridge, W.G. and Redding, D.C., "SRE Closed-Loop Error Analysis," JPL Internal Report EM-343-1049, January 1987.

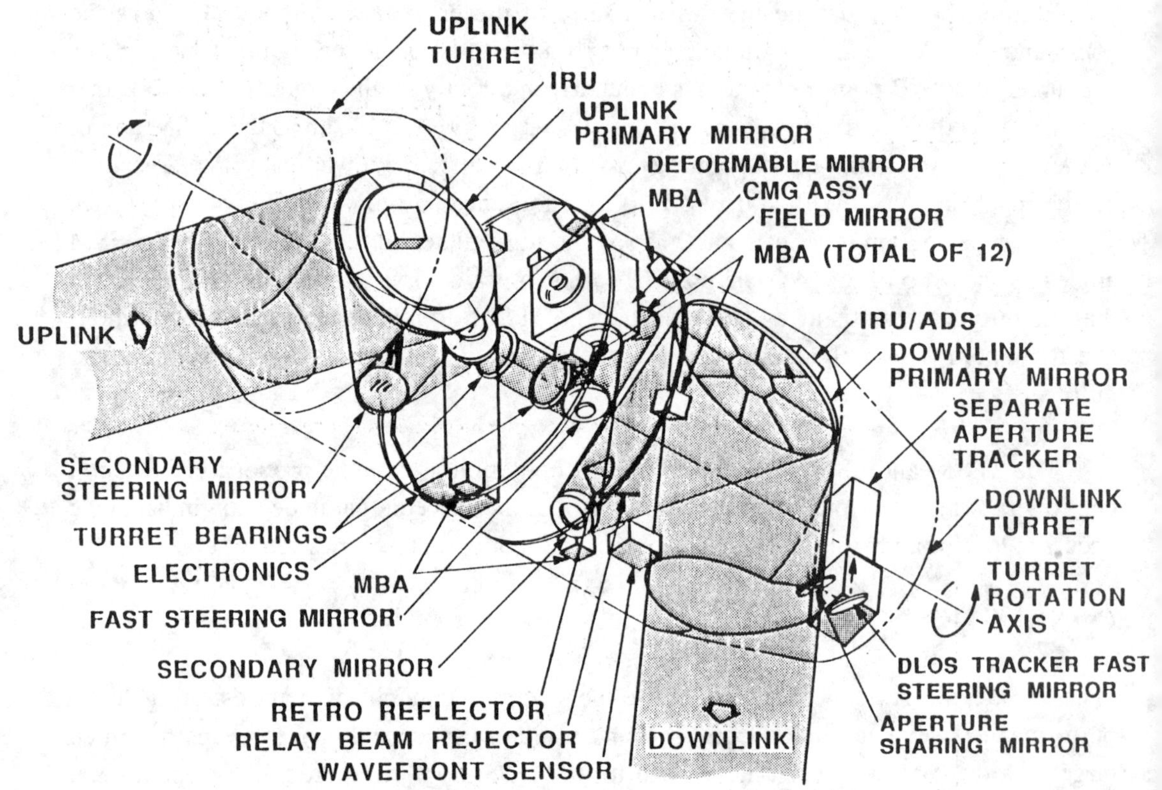

Figure 1. SAB Configuration

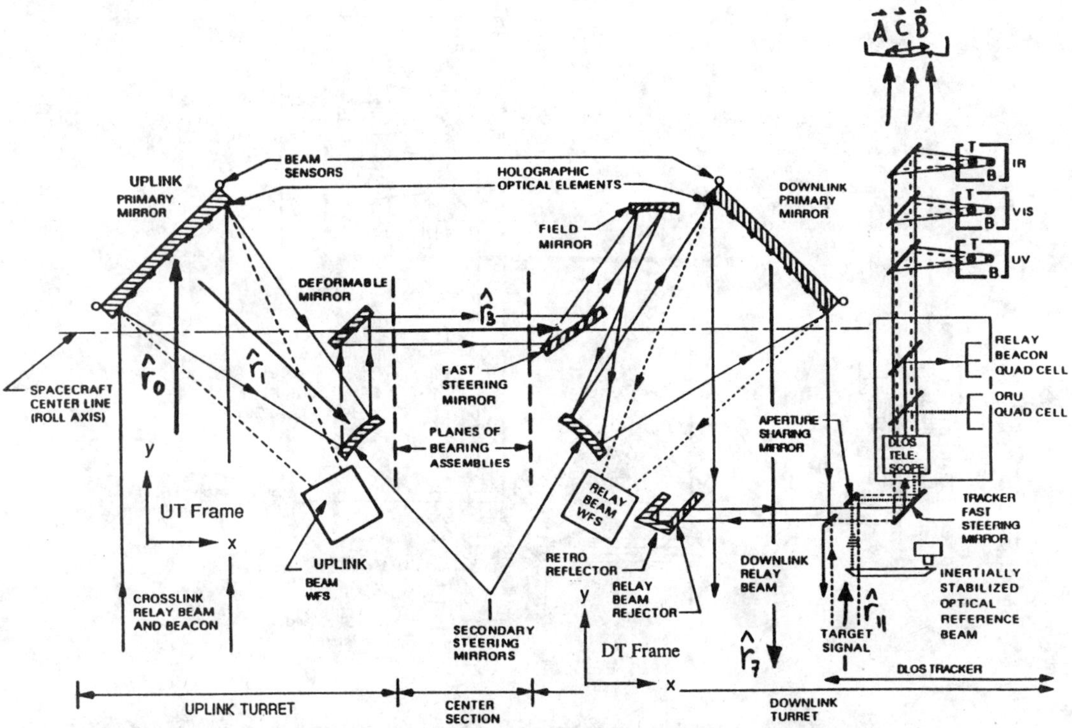

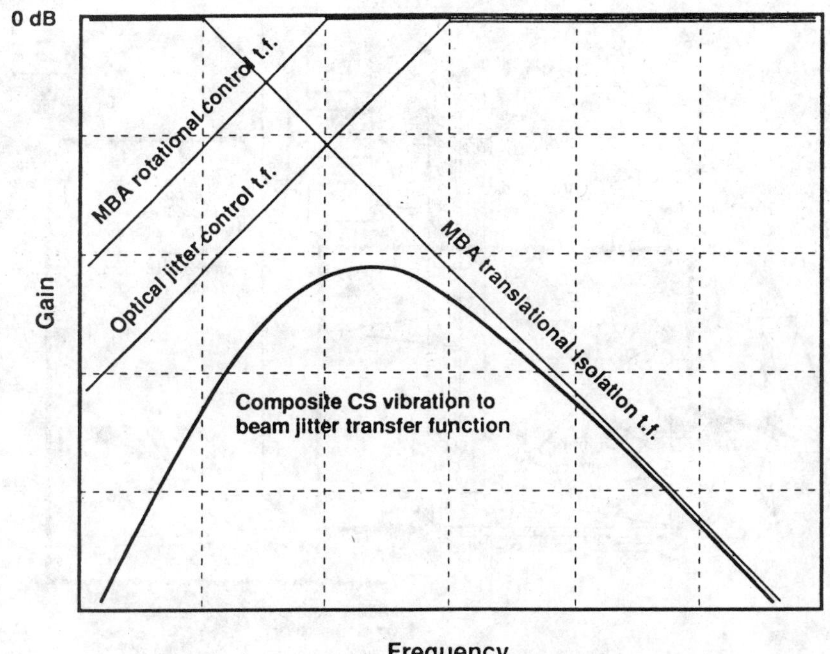

Figure 2. SAB Optical Schematic

Figure 3. Conceptual Center-Section vibration to downlink beam error transfer function

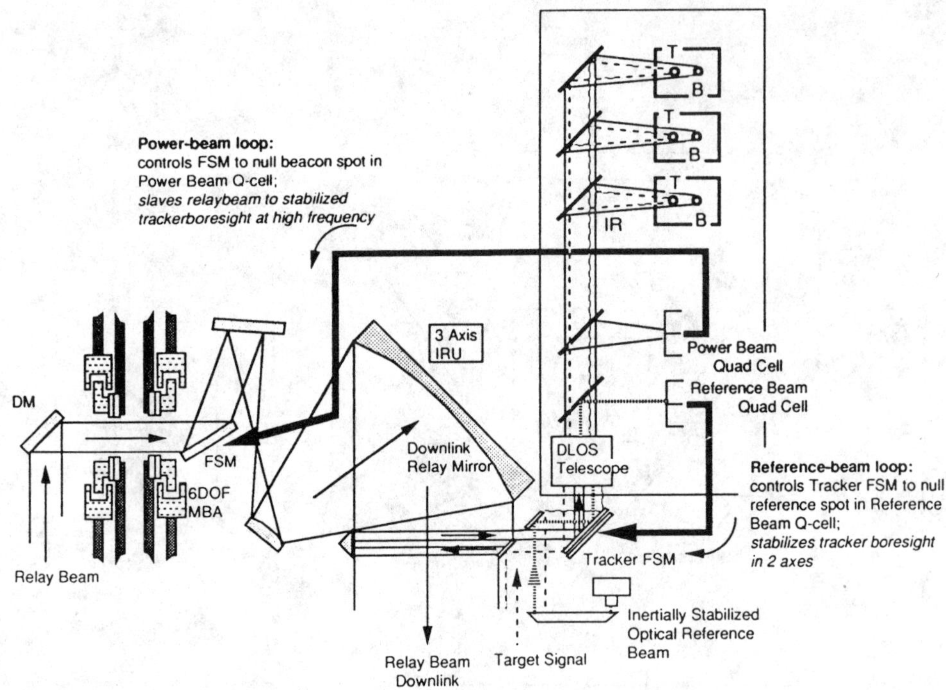

Figure 4. Reference-beam and power-beam control loops.

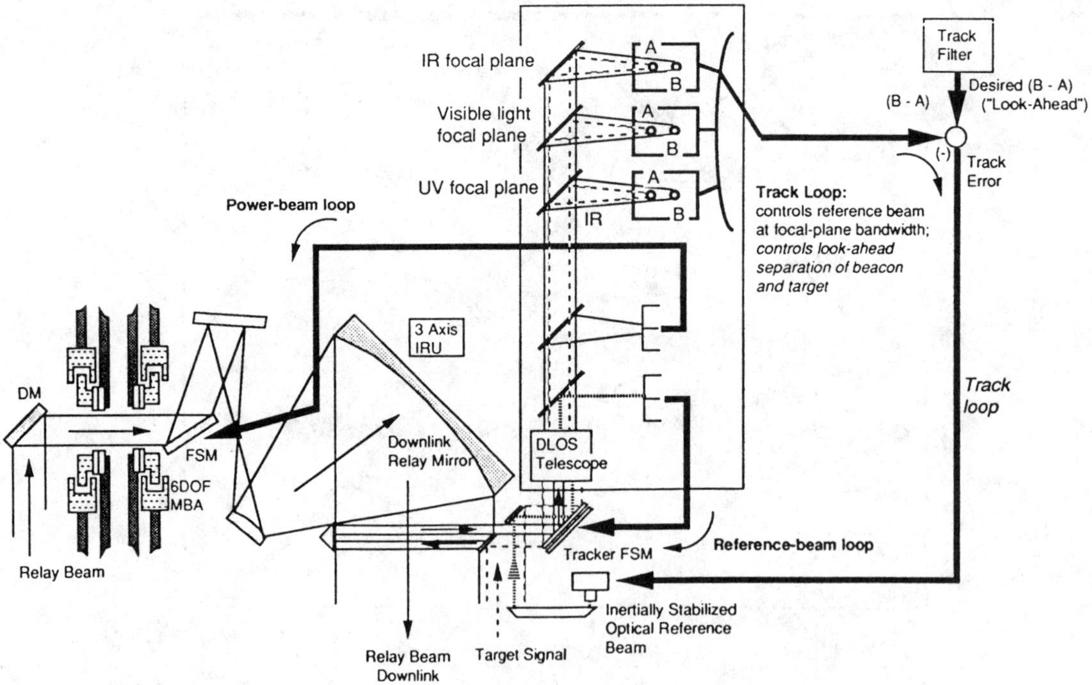

Figure 5. Target track control loop

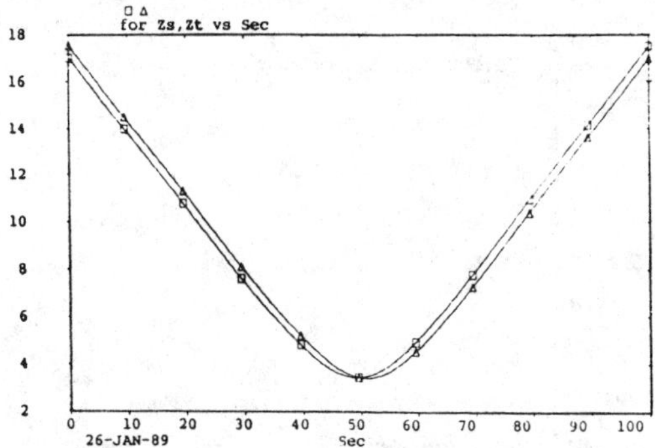

Figure 6. Zenith angles from the source and target ground sites to the spacecraft

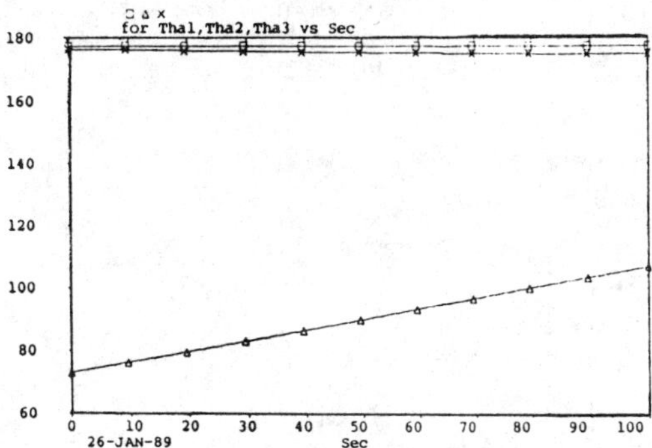

Figure 7. Spacecraft attitude

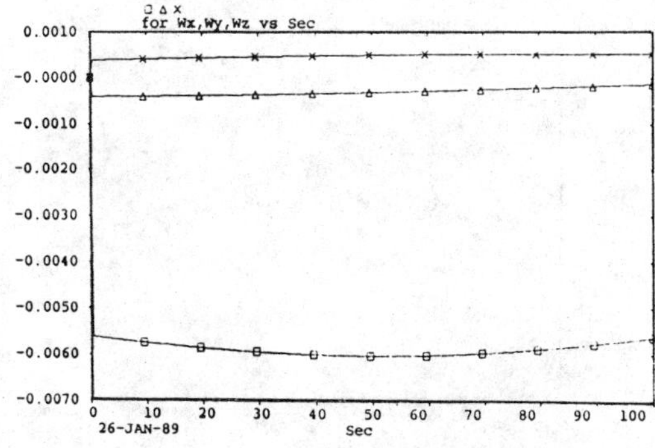

Figure 8. Spacecraft body rates

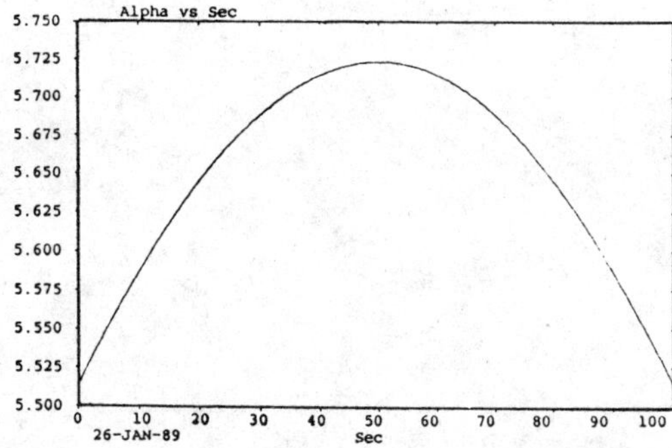

Figure 9. Gimbal angle

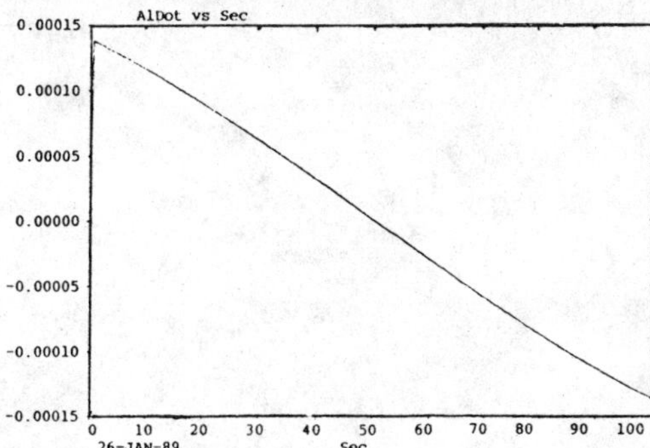

Figure 10. Gimbal rate

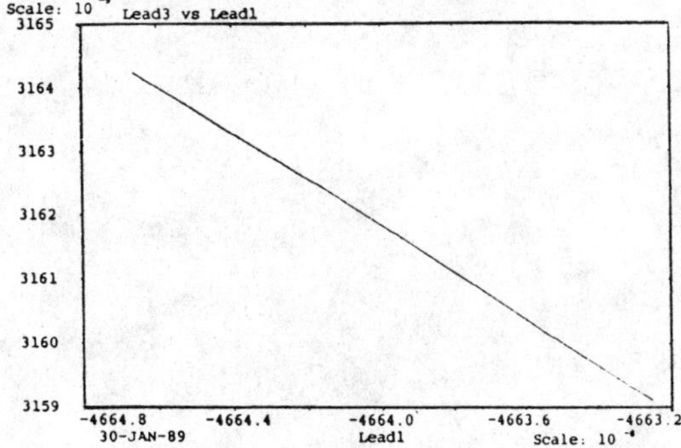

Figure 11. Look-ahead aimpoint on the focal plane

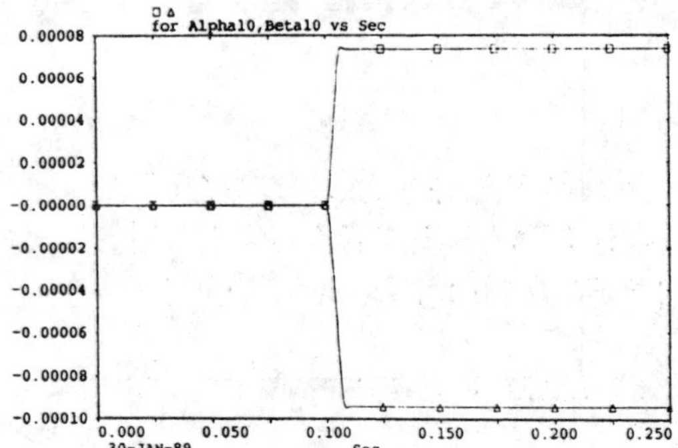

Figure 12. Tracker steering mirror angles

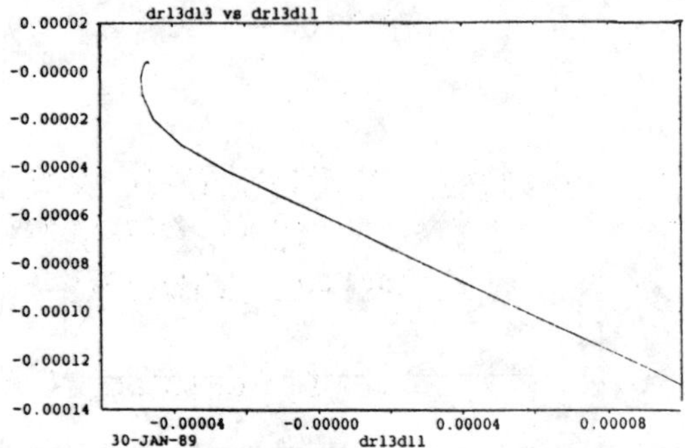

Figure 13. Target spot motion on the focal plane

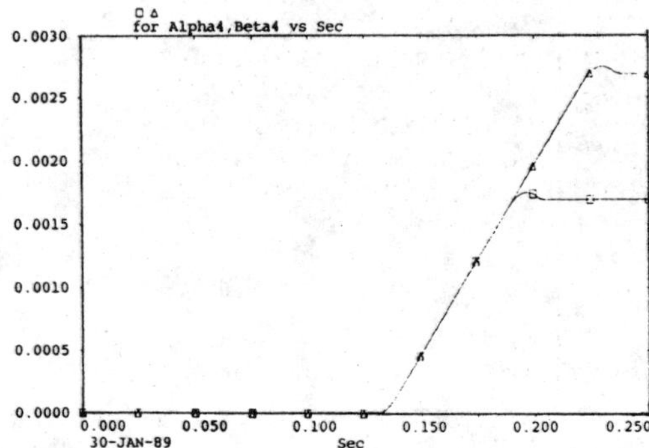

Figure 14. Power-beam steering mirror angles

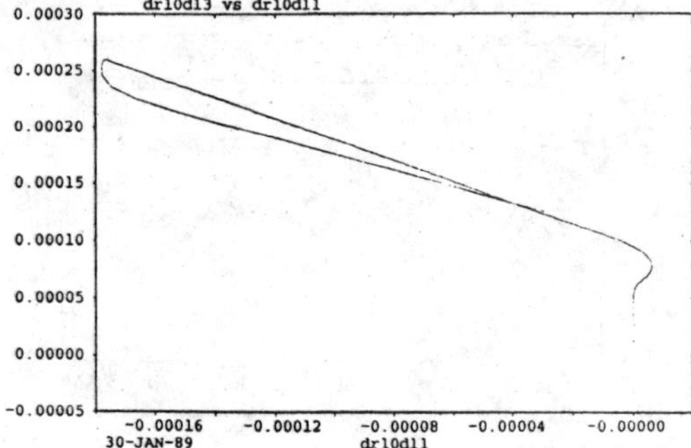

Figure 15. Power-beam spot motion on the focal plane

THE CIRCUMSTELLAR IMAGING TELESCOPE IMAGE MOTION COMPENSATION SYSTEM: ULTRA-PRECISE CONTROL ON THE SPACE STATION PLATFORM

G. E. Sevaston, M. M. Socha and A. Eisenman[*]

The Circumstellar Imaging Telescope (CIT) is a 1.9 m visible wavelength Cassegrain instrument whose primary mission is to locate extra-solar planets by direct imaging. The current reference concept calls for it to be deployed as a space station attached payload some time in the late 1990s. Mission and signal processing demands impose the following attitude control requirements: pointing accuracy - 250 nrad RMS, pointing stability - 50 nrad RMS over 30 minutes, roll accuracy - 1 mrad RMS, roll stability 1 mrad RMS over 30 minutes. This paper reports on the results of a recent design study in which it was determined that these requirements can be met on the space station by a three tiered control system consisting of a two degree of freedom mechanical gimbal for course pointing, a magnetically suspended roll bearing for roll accommodation, translational isolation and intermediate pointing control, and articulation of the secondary mirror for fine pointing.

INTRODUCTION

The Circumstellar Imaging Telescope (CIT) is a 1.9 m visible wavelength Cassegrain instrument whose primary mission is to locate extra-solar planets by direct imaging. Since extra-solar planets, if they exist, are expected to be very dim targets close to relatively bright sources (i.e., stars about which the planets orbit), the telescope faces unprecedented signal to noise challenges. The required detection performance will be achieved by employing coronagraphic principles (for control of diffracted light) in conjunction with a super smooth primary mirror and chopping (for scattered light control).

The signal detection challenges carry with them corresponding challenges on image control and stabilization. The image of the central star must be aligned with the coronagraphic mask to within 50 nrad RMS over the minimum time interval between observation breaks (approximately 30 minutes). Since the subject (i.e., the planet) is off axis, the image must be stabilized in roll as well as translation. The mission set is scoped such that the roll axis must be stable to within 1 mrad RMS over 30 min. Target acquisition dictates that image

* Jet Propulsion Laboratory, California Institute of Technology, 4800 Oak Grove Drive, Pasadena, California 91109.

translation must be accurate to within 250 nrad RMS, and
image rotation must be accurate to within 1 mrad RMS.

The control and stabilization problem is compounded by the
fact that CIT is being envisioned as a space station attached
payload. Thus, the situation is one of attempting to meet
some of the toughest control requirements ever levied within
one of the most troublesome disturbance environments ever
imagined.

This paper is about the baselined solution to the image
control and stabilization problem. The baselined system
employs a three tiered control architecture incorporating the
Space Station supplied Payload Pointing System (PPS, a
mechanical gimbal), a magnetically suspended roll bearing,
and a piezo-electrically actuated and momentum compensated
secondary mirror. Independent translation and roll charge
coupled device (CCD) array detectors serve as the primary
sensors. The paper contains a complete description of the
baselined system, supporting analyses, and simulation
results. Potential modifications to the baseline will also
be discussed.

SYSTEM DESCRIPTION

An artist's concept of CIT operating on the space station is
shown in Figure 1. As the figure suggests, CIT is a payload
of substantial dimensions. Some key telescope parameters are
listed in Table 1.

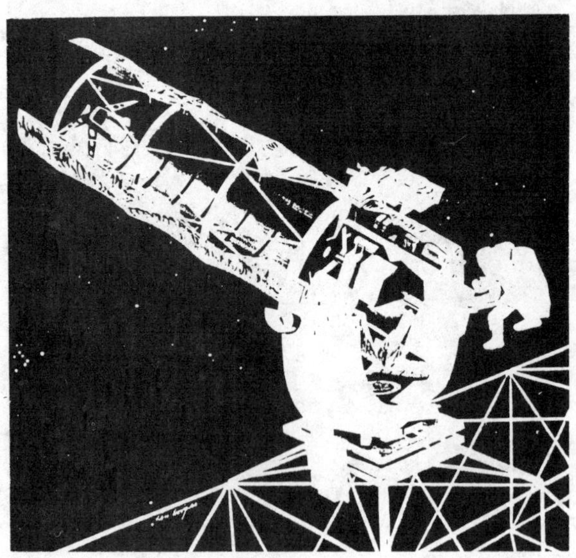

Figure 1. Artist's Concept of CIT Attached to the Space
Station

Table 1. Key CIT Parameters

Length	8.5 m
Outside Diameter	2.0 m
Mass	1,840 kg
Transverse Inertias	4,300 kg-m^2
Roll Inertia	550 kg-m^2
Plate Scale (at first focus)	60 μm/μrad
Primary-Secondary Mirror Distance	5 m
Primary Mirror Diameter	1.9 m
Primary Mirror Average Thickness	0.1 m
Secondary Mirror Diameter	0.4 m
Secondary Mirror Average Thickness	0.1 m

A schematic view of CIT, looking at it down along its boresight, is given in Figure 2. Although the PPS, as it is currently envisioned [1], will be an azimuth/elevation/x-elevation active gimbal, CIT will make use of only its first two degrees of freedom. Roll control will be provided by a magnetically suspended bearing with an integrated motor.

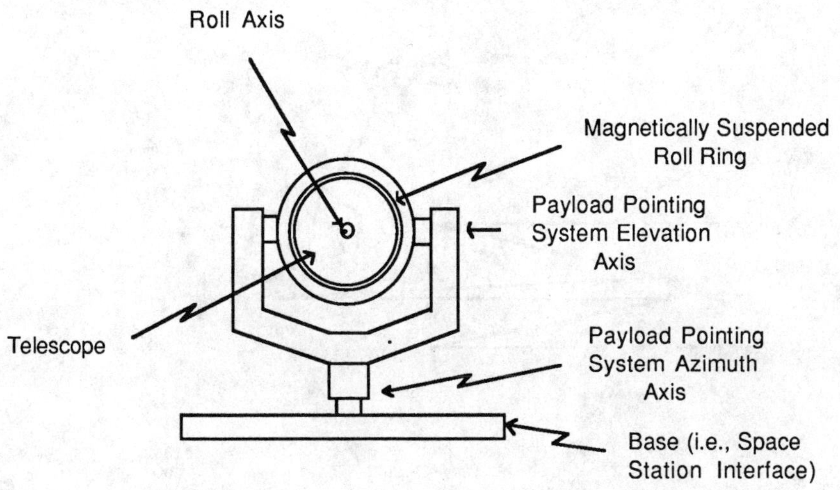

Figure 2. Schematic Diagram of CIT within the PPS

Roll bearing suspension forces and torques will be developed by electromagnets located along the periphery of the device, while roll control will be made possible by an integrated

motor. Proximity sensors will supply the measurements needed
for the suspension loops. Draped cable will transfer
signals and power across the suspension gap.

Notice that a magnetically suspended roll bearing is also
being investigated for use in the Astrometric Telescope
Facility (ATF) [2]. The main difference between that device
and the one envisioned for CIT, apart from load carrying
capacity, is the use of cabling for signal and power
transfer.

The roll bearing suspension actuators will also be used for
medium bandwidth telescope pointing. The PPS will be
controlled so that it simply tracks the telescope at low
bandwidth. This will be accomplished by driving the PPS in a
way that maintains the roll bearing gap near is nominal
configuration. Fine image motion control (equivalent to fine
pointing) will be performed by articulating the secondary
mirror within a high bandwidth loop.

Both pointing and roll stabilization will be image
referenced. As shown in Figure 3, independent CCD's will
detect pointing and roll errors. A third CCD will function
primarily as the science detector, but will also be used
during the target acquisition process.

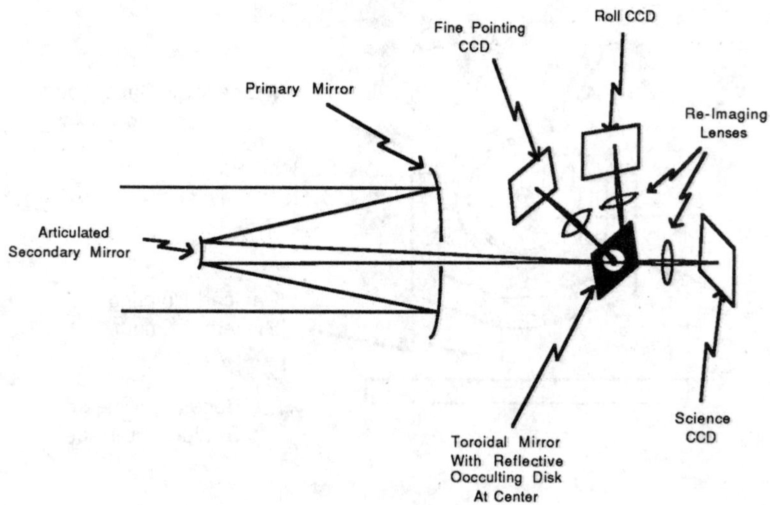

Figure 3. Schematic Diagram of CIT Optical System Layout.

All three CCD's will be located at separate focal planes.
The pointing and roll detector images will be obtained by
sampling the first telescope focal plane using mirrors in the
arrangement depicted in Figure 4. The large outer mirror is
transparent over the science field. The small inner mirror

also serves as the occulting disk for the CIT coronagraph.
The field between the mirrors is re-imaged onto the science
CCD. The structure of the telescope field is shown in Figure
5. The characteristics of re-imaging optics and CCD's are
summarized in Table 2.

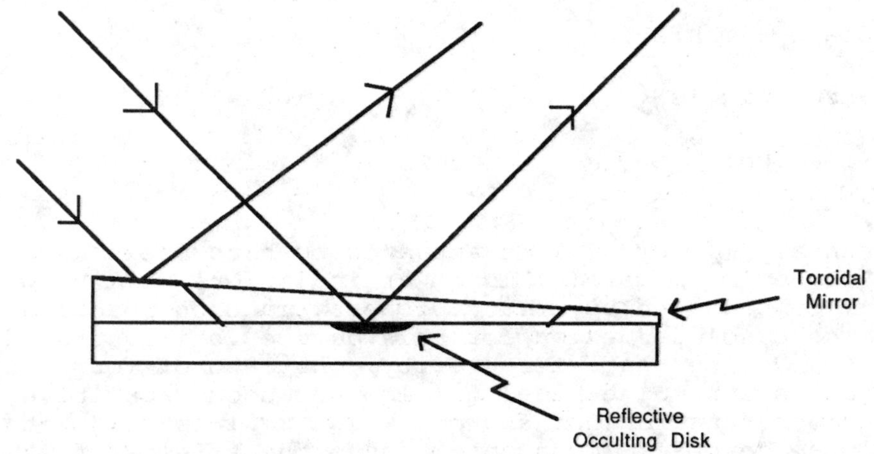

Figure 4. Sampling Mirrors

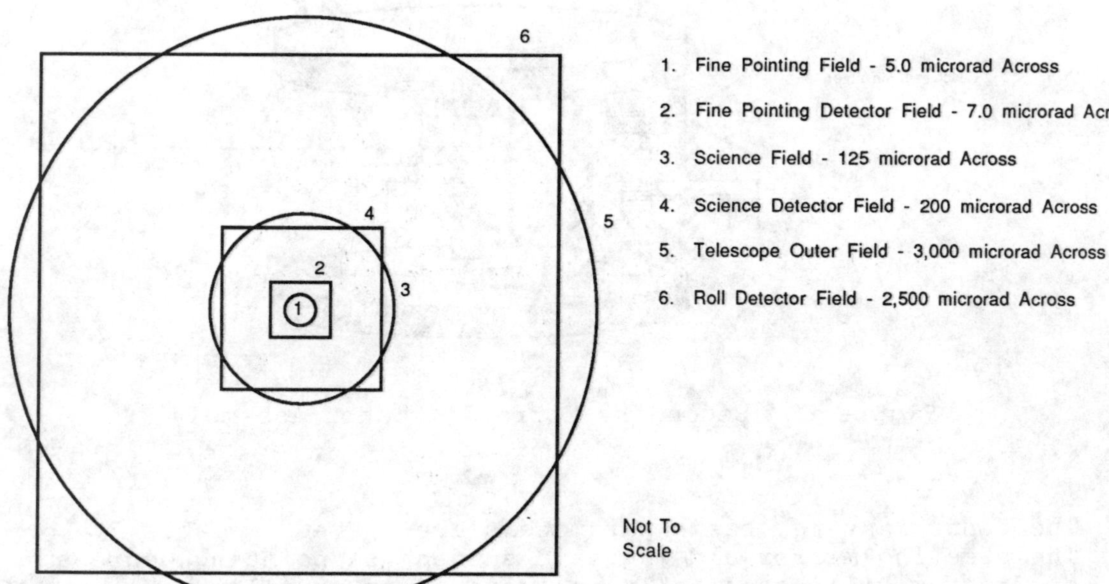

Figure 5. Telescope Field Structure.

Table 2. Re-Imaging Optics and CCD Characteristics.

Characteristic	Fine Pointing Detector	Science Detector	Roll Detector
Optics Magnification	2:1	1.5:1	1:18
Plate Scale (μm/μrad)	120	90	3.3
Array Size (pixels)	32x32	1024x1024	512x512
Pixel Size (μm)	24x24	15x15	15x15

The secondary mirror will be actuated by hard piezo-electric
(i.e. PZT) prime movers. Mirror articulations will be
momentum compensated by a mass property matched reaction
mass. The baseline design for this device is illustrated in
Figure 6. Notice that, for clarity, only two of four prime
movers are shown. The baseline also includes capacitive
displacement sensors (not shown), which will be used within
tight local feedback loops to mitigate the effects of PZT
non-linearities (e.g., hysteresis).

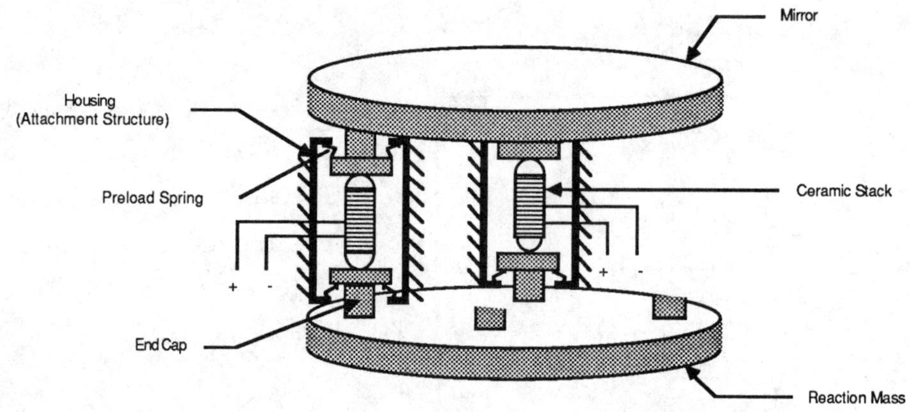

Figure 6. Secondary Mirror Actuator Assembly

The control system basic parameters are listed in Table 3.
The steering mirror and magnetic suspension pointing degrees
of freedom will essentially be operated in parallel;
bandwidth separation plus feedforward (of mirror commands to
the magnetic pointer) de-coupling will preclude de-
stabilizing interactions. As mentioned above, the PPS will
be operated as a suspension gap follower.

Table 3. Control System Basic Parameters

```
PPS Control Bandwidth                        0.5 Hz
Suspension Roll Off                          0.1 Hz
Magnetic Pointing Bandwidth                  2 Hz
Roll Control Bandwidth                       2 Hz
Steering Mirror Control Bandwidth            10 Hz
Fine Pointing CCD Frame Rate                 100 Hz
Roll CCD Frame Rate                          15 Hz
Science CCD Acquisition Frame Rate           4 Hz
```

Acquisition will be accomplished as follows:

1. Using PPS attitude data (accurate to 36 arcsec 3σ [1]), plan a maneuver that brings the target star to rest within the roll detector field of view.

2. Execute the maneuver using the magnetic pointer and the PPS (accurate to 60 arcsec 3σ [1]).

3. Using data from the roll CCD, the magnetic pointer and the PPS, stabilize the image of the target.

5. Using data from the roll CCD, plan a maneuver that brings the target star to rest within the science detector field of view.

6. Execute the maneuver using the magnetic pointer and the PPS.

7. Using data from the science CCD, the magnetic pointer and the PPS, stabilize the image of the target.

8. Using data from the science CCD, plan a maneuver that brings the target star to rest within the fine pointing field.

9. Execute the maneuver using the magnetic pointer and the PPS.

10. Using data from the fine pointing CCD, the steering mirror, the magnetic pointer and the PPS, stabilize the target image.

11. Select a bright reference star within the roll CCD field of view.

12. Using data from the roll CCD and the roll motor, stabilize the image of the roll reference star.

Collection of science data can commence at the completion of
step 12. Notice that the speed of this process is limited by
the capacity of the suspension electromagnets (capable of a
90 degree slew in less than 1 minute).

A functional block diagram of the control system is shown in
Figure 7. "Engineering CCD's" refers to both fine pointing
and roll. "Tracker Electronics" represents the hardware that
services the detectors and calculates targeting errors.
Estimated system mass and power breakdowns are given in
Tables 4 and 5 respectively.

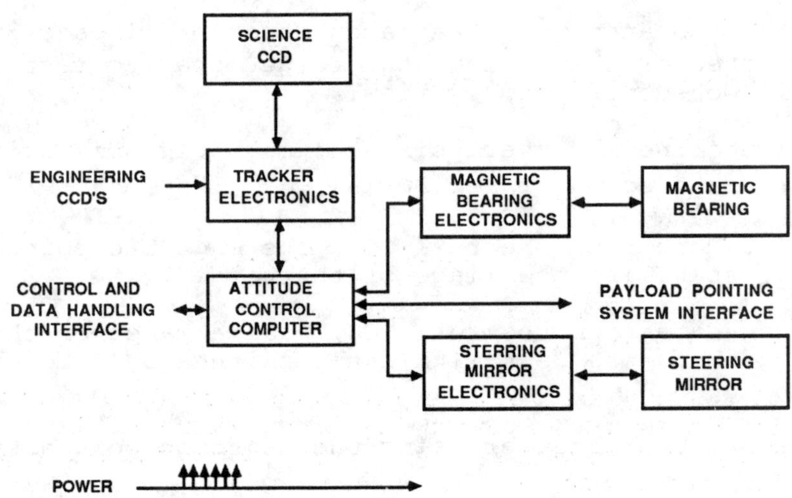

Figure 7.   Control System Functional Block Diagram

Table 4.   Control System Mass Estimate

<u>Assembly</u>	<u>Mass (kg)</u>
Roll Bearing (including Electronics)	250
Secondary Mirror Actuator (including mirror, reaction mass and electronics)	75
Central Computer	10
Payload Pointing System Interface	5
Tracker (including detectors, relay optics and electronics)	15
Total	355

Table 5. Control System Power Estimate

Assembly	Power (W)	
	Peak	Nominal
Roll Bearing (including electronics)	200	75
Secondary Mirror Actuator (including electronics)	0.01	0.002
Central Computer	5	5
Payload Pointing	5	5
Tracker (including electronics)	20	20
Total	230.01	105.002

DESIGN TRADES

The baseline architecture was arrived at as a result of the following trades:

1.   (a) Mechanical roll bearing and secondary mirror actuation vs. (b) magnetically suspended roll bearing and secondary mirror actuation vs. (c) magnetically suspended roll bearing alone option (b) was selected for superior performance (with respect to both stabilization and acquisition), minimum performance risk, minimal control-structure interaction, no high voltage requirement, and no contactless coupling requirement (signals, power and even coolant can be transferred through high compliance couplings).

2.   (a) Separate aperture trackers vs. (b) shared aperture trackers; option (b) was selected for better performance (i.e., more light gathering capacity) and minimal control-structure interaction.

3.   (a) Telescope mounted gyros, low rate trackers and high performance relative angle sensors vs. (b) telescope mounted gyros, low rate trackers and secondary mirror mounted high performance gyros vs. (c) high rate trackers alone; option (c) was chosen for superior performance, minimal performance risk, minimum cost and minimum schedule risk.

Notice that the PPS performance specifications (60 arcsec 3σ accuracy, 30 arcsec peak-to-peak stability over 30 minutes and 15 arcsec peak-to-peak jitter over 1 sec) presuppose the use of a Space Station provided attitude determination system, with its constituent star trackers, sun sensors, earth sensors and inertial reference unit [1]. Thus, it was immediately clear that standard inertial stabilization techniques would not suffice.

PZT's were selected as prime movers for the secondary mirror actuator for their bandwidth, stiffness, dimensional stability and flight worthiness. With this choice, the following benefits are realized:

1. The telescope can be roughly aligned on the ground, with the expectation that the alignment will be preserved through launch and deployment.

2. The mechanism can survive launch loads, without an external latching device.

3. The actuator will hold a position without power for substantial periods of time.

4. The resonant frequency of the loaded actuator is on the order of kHz.

Notice that the necessary range of motion (about 1 μrad) is such that a high voltage power supply is not required; the baseline power supply is required to develop only 60 V.

ANALYSIS

Fine Pointing Detector Sensitivity.

The CIT science goals are such that the ability to track magnitude 12 objects is highly desirable. The requisite frame rate can be estimated as follows. Assuming shot noise limited performance, the detector accuracy is approximated by

$$\sigma_{\epsilon\mu} \cong \sqrt{\frac{\sigma_\phi^2 + \mu_\phi^2}{S}}$$

where S is the total number of received photons, σ_ϕ is the

point spread function dispersion and μ_ϕ is the point spread

function true centroid position offset. For a centered, 1 μrad Airy disk, better than 25 nrad accuracy requires a signal of more than 1,600 photoelectrons. This can be translated into exposure time by using the approximation

$$S \cong \frac{E\lambda F_0 AT}{hc}(2.51)^{-m}$$

where E is the system efficiency, λ is the peak emittance wavelength, F_0 is the reference irradiance, A is the collection area, T is the exposure time, h is Planck's constant, c is the speed of light, and m is star magnitude. Assuming a peak emmitance wavelength of 0.6328 μm, a reference irradiance of 27.326 nW/m^2, a system efficiency of 10%, a signal of more than 1,600 photoelectrons from a magnitude 12 star requires an exposure time of at least 6.6 msec. A time of 10 msec was chosen as the baseline.

Fine Pointing Bandwidth Analysis

A simplified block diagram of the pointing control system is shown in Figure 8. The basic control system error signal is given by

$$\varepsilon\theta = \theta_T - 2(\theta_M - \theta_T)$$

where θ_T is the telescope inertial attitude angle, and ($\theta_M - \theta_T$) is the secondary mirror to telescope relative angle. The feedback input to the first summer is an estimate of θ_T constructed from $\varepsilon\theta$ and a measurement of ($\theta_M - \theta_T$). τ_d, $\delta\theta$ and ($\theta_S - \theta_T$) represent telescope disturbances (e.g., those transmitted across the cable), mirror actuator positioning errors (e.g., quantization) and mirror mount deformations respectively.

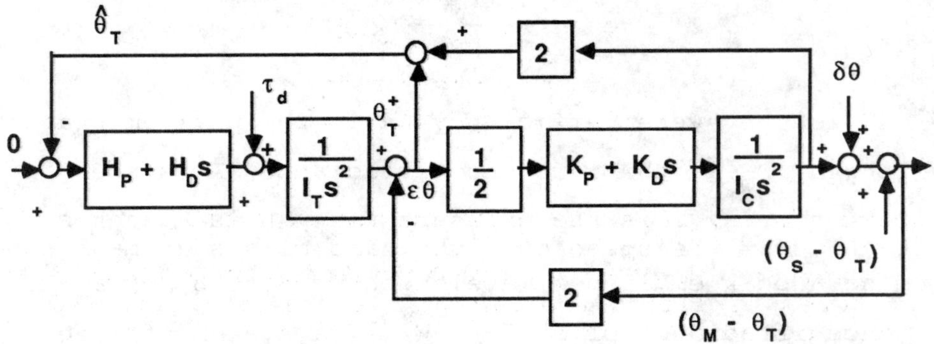

Figure 8. Bandwidth Analysis Model Block Diagram

For simplicity, only proportional plus derivative (PD) feedback is assumed for the suspension pointing controller. The input to that controller is estimated telescope attitude error. The mirror actuator controller, which is integral plus integral squared (II^2), is represented by by the PD block in cascade with a dummy rigid body.

In terms of the disturbances, the error is given by

$$\varepsilon\theta = \frac{\dfrac{1}{I_T}s^2\,\tau_d - 2\,s^4\,\delta\theta - 2\,s^4\,(\theta_S - \theta_T)}{\left(s^2 + \dfrac{H_D}{I_T}s + \dfrac{H_P}{I_T}\right)\left(s^2 + \dfrac{K_D}{I_C}s + \dfrac{K_P}{I_C}\right)}$$

With a 10 msec sampling rate, the control system bandwidth is effectively limited to about 10 Hz. If there were no isolation between the telescope and the base, base disturbances would translate into error according to

$$\frac{\varepsilon\theta}{\theta_T} = \frac{s^2}{s^2 + \dfrac{K_D}{I_C}s + \dfrac{K_P}{I_C}}$$

Assuming a 12.5 μrad rotational disturbance (PPS specification - [3]) occurring at 2 Hz (approximate treadmill fundamental frequency - [4]), a 10 Hz bandwidth limits performance to approximately 500 nrad. The magnetic bearing isolation modes, described by

$$\frac{\theta_T}{\theta_B} = \frac{\dfrac{J_D}{I_T}\left(s + \dfrac{J_P}{J_D}\right)}{s^2 + \dfrac{J_D}{I_T}s + \dfrac{J_P}{I_T}}$$

provide the required disturbance attenuation.

To achieve better that 20 dB attenuation at 2 Hz, the
isolator must roll off below 0.2 Hz. A roll off of 0.1 Hz is
baselined.

The magnetic bearing pointing control gain must be high
compared to the cable torsional stiffness. Assuming a cable
torsional bandwidth of 0.02 Hz (reasonable for a long cable
loaded by CIT), the magnetic bearing pointing bandwidth must
exceed 0.2 Hz. A pointing bandwidth of 2.0 Hz was baselined.

<u>Fine Pointing Performance Analysis</u>

Fine pointing performance was verified by simulation. The
simulation model topology is shown in Figure 9.

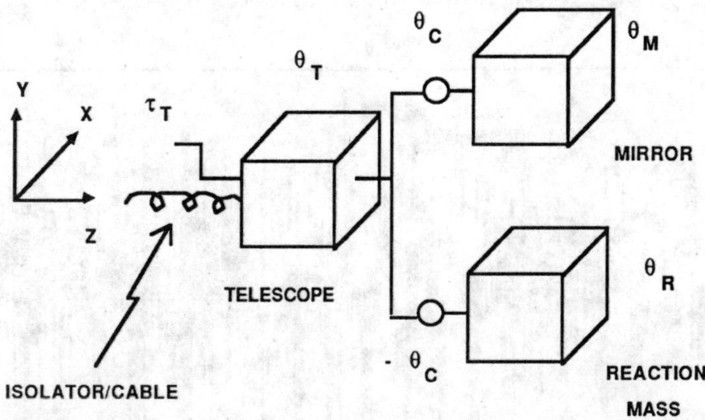

Figure 9. Simulation Model Topology

Motions were restricted to translations along the x and y
axes and rotations about the z axis. Base disturbances were
introduced as prescribed motions at the free end of the
spring, which represents both the cable and the isolation
modes of the magnetic bearing. Magnetic bearing pointing
control actuators were represented by a torque source. The
PPS was not modelled.

303

The simulation results presented below reflect the following conditions:

1. The base disturbances are characterized by the discrete spectrum described in [3].

2. The telescope center of mass offset from the pointing axes is 1.0 cm.

3. The rotational and translational isolation rolls off at 0.1 Hz and is 50 % damped.

4. Mirror motion is 95 % momentum compensated.

5. The tracker error samples are uncorrelated and have a standard deviation of 25 nrad.

6. The mirror actuator positioning error is negligible.

7. The telescope pointing control loop has a bandwidth of 2.0 Hz, 50 % damping and a 0.2 Hz integral pole.

8. The mirror actuator control loop has a bandwidth of 10 Hz and 50 % damping.

9. The sampling rate is 100 Hz.

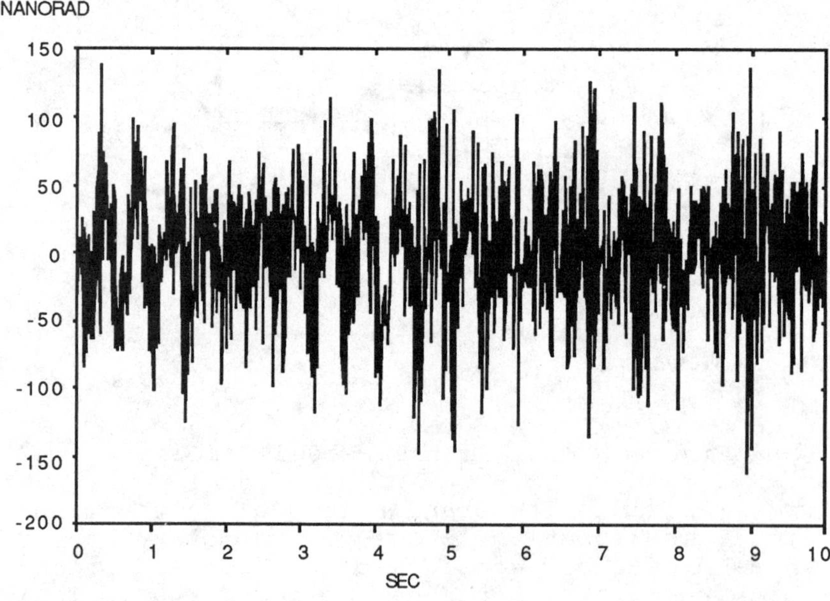

Figure 10. Translational Image (i.e., Fine Pointing) Error

Figure 10 shows that the required translational image stability (i.e., fine pointing) of 50 nrad RMS is indeed

304

achieved. Notice that although the graph depicts only a 10
sec record of performance, the virtual absence of any slow
drift makes the result representative.

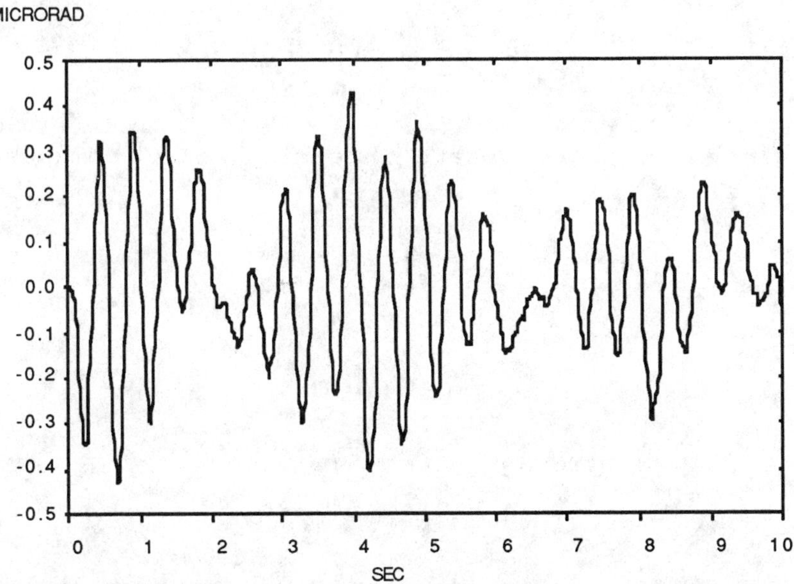

Figure 11.   True Telescope Pointing Angle

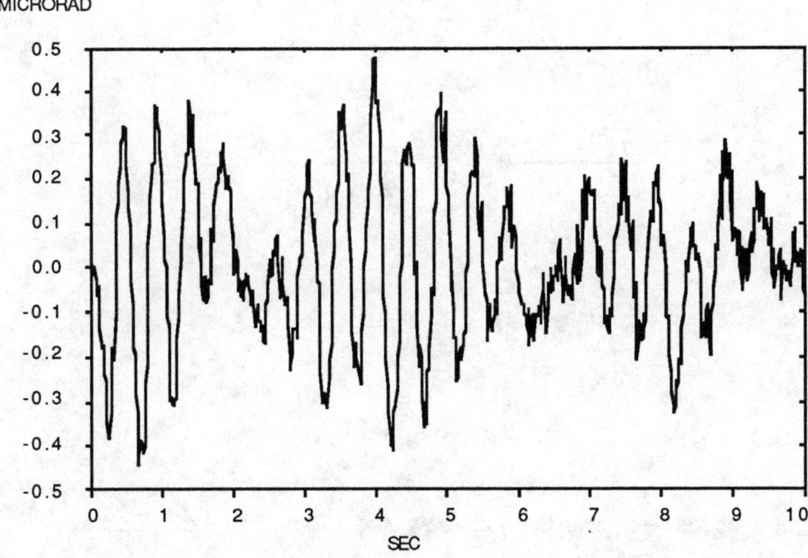

Figure 12. Filtered Estimate of Telescope Pointing Angle

The simulation included both sampled data and quantization
effects. The sampled data error signal was computed
according to

$$\varepsilon\theta_k = \frac{1}{\Delta t}\int_{t_{k-1}}^{t_k} \varepsilon\theta(t)\,dt$$

where Δt is the tracker update interval. The commanded
mirror angle was

$$\theta_{ck} = H_p\eta_k + H_D\gamma_k$$

where

$$\gamma_k = \gamma_{k-1} + \Delta t\,\varepsilon\theta_k$$

is the first integral and

$$\eta_k = \eta_{k-1} + \Delta t\gamma_k$$

is the second. The telescope pointing torque was computed
according to

$$\tau_{Tk} = -K_p\hat{\theta}'_{Tk} - \frac{K_D}{\Delta t}(\hat{\theta}'_{Tk} - \hat{\theta}'_{Tk-1}) - K_I\hat{\gamma}'_k$$

where

$$\hat{\theta}'_{Tk} = \frac{\frac{2}{\Delta t} - \omega_{LP}}{\frac{2}{\Delta t} + \omega_{LP}}\hat{\theta}'_{Tk-1} + \frac{\omega_{LP}}{\frac{2}{\Delta t} + \omega_{LP}}(\hat{\theta}_{Tk} + \hat{\theta}_{Tk-1})$$

$$\hat{\gamma}'_k = \hat{\gamma}'_{k-1} + \Delta t\,\hat{\theta}'_{Tk}$$

and

$$\hat{\theta}_{Tk} = \varepsilon\theta_k + 2\theta_{ck}$$

Notice that the estimate of θ_T used in the computation of τ_{Tk}

was low pass filtered. Notice also that, unlike in the
simplified system of Figure 8, the suspension controller
modelled here includes integral feedback.

The pointing angle of the telescope itself is shown in Figure
11, and the filtered estimate of that angle appears in Figure
12. Notice that the estimate is quite accurate.

The transient response characteristics of the system are
illustrated in Figures 13 through 15. The curves show the
results of a 10 μrad initial telescope pointing error, with
all disturbances and measurement noise disabled. The
responsiveness of the system is clearly demonstrated.

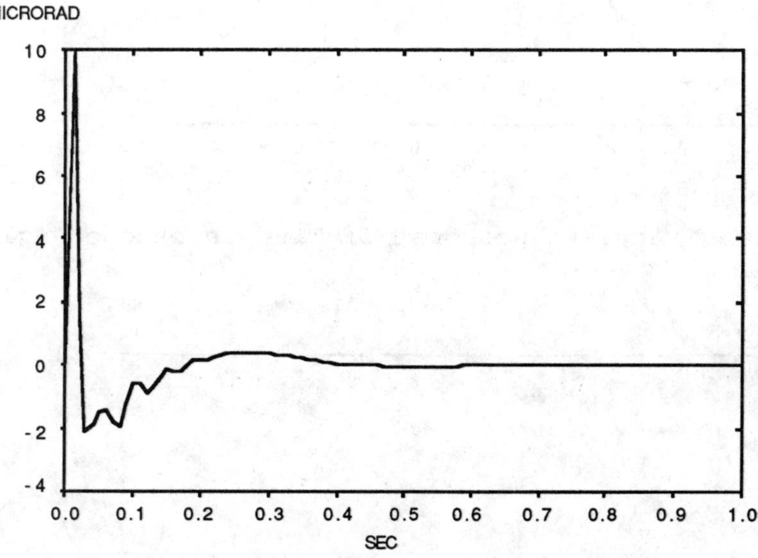

Figure 13. Transient Response of Translational Image Error

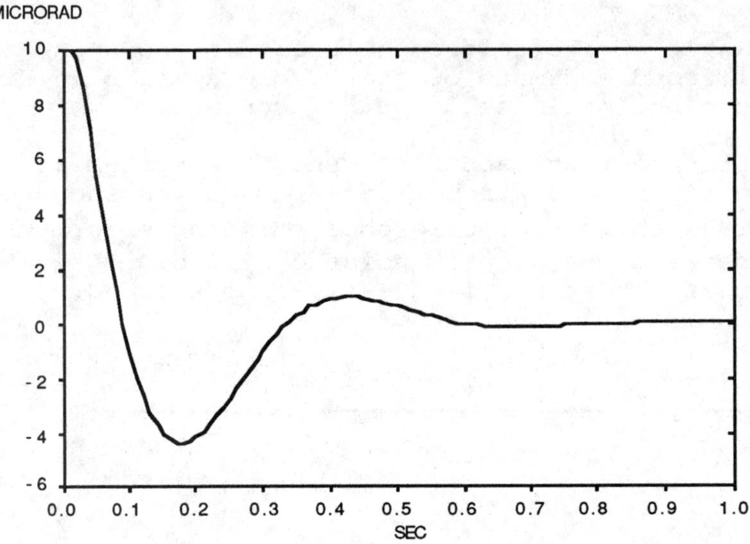

Figure 14.   Transient Response Of True Telescope Angle

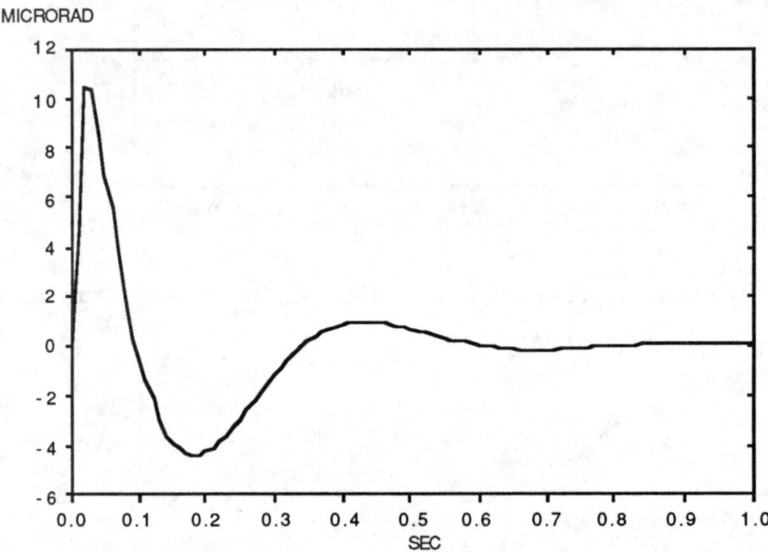

Figure 15.   Transient Response of Filtered Telescope Angle
Estimate

CLOSING REMARKS

It has been shown that milli-arcsec level image stabilization
can be achieved in a Space Station attached payload.   The

control architecture that allows this is a three level, image referenced system, which includes a piezo-electrically actuated secondary mirror, a magnetically suspended roll bearing and a Space Station PPS.

The system design described in this paper represents a baseline concept as of the time of publication. Refinement of the design is currently under way. Some of the alternatives being considered are adoption of an offset acquisition scheme (to eliminate participation of the science detector) and addition of an external star tracker (to eliminate the roll detector).

Related studies at the Jet Propulsion Laboratory (see, e.g., [4-5]) are considering the use of a passive base isolator and mechanical roll bearing for Space Station attached payloads with precision pointing and roll requirements. Results are expected in the near future, and will be reviewed from the standpoint of CIT as they become available.

ACKNOWLEDGEMENTS

This research was performed at the Jet Propulsion Laboratory (JPL), California Institute of Technology, under support from the National Aeronautics and Space Administration. Christ Ftoclas, of Perkin-Elmer, Inc. developed the telescope parameters and the pointing requirements. He also contributed many valuable insights as did Richard J. Terrile and Terrence H. Reilly, both of JPL.

REFERENCES

1.   "DRAFT - Space Station Work Package 3 Attached Payload Accommodation Equipment User Handbook," DR - UID02, Goddard Space Flight Center, Valley Forge, PA, August 1988.

2.   "Astrometric Telescope Facility Isolation and Pointing Study," W. Hibble, T. Allen, L. Jackson, J. Medbery and R. Self, NASA Contractor Report No. 177473, Ames Research Center, Moffet Field, CA, January 1988.

3.   "Development Specification for the Payload Pointing System," Spec. No. SVS-11376, General Electric Co., Astro Space Division, January 1987.

4.   "Space Science/Space Station Attached Payload Pointing Accommodation Study: Pointing Performance Analysis White Paper," R. Laskin, J. Spanos, C. Satter, L. Needels and A. Hines, JPL Internal Document D-5249, Jet Propulsion Laboratory, Pasadena, CA, January 1988.

5. "Space Station Attached Payload Pointing System Design
and Performance Analysis," K. Smith, J. T. Spanos, J.
Lilienthal, D. Rathbun, R. Miyake, E. Wong, R. A. Laskin and
R. Norton, Jet Propulsion Laboratory, in preparation.

ZENITH STAR: A CONTROLS CHALLENGE

Louis A. Morine[*]

This paper discusses the Zenith Star control system issues and how they are resolved to affect the agile performance requirements. A brief description of the hardware and mission will be presented. The critical agile control requirements and requirements rational are discussed along with the resulting control architecture necessary to achieve the described performance. Simplified models will be used to isolate and illustrate how the control system functions are accomplished (such as large precision structural control, two body control, alignment control, and acquisition, tracking control etc.). The control system hierarchical architecture will be integrated and the orchestration of the resultant architecture will be developed. While not quantitative the presentation will attempt to provide and insight to the performance potential of this control system.

---

[*]   Martin Marietta Aerospace Group, P.O. Box 179, Denver, Colorado 80201.

INTRODUCTION

The Zenith Star experiment (Figure 1) is designed to demonstrate and evaluate the performance of a laser in space to answer critical issues relevant to SDI. This experiment is fully compliant with the restrictive interpretation of the 1972 Anti Ballistic Missile (ABM) Treaty. As such it does not directly perform all of the functions of a defensive system nor to the level required by a operational system. Its results however, do provide a measure of the potential of the operational systems by applying the appropriate scaling from the benchmarks achieved by it in space.

The experiment (Figure 2) consists of a series of high power evaluations of beam control and a series of low power evaluations of the tracking and pointing functions of the system.

The high power experiments evaluate the beam control by direct measurement of the far field beam performance with a high power target board. Both space propagation and upper atmospheric effects ar measured.

The low power experiments evaluate the tracking and pointing function performance while tracking a booster throughout its boost phase flight. The Agile Control Performance is evaluatd by performing structured characterization and large and small angle repointing of the system against a star field, small test objects (carried on board) and multiple boosters to exercise the system under multiple conditions.

ZENITH STAR SPACECRAFT

The basic hardware for the Zenith Star Star experiment is shown in Figure 3. It consists of a chemical laser of the class of the alpha program, a beam expander that utilizes the segmented LAMP mirror for the primary optical element, an actuator for pointing the beam expander and isolator for attenuation the laser noise from the beam expander. The latter two are combined into one subsystem called the actuator/isolator. The laser energy is directed through the the aft body to the beam expander by a series of transfer optics and steering mirrors (beam control transfer assembly) on the aft body. A capture track system (consisting of a suite of sensors) is utilized to point the beam expander and optical train for tracking a series of test objects. The remainder of the equipment is a set of standard spacecraft subsystems that allow it to be in orbit as a free flyer that is commanded by ground operations personnel.

The system is deliver into orbit by two Titan IV launch vehicles. The forward spacecraft is launched first and check out completely. Then the aft spacecraft consisting of the Alpha laser and spacecraft support subsystems is launched into the same orbit as the first, orbit phased, and remotely operated from the ground for rendezvous and docking.

CONTROL SYSTEM ARCHITECTURE

The control architecture for the space based laser is derived from a series of stringent tracking and pointing requirements depicted in Figure 4 and the resulting interactive implications lead to a complex hierarchical control architecture. Tight accuracy and jitter requirements combined with the need for rapid repointing of the line-of-sight from one object to another necessitates isolation and suppression of disturbances to the large beam expander. The Zenith Star control system is designed to duplicate this architecture so that the experiment results can be directly related to the SBL performance.

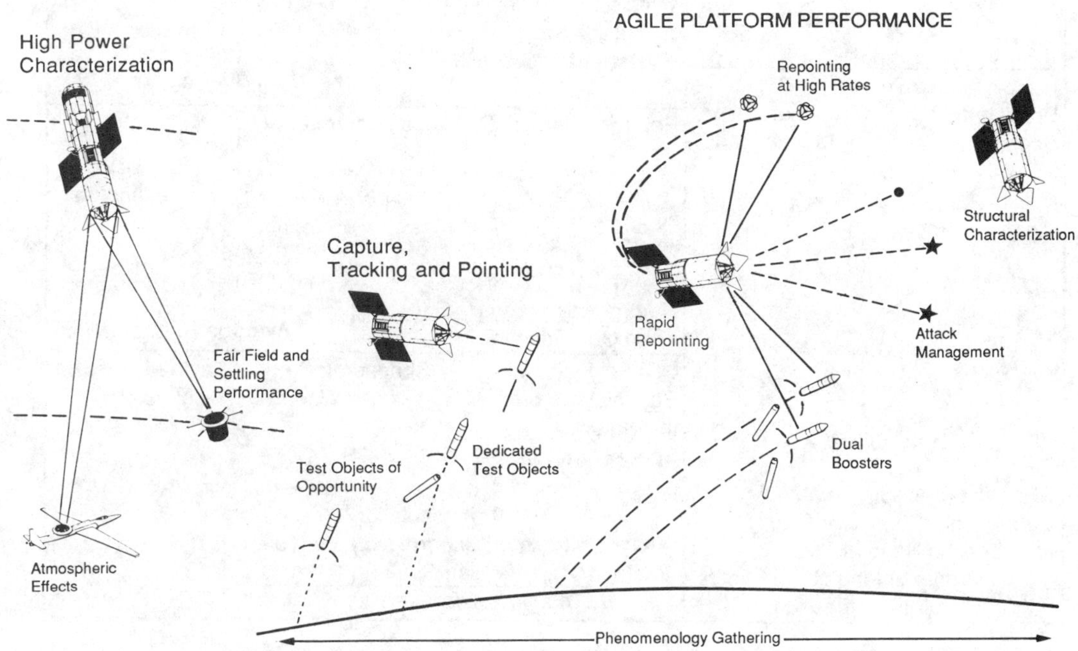

Figure 1 Zenith Star Experiment

Figure 2 Zenith Star Objectives

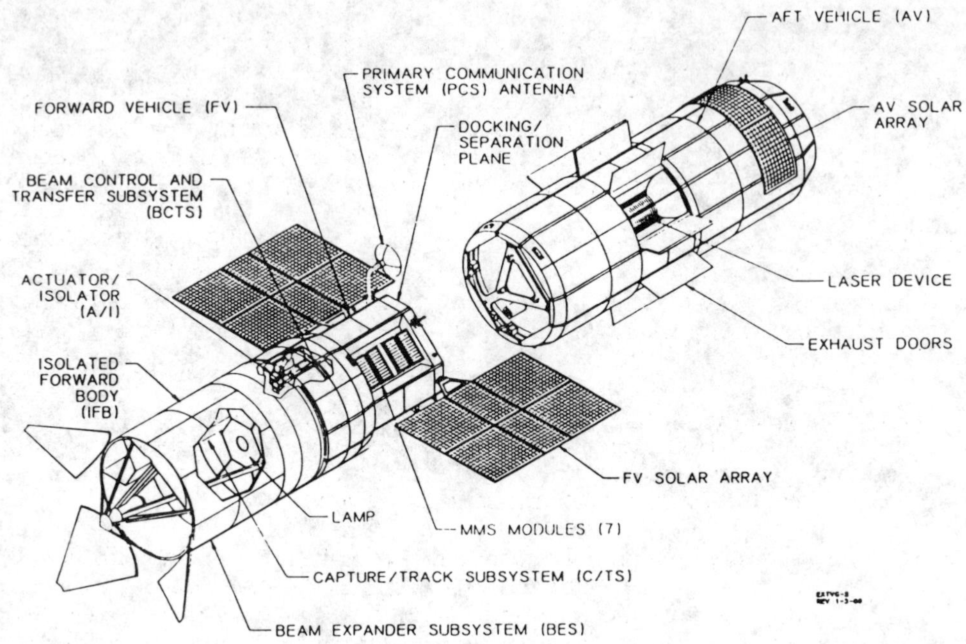

Figure 3 Zenith Star Space Vehicles

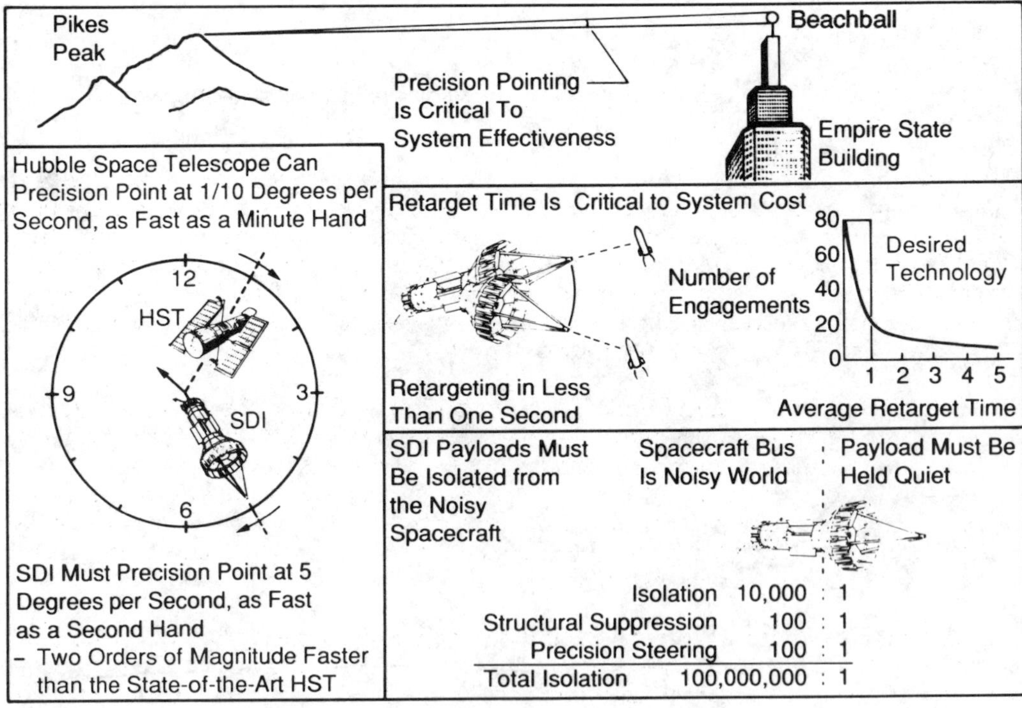

Figure 4 Space Based Laser Control Requirements

The precision and jitter are analogous to hitting a beachball on the Empire State Building in New York from Pike's Peak in Colorado. It must accomplish this while tracking objects at angular rates more than an order of magnitude higher than the capability of the Hubble Space Telescope. To accomplish this, the Line-of-sight must be isolated from disturbances by as much as 100 million to one and yet be able to repoint from one object to another in less than one second so that the system effectiveness can be high.

While the structure is made as stiff as possible, there is sufficient deformation (Figure 5) of the beam expander structure and optical geometry resulting in line-of-sight disturbances to require isolation of aft body noise from the beam expander. The are other self induced beam expander disturbances such as fluid flow and rapid repointing that require structural disturbance suppression on the beam expander itself.

Either function can be readily handled without two body interaction but when combined an actuator/isolator is required between the bodies. This actuator/isolator must provide six degrees of freedom operation which introduces other control issues, such as translation and beam walk control, that further complicate the controls problem.

In order to ease the burden of pointing the system line-of-sight, a precision pointing set of controllers is introduced to provide beam expander off axis pointing and stabilization so that the structure control can be relaxed within a small field of view and as shown by the shaded area in Figure 6. So long as the line-of-sight disturbance is within the range of the precision pointing controller authority the beam expander controller requirements are eased. In other words, the settling time is satisfied when the beam expander line-of-sight is within this band.

VEHICLE CONTROL

The formulation of the control architecture for the beam expander can be described as follows in the next series of figures in Figure 7.

An easy method of isolation (Figure 7a) of the aft body disturbances from the beam expander is to provide a gap between the forward and aft bodies and control the beam expander to point to the desired object from on-board sensor data by external torques (such as control moment gyros) and control the aft body to follow this motion by external forces and torques to maintain the desired gap within some tolerance. This is ideal isolation since there is no actuation between the bodies to force alignment of the two bodies, hence there is no transfer of disturbances from one body to the other.

Since each body tends to rotate about its own center of mass there will be large translational displacements (Figure 7b) at the optical interface between the beam expander and aft body. Also since the beam expander disturbances are to be minimized the aft body must be translated as well as rotated by external forces and torques to maintain the proper separation. This is not practical for a highly agile control system because of the large heavy aft body and the fact that the gap must be small on the order of centimeters. Consequently an actuator between these bodies is required.

This actuator introduces a disturbance coupling path from the aft body to the beam expander which then requires an isolator between the bodies (Figure 7c). This actuation and isolation must be combined into one subsystem because of this interaction. This

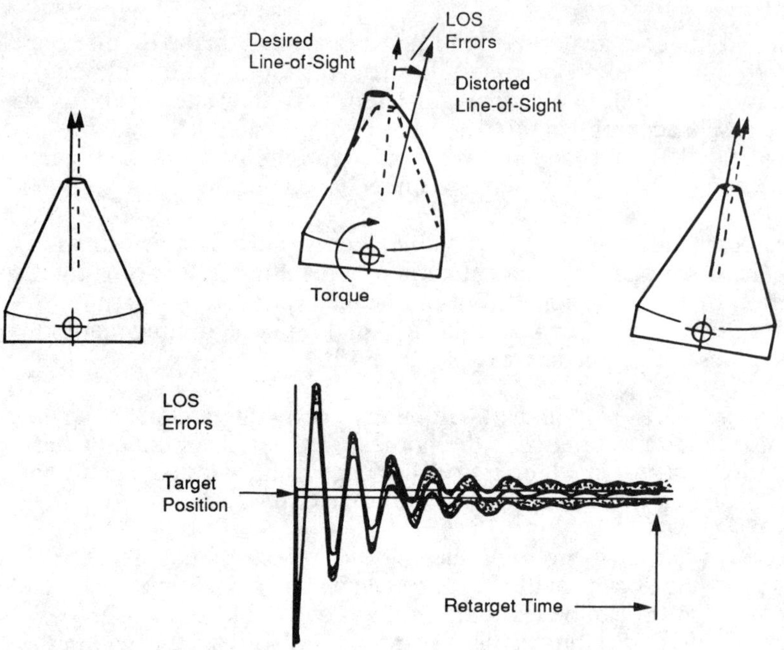

Figure 5 Beam Expander/Optical Distortions

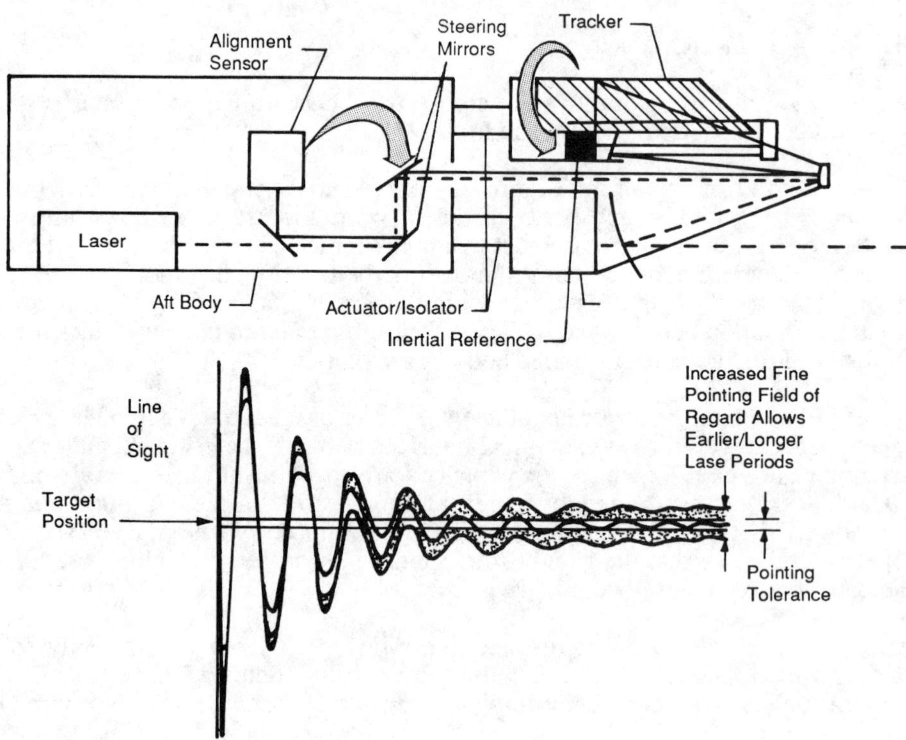

Figure 6 Precision Pointing Control System

subsystem is called the actuator/isolator subsystem and it must minimize this coupling while producing the desired pointing control forces and torques. This function is non trivial even for the baseline magnetic isolator because of non linear magnetic forces and Eddie currents which must be cancelled.

Self induced disturbances on the beam expander arising from fluid flow and rapid pointing must be dissipated through damping in the structure or transferred to non critical structural motion (non critical modes) or transferred off of the beam expander to the aft body (Figure 7d). The incorporation of the actuator/isolator allows this energy to be transferred to the aft body which can then remain isolated. Hence the beam expander line of sight can be stabilized while still tracking objects.

The pointing of the beam expander causes severe disturbances directlyinto the beam expander. In order to move the line-of-sight from one object to another (rapid repointing) it is desirable to make maximum use of the available torque from the actuator/isolator. Infact, the optimal repointing for a rigid body is a bang-bang command. This however, causes severe disturbances to the line-of-sight.

The severity is dependent on the relationship of the angle to be repointed (frequency of the bang-bang torques) and the structural frequencies. Figure 8 shows the effects of a single structural frequency of 4 hz and 8 hz separately as a function of repointing angle. The time to handover is the time that the line-of-sight error takes to settle to within the field of reguard of the beam expander where the fine off-axis steering takes over. The rigid body response is included since it represense the lower bound of maneuver time for the system.

When all structural modes are considered the picture is not quite so easily displayed because of the relative effects on the line-of-sight are intermixed. An envelope of these effects are indicated in Figure 9 where the lower bound is limited by the rigid body response and the upper bound depends on advanced structural controllablity.

The region of interest for structural control is the torque limited and rate limited regions. The algorithm limited region is the responsiveness of the precision off-axis control system for scene interpretation and control.

Figure 10 shows the improvement in repointing time that can be made by a simple modulation the technique based on the relationship of repointing angle and knowledge of the structural frequencies of the beam expander. By properly commanding or modulating the torque commands, disturbances can be minimized as shown in the figure for one technique called modulated bang-bang control.

Experiments such as structural identification and modal surveys are also planned for in the experiment objectives.Utilization of other techniques for controlling the structure such as distributed actuator structural control are not currently available on Zenith Star Star but maybe in the future depending in the interest within the community and the risk to the other Zenith Star Star objectives.

PRECISION POINTING CONTROL

The vehicle control techniques described above dominate the torque limited and rate limited regions identified in Figure 9. Since the vehicle control is only intended to bring the line-of-sight within the off axis pointing capability, the precision off axis controller must take

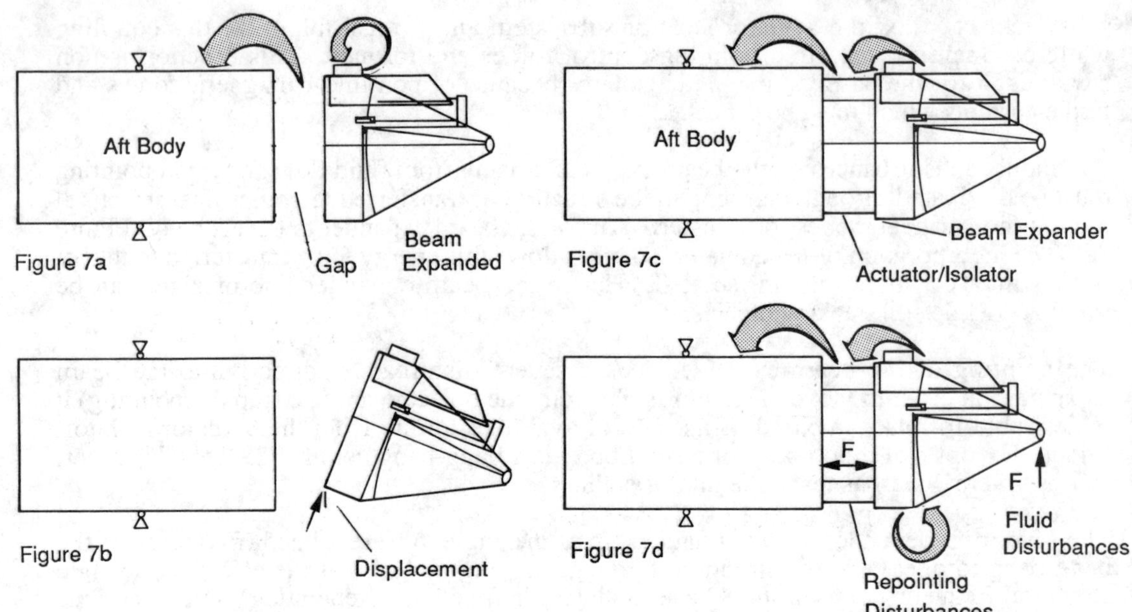

Figure 7 Vehicle Control Architecture Evolution

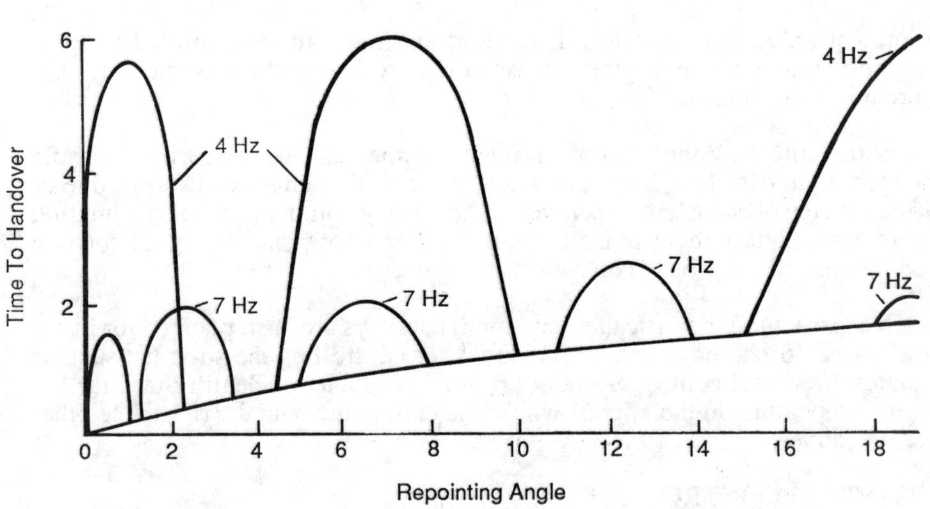

Figure 8 Effects of Structural Frequencies on Repointing Time To Precision Track Handover

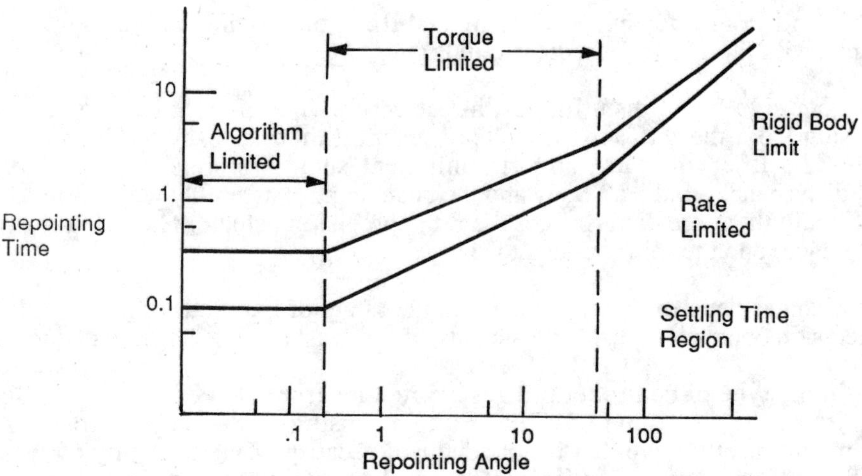

Figure 9 Repointing Characteristics

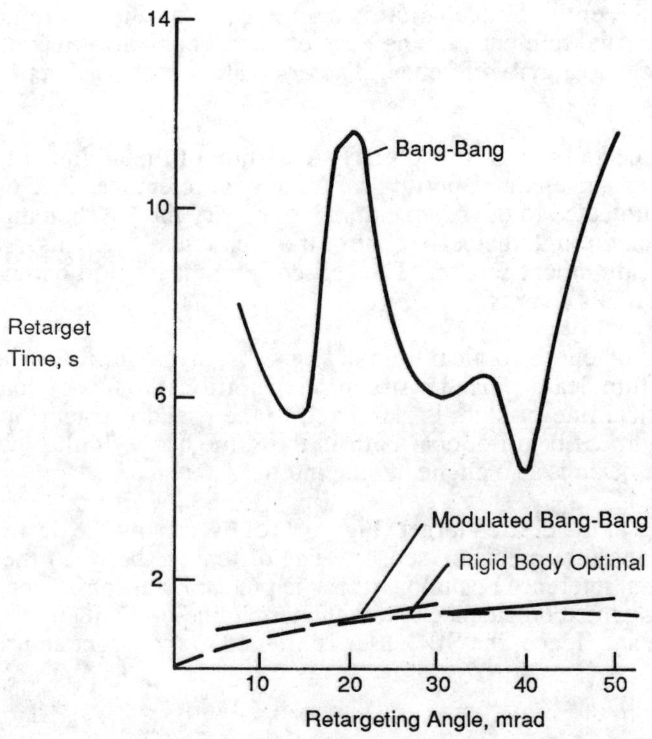

Figure 10 Multiple Mode Controller Characteristics

the handoff and continue the process of stabilization of the line-of-sight. This is accomplished with the controller indicated in Figure 6.

There are other significant disturbances that affect the line-of-sight such as the aft body motion arising from the thruster firings and reaction motion resulting from the vehicle controller. The laser and its support equipment such as pumps and turbines cause significant disturbances to the aft body and the beam control assembly which affect the line-of-sight. Finally there are disturbances from the precision actuators (steering mirrors and inertial reference) and the mirror coolant flow.

These disturbances have frequency content in the range of hundreds of Hertz . Hence the off axis precision controller must have a bandwiths at least as wide as these disturbances.

The most simple type of controller utilizes a direct measurement of the line-of-sight and direct control of the same line-of-sight. Simple sensing of the target through the high power beam path is attractive but is not adequate because of the difficulty of sensing the object with high power on the optics. Moreover, bandwiths in the hundreds of Hertz are not achievable by this method because of the transit time from the SBL to the object.

As a result of the above considerations, the SBL precision off-axis steering utilizes the controller described earlier. This controller is developed as follows for Figure 11. The geometry is shown in Figure 11a.

The line-of-sight through the entire optical train is stabilized to an inertial reference as shown in Figure 11b. An inertial reference in the form of an inertial platform containing a small laser source is stabilized to a gyro reference. The orientation of this reference will be discussed later.

This alignment beam is projected to a beam splitter. A portion of the alignment beam is projected to the precision sensor separate aperture for alignment reference. The other part of this alignment beam is projected to the beam expander primary mirror, then on through the beam expander and back through the beam control transfer assembly. This alignment beam is intercepted by the alignment sensor. The sensed error is used to command the beam steering mirrors to null this error out.

Hence, the optical path throughout the optical train is aligned. Any disturbance (including line-of-sight disturbances from beam expander structural distortions) to this optical path is corrected. That is, the optical line-of-sight is stabilized to the pseudo target (the inertial reference). Since the alignment beam optical path is the same as the high power laser optical path, the high power beam is also aligned to the inertial reference.

The pseudo target is aligned to the desired target (Figure 11c) by sensing the desired target position with the separate aperture precision sensor. The difference between the desired target position and the inertial reference position within the precision sensor is used to drive the inertial reference position into coincidence. Now the target line-of-sight and the optical line-of-sight are in alignment. Hence, the SBL laser is aligned to the target and ready for firing.

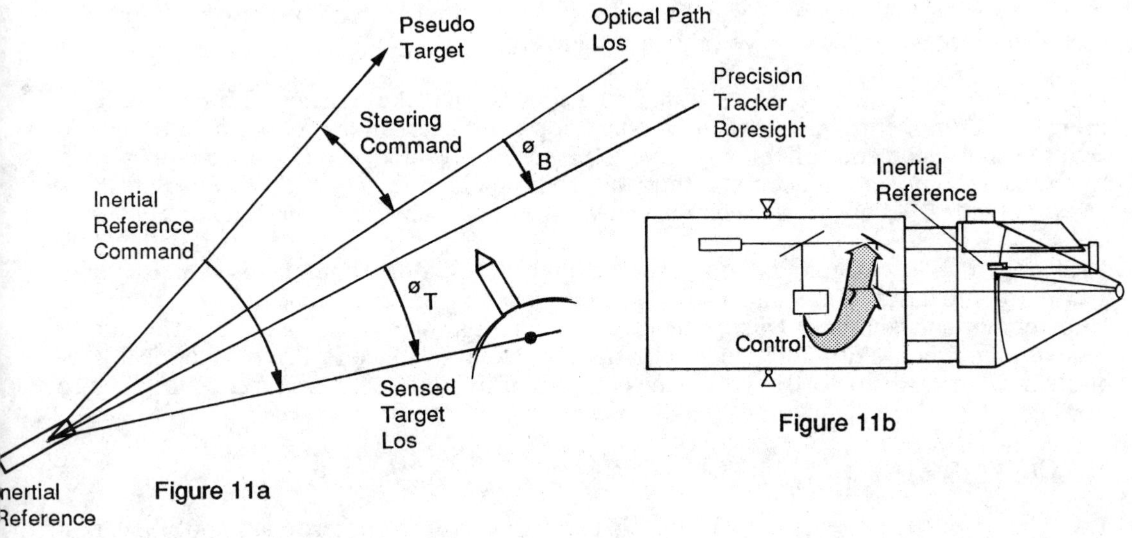

Figure 11a

Figure 11b

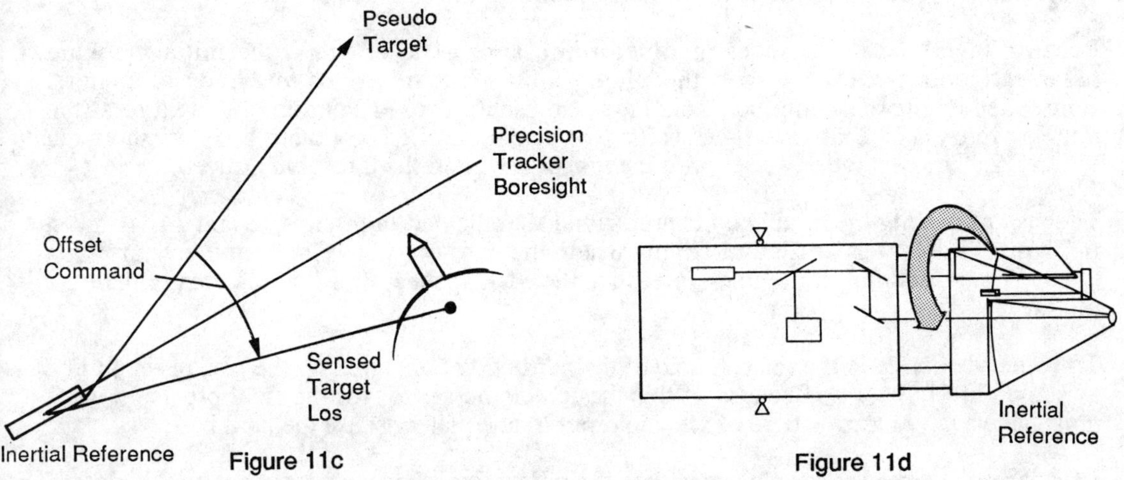

Figure 11c

Figure 11d

Figure 11 Off Axis Precision Pointing System

INTEGRATED CONTROL

This hierarchical control architecture separates the control issues so that they can be designed and stabilized separately. Accuracy and jitter analyses can be performed for near steady state conditions. However, that is not enough.

These controllers must be orchestrated as a unit so that the SBL system can meet its stringent requirements. Consequently, open loop commands in the proper time frame are required so that the controllers only have to respond to residual errors from the open loop commands. The more accurate that the commands, the faster the control task is accomplished. This obviously requires knowledge of the subsystem characteristics.

These open loop commands sometimes called anticipatory commands should make maximum utilization of the information such as beam expander motion and vibrations and aft body motion. They must also take advantage of alignment offsets and target motion to decrease the time for aligning the line-of-sight to the target. That is, drive the optical line-of-sight to the target by offset commands instead of moving the inertial reference (Figure 11d).

PRECISION TARGET SENSING

The sensing of the target is not part of this paper however the interpretation of the target scene is required for the control systems to function properly particularly since the control system must provide the handover from the plume to the missile hard body.

Because of the large distances involved (thousands of kilometers) the initial tracking (coarse tracking) can only see the plume which extends over hundreds of meters. Consequently, the most common technique is to use the sensed centroid of the plume as the tracking reference. This reference is aft of the actual missile body (hard body) as shown in Figure 12. The separation varies with the sensor used and the threshold utilized.

The task of finding the hard body is non-trivial as indicated in Figures 13 and 14. As seen in Figure 13, it is very difficult to know where the hard body is. One cannot even rely on the apparent direction of the track rate since the SBL spacecraft motion is greater than the missile.

The other difficulty is the relative size of the hard body compared to the size of the plume as shown in Figure 14. Because of this the precision sensor for the final precision track must have a very narrow field of view in order to adequately sense the hard body.

All of these parameters must be considered in order to provide the proper and rapid handoff from the plume to the hard body.

ZENITH STAR CONTROL CHALLENGE

The Zenith Star experiment is designed to provide all of the control functions of the SBL system described above and to be a direct analogy. When all of the major control functions are combined the picture looks like that shown in Figure 15. The explanation of these functions are the same as described in the paragraphs above. The Zenith Star control system is indeed a challenge.

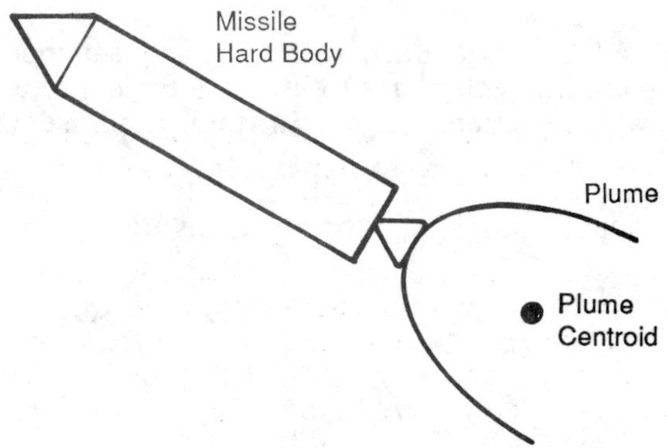

Figure 12 Missile/Plume Scene Relationship

Figure 13 Typical Plume Image

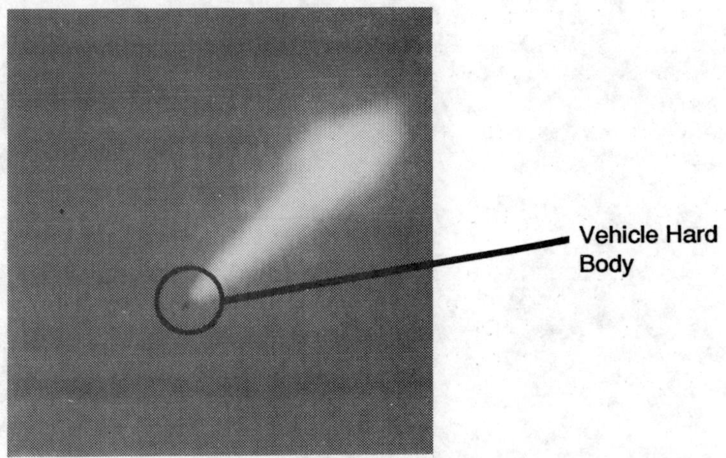

Figure 14 Typical Plume Image with Vehicle Hard Body Position

323

Because of this challenge the Zenith Star experiment offers an excellent opportunity for the controls community to develop control techniques to improve the expected SBL system performance and evaluate them with the Zenith Star experiment in a space environment with no scaling involved.

As the program progresses, opportunities will be made available for the controls community to participate.

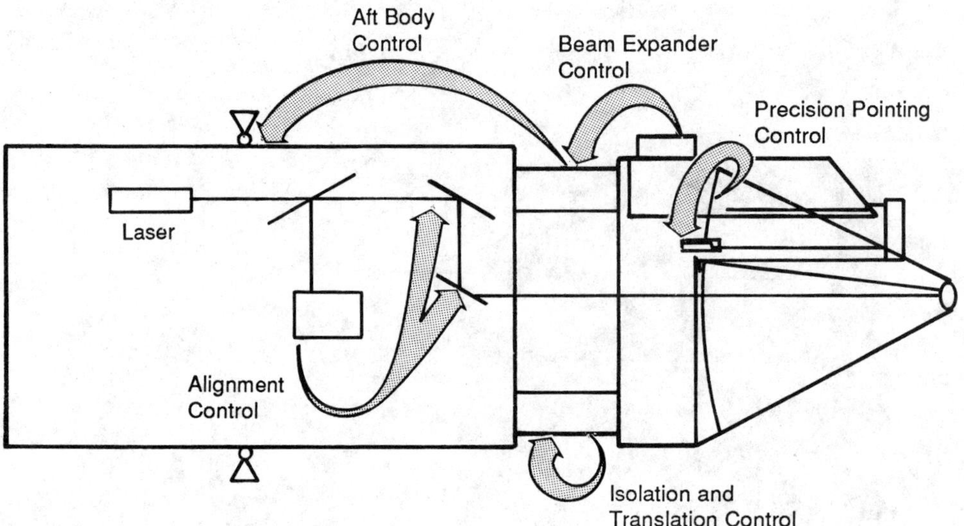

Figure 15 Zenith Star Control Challenge

PRECISION POINTING AND INERTIAL LINE-OF-SIGHT STABILIZATION USING FINE-STEERING MIRROR, AND STRAP-DOWN INERTIAL SENSORS

Avanindra A. Gupta[*] and Lawrence M. Germann[†]

In many space-based pointing and tracking applications, the pointing system must be stabilized using inertial references such as gyros, accelerometers, and star trackers. In these cases, the optical feedback signal is either unavailable or too low bandwidth, due to uncooperative targets or background clutter, to provide position reference to the pointing system.

This paper presents results of the analyses and simulations for a pointing system configuration based on the body-fixed telescope concept to show that a microradian-pointing, jitter system can be designed using inertial references from star trackers, accelerometers, and gyros for directed-energy weapons, surveillance, optical seekers, and laser communication.

INTRODUCTION

New focal-plane technology has given rise to submicroradian pointing-jitter systems. As a result, several space-based telescopes have been designed, including space infrared telescope sensors (SITS) and the Hubble Telescope, that have submicroradian pointing jitter. These systems have been specially designed to be extremely quiet with virtually no base motion above 0.5 Hz. The inertial sensor suite for these systems have low-noise gyros, accelerometers, and optical feedback from the target. The optical sensor provides a low-drift position reference to the other two sensors. These systems rely critically on the availability of this optical signal to provide the position reference. However, there are many space-based pointing, acquisition, and tracking (PAT) systems in which optical information is not available from the target. For those applications, the pointing systems must be stabilized using inertial references from the star trackers, gyros, and accelerometers.

Various sources suggest that the base motion of most space platforms is in the range of 20 to several hundred microradians, due to circulating pumps, crew motion, and other sources. Studies show that to stabilize inertial line of sight (LOS) to submicroradian levels in that base-motion environment, the pointing systems require 40

---

[*] Senior Engineer, Ball Aerospace Systems Group, P.O. Box 1062, Boulder, Colorado 80306.

[†] Senior Technical Manager, Ball Aerospace Systems Group, P.O. Box 1062, Boulder, Colorado 80306.

to 60 dB of base-motion disturbance rejection in the 1 to 10 Hz range. Some rejection capability may also be required in the 100 Hz region.

The classic approach for the inertially referenced pointing system is to mount the complete pointing system on a gyro-stabilized gimbal. But the servo for a large space-based gimbal can not have more than 10 dB base-motion rejection in the 1 to 2 Hz range limited by the structural modes of the gimbal. Even adding passive isolation stage will not achieve the required 40 to 60 dB base-motion rejection needed in the 1 to 10 Hz range. Pointing-jitter performance of a gyro-stabilized gimbal is also limited by the bearing friction. Further, it is not particularly suitable for cooled payloads, due to utility transfer problems across the two-axis gimbal.

Several alternative approaches exist, based on

o   Body-fixed telescope with an external or internal flexure-mounted gimballed flat,

o   Magnetic suspension isolation and pointing system,

o   A 6 degree-of-freedom flexure isolation and pointing system.

Advantages and disadvantages of the above systems are discussed in detail in Ref. 1. The body-fixed telescope concept offers many system-level advantages in terms of low cost, power, weight, high reliability, and superior performance.

The body-fixed telescope concept has a fine-pointing mechanism, in this case a fine-steering mirror (FSM), hard-mounted to the spacecraft. The strap-down inertial sensors measure the spacecraft base motion and their signal is used to drive the FSM in such a way that it stabilizes the LOS. In this configuration, the spacecraft is used for coarse pointing. The FSM provides fine pointing while compensating for base-motion disturbances.

To provide submicroradian pointing-jitter in a space platform environment, the body-fixed telescope concept requires high performance from the FSM and the inertial sensors. The inertial sensors are required with noise in submicroradian levels and bandwidth in the range of several hundred Hz. The FSM must have disturbance-rejection capability in the 100 Hz region to correct for base motion components at that frequency.

Only recently have high-bandwidth inertial sensors and space-qualified FSMs, with disturbance rejection up to several-hundred Hz, been available. Low-noise, inertial-grade accelerometers, with bandwidths as high as 1 kHz, and low-noise gyros with bandwidths as high as 100 Hz, have recently become available as well. With the availability of these components, it is now possible to design precision pointing systems based on the body-fixed telescopes with superior performance in terms of low jitter, cost, power, and high reliability.

Due to the size and weight of the rate sensors, inertial stabilization of the FSM in rate mode is not feasible; and due to the large dynamic range required from the rate signal, a derived rate signal also cannot be used. Therefore, the FSM has a high-control bandwidth position loop, which provides disturbance rejection in the 100 Hz region. To correct for base-motion disturbances, the FSM servo requires base-motion information in position format. Because inertial position sensors, with submicroradian noise and bandwidth from 0.01 to 100 Hz, are not currently available, position information must be derived by integrating gyro or accelerometer signals. Both of these sensors are available with low noise levels and several hundred Hz bandwidth. However, integration of the gyro offset or accelerometer noise creates a position drift, which must be corrected with a low-bandwidth, inertial-position sensor, such as a star tracker, sun sensor, or horizon sensor.

It is obvious that either phase lag or phase distortion in the base disturbances signal would limit the base-motion rejection capability of the LOS stabilization system. Therefore, a mixing filter for the low- and high-bandwidth sensors must have minimum phase distortion at the mixing frequency to use the maximum disturbance rejection available from the FSM servo.

Determining the mixing frequency is critical. Noise contributions from the low- and high-bandwidth sensors can become dominating factors if not properly balanced. Raising the mixing frequency makes the noise contribution dominant due to the low-bandwidth sensor. Conversely, lowering the mixing frequency will make high-bandwidth sensor noise dominant. Therefore, the mixing frequency must be designed to balance out the noise from both sensors.

The remainder of this paper addresses the main issues regarding the use of inertial sensors in the open-loop, feed-forward configuration. Results of simulations and analyses are presented to show that a submicroradian pointing-jitter system can be designed for moderate base-motion environments with today's technology.

SYSTEM DESCRIPTION

Fig. 1 shows a functional block diagram of a LOS-stabilization system based on a body-fixed telescope and FSM concept. In this example, the spacecraft is used as a coarse-pointing mechanism. The FSM is used as a fine-pointing mechanism. The spacecraft has strap-down, linear accelerometers to measure high-frequency components of base motion. A low-bandwidth star tracker is used for correcting the low-frequency noise (drift) of the accelerometer. The mixing filter combines the signals from the low- and high-bandwidth inertial sensors to produce a composite signal, which is used -- through coordinate transformation -- to drive the FSM and stabilize the LOS. In this configuration, the FSM is not inertially stabilized. Instead, the LOS is stabilized by moving the mirror approximately one half the base-motion angle.

Because the high-performance FSM must be flexure-mounted, its practical travel is limited to ±5 deg, which translates to approximately ±10 deg of LOS motion. In this configuration, the FSM is used primarily for fine pointing; however, if larger LOS coverage is required for the FSM, it can be accommodated by mounting the FSM on a two-axis bearing gimbal, thereby increasing the travel for low-frequency motion.

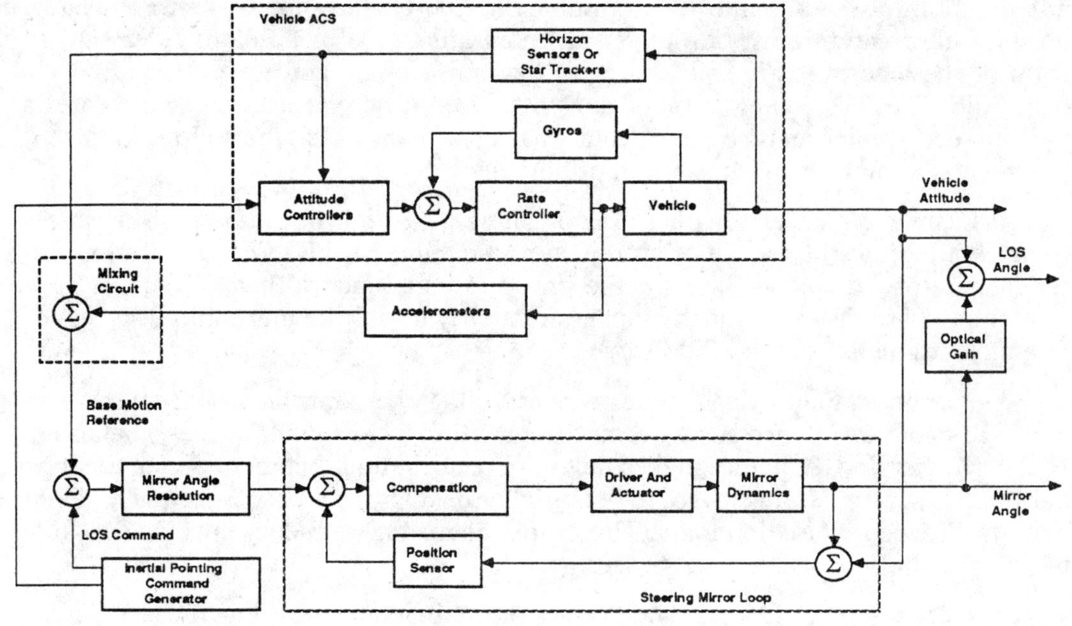

Fig. 1   The body-fixed telescope servo measures base motion with accelerometers and star trackers and uses a position-referenced FSM loop to correct for it

TEST FACILITY

Fig. 2 depicts the test facility currently used to develop the body-fixed telescope concept. It shows a three-axis (one lateral and two rotational) motion table. In this configuration, the motion table simulates the spacecraft, on which a pointing system is mounted. The fine-pointing system consists of FSM mounted to a motion table. A collimated laser beam, the source of which is mounted off the motion table, simulates an incoming beam in inertial space. It is bounced off the FSM and focused on a lateral-effect cell (LEC) to simulate the receiver and transmitter. The motion table has linear accelerometers mounted in differential pairs to measure angular motion. The position sensors of the motion table act as low-bandwidth inertial position sensors; the optical signal from the LEC measures residual LOS motion.

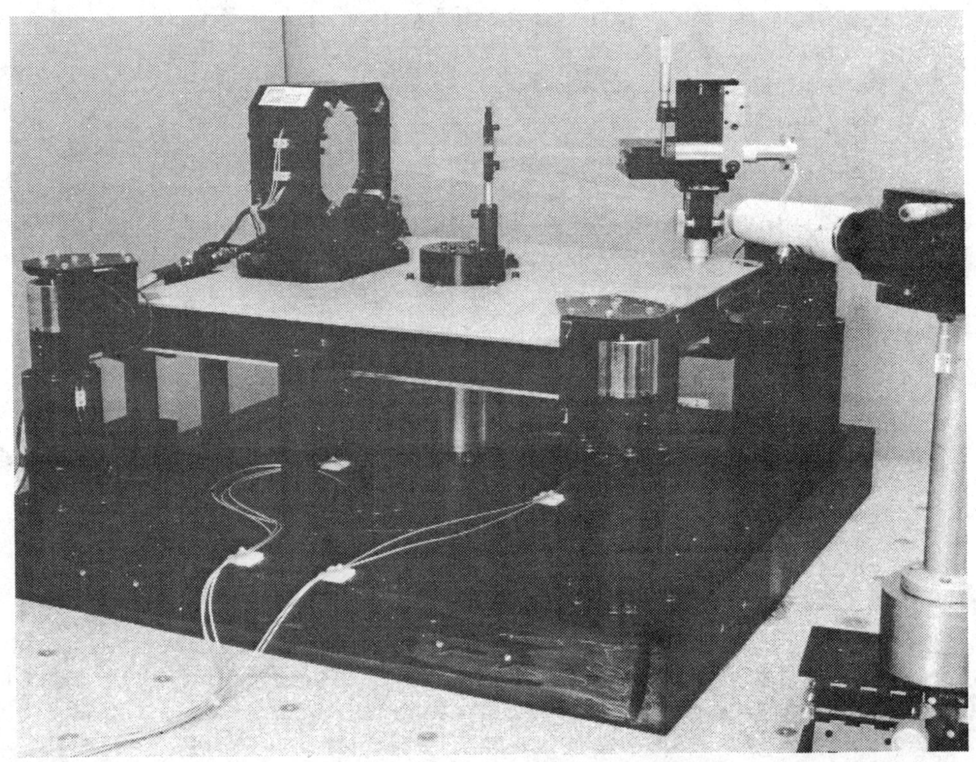

Fig. 2   Test facility for body-fixed telescope concept verifies performance

INERTIAL SENSORS

The major source of LOS jitter is noise due to inertial sensor and residual base motion. Residual base motion depends on disturbance-rejection characteristics of the FSM, and linearity and signal bandwidth of the inertial sensors.

State-of-the-art inertial sensors have 1.0 to 0.1 percent linearity which limits base-motion rejection capability to 40 to 60 dB.

The bandwidth of the inertial sensor defines the phase lag at high frequencies, therefore, inertial sensors must have bandwidth in at least the several-hundred Hz range to reject 100 Hz components of base motion.

Inertial sensor noise also contributes to system jitter. The submicroradian pointing system must have either micro-g noise accelerometers or sub-microradian noise gyros.

There are numerous low-bandwidth inertial position sensors, such as star trackers, sun sensors, and base-reference measuring systems. The choice of sensors depends on other system factors not discussed in this paper. Only star trackers are available with submicroradian accuracy and 1 Hz update rates.

A whole array of high-bandwidth inertial sensors exists. The list includes differential pairs of linear accelerometers, gas-bearing gyros, ring laser gyros, and angular accelerometers. Table 1 shows the noise and bandwidths for several state-of-the-art inertial sensors.

Table 1

**CONTROL CONCEPT PLACES STRINGENT REQUIREMENTS
ON INERTIAL SENSOR PERFORMANCE**

INERTIAL SENSORS	RMS NOISE	BANDWIDTH (Hz)
System Donner (Angular Displacement Sensor)	10 nrad	2 to 500 Hz
Northrop Gas-Bearing Gyros	76 nrad	100 Hz
QA-2000 Sunstrand Accelerometer	20 μg rms	300 Hz
Applied Technology Associates Magneto-Hydrodynamic Angular Rate Sensor	28 nrad	1 to 1000 Hz

The limitations of low-noise, gas-bearing gyros spurred the development of solid-state devices for some applications. Investigations are continuing in the use of low-noise accelerometers in differential pairs to measure angular acceleration. They are available with higher bandwidths (up to 1 kHz), may cost less, and are more reliable than gyros. A preliminary analysis on the noise power-spectral density (PSD) of the Sunstrand QA 2000 accelerometer shows that a microradian accelerometer noise level can be achieved when these units are used in differential pairs with 10 in. of separation over the 0.1 to 1 kHz frequency range.

SIMULATION/ANALYSIS RESULTS

Two principal sources of effort in a body-fixed telescope concept are inertial sensor noise and residual base motion.

The results of the simulations and analyses presented in this section show that sub-microradian pointing-jitter systems can be designed with current technology.

Sensor Noise

For purposes of these analyses, we used Sunstrand QA 2000 accelerometers and Northrop GIG6 gyros representative of low-noise, state-of-the-art inertial sensors.

Pointing-jitter was calculated for QA 2000 linear accelerometers used in differential pairs with a separation of 10 in. Fig. 3 shows the noise PSD plot (Ref. 2). Few data are available for the low-frequency region, so a flat PSD has been assumed below 0.1 Hz. Because the trend of the spectrum is toward low noise at low frequencies, this may be a conservative estimate.

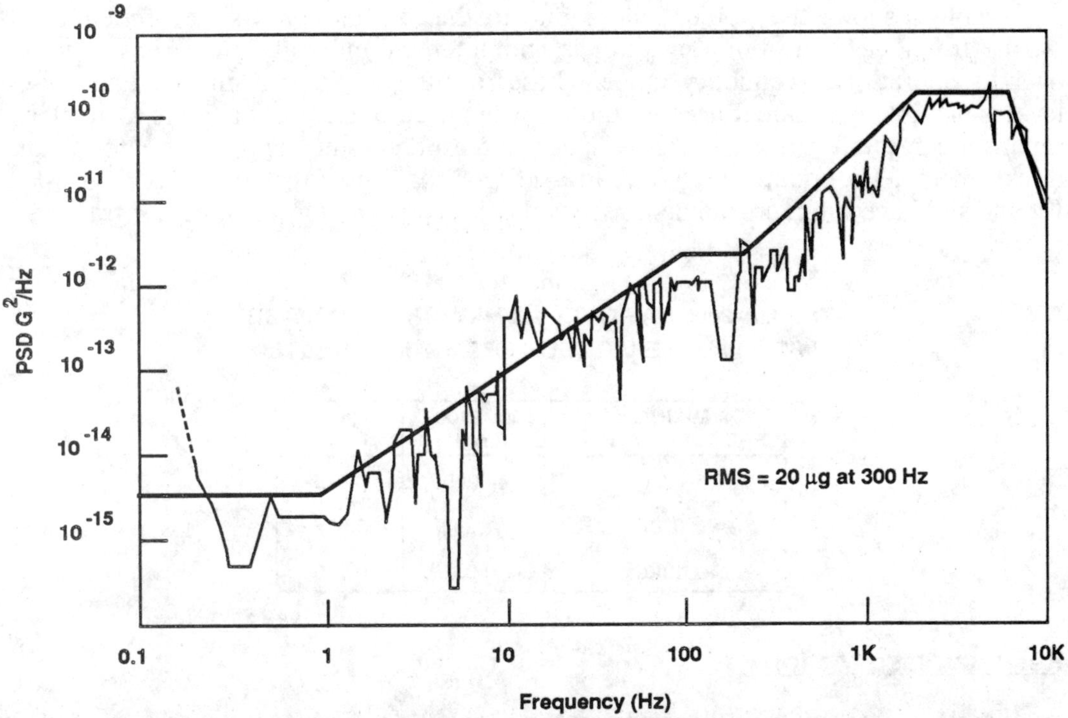

Fig. 3   Noise model for QA-2000 accelerometer is based on test data

The plot in Fig. 4 shows the noise PSD for the Northrop GIG6 gyro.

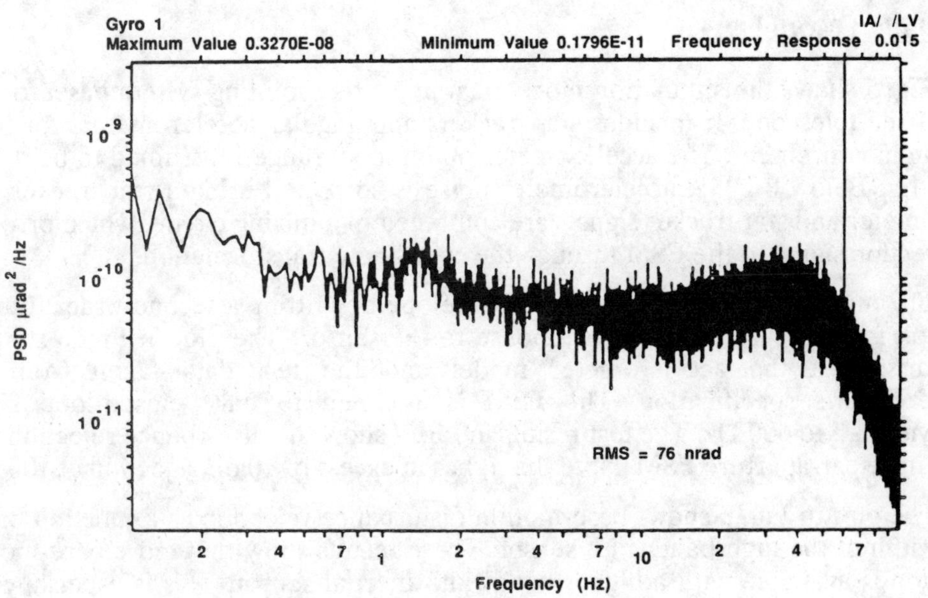

Fig. 4   Noise model for GIG6 is based on test data

Table 2 shows the pointing jitter due to the inertial sensors as a function of mixing frequency. The pointing jitter is computed by integrating the sensor noise over its operational frequency range. These results show that a microradian noise level is possible with linear accelerometers when used in differential pairs with 10 in. separation in the 0.1 to 1 kHz range. This is a reasonable mixing frequency for a star tracker with 1 Hz update rate. With low-noise gyros, it is possible to mix below 0.1 Hz and still have submicroradian noise levels, thereby allowing a slower star tracker.

Table 2

THE QA-2000 HAS ONLY 1.3 MICRORADIAN NOISE

WITH 10 IN. SEPARATION FROM 0.1 TO 1000 Hz

MIXING FREQUENCY (Hz)	QA-2000 (μrad)	GIG6 (μrad)
0.01	44.000	0.076
0.10	1.390	0.076
1.00	0.057	0.076

Residual Base Motion

Residual base motion depends on the magnitude of the base motion and the disturbance-rejection characteristics of the pointing system. This section presents the simulation results of the total pointing system, using both low- and high-bandwidth sensors, mixing circuit, and FSM servo.

Simulation Model

Fig. 5 shows the simulation block diagram of the pointing system based on the body-fixed telescope. It includes star trackers and angular accelerometers for base-motion measurement. The accelerometer operational range is assumed to be 0.1 Hz to 1 kHz. Below 0.1 Hz; accelerometer noise is corrected using a star tracker. Accelerometer and star tracker signals are combined in a mixing circuit, which provides a correction signal to the FSM through the coordinate transformation.

The model of the accelerometer is developed by fitting a second-order transfer function to the actual frequency response to the sensor. The plot in Fig. 6 shows a comparison of the accelerometer model and the test data taken from the manufacturer's specification. The FSM is assumed to have closed-loop signal bandwidth up to 600 Hz. The test results in Fig. 7 show the disturbance rejection plot of a 5 in., clear-aperture FSM. Note that it has in excess of 10 dB rejection at 100 Hz.

The plot in Fig. 8 shows base-motion disturbance rejection as a function of the bandwidth of the high-bandwidth sensor. These results show that more base-motion rejection is available with a higher bandwidth inertial sensor. This is especially important in the 100 Hz range.

The plot in Fig. 9 shows the residual base motion as a function of sensor bandwidth for 100 microradians of base motion. Note that there is no peak in the residual jitter at the mixing frequency (0.1 Hz) as would be seen if the mixing circuit had any phase distortion.

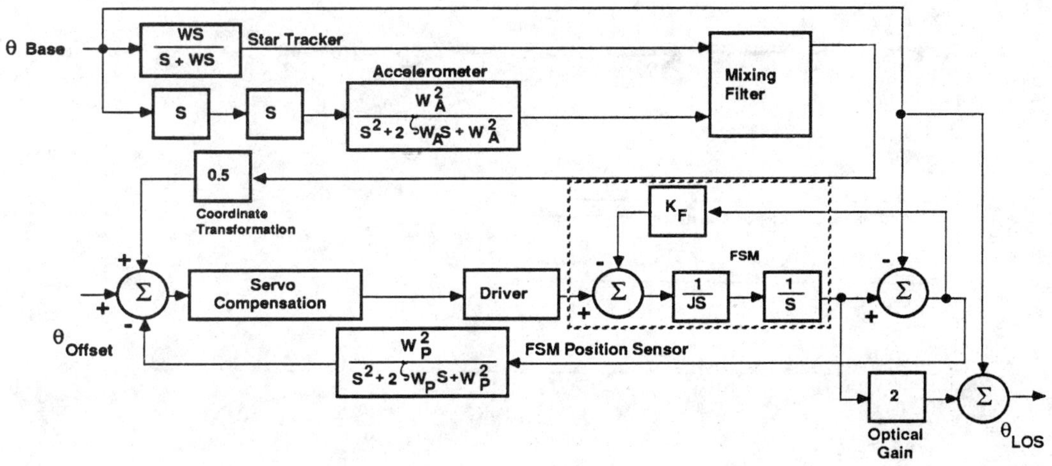

Fig. 5   Analysis of base-motion rejection uses accurate model

Tables 3 and 4 show the combined effect of residual base motion and sensor noise.

These results show that a microradian pointing system can be designed based on the body-fixed telescope concept for a space program system in a moderate base-motion environment.

SUMMARY

For a body-fixed telescope concept, the main sources of pointing-jitter are inertial sensor noise and residual base motion. Simulations and analyses have been performed to calculate the total LOS jitter due to these sources. These results are based on available inertial sensor test data and the best estimates of the base motion of a space-based platform. The results show that a microradian pointing-jitter system can be configured using available components.

ACKNOWLEDGEMENTS

The authors wish to express thanks to Sunstrand and Northrop for their cooperation in allowing the use of their test data, and to Dr. R. Synder of Jet Propulsion Laboratory (JPL) for introducing a class of mixing circuits without phase distortion.

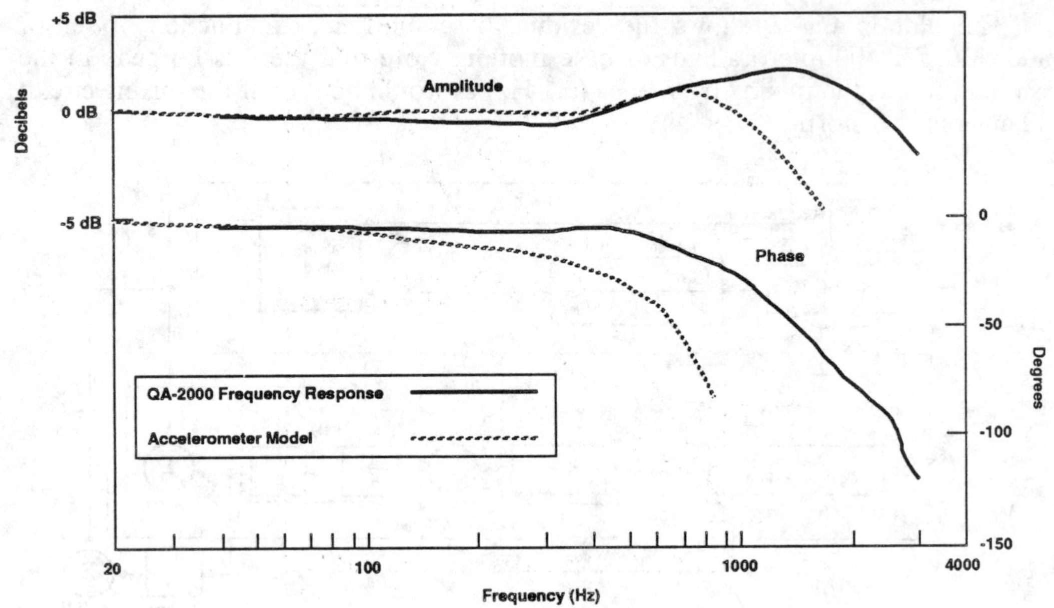

Fig. 6   Conservative simulation model is used for the QA-2000 accelerometer

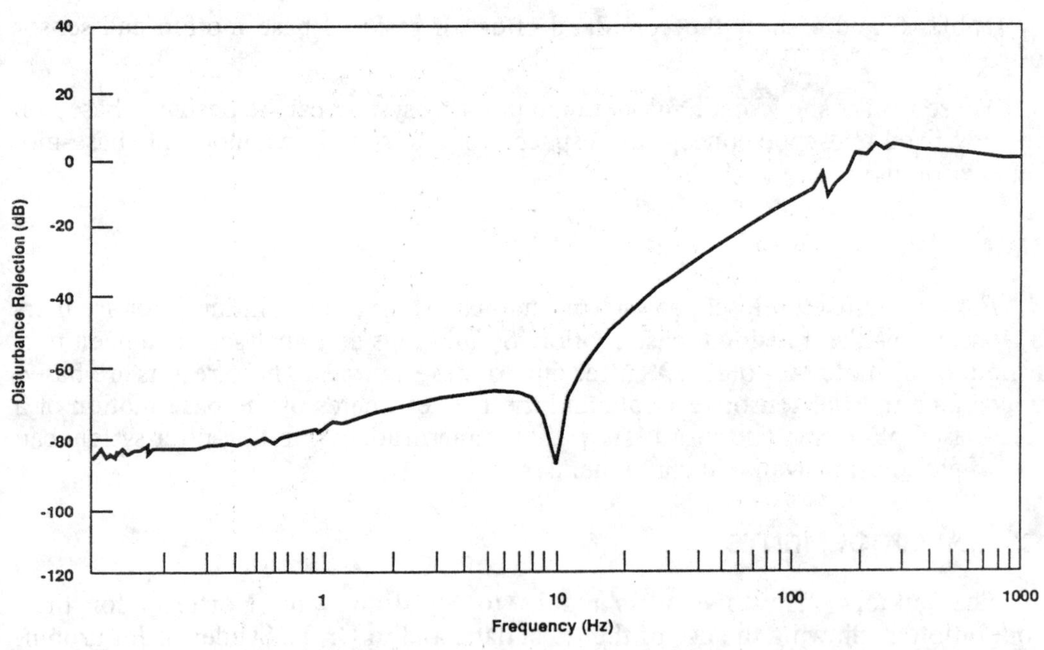

Fig. 7   Test data shows disturbance rejection in excess of 10 dB at 100 Hz for a 5 in. clear-aperture FSM

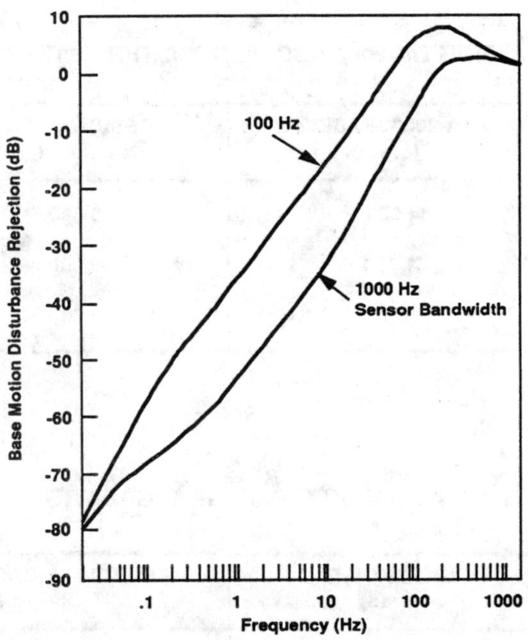

Fig. 8   Analysis shows the relationship between sensor bandwidth and rejection

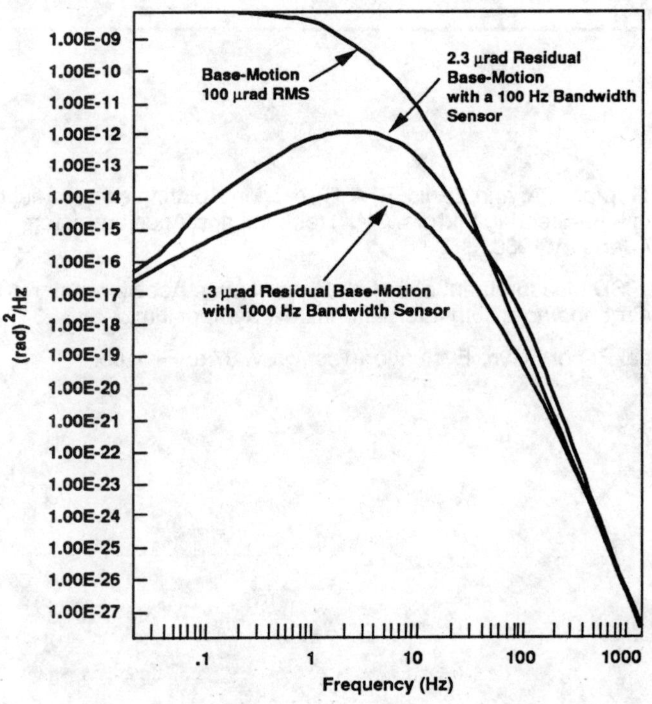

Fig. 9   Residual base-motion is a function of sensor bandwidth

335

Table 3

TOTAL LOS JITTER USING QA-2000 ACCELEROMETERS WITH 10 IN. SEPARATION

MIXING FREQUENCY (Hz)	QA-2000 NOISE (μrad)	RESIDUAL BASE-MOTION (μrad)	TOTAL (μrad)
0.01	44.000	0.300	44.300
0.10	1.390	0.300	1.690
1.00	0.057	0.300	0.357

Table 4

TOTAL LOS JITTER USING GIG6 GYRO

MIXING FREQUENCY (Hz)	GIGG NOISE (μrad)	RESIDUAL BASE-MOTION (μrad)	TOTAL (μrad)
0.01	0.076	2.3	2.373
0.10	0.076	2.3	2.373
1.00	0.076	2.3	2.373

REFERENCES

1. Germann, L. M., Gupta, A. A. and Lewis, R. A. "Precision Pointing and Inertial Line-of-Sight Stabilization using Fine-Steering Mirrors, Star Trackers, and Accelerometers." in *Proceedings: SPIE Volume 887 January 1988.*

2. Peters, R. "Noise PSD Measurement of a High-Performance Accelerometer with Motion Cancellation." in *Proceedings:* 30th International ISA Symposium, May 1984.

3. Northrop Technical Report, Gyro Evaluation Test, Nov. 87 to Feb. 88.

Section IV
GUIDANCE, NAVIGATION AND CONTROL FOR SPECIALIZED MISSIONS

SESSION IV

Chairperson:	Harold Scofield
	NASA Marshall Space Flight Center
Co-Chairperson:	Major Wiley Larson
	U.S. Air Force Academy
Local Chairperson:	John Eterno
	Ball Aerospace Systems Group

The following paper was not available for publication:

AAS 89-042

Thw following paper numbers were not assigned:

AAS 89-048 to -049

THE NATIONAL AERO-SPACE PLANE, THE GUIDANCE AND CONTROL ENGINEER'S DREAM OR NIGHTMARE?

Felix Sanchez[*]

The National Aero-Space Plane (NASP) is the most challenging and complex aircraft/spacecraft ever attempted. The purpose of this paper is threefold: first, to introduce the NASP Program to the American Astronautics Society (AAS) community; second, to discuss the major technical challenges associated with the program and the challenges of the National Aero-Space Plane from a controls perspective; and finally, to stimulate the AAS members into thinking about how we might meet the guidance and control challenges associated with the National Aero-Space Plane.

The major technical challenges which are briefly addressed are the propulsion system, the structural materials which must withstand the environment, and computational fluid dynamics which is the only mechanism to predict the aerodynamics past Mach 8. The primary discussion is on those significant challenges in the guidance and control fields relating to vehicle management systems, integrated propulsion/flight control, optimal vehicle trajectory control, and challenges in the associated fields of instrumentation and information systems. The intent of this paper is to provide insight into the complexity of the problem and to highlight the important role which the guidance and control community will play in achieving the goals of the NASP Program.

INTRODUCTION

The National Aero-Space Plane will be able to takeoff on a conventional runway, accelerate to Mach 25, fly into orbit, return to earth, and within days (not months) return to space. As in any adventure that has not been undertaken, there are always many who are skeptical and will dwell on all the reasons why it cannot be done. To those of us tasked to make this project work, the challenge of NASP is analogous to the challenge the Wright Brothers faced. It can be done!

PROGRAM GOALS

The X-1 was the first aircraft to break the sound barrier. The X-15 increased that speed six-fold. Now the X-30, as the National Aero-Space Plane is designated, will fly 25 times the speed of sound (Mach 25). The X-30 will have a horizontal takeoff capability, fly airbreathing into low earth orbit, return to a conventional run-

---

[*] Lt. Colonel, U.S. Air Force and Program Manager for Airframe Development, National Aero-Space Plane, Joint Program Office at Wright-Patterson Air Force Base, Ohio 45433-5000.

way, make an approach, and be able to go around the landing pattern prior to landing. The single-stage-to-orbit is the primary goal of the program. Other goals include hypersonic cruise, loiter capability, and subsonic ferry.

The ability to operate our future space vehicles in an aircraft-like operation will provide our nation with a capability that will reduce the cost of putting payloads into orbit by an order of magnitude. Additionally, the technology that will be developed in the process will allow us to maintain preeminence in aerospace into the 21st century. This technology base will impact all areas associated with the initiation of the design, analyses, fabrication, and flight test of two experimental flight vehicles that will fly in 1994 to demonstrate these goals.

Why Now?

A frequent question asked is, "Why do we think that we can now accomplish this feat which is technically more complex than putting a man on the moon?" There have been technological breakthroughs in three areas: airbreathing propulsion systems (ramjets and scramjets), lightweight structural materials and structures that can withstand extremely high temperatures, and super computers that allow us to solve the complex Navier-Stokes Equations that predict the fluid dynamics of air molecules. This latter capability, referred to as computational fluid dynamics (CFD), is the key link between predictions in those flight regimes where we do not have the wind tunnels to adequately simulate flight (above Mach 8) and the reality of flight test.

Recent advances made in these enabling technologies have given us the confidence that we can attain the program goals. However, these three areas, as will be discussed later in this paper, remain the major technical challenges that we must meet.

PROGRAM STATUS

Management

The National Aero-Space Program is a national effort with NASA, Air Force, Navy, DARPA, and Navy as participants. It is managed by the National Aero-Space Plane Joint Program Office at Wright-Patterson AFB, Ohio. Providing broad programmatic direction and policy guidance is a Steering Group with the Undersecretary of Defense for Acquisition as the Chairman, the NASA Associate Administrator for Aeronautics as Vice Chairman, and key acquisition executives from the Navy, Air Force, DARPA, and the Strategic Defense Initiative Organization as members of the Steering Group. The program is in Phase 2 of a three-phase program as shown in Fig. 1 (NASP Program Schedule).

Phase 2, a technology demonstration phase, consists of a competition among three airframe contractors (General Dynamics, McDonnell Douglas, and Rockwell International) and two engine contractors (Pratt & Whitney and Rocketdyne) to design the X-30 which will be built and flight tested in Phase 3. The decision on whose concept will be pursued will be made in September 1990. One of the options being studied as the acquisition strategy for Phase 3 is a consortium of these players instead of the classical airframe prime contractor and engine subcontractor. The goal for the first flight is early to mid fiscal year 1995.

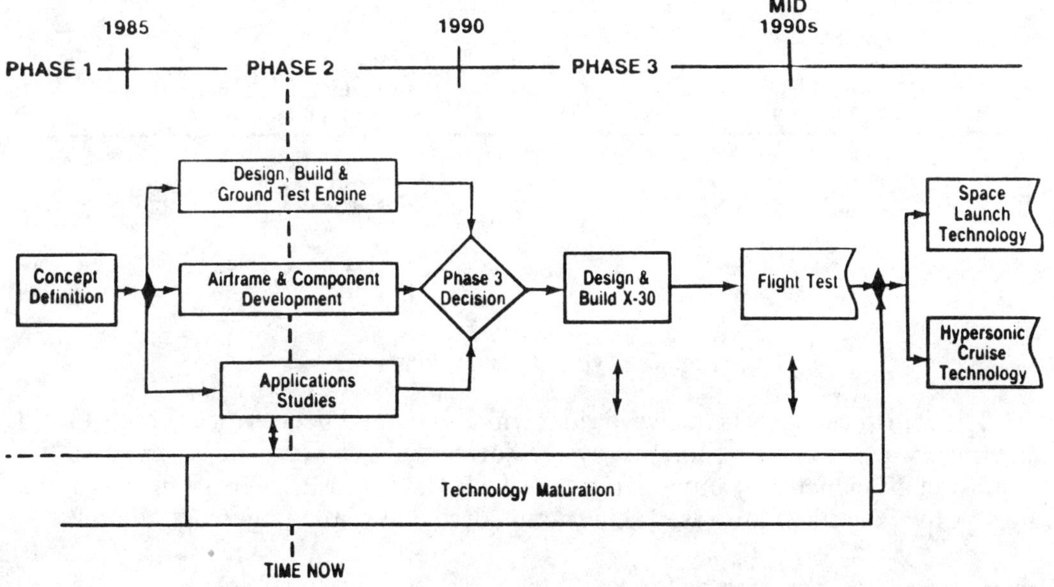

Fig. 1   NASP Program Schedule

Configuration Designs

What will the X-30 Aero-Space Plane look like? The contractors have investigated four classes of configurations for the NASP vehicle. Fig. 2 shows the four representative categories of designs. (The actual configurations are classified and competition sensitive.)

The advantages of the conical shaped vehicle is the high propulsion efficiencies for the ramjet/scramjet engines, because compression on all surfaces is put to use to produce thrust. However, packaging "minor" incidentals, such as landing gear and other subsystems, tends to be a major task. Volumetrically, the design is not as efficient as other types of configurations.

The classical fuselage/wing vehicle was the government baseline in the initial feasibility studies. The advantages are the ease of fabrication due to the cylindrical tank for the hydrogen fuel and the minimum skin friction drag.

The blended wing body provides volumetric efficiency at the expense of an increase in skin friction drag. The construction of the tanks to conform to the external shape becomes a major challenge. Engine efficiency optimization is critical to take advantage of the improved volumetric efficiencies.

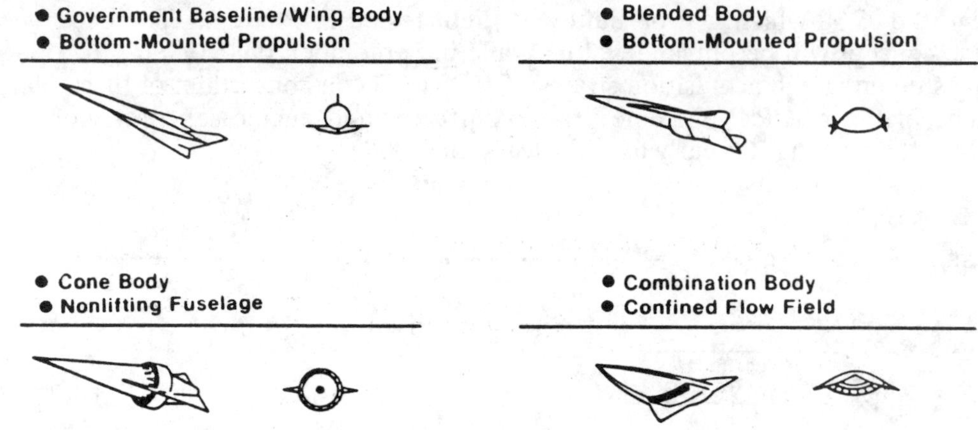

Fig. 2 Typical Configurations

The fourth category is the wave rider which is named because it "rides just inside the shock wave" that is generated by the forebody. The wave rider provides high propulsion efficiencies by minimizing the shock losses, has higher lift to drag ratios, but pays the penalty of increased skin friction drag due to the larger wetted areas.

TECHNICAL CHALLENGES

Propulsion

The engine and the airframe are indistinguishable. The propulsion system is in essence the underside of the vehicle. As shown in Fig. 3, the forebody of the vehicle is the area where compression of the air takes place (the inlet). In the ramjet mode, the air has been decelerated to subsonic flow through a normal shock prior to the addition of hydrogen fuel in the combustor. In the scramjet mode, the combustion takes place in supersonic flow (supersonic combustion ramjet). We have conducted numerous tests that have increased our confidence that scramjets will work.

The aft body of the vehicle is the nozzle of the engine where the flow is expanded and thrust is produced. The nozzle has turned out be one of the major drivers in the design of the configurations and an efficient expansion of the combustor gases is required to minimize the gross weight of the vehicles.

A major challenge is the limited facilities to test these engines above Mach 8. For Mach 8 and above, only short-duration pulsed facilities are available. The run time is usually only several milliseconds. However, since combustion in supersonic flow takes place in one-tenth of a millisecond, "steady state" conditions can be obtained. (Working on this program required a cultural acclamation because we calmly refer to velocities below Mach 3 as low speed and then we get excited to obtain 10-20 milliseconds of wind tunnel test data.)

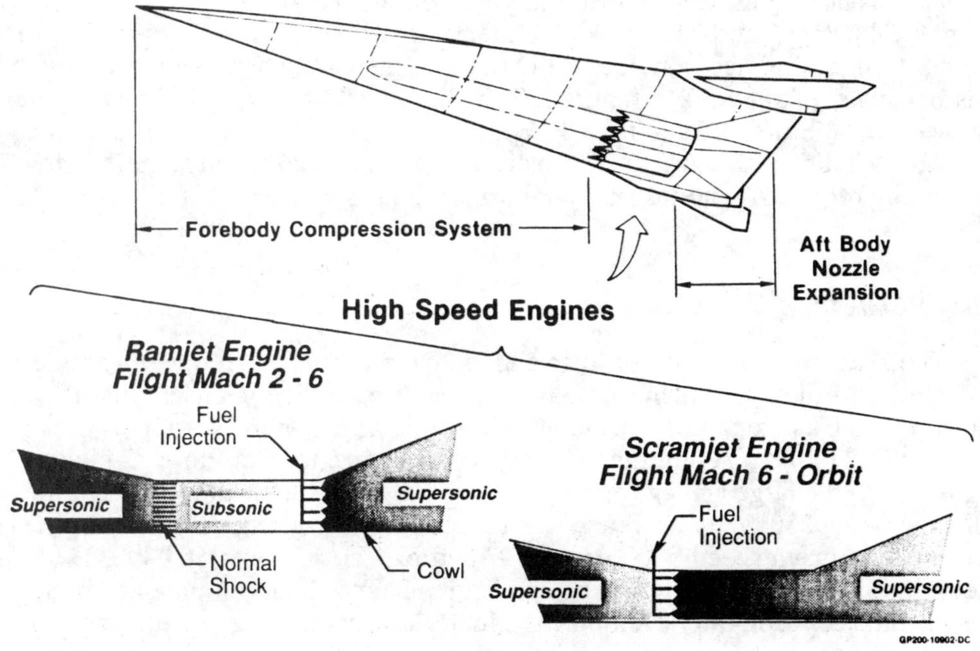

Fig. 3   Integrated Engine/Airframe

Under NASP funding, Marquardt and Aerojet Techsystems are building facilities that can test full-sized engines to Mach 8. Above Mach 8, we will be able to test the inlets, nozzles, and combustors, separately or in combinations of two. The full-scale engine will only be tested above Mach 8 during flight test on the X-30. In essence, the X-30 will be the "full-scale wind tunnel" facility. We are planning to bridge the gap from Mach 8 to Mach 25 with computational fluid dynamics.

Computational Fluid Dynamics

The advent of super computers like NASA's Numerical Aeronautical Simulation (NAS) facility, which uses the Cray II computer to process the applicable governing equations for compressible fluids, has made it possible to design and develop predictions in the regimes where we cannot test on the ground. There is a large hierarchy of simulation codes from steady, inviscid Euler codes to fully-integrated unsteady, viscous Navier-Stokes codes. The more complex, the longer the run time. For a full-body nose-to-tail simulation, a full Navier-Stokes code could take 40 hours or more to converge to a solution. What is usually done is to develop special codes for each region of the flow so they are simplified as much as possible for each region. The NASP Program currently is using half of the nation's Cray computers to conduct CFD.

343

CFD is not a panacea. In order to have confidence in the predictions, the codes must be calibrated/validated by wind tunnel or flight test. To design a vehicle for the Mach 8 + flight regimes, you need CFD to fill in the gaps where data is not available. This becomes a "Catch 22" situation. The flight test program will be structured so that we can calibrate/validate the CFD predictions as we expand the flight envelope. A related challenge is how do we instrument to gather the information the CFD community requires? The instrumentation challenges will be mentioned later in this paper.

Structures/Materials

The X-30 engines prefer to fly in a high dynamic pressure trajectory accelerating at "lower" altitudes (less than 120,000 ft). Rocket propelled vehicles (like the Space Shuttle) fly a trajectory that tries to get out of the atmosphere as soon as possible as shown in Fig. 4. As a result, the NASP will experience temperatures as high as 2500 degrees F on the large acreage areas. The nose cap and the leading edges of the wing will experience 3500 degrees F and will require active cooling. The the engine cowl leading edge may exceed 5500 degrees F and experience over 50,000 BTU/Ft^2-sec due to the shock impingement. The challenge is not only developing an efficient actively cooled concept, but developing the materials that will survive the environment. We are making progress in both areas.

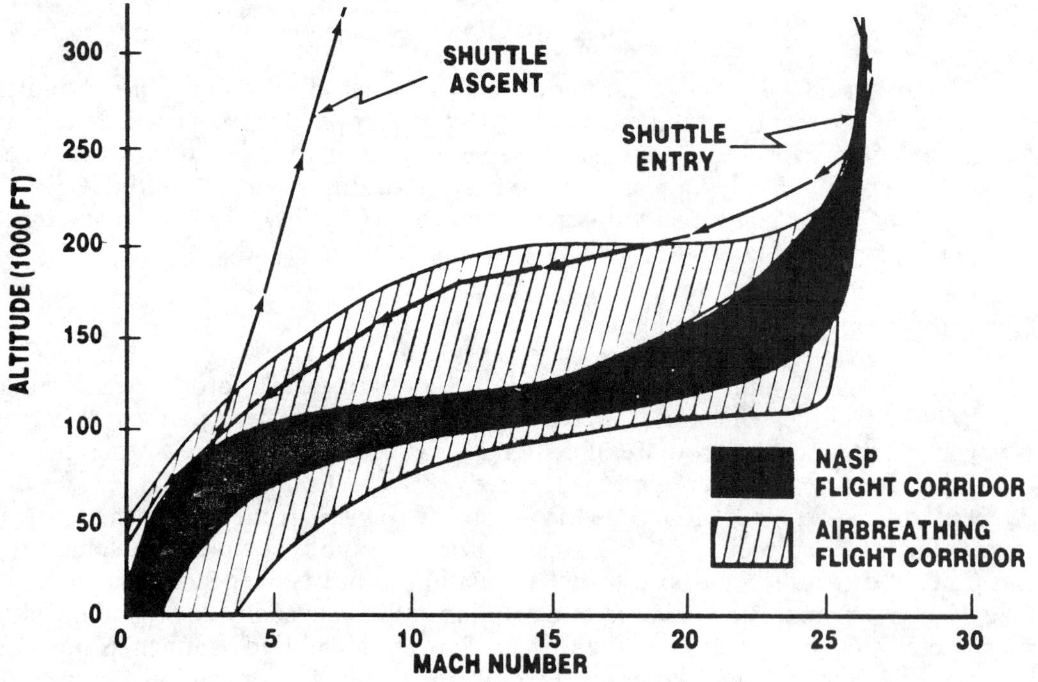

Fig. 4   Trajectory Comparison

Fig. 5 illustrates the characteristics of existing materials and the properties of those materials that we need to meet our requirements. The leading contenders are carbon/carbon, metal matrix composites (a high temperature metal such as titanium which contains strengthening fibers such as silicon carbide fibers), and ceramic materials if the brittleness disadvantages can be solved.

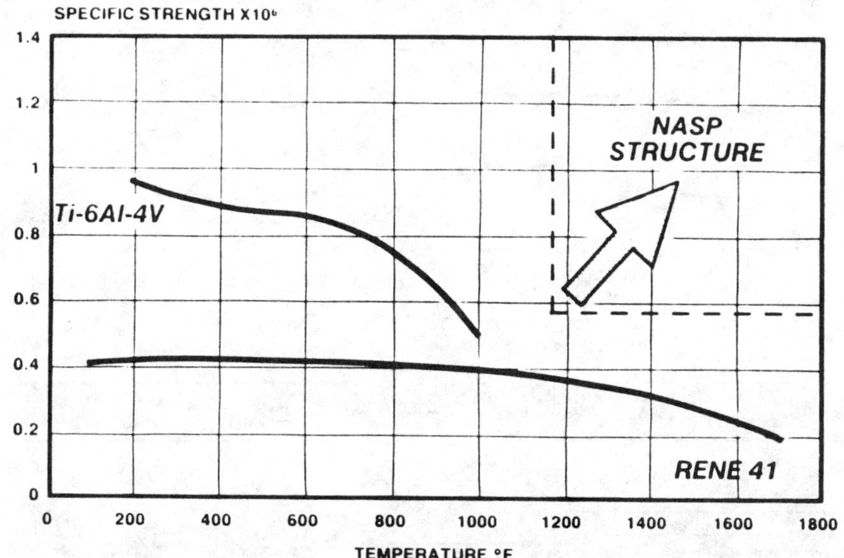

Fig. 5   NASP Structure Requirements

The NASP program has instituted an innovative approach in cooperation among the five major defense contractors working on NASP. A materials consortium was formed in 1987 to pool national technology efforts to develop the materials that both the engine and the airframe companies will need to make NASP a reality.

GUIDANCE AND CONTROL CHALLENGES

No less a challenge and of particular interest to the guidance and control community are those challenges relating to integrated propulsion/flight control, optimal vehicle trajectory control, and the associated fields of data acquisition, instrumentation and information systems. The National Aero-Space Plane is the most complex aero-propulsion interactive vehicle ever conceived. Depending on your perspective, this could be a guidance and control engineer's dream or it could be his/her worst feared nightmare. Clearly, the guidance and control technology joins the propulsion, structures/materials, and CFD as an enabling technology for the National Aero-Space Plane.

Vehicle Management System

The fully integrated engine/airframe, as illustrated in Fig. 6, is a complex system which requires the efficient combination of aerodynamics, propulsion, structures, and thermal control. In simple terms, this is a systems integration task unparalleled to date. The aerodynamics of the vehicle and the propulsion system are inseparable. The thermal control of the X-30 is dependent on the aerothermodynamics of the total system and the heat generated by the engines. One system, designated the Vehicle Management System, will be responsible for ensuring the active control of thermal, flight and engine parameters, the mission data processing, and communication and navigation functions.

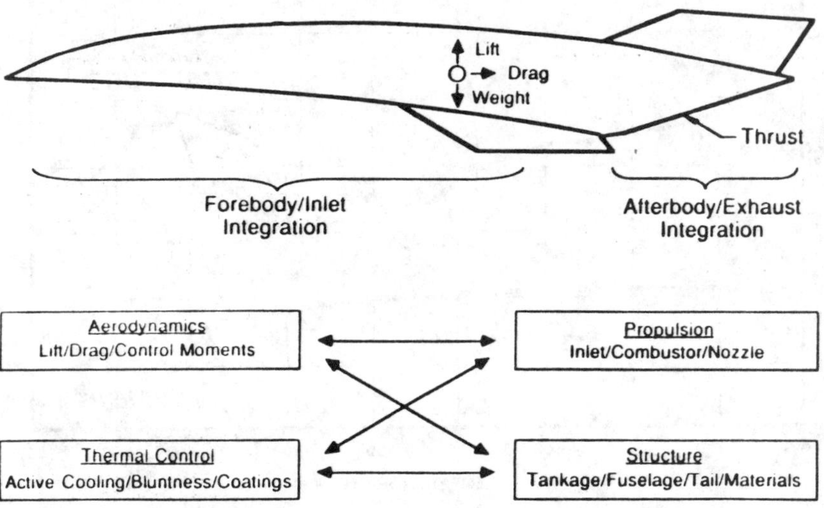

Fig. 6   NASP System Integration

The integrated avionics system, illustrated in Fig. 7, will require innovative technology such as ring laser gyros and global positioning system for vehicle navigation. Integrated, redundant digital flight and propulsion control may use high speed, high capacity fiber optics multiplexing.

Integrated Propulsion/Flight Control

The integrated flight control system must deal with a vehicle that is expanding the current flight envelope by a factor of eight. It must control many engine modes, transition between modes (i.e., ramjet to scramjet), multiple engine modules and the associated location effects on the air flow for each module. This system must be capable of coping with engine unstart, minimizing thrust asymmetries, and maintaining reasonable operating conditions for the vehicle and the engine.

To illustrate the integrated control problem, let's assume that as the test pilot I need to increase the angle of attack by 2 degrees to stay on a particular trajectory. This increase in angle of attack reduces the mass flow rate into the scramjet which decreases the thrust of each engine module. The reduction in thrust from each engine module changes the aerodynamic trim requirements on the vehicle. (Trim can be provided by either the control surfaces, the engine thrust vector, rocket thrust or a combination of the three.) The change in trim mechanism or the change in vehicle attitude can change the trim drag of the vehicle and the boundary layer transition point. When the boundary layer transitions from laminar flow to turbulent flow sooner, the heating rates and the associated skin friction drag increase substantially. This requires the thermal control system to increase the flow of hydrogen to handle the increase heat flux. This increases the volume and the temperature of the hydrogen that will be injected into the engine which produces an increase specific impulse from the engine. The change in specific impulse begins a different perturbation.

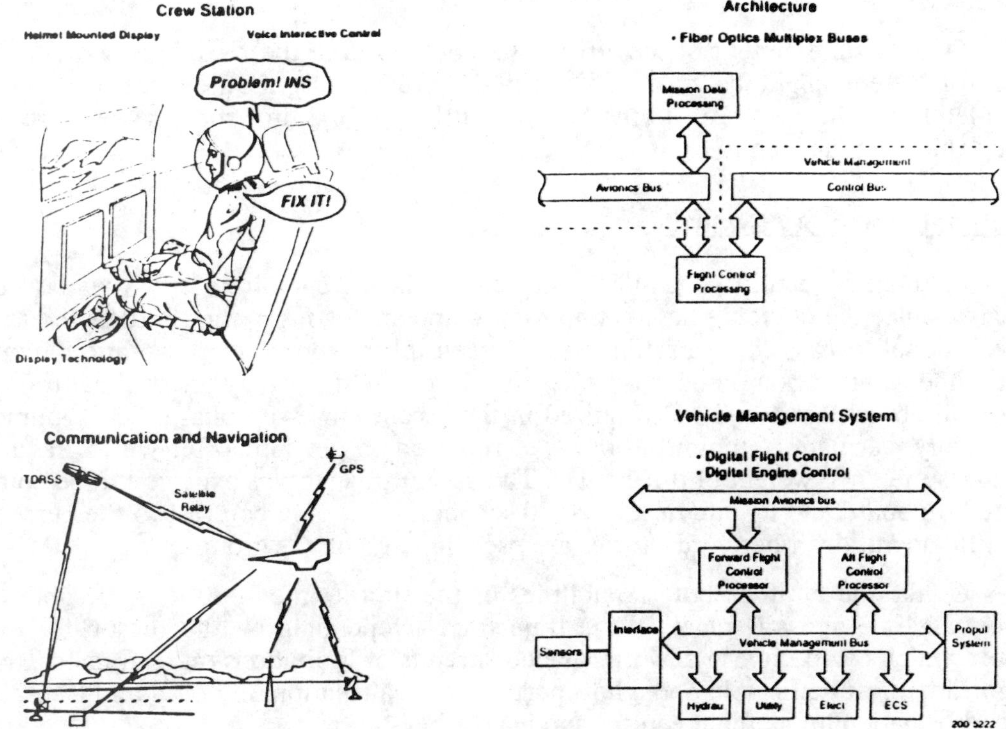

Fig. 7   Integrated Avionics System

The integrated flight and propulsion system must properly account for all these perturbations resulting from a change of in numerous control variables. One solution proposed by guidance and control engineers but rejected by management is to instruct the pilot to "keep his hands off of the stick". There may be some merit to this tongue-in-cheek solution for special situations such as optimal flight path control which will be discussed later.

While highly reliable fly-by-wire control systems are now standard, the level of flight, aerothermoelastic, and propulsion controls required pose a significant new dimension to the problem. The control system architectures for these vehicles could be driven by the simultaneous requirements for tight local control of high frequency phenomena and coordinated action across the entire system to assure appropriate control integration. This could drive the design to distributed hierarchical systems with high levels of reliability and integrity.

The uncertainties associated with the predicted environment compounds the problem. For example, the internal and external aerodynamics will be based on CFD codes that may not have been validated until the X-30 flies. The engine dynamics coupled with the aerothermoelastic response of the air vehicle make the task of a robust flight control system a challenging one. The design process is also hampered by the lack of a Flying Qualities Standard for hypersonic aircraft similar to Mil Spec 8785C for subsonic and supersonic aircraft. Again we must wait for the flight test of the X-30 to establish the desired handling qualities for an aircraft at hypersonic speeds.

One possible benefit of the aircraft-like operation of the X-30 is that the flight control system might be modified/adjusted incrementally as the envelope is expanded from Mach 1 to 25. In the Space Shuttle, the first time the engines were lit, the vehicle was headed to orbit.

Optimal Trajectory Control

Optimal trajectory control is required in the ascent phase and the descent phase. Since the descent phase is unpowered and analogous to the shuttle's reentry, the national team is concentrating on the ascent phase requirements. Autonomous real-time computation on board the aero-space plane of not only the optimal trajectory but abort trajectories are required in the ascent power-on phase. The optimal trajectory calculation minimizes the fuel required to get into orbit which in turn reduces the gross weight of the vehicle. The ascent trajectory is expected to be computed on board and updated every 20-30 seconds. This is in contrast to the current Shuttle operations where trajectories are precomputed on the ground.

Unlike the limited abort capabilities of the shuttle, the X-30 must be able to abort anytime and any place in the trajectory. Depending on the reason for the abort, which could impose new thermal constraints or limitations on load factor, the algorithm must be flexible enough to perform the calculation on board and transmit the information to the flight control and navigation system.

The computation power to accomplish this task in addition to the integrated propulsion/flight control task discussed earlier is of concern. Initial projections for total throughput requirements for the on-board information system range from 20-30 million instructions per second (mips). With a distributed computation system, a reasonable range of 3 to 10 mips could be obtained. Since the performance and the safety of the vehicle will be totally dependent upon this on board processing, an unprecedented reliability requirement on this system will be necessary.

Precise Trajectory Control

The precision desired for angle of attack and sideslip angles to optimize engine performance is in the .5 to 1 degrees range. This precision will be affected by the lower frequency structural and slosh modes. This precision can be accomplished if accurate information (airspeed, density, temperature, etc.) is available. This introduces another challenge for the community: nonintrusive instrumentation and sensors that can withstand the environment.

Sensors

Air Data. Of most interest to the NASP contractors is an accurate, nonintrusive air data sensor from Mach 3-25. The air measurement system must be able to operate within the stressing operational environment expected at the vehicle surface and within the engine. The system must be able to provide rapid update rates and in some cases spatially resolved measurements. This challenge is so tough that we have decided to take it out of the competitive arena and form a subsystem consortium among the NASP contractors to solve the problem.

The X-30 may have a full inertial navigation system with external updating using, for example, Global Positioning System (GPS). With the inertial system providing spatial position and velocity data, partial synthesized air data techniques can provide estimates of pressure altitude, temperature, and density. Angle of attack and sideslip can be defined in terms of body axes relative to spatial velocity vectors. If the local wind direction and velocity can be identified, the angles can be defined relative to the air mass. Thus, the use of computed air data may reduce the update rate requirements for a direct measurement system.

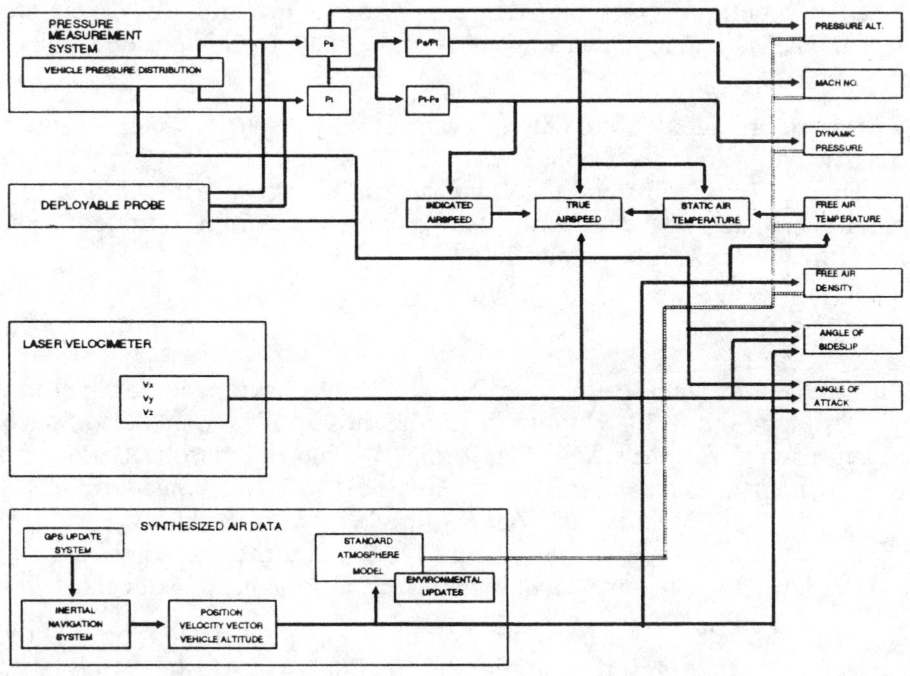

Fig. 8   Proposed Air Data System

The invention or innovation of a single direct or computed air measurement system may not be possible. One suggested approach is to use a four stage hybrid system (Fig. 8). At lower Mach numbers, a pitot-tube air data system similar to the Shuttle's would be used and retracted at around Mach 3. Past Mach 3, flush ports would record pressures and temperatures, and the airspeed would be computed real time.

Beyond Mach 7, a laser velocimeter could be used to map out the flow field around the X-30 vehicle. The laser velocimeter relies on propagation of a laser beam into the flow with subsequent scattering from the flow constituents. The scattered photons undergo a frequency shift due to the Doppler effect. The Doppler-shifted frequency is proportional to the scatterer velocity and may be used as a measure of the flow velocity (Ref. 1). Laser velocimeters have been flight tested within many vehicles including supersonic fighter aircraft. This method would require a small window for the laser. To design and cool this electro optical window will be as big a challenge as the measurement of air data. Once we reach the flight regime where there is insufficient particles to energize, we would then rely on a "precomputed guess".

Thermal/Structural. The thermal environment on the Aero-Space Plane will range from -420 degrees F near the hydrogen tank to 2500 degrees F on the majority of the external structure. The leading edges of the wing and the nose cap will be greater than 3500 degrees F. Contrast this environment with that of the Shuttle's which only experiences temperatures of 1500 degrees F for a few minutes. New structure materials and active cooling of various parts of the aircraft will be used to survive this environment.

In order to provide active cooling to various parts of the vehicle, sensors and cabling that can withstand this hostile environment are non existent. Once you develop a strain gauge that can withstand the environment, how do you attach it to the external structure? An aggressive technology development program is funded by the Joint Program Office to answer these questions.

FLIGHT TESTING THE X-30

The true test of whether the previous challenges have been conquered will be the flight testing of the X-30. Happiness will be the X-30 National Aero-Space Plane on final approach to Edwards AFB, California. We have numerous Test Pilot School graduates in managerial positions to ensure the flight test considerations are included in the vehicle. Just like in Tom Wolfe's book, *The Right Stuff,* we in the Air Force and the Navy are still debating with NASA and the contractors the age old question of whether we need real windows or artificial means of external vision.

The ground track for a hypersonic aircraft in Fig. 9 illustrates some of the challenges in gathering flight test data at high Mach Numbers. At Mach 10 it takes all of Kansas to turn around. For a 180 degree turn at Mach 15, Illinois is a checkpoint. (At least you don't have to "change in Chicago".)

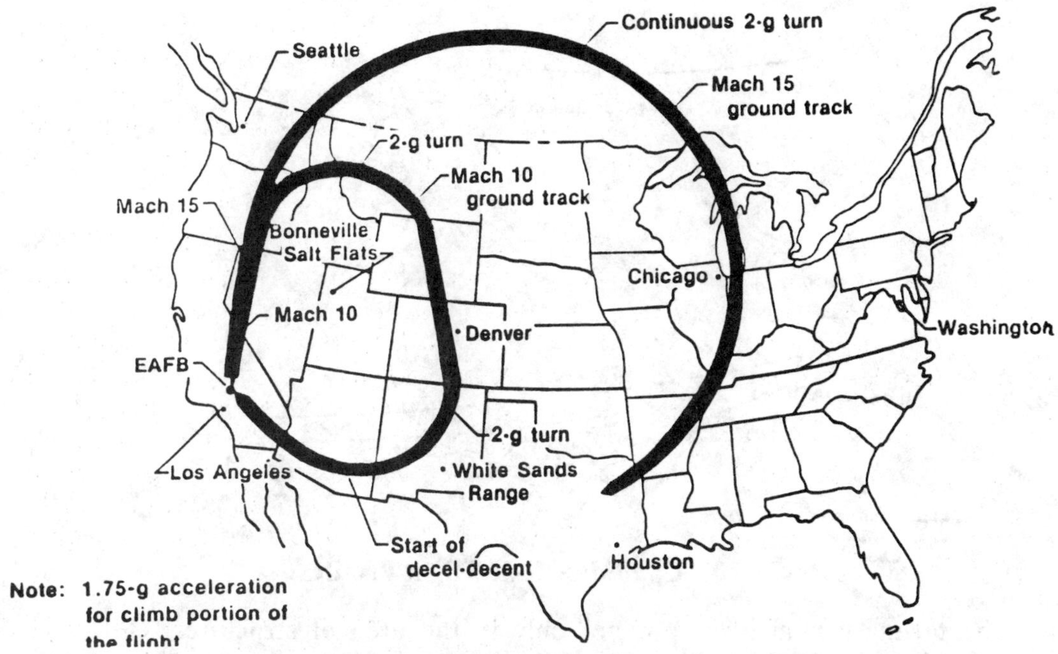

Fig. 9   Ground Track for a Hypersonic Vehicle

The typical flight envelope expansion profile might be to cruise out at a benign Mach Number and altitude above normal air traffic. At a predetermined range (based on the maximum planned Mach number and aircraft glide capability), the aircraft would be turned towards Edwards AFB and then accelerated to the test Mach number (Fig. 10). The data would be acquired at a stabilized condition and then the engines would be shut down to a dead stick landing at Edwards. This recovery profile is being considered because both the X-15 and the Shuttle flights demonstrated that unknown thermal damage can occur in this environment. This would subject the aircraft to minimum stress after reaching the high heat flux environment until the aircraft could be inspected visually. (Ref. 2)

This technique would also allow the maximum usage of our ground range capability. The ground range data acquisition will probably be complemented with the satellite relay of data, especially for orbital flights.

SUMMARY

The United States is in pursuit of a "Kitty Hawk" dream. In close trail are our economic competitors: Japan, France and Germany. The Soviet Union no doubt is following the progress closely. Whether we maintain a preeminence in aerospace depends on the success of the NASP Program. We in the NASP Program are confident that hypersonic cruise is achievable today. Single-stage-to-orbit is the formidable challenge. We are dedicated to make this dream a reality.

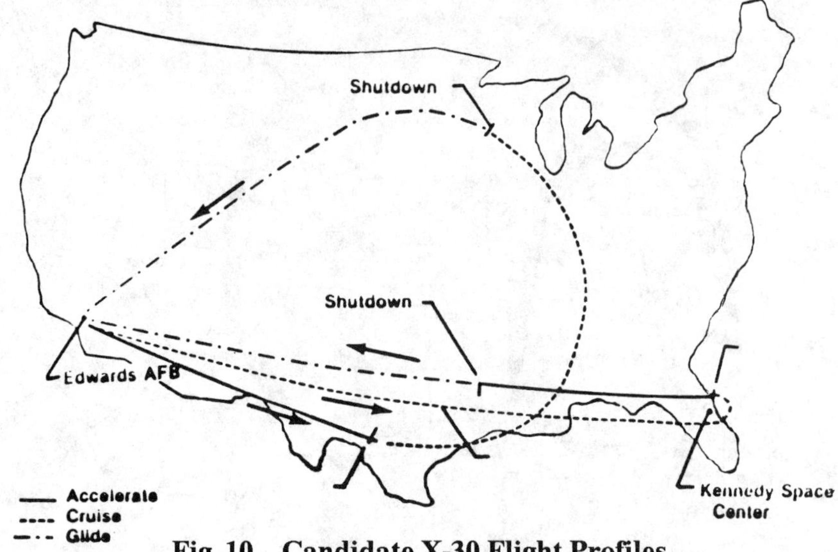

Fig. 10   Candidate X-30 Flight Profiles

The technical challenges are not only in the area of structures, CFD, and propulsion but also in the guidance and control arena. The answer to the question posed in the title of this paper (A guidance and control engineer's dream or nightmare?) depends on the individual engineer's perception of the viability of the National Aero-Space Plane given the technical challenges identified in this paper. Most likely, the answer will depend on his/her affinity to "reach for the stars".

REFERENCES

1. National Aero-Space Plane Technology Maturation Report, "X-30 Air Data System Implementation", Final Report: SRC Report No. 12345, Honeywell Inc., Minneapolis, Minn. August 17, 1988.

2. S.O. Schmitt, T.J. Wierzbanoski, and J. Johnson, "The Challenge of X-30 Flight Test", Society of Experimental Test Pilots XXXI Symposium, September 26, 1987.

AIR-LAUNCHED ORBITAL BOOSTER GN&C REQUIREMENTS: THE PEGASUS DESIGN EXPERIENCE

Daniel Rovner[*] and Antonio Elias[†]

Air launch is currently being exploited as a means of reducing the cost and improving the performance of small space boosters. The use of a carrier aircraft as a launch platform both greatly increases operational flexibility and reduces operational costs. Performance advantages accrue from a combination of effects. However, the use of this launch mode imposes a combination of requirements on the booster guidance, navigation, and control systems that have not heretofore been encountered. To provide the desired operational flexibility the navigation system must be capable of maintaining high accuracy during extended carry flights to the chosen point of drop. The autopilot must have the capabilities of both an aircraft and a launch vehicle control system. The guidance algorithm must be capable of nulling errors introduced by reliance on aerodynamic forces that are subject to significant uncertainties. This paper relates the design process that led to the current configuration of the Pegasus air-launched orbital booster vehicle scheduled for launch in the near future.

AIR LAUNCH: HISTORY AND MOTIVATION

The historical motivation for the development of the air-launched Pegasus booster were purely economical. The goal was to provide a dedicated launch for a "useful" payload at the lowest possible total mission cost, rather than achieve a certain payload capacity or cost per pound. Given recent advances in low-power microelectronic and Microwave Monolithic Integrated Circuits (MMIC's), it was decided that a payload in the 500 lbm range in a 250 nm polar orbit (giving a 1 to 3 year average life time) could be considered "useful". Thus, a very small booster was called for.

Unfortunately, several key items in a launch system do not scale gracefully with reduced size, such as mass fraction and some flight operations costs. Mass fraction, especially in the critical last stage, does not scale with size both because of structural efficiency scaling and because of the fixed overhead of the necessary guidance, control, telemetry, and other subsystems whose mass is essentially independent of vehicle size. Flight operations costs also have fixed elements, such as range support costs, which are largely independent of vehicle size.

These two factors alone would tend to make smaller launch vehicles uneconomical, unless means could be found to break the scaling laws in both mass fraction and flight operations costs.

---

[*] Guidance and Control Engineer, Orbital Sciences Corporation, 12500 Fair Lakes Circle, Fairfax, Virginia 22033.

[†] Chief Engineer, Orbital Sciences Corporation.

Air Launch Performance Advantages

Fortunately, air-launch contributes to both. On the performance side, we have five important performance-improving effects:

- The reduced outside air pressure, which contributes both directly (through reduced pA losses) and indirectly (through larger first stage expansion ratios, which on Pegasus add 20 seconds of specific impulse).

- The shifted velocity-density profile, which affects products of p and V such as dynamic pressure and heating rate, reducing both drag losses and structural and thermal protection weight.

- The flatter overall trajectory, similar to the Apollo Lunar Excursion Module ascent trajectory from the airless moon, with reduced gravity and turning (thrust direction) losses.

- The lift produced by the wing (an almost essential artifact of air-launch) that enables the above-mentioned flatter trajectory and that, even at a low lift to drag ratio, is more energy-efficient than lift from rocket thrust, thus contributing to reduced gravity losses.

- The kinetic and potential energy imparted by the carrier aircraft at drop.

While it is difficult to determine the exact contribution of each of these effects in the general case, single point simulations indicate that for a technology level similar to Pegasus (three stages, 290 seconds vacuum Isp, 0.9 propellant mass fraction, medium-weight single-string avionics, etc.) the air-launched booster has a 10% to 15% ΔV advantage over an identical ground-launched one. When carried over through a multi-stage rocket equation, this ΔV advantage equates to about a 40% improvement in final accelerated mass to orbit. For a Pegasus-sized booster, about half of this mass is payload, so that a 40% increase in mass to orbit represents a near-doubling of the payload to orbit. Thus an efficient air-launched vehicle can be made substantially smaller - and therefore cheaper - than a ground-launched vehicle with similar payload capability.

Operational Advantages

From a simple cost standpoint, air launch offers the potential of simplified, even autonomous range operations with enhanced range safety. No pad refurbishment is required, outside of refueling the carrier aircraft. Finally, air-launch imposes a discipline limiting the number of people that can physically be involved in the operations, commonly a major cost item.

There are, however, other non-economic advantages: most candidate carrier aircraft are also long-range, so that operations from U.S.territory to an equatorial launch point are feasible. Also possible are flights to launch points chosen to achieve orbital or suborbital conditions required by special scientific observations or experiments without the need for portable launch equipment and costly campaigns to remote locations. Weather can be avoided en route to a 33,000 ft drop altitude.

Finally, there are features of military interest, such as the potential for first pass overflight of any location on earth by choosing the appropriate launch longitude, the possibility of extremely secure inland operations, and a surge capacity unconstrained by ground facilities.

The Pegasus Implementation of the Concept

Pegasus, conceived in early 1987, is the first example of what could be a new family of orbital boosters. Because it was designed to strike a balance between development and recurring costs, it does not represent the ultimate air-launched booster even with our present limited knowledge and experience in designing this kind of vehicle.

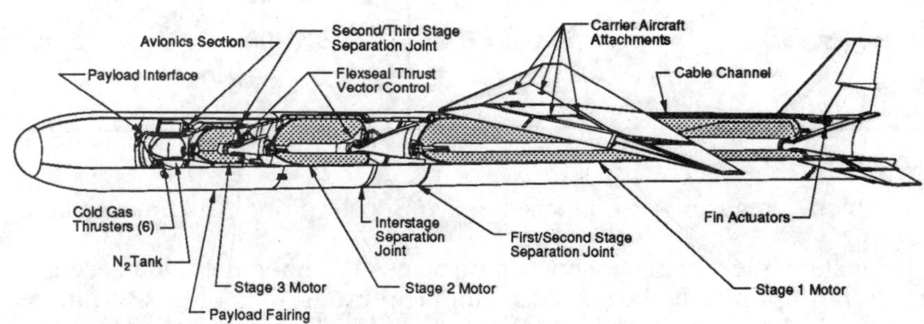

Fig. 1   Pegasus Air-Launched Space Booster

As seen in Fig. 1, Pegasus is a three-stage, all solid-propellant vehicle. Table 1 summarizes its principal dimensions. While a number of points made in this paper appear to apply specifically to Pegasus rather than to a generic air-launched booster, we have opted with particularizing them to Pegasus for two reasons: it makes it easier to explain them; and it is likely that most if not all early air-launched boosters will resemble Pegasus in its choice of implementation alternatives.

While the original Pegasus design did not have a wing, it became apparent in very early simulations that the use of a wing to achieve a positive flight path angle after drop is an absolute necessity: turning the vehicle around with thrust alone resulted in large altitude losses, high q-alpha-induced structural and control authority stresses, large thrust direction and gravity losses, or all of the above. While complete parametric runs were not made, it appeared that a wing loading between 200 and 250 psf yielded efficient initial ascent trajectories, depending on the drop altitude. A delta planform was chosen to minimize the shift in wing center of pressure with Mach number, as will be seen later. Similarly, a 45 degree sweep angle was deemed a good compromise between low-Mach and high-Mach drag. Thus the aspect ratio, and consequently the span, becomes a function of Gross Drop-Off Weight (GDOW), the latter relationship being approximately $span(ft) \approx 0.13\sqrt{GDOW(lbs)}$.

Motor Characteristics	1st Stage	2nd Stage	3rd Stage
Inert Weight (lb)	2,780	800	277
Propellant Weight (lb)	26,790	6,670	1,725
Burn Time (sec)	72.3	71.4	64.6
Max Pressure (psia)	1,088	1,003	749
Avg Pressure (psia)	797	793	637
Max Vac Thrust (lbf)	131,244	30,912	9,065
Avg Vac Thrust (lbf)	109,419	27,605	7,772
Vac Impulse (lbf-sec)	7,911,000	1,971,000	502,100
I_{sp} Vac (sec)	295.3	295.5	291.1

Table 1   Pegasus Air-Launched Space Booster Principal Dimensions

Unfortunately, the clearance between fuselage and inboard engine nacelles of the B-52 carrier aircraft used in the Pegasus development program is only 22 ft, increasing the actual wing loading to a suboptimal 300 psf. This fact influences the trajectory-derived numbers shown in this paper as well as the performance advantage actually achieved.

Fig. 2 summarizes a typical Pegasus trajectory to a 250 nm circular trajectory.

GUIDANCE, NAVIGATION, AND CONTROL REQUIREMENTS

The next sections will discuss the requirements that the Pegasus GN&C system was designed to satisfy, and the design choices that were made to satisfy them. The fundamental requirements imposed on the design can be simply stated: to insure that the vehicle places the largest achievable payload into the most accurate orbit possible, in the face of the widest range of ambient conditions expected to be encountered, at the lowest possible cost, and with the minimum of operational complexity.

NAVIGATION SYSTEM REQUIREMENTS

In common with most launch vehicles, Pegasus relies on inertial navigation to determines its position, velocity, and attitude during flight. Since (as will be discussed below) the decision was made to implement a closed-loop guidance strategy, a full inertial navigator is required as opposed to a simple attitude reference platform. However, the air launch mode of operation again imposes unusual requirements on the inertial system.

The salient aspect of Pegasus' operational scenario is the (potentially) long captive flight to the point of drop. This flight is the key to the operational flexibility of Pegasus, but it presents a significant gyro drift problem. Pegasus must be a low cost vehicle if it is to accomplish its objectives, and low cost inertial measurement units do not maintain high accuracy when operated in free inertial mode for extended periods of time.

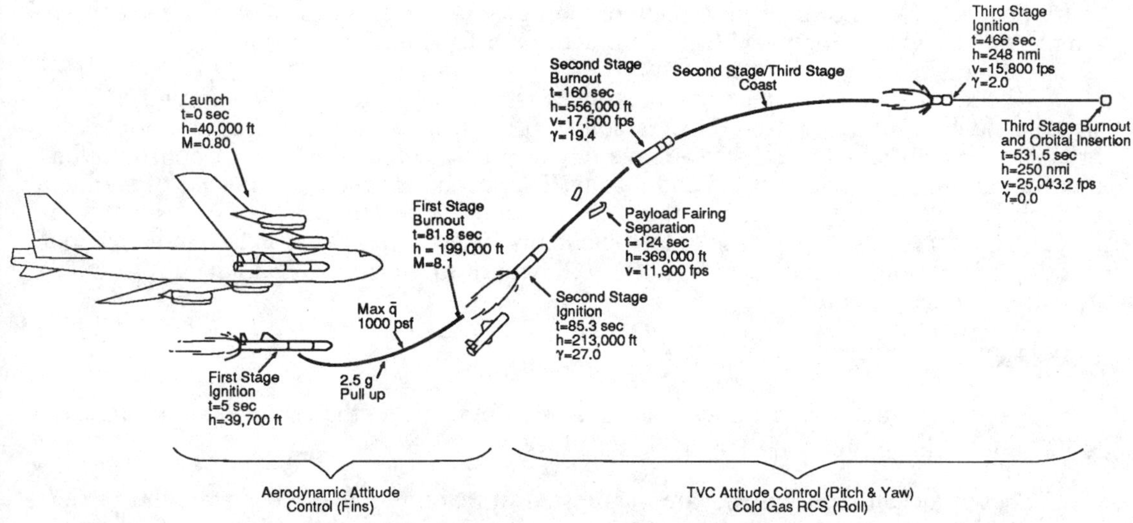

Fig. 2   Pegasus Ascent Trajectory Summary

The solution is to provide external aids to the inertial navigator. The baseline Pegasus operational scenario for the captive flight requires velocity updates to be continuously provided to the vehicle INS from an outside source. A Kalman filter on board the vehicle uses these velocity updates to infer the attitude of the gyro triad. This is possible because the estimation of attitude is correlated with the estimation of inertial acceleration. Thus vehicle attitude misalignment is "observable" from measured velocity residuals.

The Kalman filter is also used to calibrate the inertial instruments, by providing estimates of a number of other error sources. The present implementation estimates (in addition to attitude and velocity) the non-random gyro and accelerometer biases. These estimates lower the effective drift of the gyros and accelerometers, so that for the non-recurring cost of the filter software, the performance of a more expensive INS can be achieved with the relatively inexpensive Pegasus unit.

For the Pegasus implementation, a four-gimbal Litton LN-39H, with 0.1 deg/hr gyro bias drift instruments, is mounted on the carrier B-52 Launch Panel Operator's console. Velocity and position vectors are transferred to the Kalman filter in a specially-modified Litton LR-81 dry tuned-rotor gyro INS. In addition, a special signal is produced by the Pegasus LPO and supplied to both INS's for clock synchronization.

This approach does somewhat complicate launch operations, however. On the ground, two separate INS's must be allowed to gyrocompass. As soon as gyrocompassing is completed, communications must be established between the two units so that filtering of velocity measurements can begin. This communication must be continued up to the time of drop. Most importantly, to provide the Kalman filter with a sufficiently information-rich stream of measurements, just prior to drop the carrier aircraft must perform a series of maneuvers. These maneuvers provide time-varying attitude and acceleration inputs to the INS's, providing a strong signal that helps to distinguish instrument bias from noise. The strength of the signal is limited by the ability of the carrier aircraft to maneuver at high load factor, altitude, and Mach number.

Two alternative approaches are possible: first, the use of a single "compromise" INS with sufficient quality to provide satisfactory orbital injection with simple pre-takeoff ground alignment. As laser devices come down in cost, weight and power consumption this alternative may become the desirable one. The second alternative is the use of another source of external measurements, perhaps GPS, to provide position and velocity updates in lieu of the reference INS. Because of the potentially longer times between INS initialization and end of mission, air-launched boosters will benefit more from GPS than conventional ground-launched boosters.

CONTROL SYSTEM REQUIREMENTS

Air launch imposes a number of unusual requirements on the vehicle attitude control system during the initial atmospheric phases of flight.

These unusual requirements are manifested first at the time of drop from the carrier aircraft. Clearly the vehicle cannot be thrusting at this time, and for reasons of safety, it cannot begin powered flight for a minimum of several seconds subsequent to drop. Therefore the air-launched vehicle must be capable of stable gliding flight. Either the vehicle must possess a high degree of static aerodynamic stability at drop, or it must have active aerodynamic control surfaces.

The design choice made for the Pegasus vehicle steers a middle course. The vehicle is in fact statically stable for both longitudinal and lateral motions. However, it has also been equipped with aerodynamic controls, for several reasons. First, a significant performance advantage is derived by beginning to pitch the vehicle up to a significant angle of attack even before motor ignition. Preventing the nose from dropping reduces thrust direction losses after ignition considerably.

In fact, as has been discussed, the air-launch imposes a requirement that the vehicle fly stably at a large angle of attack for a significant portion of the first stage burn, as seen in Fig. 3. This is a unique aspect of the Pegasus trajectory. Previous air-launched space boosters (such as the Air Force anti-satellite missile) have relied on the carrier aircraft to provide a high initial flight path angle. In contrast, the size of the Pegasus carrier aircraft makes a horizontal drop imperative. It is thus necessary that Pegasus be controllable at relatively high angles of attack. Aerodynamic controls have traditionally been used for this purpose.

Since aerodynamic controls are desirable to confer a performance advantage early in the flight, it clearly makes sense to use them throughout the duration of the first stage burn. The additional weight and expense of a thrust vector control system is then avoided. Providing the substantial pitching moment required to maintain angle of attack as the burn progresses (and the static stability margin grows) aerodynamically also saves further on thrust direction losses compared to a TVC system. It is especially appropriate to provide the pitching moment aerodynamically, since the required moment and the available control authority both scale similarly with dynamic pressure.

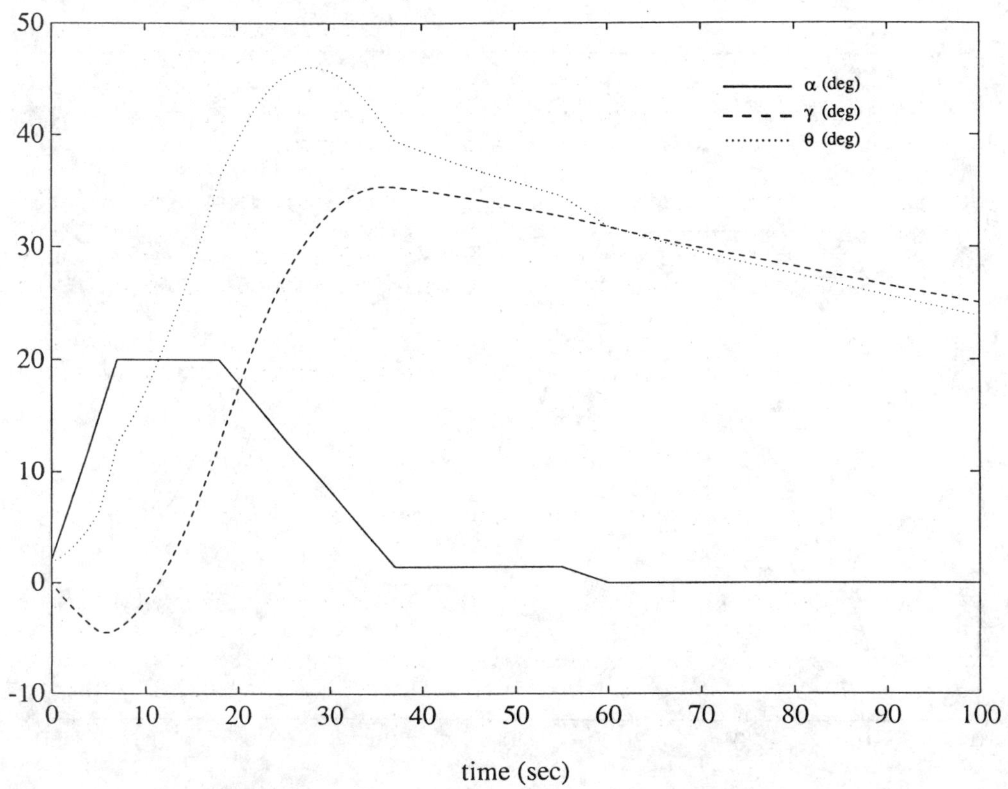

Fig. 3   Time History of Angle of Attack, Flight Path Angle, and Pitch
Angle During Atmospheric Phase of Ascent

Nevertheless, reliance on aerodynamic control surfaces leads to a serious dilemma towards the end of the first stage burn. At this time the dynamic pressure is falling rapidly, while the rocket engine thrust is still near its maximum value. Because it is physically impossible to align the thrust vector perfectly through the vehicle center of mass, the control system must be able to react a relatively constant thrust-induced overturning torque at all times. This need must be reconciled with the rapidly diminishing control authority available due to decreasing dynamic pressure.

The solution to this problem chosen for the Pegasus vehicle relies on augmentation of aerodynamic control authority through the use of a special reaction control system.   The system consists of a number of small solid-propellant rocket motors that are mounted in the aerodynamic control surfaces and ignited late in the first stage burn. This approach has a number of desirable features as a solution to the problem of maintaining control authority. First, the torques applied by the fin rocket motors are proportional (as opposed to on-off) and are controlled by the fin actuators, so that no additional actuation elements are required. Second, even in the worst case situation, a significant fraction of the thrust from the fin-rocket system is directed axially, adding to vehicle ΔV. The results of simulations conducted to evaluate the performance of the fin-rocket system are shown in Figs. 4 and 5.

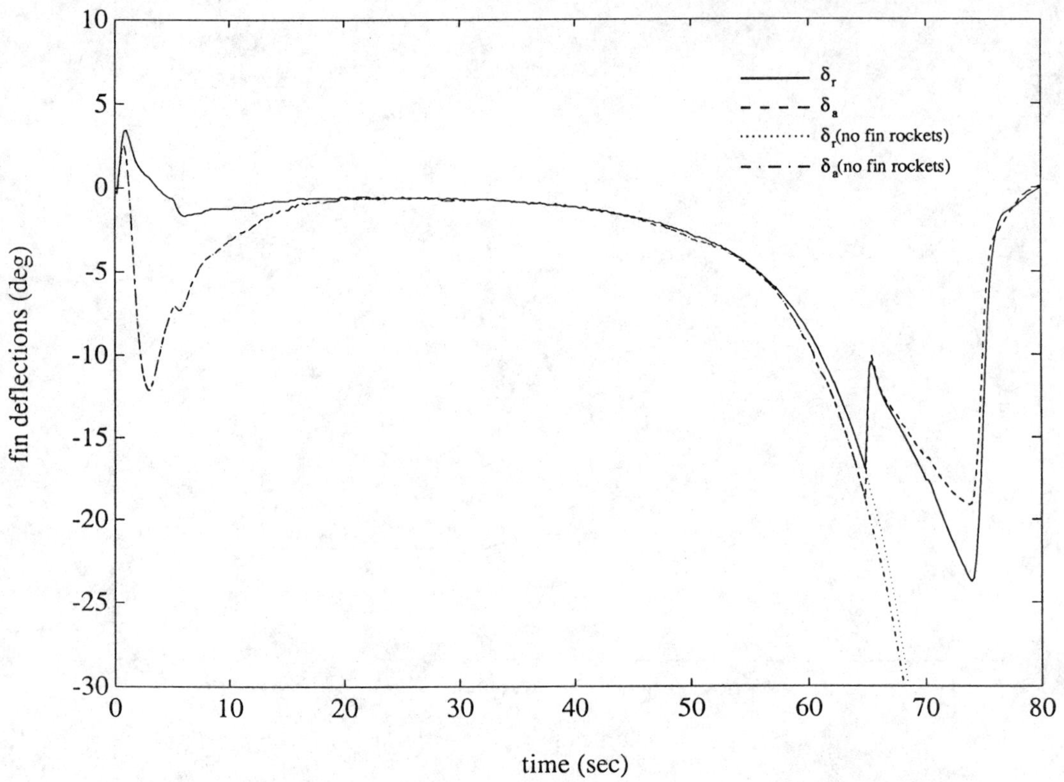

Fig. 4   Time History of Fin Deflection Angles Generated to Counter 3σ
Thrust Misalignment Torque

The effects of a (3σ worst case) 0.18° thrust yaw misalignment are shown. Without the fin
rocket system, vehicle attitude diverges as the control surfaces reach their limits (at t ≈ 68
sec, Fig. 4) without being able to provide sufficient control moment to react the thrust
disturbance torque (Fig. 5). In contrast, the corresponding curves showing fin deflections
and control torques with the fin-rocket system operational reflect the containment of yaw
attitude error to a small value. (This simulation also shows the effects of an initial 8° error
in drop heading, causing the large initial rudder excursions in Fig. 4.)

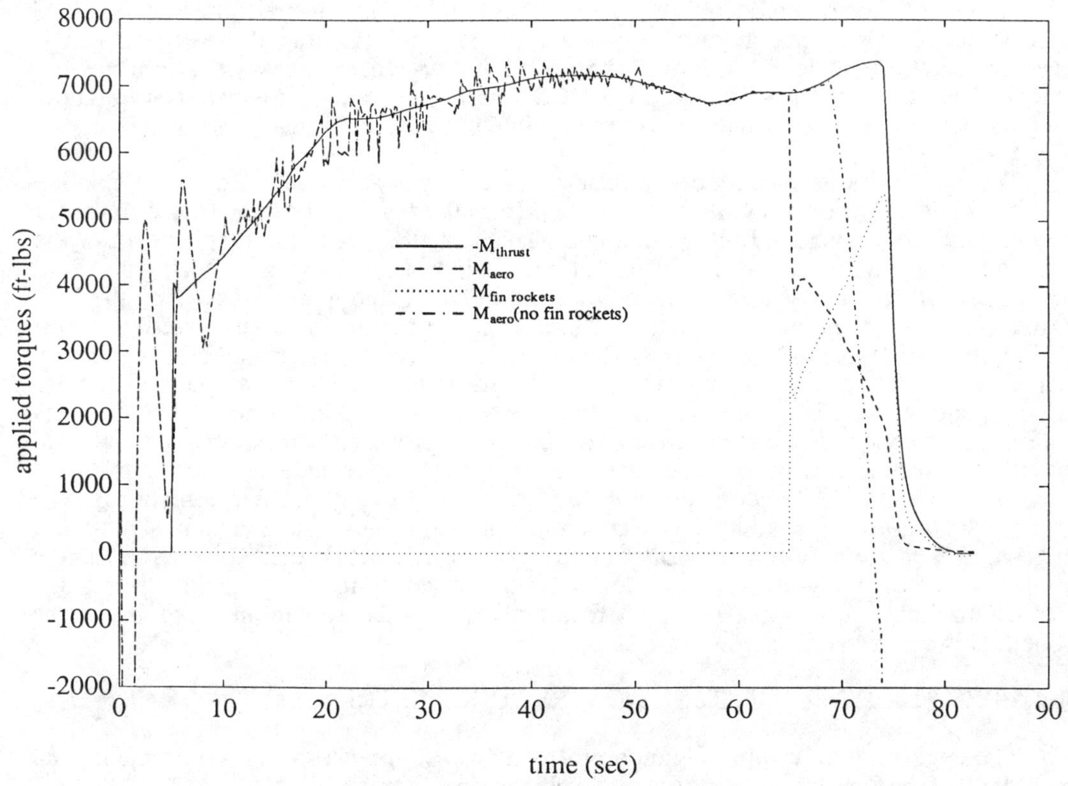

Fig. 5   Time History of Torques Applied to Vehicle with 3 Sigma Thrust
Misalignment

Paradoxically, despite the fact that during the later period of the first stage burn the lack of dynamic pressure leads to serious control problems, the continued (albeit somewhat tenuous) existence of aerodynamic forces following first stage burnout creates new problems following staging. The shape and mass properties of the upper stage stack virtually guarantee a center of mass location that is aft of the center of pressure. The coast control system minimum authority requirement imposed by the consequent aerodynamic instability is reduced by targeting for low ambient dynamic pressure. Unfortunately this approach is of course in direct conflict with the requirement to maximize the control authority available during powered first stage flight.

Two options were considered to provide control during coast. The first, adding fixed aerodynamic surfaces, was rejected due to the size and weight of the fins required to confer stability. The second was to rely on the cold gas reaction control system mounted just behind the payload shelf. This system, as will be further described, also provides roll control during powered second and third stage flight, three axis control during the exoatmospheric second/third stage coast, and payload pointing following third stage burnout.

The final control configuration strikes a somewhat delicate balance between the requirements of the powered first stage flight phase and the unpowered first/second stage coast phase. To insure that a reasonably-sized hot gas RCS system will be effective, it is necessary that the vehicle not endure a long period of low-dynamic pressure before staging. To insure that a reasonably-sized cold gas RCS system is effective, it is necessary that the vehicle reach as low a dynamic pressure as possible just before staging.

The exoatmospheric phase of the Pegasus trajectory is very similar to that of existing launch vehicles. The only significant departure from the norm that can be traced to the air launch is the lower average flight path angle. This is actually a significant benefit for the guidance algorithm, but it has little effect on the design of the autopilot. During powered second and third stage flight, attitude control in pitch and yaw is accomplished by gimbaling the rocket engine nozzle. Thrust vector control was preferred to spin-stabilization for reasons of operational flexibility. Roll control is provided by a reaction control system. Cold gas was chosen for the propellant due to the small amount of impulse required, and the ease of ground handling compared to a hydrazine system. The cold gas system also provides pitch and yaw control during unpowered coasts between first and second, and second and third stage burns. As has been described, the maximum thrust requirement for the RCS occurs just after the first staging event. Luckily, this is also the time at which the RCS has the largest moment arm about the vehicle center of mass. A trade off was made between providing control authority for this period of aerodynamic instability, and avoiding providing too much control authority in roll during the third stage burn. This latter situation arises as a result of diminished roll moment of inertia as propellant is consumed.

DETAILED FLIGHT CONTROL SYSTEM REQUIREMENTS

The Pegasus aerodynamic flight control system design must satisfy requirements that are a blend of those characteristic of launch vehicles and aircraft.

Like most aircraft, Pegasus must be controlled stably at significant angles of attack using movable surfaces. Design of the autopilot must take into account vehicle dynamics that are significantly different in character for motions in and out of the plane of symmetry. In particular, aircraft control experience at high angles of attack has shown the importance of minimizing sideslip angle to avoid serious stability problems.

As previously described, Pegasus is longitudinally stable, and so displays longitudinal dynamics that are familiar to aircraft designers. Both phugoid and short period modes are predicted to exist. Unlike most aircraft, however, the frequencies of these modes (where a quasi-steady assumption about the dynamics is of course being made) shift significantly within a time span of 60 seconds (Fig. 6). The changes are due to a change in altitude of about 200,000 feet, a change in Mach number from about 0.8 to about 8.3, and a change in dynamic pressure from about 180 psf, to 1000 psf, and then back to about 10 psf.

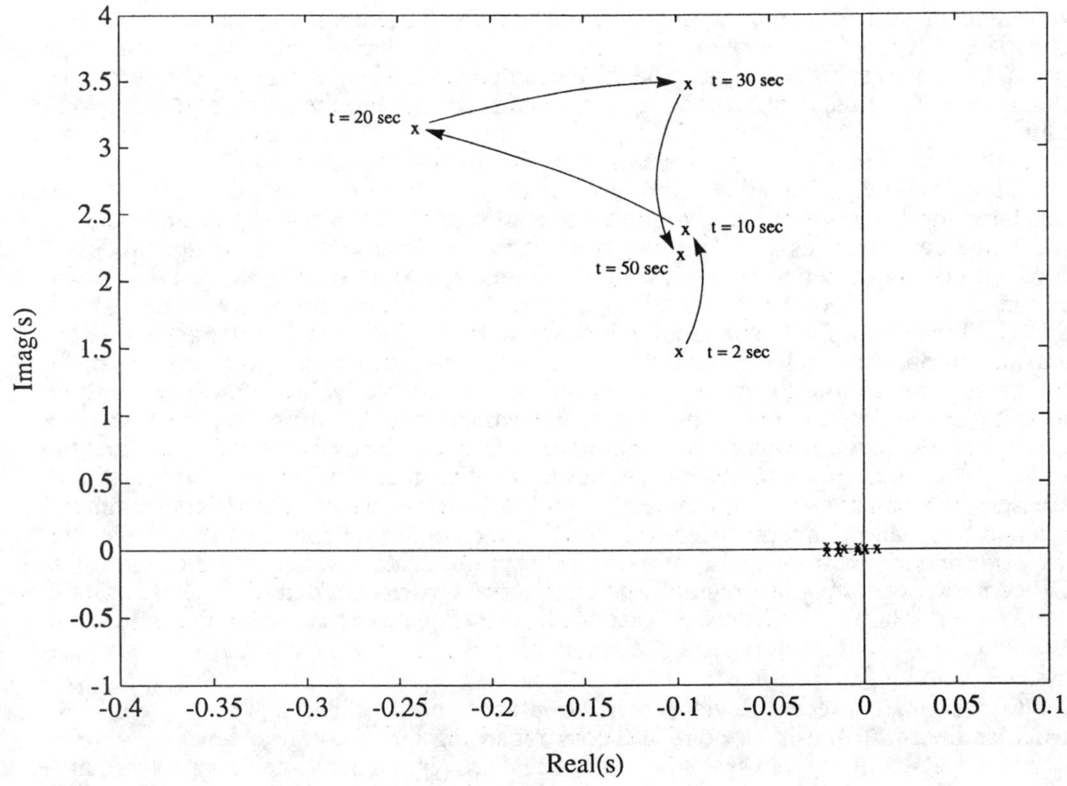

Fig. 6   Time History of Pegasus Longitudinal Modes During First Stage
Flight

Pegasus is also predicted to behave like a (fairly) typical airplane for small motions out of the plane of symmetry. There are typical spiral and roll subsidence modes, and an oscillatory rolling motion that is similar to a Dutch roll, though the great disparity between the moments of inertia in yaw and roll tends to eliminate any yawing motion. Again there are significant shifts in frequency during the course of the first stage burn. However, in contrast to the longitudinal dynamics, the lateral oscillatory mode tends to become unstable at high Mach numbers as the vertical tail loses effectiveness.

Because Pegasus is a rocket-powered launch vehicle, the design of the aerodynamic autopilot is driven by some requirements that are very unusual for winged vehicles. One of the most striking of these is the necessity of reacting large thrust-induced moments about the vehicle pitch and yaw axes. Heading angle feedback must be used to prevent potential thrust moments from spinning Pegasus about its vertical axis. Pegasus also does not necessarily always fly under conditions of positive lift. This makes flying coordinated turns problematic, as the existence of a lift vector to tilt in the direction of the turn is not guaranteed. Luckily, a space launch vehicle has no need of executing significant lateral turns, so that no capability for turn coordination is in fact planned. As noted above, though, it is very desirable that sideslip be kept to a minimum, and here Pegasus is aided by the dihedral stability conferred by the vertical tail and the small roll moment of inertia. A strong lateral wind disturbance induces a rolling response that kinematically converts any sideslip angle into angle of attack. The lift generated by this angle of attack accelerates the vehicle away from the gust, canceling its effects and allowing the vehicle to roll upright again.

Three different (linear constant coefficient) control design methodologies were considered for the design of the autopilot, modern linear quadratic, Q-design, and classical. Linear quadratic synthesis has a number of attractive features. The design process is simple and is guaranteed to result in a stable controller. Additionally, using the "pincers" method, the closed loop poles can be located as far to the left of the imaginary axis as desired. Finally, if full state feedback is available, LQR controllers have attractive robustness properties, including at least 60° of phase margin and infinite gain margin. For the Pegasus application, very nearly the full state is available, velocity, position, attitude and attitude rate from the inertial navigator, and actuator position from the actuator control electronics. However, it was decided not to use the actuator position feedbacks since this would have made the talkback electronics critical to mission success. It was found necessary to estimate the actuator states to obtain stable LQR controllers under this circumstance. Unfortunately, once an estimator is included in a linear-quadratic design, the previously-mentioned robustness properties disappear. Additionally, the behavior of an LQR controller operating in a regime where the actuators are saturated is difficult to predict or understand intuitively. For these reasons, the linear quadratic approach was rejected as unsuitable.

Q-design[1] is a recent development in optimal control theory. The 'Q' refers to a particular finding from linear control theory regarding the structure of *every* controller capable of stabilizing a given system. It has been discovered that all such controllers can be expressed very simply using a transfer function realization known as the 'Q-parameterization'. This realization in turn has led to a technique for posing controller design as an optimization problem, where the quantities being optimized have more direct physical meaning than those of linear quadratic synthesis. For example, the controller can be constrained to have a step response that falls between certain specified bounds, or a frequency response characteristic that falls below a specified value at a specified frequency (i.e. direct specification of bandwidth). Unfortunately, controllers designed by this method must be realized by finite-impulse-response filters. The resulting controllers generally are of very high order, typically with many tens of states. It was decided that this would not be implementable on the chosen flight computer.

To maintain stable control under this broad envelope of conditions, the choice has been made to use a controller design developed using classical methods. In particular, attitude error, attitude rate, and (in the case of the pitch axis) the integral of attitude error will be used to construct control surface commands for each axis. The stability and performance of classical PID feedback systems are typically very insensitive to changes in system dynamics. In addition, effective techniques of dealing with control surface saturation that have clear intuitive meaning are available. Finally, PID gains can easily be scheduled to compensate for predicted changes in the flight envelope. In fact, since the parameters of a launch vehicle ascent trajectory must necessarily be a pre-determined function of time, it is possible to schedule controller gains purely as a function of time.

A block diagram of the longitudinal aerodynamic autopilot is shown in Fig. 7. The command prefilter is a simple second order low-pass that is used to help avoid undesirable control surface saturation. If saturation is to occur, however, it will be applied by the autopilot software as shown. The control signal saturation scheme can be intuitively understood as a combination of rate command limiting and acceleration command limiting. Simulation experience has shown that this "double saturation" helps to maintain stable autopilot operation when large external disturbances are encountered. An anti-windup switch is incorporated in the (digital) integrator. The control signal actually applied to the actuators is constructed by applying a zero-order hold to the digital command. Finally, the pitch rate is sensed directly by a strap-down tuned-rotor gyro, while the pitch attitude is obtained through a kinematical integration performed by a processor incorporated in the INS package.

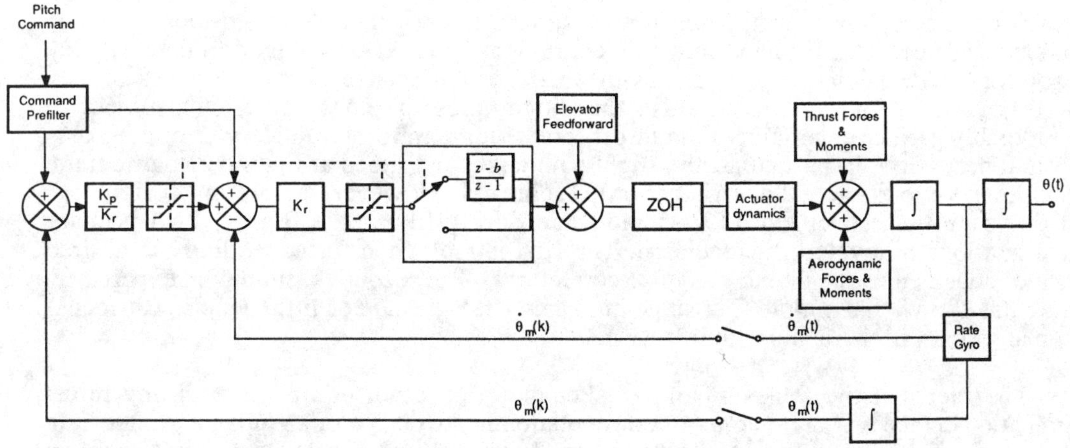

Fig. 7   Block Diagram of First Stage Longitudinal Aerodynamic Autopilot

The lateral autopilot is very similar in design to the longitudinal. The roll and yaw channels are partially decoupled by defining a "logical" set of controls that diagonalize the control distribution matrix (using a state-space realization that corresponds to the standard definition of aircraft lateral state[3]). The two partially decoupled loops are then structured almost identically to the longitudinal channel.

GUIDANCE SYSTEM REQUIREMENTS

The fundamental requirement on the guidance system is naturally to place the payload in the desired orbit within an error tolerance acceptable to the customer. The Pegasus design accuracy is tailored to the baseline mission previously described of a 250 nm altitude nominally-circular orbit.

Altitude accuracy is a major concern for low-earth orbiting satellites. This is because a trade must be made to insure adequate lifetime of the orbit. A high nominal altitude insures that the perigee altitude will be large enough to reduce atmospheric drag to an acceptable level. However, increasing nominal altitude reduces payload capability.

The accuracy with which a desired orbital inclination is achieved is also of importance, e.g. for customers desiring sun-synchronous orbits. Here there is not a significant accuracy/payload trade. Inclination accuracy is determined essentially by the accuracy with which the vehicle thrust vector can be pointed.

The fundamental choice to be made for the Pegasus vehicle was whether a closed or open-loop guidance policy should be adopted. Here an open-loop policy is defined by the characteristic of providing attitude commands to the autopilot without incorporating corrections derived from navigation data. This is the simplest imaginable scheme. It also has reliability advantages, since it reduces software complexity and makes accelerometer data optional. Currently, this is the approach used by the Scout small launch vehicle.

For Pegasus, however, the decision was made to implement a closed-loop guidance strategy that updates attitude commands continuously based on navigation data. A key reason for this decision was provided by the air-launch. Because the vehicle relies heavily on aerodynamic forces, and because the magnitudes of those forces are subject to considerably greater uncertainty than thrust forces, some form of trajectory error control is a virtual necessity. In particular, the flight path angle and speed at first stage burnout are quite sensitive to the actual coefficient of lift achieved. However, the present design does not envisage the use of closed-loop guidance during first stage flight. Two primary considerations motivated this decision. The first is the danger of exceeding the desired angle of attack envelope due to guidance corrections. The second is simulation experience, which has shown that simple first stage guidance strategies are of little benefit, especially when an exoatmospheric algorithm is used.

The fact that the Pegasus solid rocket motors are not equipped with any thrust termination device was another major consideration in the choice of a guidance strategy. In fact impulse uncertainty was a major driver towards the choice taken to implement an explicit (as opposed to a linearized) guidance algorithm.

The algorithm currently baselined for Pegasus is an adaptation of Powered Explicit Guidance[2]. This algorithm uses a predictor-corrector scheme. Trajectory end conditions are predicted in real time based on current navigation state and a projected thrust direction time function. The parameters defining the thrust direction history are then adjusted as a function of the difference between the predicted and desired end conditions so as to drive the error to zero.

The basic form of thrust direction time function is derived from optimization theory. Its salient characteristic is a pitch angle whose *tangent* is a linear function of time. By choosing this simple, and nearly fuel optimal, steering law, the (in-plane) trajectory of the vehicle is determined by just two variables: the initial direction of the thrust vector and its rate of turn. These two variables allow two (in-plane) trajectory constraints to be satisfied, for example altitude and flight path angle. Given a method of thrust termination, a third constraint, typically burnout speed, could also be satisfied.

The calculation of the values of the two steering variables that satisfy the end constraints proceeds iteratively. An initial guess is made of the two values, and is used to find the contribution of thrust to the change in vehicle position and velocity state. Gravity contributions are then approximated by constructing a free-fall trajectory that comes close to the actual powered trajectory. The end condition velocity error vector is then used to correct the mean direction of velocity gained due to thrust. The iteration generally converges within four or five cycles.

For the Pegasus application, the attractive feature of this algorithm is its ability to handle large trajectory perturbations with ease. A new trajectory is calculated in real time that is nearly minimum ΔV and satisfies a range of end condition constraints. However, it has been necessary to make modifications to allow PEG to work with motors whose total impulse is fixed. These modifications remove one of the end condition constraints. For the stage 2 burn, the constraint on burnout speed is removed. This is done by correcting only the direction of the mean velocity-gained vector, and not its magnitude. The burnout velocity vector is pointed in the desired direction, but its magnitude is in error due to any error in stage total impulses. For stage 3, the constraint on speed is traded for a constraint on altitude. This change in constraints is effected by making the thrust direction turning rate a function of the speed error instead of the desired altitude. It has been found that this combination of end condition constraints gives the best overall accuracy of orbital injection.

A new feature has been added to the guidance algorithm that makes it possible to compensate for the inability of PEG to satisfy all end constraints due to the lack of thrust termination capability. The powerful predictor built into the PEG algorithm makes it possible to calculate the solution to a non-linear programming problem during the long coast between second and third stage burns. The objective function minimized is the predicted stage 3 burnout altitude error, with the control variable being the time duration of the coast. Simulations have shown that the burnout altitude is quite controllable by the coast time, while the computational burden is well within the capability of the on-board flight computer.

By designing the nominal trajectory so that sufficient ΔV capability exists to reach a higher than desired altitude, it is insured that a low-performing vehicle can still reach the desired orbital apogee. On the other hand, nominal-or-better performing vehicles can be guided to the nominally-desired altitude by adjusting the time of coast in flight.

If sufficient velocity margin is allowed for at launch, the primary source of injection errors are off-nominal conditions during the third stage burn. There are several sources of dispersions. Perhaps the most important are INS velocity estimation errors, as these result in essentially one-for-one errors in injection conditions.

Errors in third stage motor total impulse and uncertainties in vehicle inert weight are also significant error sources. The latter are aggravated by the relatively small size of the Pegasus vehicle. Surprisingly, an error in knowledge of third stage weight of as little as 2 lbm can cause a 10 ft/sec error in injection speed.

A number of other error sources contribute second-order effects by causing the direction of thrust at any time to be pointed in a direction that is not precisely aligned with the desired direction. Since the nozzle is gimbaled, the autopilot will on the average point the thrust vector through the vehicle center of mass. Therefore, it is important that the geometric relationship between two vectors be known as exactly as possible in order that the direction of thrust be aligned correctly. The first vector is defined by a line passing through the thrust action point and the center of mass. The second vector is aligned with the instrument axes of the INS.

Inevitably, the speed dispersion caused by uncertainty in third stage motor impulse represents an undesirable limitation on achievable insertion accuracy. Thus a means of trimming the final speed would be an attractive addition to Pegasus. Since the amount of ΔV required is small, on the order of 10-20 ft/sec, a cold gas system could conceivably provide the necessary impulse. Such a system is being considered as a future Pegasus upgrade.

CONCLUSIONS

Air-launched space boosters may usher in a new era of low-cost orbital payload delivery with unprecedented operational flexibility. At the same time, they present new challenges in several areas: autonomous inertial and hybrid navigation, control strategy and design in the face of significant plant uncertainties, and (especially in the case of pure solid propulsion) guidance strategies that lead to orbital insertion accuracies limited only by navigational performance.

After the first two flights scheduled to take place in mid 1989, the current Pegasus implementation of the air-launch concept will be reviewed. The effectiveness of the chosen design solutions will be evaluated in the light of the extensive flight data that will be accumulated by the Pegasus telemetry system.

REFERENCES

1. R. L. McHenry *et. al.*, "Space Shuttle Ascent Guidance, Navigation, and Control," *J. of the Astronautical Sciences*, Vol. 27, No. 1, Jan.-March 1979 pp. 1-38

2. S. P. Boyd *et. al.*, "A New CAD Method and Associated Architectures for Linear Controllers," Information Systems Laboratory, Stanford University, TR L-104-86-1, December 1986

3. Duane McRuer, Irving Ashkenas, and Dunstan Graham, *Aircraft Dynamics and Automatic Control*, Princeton University Press, Princeton, New Jersey, 1973.

DEMONSTRATING SYSTEM FOR SERVICING SATELLITES

Lyle M. Jenkins[*]

NASA has initiated a program to demonstrate a Satellite Servicer System prototype in a series of flight tests with the Space Shuttle. When the commitment to an operational system has been demonstrated, it is anticipated that a variety of missions will develop that can productively use the servicing capability. Rendezvous and proximity operations contain key enabling technologies needed for an effective operational system. Guidance, navigation and control hardware and operational techniques must be integrated into the program. Supporting sensor technology may limit or enhance the capability of the system depending on its readiness. Requirements for docking the spacecraft in the system concept development and technology requirements will be reviewed. Examples of the initial assessment of tehnology readiness and applications opportunities will be discussed. The degree of autonomy is of particular interest. Necessary elements for an effective servicing system can be combined in a worthwhile system flight demonstration that is the basis for and operational system.

INTRODUCTION

NASA has defined satellite servicing as any activity performed on orbit to assemble, maintain, repair, resupply, upgrade, deploy, retrieve or return various spacecraft and/or facilities. In a report to congress, "On-orbit servicing is a capability in an early stage of evolution. To date, such servicing has been entirely dependent on the presence and support of humans in space. Dependence on man

---

[*]   NASA Johnson Space Center, Houston, Texas 77058.

has allowed great flexibility in the servicing tasks attempted since human beings are inherently inventive and dexterous. However, it has also severely limited the number of programs that have used or can use such a capability, since such programs must be associated with, or have the capability to achieve, a close proximity with the manned space vehicle."[1] Some of the potential servicing missions will continue to involve the space shuttle as well as expendable launch vehicles, unmanned spacecraft and the space station. The plans for implementing an operational capability for servicing satellites have been stalled for a number of years by the circular argument of"what is your capability to service satellites?" and "What are your requirements for servicing?","we can't design for servicing is there is no equipment for doing servicing", "but we can't get funding for equipment if there are no requirements for servicing." The NASA Office of Space Transportation has taken the initiative to break the cycle by planning a satellite servicer demonstration program. The project will perform activities that could lead to an operational capability. A major element is rendezvous and proximity operations including docking or berthing.

For many years, the Russians have been performing remote or automatic rendezvous and docking operations. The U.S. has not demonstrated this remote capability. Manual piloting has been at the heart of the numerous operations in the nation's experience The next phase of the development of a satellite servicing system for the nation may produce the needed demonstration of a remote or autonomous capability to meet and join spacecraft. There is a concensus that the necessary hardware and operational techniques are at an adequate state of readiness. When the appropriate objective is defined, there is little risk in being able to perform a satisfactory rendezvous and docking operation at a remote site.

The categories of technology to perform the rendezvous and proximity operations can be defined as sensors, actuators and guidance, navigation and control. These technologies can be combined to demonstrate the readiness to develop and operational system. The key enabling technology can be described in three phases; rendezvous for operations under inertial navigation, approach for navigation relative to the objective spacecraft, and docking for physically acquiring the objective spacecraft.

---

[1]"Satellite Servicing: A NASA report to Congress", March 1, 1988.

The demonstration program will define a system and operations necessary to achieve a level of confidence in a protoflight system that will permit interaction between the servicee and servicer to initiate development of capability and compatibility for operational utilization of satellite servicing. Specifically in the area of rendezvous and docking, the mission requirements, system requirements and design requirements for a variety of missions can be synthesized to establish critical issues to be addressed in the demonstration.

DEMONSTRATION PROGRAM OBJECTIVES

In order to break the cycle of inaction on satellite servicing, the demonstration of a basic capability for performing on orbit servicing is fundamental in a program development process.   On orbit servicing encompasses activities of assembly, maintenance, resupply, repair, upgrade, deployment, and retrieval.   Assembly is needed when the final configuration exceeds the capacity of the transportation system.   Maintenance assures that operational life is not limited by component parts or deterioration that is time dependent.   Resupply of expendables extends operational life. Repair determines cause of failure or damage in primary or backup systems and repairs or replaces units to return to the satellite to original function.   Other orbital replacement includes the upgrading of systems to expand performance with newer technology or different instrument packages.   Retrieval involves transporting the satellite from its operational orbit to perform other servicing or to return to earth.   The demonstration concept should accommodate as many servicing functions as is practical within time and funding limitations.

SERVICING SYSTEM DESCRIPTION

The specific servicing mission will determine the system configuration.   Pragmatic considerations of funding and schedule limit the development of new and specialized servicing system elements.   The building blocks of the servicing system will undoubtably include the Orbital Maneuvering Vehicle(OMV) and the Flight Telerobotic Servicer(FTS) that are under development.   An additional element termed the Carrier will provide functions not currently planned for the OMV and FTS.

The OMV capabilities include primary propulsion, attitude control, auxiliary propellant supply, primary communication and primary power supply. Capabilities that the FTS brings to the system are manipulative functions to accomplish Orbital Replacement Unit(ORU) exchange, tank exchange, umbilical connection, and to aid in docking. Sensors and system control functions may also be within the system capability. The Carrier serves to integrate the elements a with interfacing structure and support for payload. It supplements communication and power supply and provides mission unique sensors. Rendezvous and proximity operations including docking are a key function of the carrier. The management of onboard systems during the mission as well as checkout and fault isolation are also considered functions of the Carrier.

DEMONSTRATION OBJECTIVES

The demonstration needs to address certain critical tasks to be effective in extending the role of servicing in the development of new satellite systems. Rendezvous and docking is key to all remote servicing tasks. The degree of autonomy is a significant issue that depends in part on requirements of future missions and in part on the commitment to reduce requirements on ground control. Resupply of fluids is another function that will provide early payoff in servicing missions. Expendables are the principal limiting factor in most spacecraft. The removal and replacement of modules is the third objective within a demonstration system capability and is useful in many servicing applications.

DEMONSTRATION SCENARIO

The demonstration plan is based on an evolutionary approach that utilizes planned flight tests in the OMV and FTS programs in a three or four flight program. A phase B study is planned for early 1989 to further define the demonstration program. At this time there are four flights in the demonstration scenario. Although not specifically part of the demonstration program, the initial test of the FTS will be conducted in the payload bay of the Shuttle Orbiter and will produce data of value for subsequent tests. It addresses issues in manipulative tasks that bear on the development of the FTS. The initial flight of the OMV is an opportunity to evaluate the rendezvous and docking capabilities of a servicing system. Current plans call for a retrieval mission to bring a satellite to the vicinity of the Orbiter. The rendezvous and docking aspects of this test supplement the flight test of the basic OMV. Resupply of fluids expands the servicing objectives that are to be demonstrated.   The next demonstration features technology options and flight test of protoflight Servicer System elements in the payload bay and close the orbiter. The final test in the series would exercise an integrated protoflight servicer system with test of features and techniques to verify the application of autonomous operation to the basic system. This mission concept includes the three elements of a servicer system in servicing a cooperative spacecraft with compatible standard interfaces.   The servicing is to be conducted in the operational orbit of the objective spacecraft. It includes the rendezvous and docking, ORU exchange and fluid transfer

RENDEZVOUS AND DOCKING

The principal issue in rendezvous and docking relative to servicing is the degree of autonomy that can be achieved consistent with safe and reliable operations. Current methods depend on flight crew participation and extensive ground support. Future servicing missions will require greater piloting skill or the move to supervised autonomy and other autonomous techniques due to the remote site. These remote sites are on missions with polar orbiting platforms, geosynchronous orbiting platforms, and lunar and interplanetary spacecraft.

Other issues that are a factor in the development of a remote rendezvous and docking capability are optimization of rendezvous trajectories, cooperative control methodologies, trajectory control requirements, techniques, and boundary conditions. Operational considerations are mass properties identification and compensation, detection of hazardous conditions, abort sequences, and application of expert systems technology to the rendezvous and docking operation. Sensors to determine relative position, relative velocity, attitude and attitude rates are critical to the operation. For joining the two spacecraft, docking mechanisms guide and attenuate relative motion as well as lock the craft together. The Magnetic Assembly Tool being developed for the RMS may very well represent new technology for the docking function.

With increased requirements for autonomous operation, the integrated system must be smarter. Response to these requirements may include expert systems to aid operators and artificial intelligence for utilization of sensor data in rendezvous and docking tasks.   These techniques may also aid in management of system redundancies and selection of operational options.   Evaluation in the space and operational environment on a demonstration flight will validate computer and laboratory simulations.

SUMMARY

The commitment to a demonstration program for the testing of a protoflight system represents a significant milestone in the development of a capability to service satellites on orbit.   The demonstration flights validate the servicing system approach particularly in rendezvous and docking, fluids resupply and ORU change out.   From this beginning, the opportunities for reaping the benefits of serviceable space systems can be realized with the eventual development of full operational capability to service on orbit..

HYPERVELOCITY ORBITAL INTERCEPT GUIDANCE USING CERTAINTY CONTROL

Salvatore Alfano[*] and Charles E. Fosha, Jr.[†]

Terminal guidance of a hypervelocity exo-atmospheric orbital interceptor with free end-time is examined. A new approach called certainty control is developed where control energy expenditure is reduced by constraining the expected final state to a function of projected estimate error. Conceptually, the constraint produces a shrinking sphere about the predicted impact point with the radius being a function of estimated error. If the predicted miss is inside or touching the sphere, thrusting is not necessary. The pursuer is modeled as a satellite with lateral thrusting capability using two-body orbital dynamics. The evader is modeled as an Intercontinental Ballistic Missile (ICBM) in its final boost phase prior to burnout. Filtering is accomplished using an eight state Extended Kalman Filter with line-of-sight and range updates. The estimated relative trajectory and variances are propagated numerically to predicted impact time and then approximated by splines, eliminating the need to repeatedly propagate new data when present conditions are varied. A search is then made for a new impact time and point that will minimize present interceptor velocity changes and final miss distance. This control strategy, which is applied to two intercept problems, substantially reduces fuel consumption.

---

* Assistant Professor, Director of Research, Department of Astronautics, U.S. Air Force Academy, Colorado Springs, Colorado 80840-5151.

† Associate Professor Attendant, Department of Electrical Engineering, Director of Space and Flight Systems Laboratory, University of Colorado, Colorado Springs, Colorado 80933. Member AIAA.

INTRODUCTION

Interceptor performance can be enhanced by using a terminal guidance law that incorporates the dynamics of the pursuer and evader plus the error knowledge of their estimates. This paper develops a guidance scheme that minimizes lateral thrusting for a hypervelocity, exo-atmospheric orbital vehicle in the final thirty seconds of flight while attempting to intercept a boosting missile.

Much work has been done in the area of air-to-air guidance that has space-to-space application. Guelman[1,2] has derived a closed form solution for pure proportional navigation. Perturbation methods have been employed by Sridhar and Gupta[3] for air-to-air gudance. Design procedures using optimal and stochastic control techniques abound[4-14]. In the works cited above, the force of gravity is assumed to act equally on the pursuer and evader and is ignored in the relative dynamics. This 'flat earth' assumption is adequate for air-to-air encounters, but not for space-to-space except at short ranges. For orbital intercepts with large initial ranges the force of gravity will affect the relative trajectory and should be included in the equations of motion.

The literature for space-to-space guidance reveals many numerical approaches for determining present velocity for future rendezvous[15-21]. To date, analytic solutions for such intercepts exist only when the pursuer's impact conditions are pre-specified[19]. These works do not address hypervelocity intercept involving seconds, but are concerned with a much slower rendezvous process involving hours or even days.

The guidance scheme presented here attempts to minimize lateral velocity changes by varying the impact conditions through the use of splines. Splines were used by Johnson[16] in presenting a possible Earth-Mars transfer guidance algorithm. Dickmanns and Wells[22], as well as Hargraves and Paris[23], have used third order polynomials for general trajectory optimization. The splines result in faster searches by eliminating the need to propagate new trajectories repeatedly when conditions are varied. This feature makes them attractive for a hypervelocity orbital intercept which requires a fast and reasonably accurate numerical search.

Target tracking is accomplished with a ranging device and line-of-sight sensors for in-plane and out-of-plane measurements. Noise corrupted data is processed through an Extended Kalman Filter with serial updates occurring every tenth of a second.

SYSTEM MODELING

Here the equations of motion for the evader and pursuer are developed along with the necessary coordinate transformation for pursuer thrusting. Atmospheric drag will not be considered in the dynamics because the interceptor is assumed exo-atmospheric. Also, due to the pursuer's lateral thrusting limitation, the longitudinal axis will be assumed to be parallel to the pursuer's initial velocity vector.

It is computationally convenient to transform the present coordinate frame to align the x axis with the pursuer's initial velocity vector. This is done by first rotating about the y axis until the z component of velocity is eliminated (see Figure 1),

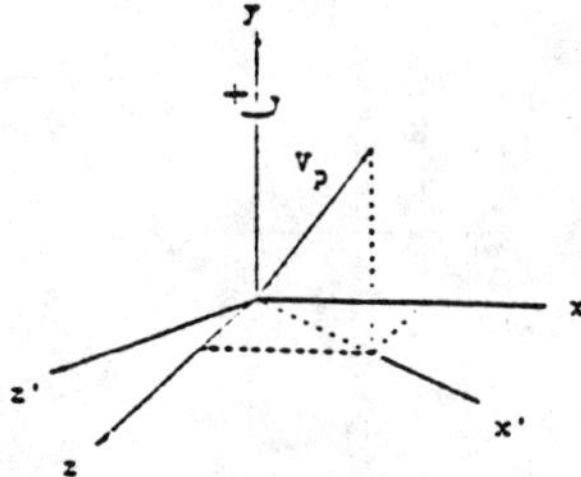

Figure 1. Rotation of coordinate frame about the y axis.

resulting in the following orthogonal transformation matrix:

$$[T_1] = \begin{bmatrix} \dfrac{\dot{x}_p}{\sqrt{\dot{x}_p^2 + \dot{z}_p^2}} & 0 & \dfrac{\dot{z}_p}{\sqrt{\dot{x}_p^2 + \dot{z}_p^2}} \\[3ex] 0 & 1 & 0 \\[3ex] \dfrac{-\dot{z}_p}{\sqrt{\dot{x}_p^2 + \dot{z}_p^2}} & 0 & \dfrac{\dot{x}_p}{\sqrt{\dot{x}_p^2 + \dot{z}_p^2}} \end{bmatrix} \tag{1}$$

The second rotation is about the new z axis (z´), eliminating the y component of velocity (see Figure 2).

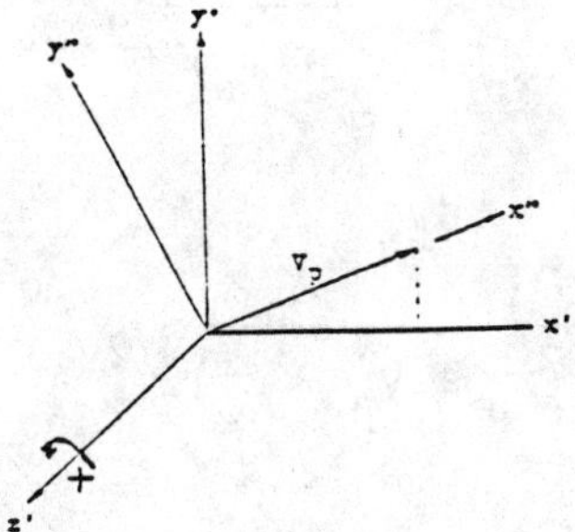

Figure 2. Rotation of the coordinate frame about the z axis.

This rotation yields the transformation matrix:

$$V_p = \sqrt{\dot{x}_p^2 + \dot{y}_p^2 + \dot{z}_p^2} \tag{2}$$

$$[T_2] = \begin{bmatrix} \dfrac{\sqrt{\dot{x}_p^2 + \dot{z}_p^2}}{V_p} & \dfrac{\dot{y}_p}{V_p} & 0 \\[2em] \dfrac{-\dot{y}_p}{V_p} & \dfrac{\sqrt{\dot{x}_p^2 + \dot{z}_p^2}}{V_p} & 0 \\[2em] 0 & 0 & 1 \end{bmatrix} \tag{3}$$

Multiplying the two matrices in the proper order produces the overall transformation matrix:

$$[T] = [T_2][T_1] \tag{4}$$

The pursuer is modeled as a satellite traveling in excess of twelve kilometers per second with lateral thrusting capability using two-body orbital dynamics. Thrusting is prohibited along the longitudinal (x) axis to prevent sensor contamination and to satisfy the structural constraints of having forward sensors and a large aft booster to achieve hypervelocity speed. The equations of motion are:

$$\ddot{x}_p = \frac{-\mu\, x_p}{(x_p^2 + y_p^2 + z_p^2)^{3/2}} \tag{5}$$

$$\ddot{y}_p = \frac{-\mu\, y_p}{(x_p^2 + y_p^2 + z_p^2)^{3/2}} + a_y \tag{6}$$

$$\ddot{z}_p = \frac{-\mu\, z_p}{(x_p^2 + y_p^2 + z_p^2)^{3/2}} + a_z \tag{7}$$

where a and a are the lateral thrust accelerations, μ is the earth's gravitational constant, and the double dots denote the second derivative with respect to time.

The evader is modeled as an Intercontinental Ballistic Missile (ICBM) in its final boost phase using two-body orbital dynamics. For tracking purposes the intercept must occur prior to burnout.

Acceleration due to thrusting is computed in the direction of the ICBM's velocity vector. The equations of motion are:

$$A = \frac{A_o}{1 - \dot{m}_o t} \tag{8}$$

$$\ddot{x}_E = \frac{-\mu\, x_E}{(x_E^2 + y_E^2 + z_E^2)^{3/2}} + \frac{A\, \dot{x}_E}{(\dot{x}_E^2 + \dot{y}_E^2 + \dot{z}_E^2)^{1/2}} \tag{9}$$

$$\ddot{y}_E = \frac{-\mu\, y_E}{(x_E^2 + y_E^2 + z_E^2)^{3/2}} + \frac{A\, \dot{y}_E}{(\dot{x}_E^2 + \dot{y}_E^2 + \dot{z}_E^2)^{1/2}} \tag{10}$$

$$\ddot{z}_E = \frac{-\mu\, z_E}{(x_E^2 + y_E^2 + z_E^2)^{3/2}} + \frac{A\, \dot{z}_E}{(\dot{x}_E^2 + \dot{y}_E^2 + \dot{z}_E^2)^{1/2}} \tag{11}$$

where A is the present acceleration, A_o is the initial acceleration, $\dot{m}_o$ is the initial mass flow rate divided by mass, and t is the time since ignition. The single dot denotes the first derivative with respect to time.

PROBLEM STATEMENT AND TRUTH MODEL

Time-to-go and pursuer velocity changes are the control parameters that must be varied to minimize miss distance and fuel expended (i.e. velocity changes). This can be done by establishing a time remaining until intercept (time-to-go), propagating the equations of motion forward, and determining the miss distance. An iterative process can then be used to find the pursuer velocity needed to bring the miss distance to zero. The difference between current velocity and that needed for intercept, known as velocity-to-go, must be minimized. To accomplish this, the time-to-go is varied and the above procedure repeated until a minimum velocity-to-go is found.

The computation of needed velocity is time consuming because the equations of motion are nonlinear and do not lend themselves to closed form solution. These equations must be propagated numerically to intercept time whenever the initial velocity is varied. The above method will serve as the basis (truth) model for this control problem using the numerical techniques found in Maron[24].

A Newton-Raphson method for solving nonlinear systems is employed to determine the proper values of the control parameters. Let

$$
\bar{\underline{u}} = \begin{bmatrix} \overline{\Delta V_y} \\ \overline{\Delta V_z} \\ \overline{t_{go}} \end{bmatrix}
\tag{12}
$$

be a solution of the nonlinear system

$$
\begin{bmatrix} f_1(\bar{\underline{u}}) \\ f_2(\bar{\underline{u}}) \\ f_3(\bar{\underline{u}}) \end{bmatrix} = \begin{bmatrix} x_E(\overline{t_{go}}) - x_p(\overline{t_{go}}) \\ y_E(\overline{t_{go}}) - y_p(\overline{t_{go}}) - \Delta V_y \overline{t_{go}} \\ z_E(\overline{t_{go}}) - z_p(\overline{t_{go}}) - \Delta V_z \overline{t_{go}} \end{bmatrix} = \begin{bmatrix} 0 \\ 0 \\ 0 \end{bmatrix}
\tag{13}
$$

where t_{go} is the time-to-go and the pursuer's velocity changes are ΔV_y and ΔV_z. The effect of small velocity changes in (13) can be considered linear because the pursuer is assumed to travel at hypervelocity, resulting in a near straight-line trajectory. Any error caused by this assumption will be accounted for in the succeeding iteration when the proposed velocity change is incorporated in the nonlinear dynamics.

The initial control values must be incrementally changed to satisfy (13). A linear approximation of the $\underline{f}$ vector for changes in u will yield approximate increments of the control parameters. The linearized system becomes

$$
[J] \begin{bmatrix} d\Delta V_y \\ d\Delta V_z \\ dt_{go} \end{bmatrix} = - \begin{bmatrix} f_1(\underline{u}) \\ f_2(\underline{u}) \\ f_3(\underline{u}) \end{bmatrix}
\tag{14}
$$

where J is the Jacobian matrix of the $\underline{f}$ vector evaluated at $\underline{u}$:

$$
[J] = \begin{bmatrix} 0 & 0 & \{\dot{x}_E(t_{go}) - \dot{x}_p(t_{go})\} \\[2em] -t_{go} & 0 & \{\dot{y}_E(t_{go}) - \dot{y}_p(t_{go}) - \Delta V_y\} \\[2em] 0 & -t_{go} & \{\dot{z}_E(t_{go}) - \dot{z}_p(t_{go}) - \Delta V_z\} \end{bmatrix} \tag{15}
$$

To determine changes in the $\underline{u}$ vector, $\underline{f}$ is multiplied by the negative inverse of J

$$
\begin{bmatrix} d\Delta V_y \\[1em] d\Delta V_z \\[1em] dt_{go} \end{bmatrix} = -[J]^{-1} \begin{bmatrix} f_1(\underline{u}) \\[1em] f_2(\underline{u}) \\[1em] f_3(\underline{u}) \end{bmatrix} \tag{16}
$$

$$
[J]^{-1} = \begin{bmatrix} \dfrac{\dot{y}_E(t_{go}) - \dot{y}_p(t_{go}) - \Delta V_y}{\{\dot{x}_E(t_{go}) - \dot{x}_p(t_{go})\}t_{go}} & -1/t_{go} & 0 \\[2em] \dfrac{\dot{z}_E(t_{go}) - \dot{z}_p(t_{go}) - \Delta V_z}{\{\dot{x}_E(t_{go}) - \dot{x}_p(t_{go})\}t_{go}} & 0 & -1/t_{go} \\[2em] \dfrac{1}{\{\dot{x}_E(t_{go}) - \dot{x}_p(t_{go})\}} & 0 & 0 \end{bmatrix} \tag{17}
$$

To find the control parameters the following procedure should be used. First, establish a time-to-go with zero velocity changes, a good choice being the time-to-go that yields the point of closest approach. This time-to-go is determined by propagating the orbits forward until a minimum relative distance is reached. Because the evader is assumed to be in its final boost phase throughout the intercept, this time-to-go will be less than or equal to time until ICBM thrust termination.

Second, propagate the dynamic equations (5 thru 11) forward to the intercept time and determine the $\underline{f}$ vector from (13). Changes to the control parameters are then obtained from (16). The velocity changes are applied to the pursuer's initial conditions and the procedure is repeated with the updated time-to-go until convergence occurs. The resulting control parameters will drive the miss distance to zero with minimum velocity changes. The difference between needed and present velocity are sufficient to determine the pursuer's thrust profile.

SPLINE APPROXIMATIONS

As discussed earlier, numerical propagation of the dynamic equations is very time consuming. It is convenient to approximate the relative trajectory by a polynomial, eliminating the need for repeated propagation. Cubic splines lend themselves well to this application[16,22,23]. The current and final states can be used to generate cubic splines along each axis of the form

$$x(t) = At_{go}^3 + Bt_{go}^2 + Ct_{go} + D \tag{18}$$

By setting the current time to zero, D and C become the current position and velocity respectively, with time-to-go being the intercept time. Changes in velocity will be reflected only in the C coefficient and the final state can be easily determined for any intercept time. With this formulation, the determination of the spline coefficients is relatively simple. The current state gives D and C with no computations:

$$D = x(o) \tag{19}$$

$$C = \dot{x}(o) \tag{20}$$

The A and B coefficients can be computed using the final states and (18) as follows:

$$x(t_{go}) = At_{go}^3 + Bt_{go}^2 + Ct_{go} + D \tag{21}$$

$$\dot{x}(t_{go}) = 3At_{go}^2 + 2Bt_{go} + C \tag{22}$$

Because there are only two unknowns in the above two equations, algebraic manipulation yields:

$$A = \frac{2[x(o) - x(t_{go})]}{t_{go}^3} + \frac{[\dot{x}(o) + \dot{x}(t_{go})]}{t_{go}^2} \tag{23}$$

$$B = \frac{3[x(t_{go}) - x(o)]}{t_{go}^2} + \frac{[2\dot{x}(o) + \dot{x}(t_{go})]}{t_{go}} \qquad (24)$$

There is an added versatility in using splines. Should the system model be changed, only the spline coefficients need be changed. A search algorithm based on the splines will remain the same, operating with the new coefficients. This is very beneficial because recomputing the coefficients is far simpler than altering an algorithm.

To ensure accuracy, new spline coefficients are computed every cycle time. To accomplish this, the truth model is propagated forward to the predicted impact time to obtain the needed final states. By using these updated final states every iteration, propagated roundoff error is eliminated in the spline coefficient computations.

CERTAINTY CONTROL

If the controls associated with cost do not affect state estimate certainty, fuel may be conserved by using that certainty to reduce the controls. By linking the controls to the certainty of the estimate, a near perfect estimate would yield the optimal control, with reduced control resulting from a poor estimate. To accomplish this, the predicted final states are constrained by a function of their variances at the final time. This form of control will be called certainty control and is implemented by establishing the cost function

$$L = \frac{\Delta V_y^2 + \Delta V_z^2}{2} \qquad (25)$$

subject to the constraint:

$$f = \frac{\hat{x}_f^2 + \hat{y}_f^2 + \hat{z}_f^2 - K[\sigma_{xf}^2 + \sigma_{yf}^2 + \sigma_{zf}^2]}{2} \leq 0 \qquad (26)$$

where K is a weighting factor. The final state estimates $(\hat{x}_f, \hat{y}_f, \hat{z}_f)$ and their deviations $(\sigma_{xf}, \sigma_{yf}, \sigma_{zf})$ are determined by running the filter (Appendix A) forward to predicted impact time without updates and then representing their time history with splines:

$$x_s = A_x t_{go}^3 + B_x t_{go}^2 + C_x t_{go} + D_x \qquad (27)$$

$$y_s = A_y t_{go}^3 + B_y t_{go}^2 + C_y t_{go} + D_y \qquad (28)$$

$$z_s = A_z t_{go}^3 + B_z t_{go}^2 + C_z t_{go} + D_z \tag{29}$$

$$\hat{x}_f = x_s \tag{30}$$

$$\hat{y}_f = y_s - \Delta V_y t_{go} \tag{31}$$

$$\hat{z}_f = z_s - \Delta V_z t_{go} \tag{32}$$

$$\sigma_{xf} = A_{\sigma x} t_{go}^3 + B_{\sigma x} t_{go}^2 + C_{\sigma x} t_{go} + D_{\sigma x} \tag{33}$$

$$\sigma_{yf} = A_{\sigma y} t_{go}^3 + B_{\sigma y} t_{go}^2 + C_{\sigma y} t_{go} + D_{\sigma y} \tag{34}$$

$$\sigma_{zf} = A_{\sigma z} t_{go}^3 + B_{\sigma z} t_{go}^2 + C_{\sigma z} t_{go} + D_{\sigma z} \tag{35}$$

Conceptually, the constraint produces a deviation sphere about the predicted impact point. If the predicted miss is inside or touching the sphere, thrusting is not necessary. If the predicted miss is outside the sphere, minimum thrusting is determined to bring the miss to the surface of the sphere. As the estimates improve, the constraint tightens and the sphere shrinks. The spline representations allow this stochastic problem to be solved deterministically. The constraint is adjoined to the cost function to form the Hamiltonian[25]:

$$H = L + \lambda f \tag{36}$$

To minimize the cost L while satisfying the constraint f, the partials of H with respect to the controls must equal zero:

$$\frac{\partial H}{\partial \Delta V_y} = \Delta V_y - \lambda \hat{y}_f t_{go} = 0 \tag{37}$$

$$\frac{\partial H}{\partial \Delta V_z} = \Delta V_z - \lambda \hat{z}_f t_{go} = 0 \tag{38}$$

$$\frac{\partial H}{\partial t_{go}} = \lambda (\hat{x}_f \dot{\hat{x}}_f + \hat{y}_f \dot{\hat{y}}_f + \hat{z}_f \dot{\hat{z}}_f -$$
$$K[\sigma_{xf} \dot{\sigma}_{xf} + \sigma_{yf} \dot{\sigma}_{yf} + \sigma_{zf} \dot{\sigma}_{zf}]) = 0 \tag{39}$$

with the dot term expansions computed in Appendix B.

Equations 26, 37, 38, and 39 constitute four equations with four unknowns, which can be reduced to two equations and two unknowns using (31) and (32). Substituting (31) into (37) yields

$$\Delta V_y = \frac{\lambda y_s t_{go}}{1 + \lambda t_{go}^2} \tag{40}$$

$$\hat{y}_f = \frac{y_s}{1 + \lambda t_{go}^2} \tag{41}$$

In a similar manner, substituting (32) into (38) yields

$$\Delta V_z = \frac{\lambda z_s t_{go}}{1 + \lambda t_{go}^2} \tag{42}$$

$$\hat{z}_f = \frac{z_s}{1 + \lambda t_{go}^2} \tag{43}$$

Equations 26 and 39 can now be solved in terms of λ and t_{go}. Once known, ΔV and ΔV can be determined from (40) and (42). The parameters λ and t can be found by numerical techniques using the Jacobian:

$$[J] \begin{bmatrix} dt_{go} \\ d\lambda \end{bmatrix} = \begin{bmatrix} -f_1 \\ -f_2 \end{bmatrix} \tag{44}$$

$$f_1 = \frac{\hat{x}_f^2 + \hat{y}_f^2 + \hat{z}_f^2 - K[\sigma_{xf}^2 + \sigma_{yf}^2 + \sigma_{zf}^2]}{2} \tag{45}$$

$$f_2 = \hat{x}_f \dot{\hat{x}}_f + \hat{y}_f \dot{\hat{y}}_f + \hat{z}_f \dot{\hat{z}}_f -$$

$$K[\sigma_{xf} \dot{\sigma}_{xf} + \sigma_{yf} \dot{\sigma}_{yf} + \sigma_{zf} \dot{\sigma}_{zf}] \tag{46}$$

with the elements of the Jacobian matrix computed in Appendix B.

Should the states be perfectly known, the σ terms will be zero. In this case, the equations for certainty control reduce to those of optimal control formulation. Should the estimate be poor, the σ terms will be large and the inequality constraint of (26) is met with very little (if any) change in velocity. This demonstrates the principle of certainty control, where the certainty of the estimate affects control energy expenditure.

COMPUTER SIMULATION

Two cases are examined with time-to-go equaling 30 seconds. Case I represents a head on, 10° out-of-plane, intercept and Case II represents a 10° out-of-plane tail chase. The pursuer is initially traveling at 12 km/s at an altitude of 750 km with a lateral acceleration range of 3-60 m/s² in each axis. The booster's initial acceleration is 3.15788 m/s² with a unitized mass flow rate of .01579 s⁻¹.

A time lag of one tenth second is used for all algorithms when computing velocity changes. It is unrealistic to assume the filter can process measurements, the controller determine thrust commands, and the thrusters respond to those commands all instantaneously. One cycle time is chosen to allow the velocity changes computed in the previous cycle to be implemented in the present cycle. The controller routines are built to take this lag into account. Also, thrusting is not permitted during the first three seconds of an intercept to account for target acquisition. This simulation, written in FORTRAN 77 to run on a VAX 8600, generates one hundred Monte Carlo runs per case.

RESULTS

The performance of the truth model versus certainty control is recorded in the tables on the following page:

Table 1. Case I Performance.
(Head On, 10° Out-of-Plane Intercept)

	MEAN MISS (METERS)	STANDARD DEVIATION (METERS)	MEAN ΔV (M/S)	STANDARD DEVIATION (M/S)
CERTAINTY CONTROL (K=.4)	.386	.191	23.21	4.18
TRUTH WITH NOISE	.545	.264	83.69	7.18

Table 2. Case II Performance.
(10° Out-of-Plane Tail Chase)

	MEAN MISS (METERS)	STANDARD DEVIATION (METERS)	MEAN ΔV (M/S)	STANDARD DEVIATION (M/S)
CERTAINTY CONTROL (K=.4)	.136	.076	29.74	9.10
TRUTH WITH NOISE	.379	.204	123.57	11.61

The effect of estimate uncertainty can be observed from the truth model in-plane thrust profiles (Figures 3 and 4). This uncertainty causes needless and often counter-productive thrusting. In contrast, certainty control requires considerably less energy expenditure (Figures 5 and 6). This result is not surprising, as the formulation of certainty control is based on reducing control energy in the presence of poor estimates. This form of control works well because filter variance is range dependent. As range decreases, the control constraint tightens and accuracy increases. Therefore, less fuel is used when range is great and estimates are poor, with refinements made as impact nears.

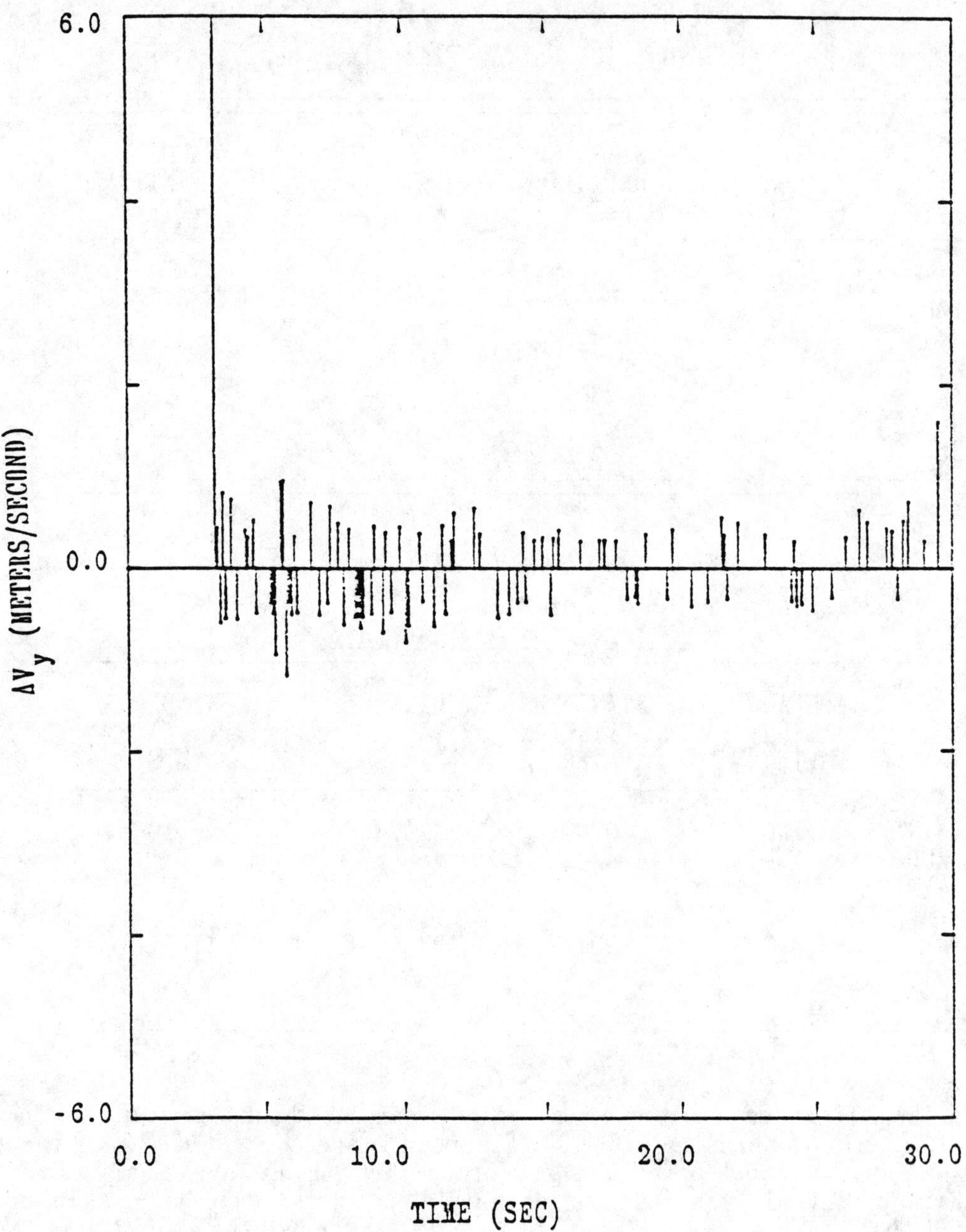

Figure 3. In-plane thrust profile of Truth Model for Case I.

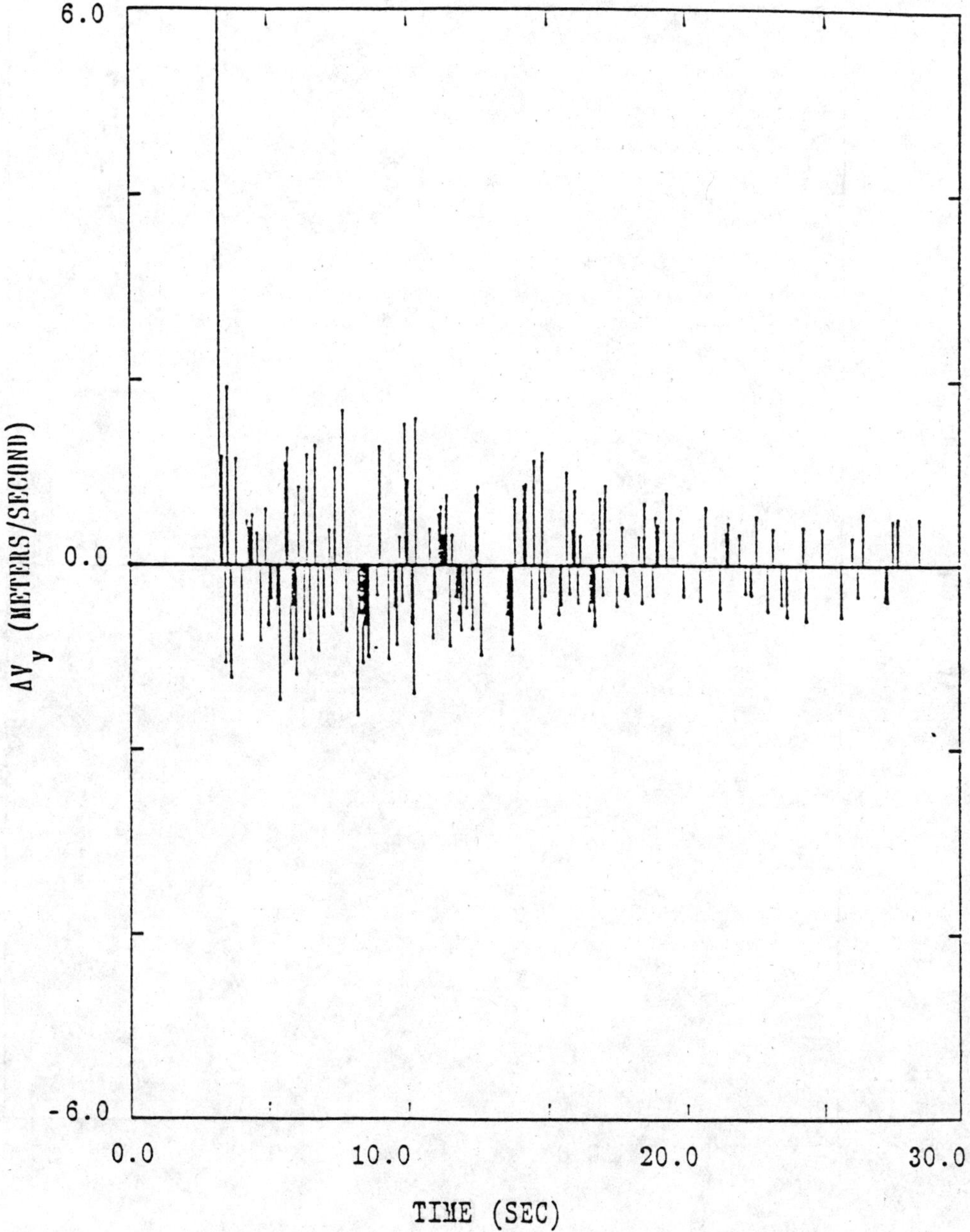

Figure 4. In-plane thrust profile of Truth Model for Case II.

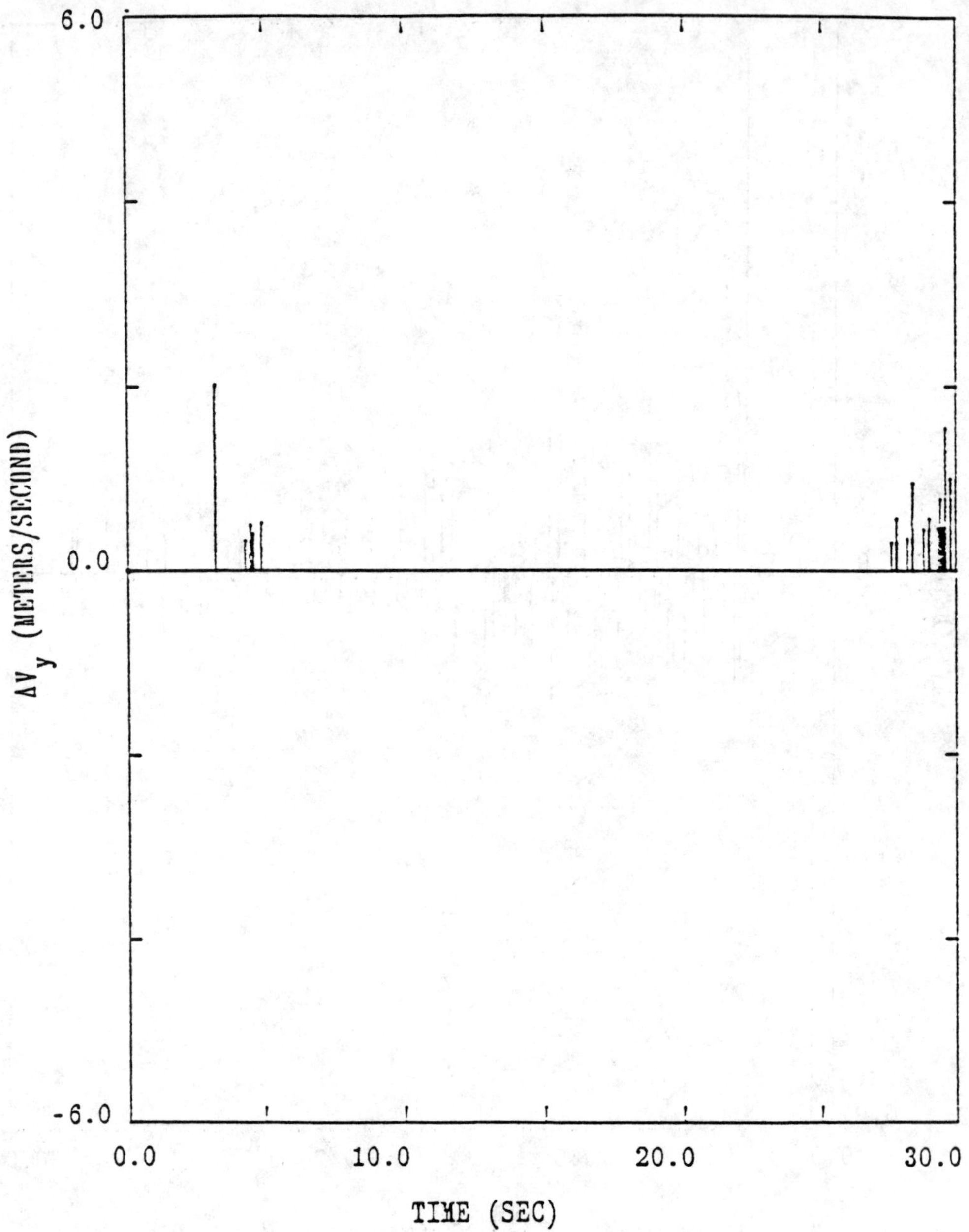

Figure 5. In-plane thrust profile of Certainty Control for Case I.

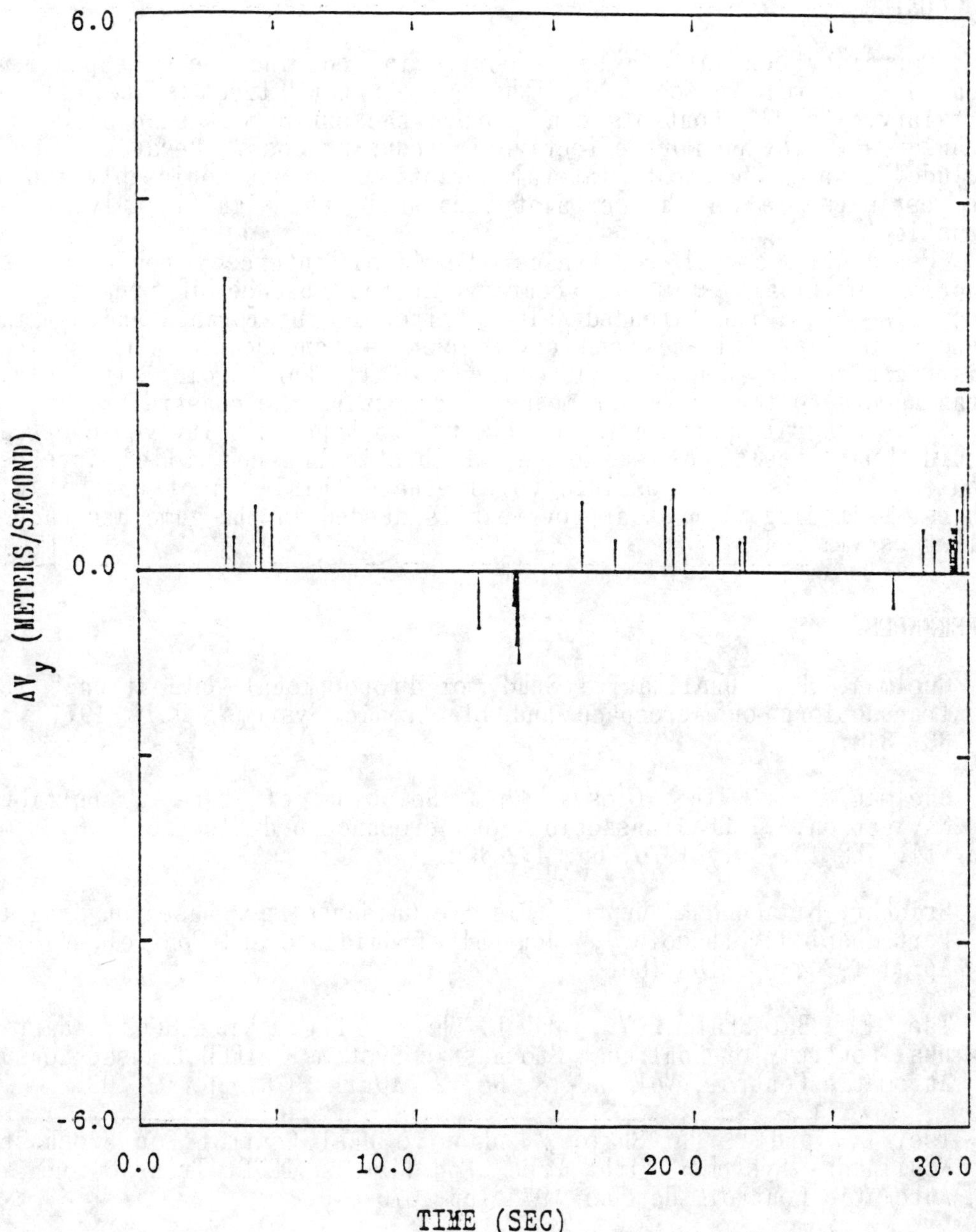

Figure 6. In-plane thrust profile of Certainty Control for Case II

391

CONCLUSION

Certainty control can be effective in reducing energy expenditure when the controls associated with cost do not affect state estimate certainty. If the controls can improve the estimate then dual control techniques[4,5] may be more effective in reducing cost. Because range is included as a measurement, lateral deviations do not noticeably improve the estimate, allowing certainty control to significantly reduce thrusting.

Certainty control constrains the final intercept position to a function of final estimator accuracy in the absence of updates. To accomplish this, the Extended Kalman Filter is run forward and the time history of the estimates and deviations is represented with splines. This technique is adequate, but computationally burdensome. It would be advantageous to find a faster means of computing the constraint.

This general approach is not limited to hypervelocity vehicles, and should lend itself well to other stochastically controlled intercept problems. It is also possible to alternate this form of control with others, switching when an improvement is needed in end-game accuracy or filter estimation.

REFERENCES

1. Guelman, M., "Qualitative Study of Proportional Navigation," IEEE Transactions on Aerospace and Electronic Systems, July 1971, pp. 337-343.

2. Guelman, M., "The Closed Form Solution of Pure Proportional Navigation," IEEE Transactions on Aerospace and Electronic Systems, Vol. AES-12, July 1976, pp. 472-482.

3. Sridhar, B. and N.K. Gupta, "Missile Guidance Laws Based on Singular Perturbation Methodology," Journal of Guidance and Control, Vol. 3, April 1980, pp. 158-166.

4. Tse, E., Bar-Shalom, Y., and L. Meier, III, "Wide Sense Adaptive Dual Control for Nonlinear Stochastic Systems," IEEE Transactions on Automatic Control, Vol. AC-18, No. 2, April 1973, pp. 98-108.

5. Tse, E., and Y. Bar-Shalom, "Adaptive Dual Control For Stochastic Nonlinear Systems with Free End-Time," IEEE Transactions on Automatic Control, October 1975, pp. 670-675.

6. Nesline, F.W., Wells, B.H., and P. Zachran, "Combined Optimal/Classic Approach to Robust Missile Autopilot Design," Journal of Guidance and Control, Vol. 4, No. 3, May-June 1981, pp. 316-322.

7. Speyer, J.L., Hull, D.G., and C.Y. Tseng, "Estimation Enhancement by Trajectory Modulation for Homing Missiles," Journal of Guidance, Vol. 7, No. 2, March-April 1984, pp. 167-174.

8. Guelman, M., and J. Shinar, "Optimal Guidance Law in the Plane," Journal of Guidance, Vol. 7, No. 4, July-August 1984, pp. 471-476.

9. Tang, Y.M., and J.A. Borrie, "Missile Guidance Based on Kalman Filter Estimation of Target Maneuver," IEEE Transactions on Aerospace and Electronic Systems, Vol. AES-20, No. 6, Nov. 1984, pp. 736-741.

10. Yeuh, W.R., and C.F. Lin, "Optimal Controller for Homing Missile," Journal of Guidance, Vol. 8, No. 3, May-June 1985, pp. 408-411.

11. Lin, C.F., and S.P. Lee, "Robust Missile Autopilot Design Using a Generalized Singular Optimal Control Techn que," Journal of Guidance, Vol. 8, No. 4, July-August 1985,pp. 498-507.

12. Ashida, S., Howe, R.M., and N.X. Vinh, "Optimal Control of Air-Launched Homing Missiles Based on Realistic Performance Indices," Proceedings of the Conference on Aerospace Simulation II, Vol. 16, No. 2, Jan. 1986.

13. Lin, C.F., and L.L. Tsai, "Analytical Solution of Optimal Trajectory-Shaping Guidance," Journal of Guidance, Control, and Dynamics, Vol. 10, No. 1, Jan.-Feb. 1987, pp. 61-66.

14. Yang, C.D., and F.B. Yeh, "Closed-Form Solution for a Class of Guidance Laws," Journal of Guidance, Vol. 10, No. 4, July-August 1987, pp. 412-415.

15. Cherry, G.W., "A General, Explicit, Optimizing Guidance Law for Rocket-propelled Spaceflight," AIAA Paper 64-638, Aug. 1964.

16. Johnson, F.T., "Approximate Finite-Thrust Trajectory Optimization," AIAA Journal, Vol. 7, June 1969, pp. 993-997.

17. Bate, R.R., Mueller, D.D., and J.E. White, *Fundamentals of Astrodynamics*, Dover Publications, New York, 1971.

18. Borisenko, I.I., and Y.P. Kulyabichev, "Algorithm for Optimization of the Solution of the Spacecraft Rendezvous Problem," Cosmic Research, Vol. 18, No. 3, May-June 1980, pp. 343-347.

19. Stuart, D.G., "A Simple Targeting Technique for Two-Body Spacecraft Trajectories," Journal of Guidance, Vol. 9, No. 1, Jan.-Feb. 1986, pp. 27-31.

20. Bhat, M.S., and S.K. Shrivastava, "An Optimal Q-Guidance Scheme for Satellite Launch Vehicles," Journal of Guidance, Vol. 10, No. 1, Jan.-Feb. 1987, pp. 53-60.

21. Menon, P.K.A., and A.J. Calise, "Interception, Evasion, Rendezvous and Velocity-to-be-Gained Guidance for Spacecraft," AIAA Paper 87-2318, Aug. 1987.

22. Dickmanns, F.D., and K.H. Wells, "Approximate Solution of Optimal Control Problems Using Third Order Hermite Polynomial Functions," Proceedings of the 6th Technical Conference on Optimization Techniques, Springer-Verlag, New York, IFIP-TC7, 1975.

23. Hargraves, C.R., and S.W. Paris, "Direct Trajectory Optimization Using Nonlinear Programming and Collocation," Journal of Guidance, Vol. 10, No. 4, July-August 1987, pp.338-342.

24. Maron, M.J., *Numerical Analysis: A Practical Approach*, Macmillan Publishing Co., New York, 1982, pp. 177-182.

25. Bryson, A.E., and Y.C. Ho, *Applied Optimal Control*, Hemisphere Publishing Corp., Washington D.C., 1975, pp. 71-75.

26. Gelb, A., *Applied Optimal Estimation*, M.I.T. Press, Massachusetts, 1986, pp. 180-228.

27. Ho, Y.C., "On the Stochastic Approximation Method and Optimal Filtering Theory," Journal Mathematical Analysis and Applications, Vol. 6, 1963, pp. 152-154.

APPENDIX A: EXTENDED KALMAN FILTERING

Optimal estimates of the pursuer and evader are needed for the search algorithms to converge properly. Due to the nature of the dynamics and sensors, the relative position and velocity must be estimated from sampled nonlinear measurements. The estimation problem for a nonlinear system having continuous dynamics and discrete-time measurements is addressed by Gelb[26]. The Extended Kalman Filter (EKF) was chosen over other estimation methods because the optimal estimate is determinate. That is, the dynamics and observations of the pursuer and evader can be well predicted in the presence of Gaussian noise. The EKF states are defined from (5) through (11) as follows:

$$x_1 = x_E - x_p \tag{A-1}$$

$$x_2 = \dot{x}_E - \dot{x}_p \tag{A-2}$$

$$x_3 = y_E - y_p \tag{A-3}$$

$$x_4 = \dot{y}_E - \dot{y}_p \tag{A-4}$$

$$x_5 = z_E - z_p \tag{A-5}$$

$$x_6 = \dot{z}_E - \dot{z}_p \tag{A-6}$$

$$x_7 = A \tag{A-7}$$

$$x_8 = \dot{m} \tag{A-8}$$

It is advantageous to process measurements one at a time. This method, called serial updating[27], eliminates the requirement to compute a matrix inverse, thereby reducing computer load and avoiding the computational problems associated with inverting an ill-conditioned matrix. Also, measurements may be skipped without reformulating the filter equations, allowing greater flexibility in examining various tracking schemes. The simultaneous components of the measurement vector (z) can be considered serially over a very short time span. The propagation (w) noise stems from using a fourth order Runge-Kutta integrator with updates every tenth of a second on a 64 bit word.

$$w_{x,y,z}(t) \sim N(0,\ 2.21516 \times 10^{-18}\ \frac{m^2}{s})$$

$$w_{\dot{x},\dot{y},\dot{z}}(t) \sim N(0,\ 5.52049 \times 10^{-20}\ \frac{m^4}{s^3})$$

$$w_A(t) \sim N(0,\ 4.29831 \times 10^{-12}\ \frac{m^2}{s^5})$$

$$w_{\dot{m}}(t) \sim N(0,\ 2.493241 \times 10^{-7}\ \frac{1}{s^3})$$

In all cases the measurement (v) noise properties associated with the filter are:

$$V_{\theta,\gamma}(k) \sim N(0,\ 1.0 \times 10^{-8})$$

$$V_R(k) \sim N(0,\ 1.0 \times 10^{-8} \times R^2\ m^2)$$

where θ is the out-of-plane line-of-sight angle, γ the in-plane line-of-sight angle and R is range. The startup variances for all runs are:

$$\sigma^2_{xx} = \sigma^2_{yy} = \sigma^2_{zz} = 100\ m^2$$

$$\sigma^2_{\dot{x}\dot{x}} = \sigma^2_{\dot{y}\dot{y}} = \sigma^2_{\dot{z}\dot{z}} = 10\ \frac{m^2}{s^2}$$

$$\sigma^2_{AA} = .1\ \frac{m^2}{s^4}$$

$$\sigma^2_{\dot{m}\dot{m}} = 2.493241 \times 10^{-6}\ s^{-2}$$

APPENDIX B: DERIVATION OF CERTAINTY CONTROL EQUATIONS

The dot terms for (39) are computed as follows:

$$\dot{\hat{x}}_f = 3A_x t_{go}^2 + 2B_x t_{go} + C_x \tag{B-1}$$

$$\dot{\hat{y}}_f = 3A_y t_{go}^2 + 2B_y t_{go} + C_y - \Delta V_y \tag{B-2}$$

$$\dot{\hat{z}}_f = 3A_z t_{go}^2 + 2B_z t_{go} + C_z - \Delta V_z \tag{B-3}$$

$$\dot{\sigma}_{xf} = 3A_{\sigma x} t_{go}^2 + 2B_{\sigma x} t_{go} + C_{\sigma x} \tag{B-4}$$

$$\dot{\sigma}_{yf} = 3A_{\sigma y} t_{go}^2 + 2B_{\sigma y} t_{go} + C_{\sigma y} \tag{B-5}$$

$$\dot{\sigma}_{zf} = 3A_{\sigma z} t_{go}^2 + 2B_{\sigma z} t_{go} + C_{\sigma z} \tag{B-6}$$

The Jacobian matrix elements for (44) are:

$$J_{11} = \frac{\partial f_1}{\partial t_{go}} = f_2 \tag{B-7}$$

$$J_{12} = \frac{\partial f_1}{\partial \lambda} = \hat{y}_f \frac{\partial \hat{y}_f}{\partial \lambda} + \hat{z}_f \frac{\partial \hat{z}_f}{\partial \lambda} \tag{B-8}$$

$$\frac{\partial \hat{y}_f}{\partial \lambda} = \frac{-y_s t_{go}^2}{(1 + \lambda t_{go}^2)^2} \tag{B-9}$$

$$\frac{\partial \hat{z}_f}{\partial \lambda} = \frac{-z_s t_{go}^2}{(1 + \lambda t_{go}^2)^2} \tag{B-10}$$

$$J_{21} = \frac{\partial f_2}{\partial t_{go}} = \hat{x}_f \ddot{\hat{x}}_f + \dot{\hat{x}}_f^2 + \hat{y}_f \ddot{\hat{y}}_f + \dot{\hat{y}}_f^2 + \hat{z}_f \ddot{\hat{z}}_f + \dot{\hat{z}}_f^2$$

$$- K[\sigma_{xf}\ddot{\sigma}_{xf} + \dot{\sigma}_{xf}^2 + \sigma_{yf}\ddot{\sigma}_{yf} + \dot{\sigma}_{yf}^2 + \sigma_{zf}\ddot{\sigma}_{zf} + \dot{\sigma}_{zf}^2] \tag{B-11}$$

$$\ddot{\hat{x}}_f = 6A_{\sigma x} t_{go} + 2B_x \tag{B-12}$$

$$\ddot{\hat{y}}_f = 6A_y t_{go} + 2B_y - \lambda(\dot{\hat{y}}_f t_{go} + \hat{y}_f) \tag{B-13}$$

$$\ddot{\hat{z}}_f = 6A_z t_{go} + 2B_z - \lambda(\dot{\hat{z}}_f t_{go} + \hat{z}_f) \tag{B-14}$$

$$\ddot{\sigma}_{xf} = 6A_{\sigma x} t_{go} + 2B_{\sigma x} \tag{B-15}$$

$$\ddot{\sigma}_{yf} = 6A_{\sigma y} t_{go} + 2B_{\sigma y} \tag{B-16}$$

$$\ddot{\sigma}_{zf} = 6A_{\sigma z} t_{go} + 2B_{\sigma z} \tag{B-17}$$

$$J_{22} = \frac{\partial f_2}{\partial \lambda} = \frac{\partial \hat{y}_f}{\partial \lambda} \dot{\hat{y}}_f + \hat{y}_f \frac{\partial \dot{\hat{y}}_f}{\partial \lambda} + \frac{\partial \hat{z}_f}{\partial \lambda} \dot{\hat{z}}_f + \hat{z}_f \frac{\partial \dot{\hat{z}}_f}{\partial \lambda} \tag{B-18}$$

$$\frac{\partial \dot{\hat{y}}_f}{\partial \lambda} = - \frac{y_s t_{go}}{(1 + \lambda t_{go}^2)^2} \tag{B-19}$$

$$\frac{\partial \dot{\hat{z}}_f}{\partial \lambda} = - \frac{z_s t_{go}}{(1 + \lambda t_{go}^2)^2} \tag{B-20}$$

LINE-OF-SIGHT STABILIZATION FOR THE DYNAMIC MAGELLAN SPACECRAFT

Clifford O. Swanson[*], Jordan Kass[†] and Alan Greiner[‡]

The evolution from the use of single-degree-of-freedom floated gyros to the dynamically tuned gyro has provided attitude reference unit hardware that can accommodate the dynamic environment of the Magellan scenario. The requirements, implementation and test results are described. Extension of these capabilities to provide for very low output noise, high bandwidths and hardening to withstand nuclear events is also described.

INTRODUCTION

Gyroscopic Attitude Reference Units (ARU's) have been employed, since early Earth and interplanetary space missions, to stabilize a Line of Sight (LOS). Performance requirements were moderate and environments benign. The Single-Degree-of-Freedom (SDF), floated gyros previously used required temperature control or low viscosity damping fluids to minimize changes in dynamic characteristics. Drift performance was specified in short-term random drift, and scale factor was not critical. Better image resolution, required in Earth-orbiting satellites, and long life prompted the use of gas-bearing-motor gyros. Lower output noise and jitter were achieved, but at the cost of increased complexity and poorer reliability.

The advent of the Dynamically Tuned Gyro (DTG) offered higher performance, simplicity, higher reliability and lower cost, factors which prompted its selection as the heart of the Dry Gyro Inertial Reference Unit I (DRIRU I) on Voyagers 1 and 2[1]. The original Voyager mission time of four years (with the potential for an extended mission time of 12 years) would have dictated the use of six SDF gyros, which would have imposed a significant increase in cost, size, weight, and power. Since 8,000 GYROFLEX® Gyros have been delivered at that point in time, it was selected as the gyro to be used in the Voyager ARU. The DRIRU I performed flawlessly through the Jupiter and Saturn fly-bys, and even beyond when, based on new requirements, torque commands to the gyros provided image-motion compensation for time exposures of Uranus' moon, Miranda. Similar techniques are again planned for the fly-by of Neptune in August, 1989.

The requirement for a very low-weight ARU launched the development of the Space-Qualified Kearfott Inertial Reference Unit (SKIRU) in parallel with the DRIRU I activity. Its characteristics and space-qualification status made it a candidate for the Magellan spacecraft ARU. Further development of the

---

* Director of Space Programs, † Member Technical Staff, ‡ Staff Engineer, Kearfott Guidance and Navigation Corporation, Wayne, New Jersey 07470.

GYROFLEX Gyro has extended its capabilities beyond that of low-rate platform stabilization to higher-rate applications and those that require precision pointing. Papers presented in previous years set forth the unique dynamic scenario of the Venus-orbiting Magellan spacecraft (to be launched in April, 1989) due to the varied mapping, data dumping, and calibration maneuvers occurring during each orbit[2]. Qualification-by-similarity modifications made to the basic SKIRU made it possible to meet these requirements.

Evolution of the Magellan ARU has continued in order to satisfy more demanding requirements.

The DTG flexure not only rotationally isolates the inertia element, the gyro rotor, but also decouples external noise sources, e.g., spin-motor bearings, from propogating into the electrical output, which could cause jitter of the spacecraft LOS. Very low jitter has been measured at three different facilities under static and dynamic conditions.

The need for higher bandwidth in some advanced systems has also been addressed. Engineering models with 50 and 100 Hz bandwidths have already been delivered and development of a 150 Hz channel is well along.

Another version of SKIRU has already been qualified to withstand a nuclear event.

THE GYROFLEX GYRO

It has been customary in papers presented in previous years to provide a description of the instrument before proceeding into a discussion of the ARU itself. An overview of the GYROFLEX Gyro will therefore be presented before presenting the Magellan requirements, implementation and results.

When company-funded development of the GYROFLEX Gyro was begun in 1958, it was considered an "unconventional gyro." Since it entered production in 1967, evolution and production have continued through the delivery of over 33,000 gyros with current production units meeting specification requirements such as $0.001°/h$, one sigma, random drift for six hours.

Very successful wide application and independent testing of the first production gyro, the MOD II, clearly identified the design elements to be addressed in further development. Changes were incorporated such as elimination of some parts, an improved pickoff, and improved processes. This resulted in the MOD IIC production gyro used in third-generation navigation systems on board the Space Shuttle, F-16, JA-37, SR-71, Pershing II, the ASAT, and SKIRU. Seventy-five percent of these gyros exhibited random drift rates of less than $0.0025°/h$, one sigma[3].

Further improvement was required, however, in order to meet the 0.2 nmi/h goal of systems such as Kearfott's High Accuracy Inertial Navigation System (HAINS). The buildup of piece-part tolerances precluded such improvement, making it necessary to incorporate some adjustments into the gyro design. The MOD IIE GYROFLEX Gyro was the result. The   capability   , has in part made the HAINS the most accurate unaided system currently available. It is being applied to the Space Shuttle upgrade, the B-1B bomber, and classified programs. Figure 1 is a histogram that depicts the random drift, trend removed, for both axes of a sample of production MOD IIE gyros. Note that the mean value for both axes is only   . The production gyro shown in Figure 2 exhibits a random drift (trend not removed) of   , one sigma, for 15 hours.

Kearfott has also successfully developed another derivative, the MOD IIE/S, which is the strapdown version of the MOD IIE. It is this gyro that was available to be applied to the Magellan ARU. The MOD IIE's torquer resistance was significantly reduced to permit driving the required torquing current with the available SKIRU 15 volt supply. In addition, the magnet material was changed in order to accommodate input rates of $15^{o}/s$. It is the MOD IIE/S that is employed in the hardware, which is covered in the remainder of the paper.

The performance of the MOD IIE/S surpasses that of the gas-bearing-motor SDF gyros available for use in space. Angle noise in the frequency range out to 100 Hz is an aspect that will be covered later in this paper. Because of particular interest for some applications, the equivalent angular error accumulated over a period of eight hours has been analyzed. Figure 3 shows the integrated angle accumulated as a function of time over a period of eight hours from the average of the hour preceding the test. The 8-hour data typically fall in the range of 5 to 30 arc-seconds. In the sample shown, the peak value of 5.8" in 8 hours is equivalent to accumulating angle error at less than one arc-second per hour; it actually works out to 0.0002"/s or 1 nrad/s. Classical Noise Equivalent Angle (NEA) tests have also been performed. Typical values for the MOD IIC, MOD IIE, as well as the MOD IIE/S lie in the range of 0.01 to 0.02 arc-seconds, peak to peak, for a series of one-hour runs.

The GYROFLEX Gyros have also been employed where long life is required. Over 4,000,000 hours have been accumulated in life test on 81 GYROFLEX Gyros, several of which have exceeded 100,000 hours with one over 165,000 hours. The latter unit did not exceed a random drift of $0.005^{o}/h$ through 100,000 hours. SKIRU-level life data will be described later in this paper.

THE MAGELLAN MISSION

The primary purpose of the Magellan Mission is to map the cloud-shrouded surface of Venus by means of a Synthetic Aperture Radar (SAR)[4]. The spacecraft will be placed in a near-polar orbit so that the SAR will map successive swaths/sectors of the surface as Venus rotates underneath the spacecraft once in 243 days. During the 400- to 460-day cruise between Earth and Venus, the spacecraft will be exercised through repetitive 48-hour calibration cycles that include a daily calibration of the ARU Acceleration-Insensitive Drift Rate (AIDR)[5]. This is followed by a special calibration sequence during the first eight days following Venus Orbit Insertion (VOI+8). The 243-day mission then consists of a series of 186-minute (3.1 hour) orbits during which mapping data are taken and transmitted to the ground, and further calibration maneuvers are performed. Some parameters can be updated during these various calibration cycles while others cannot. The stabilization and maneuver requirements, and the capability to update or not, were factors which established the ARU performance specifications. This was discussed in detail in a paper by Huang[6].

Figure 1   MOD IIE Random
Drift/Vertical Gyro

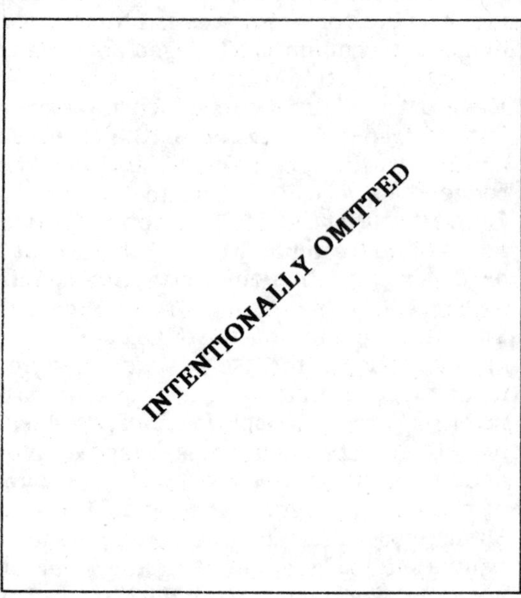

Figure 2   MOD IIE Random
Drift, 15 Hours

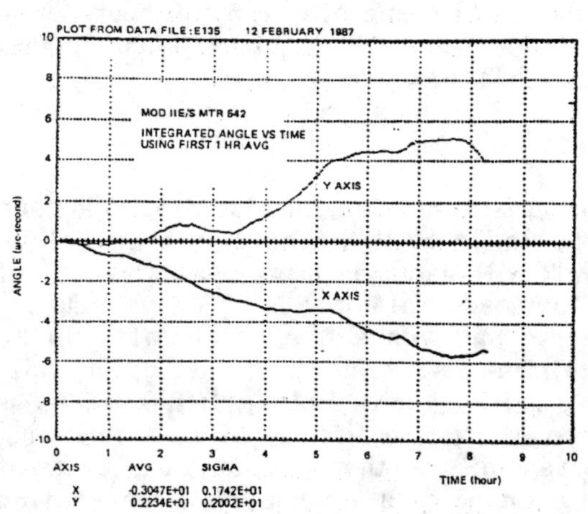

Figure 3   Integrated Angle –
MOD IIE/S

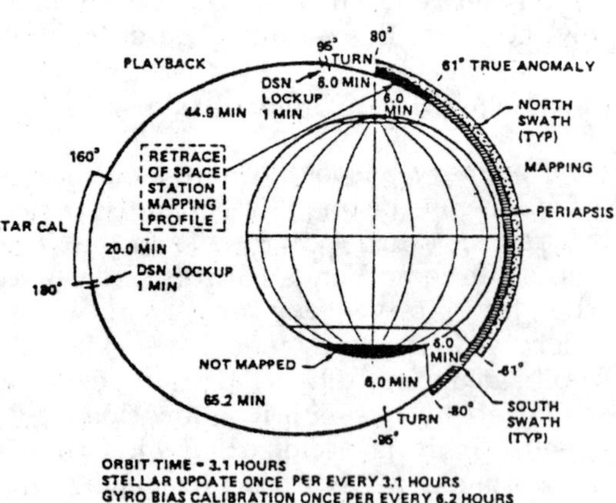

Figure 4   Magellan Mapping
Orbit Operation

The mapping orbit activities and time line are shown in Figure 4[7]. Thirty-seven minutes or radar data are collected during each 43-minute mapping window near periapsis. During this period the spacecraft-fixed High-Gain Antenna (HGA), see Figure 5[8], is pointed to the Venus surface. The spacecraft is then rotated to point the antenna at the Earth to transmit the stored data; this requires 111 minutes. Thirty-two minutes remain during each orbit for other activities such as a star-scan sequence to update the ARU (the only attitude update available during the 3.1-hour orbital period), and the pre- and post-mapping reorientation maneuvers.

Most Earth-orbital and interplanetary spacecraft place a premium on stability of the AIDR because of its impact on LOS stabilization. On the Magellan spacecraft, ARU scale factor is also critical because of the dynamic scenario—the many turns to be commanded and measured in each orbit.

ARU REQUIREMENTS

The ARU, the SKIRU, is a high-performance, three-axis strapdown package containing two GYROFLEX DTG's. Four (three orthogonal and one redundant) axes of digital, incremental-angle information are provided from each of two block-redundant ARU's, which are aligned along the body axes of the spacecraft. Each ARU has two functional modes: a low-rate mode with an input range of $\pm 1.0°$/s, and a high-rate mode with an input range of $\pm 8.0°$/s.

Short-term random drift and output noise requirements have been established by the performance requirements during the mapping and data transmission sequences. Long-term instability is dictated by whether or not a parameter can be updated during the calibration sequences. Table 1 summarizes these capabilities[9]. It can be seen that long-term scale factor stability and temperature sensitivity of the scale factor, misalignment and AIDR are critical since Factory Acceptance Test (FAT) calibrations must be relied upon. In addition, misalignment stability for a period of 243 days following VOI+8 is also critical. The mission-life temperature variation is limited to $\pm 10°C$ while the temperature variation over one orbit is limited to $\pm 1.5°C$. This gives rise to the ARU performance requirements shown in Table 2[10].

The low-range requirements are somewhat more stressful than immediately apparent since the low-range is scaled for a rather high 1°/s, i.e., electronics-related errors become larger in terms of drift rate as the input range increases.

ARU/SKIRU IV-M TEST DATA

The Space-Qualified Kearfott Inertial Reference Unit IV-M (SKIRU IV-M) was the ARU selected for the Magellan mission. The evolutionary modifications made to tailor the SKIRU capabilities to meet the Magellan requirements will be described later in the paper. They were of such a nature as to permit qualification by similarity. Two SKIRU IV-M's were built, tested and delivered to the program.

Actual FAT data for S/N 1001 are summarized in Table 3 for some of the key requirements that were listed in Table 2. Data on S/N 1002 were comparable. The low-range results are of particular interest since this is the in-orbit operating mode.

Table 1
GYRO PARAMETER CALIBRATION

ARU Parameter	Calibration Conditions and Requirements
Misalignment	A. Manufacturing/post–environmental test knowledge *B. Calibrate 2 times in cruise, one time during VOI+8 about X, Z for verification, low range only
Scale Factor (SF)	A. Manufacturing knowledge *B. Calibrate low ranges 2 times in cruise, one time during VOI+8 C. During orbit, rely on long-term stability and temperature compensation
AIDR	A. Manufacturing knowledge *B. Calibrate high and low ranges 2 times during cruise, low range one time during VOI+8 **C. Calibrate daily during cruise **D. Calibrate every 6.2 hours on orbit

* Special maneuver calibration - ground processed
**Nonspecial maneuver calibration - onboard autonomous

Table 2
ARU CHARACTERISTICS

Parameter	Specification Value	
SCALE FACTOR		
- Stability at temperature	40 ppm/8h	
- Asymmetry instability	50 ppm	(from FAT)
- Nonlinearity	50 ppm	
- Temperature sensitivity	150 ppm/5°C	
RSS	170 ppm	(over 5°C)
- Long-term stability	400 ppm	(240 days)
- Long-term asymmetry	140 ppm	(240 days)
MISALIGNMENT		
- Instability at temperature	15"/243 days from VOI+8 update	
- Nonorthogonal instability	15"/243 days from VOI+8 update	
- Temperature sensitivity	20"/5°C	
RSS	29"	(over 5°C)
AIDR		
- Instability at temperature	0.009(°/h)/8h, 3 sigma	
- Temperature sensitivity	0.018(°/h)/5°C	
RSS	0.020°/h	(over 5°C)
- Long-term instability	0.200 °/h	(240 days)

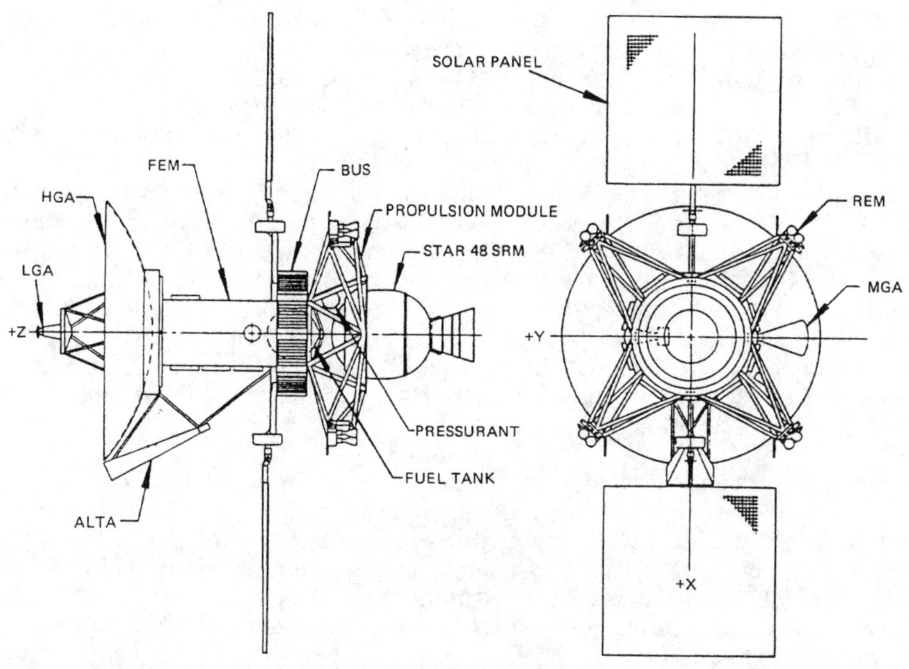

Figure 5 Magellan Spacecraft

Table 3
SKIRU-IV-M FAT DATA (S/N 1001)

Description	Range	Specification	IRU	Initial Performance 5/23/86				Final Performance 7/11/86			
				Pitch	Roll	Yaw 1	Yaw 2	Pitch	Roll	Yaw 1	Yaw 2
WEIGHT		8.4 lb maximum	8.2								
INPUT POWER											
Both gyros run	Low	26 W max (Rate: <0.0005º/s)	19.3								
SPIN MOTOR											
Run-up Time		Gyro 1: 60 s maximum	8.9								
		Gyro 2: 60 s maximum	9.5								
DIGITAL PERFORMANCE AIDR											
Instability	Low	0.0009"/s, 3 sigma, 8 h		0.0036	0.0013	0.0026	0.0015	0.0070	0.0014	0.0033	0.0014
	Low	0.027"/s, 3 sigma, 24 h		0.0057	0.0039	0.0043	0.0043	0.0072	0.0034	0.0037	0.0052
	High	0.200"/s, 3 sigma, 24 h		0.0002	0.0065	0.0002	0.0086	0.0145	0.0084	0.0105	0.0086
Temperature Sensitivity	Low	0.0035 ("/s)/ºC, maximum		0.0013	0.0000	0.0019	0.0016	--	--	--	--
Total, rss	Low	0.020"/s, 3 sigma (8 h instability & temperature sensitivity over 5ºC)		0.0038	0.0046	0.0053	0.0006	0.0082	0.0040	0.0038	0.0058
SCALE FACTOR											
SF Instability	Low	+40 ppm, 8 h, maximum		25	9	9	10	8	12	7	1
	Low	Initial-to-final		--	--	--	--	0	18	0	35
	High	Ref. only, 8 h		4	0	19	25	15	4	12	6
SF Nonlinearity	Low	± 50 ppm		8	8	25	13	12	6	10	16
	High	± 2000 ppm		140	140	100	90	260	140	110	90
SF Asymmetry	Low			12	26	34	38	13	19	22	30
Asym. Instability	Low	± 50 ppm		--	--	--	--	1	7	12	8
Temperature Sensitivity	Low	30 ppm/ºC, maximum		20	7	20	20`	--	--	--	--
Total, rss	Low	170 ppm, maximum (SF instability, nonlinearity, asymmetry and temperature sensitivity over 5ºC)		103	32	101	101	101	32	101	101

Against a low-range AIDR specification requirement of 0.009°/h, 3 sigma, the 8-hour instability typically fell in the range of 0.0015 to 0.0035°/h; and against a requirement of 0.027°/h, 3 sigma, the 24-hour instability ranged between 0.0034 and 0.0072°/h. The total instability--the rss of the 8-hour drift and temperature sensitivity over 5°C--was less than 0.0082°/h versus a specification of 0.020°/h.

Similarly, the low-range scale factor instability over 8 hours was typically 8 to 12 ppm, and over 7 weeks was less than 35 ppm; nonlinearity was typically 8 to 16 ppm; asymmetry was typically 15 to 25 ppm, while asymmetry instability over 7 weeks was less than 12 ppm. The total instability--the rss of the scale-factor instability at a given temperature, asymmetry instability, nonlinearity, and the temperature sensitivity over 5°C--was less than 103 ppm against a specification of 170 ppm. It can be seen by inspection that it is the temperature sensitivity that predominates the total error. A better trim of this sensitivity would have resulted in a lower total error.

The scale-factor temperature sensitivity must be modeled in order to meet the Magellan system requirements. Figures 6 and 7 show the scale factor modeled against temperature and the temperature-sensor output voltage. The initial and final FAT data are shown, 61 days apart. In between, the unit had been exposed to the full FAT complement of environmental tests including random vibration and thermal-vacuum testing. Repeatability of the model across the temperature range was better than 16 ppm.

The alignment instability over the 7-week period ranged between 16 and 35 arc-seconds. However, the results are significantly corrupted by the errors encountered in dismounting and remounting the unit on the test table; these are estimated to be as much as 30 arc-seconds since survey techniques were not employed. This factor is not present in the spacecraft installation where the unit is fixed in place.

The FAT results on S/N 1002 were comparable; in fact the low-range AIDR results were even better. The 8-hour instabilities were typically 0.001 to 0.002°/h, the 24-hour values 0.0012 to 0.0035°/h, and the total instabilities less than 0.0046°/h.

These two units were delivered to Martin Marietta Astronautics Group where they were integrated into the Magellan spacecraft and subjected to all the spacecraft-level environmental and functional testing. This included a minimum of 600 operating hours on each unit. Since long-term stability of several coefficients is a critical requirement to mission success, the two units were removed from the spacecraft as planned, and, while the spacecraft was enroute to the Kennedy Space Center, they were returned to Kearfott for Parameter Verification Testing (PVT), a recalibration of certain coefficients. A series of functional tests were performed during September 1988, 26 months after S/N 1001 was shipped, and 21 months after S/N 1002 was shipped. The data on key parameters are summarized in Table 4.

On S/N 1001, the low-range AIDR instabilities are comparable to those obtained during FAT. (On S/N 1002, the 24-hour data are again better, ranging between 0.0011 and 0.0039°/h, 3 sigma). Except for one axis, the total AIDR change over 26 months was less than 0.016°/h. Interpolating to 243 days, this works out to less than 0.005°/h, well within the requirement for the FAT and one to two orders of

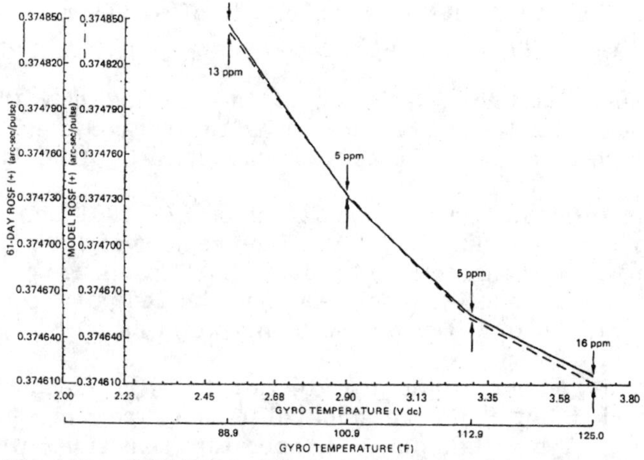

Figure 6   Magellan S/N 0001: Pitch Axis ROSF (+) Versus Temperature

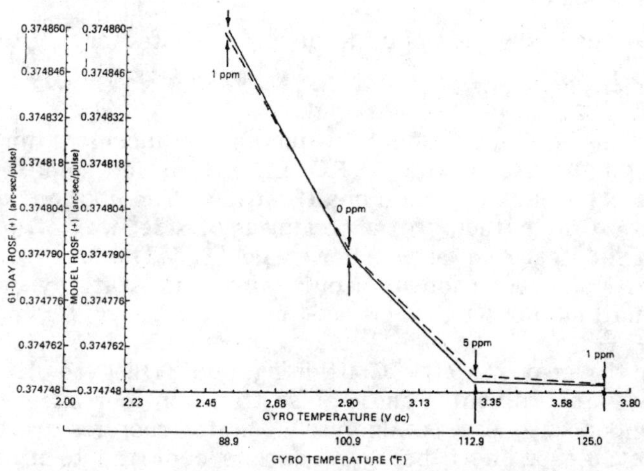

Figure 7   Magellan S/N 0001: Roll Axis ROSF (+) Versus Temperature

Table 4
SKIRU IV–M PVT DATA (S/N 1001)

Description	Range	Specification	IRU	PVT Performance 9/7/88				Interpolated Performance to 243 Days			
				Pitch	Roll	Yaw 1	Yaw 2	Pitch	Roll	Yaw 1	Yaw 2
DIGITAL PERFORMANCE AIDR											
Instability	Low	0.009"/s, 3 sigma, 8 h		0.0059	0.0065	0.0049	0.0023	--	--	--	--
	Low	0.0027"/s, 3 sigma, 24 h		0.0003	0.0072	0.0047	0.0029	--	--	--	--
Long-Term, FAT to PVT	Low	"/s, 26 mo		0.0157	0.0040	0.0482	0.0130	--	--	--	--
Long-Term Instabl.	Low	("/s)/mo		0.0006	0.0002	0.0019	0.0005	--	--	--	--
243-Day Interpolation	Low	"/s		--	--	--	--	0.0048	0.0016	0.0150	0.0004
SCALE FACTOR											
SF Instability	Low	± 40 ppm, 8 h									
Long-Term, FAT	Low	ppm, 26 mo		107	187	160	213	--	--	--	--
Long-Term, Instability	Low	ppm/mo		4	7	6	8	--	--	--	--
243-Day Interpolation	Low	ppm		--	--	--	--	32	56	48	64
SF Nonlinearity	Low	± 50 ppm		16	14	18	10	--	--	--	--
SFA Asymmetry	Low	ppm		39	0	59	2	--	--	--	--
Asym. Instabl. FAT to PVT	Low	ppm, 26 mo		26	19	37	28	--	--	--	--
Asym. Instabl.	Low	ppm/mo		1	0.7	1.4	1.1	--	--	--	--
243-Day Interpolation	Low	ppm		--	--	--	--	8	14	11	9

magnitude lower than the 240-day requirement of 0.200°/h. (On S/N 1002, all four axes would fall within 0.005°/h.)

Interpolation assumes a linear variation of the parameter with time. It also prorates the measurement uncertainties so as to average them over the entire time interval, which tends to minimize their impact on the true data.

The greatest change in low-range scale factor on S/N 1001 over 26 months (790 days) was 213 ppm, well within the 240 day requirement of 400 ppm. Interpolating to 243 days, the worst change would be 64 ppm. The change in asymmetry was less than 37 ppm for 26 months, much less than the 140 ppm requirement for 240 days; again interpolating to 243 days results in a worst case of 14 ppm.

Alignment-instability values would prorate to 0 to 10 arc-seconds for some axes over a 243-day period, but again the mount/dismount errors are suspected to have corrupted the data. Typical alignment temperature sensitivity was determined to be in the range of 1 to 2"/°C.

The FAT and PVT results were judged to be very satisfactory by the customer.

EVOLUTION TO THE SKIRU IV-M

The SKIRU product line was launched when the requirement for the low weight achievable through the use of the GYROFLEX Gyro and some hybrid electronic modules overcame the lack of space qualification. The application advantages of the DTG over the SDF floated gyro, in the areas of size, weight, power, reliability and cost, were set forth in an earlier paper[11]. The use of hybrid modules permitted a significant reduction in circuit-card real estate by compressing large functions into small modules.

The SKIRU is a three-axis ARU. Containing two TDF gyros, it provides analog rate output data and incremental-angle pulse trains on four axes. Thus, redundant outputs are provided on one axis. Individual motor supplies are provided for the two gyros so that one or the other gyro can be operated to provide single-axis, redundant outputs for those applications requiring only single-axis control in orbit. With 28 V dc power supplied, the SKIRU generates all voltages and frequencies required for its operation. SKIRU's I through IV are unheated units; hard-wired temperature compensation of the AIDR and scale factor is supplemented by temperature modeling where required. Several hybrid modules had been fully developed and purchased in quantity for Kearfott's production, third-generation navigation systems. These were qualified to the intent of SAMSO 73-2C and have since been qualified and procured to full Class S requirements. No hybrid anomaly, much less a failure, has ever been experienced on a delivered SKIRU.

Through the uses of two GYROFLEX Gyros (which together weigh less than 1.4 lb), hybrid modules, and advanced packaging techniques, the weight of the SKIRU was held to less than 8.5 lb. With enhancements described later, the SKIRU IV has grown to 9 lb maximum. The body dimensions are 7" by 9" by 3.75". Typical input power has been lowered to 16 to 17 W on the new-technology SKIRU IV from the SKIRU III typical values of 19 to 20 W.

The SKIRU II was a moderate performance, analog output, 1°/s input range package. Operation at quarter speed increased the range to 4°/s for short periods of time. The Voltage-to-Frequency Converter (VFC), which had been fully space qualified on the Voyagers 1 and 2, was repackaged and installed in the SKIRU II without increasing the outline dimensions. The result was the analog- and digital-output SKIRU III. Qualification of SKIRU II and III included random vibration at 22 $\underline{g}$ rms, pyro shock at levels up to 2,500 $\underline{g}$, and thermal-vacuum exposure for periods up to 900 hours. Both are currently operational.

The Magellan ARU, the SKIRU IV-M, is an enhancement of the SKIRU III. The 15°/s MOD IIE/S gyro replaced the MOD IIC to provide the 8°/s input range capability. Drift stability had been limited by the contribution of VFC voltage instabilities in low range. The voltage developed across the single-readout resistor in the capture loop, proportional to capture current and input rate, was applied to the VFC in the SKIRU III. Full-scale high range developed 2 V across this resistor while full-scale low range developed only approximately 0.2 V. Dual-readout resistors were used in the SKIRU IV-M so that the full voltage would be developed in both high and low ranges, thus suppressing the VFC voltage-induced drift contributions in low range by approximately a factor of 10. Switching readout-resistor values in the capture loop made it necessary to change from voltage to current feedback. The next step was to increase the full-scale voltage to 3.5 V for a potential total factor 17.5. Thus, while low-range full scale was increased from 0.1°/s to 1°/s in going from the SKIRU III to the SKIRU IV-M, the VFC drift contribution was actually decreased and performance improved. These performance results have already been detailed earlier in the paper.

The functional block diagram shown in Figure 8 supports the descriptions provided above.

The two compact subassemblies of the SKIRU IV-M are shown in Figure 9. After being individually tested, the four Circuit Card Assemblies (CCA's) are stacked on the cover and tested as an electronics assembly. The gyro/trim board/housing assembly accepts the inverted electronics assembly. The three heavy magnetics, restrained on the electronics assembly, are fastened down to the housing by screws inserted from below so that their weight is borne by the housing, not by the CCA. RF gaskets are installed under all covers. The resulting SKIRU is shown in Figure 10 where the optical reference cube is hidden by a protective cover.

EVOLUTION BEYOND MAGELLAN

Several factors combined at the right time to drive a marked improvement in capability. New programs required the ability to withstand pyro shock pulses with amplitudes up to 5,000 $\underline{g}$. In addition, the piece part junction temperature was to be limited to a maximum of 100°C (part case temperature had previously been limited to 100°C). In contrast to so many engineering issues where a compromise must be drawn between opposing requirements, in this case both requirements were satisfied by the lamination of an aluminum layer to the Printed-Wiring Board (PWB). This layer stiffened the PWB and permitted better heat sinking of the parts. At the same time Kearfott was informed by microcircuit suppliers that Class S parts would no longer be provided in the flatpack configuration; only Dual-In-Line Packages (DIP's) would be available in the future. Incorporating all these changes also afforded the opportunity to update some 10-year-old-technology parts, which were becoming increasingly difficult to obtain. The result was an

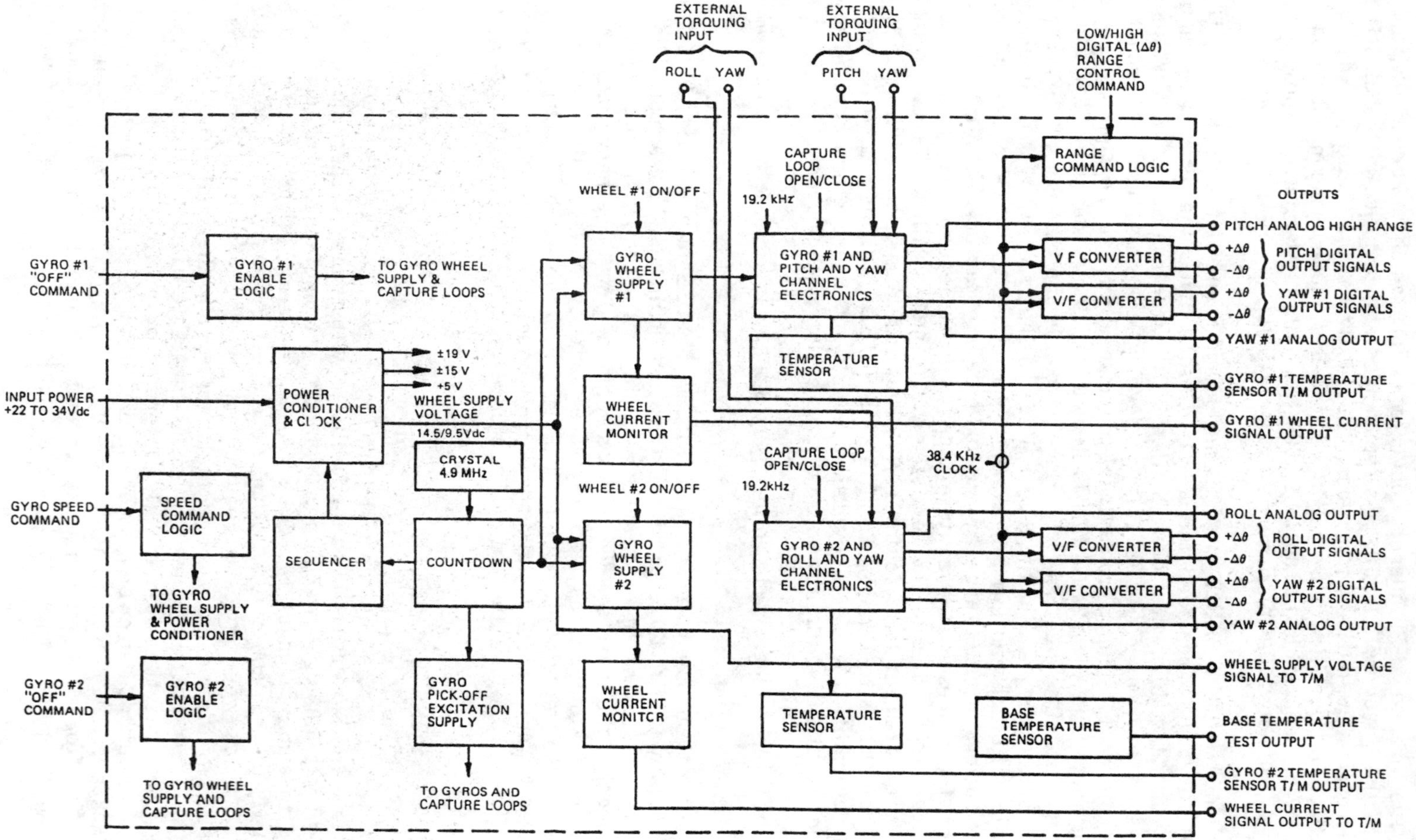

Figure 8 SKIRU IV Functional Block Diagram

Figure 9 SKIRU IV-M Gyro/Housing and Electronics Assemblies

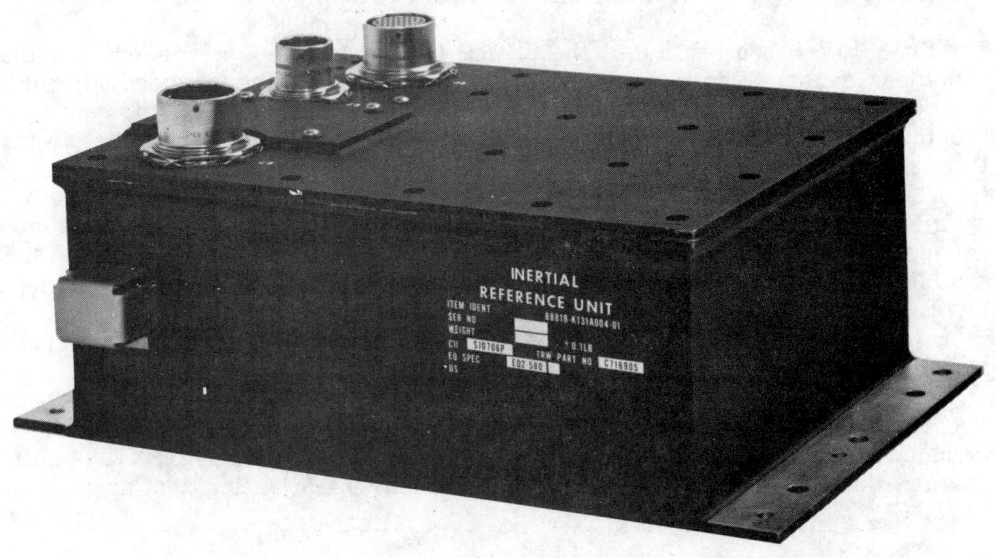

Figure 10 SKIRU IV-M

evolution of the Magellan SKIRU IV-M, which as pointed out earlier was a modified SKIRU III, to the new-technology SKIRU IV.

The predecessor SKIRU IV-M Gyro Wheel Supply (GWS) CCA is shown in Figure 11, i.e., prior to the changes outlined above. The crystal clock can be seen in one corner in its shock mount, as well as the two GWS hybrids in the center, the adjacent countdown hybrid that develops the motor and pickoff excitation frequencies, the pickoff-excitation power amplifier hybrid, and a series of flatpack microcircuits. The PWB is subdivided into small areas by a series of posts to limit deflections perpendicular to the board surface under shock and vibration. These posts also serve as thermal busbars to shunt heat out to the housing surfaces. This design was highly successful.

The new SKIRU IV GWS CCA is shown in Figure 12. The DIP's and the aluminum layer are immediately obvious. The crystal has been replaced by a crystal oscillator, and the countdown hybrid module has been replaced by microcircuits, which were not available at the time of the original design. Highlighted areas in the figure are reflections from the conformal coating. Spot shielding to withstand particle radiation dose of 10 megarads is not shown in this view.

The SKIRU IV was successfully qualified in 1988. It withstood pyro shock levels up to 5,000 g and random vibration of 21.9 g rms. The part case temperatures were reduced by 9 to 15°F; the resulting lower failure rates that were then permitted raised the predicted Mean Time Between Failures (MTBF's) to 297,425 hours (70°F baseplate). The new technology also reduced the input power to between 16 and 17 W over the full temperature range and input voltage range of 22 to 34 V.

A snapshot of significant data from the SKIRU IV qualification unit is shown in Table 5. The 24-hour AIDR instability was less than 0.0032°/h, 3 sigma, and the change from before-to-after pyro shock was less than 0.029°/h.

The SKIRU IV-H is a version of the SKIRU IV that has been hardened to withstand a nuclear event. Photocurrent burnout resistors have been added internally to protect circuits where necessary. In addition, a Faraday Cage has been added to the top of the unit as shown in Figure 13; it contains zener diode/resistor/feed-through-capacitor filter networks to prevent the Electromagnetic Pulse (EMP) from entering the unit. The cage, the optical reference cube, the heavier mounting flange that is typical of the SKIRU IV and IV-H, and the temporary connector savers can be seen in the picture of the SKIRU IV-H, Figure 14. Spot shielding was utilized throughout the unit to reduce radiation levels to acceptable levels at sensitive parts. The Engineering Model GWS CCA (with modifications) is shown in Figure 15, with shields in place, for comparison with the unshielded CCA already shown in Figure 12.

The SKIRU IV-H was successfully subjected to a protoqualification test, which included pyro-shock exposures up to 4,200 g, Flash X-Ray (FXR) testing at operate-through and threat levels, and also current-injection testing to simulate the EMP.

Both the SKIRU IV and IV-H have been designed to operate for ten years.

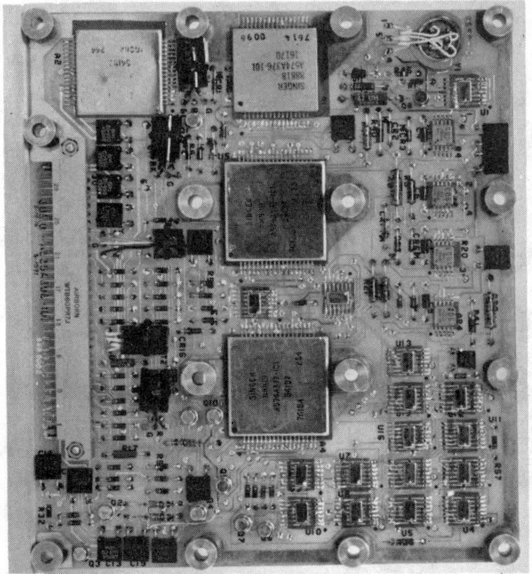

Figure 11 SKIRU IV-M GWS Board

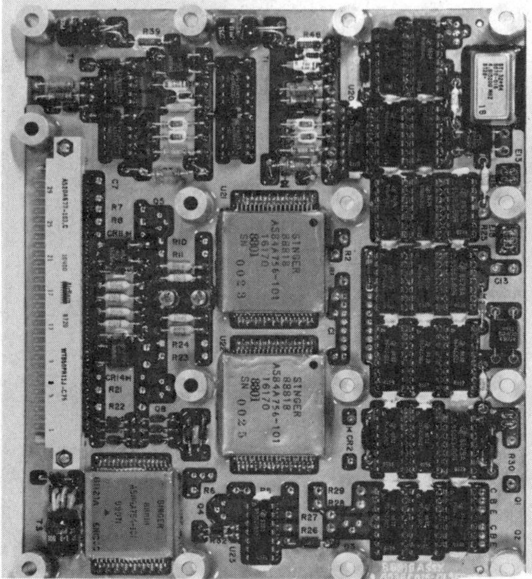

Figure 12 SKIRU IV GWS Board

Table 5
SKIRU IV (S/N 0001) DATA

Description	Range	Specification	IRU	Pitch	Roll	Yaw 1	Yaw 2
WEIGHT	--	9.0 lb maximum	8.6	—	—	—	—
INPUT POWER Both Gyros Run	Low	26 W max (Rate: $<0.005°/s$)	16.1	—	—	—	—
AIDR Stability	Low	0.027"/s, 3 sigma, 24 h	—	0.0032	0.0008	0.0012	0.0005
	Low	Before to after 5000 g shock	—	0.008	0.002	0.004	0.002
ROSF Asymmetry	Low	ppm	—	2.6	2.5	0.1	5.4

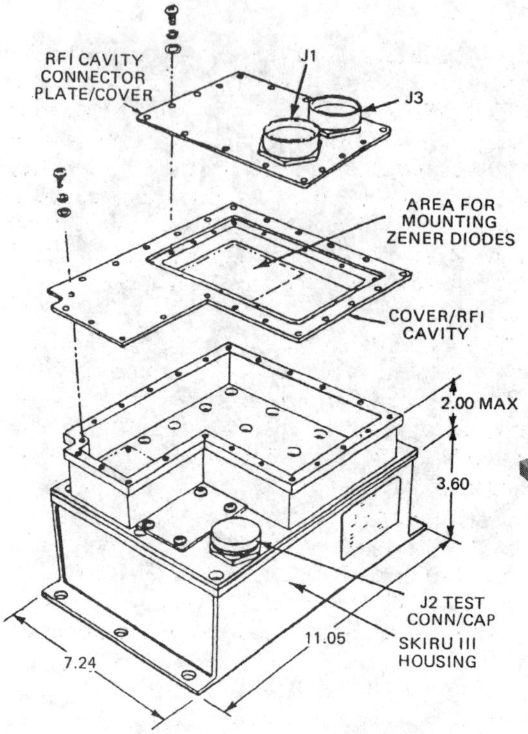

Figure 13 SKIRU IV-H Exploded View

Figure 14 SKIRU IV-H

Figure 15 SKIRU IV-H GWS CCA

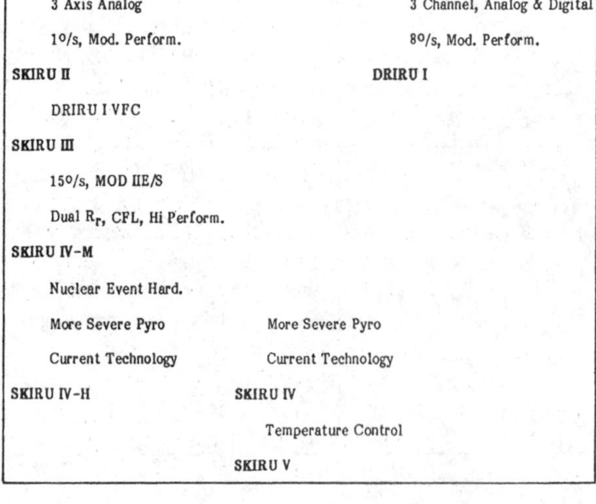

Figure 16 SKIRU Tree

OVERVIEW

The SKIRU Tree, Figure 16, recaps the evolution starting with the analog moderate performance SKIRU II; the addition of the DRIRU I VFC for the SKIRU III; the change to the MOD IIE/S gyro and the dual-readout resistor/Current Feedback Loop (CFL) for the SKIRU IV-M; the technology upgrade to the SKIRU IV; and the addition of nuclear-event hardening for the SKIRU IV-H. The SKIRU IV is currently being modified by adding temperature control and heaters to certain elements in order to achieve very high performance. The resulting SKIRU V will meet an AIDR stability requirement of 0.002°/h, 3 sigma, for a period of 8 hours over a temperature range of 40°F.

Over 30 SKIRU's have been delivered to date. The twelve of these units that are in their operating environment have accumulated over 325,000 hours at the package level and over 600,000 hours on the gyros without failure. The oldest SKIRU has passed 80,000 hours of continuous operation without anomaly. The DRIRU I units on board Voyagers 1 and 2 have been in space for over 11 years but operate on approximately a 20% duty cycle. Several more units will become operational this year such as those on Magellan and Galileo. Residual DRIRU I hardware from the Voyager program has been installed on the Galileo spacecraft. SKIRU hardware will also be delivered to General Electric for the DSCS III Integrated Apogee Boost System (IABS); to TRW for the TDRS7, which will replace the spacecraft that was lost on the Challenger; and to a number of sensitive applications.

Kearfott's participation in space programs to date is summarized in Table 6.

PRECISION POINTING

Precision pointing is of increasing importance for space applications. As mentioned earlier, gas-bearing-motor floated gyros have for years been the standard for those applications requiring low LOS jitter. However, some theorized in the mid 1970's that the flexure of the DTG could not only rotationally isolate the inertia element but also decouple it from disturbances generated on the drive side of the flexure. If true, a lower level of noise would be present in the output signal. Further, this noise level could be lower than even that of the gas-bearing gyro since the latter still contains a heavy mass rotating on a bearing inside the inertial element; though the gas bearing is quieter than a ball bearing, it is still a bearing. Some of the earliest testing that demonstrated the low output noise levels inherent in the DTG was performed at the Central Inertial Guidance Test Facility and reported upon by Craig Price in 1976[12]. Later work demonstrated the potential superiority of the DTG over the gas-bearing gyro[13].

More recently a series of tests have been performed on the MOD IIE/S GYROFLEX Gyro described earlier in the paper. All testing was performed with a Kearfott-designed, low-noise, high-gain capture loop with a bandwidth of 30 Hz. Two gyros were used to determine gyro performance in imperfect environments by using common-mode rejection of cultural noise.

A measuring technique called Equal Input Noise (EIN) was used to determine gyro performance under constant rates and accelerations. (Though also used for static testing, its effect is less noticeable since the cultural noise levels in the static

Table 6
KEARFOTT PARTICIPATION IN SPACE PROGRAMS

No Flight Failures	
FLOATED GYROS (ALPHA & KING)	
AEROBEE (HAE I, II) AGENA GRA APOLLO SURVEYOR (1 though 7) LUNAR ORBITER (3 & 4) MARINER (2 through 10)	NIMBUS (2 through 4) OAO II SKYLAB VELA HOTEL VIKING ORBITER (1 & 2) CENTAUR
DYNAMICALLY TUNED GYROS (GYROFLEX GYRO)	
SPACE SHUTTLE—IMU SKIRU II—ARU	SKIRU III—ARU VOYAGER (1 & 2)--DRIRU I

Table 7
STATIC NOISE DATA
Equivalent–Angle Output Data
0.25 to 100 Hz

Equivalent Angle	FACILITY 1 (nrad)	FACILITY 2 (nrad)	FACILITY 3 (nrad)
Gyro 1	5.7	7.4	8.2
Gyro 2	8.4	6.1	7.9

Table 8
DYNAMIC NOISE DATA
Equivalent–Angle Output Data
0.25 to 100 Hz

Facility 1 - Constant Rate Date--Table R		
RATE (^{o}/s)	GYRO 1 (nrad)	GYRO 2 (nrad)
0.3	35	35
1.0	44	44
3.0	50	50

Facility 2 - Constant Rate Date--Table S		
RATE (^{o}/s)	GYRO 1 (nrad)	GYRO 2 (nrad)
0.3	62	62
1.0	43	43
3.0	68	68

Facility 2 - Constant Acceleration Data--Table S		
ACCELERATION (^{o}/s)	GYRO 1 (nrad)	GYRO 2 (nrad)
0.01	49	49
0.03	50	50
0.05	51	50

416

environment are so much lower.) The EIN estimating technique operates under the theory that when two independent sensors are subjected to an identical external noise environment, the noise which is common to both can be removed mathematically, thus yielding estimates of sensor noise levels for each of the gyros.

Implementation of the EIN technique was accomplished by utilization of an HP3562A Dynamic Signal Analyzer.

The tests were performed at three different facilities. At each, special "quiet" rooms had been constructed to achieve minimum cultural noise environments. The data that follow have been drawn from unclassified material though the sources are not identified.

Static testing was performed at all three facilities to determine quiescent jitter. Equivalent-angle output data were similar at all three over the frequency range of 0.25 to 100 Hz. The data obtained are less than 10 nanoradians at each facility; sample data are provided in Table 7. The values shown were not the lowest levels obtained. The EIN technique was not used at one of the facilities; the fact that comparable data were obtained at all three supports the previous statement that it is perhaps not as important a tool for static testing in very quiet facilities as it is for dynamic testing.

Dynamic testing was performed at two facilities. Constant-rate testing was performed at Facility 1 while constant-rate and constant-acceleration testing were both performed at Facility 2. Performance was measured at input rates of 0.3, 1.0, and 3.0°/s and accelerations of 0.01, 0.03, and 0.05°/s^2.

Accelerations were limited to the test cable length. At one facility, data were limited to the frequency range of 0.25 to 100 Hz due to time constraints also caused by cabling. The EIN equivalent-angle data are shown in Table 8. The values shown for each gyro are the average for the clockwise and counterclock-wise results.

Equivalent-angle data seem to be independent of <u>acceleration</u> inputs over the range measured. There is a slight <u>rate</u> dependency for equivalent-angle data on Table R at Facility 2, while data are random on Table S for rates up to 3.0°/s.

The dynamic data are higher than those obtained during static testing. Were common-mode rejection (EIN technique) not used, the output noise data obtained during dynamic testing could be 100 to 1,000 times worse. The problem is still a technical challenge since one must be able to reject a large dc value in order to see the small ac data. In addition to structural bending and resonances, the drive tables also introduce noise from cogging in the drive mechanization and from speed corrections that are made many times during a revolution of the table. That noise is eliminated by the EIN technique to the degree that the gyros react equally to these inputs. Obviously the data represent the sum of the gyro and drive table noise; it seems fair to assume that the true gyro noise is less than the total.

HIGH BANDWIDTH DEVELOPMENT

Precision-pointing space applications for over a decade have used DTG's in low-to-moderate-bandwidth capture loops to take advantage of their low noise, high reliability and long life. A nominal 12 Hz bandwidth was suitable for attitude control and image-motion compensation on board the Voyager 1 and 2 spacecraft. The Magellan closed-loop bandwidth of 4 to 7.5 Hz is adequate for mapping, data dumping, and calibration maneuvers. Wide bandwidth is not required for these relatively benign applications and could be detrimental by permitting spurious frequencies to enter the control loops from the structure or other sources.

Many new, dynamic systems require a much wider bandwidth. While Ring Laser Gyros (RLG's) can satisfy this requirement and also provide excellent AIDR and scale-factor stability, they have as a class been too noisy and heavy to be considered for precision-pointing applications. Not only is there considerable noise in the output signal, but also the dithered instruments introduce mechanical vibrations into the optical bench to which they are attached. On the other hand, DTG's can be used in appropriately designed closed control loops to provide bandwidths of 100 to 150 Hz while retaining their low-noise characteristics for LOS stabilization.

100 Hz Channel

One such system offered a selectable choice of two bandwidths, 50 and 100 Hz. The Stable Two-Axis Reference (STAR) channel used a MOD IIC GYROFLEX Gyro in an unheated, analog capture loop with subsequent digitization of the analog output information. The transfer function was trimmed to meet the tight phase requirements of the control loop in which it was applied. It is well known that, unlike fluid-filled floated gyros, which experience large viscosity change and therefore wide variation in their dynamic characteristics as temperature is varied, the DTG transfer function is very stable. The transfer functions measured on one axis at 50 and 100 Hz for this system are shown in Figures 17 and 18. As noted in the next section, this is achieved without special filtering. The low-noise Power Spectral Density (PSD) plots for the 100 Hz bandwidth mechanization are shown in Figures 19 and 20. Levels of $10^{-7}(°/h)^2/Hz$ or better were achieved at frequencies between 0.01 and 10 Hz, extending to 3.3×10^{-7} at 0.00125 Hz.

150 Hz Channel

Nutation of the gyro sets an upper limit on the closed-loop bandwidth achievable with a DTG. Since the nutation frequency is somewhat less than twice the tuned-rotor speed of the gyro, the 240 Hz rotor speed of the GYROFLEX Gyro allows a 150 Hz response to be achieved without requiring nutation damping, commutating filters, or cross-axis compensation; the latter has the side effect of increasing the high-frequency noise.

Another wide-bandwidth gyro loop is under development at Kearfott as part of an R&D program. It uses the MOD IIE/S gyro in a high-resolution, digital capture loop. The transfer function of this gyro channel is shown in Figure 21. The computed response is -3 dB at 150 Hz, where the phase lag is 135°, and is flat to 50 Hz. It can be seen that the measured response closely tracks the design values. Very low random drift has been achieved as can be seen in Figure 22 where the

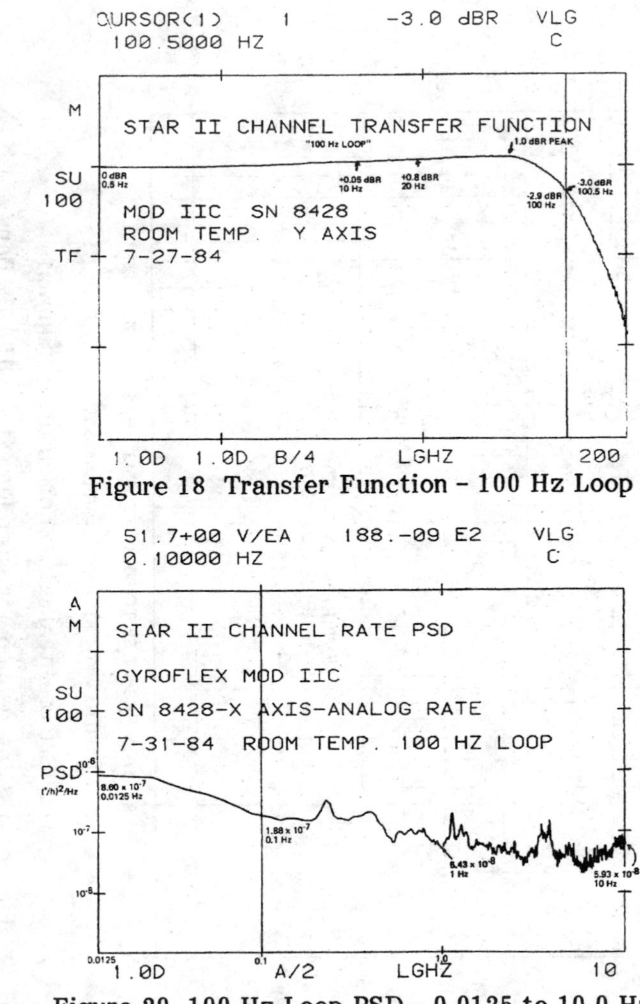

Figure 17 Transfer Function – 50 Hz Loop

Figure 18 Transfer Function – 100 Hz Loop

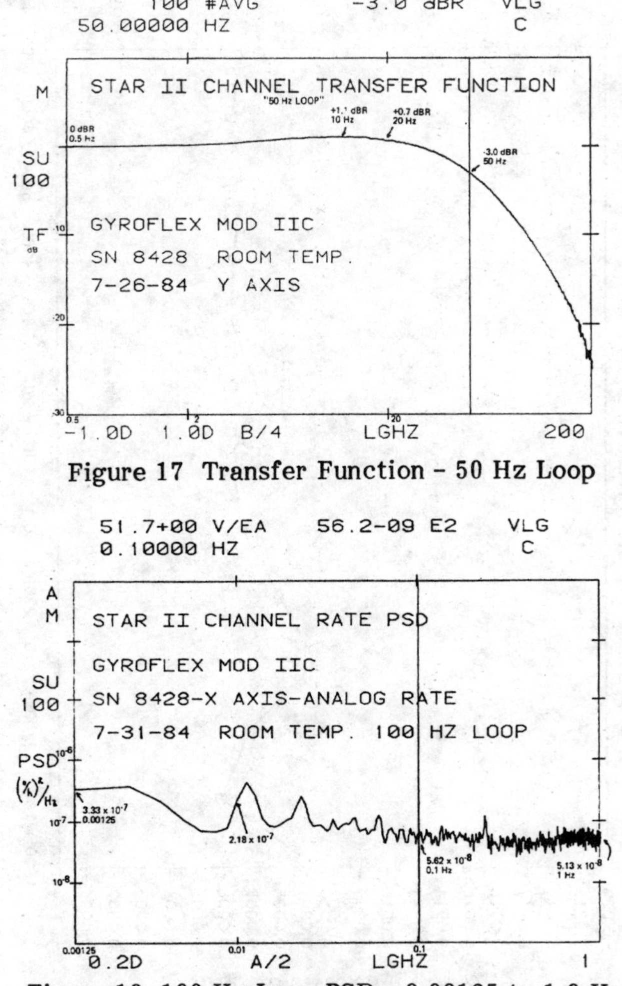

Figure 19 100 Hz Loop PSD – 0.00125 to 1.0 Hz

Figure 20 100 Hz Loop PSD – 0.0125 to 10.0 Hz

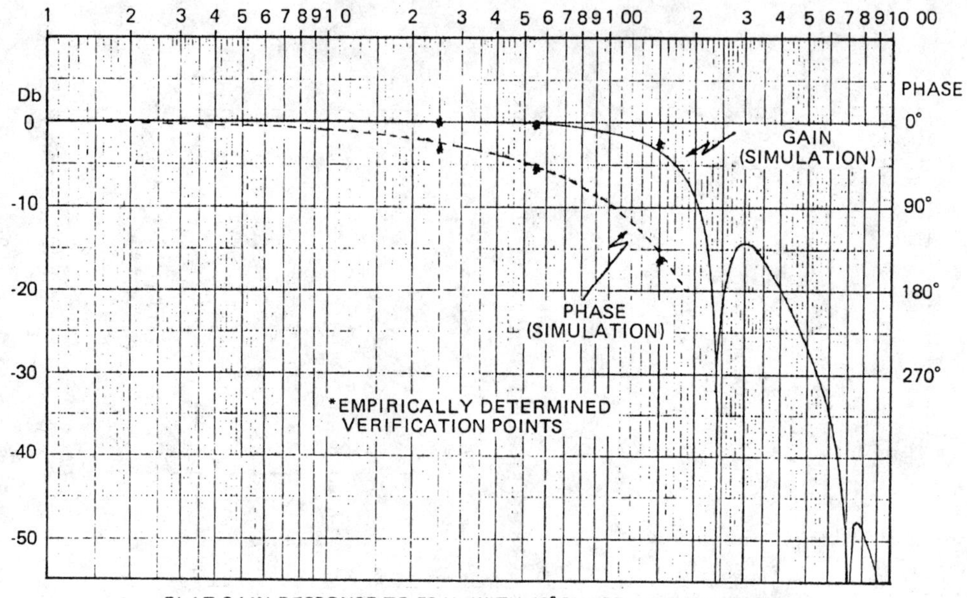

• FLAT GAIN RESPONSE TO 50 Hz WITH 43° PHASE AT THAT POINT

• -3 Db AT 150 Hz WITH 135° OF PHASE LAG

CLOSED-LOOP TRANSFER FUNCTION - 150 Hz BANDWIDTH

Figure 21 Closed–Loop Transfer Function – 150 Hz Bandwidth

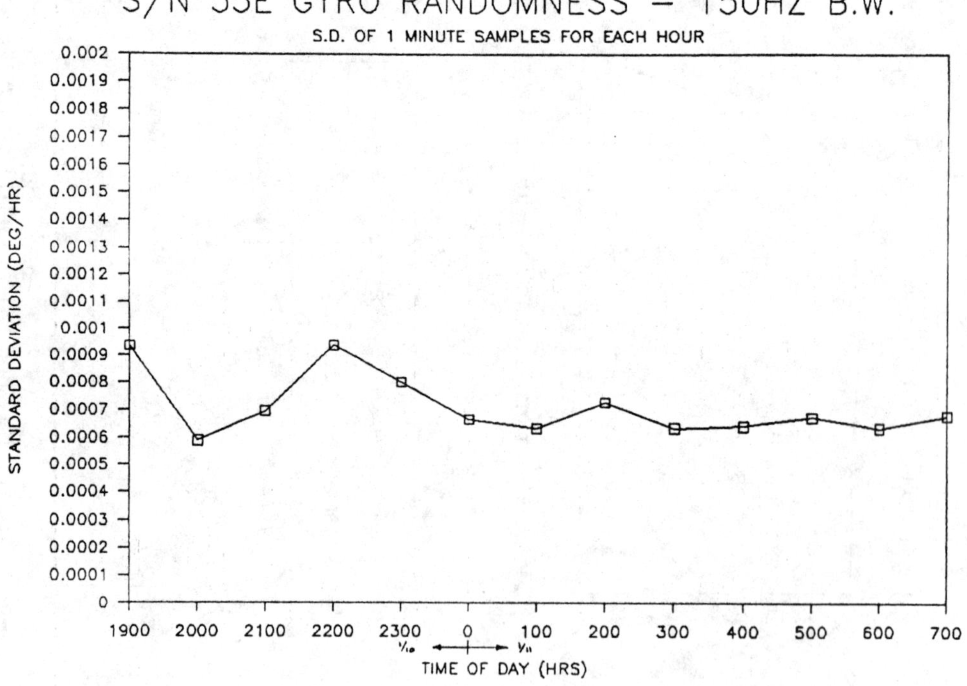

Figure 22 S/N 55E Gyro Randomness – 150 Hz Bandwidth

values range between 0.0006 and 0.0009°/h, one sigma. Each plotted point represents the one sigma value of 60 one-minute points during a one-hour interval.

SUMMARY

Because of its proven performance and qualification status, the SKIRU was well positioned to meet the requirements of the Magellan spacecraft ARU. Test data taken over a period of time longer than the Magellan mission gives confidence that mission performance will be met. The design has continued to evolve and has shown that it can satisfy advanced LOS requirements.

REFERENCES

1.   C.O. Swanson, "DRIRU I/SKIRU--The Application of the DTG to Spacecraft Attitude Control," AIAA Guidance and Control Reference, San Diego, California, AIAA 82-1624, August 9-11, 1982

2.   W.H. Huang and N.S. Reddy, "Magellan Spacecraft Attitude Determination, Updates, and Gyro Parameter Calibration," AIAA Guidance, Navigation, and Control Conference, Williamsburg, Virginia, AIAA 86-2042, August 18-20, 1986

3.   Swanson, p. 3

4.   S.B. Johnson, "Fault Protection Design for Unmanned Interplanetary Space-craft," AAS Guidance and Control Conference, Keystone, Colorado, AAS 87-001, January 31-February 4, 1987, p. 2

5.   Huang, p. 2

6.   Ibid., p. 4

7.   Ibid.

8.   Johnson, p. 3

9.   Huang, p. 8

10.   Ibid, p. 5

11.   Swanson, p. 4

12.   C.H. Price, "Gyro Noise Testing in the Low Frequency Range Using PSD Analyses," AIAA Guidance and Control Conference, San Diego, California, AIAA 76-1922, August 16-18, 1976

13.   G.J. Bukow and H. Musoff, "Fundamental Limitations of Rotational Motion Sensing for Precision Pointing Applications," AAS Rocky Mountain Guidance and Control Conference, Keystone, Colorado, AAS 85-052, February 2-6, 1985, pp. 12-17

PINPOINT LANDING CONCEPTS FOR THE MARS ROVER SAMPLE RETURN MISSION

Allan R. Klumpp[*]

Landing hazards on Mars were not well understood prior to the Viking landings. Rocks larger than the maximum size upon which the Viking lander was designed to land were prevalent in the vicinity of the landing sites. From knowledge of hazards (blocks, scarps, and slopes) gained following Viking, many people associated with the Mars Rover Sample Return (MRSR) mission have concluded that the Viking approach should not be repeated.

A variety of approaches are being pursued for making the MRSR landing safer. The most desirable approach involves a pinpoint landing capability. With a pinpoint landing, a single landing site would be certified in advance of initiating the descent orbit, using pictures transmitted to Earth from the orbiter. Then using onboard instruments, the lander would follow mapped terrain to the selected site. If navigation proves insufficiently accurate to reach a single preselected site, a small number of scattered sites would be certified, and the lander would select the most accessible one in real time.

Viking's landing footprint was an ellipse over 100 km long. Without terrain following, the MRSR footprint would measure tens of kilometers. Certifying a footprint this large would be a prodigious task for the mission support team. With a pinpoint landing capability, the size of the footprint could be reduced to about 10 m in diameter.

This paper describes some early concepts for certifying sites and navigating the lander. Some of the key problems to overcome are how to recognize hazards that are too small to be seen in pictures transmitted to Earth, how to represent terrain features in the flight computer, how to overcome errors in inertial measurements during a landing of over 12 hours duration, how to find the selected site in time to get there, and how to navigate to the selected site once it is found.

INTRODUCTION

The Mars Rover Sample Return (MRSR) is an unmanned mission planned for launch in the late 1990s. Two vehicles, launched separately, convey to Mars a mapping and communications orbiter, a Mars rover, a Mars ascent vehicle, and an Earth return vehicle. In preparation for landing on Mars, a large telescope on the communications orbiter surveys candidate landing sites to a resolution of 1 m or better and transmits the pictures to Earth. Based on the pictures, one or more landing sites

* Member of the Technical Staff, Jet Propulsion Laboratory, California Institute of Technology, 4800 Oak Grove Drive, Pasadena, California 91109.

are certified as safe for landing. A rover and an ascent vehicle then descend and land, either as a single unit or separately at sites close to one another.

The rover explores the surface for more than half a year, over a region of several tens of km, returning several collections of samples to the ascent vehicle. The ascent vehicle ascends and transfers the samples to the Earth return vehicle, still parked in Mars orbit. During and following the return to Earth, the rover may continue to explore the surface, returning its findings by radio.

The period of the orbit of the lander or landers prior to descent orbit insertion (DOI) may be as short as 2 hours or as long as 24.6 hours (one Mars day). In any case, the periapsis of the descent trajectory is close to the surface so that the descent ranges from just over 1 hour to about 12 hours. From a navigation standpoint, the most difficult case is the 12-hour descent, because of the propagation of initial errors and the drift of inertial instruments. The paper emphasizes this case.

SCENARIO OF A MARS LANDING

The scenario begins with the lander in Mars orbit, still attached to its orbiter.

The landing comprises four phases. DOI initiates the deorbit phase of up to 12 hours, during which the lander coasts more than a quarter orbit around Mars. The aerobraking phase begins at entry interface (EI, about 125 km altitude), and lasts for about 33 minutes. A parachute is deployed at 5 to 10 km altitude to begin the parabraking phase of about 38 seconds. The parachute is jettisoned and the descent engine is ignited at 2 or 3 km altitude to begin the powered descent phase of one or two minutes.

Navigation uses a variety of instruments, introduced in the following steps.

1. Prior to DOI, pictures of landmarks along the track of the descent orbit are transmitted to Earth and processed to obtain landmark images, as they will appear to the lander, starting early in deorbit and ending in the parabraking or powered descent phase. These landmarks converge on the selected landing site and are accurately mapped relative to the landing site.

2. The gimbals of a scan platform, used for articulating the camera aboard the lander, are calibrated prior to DOI.

3. The lander's inertial measurement unit (IMU) is realigned near the end of the deorbit phase, using celestial observations from the lander's camera. This overcomes most of the gyro drift occurring during deorbit. Realignment supplements, rather than replaces, the normal IMU alignment prior do DOI.

4. Landmarks selected on Earth prior to DOI are tracked by the lander during deorbit. Differences between predicted and observed directions to landmarks are used to update the lander's state vector.

5. During aerobraking, landmark tracking is probably infeasible because of atmospheric refraction and camera heating. Also, roll modulation required for guiding the lander would interfere with landmark tracking. No tracking is planned; navigation is inertial.

6. As soon after parachute deployment as possible, a landmark is found, identified, and used for computing the location of the lander relative to the preselected landing site or sites.

7. During powered descent, the lander is guided to the preselected site or to one of the sites. Either the site itself or a nearby landmark is tracked to enable the lander to be guided precisely to the site.

Based on preliminary estimates of mapping capabilities and instrument performance, a pinpoint landing appears feasible. The following sections present the rationale.

ASSUMPTIONS

A number of unknowns affect the performance of a pinpoint landing system. In some cases we have made assumptions, and in other cases we have considered more than one alternative. Here are the key assumptions and alternatives.

1. In every mission configuration, the lander will have an IMU.

2. The orbiter from which the lander separates may or may not have an IMU, but it will have, at least, gyros.

3. The lander's IMU will be aligned by observing the directions to stars, using a camera aboard the lander.

4. The lander's camera is mounted on a scan platform to permit articulating the camera without maneuvering the spacecraft.

5. The same camera can be used to observe the direction to stars and landmarks (landmark tracking) during deorbit.

6. Landmark tracking during deorbit can continue to 150 km altitude.

7. Landmark tracking is not feasible during aerobraking.

8. Radio ranging from the orbiter is available during deorbit.

9. Radio ranging from the orbiter is not available during aerobraking.

10. Landmark tracking is feasible during parabraking and/or powered descent.

11. Wind velocity during parabraking is 40 m/sec, three sigma.

12. Lander IMU errors, three sigma, are: accelerometer bias 51 micro g, accelerometer scale factor 210 parts per million, gyro drift 0.069°/hr.

NAVIGATION CAMERA

The navigation camera is used for calibrating the scan platform, aligning the IMU, and imaging landmarks during the deorbit, parabraking, and powered descent phases.

The sensing element proposed is a 1024 x 1024 charge coupled device (CCD) with a resolution of about 0.5 milliradian per pixel. The field of view is therefore 512 milliradians or 29°.

For measuring the directions to stars, images are defocused, illuminating more than one CCD element. This enables star direction to be measured to an accuracy of 0.1 pixel, three sigma. This accuracy is relative to the camera body and does not include errors due to articulating the surface on which the camera is mounted.

For measuring the directions to landmarks, images are focused. Thus landmark directions can be measured to an accuracy of about one pixel, three sigma, relative to the camera body.

The camera can be defocused by means of an extra-thick glass in one station of a filter wheel, if a wheel is required for other reasons. Otherwise a defocusing lens can be inserted by solenoid, similar to the way shutters are often actuated on flight cameras.

LANDMARK REPRESENTATION AND TRACKING

The method proposed for representing and tracking landmarks is based on a method already developed and in use at JPL for identifying cartographic control points on Io[1,2]. The essence of this method is that it places very little burden on the flight computer in terms of both storage and processing. This method is a departure from traditional correlation tracking techniques, which can be very expensive in storage and processing.

A landmark is an easily identifiable pattern of features, a feature being an area of increased or decreased brightness relative to its surroundings. The Io study has shown that a pattern of features represented by only a few numbers (perhaps 20) can locate a control point to less than one pixel, three sigma. As an example, each feature in a pattern might be defined by its brightness, size, and location.

Days or weeks before DOI, pictures from orbiter are processed on Earth to represent the major landmarks under the descent trajectory planned for the lander. Instead of defining all of the pixels in the image to be seen from the lander, only the major features are represented, as described above.

Landmarks are represented in the flight computer in the way they will be seen by the lander at the point in the descent trajectory that they are to be observed. Lighting conditions that will exist at that time are accounted for in ground processing. Generally, a landmark appears differently during landing than when photographed by the orbiter.

Each landmark is represented by a landmark dataset. A landmark dataset represents each feature in the landmark by, typically, four numbers, defining the feature's brightness, diameter, and X and Y coordinates with respect to the landmark center. In addition, a landmark dataset contains the coordinates of the landmark center with respect to Mars, and a time tag identifying the point in the descent at which the landmark is to be observed.

Landmark datasets are transmitted to the lander before DOI. They become part of the descent sequence. Each landmark is observed at the time specified by the time tag. First, the scan-platform gimbal angles required to view the landmark are computed in real time, using the coordinates of the landmark center and those of the lander, thus accounting for any deviations from the nominal trajectory. Then the camera is pointed and a picture is taken. The flight computer identifies the landmark in the picture by matching the most prominent features seen with those defined in the dataset. It is not necessary to search for a given landmark using multiple pictures because of the techniques described in the deorbit and parabraking sections of this report. Our experience shows that, with this method, feature identification and tracking will place little burden on the flight computer.

LANDING SITE CERTIFICATION

A landing site is certified by ensuring, with high confidence, that there are no rocks, scarps, or holes with a vertical dimension exceeding a value to be determined (perhaps one meter), and that the slope does not exceed a value to be determined (perhaps 15° over a 10 m baseline). Landing site certification requires measuring the size of obstacles and measuring the average slope.

Measuring the Size of Obstacles

If the lander could negotiate rocks, scarps, and holes with a vertical dimension no greater than one meter and the resolution of the orbiter camera were also one meter, then it would not be possible to see hazardous objects directly. One could see directly only objects at least five meters in size.

Objects down to the one meter height are detected by seeing their shadows. The landing site must be illuminated with the sun close to the horizon, thus amplifying the size of the shadow, and the site must be viewed from as nearly overhead as possible.

A scarp casts a shadow only when the sun is behind it. Therefore the landing site must be illuminated with the sun close to the horizon in more than one direction in order to ensure that there are no dangerous scarps.

Measuring the Average Slope

To measure the average slope across a smooth landing site, the site needs to have at least three surrounding objects of sufficient size to see them or their shadows.

SCAN PLATFORM CALIBRATION

The scan platform is calibrated prior to aligning the IMU in preparation for DOI.

The scan platform contains two gimbals: clock and cone. Calibration consists of pointing the camera at a succession of stars such that each gimbal is articulated over its range of travel in increments of about 30°. The observed directions to stars are compared with predicted directions, using the IMU as an attitude reference. The differences are used to calibrate the gimbal angles as a function of the gimbal readouts.

Calibration takes on the order of one hour, and enables the gimbal angles to be read to an accuracy of about one milliradian, three sigma.

IMU ALIGNMENT AND REALIGNMENT

The IMU is aligned prior to DOI and realigned near the end of the deorbit phase.

IMU alignment consists of pointing the camera at each of two or three known stars in directions close to normal to one another. Attitude orientation is calculated from the observed directions to the stars. Gyro drift is calculated as the apparent angular motion of the stars. Accelerometer bias is calculated as the apparent contact acceleration of the lander when in free fall.

The IMU can be aligned to the calibration accuracy of the scan platform, and the gyro drift rates can be determined to about $0.069°/hr$, three sigma. Alignment takes about two hours, mostly for calibrating gyro drift and accelerometer bias.

DEORBIT

With a pinpoint landing capability, navigation is with respect to a mapping coordinate frame whose origin is at the landing site. The navigation task is to maintain knowledge of the location of the spacecraft in this frame. The guidance task is to deliver the spacecraft to the origin of the frame.

Knowledge of spacecraft location relative to the landing site is maintained by tracking landmarks that are accurately mapped relative to the site. A preliminary estimate for the mapping error is 300 m, three sigma, in both the downrange and crossrange directions. This mapping accuracy must be achieved over a Mars central angle of nearly $90°$ back from the landing site.

By storing a convergent sequence of landmarks, that is, a sequence beginning early in the deorbit phase and zooming in on the subsatellite point at the end of deorbit, it should never be necessary to search more than one picture to find the current landmark. Landmarks are observed at the subsatellite point because this direction minimizes errors in determining lander position.

Based on current estimates of gyro drift, the IMU will have to be realigned during deorbit, perhaps more than once. The final alignment should immediately precede the final landmark observation.

With a drift rate of $0.069°/hr$, the IMU would be misaligned by $0.828°$ or 14 milliradians after 12 hours. Then computing the lander position from an altitude of 150 km by observing the directions to landmarks at the subsatellite point would result in an error of 2.2 km, three sigma, just due to platform misalignment. This error alone is probably too large to be corrected by guidance during the powered descent phase. By aligning the platform prior to observing landmarks, the pointing error can be reduced to close to that of the scan platform, i.e., one milliradian. Then observing landmarks from 150 km altitude yields a position error of only 150 m, three sigma.

In addition to the pointing error due to the scan platform gimbals, the gyros will drift in the interval between IMU alignment and observing the landmark. Assuming the gyros drift for five minutes at 0.069 degrees per hour, the pointing error will be 0.1 milliradian and the position error 15 m.

AEROBRAKING

Assuming neither landmark tracking nor radio ranging, a position error will accumulate during aerobraking due to accelerometer bias, accelerometer scale factor, gyro drift, and imperfect guidance. The first three are knowledge errors and the fourth is a control error. The knowledge errors cause the lander to be guided to the wrong location.

The error due to an accelerometer bias of 51 micro-Earth-gravity during a 33.4-minute aerobraking phase is 1004 m, three sigma.

The error due to an accelerometer scale-factor error of 210 parts per million and a velocity change of 3070 m/sec is 0.65 m/sec. Assuming this velocity error acts during half the 33.4-minute aerobraking phase, the position error is 646 m, three sigma.

The error due to a gyro drift of $0.069°/hr = 3.34E\text{-}7$ rad/sec results from sensing the braking acceleration in the wrong direction. Assuming a velocity change of 3070 m/sec at a constant rate for 33.4 minutes, the acceleration is 1.5 m/sec2. The jerk is $1.5\ \text{m/sec}^2 * 3.34E\text{-}7\ \text{rad/sec} = 5.12E\text{-}7\ \text{m/sec}^3$. The resulting position error is 687 m, three sigma. This error is normal to the average velocity vector, so it is being conservative to include it with the other errors (see tables 1 and 2).

The control error of the aerobraking phase has been estimated from simulations to be about 400 m, three sigma.

If landmarks are tracked during aerobraking, then the IMU errors are virtually eliminated. Furthermore, the mapping error is substantially reduced (conservatively, 100 m, three sigma) because landmarks at the subsatellite point near the end of aerobraking are much closer to the landing site than those at the end of deorbit.

PARABRAKING

As soon as the lander is stabilized under the parachute, the camera is redeployed to find landmarks. Landmark tracking corrects almost all of the knowledge error resulting from the preceding phases, but not the control error: the parabraking phase is unguided.

Landmark tracking in the parabraking phase is difficult because of the problem of acquiring a preselected landmark. With the parachute deployed at about 6 km altitude and a lander position error of 1.9 km (see table 1), the landing footprint subtends an angle of $35°$. This compares with the camera field of view of $29°$. Although the field of view could be increased without substantial penalty, other issues must be resolved first. One issue is how soon after parachute deployment the lander attitude is sufficiently quiescent to make the camera useful. Another issue is the size of dispersions in parachute deployment altitude.

One way to resolve these issues would be to store several landmarks within the footprint. The landmark tracking algorithm would then identify whatever landmark is first visible. It remains to be demonstrated that this is feasible, given the size of the visible footprint, the field of view of the camera, the implementation of the tracking algorithm, the time available in flight, and the speed of the computer.

The dominant control error in the parabraking phase is due to wind. Based on 40 m/sec wind velocity, the dispersion has been estimated at 1.2 km downrange and 1.1 km crossrange. It may be possible to estimate the wind velocity during aerobraking phase and use the estimate to adjust the terminal state in order to compensate for parabraking dispersions due to wind.

POWERED DESCENT

Position error estimates of the lander at the beginning of powered descent are shown in tables 1 and 2 for various assumptions stated in the table titles. The RSS position error is a control error, resulting from both control and knowledge errors of the preceding phases. The knowledge errors of the preceding phases have been corrected by landmark tracking during the parabraking phase, so that the knowledge error at the start of powered descent is only about 5 m, three sigma, due to a camera pointing error of 1 milliradian at 5 km altitude. Landmark tracking may be continued during powered descent, if necessary and feasible, in which case the knowledge error can be driven to near zero.

Table 1

POSITION ERROR TO BE CORRECTED BY POWERED DESCENT GUIDANCE, GIVEN NO RADIO RANGING ANYWHERE, AND NO LANDMARK TRACKING IN THE AEROBRAKING PHASE

Source	Value (meters)
Mapping error, deorbit phase	300
Camera pointing from scan-platform gimbals, deorbit phase	150
Camera pointing from IMU misalignment, deorbit phase	15
Accelerometer bias, aerobraking phase	1004
Accelerometer scale factor, aerobraking phase	646
Gyro drift, aerobraking phase	687
Control error, aerobraking phase	400
Control error, parabraking phase	1200
RSS position error, meters	1900

Note: Errors are three sigma, at the start of powered descent.

Table 2

POSITION ERROR TO BE CORRECTED BY POWERED DESCENT GUIDANCE, GIVEN NO RADIO RANGING ANYWHERE, BUT WITH LANDMARK TRACKING IN THE AEROBRAKING PHASE

Source	Value (meters)
Mapping error, aerobraking phase	100
Control error, aerobraking phase	400
Control error, parabraking phase	1200
RSS position error, meters	1269

Note: Errors are three sigma, at the start of powered descent.

COVARIANCE ANALYSES

The C.S. Draper Laboratory has been investigating navigation techniques and measurement sources for MRSR[3]. As part of that effort, covariance analyses were run on several deorbit and entry scenarios. Advantages of covariance analyses over the simple calculations presented above are that the analyses account for radio tracking between the lander and the orbiter and yield the statistical accuracy, accounting for the weighting between various measurements.

Depending upon the initial orbit, initial state vector errors, IMU errors, and whether there are range and Doppler measurements with respect to the orbiter, the covariance analyses show that the position error relative to the landing site grows during deorbit to about 1.5 to 4.5 km, then drops when landmark tracking begins just prior to entry interface to 0.3 to 0.9 km. The errors grow again during aerobraking and drop to less than 5 m when landmark tracking resumes at parachute deployment. All numbers are 3 sigma.

ACKNOWLEDGMENTS

The following individuals contributed ideas or concepts upon which this paper is based: Roger Bourke, Bill Breckenridge, Andy Collins, Bob Gaskell, Peter Halamek, Faith McCreary, Mac McEneaney, Steve Synnott, and Robin Vaughan of JPL; Ken Spratlin of the C.S. Draper Laboratory; and Bill Willcockson of Martin Marietta, Denver.

The work described in this paper was carried out by the Jet Propulsion Laboratory, California Institute of Technology, under a contract with the National Aeronautics and Space Administration.

REFERENCES

1. R. W. Gaskell, "Digital Identification of Cartographic Control Points", Photogrammetric Engineering and Remote Sensing, Vol. 54, No. 6, Part 1, pp. 723-727, 1988 June.

2. R. W. Gaskell, S. P. Synnott A. S. McEwen, and G. G. Schaber, "The Large-Scale Topography of Io: Implications for Internal Structure and Heat Transfer", submitted to Geophysical Research Letters, Vol. 15 No. 6 pp 581-584, 1988 June.

3. T. J. Brand et al., "Mars Rover Sample Return Mission Navigation Systems Studies", CSDL-P-2828, C. S. Draper Laboratory, 555 Technology Square, Cambridge, MA, 02139, 1988 October.

A COMPARISON OF INTERCEPT ALGORITHMS

Salvatore Alfano[*] and Charles E. Fosha, Jr.[†]

Various intercept algorithms are applied to the terminal guidance of an interceptor with free end-time. Proportional navigation, optimal control using certainty equivalence, dual control, and control with optimum thrust spacing are all examined, along with a new approach called certainty control. This new algorithm constrains the final state to a function of projected estimate error to reduce control energy expenditure. Conceptually, the constraint produces a shrinking sphere about the predicted impact point with the radius being a function of estimated error. If the predicted miss is inside or touching the sphere, thrusting is not necessary. The pursuer is restricted to lateral thrusting and the evader is modeled as a boosting missile. Filtering is accomplished using an eight state Extended Kalman Filter with line-of-sight and range updates. The estimated relative trajectory and variances are propagated numerically to predicted impact time and then approximated by splines, eliminating the need to repeatedly propagate new data when present conditions are varied. A search is then made for a new impact time and point that will minimize present interceptor velocity changes and final miss distance. These control strategies are applied to two intercept problems. Certainty control is shown to substantially reduce fuel consumption.

---

[*]   Assistant Professor, Director of Research, Department of Astronautics, U.S. Air Force Academy, Colorado Springs, Colorado 80840-5151.

[†]   Associate Professor Attendant, Department of Electrical Engineering, Director of Space and Flight Systems Laboratory, University of Colorado, Colorado Springs, Colorado 80933. Member AIAA.

INTRODUCTION

The purpose of this paper is to compare guidance schemes for an intercept vehicle in the final thirty seconds of flight that minimize lateral thrusting while attempting to intercept a boosting missile. Several air-to-air and space-to-space intercept algorithms are simulated, along with certainty control[1].

Much work has been done in the area of air-to-air intercept guidance. Guelman[2,3] has derived a closed form solution for pure proportional navigation. Perturbation methods have been employed by Sridhar and Gupta[4]. Design procedures using optimal and stochastic control techniques abound[5-15]. In the works cited above, the force of gravity is assumed to act equally on the pursuer and evader and is ignored in the relative dynamics. This 'flat earth' assumption is adequate for air-to-air encounters, but not for space-to-space. For orbital intercepts with large initial ranges the force of gravity will affect the relative trajectory and should be included in the equations of motion.

The literature for space-to-space guidance reveals many numerical approaches for determining present velocity for future rendezvous[16-22]. To date, analytic solutions for such intercepts exist only when the pursuer's impact conditions are pre-specified[20]. These works do not address intercepts involving seconds, but are concerned with a much slower rendezvous process involving hours or even days.

Certainty control[1] was developed to minimize control energy expenditure in the presence of poor estimates. This form of control lends itself well to the problem at hand because the estimator's accuracy tends to be range dependent. This dependency stems from the limitations of the target tracker.

Target tracking is accomplished with a ranging device and line-of-sight sensors for in-plane and out-of-plane measurements. Noise corrupted data is processed through an eight state Extended Kalman Filter with serial updates occurring every tenth of a second.

SYSTEM MODELING

The equations of motion for the evader and pursuer, developed in our previous paper[1], are repeated here for the reader's convenience.

The pursuer is traveling in excess of twelve kilometers per second with lateral thrusting capability using two-body orbital dynamics. Thrusting is prohibited along the longitudinal (x) axis to prevent sensor contamination and to satisfy the structural constraints of having forward sensors and a large aft booster to achieve intercept speed. The equations of motion are:

$$\ddot{x}_p = \frac{-\mu\, x_p}{(x_p^2 + y_p^2 + z_p^2)^{3/2}} \tag{1}$$

$$\ddot{y}_p = \frac{-\mu\, y_p}{(x_p^2 + y_p^2 + z_p^2)^{3/2}} + a_y \qquad (2)$$

$$\ddot{z}_p = \frac{-\mu\, z_p}{(x_p^2 + y_p^2 + z_p^2)^{3/2}} + a_z \qquad (3)$$

where a and a are the lateral thrust accelerations, μ is the earth's gravitational constant, and the double dots denote the second derivative with respect to time.

The evader is modeled as a boosting missile using two-body orbital dynamics. For tracking purposes the intercept must occur prior to burnout. Acceleration due to thrusting is computed in the direction of the missile's velocity vector. The equations of motion are:

$$A = \frac{A_o}{1 - \dot{m}_o t} \qquad (4)$$

$$\ddot{x}_E = \frac{-\mu\, x_E}{(x_E^2 + y_E^2 + z_E^2)^{3/2}} + \frac{A\,\dot{x}_E}{(\dot{x}_E^2 + \dot{y}_E^2 + \dot{z}_E^2)^{1/2}} \qquad (5)$$

$$\ddot{y}_E = \frac{-\mu\, y_E}{(x_E^2 + y_E^2 + z_E^2)^{3/2}} + \frac{A\,\dot{y}_E}{(\dot{x}_E^2 + \dot{y}_E^2 + \dot{z}_E^2)^{1/2}} \qquad (6)$$

where A is the present acceleration, A_o is the initial acceleration, $\dot{m}_o$ is the initial mass flow rate divided by mass, and t is the time since ignition. The single dot denotes the first derivative with respect to time.

Numerical propagation of the dynamic equations is very time consuming. It is convenient to approximate the relative trajectory by a polynomial, eliminating the need for repeated propagation. The current and final states are used to generate cubic splines along each axis[1] of the form

$$x(t) = At_{go}^3 + Bt_{go}^2 + Ct_{go} + D \tag{8}$$

By setting the current time to zero, D and C become the current position and velocity respectively, with time-to-go being the intercept time. Changes in velocity will be reflected only in the C coefficient and the final state can be easily determined for any intercept time.

To ensure accuracy, new spline coefficients are computed every cycle time. To accomplish this, the truth model[1] is propagated forward to predicted impact time to obtain the needed final states. By using these updated final states every iteration, propagated roundoff error is eliminated in the spline coefficient computations.

OPTIMAL CONTROL FORMULATION USING CERTAINTY EQUIVALENCE

Changes in pursuer lateral velocity will affect final position, velocity and time. The optimal control problem is to find the intercept time that minimizes changes in pursuer velocity while ensuring a hit. To solve this problem, a relative spline equation is formed for each axis and a cost function is established. The cost function (L) incorporates velocity changes and miss distance multiplied by a weighting factor (K) and is represented as

$$L = \frac{K(x_1^2 + x_2^2 + x_3^2)}{2} + \frac{(\Delta V_y^2 + \Delta V_z^2)}{2} \tag{9}$$

where the final relative state vector is determined from the spline equations:

$$\begin{bmatrix} x_1 \\ x_2 \\ x_3 \end{bmatrix} = \begin{bmatrix} x(t_{go}) \\ y(t_{go}) \\ z(t_{go}) \end{bmatrix} = \begin{bmatrix} A_x t_{go}^3 + B_x t_{go}^2 + C_x t_{go} + D_x \\ A_y t_{go}^3 + B_y t_{go}^2 + (C_y - \Delta V_y)t_{go} + D_y \\ A_z t_{go}^3 + B_z t_{go}^2 + (C_z - \Delta V_z)t_{go} + D_z \end{bmatrix} \tag{10}$$

The cost function must now be minimized with respect to the control vector $\underline{u}$:

$$\underline{u} = \begin{bmatrix} u_1 \\ u_2 \\ u_3 \end{bmatrix} = \begin{bmatrix} t_{go} \\ \Delta V_y \\ \Delta V_z \end{bmatrix} \tag{11}$$

As stated in Bryson and Ho[24], it should be possible to find a set of controls such that

$$\frac{\partial L}{\partial \underline{u}} = \underline{0} \tag{12}$$

Three equations arise from (12) with three unknowns, expressed here in vector form as

$$\underline{h} = \begin{bmatrix} h_1 \\ h_2 \\ h_2 \end{bmatrix} = \begin{bmatrix} K(x_1\dot{x}_1 + x_2\dot{x}_2 + x_3\dot{x}_3) \\ \Delta V_y - Kx_2 t_{go} \\ \Delta V_z - Kx_3 t_{go} \end{bmatrix} = 0 \tag{13}$$

with $\underline{\dot{x}}$ being

$$\underline{\dot{x}} = \begin{bmatrix} \dot{x}_1 \\ \dot{x}_2 \\ \dot{x}_3 \end{bmatrix} = \begin{bmatrix} 3A_x t_{go}^2 + 2B_x t_{go} + C_x \\ 3A_y t_{go}^2 + 2B_y t_{go} + C_y - \Delta V_y \\ 3A_z t_{go}^2 + 2B_z t_{go} + C_z - \Delta V_z \end{bmatrix} \tag{14}$$

A Newton-Raphson method from Maron[23] is used to solve (13). It is important to note that this formulation incorporates a weighting factor that allows a trade-off between miss distance and velocity changes. Zero miss distance is associated with infinite K, while zero K produces no velocity change. Also, the use of splines eliminates the need for repeated trajectory propagation, significantly reducing control parameter search time. Changes in the control vector are determined by

$$
d\underline{u} = \begin{bmatrix} dt_{go} \\ d\Delta V_y \\ d\Delta V_z \end{bmatrix} = -[J]^{-1} \begin{bmatrix} h_1 \\ h_2 \\ h_3 \end{bmatrix} \tag{15}
$$

where J is the Jacobian of (13) with respect to (11).

To execute this procedure, initialize time-to-go (preferably to the point of closest approach) and determine the spline coefficients for this initial trajectory. Compute the $\underline{x}$, $\underline{\dot{x}}$, and $\underline{h}$ vectors, in that order. Update the $\underline{u}$ vector using (15) and test for convergence. If convergence is not achieved, recompute the above vectors and test again. This algorithm will be referred to as Plan A.

The next formulation uses splines to determine the control parameters for a zero miss solution. This is a specialized version of Plan A where the weighting factor is set to infinity $(K=\infty)$. Because the only control for miss in the longitudinal direction is time-to-go, x_1 in (10) is set equal to zero,

$$
x_1 = A_x t_{go}^3 + B_x t_{go}^2 + C_x t_{go} + D_x = 0 \tag{16}
$$

and t is solved using numerical techniques. With time-to-go established, (10) is again used with x_2 and x_3 equal to zero, yielding equations for the velocity changes:

$$
\Delta V_y = A_y t_{go}^2 + B_y t_{go} + C_y + D_y/t_{go} \tag{17}
$$

$$
\Delta V_z = A_z t_{go}^2 + B_z t_{go} + C_z + D_z/t_{go} \tag{18}
$$

This plan, called Plan B, is computationally less burdensome than Plan A because the complexity of the search is reduced.

Within a few seconds of intercept the acceleration due to gravity will be nearly identical for the pursuer and evader. Also, the evader will travel in a near straight line along its velocity vector. Ignoring gravity terms in the relative dynamics leads to a simpler and faster solution, reducing the guidance to proportional navigation. The relative trajectories are expressed as

$$
D_A = \frac{A_o}{\dot{m}_o^2} \sum_{i=2}^{\infty} \frac{(\dot{m}_o t_{go})^i}{i(i-1)} \tag{19}
$$

$$x(t_{go}) = \left\{ [x_E(0) - x_p(0)] + [\dot{x}_E(0) - \dot{x}_p(0)]t_{go} \right.$$

$$\left. + \frac{D_A\dot{x}_E(0)}{\sqrt{\dot{x}_E^2(0) + \dot{y}_E^2(0) + \dot{z}_E^2(0)}} \right\} \qquad (20)$$

$$y(t_{go}) = \left\{ [y_E(0) - y_p(0)] + [\dot{y}_E(0) - \dot{y}_p(0) - \Delta V_y]t_{go} \right.$$

$$\left. + \frac{D_A\dot{y}_E(0)}{\sqrt{\dot{x}_E^2(0) + \dot{y}_E^2(0) + \dot{z}_E^2(0)}} \right\} \qquad (21)$$

$$z(t_{go}) = \left\{ [z_E(0) - z_p(0)] + [\dot{z}_E(0) - \dot{z}_p(0) - \Delta V_z]t_{go} \right.$$

$$\left. + \frac{D_A\dot{z}_E(0)}{\sqrt{\dot{x}_E^2(0) + \dot{y}_E^2(0) + \dot{z}_E^2(0)}} \right\} \qquad , \qquad (22)$$

where D_A is the distance associated with thruster acceleration in the direction of booster velocity. As in Plan B, time-to-go is computed for zero miss on the x axis using (20), and then the velocity changes can be found from (21) and (22). This plan will be known as Plan C.

It should be noted that all these techniques use the principle of certainty equivalence (CE), where expected values from a state estimator are substituted for random variables[26]. The pursuer's states are assumed known, but because the evader's states must be estimated, the resulting system is stochastic. Optimal control formulation is based on a system that is deterministic. In applying the certainty equivalence principle, the stochastic system is replaced by a deterministic one, using the expected values of the random variables from the estimator.

There is a drawback to this technique in the sense that imperfect knowledge of the present state produces needless thrusting. Any errors in the present state estimate cause errors in the predicted final state. This results in the computation of velocity changes based on the incorrect final state. Future iterations produce similar results requiring the pursuer to thrust excessively.

STOCHASTIC CONTROL FORMULATION

The excessive thrusting just mentioned can be reduced using stochastic control techniques. Three formulations are examined in the following paragraphs. The first determines the optimum spacing of corrective thrusts for Plan B. The second uses dual control methods based on predicted error knowledge, such as filter covariance. The third constrains the miss distance to a function of predicted error knowledge, at the expense of accuracy.

Corrective thrusting in the presence of state estimate errors can be optimally spaced to reduce fuel[25]. A control effectiveness ratio (ρ) is established to determine the spacing between thrusts. This ratio directly yields thrust times when control effectiveness is a linear function of time.

For this formulation, the number of corrective thrusts (N) must be chosen to minimize the sum of thrusts (S_N), which is total ΔV. The behavior of ΔV and miss distance as a function of ρ can be produced through digital computation and is done as part of the simulation to determine the best value of ρ for Plan B.

To enhance understanding this technique, assume the control effectiveness ratio is two ($\rho=2$). This implies that corrective thrusting should take place when the control has half ($1/\rho$) the effect of the previous corrective thrust. If control effectiveness is a near-linear function of time, as it is in this case, then it will be halved at about half the time to impact since the last thrust. Thrusting will take place at the start of the intercept, at one-half time-to-go, one fourth time-to-go, one-eighth time-to-go and so on. With $\rho=3$ the optimum thrust timing always occurs at a third of the time-to-go since the last correction. When the spacing is less than the estimator's cycle time, impact is imminent and thrust is terminated.

Optimal control solutions require perfect knowledge of the states, but in reality the information provided to the controller is only an estimate. As stated by Aoki[26], a theory of control should take into account the 'imperfectness' of information. This explains the need to incorporate statistical decision theory in control formulation. A solution that uses imperfect information will be sub-optimal, but it is desirable for such a solution to have the intrinsic characteristics of optimality[6]. Recognition that the control affects not only the state but also its uncertainty leads to a form of stochastic control known as dual control. This method not only drives the system to some final state, but attempts to reduce state uncertainty along the way. The result is often greater accuracy and/or reduced fuel consumption.

A dual control method for controlling stochastic nonlinear systems with free end-time was developed by Tse and Bar-Shalom[6]. This method differs from the optimal control formulations previously presented. Instead of minimizing the cost function L of (9), the expected value of the cost function ($E\{L\}$) is minimized. To accomplish this, the final states and their covariances must be computed. This can be done by running the Extended Kalman Filter[1] forward to predicted intercept time, as suggested by Tse, Bar-Shalom and Meier[5].

The solution involves establishing an expected cost function consisting of miss distance and covariance of each axis, along with the control. Assuming the estimates of the filter are Gaussian, the expected value of (9) is:

$$E\{L\} = K\left[\frac{\sigma_{xf}^2 + \sigma_{yf}^2 + \sigma_{zf}^2}{2} + \frac{\hat{x}_f^2 + \hat{y}_f^2 + \hat{z}_f^2}{2}\right] + E\left\{\frac{\Delta V_y^2 + \Delta V_z^2}{2}\right\} \tag{23}$$

The expected cost of (23) is conditioned on the controls. Two cases must be examined: the cost associated with the CE solution (Plan A) and the cost of deviating from that solution to improve the estimate. In this manner, the approximate best cost-to-go includes both estimation and control performance.

The expected cost of the CE solution is easily computed by determining the controls from Plan A and then running the Extended Kalman Filter forward to predicted impact time assuming measurement updates. The final filter data is then inserted into (23) to find the expected cost.

Finding the expected cost of deviating from the CE solution is computationally burdensome. The thrust direction that yields the greatest estimate improvement must first be determined. Thrusting in this direction will cause the expected miss distance to grow due to departure from the nominal CE path. It is therefore necessary to determine a new nominal path based on the deviation and include the control energy required for this path in the deviation cost estimate. Failure to do so may result in large expected miss distances that erroneously inflate the cost associated with deviation, causing the CE control of Plan A to always be chosen.

If the controls associated with cost do not affect state estimate certainty, fuel may be conserved by using that certainty to reduce the controls. By linking the controls to the certainty of the estimate, a near perfect estimate would yield the optimal control, with reduced control resulting from a poor estimate. To accomplish this, the predicted final states are constrained by a function of their variances at the final time. This form of control, called certainty control, is derived in its entirety in our previous paper[1]. Conceptually, the constraint produces a deviation sphere about the predicted impact point. If the predicted miss is inside or touching the sphere, thrusting is not necessary. If the predicted miss is outside the sphere, minimum thrusting is determined to bring the miss to the surface of the sphere.

As the estimates improve, the constraint tightens and the sphere shrinks.

COMPUTER SIMULATION

Two cases are examined with time-to-go equaling 30 seconds. Case I represents a head on, 10° out-of-plane, intercept and Case II represents a 10° out-of-plane tail chase. The pursuer is initially traveling at 12 km/s at an altitude of 750 km with a lateral acceleration range of 3-60 m/s² in each axis. The booster's initial acceleration is 3.15788 m/s² with a unitized mass flow rate of .01579 1/s.

A time lag of one tenth second is used for all algorithms when computing velocity changes. It is unrealistic to assume the filter can process measurements, the controller determine thrust commands, and the thrusters respond to those commands all instantaneously. One cycle time is chosen to allow the velocity changes computed in the previous cycle to be implemented in the present cycle. The controller routines are built to take this lag into account. Also, thrusting is not permitted during the first three seconds of an intercept to account for target acquisition. Filtering is done through an eight state Extended Kalman Filter[1]. This simulation, written in FORTRAN 77 to run on a VAX 8600 using double precision, generates one hundred Monte Carlo runs per case.

RESULTS

The performance of all algorithms is recorded in the following tables:

Table 1. Case I Performance.
(Head On, 10° Out-of-Plane Intercept)

	MEAN MISS (METERS)	STANDARD DEVIATION (METERS)	MEAN ΔV (M/S)	STANDARD DEVIATION (M/S)
PLAN A (K=10) (CE OPTIMAL)	.502	.224	83.82	6.99
PLAN B (K=∞) (CE OPTIMAL)	.360	.171	90.39	7.24
PLAN C (PRO NAV)	.360	.171	93.07	7.39
OPTIMUM SPACING (ρ=1.75)	.361	.171	37.19	8.50
DUAL CONTROL (K=10)	.502	.224	83.82	6.99
CERTAINTY CONTROL (K=.4)	.386	.191	23.21	4.18
TRUTH WITH NOISE	.545	.264	83.69	7.18
TRUTH WITHOUT NOISE	0	NA	7.54	NA

Table 2. Case II Performance.
(10° Out-of-Plane Tail Chase)

	MEAN MISS (METERS)	STANDARD DEVIATION (METERS)	MEAN ΔV (M/S)	STANDARD DEVIATION (M/S)
PLAN A (K=10) (CE OPTIMAL)	.190	.100	129.61	13.29
PLAN B (K=∞) (CE OPTIMAL)	.126	.061	132.65	13.32
PLAN C (PRO NAV)	.126	.061	129.47	12.26
OPTIMUM SPACING (ρ=1.75)	.126	.059	39.96	12.85
DUAL CONTROL (K=10)	.190	.100	129.61	13.29
CERTAINTY CONTROL (K=.4)	.136	.076	29.74	9.10
TRUTH WITH NOISE	.379	.204	123.57	11.61
TRUTH WITHOUT NOISE	0	NA	9.52	NA

As can be seen, the dual control's performance is no better than the certainty equivalence formulation of Plan A. This is due to the fact that range is included as a measurement, causing the control to have virtually no effect on improving filter variance. Plan B is more accurate than Plan A, but more costly in energy. Again, this result is expected because the formulation of Plan B is based on infinite miss

444

penalty (K=∞) for Plan A. By optimally spacing the thrusts of Plan B, energy expenditure is considerably reduced with little or no sacrifice in accuracy.

Plan C is just as accurate as Plan B, with slightly greater cost resulting from large initial intercept range. This extra cost is attributed to the negligible gravity assumption used in the formulation of Plan C. For the smaller ranges associated with a tail chase, Plan C was actually less costly than Plan B.

In both cases, certainty control yields the least energy expenditure. This result is not surprising, as the formulation of certainty control is based on reducing control energy in the presence of poor estimates. This form of control works best because filter variance is range dependent. As range decreases, the control constraint tightens, and accuracy increases. Therefore, less fuel is used when range is great and estimates are poor, with refinements made as impact nears. The last two entries (truth with and without noise) are included as a baseline reference of performance.

CONCLUSION

In this paper, six guidance schemes were examined to determine their capability to minimize lateral velocity changes of an intercept vehicle. Optimal control using certainty equivalence (Plans A and B), proportional navigation (Plan C), control with optimum thrust spacing, dual control, and certainty control were all implemented for two cases. Certainty control was shown to be the most energy efficient.

Certainty control constrains the final condition to a function of final estimator accuracy in the absence of updates. This general approach is not limited to the specific cases presented, and would suggest other applications of this form to control intercepts stochastically.

This control requires a measure of final estimator accuracy which was achieved by running the Extended Kalman Filter forward to intercept time without updates. This time consuming process could be eliminated if filter variances could be estimated by some function (polynomial or otherwise). Also, the constraint multiplier was assumed constant for this formulation. A future area of research is to develop a multiplier that is range or time dependent to further reduce interceptor thrusting.

In summary, the approach identified by this research not only improves the efficiency of intercept, but can be applied to a broad range of stochastic problems where control energy does not improve filter accuracy. It is also possible to combine the effects of dual and certainty control in certain cases by initially using dual control to improve estimator accuracy and then switching to certainty control. End-game accuracy may be improved by switching from certainty control to a certainty equivalence formulation just prior to impact.

REFERENCES

1. Alfano, S. and C.E. Fosha, Jr., "Hypervelocity Orbital Intercept Guidance Using Certainty Control," AAS Paper 89-044, Feb. 89.

2. Guelman, M., "Qualitative Study of Proportional Navigation," IEEE Transactions on Aerospace and Electronic Systems, July 1971, pp. 337-343.

3. Guelman, M., "The Closed Form Solution of Pure Proportional Navigation," IEEE Transactions on Aerospace and Electronic Systems, Vol. AES-12, July 1976, pp. 472-482.

4. Sridhar, B. and N.K. Gupta, "Missile Guidance Laws Based on Singular Perturbation Methodology," Journal of Guidance and Control, Vol. 3, April 1980, pp. 158-166.

5. Tse, E., Bar-Shalom, Y., and L. Meier, III, "Wide Sense Adaptive Dual Control for Nonlinear Stochastic Systems," IEEE Transactions on Automatic Control, Vol. AC-18, No. 2, April 1973, pp. 98-108.

6. Tse, E., and Y. Bar-Shalom, "Adaptive Dual Control For Stochastic Nonlinear Systems with Free End-Time," IEEE Transactions on Automatic Control, October 1975, pp. 670-675.

7. Nesline, F.W., Wells, B.H., and P. Zachran, "Combined Optimal/Classic Approach to Robust Missile Autopilot Design," Journal of Guidance and Control, Vol. 4, No. 3, May-June 1981, pp. 316-322.

8. Speyer, J.L., Hull, D.G., and C.Y. Tseng, "Estimation Enhancement by Trajectory Modulation for Homing Missiles," Journal of Guidance, Vol. 7, No. 2, March-April 1984, pp. 167-174.

9. Guelman, M., and J. Shinar, "Optimal Guidance Law in the Plane," Journal of Guidance, Vol. 7, No. 4, July-August 1984, pp. 471-476.

10. Tang, Y.M., and J.A. Borrie, "Missile Guidance Based on Kalman Filter Estimation of Target Maneuver," IEEE Transactions on Aerospace and Electronic Systems, Vol. AES-20, No. 6, Nov. 1984, pp. 736-741.

11. Yeuh, W.R., and C.F. Lin, "Optimal Controller for Homing Missile," Journal of Guidance, Vol. 8, No. 3, May-June 1985, pp. 408-411.

12. Lin, C.F., and S.P. Lee, "Robust Missile Autopilot Design Using a Generalized Singular Optimal Control Technique," Journal of Guidance, Vol. 8, No. 4, July-August 1985, pp. 498-507.

13. Ashida, S., Howe, R.M., and N.X. Vinh, "Optimal Control of Air-Launched Homing Missiles Based on Realistic Performance Indices," Proceedings of the Conference on Aerospace Simulation II, Vol. 16, No. 2, Jan. 1986.

14. Lin, C.F., and L.L. Tsai, "Analytical Solution of Optimal Trajectory-Shaping Guidance," Journal of Guidance, Control, and Dynamics, Vol. 10, No. 1, Jan.-Feb. 1987, pp. 61-66.

15. Yang, C.D., and F.B. Yeh, "Closed-Form Solution for a Class of Guidance Laws," Journal of Guidance, Vol. 10, No. 4, July-August 1987, pp. 412-415.

16. Cherry, G.W., "A General, Explicit, Optimizing Guidance Law for Rocket-propelled Spaceflight," AIAA Paper 64-638, Aug. 1964.

17. Johnson, F.T., "Approximate Finite-Thrust Trajectory Optimization," AIAA Journal, Vol. 7, June 1969, pp. 993-997.

18. Bate, R.R., Mueller, D.D., and J.E. White, *Fundamentals of Astrodynamics*, Dover Publications, New York, 1971.

19. Borisenko, I.I., and Y.P. Kulyabichev, "Algorithm for Optimization of the Solution of the Spacecraft Rendezvous Problem," Cosmic Research, Vol. 18, No. 3, May-June 1980, pp. 343-347.

20. Stuart, D.G., "A Simple Targeting Technique for Two-Body Spacecraft Trajectories," Journal of Guidance, Vol. 9, No. 1, Jan.-Feb. 1986, pp. 27-31.

21. Bhat, M.S., and S.K. Shrivastava, "An Optimal Q-Guidance Scheme for Satellite Launch Vehicles," Journal of Guidance, Vol. 10, No. 1, Jan.-Feb. 1987, pp. 53-60.

22. Menon, P.K.A., and A.J. Calise, "Interception, Evasion, Rendezvous and Velocity-to-be-Gained Guidance for Spacecraft," AIAA Paper 87-2318, Aug. 1987.

23. Maron, M.J., *Numerical Analysis: A Practical Approach*, Macmillan Publishing Co., New York, 1982, pp. 177-182.

24. Bryson, A.E., and Y.C. Ho, *Applied Optimal Control*, Hemisphere Publishing Corp., Washington D.C., 1975, pp. 71-75.

25. Lietmann, G., *Optimization Techniques*, Academic Press Inc.New York, 1962, Breakwell, J., Ch. 12, pp. 353-357.

26. Aoki, M., *Optimization of Stochastic Systems*, Academic Press Inc., New York, 1967.

Section V
RECENT EXPERIENCES

SESSION V

Chairperson:	Brian McGlinchey Jet Propulsion Laboratory
Co-Chairperson:	Bob Williamson The Aerospace Corporation
Local Chairperson:	Kirk Sterling Honeywell, Inc.

The following paper was not available for publication:

AAS 89-053   Result of the Propulsion and Attitude Control Systems on the AMSAT-OSCAR-13 Spacecraft, K. Meinzer, J. King, R. Daniels, The Radio Amateur Satellite Corporation

The following paper numbers were not assigned:

AAS 89-056 to -059

AAS 89-061 to -099

DELTA 181 POINTING AND TRACKING EXPERIENCE

T. S. Englar and J. C. Ray[*]

The SDIO Delta 181 mission was flown in February 1988 to obtain phenomenology data in support of the SDI midcourse phase. The system consisted of a MDAC Delta 3910 second stage, which was used in orbit as a platform to point at a variety of objects using a suite of seven primary sensors mounted on the platform. Objects carried into orbit with and deployed from the Delta included test objects (TO's) as well as rocket motors for plume characterization data; one three-stage sounding rocket was launched from the ground. Other pointing activities included viewing of the TO container after all deployments and measurement of various backgrounds (limb/earth/space).

The pointing was accomplished by interfacing three of the sensors to the existing Delta attitude control system, via an on-board multi-sensor/multi-object navigation filter. The tracking sensors were a pulsed LADAR, a ku band Doppler radar, and an IR imager. About 100 different acquisition and track sequences were attempted, and possibly as many as half of these achieved less than solid track. Although tracking exceeded mission data gathering requirements, as a pointing and tracking experiment it inflicted some valuable lessons. Specifically, tracking was adversely affected by sensor boresight errors, disappointing sensor performance, dependence of navigation algorithms on dynamic models, and ambitious estimates of settling and acquisition time.

INTRODUCTION

The Delta 181 mission objectives centered around the gathering of SDI phenomenology data. This included characterization of a variety of test objects (TO's) and plumes against various backgrounds, characterization of the background itself, and maneuvering, pointing, and tracking to acquire and view targets. The mission required nearly continuous maneuvering to view multiple test objects under various conditions of lighting, background, range, and aspect angle. Accelerating objects, including a ground-launched rocket, were also to be tracked and observed.

---

[*] The Johns Hopkins University Applied Physics Laboratory, Johns Hopkins Road, Laurel, Maryland 20707.

The mission was launched on 8 February 1988. Data collection was accomplished over the next 12 hours and retrieved from the onboard recorders during the following 10 days. Several complete data playbacks were accomplished, assuring redundant copies of the data on the ground.

Though many parts of the mission were executed nominally, there were anomalies and these will be described in some detail, both as a matter of interest and for their lesson content.

SPACECRAFT

The launch vehicle was a Delta 3910, with an augmented second stage which became a part of the "satellite". The seven instruments were mounted around the outside of the front end, all looking forward. The entire stage was maneuvered to point them at desired targets. Inside was the Canister Cluster which housed the test objects and SP5 (a gas release experiment). (The Canister Cluster was ejected following test object releases and became one of the test objects.) Buried further inside was SPX, a Star 13 solid motor which was ejected and fired as a plume generator.

The seven primary instruments were fixed relative to the vehicle centerline and had fields of view of order 5 degrees or less; some were gimbaled giving them somewhat larger fields of regard. The instruments required up to about 30 seconds to cycle through various modes, so viewing ("scan") times were the order of 35 seconds for most test objects. The instruments used in Pointing and Tracking were:

      I5 -   An IR imager built by Aerojet ElectroSystems Company,
      L1 -   A pulsed diode-laser range and angle tracker built by GTE Government Systems,
      R -   A radar built by Teledyne Ryan Electronics.

The instruments were aligned utilizing optical cubes prior to launch to an accuracy of better than 2 arc minutes, including expected changes in orbit. (Parallax was accounted for in the reference axes rotations.)

The TO's included eight objects carried into orbit with and deployed from the Delta, four reference spheres for sensor calibration (two each emissive and reflective), and rocket motors fired in space for plume characterization data. The plumes were generated by four solid rockets carried into orbit and deployed, and one 3-stage sounding rocket launched from the ground. The TO canister was ejected and viewed, simulating a Post-Boost Vehicle.

GUIDANCE AND CONTROL

The pointing was accomplished by interfacing three of the sensors to the existing Delta attitude control system, via an on-board multi-sensor/multi-object navigation filter. The tracking sensors were the pulsed LADAR (L1), the Doppler radar (R), and the IR imager (I5). The Delta attitude system consists of the RCS, the DRIMS, and the DGC. The Reaction Control System (RCS) provides maneuvering torque by a set of GN_2 reaction control jets, and the Delta Redundant Inertial Measurement System (DRIMS) uses three two-degree-of-freedom strapdown gyros for attitude reference. The control laws are implemented in the Delta Guidance Computer (DGC), which although relatively small and slow by today's standards has proven 100% reliable in flight. The Delta attitude control system was designed for a launch vehicle role, and normally operates no more than about 90 minutes. Delta 181 was controlled continuously for about twelve hours, an order-of-magnitude extension in life. The engineering changes to accomplish this included extra GN_2 and batteries, and logic to bound gyro drift with sun sensors.

The interface into the DGC and hence the Delta control system was not nearly so straightforward. A digital interface ("InterFace Box", IFB) had been developed for Delta 180 and was used essentially unmodified, but the three tracking sensors were all new. A computer ("Flight Processor", FP) on the APL Sensor Module served as interface with the sensors, as well as running one of the navigation filters to keep track of the TO's. The FP, an SC-1E produced by Southwest Research Institute and flight qualified, was based on the 80C86 microprocessor and had computing power roughly comparable to an IBM PC. It's 512K bytes of memory and high-level-language programming (Pascal) contrast to the DGC's 8K and machine language. It was decided early on to "let the Delta drive the mission", i.e. it would navigate all TO's and point to them even if all sensors and the FP failed. The FP nav filter (called the "Mapper") controlled pointing of gimbaled sensors, and pre-processed the sensor data to fit the IFB and DGC constraints. Logic to allow the FP to override the DGC under some conditions was included to take advantage of the reliability inherent in two computers.

The vehicle was maneuvered by time sequence in the DGC to point at a predetermined object, for a preset time. The location of the object, however, was not sufficiently predictable and on-board estimates from the two navigation filters were used (DGC first). During the scans, the time to dwell on a given object before moving on to the next was based on pre-launch best estimates of the object dispersion and Delta slew rates. Larger than expected dispersions led to trouble, in that sometimes the maneuver to an object had not completed, or at least acquisition transients not settled, before it was time to move on. Lighting conditions and geometry for the various viewings was a big driver in laying out the sequence; in fact SP4, the Strypi sounding rocket from Hawaii, set the launch window. Off-axis sunlight rejection capability of the sensors also was a constraint. All in all, the planned maneuver sequence was very tight and ambitious, and we expected we'd probably miss some of the data. The timeline had plenty of redundancy built into it, which proved fortunate.

MISSION TIMELINE

We summarize the timeline, then describe interesting pointing & tracking events from each phase. The maneuvers, deployments, and data gathering evolved essentially nominally. Data retrieval proceeded within expected deviations and achieved at least 95% recovery of the data. We will discuss only the anomalies, which are more interesting and educational. None of the non-nominal situations prevented overall mission accomplishment, although we weren't so sure this would be the outcome during the operation.

Liftoff was 1988 Feb 8, 22:07:00.475 UT, and the Delta placed the payload in its usual excellent orbit. Following instrument calibration, the first group of TO's was deployed and observed. This group of six included two reference spheres. The TO's were deployed upward (away from earth), at about 2-3 m/s relative velocity. For a purely radial deployment, the TO is in a neighboring orbit with the same period as the spacecraft, and (neglecting drag) describes an elliptical path relative to it in the orbit plane. The idealized TO path would bring it back to the spacecraft location every orbit. Maximum range was set to about 10 km, based on sensor characteristics. We tailored specific deployment velocities and directions to maximize subsequent viewing opportunities. The reference spheres were given a deliberate retrograde component to get them out of the way, after viewing. Some TO's had higher area/mass than the spacecraft, so in general were expected to be (and were) too far away for viewing after the initial deployment. Others could and did return for several re-visits later in the mission.

Each TO was viewed continuously for about 300 sec after deployment, i.e. until about 1 km away. The entire group was then "scanned", meaning the sensor suite was pointed at each object in turn for about 30 seconds. The Group 1 TO's had 8 scans with sky background, followed by four more scans of some objects at max range with earth/limb background. These scans were in both daylight and darkness. The first orbit activities finished with longer (about 200 sec) views of three objects to update the onboard navigation (NAV) filters (described below).

The second orbit was similar to the first, in that a second group of TO's ("Group Two") was deployed, observed, and scanned. There were 8 scans in daylight and darkness, and a protracted view of one TO as it crossed the earth limb. This orbit ended by tracking some TO's from Group 1 again for Nav update (we needed to keep track of them for later viewing). Both Groups 1 & 2 began with deployment as the spacecraft came over the Pacific, in full sun.

The third orbit was devoted largely to the Test Object Exciter ("TOE") experiment, involving views of five TO's. Following the TOE experiment, a final track of two TO's for Nav update was performed. The TO canister was then deployed and tracked, and five gas samples were released from it to investigate their signatures. This orbit concluded by viewing the plumes of three small (Thiokol Star 6) solid rockets with all sensors. These rockets, contained in TO's as plume generators, were fired in directions to assure the TO's were in safe orbits.

The next several hours were dedicated to gathering background and phenomenological data. There were three earth limb scans, two "flat spins" and a "cartwheel" maneuver, where the spacecraft was rotated slowly through 360 degrees. Purpose of these was to investigate the so-called "shuttle glow" phenomenon.

About eight hours after launch a three stage STRYPI sounding rocket ("SP4") was launched from Hawaii at a carefully coordinated time to enable viewing of its plumes from space. The closest approach to the spacecraft was about 70 km during 3rd stage (Star 27) burn. (Unfortunately a timing error resulted in loss of 3rd stage track. Second stage was tracked nominally.) Tracking of this rocket was aided by a radar beacon on the STRYPI and the radar sensor on the spacecraft. On the next orbit, a Star 13 rocket carried aboard the spacecraft ("SPX") was deployed over the Indian Ocean and fired over Hawaii. This event, which went off without a hitch, was at a range of a little over 10 km from the sensor platform and produced some of the best plume data from the mission.

Having completed the Data Collection phase, the spacecraft was placed in its final orbit by two restarts of the Delta engine. First the orbit was circularized at about 330 km altitude, then the engine burned to fuel depletion over Kwajalein, which increased the orbit inclination to 30 deg. The Data were then recovered by several complete tape playbacks over about the next ten days.

POINTING AND TRACKING PERFORMANCE

The mission was designed to execute an ambitiously large number of maneuvers in a short time; most scans were less than 40 seconds with only about 15 seconds to maneuver between scans. In all, there were about 100 different acquisition & track sequences attempted, and possibly as many as half of these achieved less than solid track. Although as a pointing and tracking experiment we learned some hard and valuable lessons, tracking was plenty good enough to meet mission data-gathering objectives. Specifically, tracking was adversely affected by sensor boresight errors, disappointing sensor performance, dependence of navigation algorithms on dynamic models, & overly ambitious underestimates of time for acquisition transients to settle.

Physically and functionally, the Guidance & Control system operated nominally. There was nominal two-way communication between the sensors and the FP, and between the FP and the DGC. Throughput speed, which had been a prelaunch concern, was nominal with only 0.3 percent of iterations taking longer than the nominal 0.5 seconds; no iterations greater than 1 second were observed. Sense (polarity) of the pointing vectors, always a nagging concern in any closed loop system, was proper. The FP command and telemetry interface hardware and software also operated nominally; DGC and FP uplinks were handled without error. All deployments and scans were properly initiated. Sun sensor processing was flawless. In summary, All FP hardware and software performed exactly as designed.

It appears that NAV filters had satisfactory estimates of TO position whenever good sensor tracking data was available. In many cases, however, information from the sensors was insufficient for either FP or DGC to obtain a TO state estimate accurate enough for reacquisition after open-loop propagation. The unexpectedly large number of overrides (in which the FP assumes pointing control from the DGC) were caused by the sensors' failure to find any object within their FOV at the predicted location.

Some explanation of the "override" logic is in order. There were navigation filters in both DGC and FP, with intent that they would complement one another rather than be strictly redundant. The DGC filter navigated all TO's using essentially the same code used to navigate the Delta into orbit on so many successful flights. As such it was intended to be reliable rather than experimental, and was straightforward and simple in design. The mapper, having the advantage of a larger computer and high level language, was allowed to be more experimental in modeling, but was not allowed to directly control vehicle attitude, so that there was no risk of an experimental algorithm destroying the mission. By default, the platform was always initially pointed to where the DGC's filter believed the target to be.

To hedge against various failures of either nav filter, a protocol was established allowing the FP to "override" the DGC if no data was received after a specified time. The intent here was to give the Mapper a chance to acquire, if the DGC's filter had "lost" the object. The time chosen was about 25 seconds, which was long enough for the DGC to get enough data for a velocity update. The DGC velocity update algorithm was an example of launch vehicle heritage providing reliability under conditions for which it was designed, but having too much inertia to be robust and adaptable to the changing environment actually encountered when multiple TO's were deployed. Note that the override time was almost as long as a typical scan; there was very little time for an override on a scan to do any good before it was necessary to move on to the next object. We were of course aware of this, but felt that as long as we were getting data both filters would converge to the same correct state, and override would only be necessary (if at all) to aid the DGC in re-acquiring after a long prediction without data (its propagation had no model for drag). However, the override logic could do nothing to provide data if the sensors themselves could not acquire, even if pointed directly at the target.

For further safety, the mission was divided into five phases for control gas usage. When the quota of RCS gas was used for a given phase, no further maneuvers would be done in that phase, so the next phase would start with its full gas allotment. The phase points and gas allotments were chosen by pre-launch analysis to maximize probability of critical mission events taking place. As it actually happened, there was satisfactory tracking during deployment, and reacquisition was successful for the first two scans. However, most of the following reacquisitions resulted in an override, caused by a failure of the system to acquire. Once a rescan was missed, of course, the probabilities continued to decrease. During the Group 1 scans, 42 FP overrides of DGC positions occurred. This large number of overrides caused Delta to expend its gas allotment for that portion of the mission. At that time, FP ability to override DGC was inhibited by ground command to conserve gas. Thus, pointing vectors to the sensors and Delta pointing for the rest of the mission were based solely on DGC navigation filter estimates (updated by sensor data when available).

DGC and FP estimates of TO location during the Groups 1 and 2 scans usually agreed to within 1 deg pointing error and 500 meter range error. When L1 data were available, Mapper/L1 residuals dropped to less than 0.5 meter rms. One noteworthy case of significant DGC/FP disagreement involved the deployment of TO6 (described below). This object experienced a large out-of-plane velocity at deployment, and L1 locked onto the trajectory and gave FP good tracking data to update its state. However, the short scan times relative to DGC's velocity update time prevented DGC from getting a velocity update on this object. Thus, in effect DGC was unaware of the large out-of-plane component, and DGC and FP disagreed on the position of this object from the outset. Throughout the rest of Group 1 scans, DGC would first try to find TO6 at its (erroneous) own estimate of position, no data would of course be obtained, the FP would override, and the Delta would begin a large and gas-wasting maneuver out of plane to find the TO. But because of the short times available, the Delta did not reach the FP's estimated position in time to get data there either, and then had to start another large gas-wasting maneuver back in plane to the next object. All these maneuvers took a lot of time as well as RCS gas, so that not just TO6 but all acquisitions in Group 1 were sorely pressed for time. The situation deteriorated until we had no choice but to inhibit the override to prevent wasting all the control gas.

Overall it appears that L1 data was not obtained on any object at a range greater than 3 km, and R data was very sporadic and was biased about 0.2 deg from the L1 readings. Furthermore the I5 data was completely unusable for tracking because a bright (hot in IR) band along one edge almost always swamped the tracked object of interest. Had we had reliable, accurate sensor data, the onboard algorithms were probably robust enough to recover from the TO6 debacle. At least, Group 2 would have started fresh and the nav tracks would have gotten sufficient data to reacquire some of the more difficult and long-range TO viewing, e.g. during TOE and the TO1, 2, & 3 orbit-adjust rocket motor events. The lesson here is fairly clear: although sensor experiments are valuable, don't rely exclusively on experimental sensors for operational tracking.

The FP included a data editor to screen wild points from the data before being sent to the DGC. The editor also included a "track correlation" algorithm, a partly experimental algorithm intended to address the problem of multiple targets in the field of view. The editor functioned essentially nominally and was observed to save some track states from being corrupted by false data. At about 12700 sec after launch, an attempted scan of TO1 (range about 8 km) was in progress when TO10 (range about 2 km) drifted into view. L1 identified the object as TO10, and the editor rejected the data three different ways - wrong ID by L1, bad position estimate from L1, and bad position estimate from R. However, the track editor was designed to control data flow to the FP and not to inhibit data flow to the DGC, which therefore used the data erroneously to update its TO1 track state. If override had not been inhibited, FP would have prevented DGC from following that object. In other instances, the editor allowed such obviously bad data as negative range measurements from R to be passed to the DGC, causing both FP & DGC to become confused. The track correlation editor shows promise as a means of handling the "two-in-view" problem, and should be developed further for future use.

SPECIFIC TRACK EXAMPLES

The H-L-C coordinate system referred to here is a local, orbit-determined frame with H vertical, C along the angular momentum, and L right-handed.

Table 1 summarizes the results of all tracking.

GROUP 1 DEPLOYMENT

The TO3 H-axis mapper covariance shows a slight drop at 4200 sec, probably due to beginning to get some R range data (Figure 1). Then at 4213 the covariance drops rapidly, indicating L1 acquisition & tracking (Figure 2). This continues through deployment view, proving L1 stays locked. This is a typical deployment track of most Group 1 TO's. Oscillations in the body frame az and el data are caused by the DELTA control system deadband. (Body-fixed elevation is in the orbit plane, for deployments and scans.)

For the reference objects, no L1 data had been planned. The R data is much sparser than for TO3 and even more noisy. This was typical of all reference object data; however we did successfully track all four reference objects.

TO10 was tracked by both L1 and R throughout the deployment viewing, with the result that there were good estimates. TO2 deployment was essentially the same as that of TO3.

The out-of-plane velocity event occurred in the TO6 deployment. Data looks very much like that in TO3 deployment, except that mapper picks up the velocity perturbation and DGC does not. Mapper & DGC agree until loss of L1 data occurs at about the time when the track mode changes from deployment to scan 1, at which time DGC continues to follow R, while mapper propagates the velocity perturbation. L1 and Mapper agree and R agrees with both (with its usual bias) until L1 drops track. Loss of track could be "normal"; even though already tracking TO6, switching to Scan mode may be like a new object ID to L1, and this may have caused L1 to break lock and search (with unfortunate result of losing the target).

GROUP 1 SCANS

TO2 and TO10, Scan 1, were successful tracks. On TO3, there is basically no data (Figure 3), and we get our first override. Maneuver is complete by 5430, but Mapper and DGC are over 2 degrees apart, and no data is obtained. After the override at 5450, Delta slews as planned over to the Mapper estimate, as seen by the mapper az & el dropping to zero. At this point (5469 sec) R acquires something, presumably TO3, but we have run out of time and shortly thereafter slew to next object begins. The range is over 5 km, which proved to be beyond the max range for L1, so it did not acquire. Note that the Ground "truth" (radar best-estimate-trajectory, BET) data here is noticeably closer to the mapper estimate than DGC, although still about a degree off. It is hard not to conclude from this that the DGC navigation estimate of TO3 was way off by this time. Post flight analysis by MDAC has discovered a software problem which introduced errors during DGC velocity updates; thus it may have been beneficial that we almost never got a DGC velocity update.

There was no override on TO2, Scan 2; L1 acquires just before the end. BET data is now becoming useful, and shows we were pointing within about 1 deg of TO3, on Scan 3. Pattern was generally the same throughout rest of Group 1 scans; we overrode on all but two after scan 1. The R range estimates seem consistently low through this period. DGC may have been using them, and thus have had a bad orbit (nav estimate) on some TO's. R data is probably not being accepted by Mapper because the range is so short the data fails the track correlation test. BET when available differs from the realtime estimates by about 1 deg. Range estimates remain biased low, due to R data bias. On TO10, mapper seems better than DGC, and we often get R data after override. On TO3, however, DGC is better and we get R data only before the override. (Figures 4 & 5 are typical)

Scan 8, TO2 (Figure 6) is a good track, no override. The variance updates at 6983 (maneuver complete 6958). BET, DGC, and mapper are within 1 deg after this update, but the R range data remains inconsistent. It is negative and grossly wrong until 6975, when DGC accepts the first point, biased high for a change. Mapper filters out the odd high range, and settles in biased low. So even with this update, from same data, mapper and DGC will now begin to diverge again. They are both wrong, because sensor data is wrong. Message: sensor biases are important and must be dealt with, by eliminating them or possibly estimating them.

On the four limb/background scans (TO's 2, 10, & 3), DGC has good estimates of TO3; it's close to the BET data and we get R data only before the override (which occurs on all scans). On 2 & 10, there is R data after the override, and mapper is closer to BET. Mapper is 15 deg off on TO3 by scan EL-4, and DGC is 4-5 deg off on TO2. Both are within 0.5 deg of TO10, thanks to an update on scan 8, although the range estimates are low, as by now expected.

Prior to the nav tracks (TO's 2,3,10, 8050-9000 sec), Delta control gas allotment for the Group 1 phase is exhausted, and the vehicle does not respond to pointing commands. There are in general large error angles to the desired object. In the TO2 time frame, R gets data, probably from TO10, which is coincidentally about 3 deg off boresight. Mapper did not accept this data but DGC may have.

GROUP 2 DEPLOYMENT AND SCANS

Group 2 Deployments were much like the Group 1 Deployments, with sparse data and limited success for the Pointing and Tracking Systems. An additional hazard in this phase was the post-deployment modification of two TO's, which apparently caused loss of track.

After Group 2 scan 3 we tried to track TO's 2 & 3 for a Nav update. On TO2 we were not close at all, apparently about 30-40 degrees away. Mapper had a good range estimate, but not good enough. DGC was closer to truth than Mapper but both were grossly wrong in their estimates. TO3 was even worse, with DGC about 90 degrees off and 5 km off in range. Mapper was wrong in range by 7 km. These views were intended to be about 30 minutes after last previous update on these TO's, but the last data used was about an orbit stale. With good updates we expected to be able to propagate for 30 minutes, but it's not surprising we couldn't find the object after 84 minutes (TO3) of open loop propagation (starting from a poor and erroneous estimate anyway). There was no search algorithm using the vehicle, because the Delta gas budget could not have accommodated one (Group 1 scans prove this true).

An entire scan of Group 2 objects was missed because the vehicle was unable to complete its maneuver back in the allotted time after these attempted Nav tracks of TO's 2 and 3.

There are next attempted reacquisitions of TO's 2,1,3,& 10 for nav update, at a so-called "hardpoint". These are positions in the relative orbital motion where the line-of-sight direction is supposed to be predictable, i.e. insensitive to initial condition and/or propagation errors. For the first of these, supposed to be TO2, L1 has a solid track (Figure 7), and covariance shows Mapper updates with it for over 50 sec. L1 is clearly on the wrong object; in fact it is TO10, which is at about 2 km instead of the 6+ of TO2. Truth data for this span on TO10 overlay the L1 data (Figure 8). Note that R was initially reporting angles and range reasonable for TO2, but as TO10 drifted in front, R switched over and tracked the closer object, TO10. Mapper is now hopelessly and irrevocably confused about the whereabouts of TO2, and DGC is probably also. TO2 is now effectively lost, due to confusion introduced by multiple objects in the tracking sensors' field of view.

Shortly thereafter the same confusion with TO10 corrupts the estimate of TO1; at this point the estimates of TO3 and TO10 are also useless.

SPECIAL CASES

Next we have SP5 (the empty TO canister) deployment and track (Figures 9 & 10). As the covariance plots show, we updated this object repeatedly over a track span of about 1000 sec. Although there is no truth data, the R data (gaps are when the tape recorders were off) is within a degree of boresight most of the time, so SP5 was probably in the field of view of the science instruments at the gas-release times. Note that the R range data are very erratic, often negative. (Note: editor was inhibited at 16725). Also, the question of why no L1 data on SP5 remains open. The first covariance update looks large enough to be L1 data, and Mapper is not following R angles but the biased-off point, suggesting L1 track at least initially on this target. Later Mapper follows R, which would be consistent with L1 losing track at longer range.

The last two pointing & tracking events were also plume viewing opportunities, SP4 & SPX. SP4, a three-stage Strypi sounding rocket launched from Hawaii, had a beacon for R to home on in angles-only track mode, like a home-on-jam mode. This worked quite well, and this phase could have been extremely successful except for a 30 sec delay in firing the third stage. Because tracking in this high angular rate situation required modeling of the motion, the delay in angular acceleration meant that continuous viewing was lost.

The final tracking event was SPX, the Star 13 motor and associated equipment to spin it up & fire it. Although last, this was the most successful tracking event (maybe our computers were learning...). The object was deployed and tracked as it moved away; it had GN_2 jets for both separation and spinup, which fired immediately after deployment. As Figure 11 shows, we had solid lock by L1 in the post-deployment time, and our by now familiar R bias. Note the very ratty and even negative R data after 31100 sec; this is believed to be the time when range and velocity are ambiguous for the doppler radar. There is no covariance plot because these telemetry locations were used for mapper/editor status words and the covariance was not telemetered. Clearly mapper had a good state vector on this object, however. The recorders were stopped to conserve tape, so the entire 3000 sec track was not recorded. As firing time in view of Hawaii approached the recorders were restarted, and we find that the last L1 data was at about 12 km range at 32840. R continued to track the object, but both Mapper & DGC had such a good nav update on it by now that pointing open-loop would be adequate. Mapper's estimate shows large velocity steps at firing, but its bandwidth is too slow to actually follow the object through its burn. This is why the mapper estimates go way off boresight at firing (Figure 12); DGC does attempt to follow the accelerating target. Data from the science instruments confirmed that in fact SPX remained in view throughout its burn, and excellent science data were obtained. This experiment (SPX) illustrates with success, as strongly as does SP4's failure, the value of having a good range measurement to track a thrusting target. In both these cases there was no effective measurement of range during firing, but since our SPX estimate was correct it is equivalent to a measurement. I. e., you can't correct for surprises without data, but if there are no surprises then prior estimates are adequate.

CONCLUSION

This paper has not been an exhaustive presentation of the Pointing and Tracking results from DELTA 181; it has shown some of the high and low points. Mission requirements for Pointing and Tracking were met by the flight system. However, the Pointing and Tracking Experiment highlighted some real system shortcomings, many of which would have been addressed if the schedule had permitted. We have noted several times that there remain some unexplained aspects of system behavior, and part of this is caused by the limited visibility into the system and the limited availability of ground truth, a particular hardship for dealing with the experimental aspects of Pointing and Tracking performance. For instance, have we missed some fundamental problem with the intercomputer communications?

One dominant feature of the P&T system is that it expected to receive data. While it could function in the absence of data, it was critically dependent on sensor data to compute a state which could be used for reacquisition, and the estimator was designed to evaluate and incorporate whatever sensor data was available. We have described how the I5 sensor was essentially out of commission and the L1 instrument was not providing data for the full range required by the Mission plan. R data was more sporadic than anticipated and had biases that affected the use of its data for longterm propagation in the absence of target viewing. For tracking following deployment, these difficulties are supportable. Openloop propagation, however, requires accurate estimation and accurate modeling to be successful. While the periods of openloop propagation were tailored reasonably for the model accuracy, they were too long for the estimation accuracy which was achieved, even by the robust P&T system. If reacquisition or high performance pointing and tracking are essential to Mission goals, then it may be necessary to include a proven sensor in the system. For Delta 181, it is not clear that the design would have been changed.

The timeline was very success oriented, in providing short intervals to slew and acquire. This was driven by Mission requirements for different viewing backgrounds and aspects, but was difficult to achieve in the sensor environment described above. For Delta 181 it is not clear how many additional reacquisitions would have been purchased by relaxing the timeline, but providing more time would have reduced the risk of missing a scan and thereby jeopardizing all subsequent reacquisitions. These possibilities were apparent during the P&T design process, but were given less weight because of a single-point failure design philosophy.

Computer coordination was also part of the design considerations on both sides. However this may not have been given the depth of analysis it deserved; excursions at deployment, such as TO6 experienced, were anticipated and were inside the sensor FOR's. The design then expected that one of the two redundant estimators would be able to find the TO. If DGC were correct, acquisition would be immediate. If FP were correct, then reacquisition would take place at the end of the scan. As stated above, the time in the scan was short to enable all of these operations to take place and to get a DGC velocity update. In the absence of such a DGC velocity update, the DGC would never obtain a correct estimate and the Override would be executed, requiring additional gas usage. A further source of confusion was the difficulty, under the ground rules of sending only raw data to the DGC, of informing the DGC that this data had been passed by the Editor as being good, but was from the wrong object.

During the design of the P&T system, the problem arose of how to handle multiple objects in the sensor FOV. Preflight analysis of the TO orbits and the Mission Timeline showed that this was not an idle concern and could be expected, but also indicated that, when multiple objects appeared, they could be distinguished on the basis of range. Therefore the system was given a Track Correlator design which would compare current sensor readings with that expected by the estimator, and associate the reading with the more likely TO. A decision was made to reject data if it appeared not to be from the object expected at that point in the timeline - realtime redefinition of the timeline was not a design objective. In addition, several objects could be directly identified by the L1 sensor and in one case this method of track correlation was used.

Delta 181 Pointing and Tracking was successful from a Mission standpoint. Features of the system which could have been improved were primarily those to which either the schedule, the instrument complement, or the Mission Timeline prevented significant changes.

Table 1 - Pointing and Tracking Summary

	Group 1: TO3	TO10	TO2	TO6	Group 2: TO1	TO4	TO5	TO11	Deploy only: E1	R1	E2	R2
Deploy	4160 LR	4570 LR	4720 LR	5090 LR	9200 LR	9600 LR	10010 R	14500 r	4385 R	4890 R	9400 R	9770 R
Scan 1	5406/5426 r (5450)	5346/5372 Lr	5253/5310 L	5215/5215 r	10260/10296 R	10160/10220 (10246)	10125/10125		SP5	SP4	SPX	
Scan 2	5688/5712 r (5734)	5632/5652 r (5674)	5551/5598 lr	5476/5516 (5538)	10493/10522 r	10405/10453	10334/10368		17400 R	27410 R	30720 LR	
Scan 3	5943/5958 R	5890/5905 R (5928)	5819/5852 r (5876)	5751/5782 (5806)	10703/10727 r	10625/10663	10560/10588					
Scan 4	6181/6194 (6216)	6130/6140? (6165)	6066/6092 r (6116)	6002/6025 (6050)	11320/Inc	11259/Inc	11200/Inc					
Scan 5	6406/6417 r (6441)	6356/6365? r (6388)	6297/6317 r (6342)	6237/6250? (6282)	11485/11505 R	11426/11446 r	11373/11390 r					
Scan 6	6624/6633 r (6658)	6574/6580? r (6606)	6518/6535 r (6560)	6460/6477 (6502)	11646/11666 lr	11589/11610	11537/11553 r					
Scan 7	6837/6845 r (6870)	6787/6796 r (6820)	6734/6748 r (6773)	6677/6694 (6717)	11804/11825	11748/11763	11697/11713 r					
Scan 8	7048/7058 r (7081)	6997/7006 r (7029)	6945/6958 R	6889/6903 (6927)	11961/11980	11906/11920	11855/11872 r					
EL Scan 1	7205/7213 r (7238)	7152/7156 r (7184)	7100/7105 r (7130)		13500/13550?							
EL Scan 2	7365/7372 r (7399)	7311/7320 r (7344)	7259/7263 r (7287)									
EL Scan 3	7528/7542 r (7564)	7472/7484 r (7508)	7419/7422 r (7447)									
EL Scan 4	7695/7708 r (7734)	7637/7653 r (7677)	7583/7588 r (7611)									
NAV 1	8050/Inc (8175)	8400/Inc (8670)	8700/Inc r(TO10?)(8945)									
NAV 2	10985/11050		10765/10860									
HP5	12850/12920	13050/13130	12250/12280 LR (TO10)		12350/12370 LR (TO10)							
TOE	15220/15320	15870/15900?	15640/15610		14765/14860						16245/16245? R (TO10?)	
NAV 4	16670/Inc		16550/16650?		16815/16750?							
OA	18800/18970		19000/19300? r?		18600/18700? LR (SP5?)							

Notes:
1. R,L means R, L1 data with update
   r,l means data but no update
2. All times seconds after liftoff
3. Deploy times: Nominal
4. Scan times: Nominal/Maneuver complete
   (Inc means maneuver not completed)
5. Override time in parens (override
   inhibited at 10318 sec)
6. If tracking wrong object, probable object
   being tracked in parens.

Figure 1

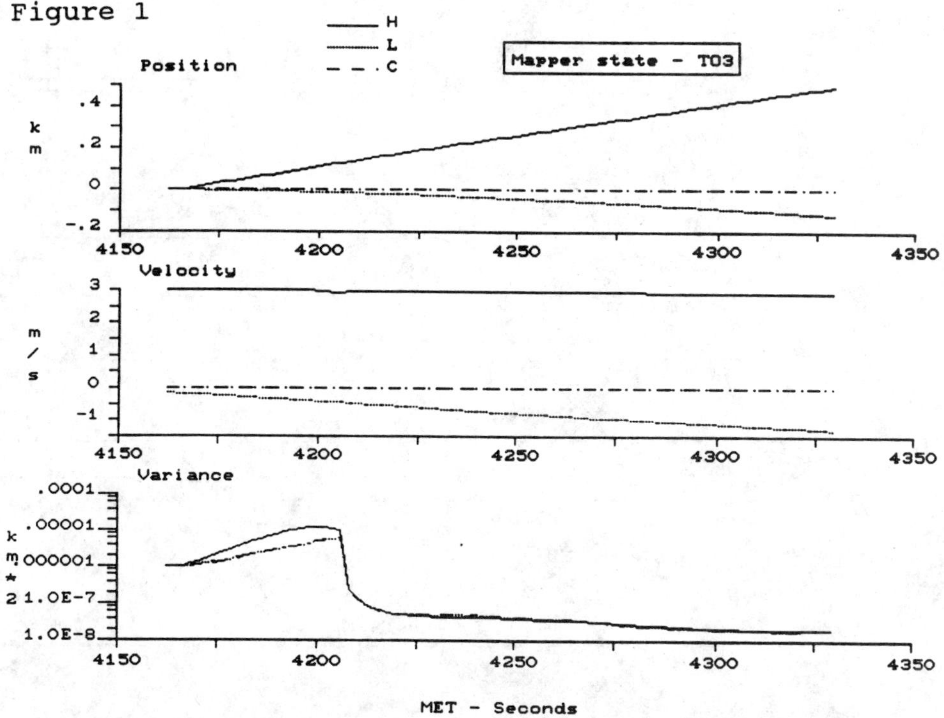

Figure 2

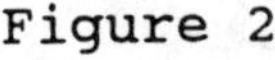

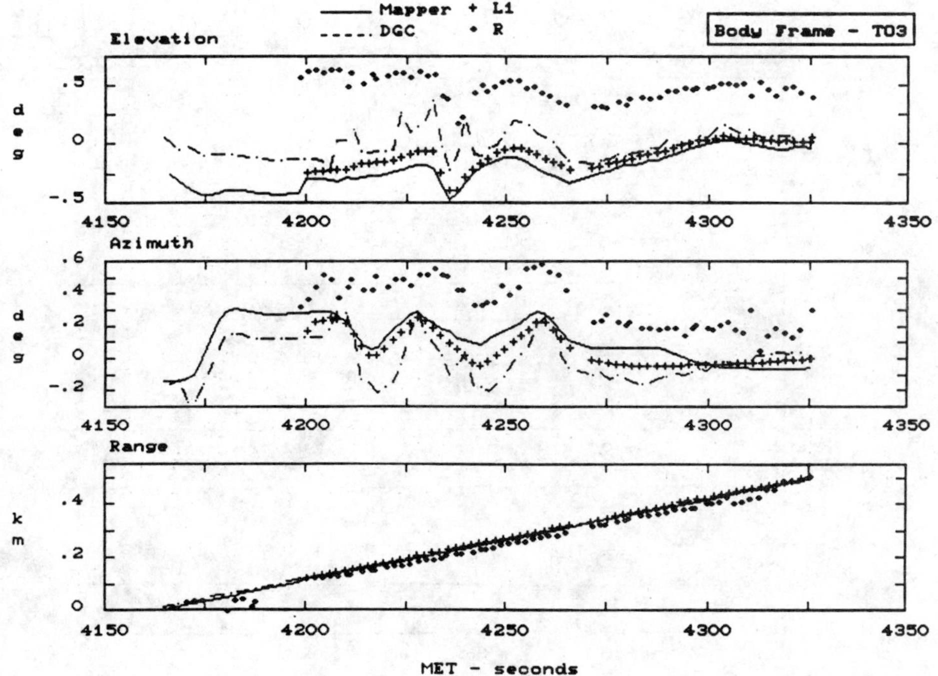

Figure 3

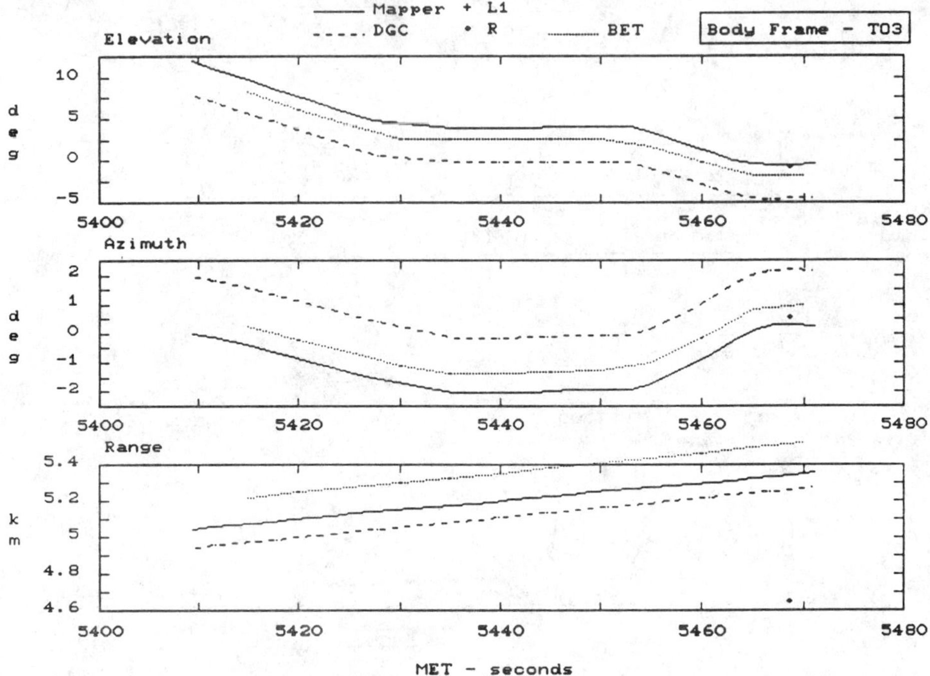

Figure 4

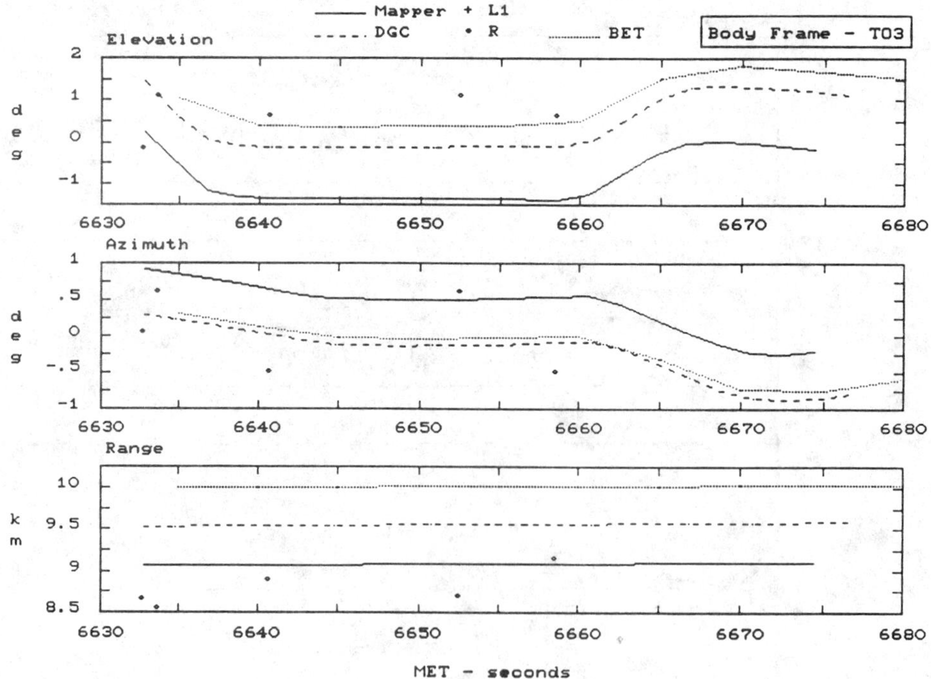

Figure 5

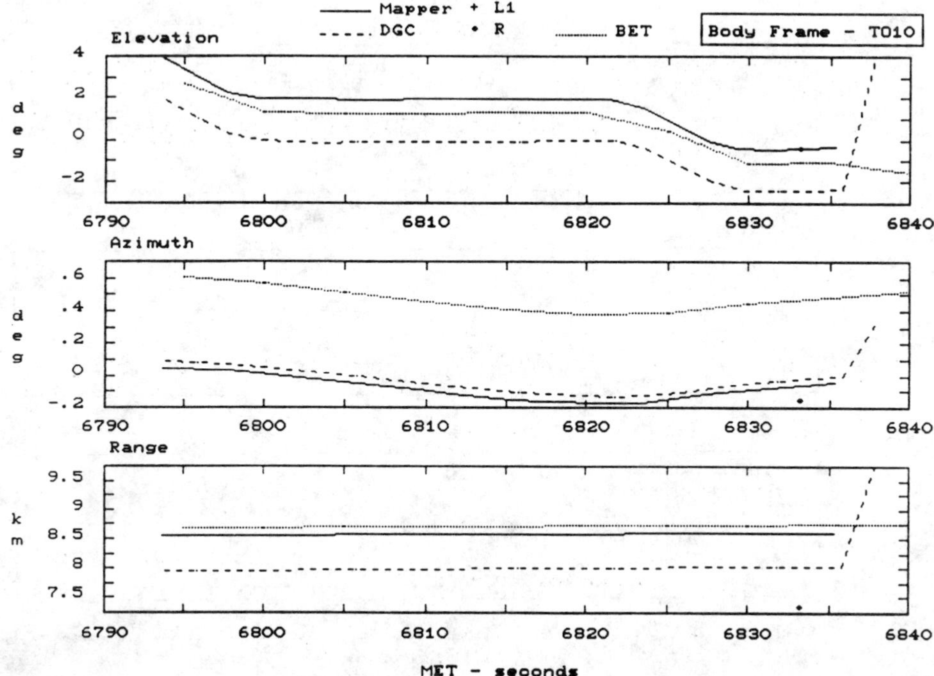

Figure 6

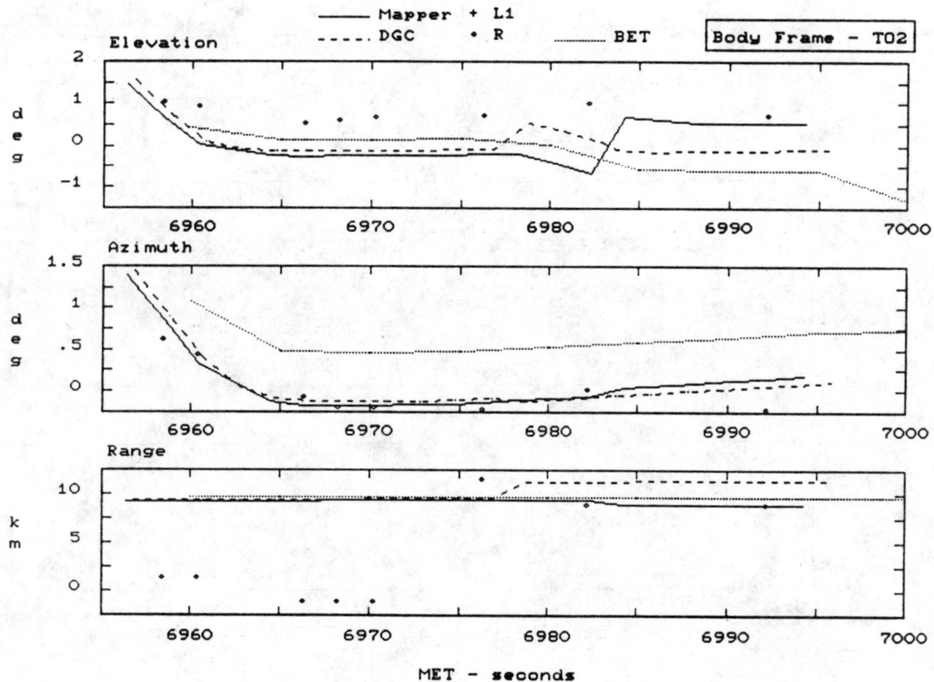

Figure 7

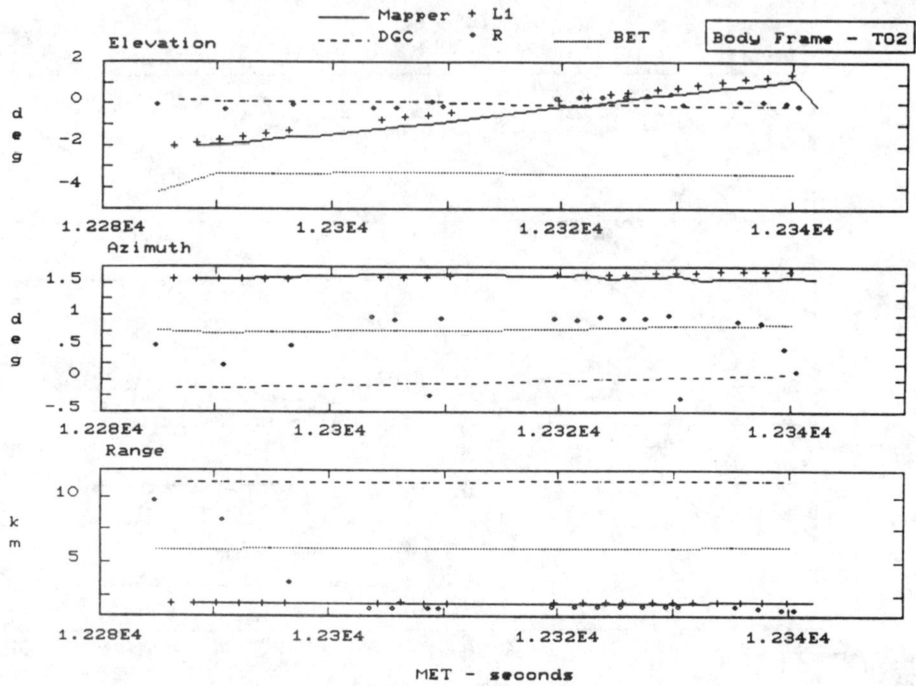

Figure 8

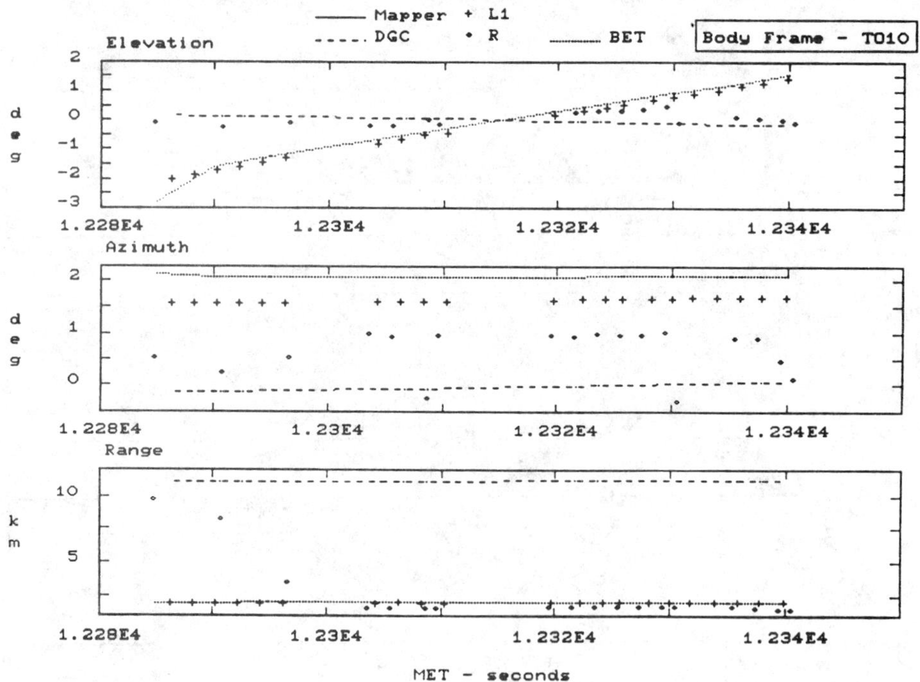

466

Figure 9

H
L
C
Position
Mapper state - SP5
k
m
4
2
0
-2
1.75E4   1.8E4   1.85E4
Velocity
m
/
s
5
0
-5
1.75E4   1.8E4   1.85E4
Variance
.0001
.00001
k
m .000001
*
2 1.0E-7
1.0E-8
1.75E4   1.8E4   1.85E4
MET - Seconds

Figure 10

Mapper + L1
DGC   • R
Elevation
Body Frame - SP5
d
e
g
1
.5
0
-.5
-1
-1.5
1.75E4   1.8E4   1.85E4
Azimuth
d
e
g
1
.5
0
-.5
1.75E4   1.8E4   1.85E4
Range
k
m
5
0
-5
1.75E4   1.8E4   1.85E4
MET - seconds

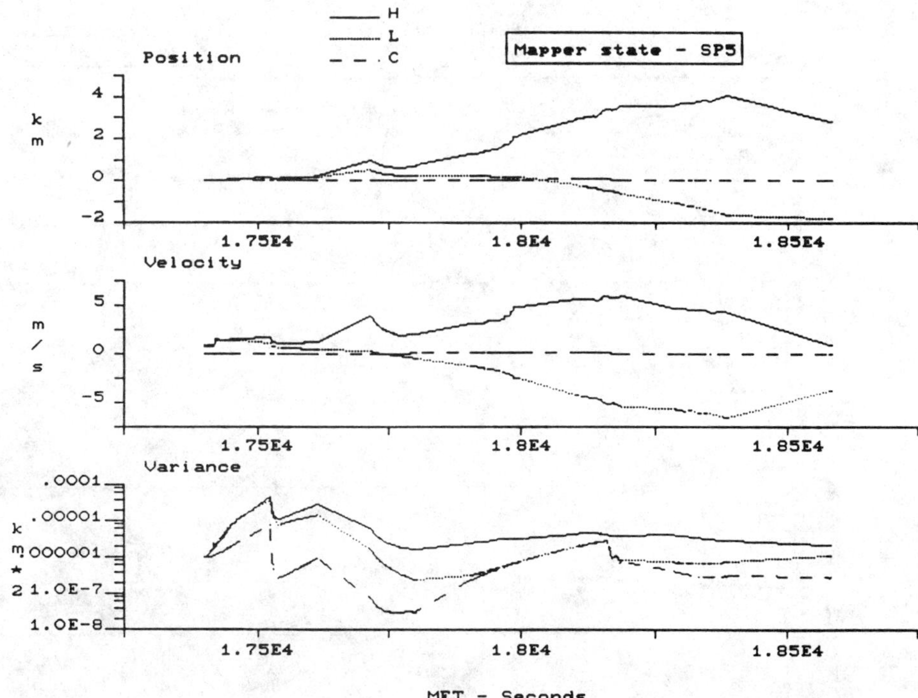

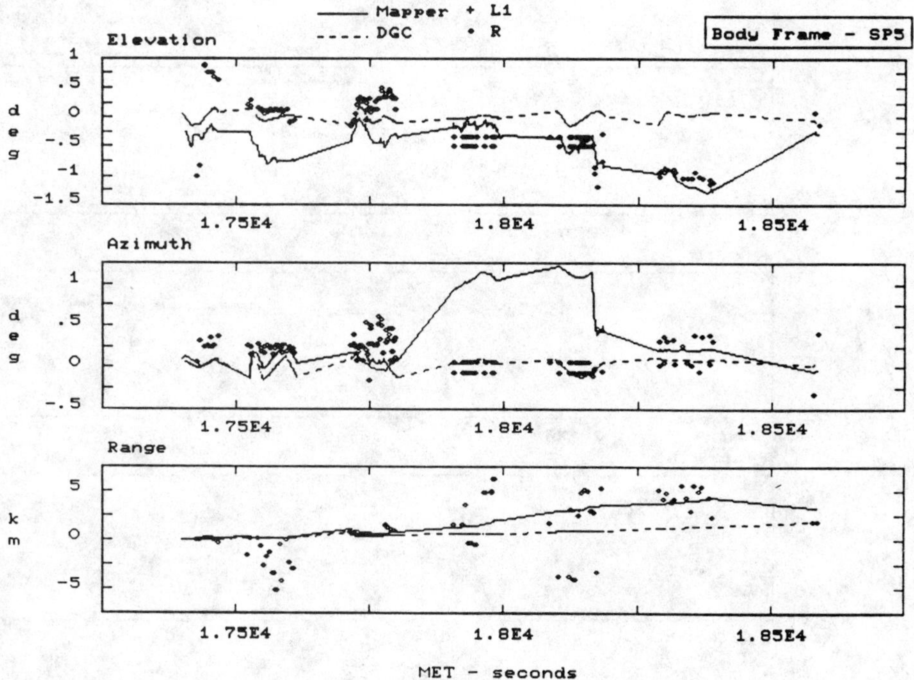

Figure 11

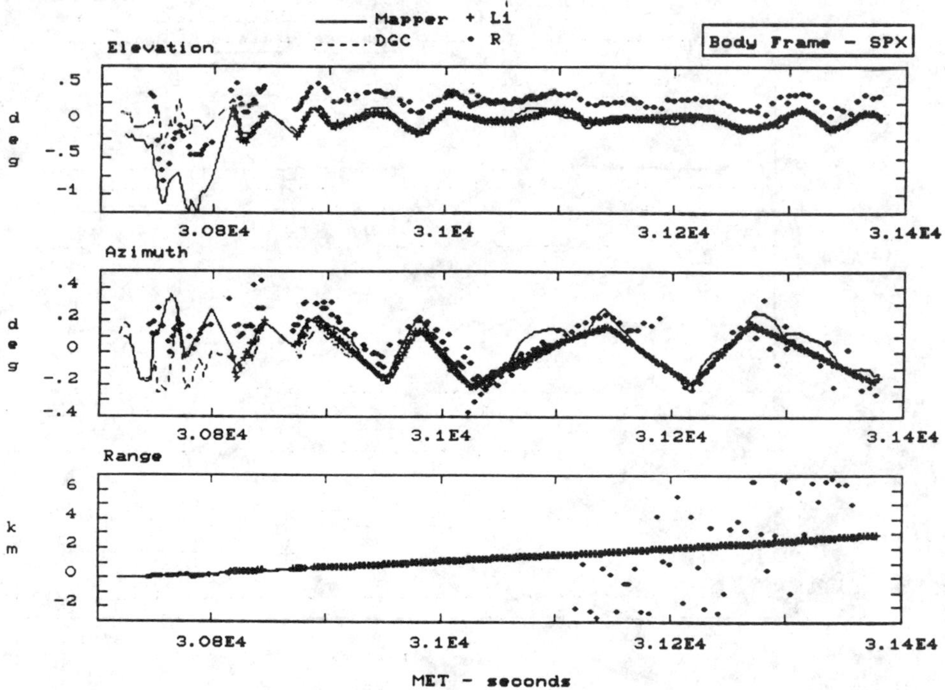

Figure 12

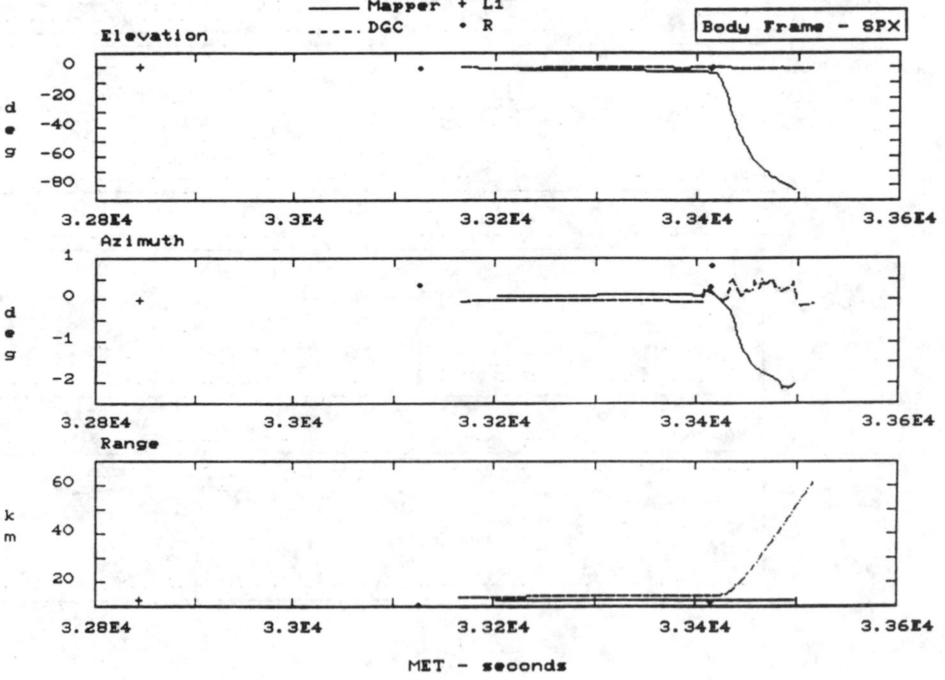

LAUNCH DELAY IMPACT ON THE GALILEO ATTITUDE CONTROL SYSTEM

J. L. Chodas[*] and D. M. Weisenberg[†]

The Galileo spacecraft was scheduled to be
launched in May 1986 aboard the Space Shuttle
and injected on a direct trajectory to Jupiter
using a Centaur upper stage. This mission
plan was severely impacted by the Challenger
disaster and subsequent cancellation of the
Centaur. Galileo is now scheduled to be
launched from the Shuttle in October 1989,
using a two-stage Inertial Upper Stage. The
trajectory had to be drastically modified to
include flybys of Venus and Earth for gravity
assists. The mission duration increased to a
total of 8 years rather than the original 4.5
years. The major impact of this new mission
plan was on the thermal control design. A
viable design was achieved, with one major
Attitude Control constraint: the spacecraft
antenna boresight must be pointed to within 14
degrees of the Sun whenever the spacecraft is
within 1 AU of the Sun, even in the presence
of faults. Changes to hardware, software and
operations strategies had to be made to handle
the new and longer mission.

INTRODUCTION

   Galileo is a dual-spin spacecraft designed to explore
Jupiter and its moons. It was scheduled to be launched in
May 1986 from the Space Shuttle using a Centaur upper stage
on a direct trajectory to Jupiter, arriving there in December
1988. This mission plan was severely impacted by the
Challenger disaster and subsequent cancellation of the

---

* Technical Manager, † Member of Technical Staff, Guidance and Control Section, Jet Propulsion Laboratory, California Institute of Technology, 4800 Oak Grove Drive, Pasadena, California 91109.

Centaur. The mission had to be completely redesigned to be
compatible with a new upper stage and a very different
trajectory. Galileo is now scheduled to be launched from the
Shuttle in October 1989, using a two-stage Inertial Upper
Stage (IUS). Because the IUS provides less energy than the
Centaur, the spacecraft must use gravity assists once at
Venus and twice at Earth in order to gain enough energy to
send it on to Jupiter. The cruise phase will last 6 years in
the new mission instead of the original 2.5. Upon nearing
the planet, the spacecraft will eject a Probe into the Jovian
atmosphere. The Orbiter will be injected into orbit around
the planet to study the Jovian system. The Jupiter orbital
phase remains essentially unchanged with a duration of about
2 years.

   This new trajectory is referred to as the Venus-Earth-Earth
Gravity Assist or VEEGA trajectory[1] (see Fig. 1). The
spacecraft will fly within 0.71 AU of the Sun on the prime
October 1989 mission or within 0.65 AU on the backup July
1991 mission, although it had originally been designed to go
no closer than 1 AU. The major impact of this new mission
was on the thermal control design. A viable design was
achieved, with one major Attitude Control constraint:
whenever the spacecraft is within 1 AU of the Sun, its
antenna boresight must be pointed to within 14 degrees of the
Sun. If the spacecraft exceeds this limit, the High Gain
Antenna (HGA) will be severely damaged by exposure to the
Sun. The impact on the Attitude Control Subsystem is that,
for this portion of the mission, the spacecraft has to be
kept Sun-pointed even in the presence of faults. New
hardware, software and operations strategies had to be
developed to handle this new mission constraint.

   Another impact of the new mission on Attitude Control is
the longer mission duration. The hardware had to be reviewed
carefully to uncover any problems with a longer shelf life
and operating life and modifications had to be made. The
change in the trajectory impacted the orientation of the Star
Scanner and exacerbated difficulties in generating acceptable
sets of target stars for the onboard attitude determination
process. The increase in the time spent on the ground before
flight meant that the hardware was handled more creating
opportunities for damage, and also that there was more time
to find and fix problems. Major changes and/or additions to
Attitude Control hardware, software and operations were
needed to accommodate the new mission.

ATTITUDE CONTROL SUBSYSTEM OVERVIEW

   The Galileo spacecraft is comprised of a large rotating
portion known as the rotor and a smaller despun section, or

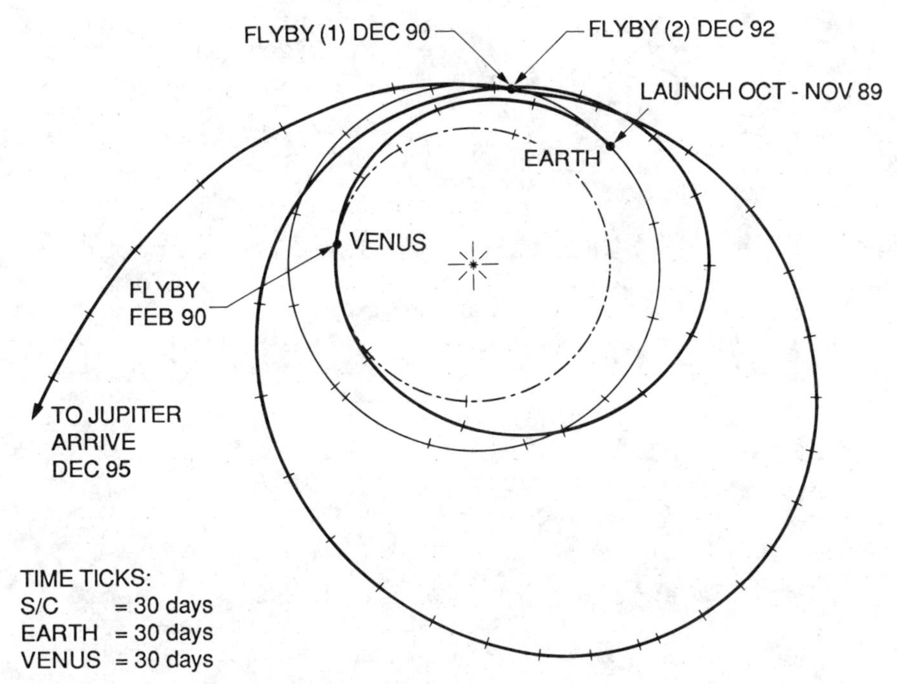

Fig. 1   Galileo VEEGA Trajectory

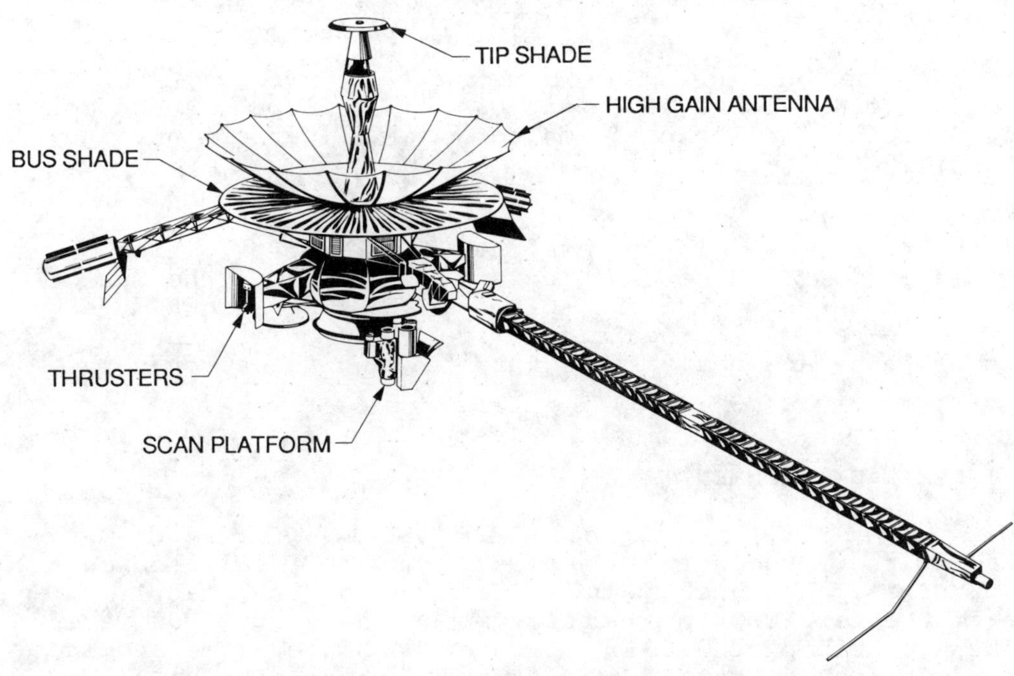

Fig. 2   Galileo Spacecraft Configuration with HGA Deployed

stator. The scan platform on which the imaging instruments
are mounted is connected to the stator. The spacecraft can
operate in a dual-spin mode, with a nominal rotor spin rate
of 3.15 rpm, or as a single spinner at either 2.9 or 10.5
rpm. The higher spin rate is used to provide greater spin
stability during firings of the large axial engine and during
the release of the Probe.

The Attitude Control Subsystem consists of an onboard
processor and several sensors and actuators[2]. The primary
sensor for attitude determination is the rotor-mounted Star
Scanner which provides information for computing the rotor
spin rate and the orientation of the spin axis. Spin rate
information is also supplied by the Acquisition Sensor which
detects Sun pulses. Control of the spacecraft attitude is
maintained by commanding the Propulsion Subsystem thrusters
and isolation valves. Galileo uses a bipropellant system,
with two isolation valves (one each for fuel and oxidizer)
upstream of each of the two redundant sets of thrusters.

The spin bearing actuator provides a rotational degree of
freedom between the rotor and the stator while the scan
platform actuator allows the scan platform to rotate with
respect to the stator. These actuators provide the platform
with two orthogonal axes of articulation. High precision
control of the scan platform is obtained by using platform-
mounted gyros to sense its inertial orientation.

Most of the Attitude Control hardware assemblies are
redundant to provide a backup in case of failure. Autonomous
Fault Protection algorithms monitor nominal performance and
swap units as necessary.

The major additions to the Attitude Control Subsystem to
accommodate the new mission are the Spin Detector, which
provides another source of spin rate information, the Sun
Gate, which detects when the spacecraft HGA boresight moves
off sunline, and the Propulsion Drive Electronics Annex,
which protects against various thruster and isolation valve
failures.

NEW THERMAL ENVIRONMENT

The most significant impact of the VEEGA mission was the
change in the spacecraft thermal environment. Galileo was
not originally designed to get closer to the Sun than 1 AU;
however the new VEEGA trajectory imposed a design requirement
on the spacecraft of 0.65 AU to cover both the prime and the
backup missions. Protection from the increased solar
intensity became a major concern.

The Thermal Control Subsystem designed a set of shades and blankets to protect sensitive spacecraft hardware. The most sensitive component was found to be the HGA, which could be severely damaged by even brief exposure to sunlight (on the order of seconds) at the closer solar distances. The solution to this problem consisted of an operational change and a hardware modification. Operationally, the HGA will not be deployed from its stowed configuration until after the spacecraft has permanently passed beyond 1 AU. Hardware modifications (see Fig. 2) included adding a shade at the base of the HGA to protect the bus hardware (the "bus shade"), placing a shade on the tip of the stowed HGA to protect the HGA itself (the "tip shade") and changing the blanket material in several areas. In the initial design of the tip shade, an HGA-Sun angle of greater than 5 degrees would result in illumination of the HGA. The spacecraft attitude would have to be maintained within 5 degrees of the Sun. The shade was later enlarged to cover up to 14 degrees.

This new thermal design had several major impacts on the spacecraft attitude control. To accommodate the new design, two hardware devices were added to the spacecraft: the Spin Detector and the Sun Gate. In addition, Attitude Control maneuver philosophy changed substantially.

<u>Spin Detector</u>

The Galileo Attitude Control Subsystem has two primary methods of determining the rotor spin rate. These methods are referred to as the Star Scanner Spin Rate Estimator (SSSRE) and the Acquisition Sensor Spin Rate Estimator (ASSRE). Each of these spin sources has its own set of idiosyncrasies. With the new requirement of remaining very closely Sun-pointed for much of the mission, these sources of spin data are inadequate for detection of stuck-open spin-down thrusters. The problem is that a spin-down thruster can fail open and cause the spacecraft to despin (a mission catastrophic event) without the fault being detected or corrected.

The SSSRE determines the rotor spin rate by collecting 60 seconds of star data (about 3 revolutions) and finding a spin period which corresponds to the time between sightings of a single star. For the algorithm to converge, the rotor spin rate must be constant.

The ASSRE determines spin rate by finding the time between passings of the Sun through the Acquisition Sensor field of view. When the spacecraft is Sun-pointed, the Sun is always in the field of view. Analysis has determined that, in order to guarantee a reliable Acquisition Sensor spin rate estimate (accounting for sensor misalignments, nutation, wobble,

473

etc.), the spacecraft must be pointed more than 7 degrees
away from the Sun.

This requirement obviously conflicts with the VEEGA thermal
requirement that the spacecraft remain pointed within 5
degrees of the Sun (based on the initial size of the tip
shade). Attitude Control Fault Protection is required to
detect stuck-open spin thrusters and respond to them before
any mission catastrophic consequence occurs. However, if a
spin-down thruster were to stick open, it would result in a
changing spin rate. The SSSRE would not function. If the
spacecraft is Sun-pointed, this leaves Fault Protection with
no source of spin information since the ASSRE is not
reliable. Thus Fault Protection is left with no way of
detecting a stuck spin thruster.

The solution to this problem was to add a new spin sensor
to the spacecraft that does not depend on the Sun cone angle
and that will operate when the rotor spin rate is changing.
This new sensor, called the Spin Detector, consists of an
accelerometer mounted radially on the rotor such that it
detects the rotor centripetal acceleration. From this
acceleration, the Flight Software can calculate the spin
rate.

The Spin Detector was designed to have a dynamic range from
0 to 14 rpm. It consists of two completely redundant
accelerometers and analog electronics. To improve the
accuracy at lower spin rates, the electronics provide two
outputs to the Attitude Control processor - a high resolution
low-rate signal, with dynamic range from 0 to 3.5 rpm, and a
lower resolution high-rate signal, with the full dynamic
range. The Spin Detector estimate is not as accurate as
either the Star Scanner or the Acquisition Sensor estimates
(accelerometer bias drift between calibrations is the major
error source), but, since it provides a spin estimate
independent of the Sun cone angle, it is adequate for Fault
Protection purposes.

<u>Sun Gate</u>

A second problem caused by the new thermal constraints also
arises from considerations of thruster Fault Protection. The
Attitude Control electronics is a redundant set. The failure
of a component on one side will be detected by the onboard
Fault Protection and a switch to the redundant unit will be
performed autonomously. There are several failures in the
electronics that could cause an attitude thruster valve to
stick open or a valve to fail closed when two thrusters are
firing as a balanced couple. If such a failure had occurred
on the original 1986 mission and caused the spacecraft
attitude to be perturbed, it would not have been mission

catastrophic since there was no requirement to stay Sun-
pointed as on the VEEGA mission. With the new Sun-pointing
constraint, however, these failures could be mission
catastrophic. Because Galileo is spin stabilized, the Sun
cone angle does not increase indefinitely - the peak cone
angle has been determined to be approximately 18 degrees.
However, an 18 degree off-Sun excursion is clearly in
violation of the initial 5 degree thermal requirement for
protection of the HGA.

At first, enlarging the radius of the tip shade was
considered. It did not prove to be feasible to make it
larger than the size that would protect the spacecraft out to
14 degrees off-Sun, which was insufficient protection against
thruster faults. The Thermal Control Subsystem engineers
insisted that any exposure of the HGA to the Sun, no matter
how brief, could be mission catastrophic at the closer solar
distances. The tip shade was enlarged, but that did not
solve the problem completely.

The final solution to this problem was to add a simple
optical sensor to the spacecraft. This sensor, mounted
parallel to the HGA boresight, will detect whether or not the
Sun is in its field of view. If not, it will send a signal
to the processor and a Fault Protection response will be
initiated to keep the spacecraft pointed to within 14 degrees
of the Sun.

For cost and schedule reasons, it was decided to use a
residual Viking sensor to perform this task - the Viking Sun
Gate. The Sun Gate has a simple cadmium sulfide sensor in
each of two redundant channels whose resistance varies with
illumination. At closer solar distances and smaller off-Sun
angles, a large amount of light hits the sensor and its
resistance is very low. At greater solar distances and
larger off-Sun angles, the resistance increases. By
measuring the voltage across a reference resistor placed in
parallel with the sensor, a coarse indication of the Sun cone
angle can be obtained.

There were many difficulties associated with adapting the
Sun Gate for use on Galileo. The sensor was originally
designed to be used from 1.0 to 1.5 AU, but Galileo intends
to use it from 0.65 to 2.4 AU, 3.5 times the dynamic range.
In addition, use of the sensor output on a spinning
spacecraft was unexpectedly difficult. For these reasons,
the Sun cone angle determined by the Sun Gate is very coarse
- recent analysis indicates that it is only accurate to
within 1.7 degrees and that this accuracy is only obtained
for a narrow range of cone angles. The Sun Gate output
remains constant for smaller cone angles (less than about 3
degrees), then falls off sharply as the cone angle is

increased. The 1.7 degree accuracy is obtained in the
regions where the output is decreasing.

   The nominal trip angle of the Sun Gate Fault Protection
response was chosen to be 5.5 degrees, because this falls
within the region in which the sensor output varies with cone
angle. However, because of the inaccuracies in the Sun Gate
cone angle, the spacecraft attitude must be maintained within
3.8 degrees of the Sun at all times to prevent an erroneous
trip of the fault response. This is a major operational
constraint which will require frequent pointing corrections
(sometimes as often as once every other day).

   The Fault Protection response to an excessive off-Sun
excursion is to abort any maneuver in progress, swap Attitude
Control hardware, and close isolation valves. The Propulsion
Drive Electronics Annex (discussed in a later section)
guarantees the termination of any thruster firings when its
isolation valves are closed.

<u>Maneuver Philosophy</u>

   The Galileo Attitude Control Subsystem has the capability
to perform maneuvers in two modes. In the first, called
"turn/burn/turn", the spacecraft is turned until it is
pointing in the direction in which a change in velocity is
desired, axial thrusters fire to create the necessary
velocity change, and then the spacecraft is turned back to
its initial attitude. The second mode, referred to as
"vector mode", does not require any turns. The axial
component of the velocity change can be obtained by firing
the axial thrusters while the lateral component is generated
by performing a lateral maneuver in which lateral thrusters
are pulsed once per revolution.

   In the original 1986 mission, there was no overwhelming
reason to do vector mode maneuvers. They were planned only
for very small velocity increments, and very little testing
of the capability took place. However, the new VEEGA
mission, with its stricter pointing requirements, made vector
mode maneuvers much more important. When turns off Sun are
prohibited, only vector mode maneuvers are possible. This
created a whole new emphasis in testing. Both the vector
mode sequence components in the Ground Software and the
corresponding onboard algorithms in the Flight Software were
tested extensively on the spacecraft during System Test to
make up for the limited testing performed previously.

INCREASED MISSION DURATION

The VEEGA trajectory results in a much longer mission than
was originally planned. The 1986 mission would have lasted
approximately 4.5 years, while the 1989 mission will last 8
years. This raises significant concerns for shelf life and
aging effects and for adequacy of spacecraft consumables.

<u>Shelf Life and Aging Review</u>

An extensive study of hardware shelf life and aging was
instigated to identify components that could not tolerate an
additional 5 years on the ground (to cover the prime launch
opportunity in October 1989 and the backup one in July 1991)
or an increase of 3.5 years in flight. Existing waivers were
reviewed to determine if they were still acceptable for the
new mission. Electronic parts were reviewed for reliability
over the life of the VEEGA mission, and worst case analyses
were reassessed with new derating parameters. This resulted
in the changeout of several types of electronic parts that
were shown to be unreliable over the mission life.

Several Attitude Control hardware devices were inherited
from previous missions such as Voyager and Viking. These
items were reviewed to ensure that they could still meet the
Galileo mission requirements despite the launch delay and the
longer mission. The materials were also reviewed for
acceptability with special attention being paid to seals and
leak rates. No concerns were found.

The most significant concern that surfaced in the Attitude
Control Subsystem involved the Light Emitting Diodes (LEDs)
in the spin bearing and scan platform actuator encoders. The
LEDs are pulsed in order to read the encoder angles and the
values are used to commutate the motors and to point the
platform. The pulse rate was set at 156 Hz in order to
minimize motor disturbance torques that would perturb scan
platform pointing. The LEDs degrade when they are on, and
since the longer mission requires them to be on for a longer
total time, questions were raised about their ability to
perform adequately later in the mission.

Tests were performed on LEDs from the same wafer as the
flight units to determine their degradation rate. Results
showed that the LEDs had only an estimated 80% chance of
surviving the mission at the 156 Hz pulse rate. The simplest
solution to this problem was to reduce the pulse rate, which
reduces the total on time but results in coarser commutation
of the motors. It was determined that a reduction in the
pulse rate from 156 Hz to 15 Hz would degrade scan platform
pointing accuracy slightly, but would eliminate any lifetime
concerns. The hardware change to accomplish this was very

straightforward and easy to test. Since the effect on
pointing performance was extremely small and did not result
in the violation of any pointing accuracy requirements, the
actuator electronics were modified to pulse the LEDs at the
lower rate.

<u>Spacecraft Consumables</u>

   The new mission called greater attention to spacecraft
consumables. Star Scanner Sun Shutter cycles and thruster
cycles are examples of this concern.

   The Star Scanner is used to provide a celestial attitude
reference and to estimate the spacecraft spin rate. The
hardware is designed to measure stars with a sensitivity on
the order of photons, so the Sun can damage the
photomultiplier tube. To protect against this possibility, a
Sun sensor is mounted on the Star Scanner baffle to detect
whether or not the Sun will enter the boresight field of
view. If so, a shutter is closed to prevent light from
striking the photocathode.

   During maneuvers, this Sun Shutter is commanded closed by
the Flight Software as a safeguard against hardware
contamination from thruster firings and against the chance
that the maneuver might sweep the Star Scanner boresight
through the Sun. In addition, when the spacecraft is at
certain attitudes, the Sun or another bright celestial body
may pass through the field of view once every spacecraft
revolution (about once every 20 seconds).

   The increased number of maneuvers required by the VEEGA
mission for attitude maintenance and trajectory corrections,
in addition to the preponderance of new bright bodies that
will be seen early in the mission (Venus, Earth and moon),
greatly increase the number of cycles that the Sun Shutter is
expected to experience. It has been estimated that,
including ground testing, the number of cycles that will be
put on the flight shutter by the end of the mission could be
as high as 75,000. However, the shutter had only been tested
to 25,000 cycles; it was not clear that the shutter could
survive three times its previously tested limits.

   Fortunately, there was a very simple solution to this
problem. A spare shutter is being requalified by forcing it
to cycle closed and then back open three times every minute.
The current intention is to continue testing until the
shutter fails; to date, more than 500,000 cycles have been
performed successfully.

   The VEEGA mission, by increasing the required number of
maneuvers, requires many more cycles of thruster valves and

isolation valves. The thruster valves have been qualified to 60,000 cycles, and the Propulsion Subsystem recommends against cycling them more than 40,000 times. The isolation valves have been qualified to 5,000 cycles, and Propulsion recommends cycling them fewer than 4,000 times.

A conservative estimation of the number of maneuvers showed that the number of cycles on the isolation valves would be less than 4,000, but without much margin. However, the precession thrusters, which are used for Sun pointing maneuvers and for certain vector mode maneuvers, are in danger of exceeding the required limit on thruster cycles by about 12,000 cycles.

This problem has not yet been entirely solved; efforts to reduce the number of thruster cycles are ongoing. Several changes to the mission plan have already been made to reduce the number of cycles. The Navigation Team redesigned the trajectory slightly to reduce the number of maneuvers that require the use of the overworked precession thrusters. Attitude Control allowed the backup thruster branch to be used as the nominal one for spin correction maneuvers in order to reduce the number of isolation valve cycles on the primary thruster branch.

Finally, the pointing strategy used to maintain Sun point was changed. For the 1986 mission, a maneuver referred to as an "HGA correction" was planned to be used for attitude maintenance. The number of thruster cycles was reduced by simply replacing HGA corrections with turn maneuvers which require fewer thruster pulses to turn the spacecraft through the same angle. Turns are slightly less accurate than HGA corrections, but the decreased accuracy was determined to be acceptable while the HGA is in its stowed configuration and lower data rates are being used with the Low Gain Antennas. After the HGA is deployed and high data rates are planned, the increased accuracy of the HGA correction maneuver will be required.

NEW VEEGA TRAJECTORY

The new VEEGA trajectory had other impacts on the Attitude Control Subsystem besides the necessity to keep the spacecraft pointed at the Sun whenever it is within 1 AU of the Sun. The optimal orientation of the Star Scanner had to be determined, the limitations of the Star Scanner to provide a celestial reference became more restrictive and hardware dependencies on distance to the Sun had to be reviewed to ensure that this trajectory would not introduce problems.

Due to the new launch date and mission, the spacecraft's arrival date at Jupiter changed. The Star Scanner orientation on the spacecraft has always been optimized to view bright, unambiguous stars during the Jupiter Orbit Insertion (JOI) time period; it had been necessary to redo this analysis for every new launch date. At the time that the decision had to be made (for hardware schedule reasons) as to how to orient the Star Scanner, the 1989 launch period spanned six weeks, which corresponded to about a one year time period for JOI (from Oct 1995 to Aug 1996). The analysis to optimize the Star Scanner boresight orientation with respect to the HGA (given that the HGA is pointed at the Earth during JOI) had to be redone for this one year time period.

This analysis had been done several times already as the Galileo mission changed over the years, so the method was well known. The analysis was redone for 27 different angles in 1 degree increments. The optimal angle enables good target stars to be seen easily by the Star Scanner during the JOI time period early during the one year arrival space and acceptably during the remainder of the arrival space. The Mission Design Team has now selected 7 December 1995 for the JOI arrival date which falls in the time period when the selected orientation angle sees the target stars most easily.

Star Set Difficulties

Due to the design of the VEEGA trajectory, two new problems arose that had not been present for the original 1986 mission. The first one is concerned with the fact that when the spacecraft flies past Venus and Earth, a large percentage of the sky is blocked out by the Sun and planets. The Star Scanner Sun Shutter closes when a celestial body is predicted to pass through the field of view to prevent the photomultiplier tube from being damaged. With the shutter closed, no stars are sensed during that portion of the revolution. As a result, fewer stars are available to be selected as target stars for the onboard attitude determination process.

For this reason, during the close planetary flybys, it may not be possible to determine the spacecraft attitude based on star information. The gyros may have to be used for extended periods of time to provide spin rate and attitude information. This configuration makes the spacecraft vulnerable to attitude changes due to gyro drift since there will be no celestial updates to null out the drift. The spacecraft will also experience sequence aborts if one of the gyros fails for any reason since there will be no other

source for attitude information. This strategy requires the
gyros to be on for a longer total time than was originally
planned. Fortunately, the gyro lifetime specification is
much greater (approximately 10,000 hours) than that needed
for the VEEGA mission under these conditions.

The second problem relates to the requirement to keep the
spacecraft pointed at the Sun within 3.8 degrees during
certain portions of the mission. Planned attitude
maintenance must allow for uncertainties in the spacecraft
attitude, nutation, etc., so the strategy is to command an
attitude update whenever the HGA has drifted more than
approximately 2 degrees from the sunline. Close to Venus,
the Sun moves with respect to the spacecraft at up to 2
degrees per day so the Operations Team must command attitude
updates as frequently as once every other day.

Each new attitude requires a new star set of target stars
to be uploaded from the ground giving the target stars that
the Star Scanner should look for at that attitude. Having to
do this once every other day greatly increases the workload
on the Operations Team. For the 1986 mission, it was
anticipated that star sets would only need to be updated once
every two weeks to a month.

Both of these problems increased the emphasis on the
Operations Team's capability to select star sets. Ground
software programs had been developed for the 1986 mission to
aid the analyst in this process. These programs were
difficult to use, however, and required a highly trained,
experienced engineer to perform the task successfully in a
reasonable amount of time. Difficulties arise when a
satisfactory star set cannot be found by the programs.
Engineering judgement has to be used then to selectively
relax some constraints without jeopardizing the onboard
attitude determination process.

For these reasons, an expert system called STARS was
developed for the VEEGA mission. This program aids the
analyst in several ways. First of all, it provides a user-
friendly interface to the ground software programs that makes
them much easier to use. It processes the data generated by
these programs and presents the information in more easily
understandable formats. Finally, it helps the analyst in the
decision-making process by recommending which constraints to
relax, based on the rules in its knowledge base. These
capabilities enable the analyst to select star sets more
easily and in less time.

The STARS expert system has already proven its usefulness
in analyzing problem star sets encountered during integration

testing, and is expected to be an important tool during
mission operations.

<u>Expanded Range for Hardware</u>

The new trajectory increased the range of distances from
the Sun over which the Attitude Control hardware had to
operate. This was taken into account when the Sun Gate was
added but it also impacted the Acquisition Sensor.
Additional testing had to be done to identify the sensor
characteristics at solar distances less than 1 AU.

The Acquisition Sensor consists of a cadmium sulfide
detector whose resistance varies with illumination. There
was concern that the output might saturate at 0.65 AU, making
it necessary to add a filter over the boresight. It was not
desirable to add a filter since it would dim the sensor
output at Jupiter (unless a retractable filter was designed,
leading to more problems). Fortunately, tests with the
flight hardware showed that the Acquisition Sensor output
does not saturate at 0.65 AU so no filter had to be added.

LONGER GROUND TIME

One of the more interesting impacts on the Attitude Control
Subsystem due to the launch delay was the longer time spent
on the ground. This manifested itself as a return to the
"development phase" of the Project. Since there was more
time before launch, more problems could surface, be analyzed,
and have hardware and/or software fixes implemented. The
most obvious example of this was the new hardware assembly
added to the spacecraft to protect against thruster-related
faults, the Propulsion Drive Electronics Annex.

<u>Propulsion Drive Electronics Annex</u>

The Galileo single point failure policy states that no
single failure can lead to a mission catastrophic event.
After the launch delay, a more detailed study was done of
possible failures in the sections of the Attitude Control
electronics that control thruster firings. This analysis
uncovered several serious concerns with the impacts of these
failures.

Failures had already been identified that could cause a
thruster valve to open without its associated isolation
valves being open. The Sun Gate had been added to the
spacecraft to detect attitude perturbations due to this type
of failure. It was later realized, though, that parts of the
spacecraft could be contaminated when the fuel and oxidizer
that are in the lines between the isolation valves and the

thruster valve spill out onto exposed spacecraft surfaces.
In addition, if a thruster valve opens with only one of its
isolation valves open, the Propulsion Subsystem engineers
became concerned that a mission catastrophic explosion could
result.

Other single point failures could cause a thruster valve to
open erroneously when its isolation valves were open for a
scheduled maneuver that used different thrusters. If this
happened, unplanned changes in attitude or velocity could
occur. Attitude errors are a serious concern during the
release of the Probe, since the Probe has no attitude control
of its own. It depends on the Orbiter to release it at the
correct orientation. Attitude errors that cause the
spacecraft to turn away from the Sun while at less than 1 AU
are a new concern for the VEEGA mission. Velocity errors
could be catastrophic if they occurred during the maneuvers
that take place just before the close planetary or satellite
flybys. Maneuvers in the vicinity of Earth were also
examined very closely for safe execution due to the
requirement of avoiding Earth impact. Unplanned velocity
errors were unacceptable. Contamination of the spacecraft is
again a concern with this failure, as is plume heating of
certain sensitive hardware on the stator or scan platform.

Finally, single point failures were uncovered that could
cause the isolation valves to be powered continuously. These
valves are normally pulsed for about 71 ms. If one is
powered continuously, the valve itself could be destroyed.
Also, the increased power draw could cause a spacecraft
undervoltage trip since the predicted power margins during
the mission are generally not large enough to handle this
type of failure. Because of the extremely large power draw
of an isolation valve, the Fault Protection responses from
the Command and Data Subsystem would not necessarily reduce
the power demand enough to terminate the undervoltage
condition.

These failures all existed for the 1986 mission but their
impacts were not completely understood. The autonomous use
of the redundancy of the electronics design was felt to be
adequate for detecting and correcting stuck open or closed
thruster valves. The seriousness of the plume contamination
and thruster explosion concerns was not appreciated.

The solution to these concerns was to add a new hardware
assembly to the spacecraft to monitor the commanded states of
the isolation valves and thruster valves and to compare them
with the actual states. If a thruster valve is opened
erroneously, the thruster will be disabled. If an isolation
valve is being powered continuously, a signal will be sent to

the Command and Data Subsystem to tell it to turn off the electronics providing power to the valve.

This new hardware device is referred to as the Propulsion Drive Electronics Annex, since it was added as a sort of "annex" to the existing Propulsion Drive Electronics. It consists of two redundant channels tied to the redundant electronics I/O channels to monitor the commanded states of the thrusters and the isolation valves. It senses the actual states of the thruster valves by monitoring outputs of the Propulsion Drive Electronics, and obtains the actual states of the isolation valves from reed switches in the Propulsion Subsystem. It has two redundant relays for each of the 12 thrusters for disabling them as necessary. It also supplies telemetry to the Attitude Control processor to provide information on any thruster disablings as well as on the health of the Annex itself. As was noted earlier, the Annex has an interface with the Command and Data Subsystem to inform it as to whether any isolation valve is being powered continuously.

It was critical to the success of the mission to implement solutions to the single point failures uncovered in the analysis of thruster-related faults. It was just as critical, however, to ensure that the thrusters <u>could</u> be fired when they needed to be. For this reason, the Propulsion Drive Electronics Annex was designed such that its disable relays could not be activated when power was removed. If the Annex were to erroneously disable a thruster, use of the thruster can be restored by simply turning off the Annex itself. This disable function had to be designed so that no single failure could simultaneously cause an erroneous thruster disable <u>and</u> prevent power from being removed.

With all these interfaces, the design of the Annex became very complicated. It had to be compatible with existing spacecraft cabling since there was not enough time to build and install new cables. The availability of spaceflight qualified parts was another major constraint on the design. However, since the Annex was the newest piece of hardware to be added to the spacecraft, there was a tendency to regard it as the easiest assembly to modify. As it turned out, there was enough room in the Annex to add one more function related to spacecraft safety while in the Shuttle payload bay.

<u>Shuttle Safety Issue</u>

Galileo has a requirement to be two-fault tolerant to failures which could jeopardize astronaut safety while the spacecraft is in the Shuttle payload bay. A scenario of concern postulated that the Command and Data Subsystem or Power Subsystem could fail and turn on the Propulsion Drive

Electronics. A second failure in the Propulsion Drive
Electronics could cause it to power an isolation valve
continuously. If this happens, a fire could start in the
Shuttle payload bay. Only one level of protection is
provided by not turning the Propulsion Drive Electronics on
at launch. A second level, however, could be obtained by
ensuring, via a hardware inhibit, that the Propulsion Drive
Electronics could not be powered on until after the
spacecraft was deployed.

This was implemented by adding redundant relays to the
Annex to interrupt power from the Power Subsystem to the
Propulsion Drive Electronics. These relays will be open at
launch so that there will be no way that the electronics can
be powered and thus fire thrusters. The relays will be
closed by the IUS when it separates from Galileo and they
will remain closed for the life of the mission.

Increased Hardware Handling

Because of changes to the spacecraft necessitated by the
VEEGA mission, the hardware was handled and tested more than
was originally planned. The spacecraft as a whole had to
undergo a second Solar Thermal Vacuum test to verify the new
thermal design. Several hardware assemblies had to have
specific parts replaced as data surfaced that implied that
the parts would not support the mission adequately. Each
time this happened, the necessary environmental and
functional retests required to ensure reliability had to be
performed.

In some instances, there had not been enough time before
the 1986 launch to perform the standard set of environmental
tests. Problems were uncovered and changes had to be made so
late that the hardware had to be delivered to the spacecraft
for final integration before completing all unit testing.
For these cases, the launch delay provided time to complete
the qualification of the hardware per the original
guidelines.

In addition to hardware-related tests, the flight hardware
had to support extensive System tests. Flight Software
programs for the Attitude Control Subsystem and the Command
and Data Subsystem experienced major modifications for the
VEEGA mission. The Ground Software programs and sequence
components also made changes to accommodate the new mission.
These changes had to be tested in an integrated manner on the
spacecraft in System Test. The flight hardware obtained more
hours pre-launch than was originally planned for the 1986
mission due to this usage.

SOFTWARE IMPACTS

<u>Flight Software</u>

   As might be expected, many of the changes discussed in the
previous sections have had significant impacts on the
Attitude Control Flight Software. Three new hardware
assemblies were added to the spacecraft for the VEEGA mission
- the Spin Detector, the Sun Gate and the Propulsion Drive
Electronics Annex. Software had to be added to process the
outputs of each of these devices to enable them to carry out
the functions for which they were designed.

   In addition, each of these new Fault Protection-related
devices were found to have mission catastrophic failure modes
of their own which required further additions to software
Fault Protection. Each device has to be monitored to ensure
that it is operating correctly. If not, the outputs have to
be ignored and some appropriate action taken.

   Since only 32K of flight memory exists and it was
essentially all used for the original 1986 mission, these new
requirements led to substantial scrubbing of the software.
Several non-essential functional capabilities were deleted.
For example, software limitations on the rate of scan
platform slews were eliminated and replaced with an
operational rule forbidding the commanding of large platform
rates. This constraint will now be checked by the Ground
Software. In addition, three major modules were recoded from
the high-level language HAL/S into assembly language. A size
reduction for these modules of over 40% was obtained by doing
this, although the software understandability and
maintainability suffered as a result.

   Finally, there were many problems remaining after the
launch version of the Flight Software for the 1986 mission
had been delivered. These had been deferred for various
reasons. Some applied to situations that the spacecraft
would not encounter until it reached Jupiter so they did not
need to be worked before launch. Some arose so late in the
software development process that decisions had been made to
not incorporate the fixes. Some were thought to be non-
problems but the details had not been worked through. The
launch delay enabled all of these problems to be worked and
incorporated into the Flight Software as necessary.

<u>Ground Software</u>

   The Attitude Control Ground Software programs incorporated
several changes after the launch delay. The new Spin
Detector needed to have calibration software to ensure valid
outputs. Some changes resulted from the fact that the

spacecraft will now fly closer to the Sun than 1 AU.
Programs dependent on solar distance such as the model of the
Acquisition Sensor output had to be updated to reflect new
data. Several new operational constraints were imposed that
required changes. For example, the restriction on maximum
scan platform slew rate that was scrubbed from the Flight
Software had to be added to the appropriate Ground Software
programs. The spacecraft configuration had changed
appreciably due to the addition of shades and blanket
modifications. This meant that, for each instrument on the
scan platform, the obscuration mask that defines what parts
of the sky are obscured by the spacecraft had to be updated.

Additional changes were needed to correct failures that
were uncovered just before the 1986 launch but were not fixed
then because they were found so late and the risk was deemed
acceptable. Several more changes resulted from using the
programs for various analyses and uncovering deficiencies.
In the process of defining the scan platform misalignment
calibration procedure, it was discovered that the
corresponding program gave little flexibility for plotting
the star fields. Changes were made to allow the user to
obtain greater resolution and to expand the range of angles
over which the plots could be generated.

The maneuver analysis program was used extensively to
support sequence development for System Test and for training
exercises. Changes were made to the program to print out
more information that can be used to verify predicted
performance against actual telemetry when a maneuver is
executed. In the process of designing maneuvers for
preventing the spacecraft from entering the Earth's
atmosphere on a flyby, the Navigation Team requested more
flexibility in the order in which maneuver types can be
combined to deliver a specified velocity change. The
maneuver analysis program was modified to accommodate this
request.

All these changes required the Attitude Control Ground
Software programs to be completely retested and redelivered
for the VEEGA mission.

CONCLUSIONS

A tremendous number of changes have been made to the
Galileo spacecraft over the last 3 years in order to respond
to the Challenger disaster and subsequent cancellation of the
Centaur upper stage. Major modifications made to the
Attitude Control Subsystem include the addition of new
hardware components, the rework of major portions of the
Flight Software and the development of new strategies for

flying the spacecraft on the VEEGA trajectory. These changes
have now been incorporated and tested and the spacecraft will
be ready to launch in October 1989.

ACKNOWLEDGEMENT

   The work described in this paper was carried out at the Jet
Propulsion Laboratory, California Institute of Technology,
Pasadena, California, under contract with the National
Aeronautics and Space Administration.

REFERENCES

1. D'Amario, L. A., Byrnes, D. V., Johannesen, J. R., Nolan,
   B. G., "Galileo 1989 VEEGA Trajectory Design", Paper No.
   AAS-87-421, AAS/AIAA Astrodynamics Specialist Conference,
   Kalispell, Montana, Aug. 1987.

2. Rasmussen, R. D. and Brown, T. K., "Attitude and
   Articulation Control Solutions for Project Galileo", Paper
   No. AAS-80-019, AAS Annual Rocky Mountain Guidance and
   Control Conference, Keystone, Colorado, Feb. 1980.

A FAST ATTITUDE RECOVERY SYSTEM FOR COMMUNICATIONS SATELLITES: EXPERIENCE OF EUTELSAT 5

L. Van Holtz[*], M. Burton[†] and R. C. Rogers[†]

Recovery of telecommunication satellite
operational attitude following an anomaly
must be achieved rapidly to minimise
service disruption.   A low cost CCD
Star Sensor has been developed for
this purpose and is being tested on
Eutelsat 5.   The paper describes the
sensor, its development and the results
of in flight tests.

INTRODUCTION

Three-axis stabilised telecommunications satellites require
a simple, robust control loop which can place the spacecraft
in a safe holding attitude in the event of an anomaly which
cannot be resolved by the on-board Fault Management System
(FMS).   Current generation European communications
satellites, such as EUTELSAT, employ an Emergency Sun
Re-acquisition (ESR) loop to implement this function.
Whilst this assures spacecraft safety it can lead to a
lengthy interruption in the communications service whilst
the normal, Earth-pointing, attitude is re-established.

Conventional recovery from Sun-pointing to Earth-pointing
requires waiting for orthogonality between the Sun and Earth
directions and then a rotation around the Sun-line until the
Earth appears in the field of view of an Infra-red Earth
Sensor (IRES).   This condition occurs only twice a day and
so an outage of up to twelve hours can ensue following
resolution of the original anomaly.

---

[*]   European Space Agency/European Space Research and Technology Centre (ESA/ESTEC), P.O. Box 299, NL-2200
AG Noordwijk ZH, The Netherlands.

[†]   British Aerospace (Space Systems) Ltd., Argyle Way, Stevenage, Herts SG1 2AS, United Kingdom.

Recently consideration has been given to ways in which this conflict between safety and continuity of operation may be alleviated. There are two approaches to the problem:

o   avoid the transition to the Sun pointing
     attitude

o   develop more rapid means of recovery to Earth-
     pointing

Reference 1 describes one example of the first approach which was demonstrated in-orbit on the European Orbital Test Satellite (OTS-2).   The technique allows the spacecraft to approach flat spin conditions following an unresolved anomaly.   It then uses a ground-based control loop which minimises the spin energy of the satellite body using the on-board momentum wheel as the only actuator.   The method is most effective for relatively lower power satellites.

The alternative strategy is the subject of this paper.   It involves the addition of a new, dedicated, coarse pointing star sensor to support the recovery process.   This sensor would be used to detect a bright star such as Canopus, while rotating about the Sun-line, and hence avoid orthogonality restrictions.   The Fast Recovery Sensor (FRS) was developed for the EUTELSAT-5 spacecraft and has been successfully commissioned in-orbit.   The following sections contain a description of the FRS and discuss the results of ground-based and in-orbit testing.

FAST RECOVERY SENSOR TECHNIQUES

Sensor Description

The Fast Recovery Sensor (FRS) is a CCD based star sensor with a wide field of view and is shown in Figure 1.   The design is based on TV camera concepts, with optics, electronics and packaging adapted for space application and compatibility with EUTELSAT power and telemetry subsystems.

The sensor optics are based on a double Gauss system in order to achieve low distortion with focal length approximately 13mm and f/2 relative aperture. The image is focused in the plane of the CCD which has a pixel size of the order of 0.1 degrees.   There are 288x384 pixels providing a field of view of 30x40 degrees.

The optical head, which is separate from the electronics, has sufficient wall thickness to protect against radiation and act as a heat sink.   It is thermally decoupled from the spacecraft in order to provide a suitable environment for the CCD which in this application does not require active cooling.

Fig. 1   The Fast Recovery Sensor (FRS)

A short baffle, comprising two knife edged vanes, rejects
direct entry of the Sun's rays under grazing angles of up to
10 degrees with an attenuation of 10-6.

An electronics unit, comprising four printed circuit boards
provides power conversion and processes the outputs from the
CCD.   This unit also provides the interface with the
Telemetry and Telecommand Subsystem.

A summary of the sensor characteristics is given in Table 1.

The sensor is mounted on EUTELSAT such that its line of
sight is orthogonal to the Sun pointing direction in ESR.
This simplifies the baffling requirements and facilitates
acquisition of Canopus, which is located close to the
southern ecliptic pole.

Table 1
FRS Characteristics

MASS   2.2 kg

POWER   3.5 W

FIELD OF VIEW   30° x 40°

RESOLUTION   0.1°

<u>Sensor Operation</u>

The FRS does not provide data to the Attitude Control
Subsystem, but interfaces directly with the telemetry and
telecommand subsystem.   A serial digital telecommand word
is provided which enables selection of

o   sensor gain

o   CCD thresholds

o   field of view

The four sensor gain settings (20, 40, 60 and 80) are used
in conjunction with a choice of 15 upper and lower threshold
levels on the CCD output to limit the sensor response to a
range consistent with the target star.   In addition areas
of the field of view can be selectively enabled to ensure
that only one bright star is processed in each scan of the
CCD outputs.   The selectable field of view areas are shown
in Figure 2.

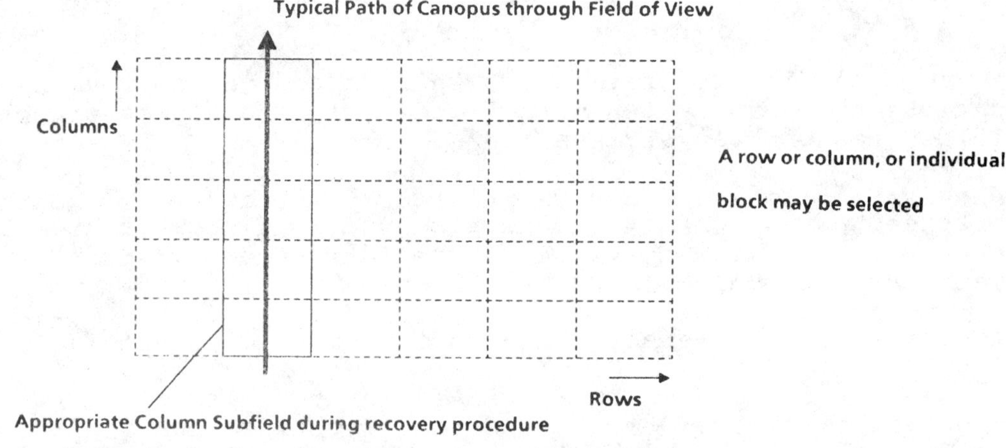

Fig. 2   Field of View Organisation

Three serial digital telemetry words are provided, one of
which is used to verify the configuration of the sensor as
set up by telecommand.   The other two give the CCD
co-ordinates of the first bright star to be encountered
during the scan of the enabled section of the field of view.
This data may then be used to determine the phase angle of
the spacecraft rotation.   The CCD is read out every 289ms,
but because of limitations on telemetry only the most recent
sighting of one star is accepted for transmission in each
19.2 second telemetry format.

The telemetry data transmitted is sufficient to define the direction of the star when observed, but while the spacecraft is rotating the time of observation is also required to determine the spacecraft rotation phase angle. It was intended to provide this by counting the number of CCD read-outs between the observation and its inclusion in telemetry. Unfortunately only after installation on the spacecraft behind a reflector (at the position of a balance mass) was this counter found not to work correctly. In view of the time scale and the experimental nature of the sensor in an operational mission it was not possible to introduce the minor wiring change necessary to correct this problem before flight.

<u>Fast Recovery Procedure</u>

The essence of the fast recovery procedure is shown in Figure 3.

This procedure developed by British Aerospace has been shown by simulation to generate re-acquisition of the nominal pointing mode in not more than one and a half hours.   This is a considerable improvement over the conventional method.

FRS DEVELOPMENT

The Fast Recovery Sensor experiment was proposed in February 1986 and the flight model was delivered for integration on the Eutelsat 5 spacecraft in February 1987.   The activities up to launch in July 1988 are indicated in Figure 4.

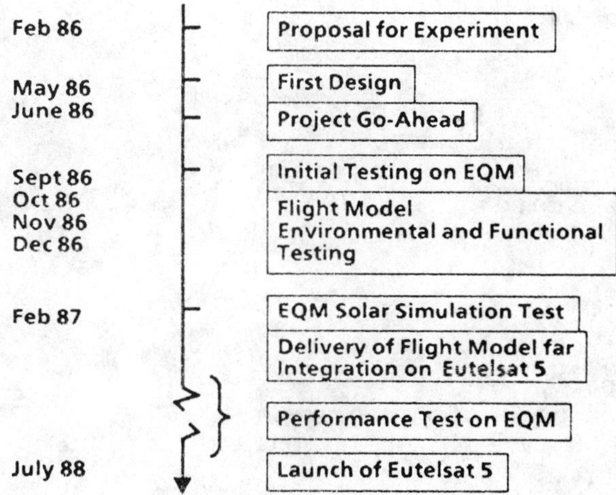

Fig. 4 Development History of the FRS

Spacecraft initially
Sun pointing.

Rotate about the Sun
Line until the FRS
sees CANOPUS.

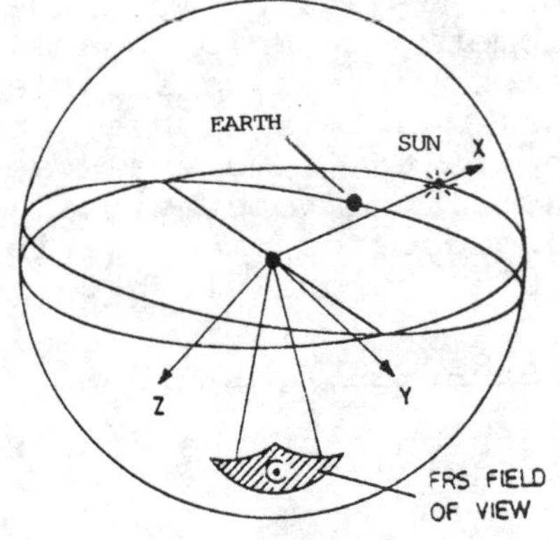

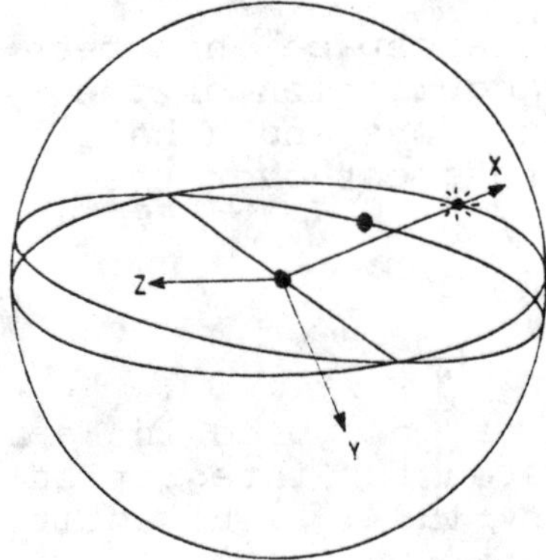

Run up wheel to give
momentum bias about
the y-axis.
Reoreint the y axis
normal to the Sun/Earth
plane.

Slew about the y-axis
until the Earth is
seen by the z-facing
IR sensor.

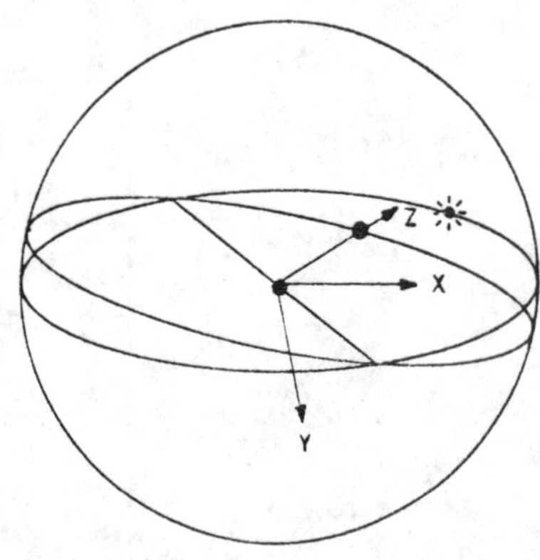

Finally, erect the y-axis
by a precession manouevre
to achieve the operational
attitude

Fig. 3 Outline Recovery Procedure

ESTEC was responsible for the overall design engineering and
integration of the sensor, the digital electronics and the
spacecraft interfaces.   The sensor optical head was
developed by Fisba Optik AG in Switzerland and Aqua TV in
Germany defined the analog electronics concepts based on
their slow-scan TV experience.

The optical system incorporates a six element lens system
made of radiation hardened glass with a fused silica window
bandpass filter.   Optical design was driven by the 25
degrees field of view half cone angle requirement, the
desired spectral range and the need to maximise collecting
area.   A key constraint however was the eight month time
scale for delivery, meaning that the design was restricted
to the radiation hardened glass types chosen at the start of
the development.
The FRS electronics using discrete, hard wired components
are shown in simplified block diagram form in Figure 5.
This simple approach is possible because of the minimal
requirements on internal processing of star observations.

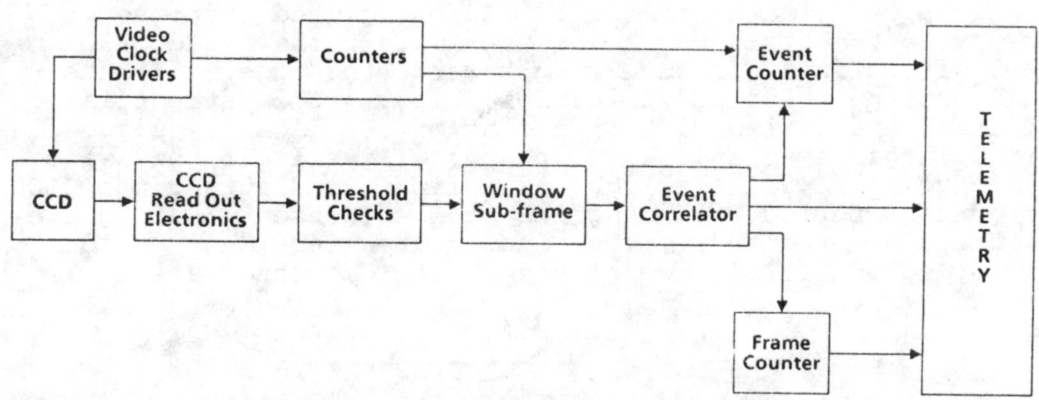

Fig. 5   FRS Functional Block Diagram

The FRS provides no active thermal control for the CCD which
is thermally isolated except for heat received through its
electrical connections and dissipated through the optics.
Resulting from thermal analysis carried out by MBB/ERNO
thermal straps were added between the CCD and the
electronics, and the CCD and a radiator surface to reduce
variations in the CCD temperature while Earth pointing.
During a recovery procedure the CCD temperature is expected
to be held between -9° and +9°C.

PRE-FLIGHT TESTS

After initially confirming the ability of the FRS to respond
to bright stars at the Metzelen Astronomical Observatory in
Switzerland with the help of Dr. Trefzger of the University
of Basel, more detailed performance testing of the sensor
was carried out in the ESTEC optical laboratory. In
addition to general investigation of the sensor performance
the implications on threshold selection were considered
because of the importance of this in ensuring that useful
data is included in telemetry.

Laboratory tests were carried out both with uniform
illumination of the Field of View and with a point source
giving rise to an image on the sensor CCD of approximately
one pixel in diameter. The intensity of this image was
made comparable to that of bright stars, as observed at
ESTEC, through the use of a calibrated detector.

Three effects contribute to the sensor response to a given
star. Firstly the maximum pixel response depends on
whether the image is centred in a single pixel or at a cross
over between pixels. This variation, for a static image,
may be up to a factor of four. Secondly, the response
depends on the position of the star within the field of
view, with a fall off occurring towards the edge. Figure 6
indicates the response variations found. This effect is
primarily associated with spreading of the light received
over a larger area in the CCD plane. Finally if the image
is moving the available light is again spread during the CCD
integration period, reducing the peak response. During the
search phase of a recovery procedure the star image would
move by approximately two pixels during one integration
period.

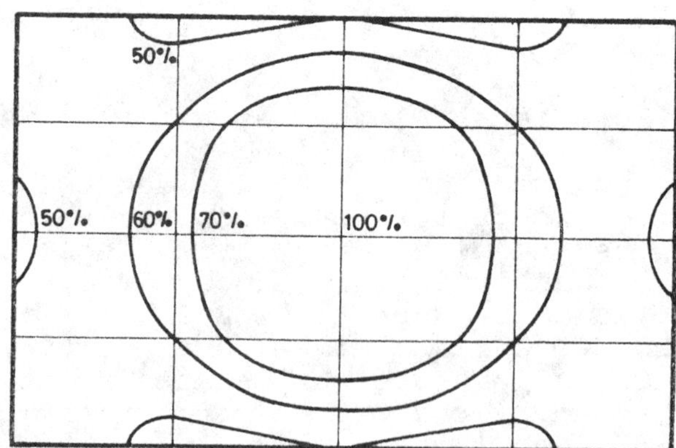

Fig. 6 Response Variation Over the Field of View

Analysis indicates that these effects are not independent
and that the reduction in response from the maximum for an
ideal stationary centred image due to all these effects
together can be less than that expected by considering each
effect separately.   This is beneficial to the operation of
the sensor as a single low threshold must be used for
detection as a star moves through the field of view.

The final stage of performance testing was carried out at
ESTEC observing real stars, focussed on individual pixels.
The results of this work were used as a basis for setting up
a catalogue of expected star responses for use during flight
tests.

DRIFT ORBIT TESTS

As part of the normal sequence of operations the Eutelsat 5
spacecraft was held in a sun pointing attitude for some time
following the apogee burn, using the same attitude control
loops as would be used in the recovery manoeuvre.   This is
a quiet period with little ground command activity and
therefore provided an excellent opportunity to test the
performance of the Fast Recovery Sensor under its intended
operating conditions.

In this situation the spacecraft is rotating at
approximately 0.8°/s about the sun line and based on
Pre-flight tests it was expected that the objects listed in
Table 2 would be visible.

Table 2
EXPECTED OBSERVATIONS

OBJECT	Expected Max Pixel Response (mV)	FRS CONFIGURATION					
		Gain 20			Gain 40		
		Thres-hold 1 (0 mV)	2 (9.8mV)	3 (25.4mV)	1 (9.8mV)	2 (17.6mV)	3 (25.4mV)
CANOPUS	19-29	**	*	*	**	**	*
ARCTURUS	16-24	**	**		**	*	
MOON		**	**	**	**	**	**
O CET	8.7-13.1	**	*		*		
♉ CRUX	6.3-9.5	**					

**   Always observed   *   Maybe observed

The tests were organised as a systematic variation of the
threshold levels, starting with maximum sensitivity and
increasing the threshold until no stars were visible.
Approximately three revolutions of the spacecraft were
allowed with each set up condition, so that responses over
subsequent revolutions could be compared and consistent
responses separated from isolated, possibly noise responses.
This exercise, as well as the matching of sensor data to
catalogued stars, was made more difficult however by the
missing timing information in telemetry.   Figure 7 shows
how this can introduce up to +/-7.5 degrees uncertainty in
the direction of a sighting.

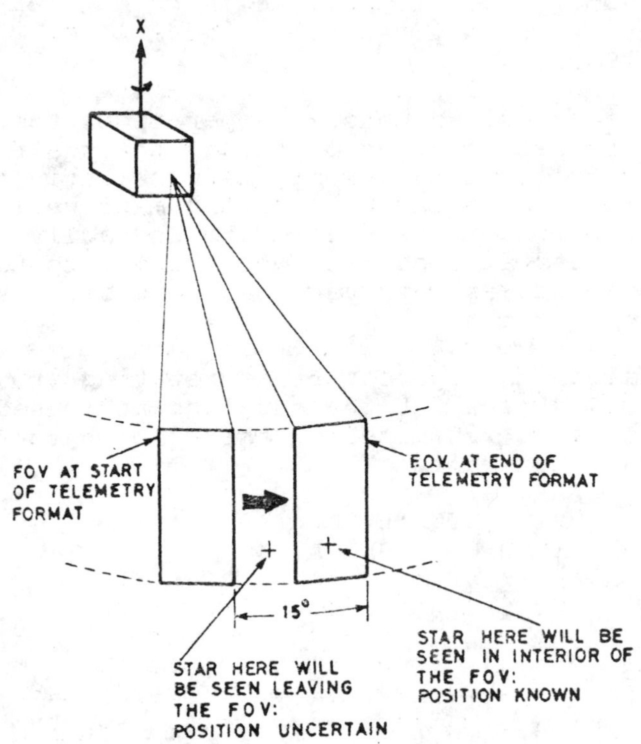

Fig. 7 Drift Orbit Operation of the FRS

Ideally when used for recovery purposes a column subfield of
the complete field of view would be used and the star of
interest would be seen travelling across the CCD while as
many other objects as possible would be rejected.
Unfortunately at the time of the test it was found that
Canopus would travel through the field of view close to a
boundary between two such columns and so this approach was
not appropriate.   A set of overlapping columns would have
been a more flexible way of organising the sensor field of
view.   Instead tests were carried out using a row subfield
so that fewer sightings of Canopus and more sightings of
other objects would be expected.

The tests were timed so that as the threshold value
approached the expected levels for useful observations the
Earth no longer entered the field of view. However the
Moon was seen, and because of its large diameter excited
many pixels. The count of the number of pixels excited is
included in telemetry in order to discriminate between large
objects and stars.

Despite the position uncertainties it was clear that when a
gain of 40 was used with lower threshold 1 or 2 Canopus,
Arcturus and the Moon were consistently detected on each
revolution of the spacecraft (Figure 8). The sightings
were all within the uncertainty level of the catalogued
positions, and were in line with expectations.

Although stars with slightly lower expected responses do not
seem to have been detected there were a considerable number
of other detections. About 10 of these appear consistent
with sightings of lower intensity stars, although only
occurring on single revolutions of the spacecraft. An
analysis of the number of apparent correspondences which
could occur by chance suggested that these sightings were
probably due to noise responses.

Excluding stars of variable intensity and those very close
to the edge of the field of view, the brightest star not
seen was Gamma Crux. The image of this would have passed
through the centre of the CCD. A limit on the sensitivity
of the Fast Recovery Sensor has therefore been found for the
recovery manoeuvre situation close to the expected sensor
performance.

Some anomalies were found in the sensor response. In
general there were many spurious responses. In particular
while with a fixed gain level an increase in the low
threshold level from 1 to 2 led, as would be expected, to
fewer explainable pixel excitations, there was an increase
in the number of unexplained responses. This indicates a
noise problem in the electronics. Also when the sensor
gain was set at 20 rather than 40 the behaviour did not
correspond to expectations, only the Moon being detectable.
Some problems of thermal sensitivity in the amplifier had
been found in ground test which it is believed explain this
problem.

Despite these anomalies however the sensor did demonstrate
its ability to detect Canopus consistently with an
appropriate selection of gain and thresholds. Had the
missing timing data in telemetry been available in order to
relate observations to precise directions in space the
sensor could have been used to perform the newly developed
recovery procedure. As installed on the Eutelsat 5

satellite however a   procedure incorporating an alternative
low speed acquisition sequence would be required in order to
reduce uncertainties in detection of Canopus.

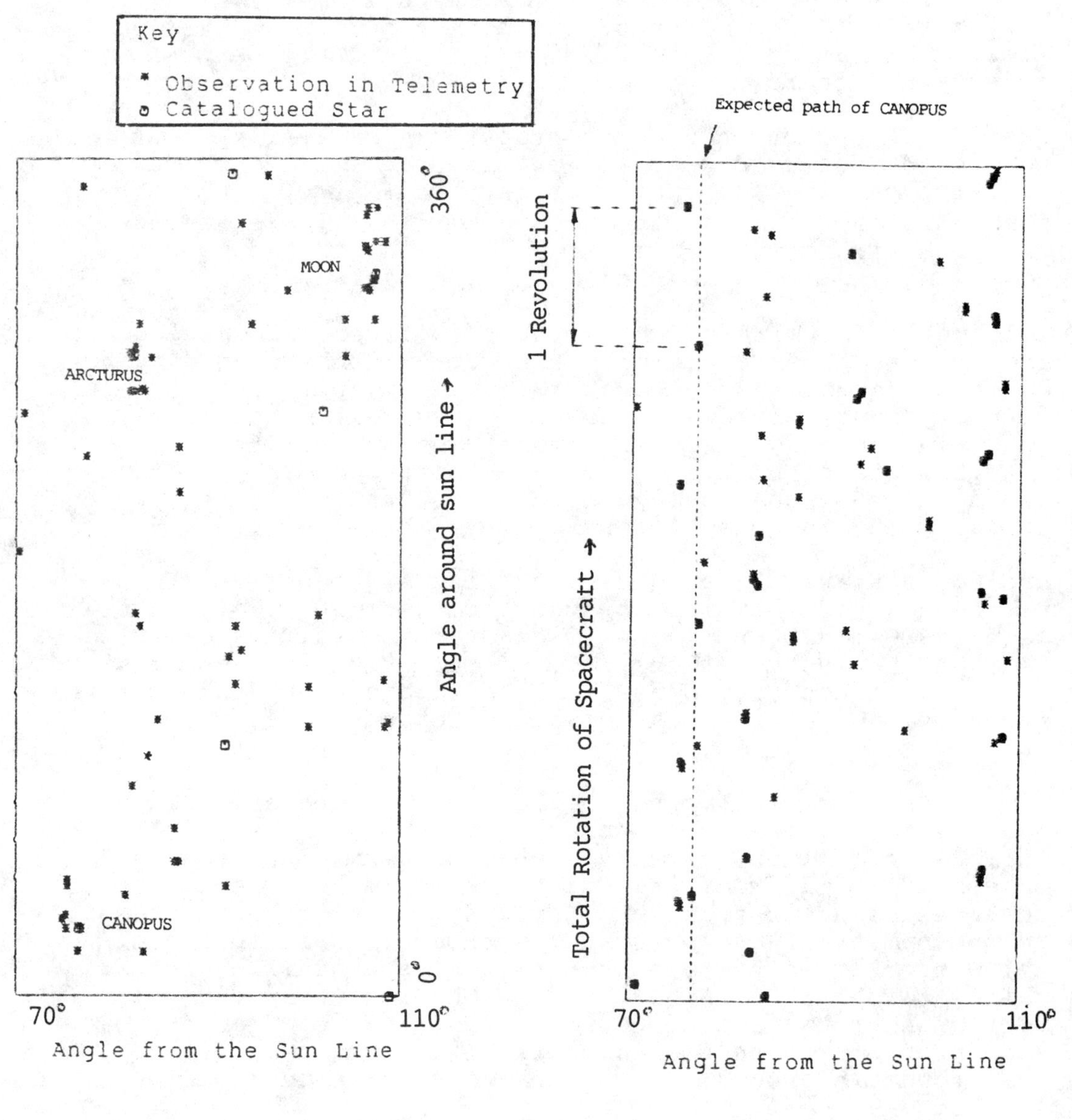

Deduced Directions of
Observed stars shown with
Catalogued Brightstars

Observations over 5
revolutions, with
possible catalogue
correspondancies

Fig. 8   Sample Drift Orbit Results

500

ON-STATION TESTS

Although the design and mounting of the FRS was optimised
for the recovery manoeuvre situation, in the nominal mission
there is no further planned opportunity to test the sensor
performance under these conditions after entry into normal
Earth pointing attitude control.   Nevertheless useful
performance data can also be obtained when the spacecraft is
Earth pointing. The sensor is then looking into the
southern hemisphere with the centre of its field of view
canted by 30 degrees from the south pointing spacecraft
y-axis toward the Earth.   Consequently the sensor sweeps
once per day over a band somewhat wider than 30 degrees
around the southern pole.

Various observations have been made starting in the
commissioning phase when the spacecraft was drifting towards
its nominal position. These measurements are part of a
continuing long term test program aiming to:

o   Measure sensor performance (sensitivity, noise
      characteristics) under stationary conditions

o   Investigate possible thermal effects on a daily
      and yearly basis

o   Identify Sun blinding effects (the sensor
      baffle not being designed for this pointing mode)

o   Monitor long term sensor degradation caused by
      radiation.

To support this test programme software has been developed
at ESTEC to ease test planning and assist in the evaluation
of the sensor telemetry data.   A test planning programme
supports the selection of suitable threshold levels and area
inhibition commands in order to detect selected stars.
Evaluation programmes process the raw telemetry data
received, identify real star observations and noise events,
generate statistical performance data and present results in
graphical form.

As an example of the results Figure 9 shows the trace of the
star Alpha pav (m_V=1.94) and Alpha tuc (m_V=2.86, a red star)
in the sensor field of view together with noise events
during an observation period of 5 hours. As threshold
settings and enabled areas have been changed during this
period interruptions to the star traces occur.

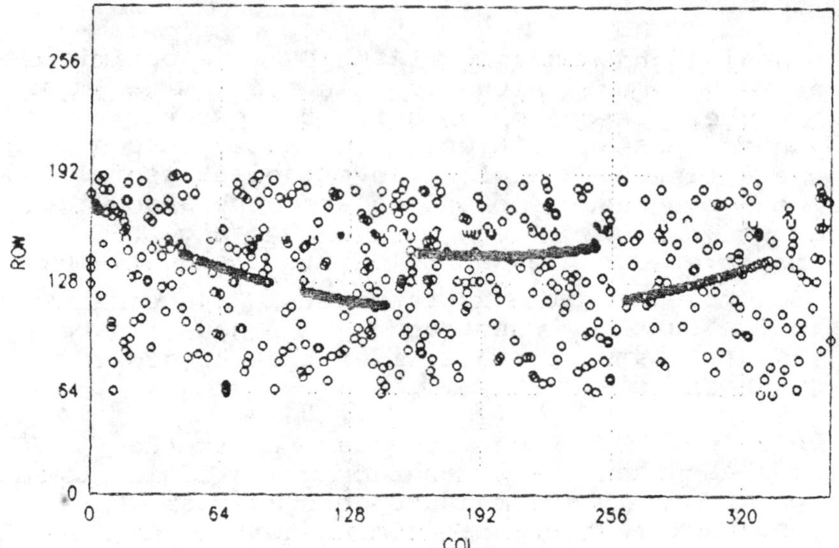

Fig. 9   Traces of ∝PAV and ∝TUC over 5 hours

Figure 10 shows the corresponding row and column address numbers versus time during the observation period. In these plots the Alpha pav events have been suppressed, therefore outside the observation time of Alpha tuc from 9.05h to 10.22h only noise events are displayed.

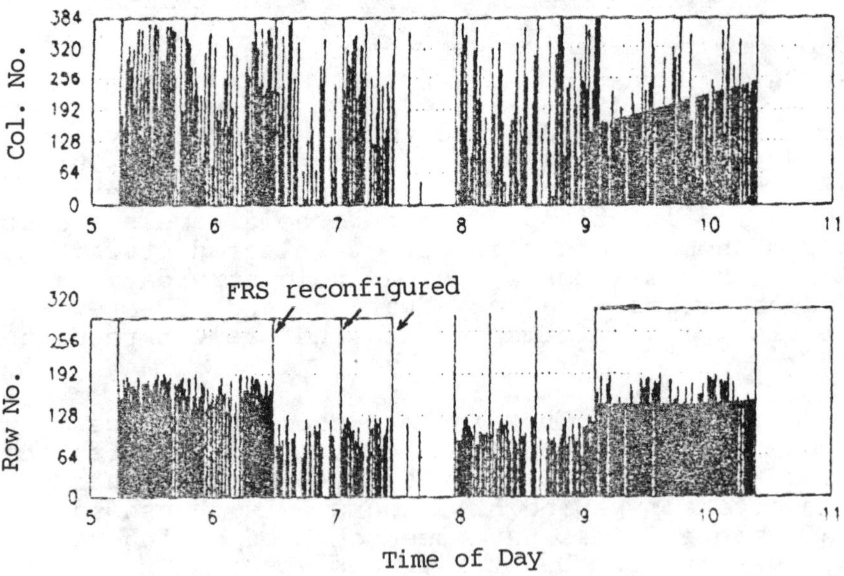

Fig. 10   Row and Column Telemetry

502

CONCLUSIONS

The results of the FRS experiment show satisfactory
performance and the viability of such a simple star sensor
in providing adequate information for attitude
determination.   The sensor as flown does have a number of
problems associated with its rapid development and its
introduction into a running commercial telecommunications
spacecraft programme.   The experience gathered since
launch, and future data on performance variation over the
Eutelsat 5 lifetime are considered to provide a valuable
basis for further development

ACKNOWLEDGEMENTS

A crash programme of flight hardware development can only be
realised through the generous cooperation of all parties
involved.   In particular the authors acknowledge the
cooperation of BAe ECS Project Team led by D. Forder and the
ESA ECS Project team led by J. Durant and G. Fuller and all
engineers at ESTEC and in industry who demonstrated that
rapid production of new, innovative flight hardware is
certainly within the competence of European cooperation.

REFERENCES

1.   C.K. Leong, N. Matthews, L.V. Holtz and A.G. Bird.
      An In-Orbit Demonstration of Recovery from Flat
      Spin for a Momentum Bias Controlled Communications
      Spacecraft OTS Proc. Second International
      Symposium on Spacecraft Flight Dynamics, Darmstadt
      FR Germany Oct. 1986.

INSTABILITY OF GRAVITY GRADIENT SPACECRAFT IN FULL SUN ORBIT: FLIGHT EXPERIENCE FROM THE POLAR BEAR MISSION

Alan Lewis[*] and Arnold Streland[†]

Several gravity gradient stabilized spacecraft have experienced instability when exposed to the full sun condition in orbit (no eclipse). The Polar Beacon and Auroral Research (Polar BEAR) Spacecraft experienced instability and eventually inverted three of the five times it was exposed to the full sun condition. The characteristic oscillations of the spacecraft have a known profile. Several possible perturbation sources and their application to the Polar BEAR spacecraft are discussed. Recommendations of design criteria for future spacecraft to avoid full sun instability, and to recover from instability and possible inversion, are presented.

INTRODUCTION

Polar Beacon and Auroral Research (Polar BEAR) mission is a joint effort between the Defense Nuclear Agency (DNA) and the Air Force Systems Command's Space Test Program (STP). The Polar BEAR spacecraft is a Navy Orbiting Satellite Communications Amatuer Radio (OSCAR) navigation satellite, extensively modified by the Applied Physics Lab (APL) of Johns Hopkins University. The spacecraft was launched on a Scout Launch Vehicle from Vandenberg Air Force Base on 13 November 1986 to a 545 nautical mile altitude (105 minute period), 90 degree inclination, circular orbit. Polar BEAR carries four Department of Defense experiments designed to study the effects of the polar atmospheric environment on communications and radar systems.

The four experiments include the Auroral/Ionospheric Remote Sensor (AIRS), the Beacon Experiment, the Magnetometer Experiment and a communications experiment. The objective of AIRS is to gather day and night multispectral auroral images. The Beacon Experiment measures the effects of the ionosphere on radio frequencies

---

*   The Aerospace Corporation, P.O. Box 92957, Los Angeles, California 90009.

†   Lieutenant, U.S. Air Force Space Division, Los Angeles AFS, P.O. Box 92960 WPC Station, Los Angeles, California 90009.

used in radar and communications. The Beacon L Band
transmitter is also the primary RF link for science
telemetry from all four experiments. The Magnetometer
Experiment measures field aligned magnetic currents in the
polar region while also providing spacecraft attitude data.
The communications experiment investigates methods of
countering the adverse effects of the polar environment on
communications.

SPACECRAFT DESCRIPTION

     The Polar BEAR spacecraft in Fig 1, shows the basic
OSCAR structure and solar panels and the significant
additional structure added to accommodate the four
experiments. The "penthouse" contains the electronics for
the communications experiment. The upper experiment deck
provides mounting area for the AIRS sensor and the Beacon
and antennas.   The additional pedestal provides mounting
area for the magnetometer, Beacon electronics, and a
momentum wheel and its associated electronics. An overview
of the spacecraft subsystems is presented below. More
details are available in Ref 1.

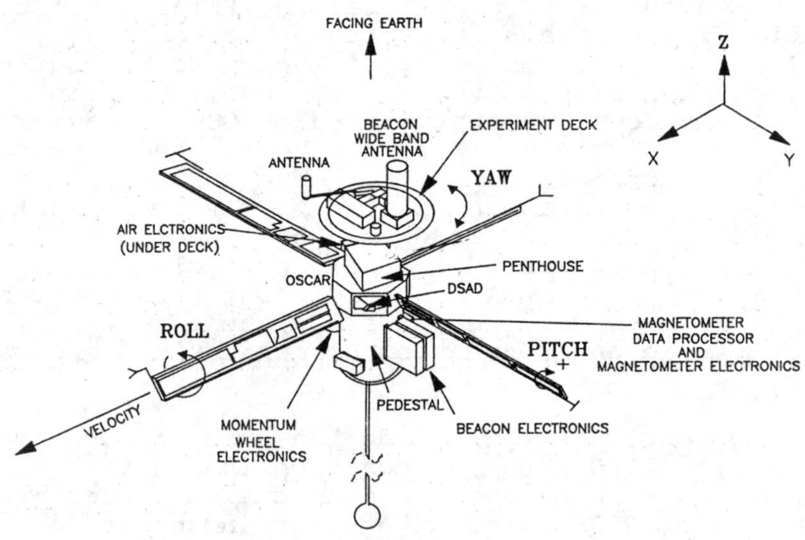

CONCEPTUAL VIEW OF POLAR BEAR SATELLITE CONFIGURATION

Fig 1 Polar BEAR Spacecraft

Four solar panels charge 8, 12 Amp Hour Nickel Cadmium
batteries which provide 35-50 W orbit average power for the
spacecraft systems and the experiments. Primary RF
communication for spacecraft health and status data is
provided by a 150 MHz transmitter. A 400 MHz transmitter is
available as a back-up for transmitting health and status
and science data.

There is no on-board data storage. All spacecraft and
experiment data must be collected and downlinked while the
satellite is in view of a ground station. Locations of the
Naval Astronautics Group (NAVASTROGRU) ground stations used
to control the satellite and the Stanford Research
Institute (SRI) ground stations operated for the Defense
Nuclear Agency, used to collect experiment data, are shown
in Fig 2.

ATTITUDE DETERMINATION AND CONTROL SYSTEM

The Polar BEAR spacecraft is three axis stabilized to
the local vertical coordinate system. Experiment operations
requirements dictate that Polar BEAR be stabilized to
within +/- 10 degrees in the roll, pitch, and yaw axes.
Roll and pitch stability is provided by a 60 foot gravity
gradient boom. The boom is made out of two interlocking
strips of beryllium-copper tape which are unwound by an
electric motor and "zippered" together to form a semi-rigid
tube. Yaw stability is provided by a constant speed
momentum wheel. The wheels spin axis is parallel to the
spacecraft Y axis. The wheel spins at 2000 rpm to provide
a momentum of 1.77 slug-ft^2/sec.

Additional passive stabilization is provided by an eddy
current damper. The eddy current damper is located at the
end mass of the gravity gradient boom. It provides a
damping coefficient of 0.0051 ft-lb-s/rad. The two
magnetic hysteresis rods, located on the -Z side of the +X
and +Y solar panels, provide magnetic damping during despin
but their contribution to normal on-orbit attitude control
is minimal.

Attitude knowledge is obtained by the triaxial
magnetometer, mentioned previously as one of the spacecraft
experiments, and a sun sensor. The magnetometer sensor is
located on the outer tip of the -X solar panel. It provides
magnetic field measurements in three axes with an accuracy
of +/- 16 millioersted. The digital solar attitude

POLAR BEAR GROUND STATIONS
DNA GROUND STATIONS
(DATA COLLECTION)
APL GROUND STATION
NAVASTROGRO GROUND STATIONS
(SPACECRAFT CONTROL)

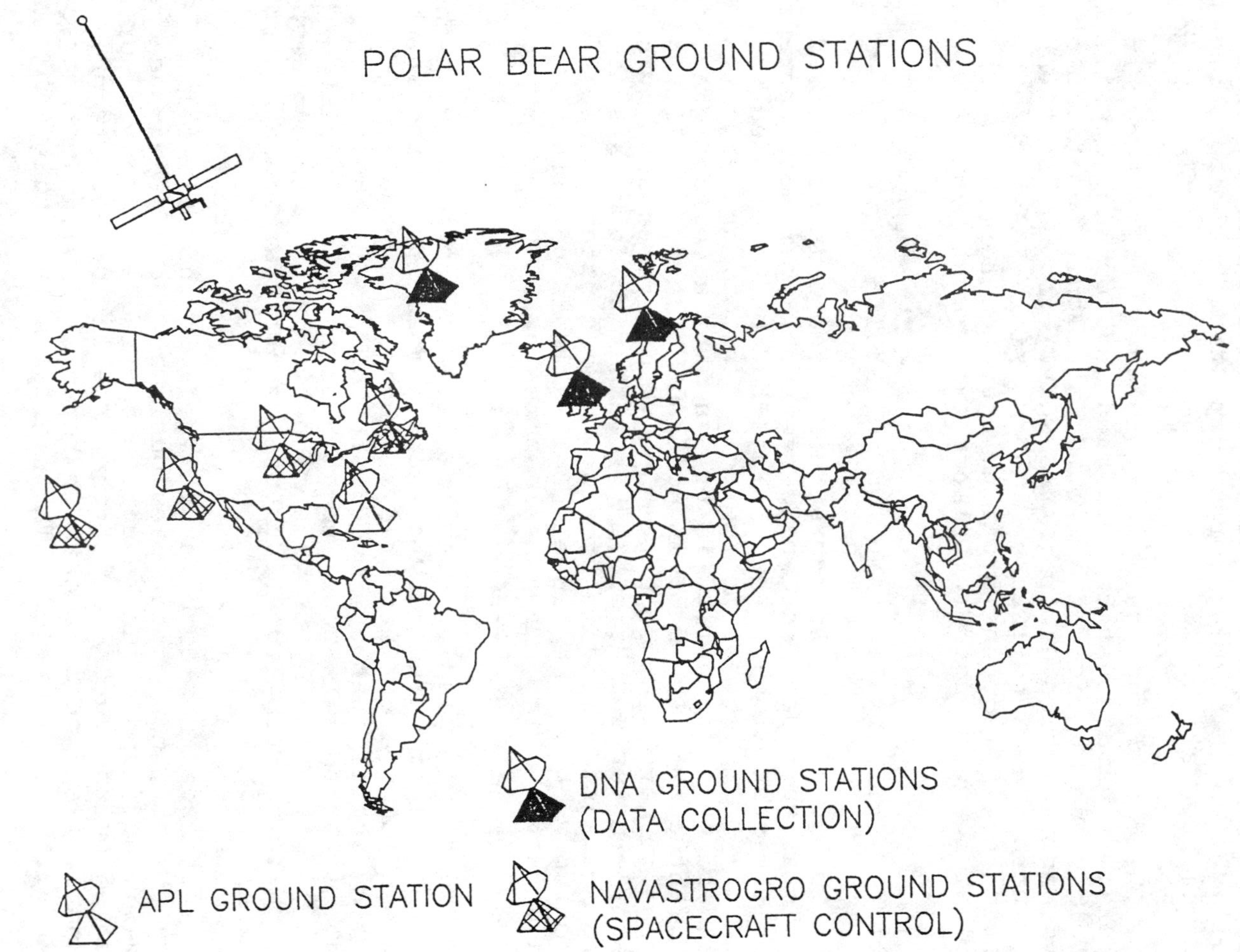

FIGURE 2

detector (DSAD) has three sun sensors which measure the
orientation of the sun with respect to the spacecraft to an
accuracy of +/- 1 degree. A mathematical description of the
attitude control system will be given later in the paper.

INITIAL ORBIT OPERATIONS AND ANOMALY EXPERIENCE

     Following the successful 13 Nov 86 launch, the
spacecraft and its orbit insertion motor were spun up to 140
rpm prior to separation from the Scout fourth stage. After
separation and orbit insertion stage burn, the spacecraft
and orbit insertion stage were despun with a yo-yo despin
system. As the yo-yo wires unwound the solar panels
deployed from their stowed launch configuration. The
spacecraft then separated from the orbit insertion stage.
Details of orbit operations and anomalous performance are
provided in Ref 2.   The events are summarized below.

     During the first pass over the APL satellite control
facility the momentum wheel was turned on. When the yaw
angle < 30 degrees and the spin rate < 1 revolution/hour the
z-coil was commanded on to initiate magnetic stabilization.
Magnetic stabilization was achieved one day after launch.
The gravity gradient boom was extended three days after
launch when the z axis pointing was within +/- 15 deg. The
z-coil was turned off once gravity gradient capture had been
achieved.

     Oscillations in all three axes were reduced to +/- 10
degrees within five days after extension. Control of Polar
BEAR was turned over to NAVASTROGRU 30 days after launch,
following a complete spacecraft and experiment system
check-out.

     Polar BEAR entered the minimum sun period of its orbit
15 days after launch. The spacecraft remained basically
stable within specification for the 55 days following entry
into the minimum sun period. During this time however two
significant anomalies were noted. First as the spacecraft
passed from eclipse into sun illumination, the x axis of the
magnetometer indicated bending of the gravity gradient boom.
Also during this time there was a steady increase in yaw
angle oscillations. By day 44 yaw angle oscillations had
become greater than the specified +/- 10 degrees.

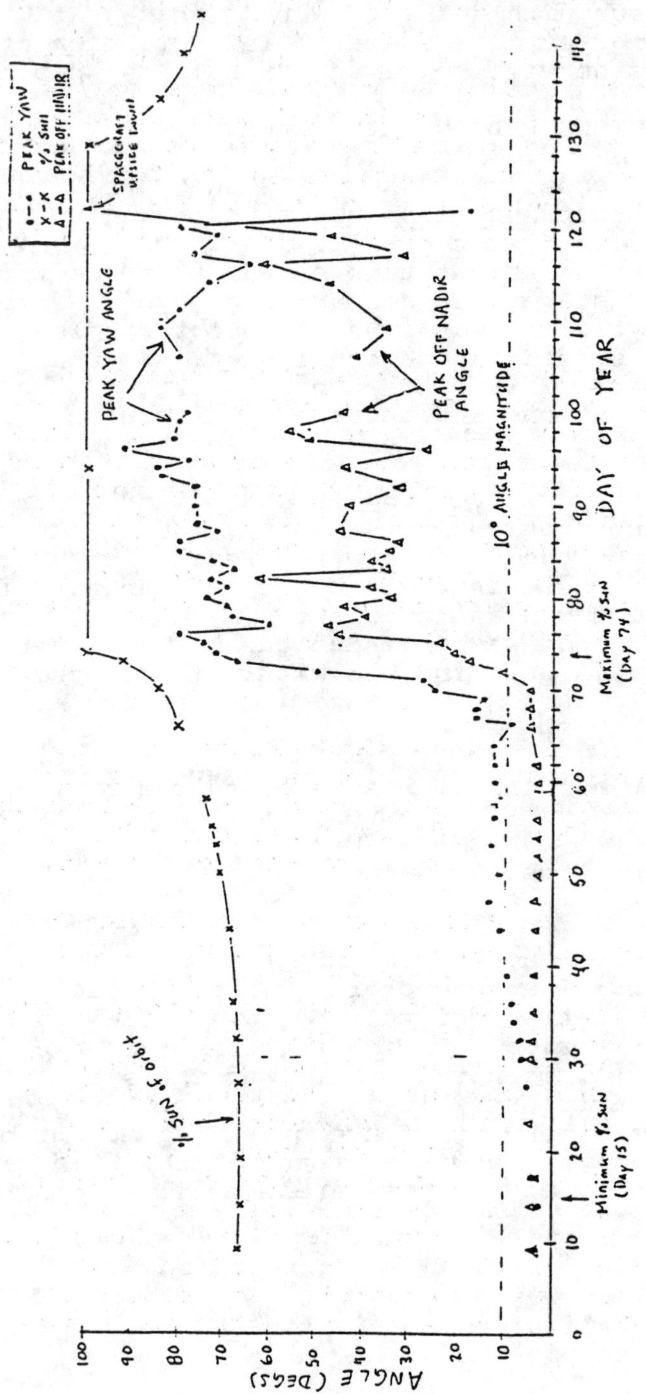

Fig 3

Spacecraft Performance Prior to Inversion (Ref 2)

The spacecraft entered the full sun portion of its
orbit for the first time on day 70. Large attitude
excursions were observed within a week of entry into the
full sun orbit. Peak excursions of 50 degrees in yaw, 30
degrees in pitch, and 10 degrees in roll were observed by
day 75. By day 121 the spacecraft appeared to be tumbling.
Exact knowledge of spacecraft behavior is limited because
its motion could only be observed in "snapshots" over the
ground stations shown in fig 2 due to the lack of on board
data storage.

<u>Inversion</u>

Fig 3 shows a diagram of Polar BEAR's anomalous behavior
prior to inversion. Between day 121 and day 123, 16 Mar 87,
the spacecraft yaw motion reduced to less than 20 degrees
but the pitch oscillations increased to greater than 90
degrees. On day 123 it was confirmed that the satellite had
achieved gravity gradient capture in the inverted position.[2]
At this point all experiments were turned off since they
were now facing away from the earth. The spacecraft was
placed in a power saving configuration with only essential
spacecraft subsystems operating.

REINVERSION

Polar BEAR remained inverted and passed into the
eclipse period of its orbit in early May 1987. During the
eclipse period the satellites attitude oscillations damped
to within original performance specifications. At this
point an attempt was made to reinvert the spacecraft.

Satellite control was transferred from NAVASTROGRO to
APL for the reinversion maneuver. The procedure for
reinversion consisted of the momentum wheel being turned off
and allowed to spin down for one orbit. On the following
pass over the APL satellite control facility the wheel was
commanded on again and began to spin up to its operating
speed of 2000 rpm. This spin up of the wheel exerts a
torque in the pitch axis which is greater than the gravity
gradient restoring torque causing the satellite to tumble
about the pitch axis.[3] This operation has a 50 percent
chance of flipping the satellite right side up since the
gravity gradient boom could capture in either the right side
up or the up side down position. The third attempt to
reinvert the satellite was successful. Polar BEAR captured
in the proper orientation on 22 May 87 after about two
months in the inverted position.

The Polar BEAR spacecraft has inverted twice since its first inversion and recovery. The second inversion occurred on 15 April 1988. Following inversion the spacecraft continued to oscillate and recaptured in the right-side-up position on 21 April. The third and most recent inversion occurred on 2 October 88. This time the spacecraft had to be reinverted by ground command. The reinversion procedure was similar to that of the first reinversion. Operations were conducted by NAVASTROGRU using procedures developed by APL.

ANALYSIS OF SPACECRAFT BEHAVIOR

Polar BEAR's attitude anomalies occurred in spite of steps taken in the spacecraft design to avoid the type of instability the spacecraft experienced indicating the exact cause of the attitude perturbations is still not certain. To understand the possible causes of the instability, first consider the equations that govern the normal motion of the spacecraft.

<u>Rigid Body Equations of Motion</u>

The Euler rigid body equations of motion were written for Polar BEAR which included the wheel momentum and gravity gradient torques. It is further assumed that the center of mass is in a circular orbit and oblate earth effects are negligible. Later, the rigid body assumption will be relaxed and other effects modeled. The equations of motion can be derived from the following momentum equation written in terms of the angular velocity vector of the body $\bar{\omega}_b$ and the system momentum $\bar{h}$, in vehicle coordinates. Vectors are noted with a bar ($\bar{h}$). Dyadics are noted with a double bar ($\bar{\bar{I}}$). Magnitudes of vectors and all other scalars are given with no bar.

$$^{d}/_{dt}[\bar{h}] = \dot{\bar{h}} + \bar{\omega}_b \times \bar{h}$$

$$= \Sigma \bar{M} \tag{1}$$

Where $\Sigma \bar{M}$ is the sum of external moments. The system angular momentum consists of the angular momentum of the body, $\bar{h}_b$, and of the momentum wheel, $\bar{h}_w$.

$$\bar{h} = \bar{h}_b + \bar{h}_w \tag{2}$$

$$\bar{h}_b = \bar{\bar{I}}_{bo} \cdot \bar{\omega}_b \tag{3}$$

$$\bar{h}_w = \bar{\bar{I}}_w \cdot \bar{\omega}_w \qquad\qquad (4)$$

$$\bar{\bar{I}}_b = \bar{\bar{I}}_{bo} + \bar{\bar{I}}_w \qquad\qquad (5)$$

$\bar{\bar{I}}_{bo}$ and $\bar{\bar{I}}_w$ are the inertia dyadics of the body and wheel respectively. $\bar{\bar{I}}_b$ is the inertia dyadic of the body including the momentum wheel. The term $\bar{\omega}_w$ is the angular velocity of the momentum wheel. Using Eqs. (2), (3), (4), and (5) in Eq. (1) yields the following differential equation.

$$\bar{\bar{I}}_b \cdot \dot{\bar{\omega}}_b + \bar{\bar{I}}_w \cdot \dot{\bar{\omega}}_w + \bar{\omega}_b \times (\bar{\bar{I}}_{bo} \cdot \bar{\omega}_b + \bar{h}_w) = \Sigma \bar{M} \qquad (6)$$

We wish to write the equations of motion with respect to an orbital reference frame, x_o, y_o, and z_o which is aligned with the local vertical and the velocity vector of the spacecraft as shown in Fig. 4. We shall define the angular velocity of the body with respect to orbital coordinates as $\bar{\omega}_{b/o}$ and the angular velocity of the wheel with respect to the body as $\bar{\omega}_{w/b}$.

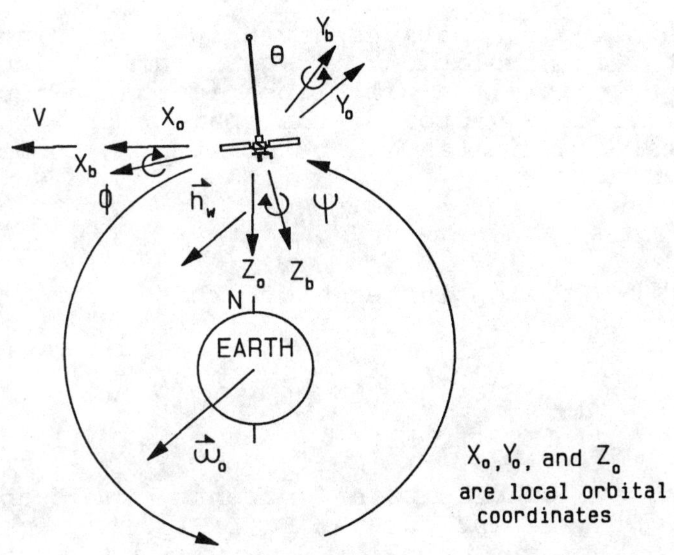

Fig. 4 Spacecraft Orbital Configuration

$$\bar{\omega}_{b/o} = \bar{\omega}_b - \bar{\omega}_o \qquad (7)$$

$$\bar{\omega}_{w/b} = \bar{\omega}_w - \bar{\omega}_b - \bar{\omega}_o \qquad (8)$$

Substituting Eqs. (7) and (8) into Eq. (6) yields the following.

$$\bar{\bar{I}}_b \cdot (\dot{\bar{\omega}}_{b/o} + \dot{\bar{\omega}}_o) + \bar{\bar{I}}_w (\dot{\bar{\omega}}_{w/b}) +$$

$$(\bar{\omega}_{b/o} + \bar{\omega}_o) \times [\bar{\bar{I}}_b \cdot (\bar{\omega}_{b/o} + \bar{\omega}_o) + \bar{\bar{I}}_w \cdot (\bar{\omega}_{w/b})] = \Sigma \bar{M} \qquad (9)$$

The term $\bar{\omega}_o$ is constant since the orbit is circular. The spin rate of the wheel, $\bar{\omega}_{w/b}$ is constant by design. Thus $\dot{\bar{\omega}}_o = \dot{\bar{\omega}}_{w/b} = 0$. The roll, pitch, and yaw angles (ϕ, Θ, and Ψ) are referenced to the vehicle coordinates as shown in Fig. 4. The x, y, and z axes are nominally the principal axes. In this configuration, the momentum vector of the momentum wheel points in the negative y direction. Gravity gradient torques (with orbital radius vector, $\bar{r}$, and earth graviy constant, μ) are given as[4]:

$$M_{gg} = 3\mu/r^5 [\bar{r} \times \bar{\bar{I}}_b \cdot \bar{r}] \qquad (10)$$

These nonlinear equations given in Eqs. (9) and (10) are used later in simulation. To gain insight into the motion of the vehicle, we will linearize these equations. The angular velocity vector of the spacecraft with respect to the orbital coordinates can be written in terms of roll, pitch, and yaw.

$$\bar{\omega}_{b/o} = d/dt[\phi \quad \Theta \quad \Psi]^T \qquad (11)$$

Eq. (10) can be written in terms of orbital rate, roll angle, and pitch angle.

$$\bar{M}_{gg} = 3\omega_o^2 [(I_z - I_y)\phi \quad (I_z - I_x)\Theta \quad 0]^T$$

$$= [M_x \quad M_y \quad M_z]^T \qquad (12)$$

Ignoring all other external moments, two stable equilibrium points exist for the Eq. (9) with moments given by Eq. (10).[5] One configuration is shown in Fig. 4 where ϕ, Θ, and Ψ are all zero. This is the configuration in which the spacecraft is designed to operate, with the antennas pointing toward the earth. The other stable configuration is given by a 180 degree rotation about the y-axis. Using small angle approximations, we can write Eqs. (9) and (10) in three scalar equations of motion.   Linearizing about the point $\Phi = \Theta = \Psi = 0$ yields the following equations. Let $\bar{h}_{w/b} = \bar{\bar{I}}_w \bar{\omega}_{w/b}$.

$$\dot{\Phi} - [(1+R_x)\omega_o - h_{w/b}/I_x]\dot{\Psi} + (-4R_x\omega_o^2 + \omega_o h_{w/b}/I_x)\Phi = 0 \qquad (13a)$$

$$\ddot{\Theta} + 3\omega_o^2 R_y\Theta = 0 \qquad (13b)$$

$$\dot{\Psi} - [(R_z-1)\omega_o + h_{w/b}/I_z]\dot{\Phi} + (R_z\omega_o^2 + \omega_o h_{w/b}/I_z)\Psi = 0 \qquad (13c)$$

where:
$$R_x = (I_z - I_y)/I_x$$
$$R_y = (I_x - I_z)/I_y$$
$$R_z = (I_y - I_x)/I_z$$

Eq. (13b) is a simple harmonic oscillator that is stable for $I_x > I_z$. Eqs. (13a) and (13c) are coupled. Their solution and stability can be found using state space formulation. Define the following state space vector $\bar{X}$.

$$\bar{X} = [\Phi \quad \Psi \quad \dot{\Phi} \quad \dot{\Psi}]^T \qquad (14)$$

Eqs. (13a) and (13c) can be expressed in state space form.

$$\dot{\bar{X}} = A\bar{X} \qquad (15)$$

where:
$$A = \begin{bmatrix} 0 & 0 & 1 & 0 \\ 0 & 0 & 0 & 1 \\ a_{31} & a_{32} & 0 & a_{34} \\ a_{41} & a_{42} & a_{43} & 0 \end{bmatrix}$$

$$a_{31} = 4R_x\omega_o^2 - \omega_o h_{w/b}/I_x$$
$$a_{32} = 0$$
$$a_{34} = (1+R_x)\omega_o - h_{w/b}/I_x$$
$$a_{41} = 0$$
$$a_{42} = -R_z\omega_o^2 - \omega_o h_w/b/I_z$$
$$a_{43} = (R_z-1)\omega_o + h_w/b/I_z$$

This system can be shown to be marginally stable for $I_y > I_x > I_z$. That is, nonzero initial conditions will cause the system to oscillate about the equilibrium. Note that this system has been derived without damping, but will be included in the augmented equations of motion. Proper damping will allow the system to reach a steady-state value at equilibrium. The solution of Eq. (15) takes the following form.

$$\phi(t) = (c_1\phi_o + c_2\dot{\psi}_o)\cos\omega_1 t + (c_3\psi_o + c_4\dot{\phi}_o)\sin\omega_1 t$$
$$+ (c_5\phi_o + c_6\psi_o)\cos\omega_2 t + (c_7\psi_o + c_8\dot{\phi}_o)\sin\omega_2 t \qquad (16a)$$

$$\psi(t) = (d_1\phi_o + d_2\dot{\psi}_o)\sin\omega_1 t + (d_3\psi_o + d_4\dot{\phi}_o)\cos\omega_1 t$$
$$+ (d_5\phi_o + d_6\dot{\psi}_o)\sin\omega_2 t + (d_7\psi_o + d_8\dot{\phi}_o)\cos\omega_2 t \qquad (16b)$$

The physical system parameters for Polar BEAR are given in the appendix. The frequencies ω_1 and ω_2 are the square roots of the imaginary part of the eigenvalues of matrix A. The coefficients c_1 through c_8 and d_1 through d_8 are found by substituting Eqs. (16a and b) into Eq. (15).

The rigid body mode of frequency $\omega_1 = 0.015/\text{sec}$, the fast roll-yaw mode, is present prior to inversion of the spacecraft. The yaw amplitude is about 5.6 times as large as the roll amplitude. This mode is sensitive to an impulse to roll. This mode can also be excited from an external moment that makes the term $a_{41}>0$ or $a_{32}<0$. Either would give the system positive roots thus making it unstable. A physical interpretation will be given later.

<u>Augmenting Equations of Motion</u>

The equations of motion given by Eqs. (9) and (10) explain much of the observed behavior, but do not predict inversion. In this section, the modeling of thermal bending and flexibility of the boom are described as well as damping from the tip mass damper. Boom bending is dynamically coupled to the nonlinear equations of motion.

The 60 foot long boom is made of two strips of 0.002 inch thick pre-stressed silver plated beryllium-copper formed in a 0.5 in diameter tube and fastened together by interlocking tabs along the edge.[6] The total weight of this boom is approximately 1 pound. When the boom is heated by solar radiation, one side of the boom heats and expands, causing the boom to bend away from the sun. These deflections occur parallel to the x-y plane in vehicle coordinates. The displacement in the z direction is negligible. The displacement as a function of time is given by the following.

$$x_t = (\hat{s}\cdot\hat{1}_v)(1- \hat{s}\cdot\hat{k}_v)D_t(1-e^{-\beta t}) \qquad (17a)$$

$$y_t = (\hat{s}\cdot\hat{j}_v)(1- \hat{s}\cdot\hat{k}_v)D_t(1-e^{-\beta t}) \qquad (17b)$$

The vector $\hat{s}$ is the unit vector from the sun to the spacecraft in body coordinates. The unit direction vectors of the body fixed coordinates are given by $\hat{1}_v$, $\hat{j}_v$, and $\hat{k}_v$. The maximum deflection D_t, and the time constant β were

determined experimentally[7]. The time constant β, was given
for a seamless boom. The value given in the appendix was
varied by two orders of magnitude with little difference
seen in simulation results given later in this paper.

Boom flexibility is modeled with the assumption that
the boom is massless. The boom is modeled in bending as a
cantilevered beam in the vehicle coordinates. The force due
to bending is dependent upon displacement and thermal
equilibrium.

$$m_d \ddot{x}_d + c_x [{}^d/_{dt}(x_d - x_t)] + k_x [x_d - x_t - k_m (\bar{M}_d \cdot \hat{i})] =$$

$$-m_d [\dot{\bar{\omega}}_b l + \bar{\omega}_b \times (\bar{\omega}_b \times l \hat{k}_v)] \cdot \hat{i} \qquad (18a)$$

$$m_d \ddot{y}_d + c_y [{}^d/_{dt}(y_d - y_t)] + k_y [y_d - y_t - k_m (\bar{M}_d \cdot \hat{j})] =$$

$$-m_d [\dot{\bar{\omega}}_b l + \bar{\omega}_b \times (\bar{\omega}_b \times l \hat{k}_v)] \cdot \hat{j} \qquad (18b)$$

Spring constant values, k_x and k_y, and the damping
coefficients, c_x and c_y, were determined experimentally[7].
The natural frequency of this system is approximately 0.22
rad/sec. The damping torque vector , M_d, is produced by a
magnetically anchored eddy current damper that is mounted at
the tip of the gravity gradient boom. A magnet inside a
conductive shell is anchored to the earth's magnetic field.
The relative rotation of the magnet produces a resistive
torque in the shell.[8]

$$M_d = -k_d [(\bar{\omega}_{tm} - \dot{\hat{B}}) - \hat{B}((\bar{\omega}_{tm} - \dot{\hat{B}}) \cdot \hat{B})] \qquad (19)$$

$\hat{B}$ is the magnetic field unit vector in vehicle coordinates.
The term $\bar{\omega}_{tm}$ refers to the angular velocity of the tip of
the boom. It is a function not only of body and orbital
angular velocities, but also of the length of the boom, l,
and twist rate, .

$$\bar{\omega}_{tm} = \bar{\omega}_{b/o} + \bar{\omega}_o + \dot{\zeta}(y_{tm}/l)\,\hat{i}_v - \dot{\zeta}(x_{tm}/l)\,j$$

$$+ [\dot{\zeta} + (\dot{x}_{tm} y_{tm} - \dot{y}_{tm} x_{tm})(x_{tm}^2 + y_{tm}^2)^{-1}]k_v \qquad (20)$$

where:
$$x_{tm} = x_t + x_d$$
$$y_{tm} = y_t + y_d$$

The magnetic field unit direction vector is given by a
spherical harmonic model[4] that uses 1985 International
Geomagnetic Reference Field (IGRF) data.[9] In a polar orbit,
the we can approximate $\hat{B} = -2\bar{\omega}_o \hat{j}$ and $\bar{\omega}_{tm} = -\omega_o \hat{j}$, ignoring boom
deflection, twist, and oscillations. The torque produced by
Eq. (19) for these approximations is $M_d = -k_d \omega_o \hat{j}$,

which is a forward pitching torque fixed in orbital
coordinates. In the linearized system given by eqn. 15, we
get the term $a_{32}=-k_d\omega_o$. This term would destablize the fast
roll-yaw mode if we did not also get the terms $a_{33}=-k_d$ and
$a_{44}=-k_d$. This linearized system is stable for all values of
k_d.

The equation of motion for the twist angle of the boom, ζ
, is given by the following scalar equation.

$$I_{tm}\ddot{\zeta} + (M_{tw} + M_{td})k + (\overline{M}_d\ \hat{k})$$

$$+ \,{}^d/_{dt}[\,(\dot{x}_{tm}y_{tm}-\dot{y}_{tm}x_{tm})\,(x_{tm}^2+y_{tm}^2)^{-1}]I_{tm} = 0 \qquad (21)$$

M_{td} and M_{tw} are the damping and restoring torques
respectively. They are modeled from emperical data[6] and are
expressed in the following equations.

$$\begin{aligned}
M_{td} &= 0 &&\text{for} \quad <0.1 \text{ rad} \\
M_{td} &= -k_{td}\text{sign}(\zeta) &&\text{for} \quad >0.1 \text{ rad} \qquad (22)
\end{aligned}$$

$$M_{tw} = A_1\zeta + A_2\zeta^2 + \ldots + A_7\zeta^7 \qquad (23)$$

Coefficients A_1 through A_7 are given in the appendix.

Boom tip displacement and twisting are dynamically
coupled to roll, pitch, and yaw angles. The products of
inertia change with tip displacement and tip motion.
Momentum due to tip displacement and twist rates are added
to the total angular momentum. Augmenting the equations of
motion given in Eq. (9) and Eq. (10) with Eqs. (17) through
(23) gives a model that can be used to simulate the
spacecraft.

Coupling between the twisting and bending induced by
thermal bending has been identified in the literature as a
source of instability for the type of gravity gradient boom
used on Polar BEAR.[10] The internal boom damping and damping
provided by the eddy current damper are much larger than
needed to prevent this type of instability as predicted by
Ref. 10. Bending and twisting remain bounded in simulation.

Nonlinear Simulation

A fourth order Runga-Kutta integrator with a 5 second
step size was used to simulate the nonlinear differential
Eqs. (9) and (10) augmented by Eqs. (17) through (23). Fig.
5 gives roll pitch and yaw angles as a function of time. In
this configuration, the spacecraft is fully exposed to the
sun and the orbit normal is making approximately a 20° angle

with the sun vector. Roll, pitch, and yaw angles and rates
were initially zero as were tip displacements and twist.
The high frequency oscillations that occur at the beginning
of the simulation is produced by the combination of thermal
bending and boom flexibility. This is the type of activity
that is seen when the spacecraft emerges from earth shadow
to sunlight. We can see that the fast roll-yaw motion is
excited by the thermal deflection. It is not apparent in
this figure, but this system is stable.

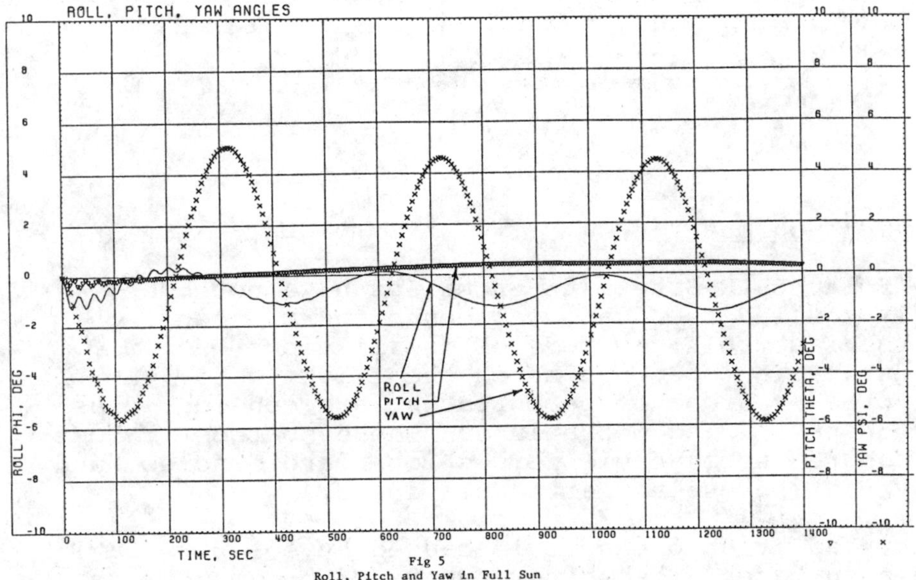

Fig 5
Roll, Pitch and Yaw in Full Sun

    To demonstrate how a small pitch torque (in orbital
coordinates) can destabilize this system, a pitch torque of
-6×10^{-34} in-lbs was added. Fig. 6 shows roll, pitch, and yaw
for this run. This pitch torque would show up as a negative
value for a_{34} in the linear analysis, giving the system
positive roots. According to the linear analysis, if the
sign of the pitch torque were changed, the fast roll-yaw
mode would be damped and the slow roll-yaw motion would be
unstable. Inversion occurs when yaw angle exceeds 90°.

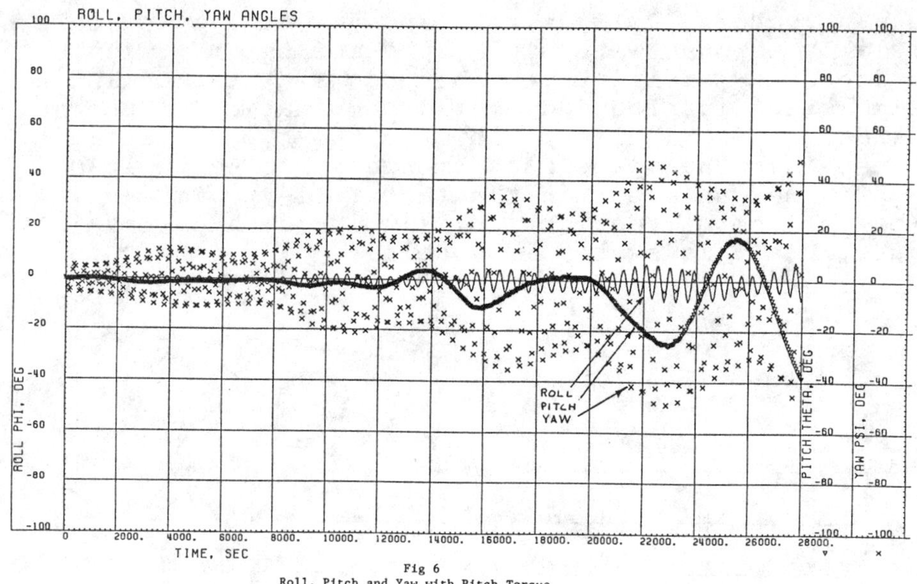

Fig 6
Roll, Pitch and Yaw with Pitch Torque

Perturbation Sources

The Polar BEAR Satellite anomalies have not been
adequately modeled thus far. Although thermal bending
excites the fast roll-yaw mode of oscillation (ω_1), this
type of excitation does not match the flight data. In this
section, other sources of perturbation are considered as
possible destabilizing mechanisms. These sources include;
Solar pressure, aerodynamic forces, on-board dipoles, and
gravity variation due to earth oblateness.

Solar radiation pressure is caused by solar radiation
striking a surface. The vector equation to compute the
moment on a body from pressure on a given surface is[4] :

$$M_{sp} = AP_r(\bar{N}\cdot\hat{s})\{(\breve{r}_{cg}\cdot\hat{s})(c_a+c_d)+2(\breve{r}_{cg}\times\bar{N})[c_d/3+(\bar{N}\cdot\hat{s})c_s]\} \qquad (25)$$

For a given surface with area A, unit normal N, surface
reflection characteristics c_a, c_d, and c_s, and moment arm to
the center of gravity $\breve{r}_{cg}$, the moment is a function of the
sun vector $\hat{s}$ in vehicle coordinates. The moment, then,
changes with orbit position and vehicle attitude. Summing
the moments of various surfaces gives a composite moment due
to radiation pressure.

Of particular interest is the radiation pressure on the
solar arrays. The configuration of the arrays (Fig. 1) is
such that a yaw torque is produced that is dependent upon
roll angle when the sun vector s has a component in the y
direction. In the linearized system, Eq. (15), a_{41} becomes
nonzero. The size and sign of a_{41} is dependent upon the
relationship of the orbit normal and sun position. When the
orbit normal is pointing to the sun, $a_{41}<0$ and the fast
roll-yaw motion is stable. When the orbit normal is
pointing away from the sun, the fast roll-yaw motion is
unstable when no damping is present. Using the model
described in the previous section, we find that radiation
pressure torques are negligible and are overwhelmed by
damping. Furthermore, attitude anomalies have occurred when
the orbit normal is roughly parallel to the sun vector,
regardless of sign. Radiation pressure from the earth
reflectance of the sun is much smaller than direct
radiation and has been ignored.

At the 545 nm (1009 km) orbital altitude of Polar BEAR,
the atmospheric density can be modeled by a Jacchia
atmospheric model.[11] This model accounts for density
changes due to solar flux variation and density variation
due to the relative position of the orbit with respect to
the earth and sun. The torque produced by aerodynamic drag
for a given surface is:

$$\overline{M}_a = (1/2)\,C_d\,\rho\,VA(\bar{r}_{cg} \times \overline{V}) \qquad (26)$$

The density, ρ, is given by the Jacchia model. For the
low density, high velocity situation produced by this
particular orbit, the drag coefficient can be approximated
by $C_d \dot{=} 2$. The area, A, is the area projected by the surface
perpendicular to the velocity vector $\overline{V}$. The velocity vector
is the vector sum of the orbital velocity and the relative
velocity of the atmosphere with respect to the orbit due to
the rotation of the earth. The velocity vector is parallel
to the x-y plane in orbital coordinates. The boom,
spacecraft body, and solar arrays were modeled for a
composite aerodynamic moment. The linearized system of Eq.
(15) reveals a fast roll-yaw instability due to the y
component of velocity acting on the solar arrays, similar to
the instability encountered with solar radiation pressure.
This effect is negligible compared to damping. Also, The
instability is only present on the half of the orbit
containing the descending node. The sign of the moment
changes on the other half of the orbit, tending to damp the
fast roll-yaw motion. The positive pitch moment produced by
the boom gives a_{32} a value greater than zero in Eq. 15,
which damps the fast roll-yaw motion. Simulation of the

nonlinear system did not reveal any instability due to
aerodynamic moments.

A magnetic dipole produced by the current in a solar
array was considered since the resulting interaction with
the geomagnetic field would produce a moment dependent upon
attitude and sun direction. This assumes that a solar panel
was miswired since the solar panels are typically wired to
avoid producing a net dipole. The moment is given by[4] :

$$\overline{M}_m = k_m(\overline{N} \cdot \hat{s})(\overline{N} \times \overline{B}) \qquad (27)$$

The coefficient k_m is the maximum magnitude of the dipole
assuming the current loops the outer edge of the array.
Although the magnitude is potentially large enough to
produce instability, no combination of array moments could
be found to produce instability under orbit conditions that
the anomalies occurred. This was confirmed by simulation.

When deriving Eq. (10), it was assumed that the earth's
gravitational field was constant. A more accurate
representation is to model the gravitational field of an
oblate spheroid and derive the gravity gradient torque from
this model. The following equation is the potential
function for a body in a gravitational field of an oblate
spheriod[12]

$$U = C_1 I_{zzo} + C_2 I_{xxo} + C_3 I_{xzo} \qquad (28)$$

where:
$$C_1 = 3\mu/2r^3[1 + fa^2/r^2(5 - 27\sin^2\lambda)]$$
$$C_2 = -3f\mu a^2/r^5\cos^2\lambda$$
$$C_3 = 12f\mu a^2/r^5\sin^2\lambda$$

The form factor f, is a dimensionless quantity that
defines the oblateness. The constant, a, is the polar
radius of the earth, λ is the geocentric latitude. The
inertia terms used in Eq. (28) are the products of inertia
about the <u>orbital</u> coordinates. Gravity gradient torques are
then given by:

$$\overline{M}_{og} = -[\frac{\partial U}{\partial \phi}(^{\cos\psi}/_{\cos\theta}) + \frac{\partial U}{\partial \theta}\sin\psi - \frac{\partial U}{\partial \psi}\cos\psi\tan\theta]\hat{i}$$

$$-[\frac{\partial U}{\partial \phi}(^{\sin\psi}/_{\cos\theta}) + \frac{\partial U}{\partial \theta}\cos\psi + \frac{\partial U}{\partial \psi}\sin\psi\tan\theta]\hat{j}$$

$$-\frac{\partial U}{\partial \psi}\hat{k} \qquad (29)$$

No instabilities were found in the nonlinear simulation[13].
Note that the linearized analysis is the same for oblate
earth gravity gradient and spherical earth gravity.

It should be noted that the most recent two inversions, in April and October of 1988, occurred during periods of high sunspot and solar flare activity.[9] The _total_ variation in solar flux is insignificant and its effect is a small change in the maximum deflection of the boom due to thermal bending. No other connection between solar activity and spacecraft performance could be found by the authors.

SIMILAR EXPERIENCE

Polar BEAR is not the first satellite to experience attitude anomalies during the full sun period of its orbit. Other gravity gradient stabilized spacecraft including Space Test Program Mission P83-1 High Latitudes Mission (HiLat), and two satellites built by the Naval Research Laboratory (NRL) have experienced significant attitude oscillations while in the full sun portion of their orbits. A documentation of several gravity gradient missions may be found in Ref. 14.

Thermal bending of gravity gradient booms has been identified in the literature as a source of instability for three axis stabilized spacecraft without a momentum wheel. The OV1-10 satellite showed large yaw excursions and subsequent inversions[15]. Also exhibiting this type of behavior is the NRL-164 spacecraft.[16]

<u>NRL Satellites</u>

A two axis stabilized (no yaw control) gravity gradient satellite built by NRL and launched in the spring of 1967 inverted five times during its orbit life.[17] All five inversions occurred at or near the full sun period of the satellites orbit. Details on the satellite behavior are somewhat limited since the spacecraft had no onboard data storage and its attitude sensor only worked when fully exposed to the sun. The spacecraft was in a 500 nautical mile, 70 degree inclination orbit.

Another NRL satellite which incorporated three axis gravity gradient stabilization experienced the same problems. The GGSE V spacecraft had two booms extending out opposite sides of the spacecraft. This is the same boom configuration as the NRL satellite mentioned earlier. GGSE V had a momentum wheel unlike the 2 axis stabilized spacecraft however. GGSE V was also unique in the fact it had on board data storage and therefore the ability to collect continuous attitude data. This data showed several spacecraft inversions all at times the spacecraft was at or near the full sun period of its orbit.[17]

The satellites were recovered by several methods once they inverted. GGSE V was reinverted by retracting and reextending the gravity gradient booms. The two axis stabilized satellite had micro thrusters which were fired in short bursts to tumble the spacecraft and then allow gravity gradient capture, hopefully in the right orientation. NRL has also maneuvered satellites using the momentum wheel in a manner similar to the Polar BEAR reinversion method.[17]

P83-1 HiLAT

Space Test Program Mission P83-1, HiLAT, was very similar in design to Polar BEAR. This spacecraft was also built by APL. It has basically the same subsystems as Polar BEAR. However the gravity gradient boom is less rigid in the deployed configuration since it does not have the interlocking zipper mechanism of the Polar BEAR boom.[18] The momentum wheel is the same size as Polar BEAR but is located higher up on the spacecraft body of HiLAT than it is in Polar BEAR. There is also no eddy current damper at the tip of the HiLAT boom as there is in Polar BEAR.

Since its launch in 1983 it has experienced several periods of out of specification attitude oscillations. These oscillations began when the spacecraft passed from eclipse into sun during orbit. The oscillations increased as the orbit normal plane moved closer to the sun.[18] Oscillations however were never greater than 40 degrees and occurred only about 20 percent of the time. No inversions of the spacecraft were observed.

CURRENT STATUS AND RECOVERY OPERATIONS PLAN

The spacecraft is now operating within specification. The next full sun period in the orbit will begin 11 February 1989. The same procedures used for the previous recoveries will be implemented by NAVASTRGRU should another inversion occur.

RESULTS/EFFECTS

The inversions have apparently not damaged the Polar BEAR spacecraft. However the ability to collect experiment data, the entire mission of the spacecraft, is lost each time the spacecraft inverts. Experiment data collection is also impaired without inversion when the attitude oscillations significantly exceed performance specifications.

The ability to continue reinversion using the momentum
wheel may also be in question because of the wheel's
reliability. Frequent on and off cycling of the wheel may
damage the device and result in a disabled spacecraft. The
momentum wheel must continue to be used for reinversion
since no viable alternative exists.

CONCLUSIONS, RECOMMENDATIONS, AND LESSONS LEARNED

It is difficult to recommend solutions to a problem
that is not fully understood. The Polar BEAR mission has
been very successful however the spacecraft's attitude
performance difficulties must be taken into consideration
when designing future gravity gradient spacecraft. Since
the spacecraft was originally designed to overcome attitude
oscillations in the full sun condition it is evident that
more research is needed to determine the exact cause of the
anomalies. This research is currently underway at APL.

Physical Layout of Spacecraft

Several steps can be taken in the physical design of
the spacecraft to minimize the potential for large attitude
oscillations, even though the exact cause of those
oscillations is not known. One step that might be taken to
reduce the problems experienced by Polar BEAR is to increase
damping, particularly in yaw. The tip mass damper sits at
the end of a boom that is very flexible in twist and has
little effect on yaw damping.

Another consideration is the design of the boom
itself. A more rigid boom could be less susceptible to
thermal bending. Materials used in the construction of the
boom could be changed to minimize thermal bending. The
recommendation made from Ref. 14 is that active damping
(feedback control) be used with three axis stabilized
gravity gradient missions.

On-Board Data Storage

One problem frequently noted in the analysis of Polar
BEAR's attitude anomalies is lack of attitude data. Only
"snap-shots" of data taken real time over ground stations
are available. Simulations can be used to interpolate
between spacecraft performance between ground station
contacts but these simulations are inaccurate. On-board
data storage is essential to allow attitude data to be
collected throughout the entire spacecraft orbit. The
ability to analyze spacecraft performance throughout its

entire orbit could greatly contribute to the understanding
of the attitude anomalies.

On-board data storage would also aid in the
reinversion process because reinversion commands could be
uplinked and stored for execution at a later time. This
would permit commanding of the spacecraft outside of ground
station contact and would present more opportunities for
reinversion, thus reducing the time the spacecraft is
inverted and not performing its mission. On-board data
storage would also benefit the mission payload by allowing
data collection and downlink to ground at any time in the
orbit. This would provide many more opportunities for data
collection and allow ground stations to be located more
conveniently. Current Polar BEAR and HiLAT stations must be
positioned in remote northern locations which greatly
increase operating cost. Ground stations location would be
much more flexible if the data could down linked at any time
rather than just during real time collection.

Data collection devices could be the standard tape
recorder although most of those are rather large for small
satellites. A better solution would appear to be solid
state memory. Radiation hardened memory chips are currently
available and large capacity, small volume solid state tape
recorders are currently in development.

Design for Reinversion Capabilities

After all the previous design considerations have been
reviewed there is still no guarantee the spacecraft will not
invert. Because of this, the user must be prepared for an
inversion. A quick recovery can be made possible by first
designing the spacecraft with the ability to recover and
then preparing the operating agencies to use that ability
quickly and effectively.

Spacecraft design should consider an active method of
inversion. The momentum wheel cycling of Polar BEAR was
conceived after launch, whereas future spacecraft should
consider a method of inversion in the conceptual design
phase. Possible solutions would be to add micro thrusters
like the NRL satellite, or to have a stronger z-coil. The
Polar BEAR z-coil was only effective with the boom
retracted. A more powerful coil could overcome the gravity
gradient restoring force and reinvert the spacecraft even
with the boom extended. A retractable boom could also be
used but this requires a more complex mechanism than booms
designed for one time deployment. Complexity and greater
cost, weight, and space requirements of a retractable boom

make a one time deployment boom more attractive for smaller inexpensive spacecraft.

Operational planning for reinversion should be done along with the spacecraft design. Ensuring that using agencies are prepared to effectively implement a standard reinversion procedure will minimize the time the spacecraft is in an improper attitude and not performing its primary mission.

Conclusion & Summary

The linearized analysis shows how the mode of oscillations observed prior to inversion can be excited, but did not identify a source of this disturbance. Simulating a nonlinear model including flexibility and damping failed to predict inversion. Some likely sources of excitation were modeled, but these also failed to predict inversion. As stated previously, the anomalous behavior of Polar BEAR and other gravity gradient satellites is a problem with as yet no clear solution. The spacecraft should be designed for reinversion since inversion cannot be confidently prevented at this time. More flight data from future missions would be very valuable to understanding the problem.

Acknowledgements

The authors would like to acknowledge Dr. Thomas Alley for developing the original numerical simulation model and for his general help and suggestions relating to the work presented in this paper. We would also like to thank the members of the Air Force Space Test Program and The Applied Physics Laboratory at Johns Hopkins University who worked on the Polar BEAR mission and provided invaluable information for this paper.

NOTATION

$\overline{h}$ - Angular momentum
$\overline{h}_b$ - Angular momentum of the vehical
$\overline{h}_w$ - Angular momentum of the momentum wheel
$\sum \overline{M}$ - sum of the external moment vectors about body
     coordinates
$\overline{\overline{I}}_b$ - Inertia matrix of the body
$\overline{\overline{I}}_w$ - Inertia matrix of the momentum wheel
$\overline{\overline{I}}_{bo}$- Inertia dyadic of the vehical sans momentum wheel

$\bar{\omega}_b$ - angular velocity vector of the body with respect to the inertial coordinates

$\bar{\omega}_w$ - angular velocity of the wheel with respect the to inertial coordinates

$\bar{\omega}_o$ - orbital angular velocity vector

$\bar{\omega}_{b/o}$ - angular velocity vector of the body with respect to orbital coordinates

$\bar{\omega}_{w/b}$ - angular velocity vector of the wheel with respect the body coordinates

ϕ - roll angle

θ - pitch angle

ψ - yaw angle

M_x, M_y, M_z - Moments about the x, y, and z body coordinates

μ - gravitation constant

$\bar{M}_{gg}$ - Gravity gradient moments

R_x, R_y, R_z - Product of inertia ratios

h_{wb} - momentum wheel momentum

A - System matrix of linear model

$\hat{X}$ - State vector for linear model

ω_1, ω_2 - Roll-yaw natural frequencies

$\bar{r}$ - orbit radius

x_t, y_t - x and y thermal displacement of the boom tip

$\hat{s}$ - unit vector from sun to spacecraft in body coordinates

D_t - maximum thermal deflection

l - length of gravity gradient boom

x_d, y_d - x and y boom displacement from thermal deflection

c_x, c_y - x and y boom damping coefficient

k_x, k_y - x and y boom spring constant

k_d - tip mass damping constant

m_d - tip damper mass

$\bar{\omega}_{tm}$ - inertial angular velocity vector for tip mass damper

I_{tm} - product of inertia matrix of the tip mass damper about center of gravity of damper

x_{tm}, y_{tm} - x and y total tip displacement

M_{td} - Moment from damper

M_{tw} - Moment from torsional stiffness

A_1, A_2, $\ldots$, A_7 - torsional stiffness coefficients

C_D - drag coefficient

M_{sp} - moment vector due to solar pressure in body coordinates

M_{sp} - moment vector due to solar pressure

A - Area exposed to sunlight

P_{sp} - Solar pressure at 1 astronautical unit

$\bar{r}_{cg}$ - distance vector from center of mass to center of pressure

$\bar{N}$ - normal unit vector of a surface

$\bar{V}$ - velocity vector

M_a - moment vector due to aerodynamic forces

M_m - moment vector due to magnetic torques

$\hat{B}$ - magnetic field vector

$\bar{U}$ - gravitational field vector
$\bar{M}_{og}$ - oblate earth gravity gradient moment vector
c_a - coefficient of absorbtion
c_s - coefficient of specular reflection
c_d - coefficient of diffuse reflection
  - density

REFERENCES

1. <u>Payload Description of the Polar BEAR Spacecraft (P87-1)</u>, The John Hopkins University Applied Physics Laboratory, April 1986.

2. W. L. Ebert and D. G. Grant, <u>Polar BEAR Attitude Anomaly Activities</u>, The Johns Hopkins University Applied Physics Laboratory, July 1987.

3. J. W. Hunt Jr and C. E. Williams, "Anomalous Attitude Motion of the Polar BEAR Satellite," <u>Johns Hopkins APL Technical Digest</u>, Vol 8, No. 3, 1987, pp. 324- 328.

4. J. R. Wertz, editor, <u>Spacecraft Attitude Determination and Control</u>, Vol. 73, D. Reidel Publishing Company, Boston, 1985.

5. R. W. Longman, "The Equalibria of Orbiting Gyrostats with Internal Angular Momenta Along Principal Axies", Proceedings of the Symposium on Gravity Gradient Attitude Stabilization, The Aerospace Corporation, 2350 El Segundo Blvd., El Segundo CA, 1968.

6. R. W. Wilder, "Polar BEAR Gravity Gradient Boom Stiffness Testing", TOR-0086A(2508-16)-1, The Aerospace Corporation, 2350 El Segundo Blvd., El Segundo CA, 1987.

7. C. L. Staugaitis and R. E. Predmore, "Thermal Static Bending of Deployable Interlocked Booms," NASA TN D-7243

8. W. W. Pettus, "Performance Analysis of Two Eddy Current Damper Systems," Proceedings of the Symposium on Gravity Gradient Attitude Stabilization, The Aerospace Corporation, 2350 El Segundo, CA, 1968.

9. National Oceanic and Atmospheric Administration, National Geophysical Data Center, 325 Broadway, Boulder CO.

10. H. P. Frisch, "Coupled Thermally Induced Transverse Plus Torsinal Vibrations of a Thin-Walled Cylinder of Open Section", NASA TR R-333, 1970.

11. R. W. Bruce, "Atmosphere Models", TOR-1001(2307)-24, The Aerospace Corporation, 2350 El Segundo Blvd., El Segundo CA, 1968.

12. R. E. Roberson, "Gravitational Torque on a Satellite Vechicle", J. Franklin Institute, Vol. 265, p.13-22, 1985.

13. T. L. Alley, "Response of Polar BEAR Satellite to Gravity Gradient Torques from an Oblate Earth", A89.5415.TLA.01, The Aerospace Corporation, 2350 El Segundo Blvd., El Segundo CA, 1988.

14. P. C. Hughes, <u>Spacecraft Attitude Dynamics</u>, John Wiley and Sons, New York, 1968.

15. G. M. Connell and V. Chobotov, "Possible Effects of Boom Flutter on the Attitude Dynamics of the OV1-10 Satellite," Journal of Spacecraft, Vol. 6, No. 1, 1969, pp. 90-93.

16. R. L. Goldman, "Influence of Thermal Distortion on Gravity Gradient Stabilization", Journal of Spacecraft and Rockets, Vol. 6, No. 1, 1969, pp.90-93.

17. F. W. Raymond, P. G. Wilheim, R. T. Beal, "Gravity Gradient Flight Experience Acquired with the Naval Research Lab Satellites". US Naval Research Laboratory, Washington DC.

18. <u>Payload Description of the HiLAT Spacecraft (P83-1)</u>, The Johns Hopkins University Applied Physics Laboratory, Feb 1983.

APPENDIX

The following physical parameters were used in the simulation of the Polar BEAR spacecraft.

$$I_{bxx} = 705.4 \text{ slug-ft}^2$$
$$I_{byy} = 707.4 \text{ slug-ft}^2$$
$$I_{bzz} = 21.6 \text{ slug-ft}^2$$
$$I_{bxy} = 0.18 \text{ slug-ft}^2$$
$$I_{bxz} = 0.33 \text{ slug-ft}^2$$
$$I_{byz} = -0.25 \text{ slug-ft}^2$$
$$I_{tm} = 1.005 \text{ slug-ft}^2$$
$$m_{tm} = 6.0 \text{ lbm}$$
$$h_{w/b} = 1.77 \text{ slug-ft}^2/s$$
$$r = 3988.9 \text{ nm}$$
$$\mu = 1.408e17 \text{ ft}^3/s^2$$
$$\omega_o = .05707 \; \hat{\jmath} \text{ deg/s}$$
$$k_d = .005 \text{ ft-lb-sec/rad}$$

$\beta = 0.1 \text{ s}^{-1}$
$k_x = k_y = 1.6e{-}05 \text{ lbf/in}$
$c_x = c_y = 3.77e{-}03 \text{ in-lb-s/rad}$
$A_1 = 2.114e{-}05 \text{ in-lb/rad}$
$A_2 = 7.418e{-}06 \text{ in-lb/rad}^2$
$A_3 = 1.190e{-}05 \text{ in-lb/rad}^3$
$A_4 = -4.762e{-}06 \text{ in-lb/rad}^4$
$A_5 = -5.357e{-}05 \text{ in-lb/rad}^5$
$A_6 = -9.807e{-}07 \text{ in-lb/rad}^6$
$A_7 = 9.674e{-}06 \text{ in-lb/rad}^7$
$k_{td} = 2.6e{-}04 \text{ in-lb}$
$c_a = 0.708$
$c_s = 0.050$
$c_d = 0.242$
$l = 720 \text{ in}$
$D_t = 18 \text{ in}$

SPACE MISSILE GUIDANCE AND CONTROL SIMULATION AND FLIGHT TESTING

David C. Ductor, E. Bryan Wallace and Charles H. Dillon[*]

This paper presents the theoretical development and experimental results of a guidance and control system for an exoatmospheric interceptor. Designed to operate from space platforms at low earth orbits, Kinetic Energy Weapons (KEW) are a critical element of an effective strategic defense, responsible for the boost, post-boost and mid-course intercept of ballistic missiles and re-entry vehicles. Earlier technology development efforts supported by the Air Force Astronautics Laboratory (AFAL) demonstrated the necessary rocket control components for the KEW guidance and control loops. Subsequently, the AFAL initiated a flight test program in Sept 1987 to validate the integrated flight control performance on a research vehicle configured to represent a baseline KEW. In parallel to the hardware integration and testing, extensive simulation and analysis was performed to characterize the expected dynamic response of the vehicle in free flight. In addition, an Extended Kalman Filter was derived and implemented to update system models using measured flight test data. This paper reports on these analytical studies, and details the observed flight trajectories of the missile.

INTRODUCTION

The pursuit of a near term strategic defensive system requires the design, development and testing of Kinetic Energy Weapons (KEW's) that would track, intercept and destroy ballistic missiles or their reentry vehicles. The KEW payload, a kinetic kill vehicle (KKV), consists of propulsion, avionics, and guidance systems that enable the KKV to track and intercept its target. The propulsion system of a KKV includes both divert engines for changing the vehicle's velocity vector, and attitude control engines to stabilize the vehicle, while keeping the target in the field of view. Because the KKV must hit the target in order to acheive a kill, it is critical that the KKV have the capability for precise control in both attitude and position. It is the accurate control of the KKV during the terminal phase of the intercept that is of the most interest and the subject of this effort.

---

* Systems Analysts, Air Force Astronautics Laboratory, Edwards Air Force Base, California 93523.

BACKGROUND

Over the past four years, the Air Force Astronautic Laboratory (AFAL) has sponsored KKV technology development efforts in support of the Strategic Defense Initiative (SDI). As part of these efforts, a program to reduce the risk of a near term flight experiment was conducted from 1985 to 1987, with the goal of demonstrating the successful development and testing of lightweight control system components for a cruciform divert, three axis stabilized rocket controlled projectile. Following the successful completion of this program in September 1987, the AFAL initiated the Kinetic Kill Vehicle Hovered Interceptor Test (KHIT) program with the objective of integrating these lightweight propulsion components into a complete vehicle and performing free flight tests in a laboratory environment. Because of a lack of data on projectiles of this kind, and the extreme cost and risk associated with space testing experimental vehicles, the KHIT program is also intended to develop the facility and a test methodology that would support the ground testing of integrated KKV's.

APPROACH

The overall performance of a KKV interceptor system can be characterized by nonlinear relationships between the various vehicle subsystems, such as the avionics implementation of the designed controller, the various sensors required by the control system, the actuators, and the integrated structure. The analysis of the performance of such a system is based on extensive simulation using models which represent the dynamic characteristics of the subsystems and their nonlinear relationships. Because the testing of integrated KKV's has the goal of better characterizing the response of the complete system to various inputs and validating the theoretical models of the system, a typical cycle of simulation analysis and hardware testing, shown in Fig. 1, is used to compare the test results to the simulation results in order to improve the system models.

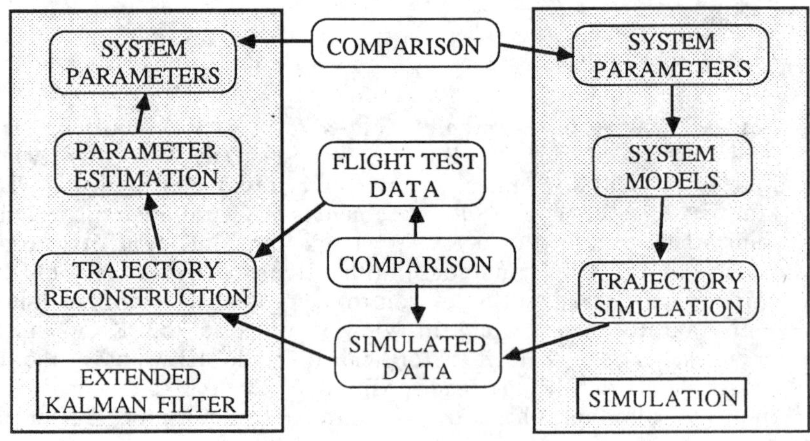

Figure 1.   Analysis And Testing Cycle

In this analysis cycle, best estimates of the system parameters of interest, such as any thruster misalignments and center of gravity (CG) offsets, are modeled in the simulation and the simulation is used to produce data that will be compared to the results of the flight test. The flight test data can then be used to produce the most likely estimates of the system parameters of interest.

The most likely estimates of system parameters from test data can be obtained by several methods. The method of utilizing engineering approximations and assumptions to calculate parameters of interest is straightforward, but this technique does not possess the fine detail that is available by other methods. The batch processing method uses techniques such as Maximum Likelyhood Estimation or Least Squares to process the entire data set before estimating the pertinent parameters. Because the batch method requires an accurate model of the system dynamics, poorly modelled dynamics can produce poor results. The sequential processing method processes the data sequentially from start to finish and uses various optimal filtering techniques such as the Kalman Filter or the Extended Kalman Filter (EKF) to estimate the parameters. The sequential method also utilizes models of the system dynamics, but allows for uncertainty in the models by incorporating noise into the dynamic equations. The sequential method, utilizing an EKF, was chosen to estimate the system parameters from the test data, because of its ability to include uncertainties in the system models.

EXTENDED KALMAN FILTER

An EKF, when applied to the problem of estimating the states and parameters of the KHIT vehicle, is essentially a discretized, minimum variance estimator applicable to nonlinear states. The code that was developed is structured to use a set of measurements and propagate the measured and estimated states until a new set of measurements becomes available. Measurements from the KHIT inertial measurement unit (IMU) are available at 100 Hz and the EKF propagates the states at 1000 Hz.

The mass and moments of inertia (MOI's) time histories for the KHIT vehicle will not be estimated directly through the use of the EKF due to the fact that the increased computational burden does not produce any improvement in estimation accuracy of these two parameters. Also, since all estimated parameters share a common gain matrix, the reduction in the number of states results in a reduction in the covariance for the estimated parameters.

Instead of using the EKF to estimate the mass of the vehicle, the mass is computed as shown below in Eq. (1), using the vehicle's original mass (m_0), the known specific impulses for the divert and attitude control engines ($I_{x,sp}$), and the total impulses produced by the respective engines ($I_{x,t}$). The total impulses delivered by each thruster are based on the recorded valve commands and an earlier analysis of data taken during static tests of individual divert and ACS engines. This analysis provided thrust time histories which, when combined with the valve commands, allow the total impulse of each engine to be estimated.

$$m = m_o - \frac{I_{a,t}}{g\, I_{a,sp}} - \frac{I_{d,t}}{g\, I_{d,sp}} \qquad (1)$$

The MOI's of the KHIT vehicle are determined using the original MOI's and the vehicle's mass time history. The MOI's are measured directly, prior to the vehicle being enabled for flight, using specialized equipment. The MOI's at liftoff are estimated by accounting for the mass of the propellants which enter the manifolds and its influence on the MOI and CG. To determine the time histories of the MOI's for the vehicle, it is assumed that the vehicle is a cylinder of uniform density, mass m, length L, diameter D, and that the MOI's are therefore directly proportional to the mass of the vehicle as shown below in Eq. (2).

$$I_x = \frac{1}{8}\, mD^2 \qquad I_y = \frac{1}{12}\, mL^2 \qquad I_z = \frac{1}{12}\, mL^2 \qquad (2)$$

We can therefore assume that any change in mass of the vehicle is also directly proportional to the the resulting change in moments of inertia as shown in Eq. (3).

$$\Delta I_x = C_x \Delta m \qquad \Delta I_y = C_y \Delta m \qquad \Delta I_z = C_z \Delta m \qquad (3)$$

The above relationships can be solved to find the constants of proportionality. With this information, the mass and MOI time histories can be determined and used within the EKF for the estimation of the remaining vehicle states.

The primary states to be estimated through the EKF are the individual engine thrust levels, divert thruster misalignments and the vehicle's CG travel. The actual vehicle states such as position and velocity are also outputs of the EKF, but they are not of interest since there are several other systems performing these measurements. The primary reason for interest in the thrust magnitudes, thruster misalignments, and the change in CG, is that these variables represent the disturbance torques to the vehicle, and the characterization of these torques is required in order to understand the performance of the system.

One of the problems associated with the estimation of various vehicle parameters is caused by the engine on-delay. These delays, obtained from static tests of individual engines, introduce errors in the EKF due to the differences between the static test results and the performance of the engines during the actual flight. If the thrust profiles delivered during the flight test do not match the models in the EKF, the estimated states such as thruster misalignments and CG travel would be affected. The solution to this problem is to make a comparison between the thrust determined by the measured accelerations from the IMU and the thrust determined by the modelled thruster profile. If the differences exceed a preset error band, the propagation of the vehicle states is temporarily stopped, but the propagation of the assumed measurement error and the state error covariance continues.

CONTROL SYSTEM DESIGN

While the design and development of control sofware to enable missiles to track and intercept targets has been well characterized, controlled flight of a three axis stabilized rocket propelled interceptor vehicle inside a building in a one-g environment presents a new and interesting technical challenge. In the process of selection of the vehicle control system, several factors need to be considered. First, the effect of gravity must be compensated for to keep the vehicle at a constant altitude during the test, whereas for an actual interceptor performing a mission in a space environment, this effect would be very small. Second, aerodynamic effects must be considered when flying within the earth's atmosphere, because if high divert velocities are reached, this could result in aerodynamic moments which may cause instabilities. Also, due to the size of the test area selected, a means of controlling the vehicle's position safely within the allowed flight volume must also be considered.

After careful consideration of these factors, a linear error plus rate control scheme based on rate-to-gain was selected for attitude control in three axis and position control in the vertical plane. This controller calculates attitude engine commands based on body angles and changes in body angles. Position control in the vertical plane is based on the vehicle's integrated position and the vehicle velocity obtained from the IMU. This control scheme was chosen for its simplicity and ease of implementation and is described in the block diagram shown in Fig. 2.

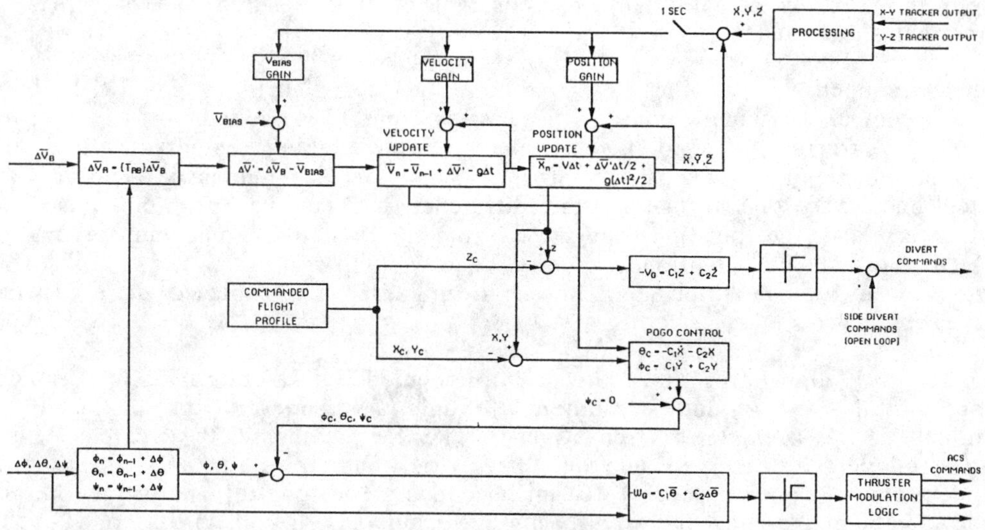

Figure 2. Control System Block Diagram

The present design of the KHIT vehicle, shown in Fig. 3, has no axial thrusters, and the lateral divert engines will not be used for position control until later in the test program.

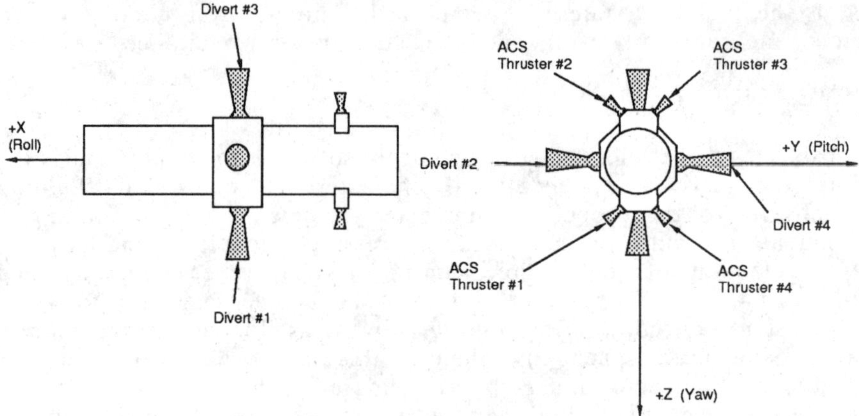

Figure 3. KHIT Vehicle

In order to control position of this vehicle in the axial (X) and lateral (Y) directions, a slightly more complex scheme had to be designed. The method of control selected combines the use of the attitude control system and the vertical position control to create the appropriate correctional forces. This system, deemed "pogo control", is a form of proportional plus integral steering which generates attitude commands in pitch and roll based on position and velocity information which will produce components of force in the X and Y directions intended to nullify any position error within a specified length of time. These commands are based on a second order differential equation dependent on the vehicles position and velocity. The length of time required to zero the error is actually the time constant of this equation and is dependent on the gains selected.

To help compensate for the fact that the IMU has drifts which will affect position control, a video contrast tracker system was added as an external reference. Correct positioning of these two tracking cameras allow the determination of the true position of the vehicle. The guidance equations are updated once a second by taking the difference between the position sensed by the trackers and the position integrated from IMU outputs and multiplying by gains to correct for position, velocity, and velocity bias. The gains were selected by a Kalman filter and chosen to minimize the expected drift between updates.

These control algorithms have been implemented on a MicroVAX™ ground computer and this ground computer, through a telemetry link, is used to control the KHIT vehicle in free flight. The IMU onboard the vehicle senses changes in velocity and attitude in three axis, and transmits this information to the ground computer every 10 milliseconds. The control algorithm in the ground computer updates the divert engine commands in 50 millisecond cycles and the ACS engine commands in 10 millisecond cycles based on position and attitude information that has been updated at 10 millisecond intervals. These engine commands are then transmitted to the vehicle every 10 milliseconds for the duration of the flight.

SIMULATION

A six degree of freedom simulation of the KHIT vehicle was constructed using the SYSTEM BUILD™ portion of MATRIXx™. MATRIXx™ is a VAX™ operating environment that allows the user to rapidly construct and analyze high fidelity models of various systems and also has the ability to allow the incorporation of the actual flight software to control the model of the vehicle. The simulation of the KHIT vehicle attempts to represent the dynamics of the vehicle by including hardware peculiarities and nonlinearities such as thruster delays, rise times, decay times, IMU measurement noise, biases, drifts, and aerodynamic effects. The simulation also includes structural dynamic effects, but because the present test program doesn't verify the predicted structural modes, these effects have not been included here.

The KHIT vehicle weighs approximately 150 lbs when loaded with propellant. The engines burn nitrogen tetroxide and monomethylhydrazine propellants using Helium to pressurize the propellant system. The divert engines produce an estimated 350 lbs of thrust and the ACS engines produce 5 lbs of thrust. The KHIT vehicle carries almost 20 lbs of propellant onboard, which is enough to fly a nominal flight profile lasting approximately 22 seconds, described in Fig. 4.

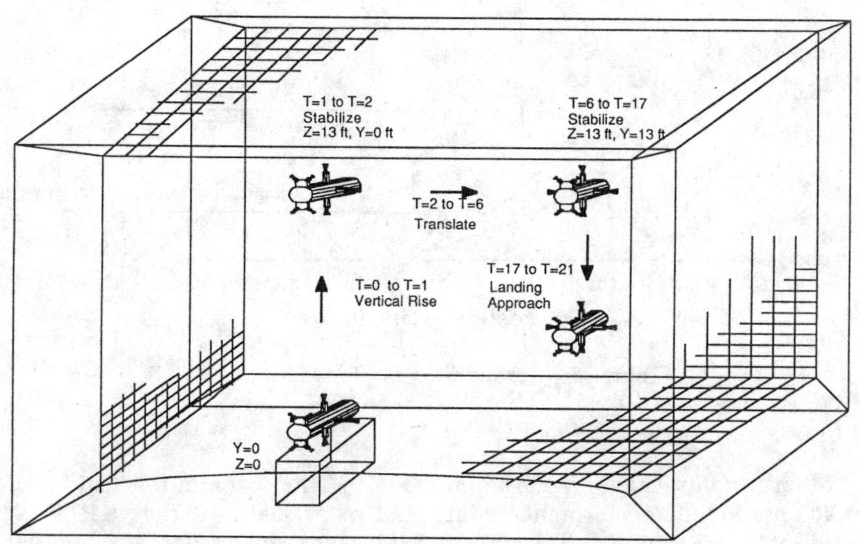

Figure 4.   KHIT Flight Profile

The initial flight segment consists of a lift off from the launch cradle with a 13 ft rise in 1 second, followed by a 1 second period at altitude for stabilization. The next 4 second segment commands the vehicle to laterally translate 13 ft, followed by 11 seconds of hovering in place. The vehicle will use the remaining 4 seconds of available flight time to gradually lower the altitude in order to minimize to free fall distance into the net for capture. While the vehicle hovers in place, the steering control loops will be opened for several seconds, to allow for the determination of the vehicle's attitude limit cycles without pogo control corruptions. Further testing may include segments to perform pitch, yaw, and roll doublet maneuvers for the accurate determination of MOI's and control effectiveness.

The CG and MOI's of the KHIT vehicle are measured prior to flight. Although these measurements are obtained from a fully loaded vehicle, some changes occur as the propulsion system is enabled immediately before liftoff. These changes are a result of firing squibs and introducing propellants into the manifolds of the vehicle. Estimates of these changes place the location of the CG prior to flight at 0.007 in. along the X-axis forward of the divert plane and the predicted CG travel is shown in Fig. 5.

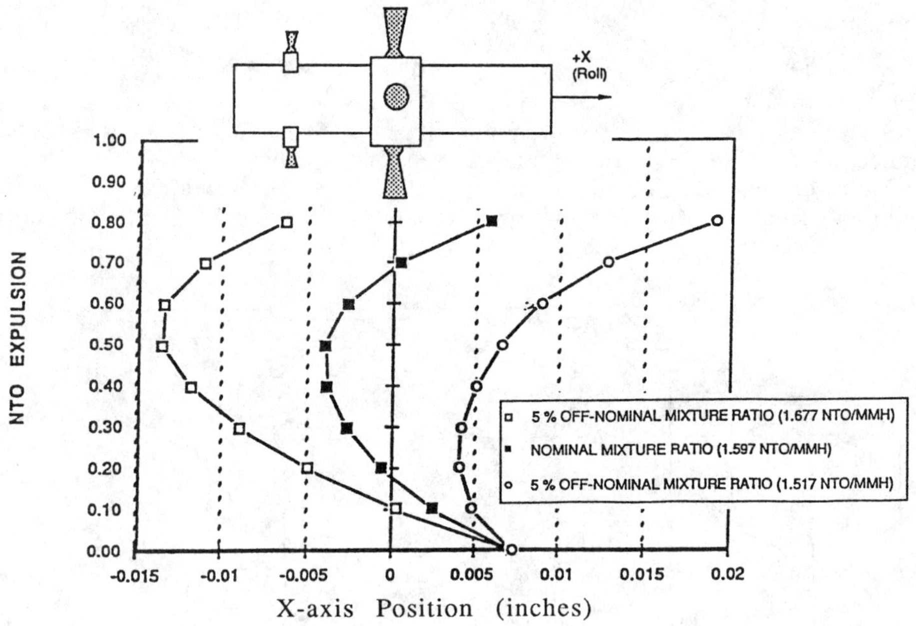

Figure 5. Predicted CG Travel

These and other inputs have been included in the simulation and selected results of the predicted performance during first flight of the KHIT vehicle are shown in Figs. 6, 7 and 8. The first 2.5 secs are shown here to highlight the expected behavior of the system, and predicted engine commands are shown in Fig. 9. These results indicate the expected performance of the system, but these results are dependant on initial estimates of several variables, such as the noise levels present in the IMU, and these estimates must wait for test results in order to be verified.

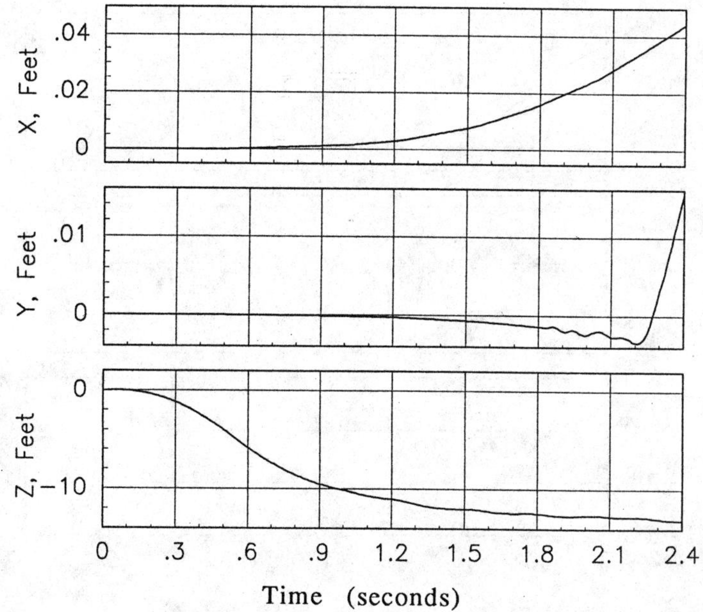

Figure 6. Predicted Position

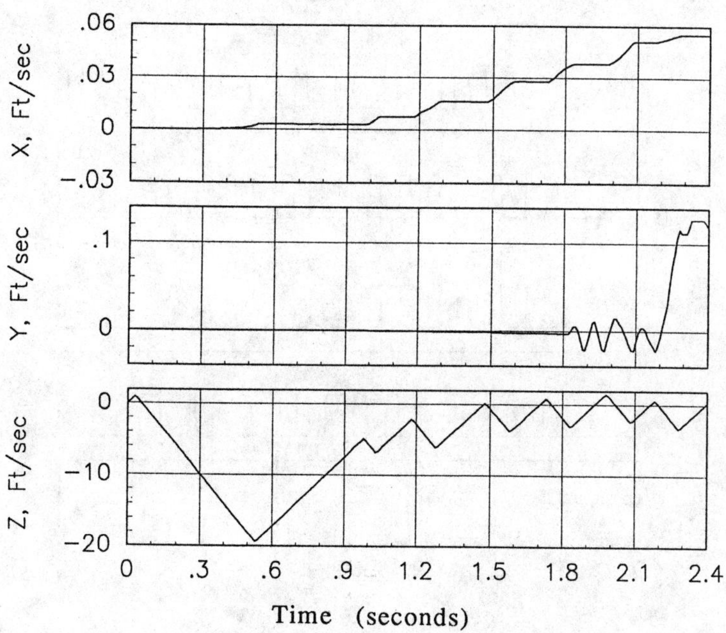

Figure 7. Predicted Velocity

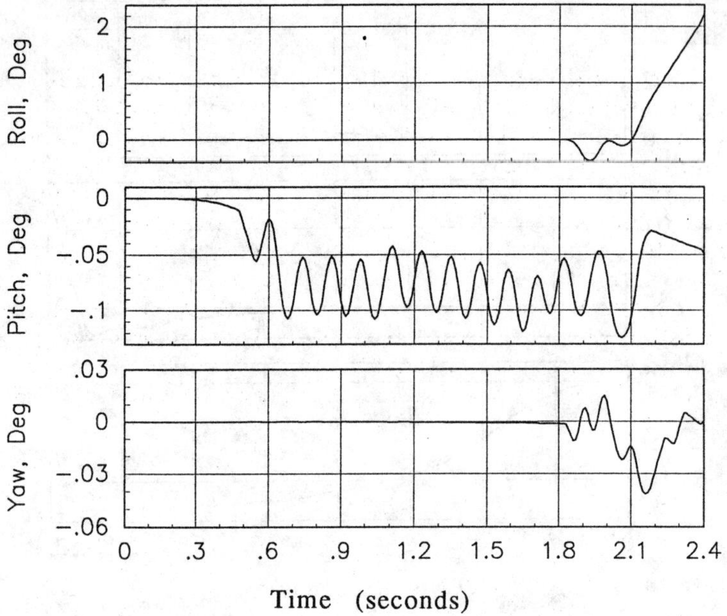

Figure 8.   Predicted Attitude

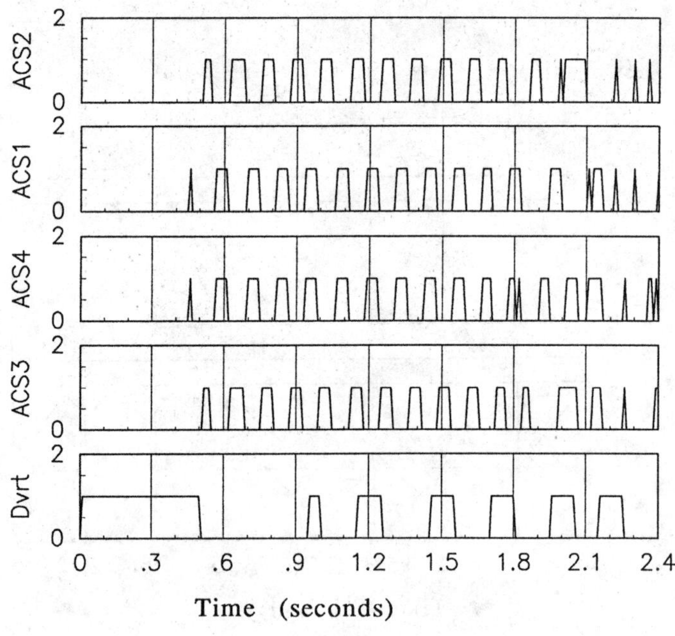

Figure 9.   Predicted Engine Commands

TESTING

Figure 10 shows the facility layout for the tests. Prior to the first free flight test of the KHIT vehicle, several static strapdown propulsion tests were performed in order to build confidence in the facility test setup, the test procedures, as well as the basic vehicle performance. Each static test built upon the information gained from, and the success of, the previous test. The first test of this static test series involved firing of each engine for a one second duration in order to verify the integrated propulsion system performance. The second test of the series was the first mission duty cycle firing of an integrated KKV propulsion system. The third and final static test of the series was a systems integration test which included the complete ground and airborne electronics controlling the vehicle through a mission duty cycle. Each of these tests increased the confidence of performing a successful free flight test.

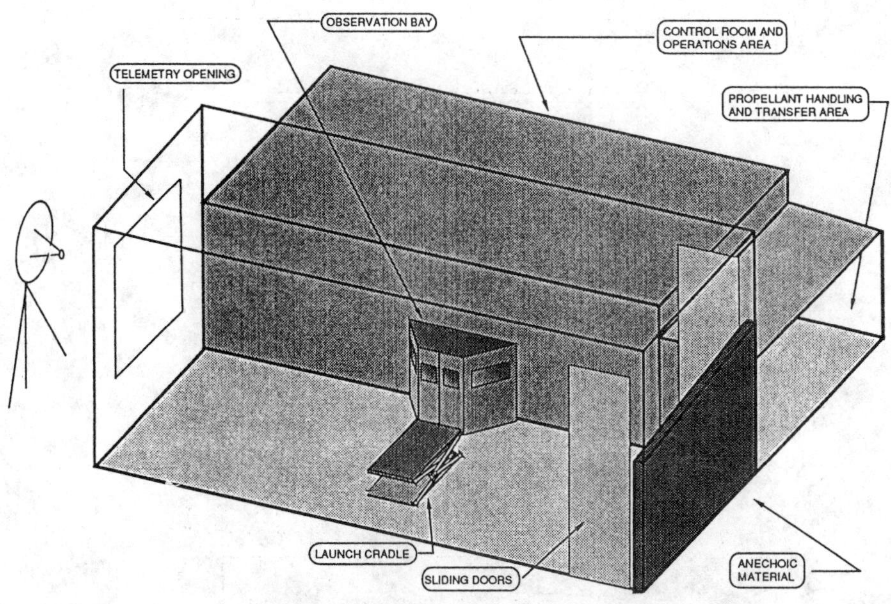

Figure 10.   Facility Orientation

The first free flight test of the KHIT vehicle occured on November 18,1988. The test duration was approximately 2 seconds, and was terminated early due to an inoperative ACS engine. Selected results of this first flight test are shown in Fig. 11 through 14. As can be seen from the engine command data presented in Fig. 14, ACS engine #3 was commanded to fire for the duration of the flight, but no change in pitch rate was detected. The range safety system terminated the flight at 2.15 seconds into the flight because the vehicle had exceeded the pitch attitude safety limit of .2 radians.

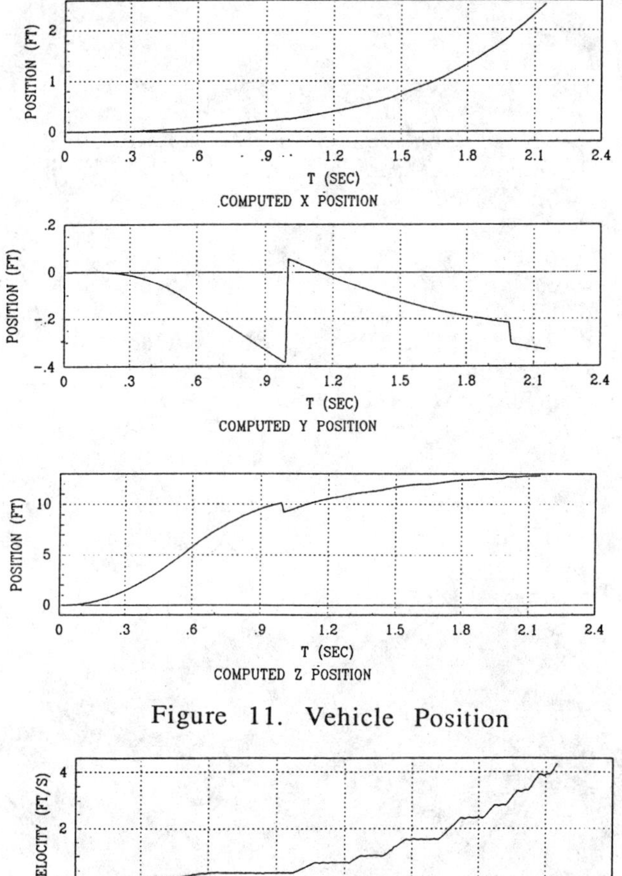

Figure 11. Vehicle Position

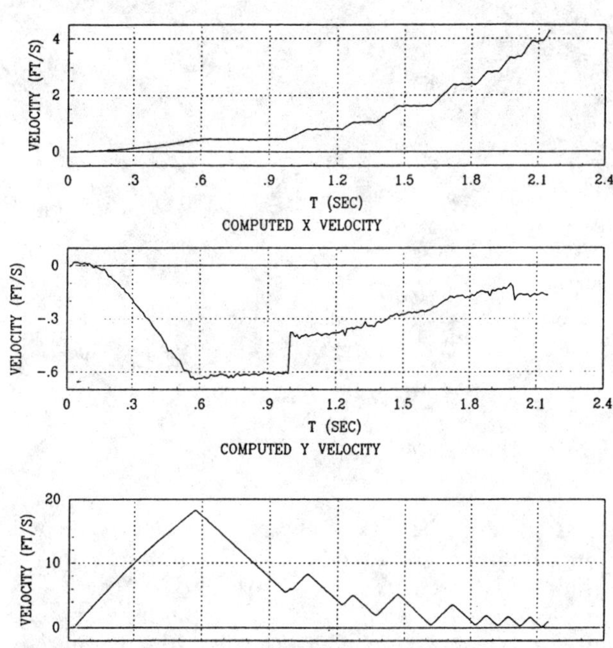

Figure 12. Vehicle Velocity

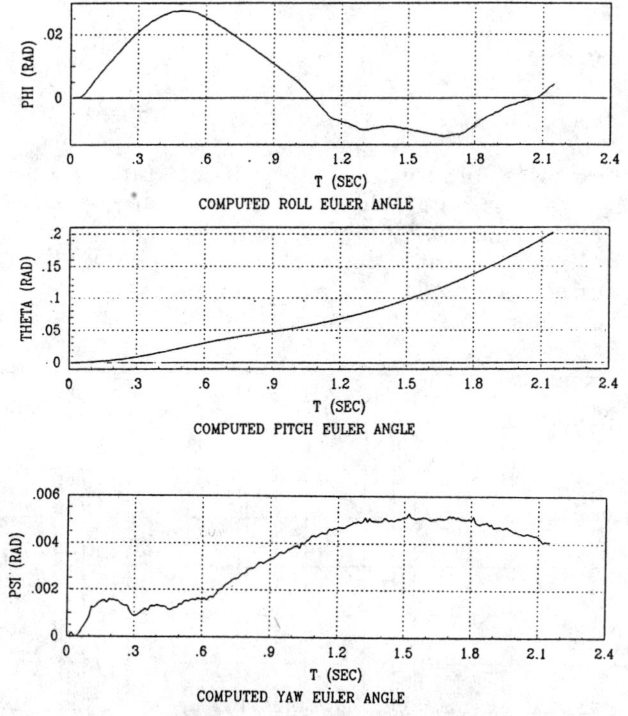

Figure 13.   Vehicle Attitude

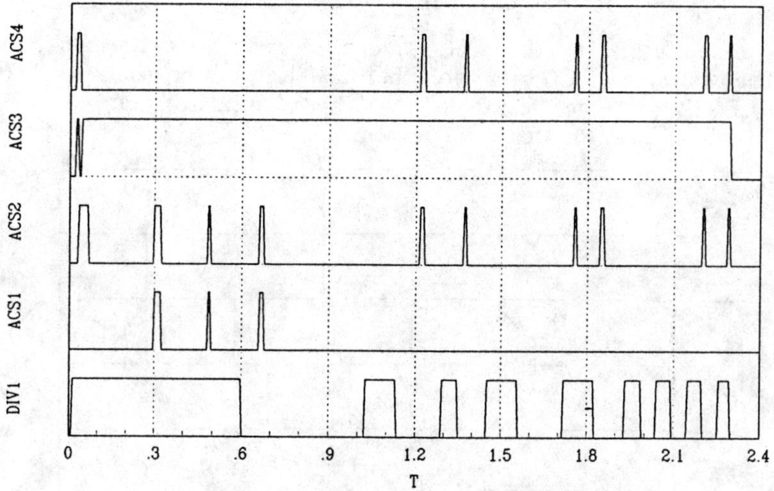

Figure 14.   Measured Valve Commands

PARAMETER ESTIMATION

The data obtained from this test were used as input to the EKF and the results obtained from this process are presented here. A significant departure from the modeled behavior was observed in the divert thruster performance. As can be seen in Fig. 15, the thrust level decays steadily during the first divert pulse from the predicted value of 350 lbf to a final value of 285 lbf. Thereafter, the thrust averages 320 lbf during each of the remaining divert pulses. This variation directly correlates to the increased pulse width observed during the first divert pulse, and the increased time required to acheive the commanded altitude.

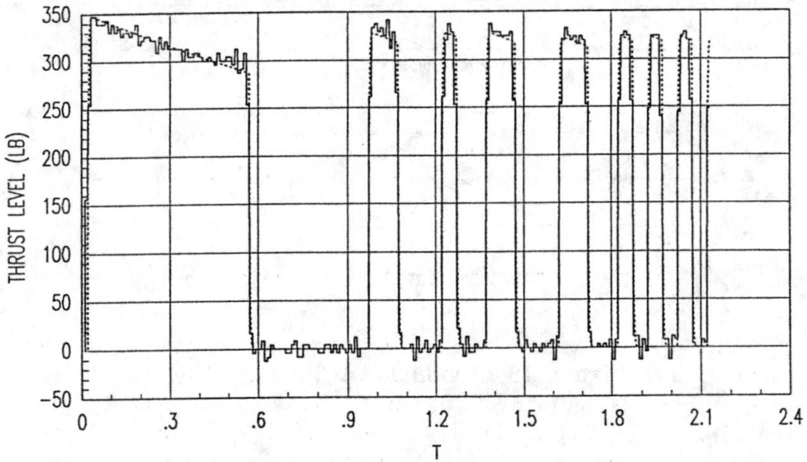

Figure 15. Estimated Divert Thrust

As can be seen in fig. 16, the CG travel in the x-axis does not behave as predicted. The estimated CG location before flight was used as the starting point for the EKF and the results shown here indicate that the CG travels forward instead of aft as the test progressed.

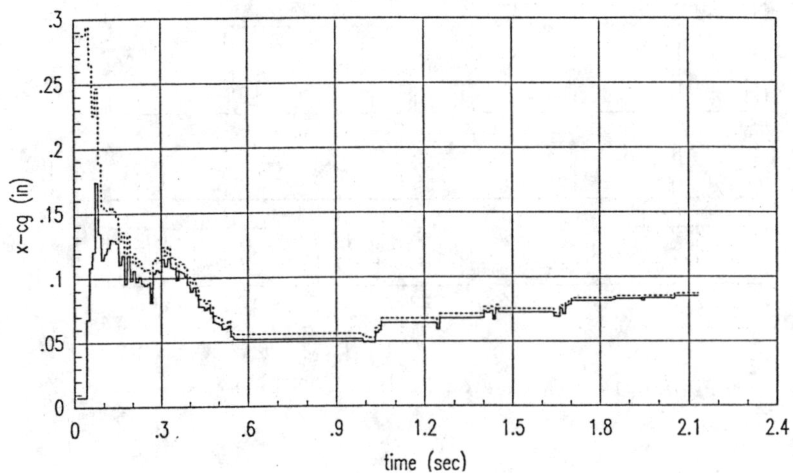

Figure 16. Estimated X-Axis CG Travel

As a check on the validity of these results, a different starting point of .285 in. forward of the divert plane was used and the EKF converged to the same solution after .3 seconds. The estimated CG travel in the Y-axis is shown in Fig. 17, and as expected, remains close to zero for the duration of the test.

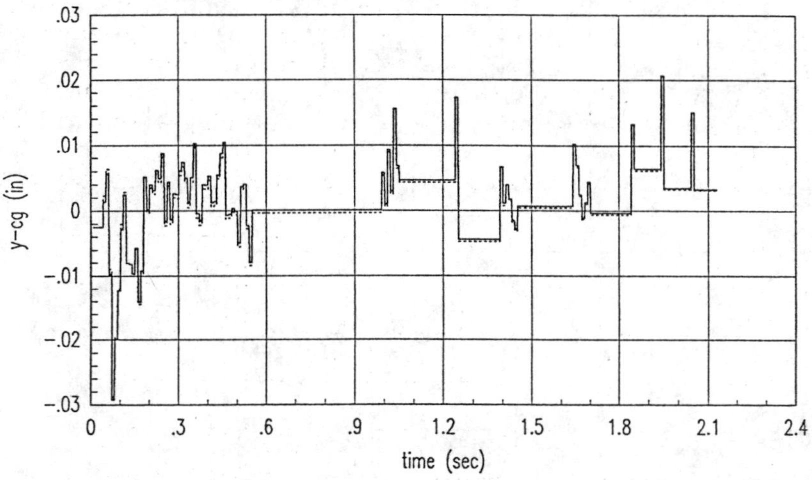

Figure 17. Estimated Y-Axis CG Travel

Both CG offsets and thrust vector misalignments cause rotational motion of the vehicle, but because any thrust misalignment will generate small forces in the axial and lateral directions, the EKF is able to attribute these forces to the misaligned thrust vector. The EKF estimates of divert engine thrust vector misalignments in the roll plane and the pitch plane are shown in Figs. 18 and 19. The estimated magnitudes of the CG offsets and thrust misalignments are within the limits established by the KHIT control system's capability, but as future KEW systems become smaller and as a result more sensitve to such disturbances, more precise manufacturing and assembly proccesses combined with more sophisticated control systems may be required.

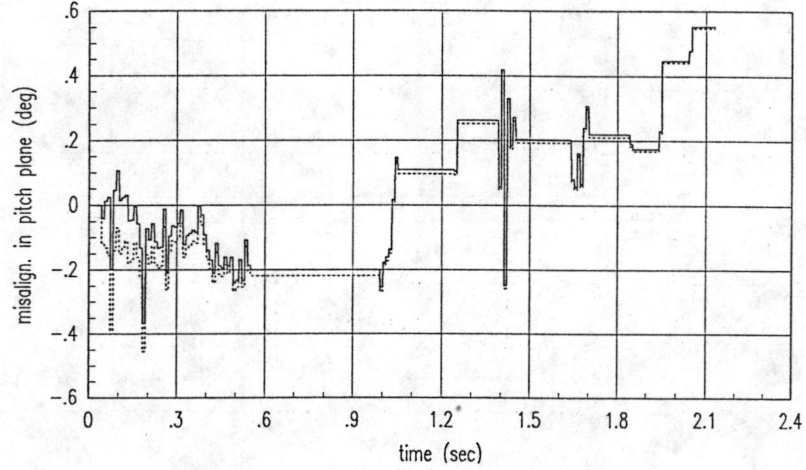

Figure 18. Estimated Thrust Misalignment in Pitch

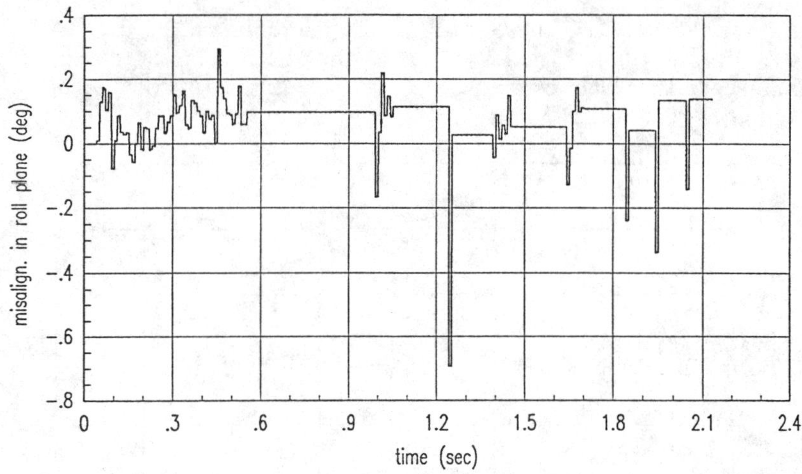

Figure 19.   Estimated Thrust Misalignment in Roll

COMPARISON

The next step was to use the output of the EKF to improve the correlation between the simulation and the actual test data, in order to validate the simulation, and to better understand the causes of the observed flight performance.   The estimates of the divert thrust variations, CG travel, and divert thruster misalignments obtained from the EKF were used in the simulation and selected results of the attempt at recreating the observed trajectory are presented in Figs. 20 through 23.

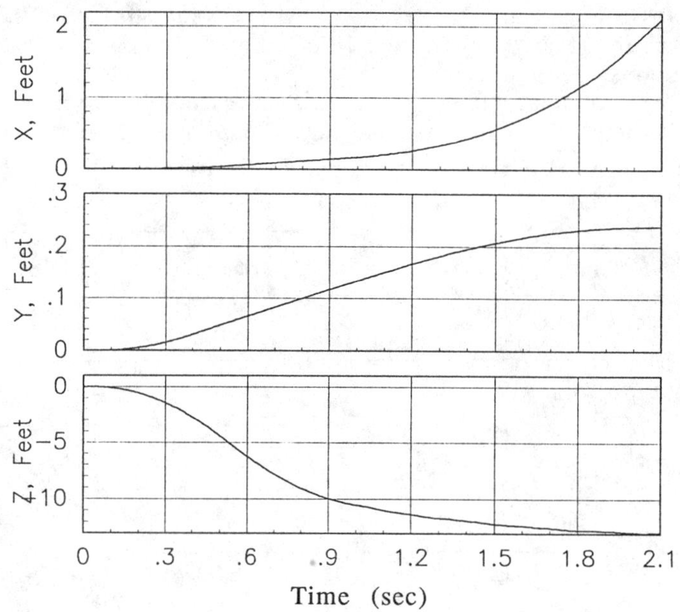

Figure 20.   Simulated Position

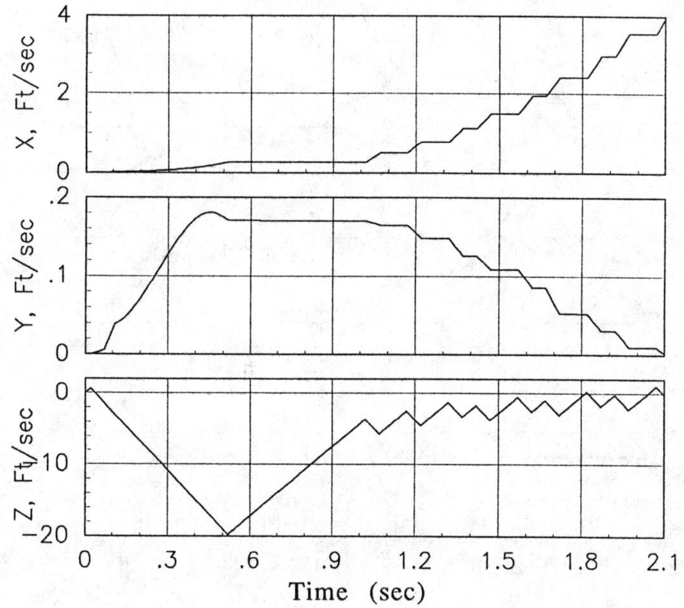

Figure 21.   Simulated Velocity

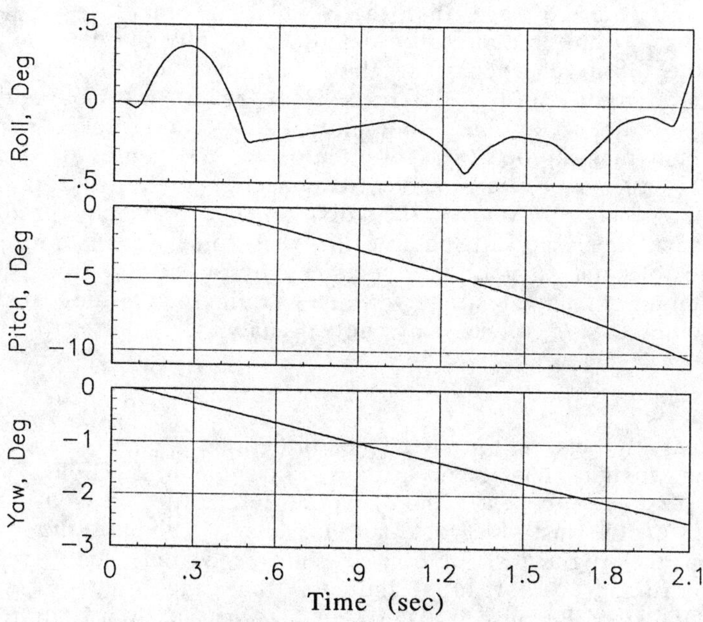

Figure 22.   Simulated Attitude

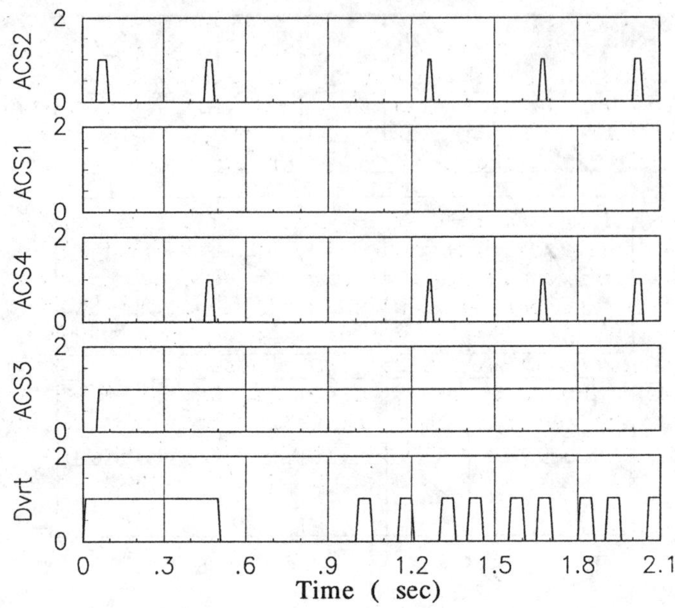

Figure 23. Simulated Engine Commands

By comparing the position integrated from the data obtained from the onboard IMU, shown in Fig. 11, with the position results obtained from the simulation, shown in Fig. 20, it can be seen that the simulation provides good agreement with the test results. The predicted pitch attitude obtained from the simulation closely follows the observed behavior also, but the velocity and remaining attitude results are significantly different from what was observed. The engine commands predicted by the simulation, shown in Fig 23, reflect the updated thruster and mass properties information as a result of the test, but when compared to the measured valve response, shown in Fig. 14, some disagreement still exists. While the differences between predicted and observed performance may be attributable to the noise levels present in the IMU, aerodynamic effects, structural dynamics, or a tipoff from the launch cradle, further simulation and testing will be required in order resolve these differences and complete the process of model analysis.

CONCLUSIONS

By using analytical models to predict the integrated response of the KHIT vehicle, significant insight has been gained into the real flight performance achieved in the first flight test, and several interesting lessons have been learned as a result of the test data obtained to date. Although the experimental data obtained from the flight test is limited and the vehicle had an inoperative attitude control thruster, the results indicate that the control system was performing as intended. Because the present design and configuration of the control system does not allow for recovery if an attitude engine fails, a slight change to the flight control and engine selection logic may help with a partially failed thruster. During the flight test, attitude engine #3 was commanded to be on for the duration of the flight because there was no

550

change in torque requirement. Post test examination of the engine valve determined that the valve occasionally required a higher pull-in current to open than was provided by the valve driver circuit. If the flight control system cycled ACS engine valves after an extended on command lasting 200 milliseconds, then the sticky attitude valve might have been coaxed into opening and the test may have been longer in duration.

The results obtained from the EKF imply that such a tool can be designed and used for system parameter identification, and the feedback of the results obtained from this tool can be used to better understand the modelling process. The thruster models are one area where performance predictions will have a significant impact on the behavior of the system, and accurate modelling is required if the predicted performance of the system is to be of any use to the system designer. Although a full understanding of the KHIT vehicle's flight performance cannot be obtained from the data because of an inoperative ACS thruster, this test provided an excellent opportunity to upgrade the simulation to better reflect what has been learned about the various systems, and their true characteristics.

ACKNOWLEDGMENT

The authors would like to gratefully acknowledge the help of Dr. Richard Bortins, Rob McEwen, Tom Trikas, and Minjea Tahk, of Integrated Systems Inc., for their effort in developing the simulation and post-flight data reduction tools discussed in this report.

APPENDICES

Applications of Modern State Space Analysis in Spacecraft Dynamics, Estimation and Control

John L. Junkins

Department of Aerospace Engineering
Texas A&M University

Tutorial Lecture
1989 AAS Guidance and Control Conference
Keystone, Colorado
February 5, 1989

OUTLINE

Session I

0700-0750am *Matrix manipulations: for the fun of it!*

0800-0850 *Attitude Dynamics and Control: Basic Formulations*

0900-1000 *Example applications: optimal open loop attitude maneuvers*
NOVA magnetic attitude maneuvers
Optimal momentum transfer maneuvers

Session II

1600-1650pm *Feedback control design*
Stability robustness measures
Eigenstructure assignment methods
Multiple criteria optimization methods in controller design

1700-1750 *Structural Identification: analytical & experimental results*
Structural ID using TAMU stereo otpical triangulation system
Minimum sensitivity output feedback control

1800-1900 *Feedback control design: nonlinear and distributed parameter systems*

Typical Structure of a Dynamic Optimal Control Problem Statement

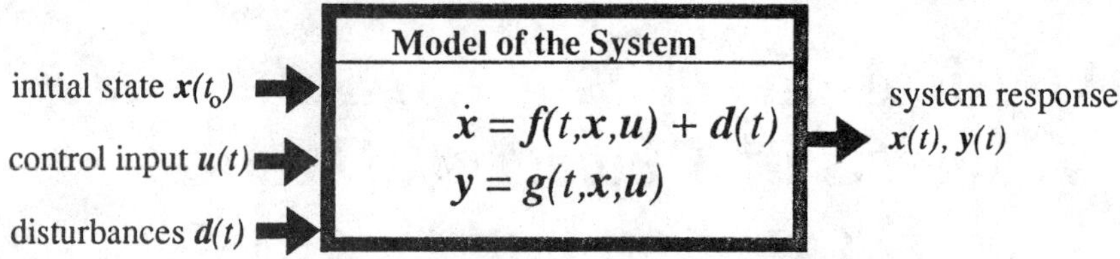

Given

- **model of the system**

- **initial state of the system**

- **specification of admissible controls**

- **specification of admissible outputs**

- **specification of performance index**

Determine

- **Optimum control $u^*(t)$, or _feedback law_ $u^*(t, x(t))$.**

A not so minor problem with the above is that all of the assumed "given" information must be <u>invented</u> by the engineer before he/she can proceed to use optimal control theory!!

Interplay of System Modeling, Control Formulation, Simulation Studies, and Implementation

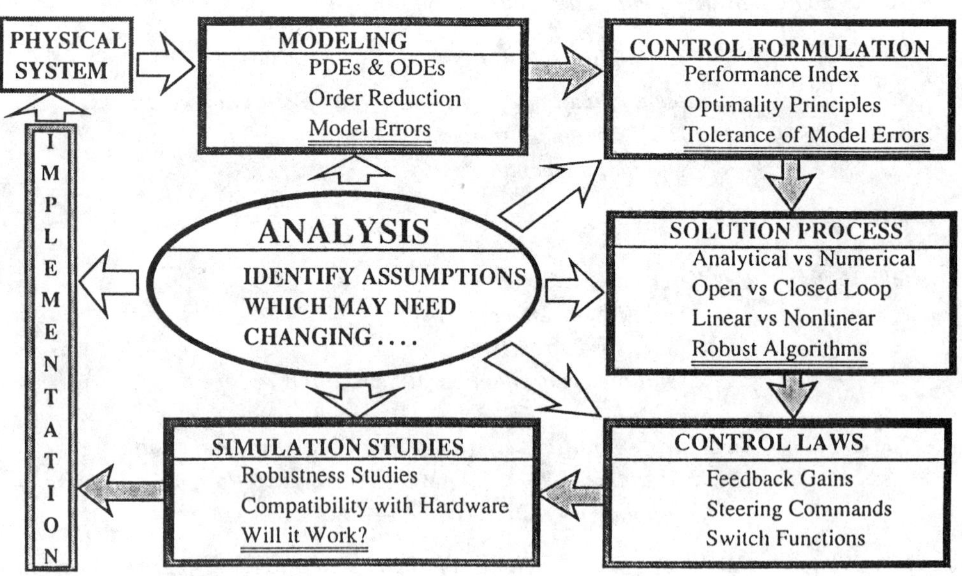

State Space Analysis ...
What is it?

It is a synthesis of topics from applied mathematics:

- *matrices and linear algebra*
- *differential equations*
- *numerical methods*
- *variational calculus*
- *random processes*
- *control theory*

➤ *Analysis of dynamical systems' behavior, especially vis-a-vis the response to disturbances, stability characteristics, estimation and error analysis, and design/performance evaluation of control laws.*

Historical Perspective

- ## State space analysis isn't really new!

  The key ideas had their 1800's genesis in the "generalized coordinates" and variational methods developed in the classical mechanics of Lagrange and Hamiltion. Also important are the classical probability & estimation concepts, and linear algebra foundations laid by Gauss et al. Since the mid 1940's, the following individuals added key additional ingredients:

  ➤ *Bliss, Pontryagin, Bellman (the Maximum Principle)*

  ➤ *Kalman, Bucy, Battin, et al (the Kalman filter)*

  ➤ *Many less monumental, but in sum decisively important contributions of analytical results and algorithms.*

- ## What's new? ... *Well, mainly, the notational unification and synthesis of a large body of material using systematic vector/matrix notations. "New" means since the mid 60's.*

Acknowledgements

Several colleagues and students have directly or indirectly made significant contributions to the analytical, numerical, and experimental results reported herein. I wish to especially thank the following individuals: D. Anderson, A. Amos, R. Battin, J. Blanton, D. Bodden, A. Browder, C. Carrington, N. Creamer, N. Hecht, I. Jacobson, G. James, J. Juang, Y. Kim, K. Lim, L. Meirovitch, S. Morgan, H. Morton, T. Pollock, Z. Rahman, D. Rew, R. Thompson, J. Turner, C. Williams, and R. Vadali.

I am pleased to acknowledge the generous institutional support received from the American Astronautical Society, and Texas A&M University to directly support preparation of this tutorial. The expert slide preparation support of Ms Becky Masters and my son J. Stephen Junkins are especially appreciated.

Some Random Thoughts

- **History Lesson:**

  "The Panama Canal could have been dug with a teaspoon, but it wasn't!!"

- **Trends:**

  ➤ *State space analysis is gaining importance as the dimensionality of dynamics & controls problems increases (# of coordinates, sensors, actuators, and other devices)*

  ➤ *Due to on-going refinements of pertinent computer software, especially vis-a-vis reliability and improved user interfaces, the computations are ever easier to carry out.*

Matrix Manipulations: for the fun of it!

- **Introduction and Basic Notions**

  - *Linear algebraic systems, least square & minimum norm solutions, partitioned algebra, vec & kronecker notations.*

  - *Eigenvalues and eigenvectors*

  - *MATLAB $\Longleftrightarrow$ "painless matrix computation"*

- **Matrix Decompositions & Other Wonders**

  - *Cholesky, Schur, Q-R, Singular-Value, and Spectral decompositions; $\Rightarrow$ Generalized inverses.*

  - *Examples*

- **Sensitivity and Conditioning Issues**
  - *Examples*

INTRODUCTION AND BASIC NOTIONS

<u>*system of m algebraic equations:*</u>

$$y_1 = a_{11}x_1 + a_{12}x_2 \; ... + a_{1n}x_n$$
$$y_2 = a_{21}x_1 + a_{22}x_2 \; ... + a_{2n}x_n$$
$$\vdots$$
$$y_m = a_{m1}x_1 + a_{m2}x_2 \; ... + a_{mn}x_n$$

$\Longleftrightarrow$

matrix equivalent:

$$y = Ax$$

where

$$y = \begin{Bmatrix} y_1 \\ y_2 \\ \vdots \\ y_m \end{Bmatrix}, \quad x = \begin{Bmatrix} x_1 \\ x_2 \\ \vdots \\ x_n \end{Bmatrix}, \quad A = \begin{pmatrix} a_{11} & a_{12} & ... & a_{1n} \\ a_{21} & a_{22} & ... & a_{2n} \\ \vdots & \vdots & & \vdots \\ a_{m1} & a_{m2} & ... & a_{mn} \end{pmatrix}$$

Standard Inverse Problem: Given y & A, determine x.

Overview of solutions of the inverse problem:

<u>Determined</u> **case:** [$m=n=$rank(A)], *exact & unique solution* $\Rightarrow$ $\quad x = A^{-1}y$.

<u>Overdetermined</u> **case:** [$m>n=$rank(A)], *no exact solution* generally exists,
the *'least square'* solution minimizing $|y - Ax|$ is $\quad \Rightarrow \; x = (A^TA)^{-1}A^Ty$.

<u>Underdetermined</u> **case:** [$n>m=$rank(A)], *an infinity of exact solutions* exist,
the *'minimum norm'* solution minimizing $|x|$ is $\quad \Rightarrow \; x = A^T(AA^T)^{-1}y$.

If the rank conditions aren't satisfied, can *use the 'generalized inverse'* $\Rightarrow x = A^{\dagger}y$.

559

Classification of Matrices $A=[a_{ij}]$

Type of Matrix	Defining Property	Nec. & Suff. Cond.
Symmetric	$A = A^T$	$a_{ij} = a_{ji}$
Hermitian	$A = A^H$	$a_{ij} = conj(\,a_{ji})$
Skew-Symmetric	$A = -A^T$	$a_{ij} = -\,a_{ji}$
Skew-Hermitian	$A = -A^H$	$a_{ij} = -\,conj(\,a_{ji})$
Orthogonal	$AA^T = A^TA = I$	
Unitary	$AA^H = A^HA = I$	
Idempotent	$A^2 = A$	
Nilpotent	$A^k = 0,\ \text{for some integer } k$	
Positive Definite (A>0)	$x^HAx > 0,\ \text{for all } x \neq 0$	all eigenvalues of $A > 0$
Pos. Semi- Def. (A ≥ 0)	$x^HAx \geq 0,\ \text{for all } x \neq 0$	all eigenvalues of $A \geq 0$

Notations for Elementary Matrix Operations $A=[a_{ij}]$

Function	Symbol	Description/Definition (computation)
Transpose	A^T	Interchange rows & columns $(i,j^{th}\text{ element of } A^T = a_{ji})$
Conj. Transpose	A^H	$(ij^{th}\text{ element of } A^H = conj(a_{ji}))$
Rank	$r(A)$	The smallest of the # of lip[*] rows and the # of lip columns (compute via SVD)
Trace	$tr(A)$	Sum of the diagonal elements (Σa_{ii})
Column Space	$C(A)$	The vector space spanned by the columns of A. (determine via SVD or QR)
Null Space	$N(A)$	The vector space containing all $x \neq 0$ satisfying $Ax = 0$. (determine via QR or SVD)

---

[*] Linearly independent (lip).

CLASSICAL LINEAR LEAST SQUARES PROBLEM
(Overdetermined Algebraic Systems)

or

$$y = Ax + e, \quad A \in R^{mxn} \quad (m \geq n) \Rightarrow \text{"more equations than unknowns"}$$

$$e = y - Ax = \text{"residual errors after the solution"}$$

Generally, no exact solution is possible $\Rightarrow$ we seek a solution with "*the smallest*" *e*.

Principle of Least Squares (Gauss): Find $\hat{x}$ to minimize

$$J = \tfrac{1}{2}\sum_j e_j^2 = \tfrac{1}{2}e^T e = \tfrac{1}{2}(y - Ax)^T(y - Ax) = \tfrac{1}{2}(y^T y - 2y^T Ax + x^T A^T Ax)$$

Gradient of *J* w.r.t $x = 0$ (necessary condition): "*Normal Equations*"

$$\nabla_x J = o = A^T y + A^T A\hat{x} \implies A^T A\hat{x} = A^T y$$

If rank (A) = rank $(A^T A)$ is *n*, then the normal equations can be solved for $\hat{x} = (A^T A)^{-1}A^T y$
More generally, we can introduce a positive definite weighting matrix *W* and minimize

$$J = \tfrac{1}{2}\sum_i\sum_j w_{ij}e_i e_j = \tfrac{1}{2}e^T We = \tfrac{1}{2}(y - Ax)^T W(y - Ax)$$

This gives the *weighted normal equations*: $\Rightarrow (A^T WA)\hat{x} = A^T Wy \quad or \quad \hat{x} = (A^T WA)^{-1}A^T Wy$

MINIMUM NORM PROBLEM
(Underdetermined Algebraic Systems)

$$y = Ax, \quad A \in R^{mxn} \quad (m \leq n) \Rightarrow \text{"fewer equations than unknowns"}$$

Typically, we expect an infinity of exact solutions, must introduce some criteria to select a particular exact solution. One often-useful choice is the "minimum norm" solution: Find $\hat{x}$ to minimize $\tfrac{1}{2}x^T x$ subject to satisfying $y - Ax = 0$.

Use Lagrange Multiplier Rule; form the augmented function: $J = \tfrac{1}{2}x^T x + \lambda^T(y - Ax)$

The necessary conditions for a constrained minimum are:
$$\nabla_x J = \hat{x} - A^T\lambda = 0$$
$$\nabla_\lambda J = y - A\hat{x} = 0$$

From which we find the minimum norm solution is obtained from: $\hat{x} = A^T\lambda, \quad and \quad (A A^T)\lambda = y.$

We can indicate the solution formally [if rank(A) = rank$(A A^T) = m$], as: $\hat{x} = A^T(A A^T)^{-1} y.$

The introduction of a weight matrix in $J = \tfrac{1}{2}x^T Wx + \lambda^T(y - Ax)$ leads to $\hat{x} = W^{-1}A^T(A W^{-1}A^T)^{-1} y$

561

FOUR PATHS TO SOLVING LEAST SQUARES PROBLEMS

1. *"Brute Force" inversion:* $\hat{x}_1 = (A^T A)^{-1} A^T y$ {MATLAB code: xhat1 = inv($A'*A$)*$A'*y$}

2. *Gaussian Elimination Solution for $\hat{x}_2$:* $(A^T A)\hat{x}_2 = A^T y$ {Gauss's Preferred Method}

3. *Pseudo-Inversion Using the Singular-Value-Decomposition* (Moore-Penrose generalized inverse):

$$Ax = y \quad , A = U\Sigma V^T \; , U^T U = UU^T = I \qquad\qquad \Sigma = \begin{bmatrix} \sigma & O \\ O & O \end{bmatrix}$$

$$A^\dagger = V\Sigma^\dagger U^T \; , Q^T Q = QQ^T = I \qquad\qquad \Sigma^\dagger = \begin{bmatrix} \sigma^{-1} & O \\ O & O \end{bmatrix}$$

$$\Rightarrow \quad \hat{x}_3 = A^\dagger y \qquad \{\text{MATLAB code: xhat3 = pinv}(A)*y\} \qquad \sigma = diag(\sigma_1, \sigma_2, \ldots, \sigma_r)$$

4. *QR Factorization:*

$$Ax = y, \; A = QR \; , Q^T Q = QQ^T = I ,$$

$$so \quad QR\,\hat{x}_4 = y$$

$$R\hat{x}_4 = Q^T y$$

$$U\hat{x}_4 = Q_1^T y \quad \Rightarrow \quad \text{solve for } \hat{x}_4 \text{ by back--substitution}$$

$$R = \begin{bmatrix} \diagdown \\ O \end{bmatrix} = \begin{bmatrix} U \\ O \end{bmatrix} \qquad Q = \begin{bmatrix} Q_1 \\ Q_2 \end{bmatrix}$$

Four Paths to Solving Least Squares Problems

Question: Does it really make any difference vis-a-vis speed & precision? Answer: Yes!

SPEED	PRECISION
fastest $O(n^2)$:	*most accurate & reliable:*
$\left\{\begin{array}{l} Gaussian\ elimination \\ QR\ factorization \end{array}\right.$	$\left\{\begin{array}{l} QR\ factorization \\ SVD\ generalized\ inverse \end{array}\right.$
slowest $O(n^3)$:	*least accurate & reliable:*
$\left\{\begin{array}{l} "brute\ force"\ inversion \\ SVD\ generalized\ inverse \end{array}\right.$	$\left\{\begin{array}{l} Gaussian\ elimination \\ "brute\ force"\ inversion \end{array}\right.$

Moral: The "brute force" inversion approach is slow, but it is inaccurate!

A FAMILAR EXAMPLE: POLYNOMIAL CURVE FITTING

$$y_j = x_1 + x_2 t_j + x_3 t_j^2 + \ldots + x_n t_j^{n-1} \, , \quad j = 1,2, \, , m$$

$$\text{or} \quad \begin{Bmatrix} y_1 \\ y_2 \\ : \\ y_m \end{Bmatrix} = \begin{bmatrix} 1 & t_1 & t_1^2 & \ldots & t_1^{n-1} \\ 1 & t_2 & t_2^2 & \ldots & t_2^{n-1} \\ : & : & : & \ldots & : \\ 1 & t_m & t_m^2 & \ldots & t_m^{n-1} \end{bmatrix} \begin{Bmatrix} x_1 \\ x_2 \\ : \\ x_n \end{Bmatrix}$$

$$\text{or} \quad y = Ax$$

Since the problem is over-determined, generally no coefficient vector x will yield an exact solution, so letting e denote the residual error vector,

$$y = Ax + e$$

Least squares solution: Find $\hat{x}$ which minimizes $J = e^T e$ as follows:

$$J = e^T e = (y - Ax)^T (y - Ax) \; = y^T y - 2y^T Ax + x^T A^T Ax$$

Necessary Condition: $\qquad \nabla_x J = O = -2A^T y + 2A^T A \hat{x} \implies \begin{array}{l} \textit{Normal Equations} \\ A^T A \, \hat{x} = A^T y \end{array}$

NUMERICAL DEMONSTRATIONS OF THREE LEAST SQUARES ALGORITHMS

$$A = \begin{bmatrix}
1 & 0 & 0 & 0 & 0 \\
1 & 1 & 1 & 1 & 1 \\
1 & 2 & 4 & 8 & 16 \\
1 & 3 & 9 & 27 & 81 \\
1 & 4 & 16 & 64 & 256 \\
1 & 5 & 25 & 125 & 625 \\
1 & 6 & 36 & 216 & 1296 \\
1 & 7 & 49 & 343 & 2401 \\
1 & 8 & 64 & 512 & 4096 \\
1 & 9 & 81 & 729 & 6561 \\
1 & 10 & 100 & 1000 & 10000 \\
1 & 11 & 121 & 1331 & 14641 \\
1 & 12 & 144 & 1728 & 20736 \\
1 & 13 & 169 & 2197 & 28561 \\
1 & 14 & 196 & 2744 & 38416 \\
1 & 15 & 225 & 3375 & 50625 \\
1 & 16 & 256 & 4096 & 65536 \\
1 & 17 & 289 & 4913 & 83521 \\
1 & 18 & 324 & 5832 & 104976 \\
1 & 19 & 361 & 6859 & 130321
\end{bmatrix}$$ (20 X 5 VANDERMONDE MATRIX)

$$x = \begin{Bmatrix} 1 \\ 1 \\ 1 \\ 1 \\ 1 \end{Bmatrix}, \quad y = \begin{Bmatrix} 1 \\ 31 \\ 121 \\ 341 \\ 781 \\ : \\ 137561 \end{Bmatrix} = Ax$$

INVERSION VIA GAUSSIAN ELIMINATION	GAUSSIAN LU REDUCTION AND BACK SUBSTITUTION	SVD COMPUTATION OF THE MOORE-PENROSE PSEUDOINVERSE
$\hat{x}_1 = (A^T A)^{-1} A^T y$	$(A^T A)\hat{x}_2 = A^T y$	$\hat{x}_3 = A^\dagger y$
$\{ \text{xhat1} = \text{inv}(A'*A)*A'*y \}$	$\{ \text{xhat2} = (A'*A) \backslash A'*y \}$	$\{ \text{xhat3} = \text{pinv}(A)*y \}$
$\text{error}_1 = \dfrac{\lvert \hat{x}_1 - x \rvert}{\lvert x \rvert} = 6x10^{-10}$	$\text{error}_2 = \dfrac{\lvert \hat{x}_2 - x \rvert}{\lvert x \rvert} = 1x10^{-10}$	$\text{error}_3 = \dfrac{\lvert \hat{x}_3 - x \rvert}{\lvert x \rvert} = 7x10^{-12}$

How do the dual curses of increasing dimensionality and ill-conditioning affect the above computations?

Accuracy of Least Square Solutions With Vandermonde Matrices

| | SOLUTION METHODS* | | |
Dimension of $A(m,n)$	Brute Force Inversion $e_1 = \dfrac{|\hat{x}_1 - x|}{|x|}$	Gaussian LU Factorization & Backsubstitution $e_2 = \dfrac{|\hat{x}_2 - x|}{|x|}$	SVD - Based Moore-Penrose Psuedo-Inverse $e_3 = \dfrac{|\hat{x}_3 - x|}{|x|}$
(20,2)	8×10^{-16}	5×10^{-16}	8×10^{-16}
(20,3)	9×10^{-14}	3×10^{-14}	4×10^{-14}
(20,4)	6×10^{-12}	2×10^{-12}	6×10^{-13}
(20,5)	6×10^{-10}	1×10^{-10}	7×10^{-12}
(20,6)	5×10^{-8}	8×10^{-9}	8×10^{-11}
(20,7)	7×10^{-6}	9×10^{-7}	4×10^{-9}
(20,8)	9×10^{-4}	6×10^{-4}	5×10^{-8}
(20,9)	1×10^{-1}	1×10^{-2}	3×10^{-6}
(20,10)	2×10^{1}	2	7×10^{-5}

All computations were performed using PCMATLAB software on an IBM AT with an 80 bit math co-processor.

Algorithm:	$\hat{x}_1 (A^T A)^{-1} = A^T y$	$(A^T A)\hat{x}_2 = A^T y$	$\hat{x}_3 = A^\dagger y$
		Solved by reducing $A^T A$ to upper triangular form and back-substituting.	$A^\dagger = V \Sigma^\dagger U^T$

Some Important Matrix Decompositions

- *Cholesky Decomposition*
- *QR Factorization*

- *Schur Decomposition*
- *Singular Value Decomp.*

- *Spectral Decomposition*

These Matrix decompositions are playing an increasingly important role in the analytical and computational aspects of dynamics, estimation, and control of aerospace vehicles. We'll consider several illustrative examples. Before concerning ourselves with the role that these ideas play in analysis of dynamical systems, it is instructive to first consider the "cleaner" applications of these ideas to solution of linear algebraic systems of equations.

Cholesky Decomposition

Consider a symmetric, positive definite matrix $A=[a_{ij}]$. The Cholesky Decomposition ("matrix square root") is

$$A = U^T U = \begin{bmatrix} u_{11} & u_{12} & .. & u_{1n} \\ 0 & u_{22} & .. & u_{2n} \\ \vdots & \vdots & & \vdots \\ 0 & 0 & .. & u_{nn} \end{bmatrix}^T \begin{bmatrix} u_{11} & u_{12} & .. & u_{1n} \\ 0 & u_{22} & .. & u_{2n} \\ \vdots & \vdots & & \vdots \\ 0 & 0 & .. & u_{nn} \end{bmatrix}$$

> MATLAB syntax
> \> U = chol(A)

Carrying thru the matrix multiplications and equating elements, we can establish the following algorithm for computing U (note U exists only if A is pos. def.):

$$a_{11} = u_{11}^2 \quad \Rightarrow \quad u_{11} = +\sqrt{a_{11}}$$

$$a_{12} = u_{11} u_{12} \quad \Rightarrow \quad u_{12} = a_{12}/u_{11}$$

$$a_{13} = u_{11} u_{13} \quad \Rightarrow \quad u_{13} = a_{13}/u_{11}$$

$$\vdots \qquad\qquad \vdots$$

$$a_{1n} = u_{11} u_{1n} \quad \Rightarrow \quad u_{1n} = a_{1n}/u_{11}$$

$$a_{22} = u_{12}^2 + u_{22}^2 \quad \Rightarrow \quad u_{22} = +\sqrt{a_{22} - u_{12}^2}$$

$$a_{23} = u_{12} u_{13} + u_{22} u_{23} \quad \Rightarrow \quad u_{23} = (a_{23} - u_{12} u_{13})/u_{22}$$

$$\vdots \qquad\qquad \Rightarrow \qquad\qquad \vdots$$

$$a_{2n} = u_{12} u_{1n} + u_{22} u_{2n} \quad \Rightarrow \quad u_{2n} = (a_{2n} - u_{12} u_{1n})/u_{22}$$

$$\vdots \qquad\qquad \vdots$$

$$a_{nn} = \sum_{i=1}^{n} u_{in}^2 \quad \Rightarrow \quad u_{nn} = +\sqrt{a_{nn} - \sum_{i=1}^{n-1} u_{in}^2}$$

Example Cholesky Decomposition

$$A = \begin{bmatrix} 5 & 10 & 30 & 100 \\ 10 & 30 & 100 & 354 \\ 30 & 100 & 354 & 1300 \\ 100 & 354 & 1300 & 4890 \end{bmatrix} = U^T U$$

$$U = \text{chol}(A) = \begin{bmatrix} 2.2361 & 4.4721 & 13.4164 & 44.7214 \\ 0 & 3.1623 & 12.6491 & 48.6991 \\ 0 & 0 & 3.7417 & 22.4499 \\ 0 & 0 & 0 & 3.7947 \end{bmatrix}$$

Applications of Cholesky Decomposition

- *Solving linear systems with positive definite coefficient matrices.*

- *The "square root form" of the Kalman filter covariance update algorithm*

- *Parameterization of any matrix for which we wish to guarantee positive definiteness (e. g., weight matrices, covariance matrices, ...)*

Diagonalization of Matrices: Spectral Decomposition

Question 1: Can $A \in R^{nxn}$ be diagonalized using a non-singular matrix U to operate on A as

$$U^{-1}AU = \Lambda = diag(\lambda_1, \lambda_2, ..., \lambda_n) ? \qquad (1)$$

This question can be re-stated {pre-multiplying (1) by U and post-multiplying by U^{-1}}, the above question can be re-stated as follows.

Question 2 ≡ Question 1: Can A be factored as: $\qquad A = U\Lambda U^{-1} ? \qquad (2)$

Some insight can be gained by rewriting (1) and/or (2) as: $\quad AU = U\Lambda \qquad (3)$

and equating columns on both sides of eq. (3), using $U = [u_1 \, ... \, u_n]$, we obtain

the eigenvalue problem: $\qquad A\,u_i = \lambda_i\,u_i\,,\ i=1,2,...,n \ (4)$

Thus diagonalization of A and existence of a non-singular U depends upon matrix A having n linearly independent eigenvectors. If the eigenvalues of A are distinct, then no problem! ... The story is more complicated otherwise.

The Singular Value Decomposition (SVD)

- *A square real* matrix $A \in R^{nxn}$ **can be factored as**

$$A = U\Sigma V^T , \quad U^T U = U U^T = I_n , \quad \Sigma = diag(\sigma_1, \sigma_2, ..., \sigma_n)$$

$$V^T V = V V^T = I_n , \quad \sigma_i \geq 0 \;\; \text{are the real singular values}$$

- **A square** *complex* **matrix** $A \in C^{nxn}$ **can be factored as**

$$A = U\Sigma V^H , \quad U^H U = U U^H = I_n , \quad \Sigma = diag(\sigma_1, \sigma_2, ..., \sigma_n)$$

$$V^H V = V V^H = I_n , \quad \sigma_i \geq 0 \;\; \text{are the real singular values}$$

- **A rectangular** *complex* **matrix** $A \in C^{mxn}$ **can be factored as**

$$A = U\Sigma V^H , \quad U^H U = U U^H = I_m , \quad \Sigma = \begin{bmatrix} \sigma & O \\ O & O \end{bmatrix}, \sigma = diag(\sigma_1, ..., \sigma_r)$$

$$V^H V = V V^H = I_n , \quad \sigma_i \geq 0, \quad \text{for } i = 1,2 ..., r, \; r = rank(A)$$

The SVD exists for *any* matrix and can usually be computed reliably using methods given in references 1-5.

An Important Relationship: $SVD \Longleftrightarrow \{A^T A v = \lambda v \; \& \; A A^T u = \lambda u\}$

To gain some insight, consider the case of a square real matrix A

$$A = U\Sigma V^T , \quad A^T = V\Sigma U^T , \quad \Sigma = diag(\sigma_1, \sigma_2, ..., \sigma_n) \qquad (1)$$

or

$$AV = U\Sigma \qquad \& \quad (2) \qquad A^T U = V\Sigma \qquad (3)$$

Denoting column vectors of U and V as $U = [u_1 ... u_n]$, $V = [v_1 ... v_n]$, *then equating columns on both sides of Eqs. (2) and (3), we obtain*

$$A v_i = \sigma_i u_i \; i = 1,2,...,n, \quad \& \quad (4) \qquad A^T u_i = \sigma_i v_i \; i = 1,2,...,n \qquad (5)$$

Substituting (4) $\Rightarrow$ (5), *and* (5) $\Rightarrow$ (4) *yields, for* $\sigma_i \neq 0$,

$$A^T A v_i = \sigma_i^2 v_i \; i = 1,2,...,n, \quad \& \quad (6) \qquad A A^T u_i = \sigma_i^2 u_i \; i = 1,2,...,n \qquad (7)$$

Conclude that the singular values (σ_i^2) *of A are the eigenvalues of* $A^T A$ & $A A^T$ *and the columns of* V, U *are the corresponding eigenvectors.*

The Moore-Penrose Generalized Inverse

● **For *any* mxn matrix A, there is a unique matrix $A^\dagger$ which satisfies the properties:**

(i) $\quad AA^\dagger A = A$ $\qquad\qquad\qquad$ (ii) $\qquad AA^\dagger = (AA^\dagger)^T$

(iii) $A^\dagger AA^\dagger = A^\dagger$ $\qquad\qquad$ (iv) $\qquad A^\dagger A = (A^\dagger A)^T$

(v) $\quad A^\dagger$ minimizes $|A^\dagger A - I_n|$, m>n, or $A^\dagger$ minimizes $|AA^\dagger - I_m|$, m<n.

● **For *any* mxn matrix, there exists a Singular Value Decomposition (SVD):**

$$A = U\Sigma V^H , \quad U^H U = U U^H = I_m , \quad \Sigma = \begin{bmatrix} \sigma & O \\ O & O \end{bmatrix}, \quad \sigma = diag(\sigma_1, ..., \sigma_r)$$

$$V^H V = V V^H = I_n , \quad \sigma_i \geq 0, \quad for\ i = 1,2\,...,\,r,\ \ r = rank(A)$$

● **For *any* mxn matrix, the Moore-Penrose Generalized Inverse can be written as**

$$A^\dagger = V\Sigma^\dagger U^H \quad , \quad \Sigma^\dagger = \begin{bmatrix} \sigma^{-1} & O \\ O & O \end{bmatrix}, \quad \sigma^{-1} = diag(\frac{1}{\sigma_1}, ..., \frac{1}{\sigma_r})$$

The fact that $A^\dagger = V\Sigma^\dagger U^H$ satisfies the first four properties can be verified by direct substitution.

Connections of the Generalized Inverse to the Least Square & Minimum Norm Problems

For an arbitrary real mxn matrix, the SVD is:

$$A = U\Sigma V^T, \quad U^T U = U U^T = I_m, \quad V^T V = V V^T = I_n , \quad \Sigma = \begin{bmatrix} \sigma & O \\ O & O \end{bmatrix}_{mxn}, \quad \sigma = diag(\sigma_1, ..., \sigma_r)$$

The Moore-Penrose Inverse is: $\qquad A^\dagger = V\Sigma^\dagger U^T , \quad \Sigma^\dagger = \begin{bmatrix} \sigma^{-1} & O \\ O & O \end{bmatrix}_{nxm}, \quad \sigma^{-1} = diag(\frac{1}{\sigma_1}, ..., \frac{1}{\sigma_r})$

<u>PROPERTIES:</u> We previously established that

I. $\quad A^T A = V S_n V^T , \quad S_n = \begin{bmatrix} \sigma^2 & O \\ O & O \end{bmatrix}_{nxn}, \quad \sigma^2 = diag(\sigma_1^2, ..., \sigma_r^2)$

II. $\quad A A^T = U S_m U^T , \quad S_m = \begin{bmatrix} \sigma^2 & O \\ O & O \end{bmatrix}_{mxm}$

Using I. and II., you can verify the following three important algebraic equivalences*:

III. $\quad$ If $m>n$ and $r = rank(A) = n$, then: $\quad A^\dagger = (A^T A)^{-1} A$ $\qquad$ (least square case).

IV. $\quad$ If $m<n$ and $r = rank(A) = m$, then: $\quad A^\dagger = A^T (A A^T)^{-1}$ $\qquad$ (minimum norm case).

V. $\quad$ If $r = rank(A) \leq min(m,n)$, then: $\quad A^\dagger = V\Sigma^\dagger U^T$ $\qquad$ (general, rank deficient case).

*Note algebraic equivalence *does not* imply computational equivalence!

The Singular Value Decomposition: Numerical Examples

$$A = \begin{bmatrix} 1 & 2 & 3 & 0 & 0 \\ 0 & 0 & 4 & 5 & 6 \\ 0 & 0 & 0 & 0 & 7 \\ 1 & 2 & 7 & 5 & 6 \end{bmatrix} = U\Sigma V^T = \begin{bmatrix} -.1303 & -.4318 & .6806 & .5774 \\ .5908 & .0686 & -.5594 & .5574 \\ .3375 & .8227 & .4574 & .0000 \\ -.7711 & -.3632 & .1212 & -.5774 \end{bmatrix} \begin{bmatrix} \sigma_1 & 0 & 0 & 0 & 0 \\ 0 & \sigma_2 & 0 & 0 & 0 \\ 0 & 0 & \sigma_3 & 0 & 0 \\ 0 & 0 & 0 & 0 & 0 \end{bmatrix} \begin{bmatrix} .0584 & -.1364 & .2708 & -.9506 & .0322 \\ .1169 & -.2729 & .5416 & .1746 & -.7669 \\ .5355 & -.6115 & .2205 & .2004 & .5005 \\ .4502 & -.2528 & -.7398 & .1604 & -.4004 \\ .72024 & .6849 & .1935 & .0000 & .0000 \end{bmatrix}^T$$

where the singular values of A are $\sigma_1 = 14.5695$, $\sigma_2 = 5.8276$, $\sigma_3 = 2.9611$.

The two "squared" matrices are: $AA^T = \begin{bmatrix} 14 & 12 & 0 & 26 \\ 12 & 77 & 42 & 89 \\ 0 & 42 & 49 & 42 \\ 26 & 89 & 42 & 115 \end{bmatrix}$, $A^T A = \begin{bmatrix} 2 & 4 & 10 & 5 & 6 \\ 4 & 8 & 20 & 10 & 12 \\ 10 & 20 & 74 & 55 & 66 \\ 5 & 10 & 12 & 50 & 60 \\ 6 & 12 & 66 & 60 & 122 \end{bmatrix}$

The eigenvalues and eigenvectors of AA^T, and $A^T A$ are related to the singular values and singular vectors of A as follows:

$$U^T(AA^T)U = \begin{bmatrix} 212.27 & 0 & 0 & 0 \\ 0 & 33.96 & 0 & 0 \\ 0 & 0 & 8.77 & 0 \\ 0 & 0 & 0 & 0 \end{bmatrix} = \begin{bmatrix} \sigma_1^2 & 0 & 0 & 0 \\ 0 & \sigma_2^2 & 0 & 0 \\ 0 & 0 & \sigma_3^2 & 0 \\ 0 & 0 & 0 & 0 \end{bmatrix}, \quad V^T(A^T A)V = \begin{bmatrix} 212.27 & 0 & 0 & 0 & 0 \\ 0 & 33.96 & 0 & 0 & 0 \\ 0 & 0 & 8.77 & 0 & 0 \\ 0 & 0 & 0 & 0 & 0 \\ 0 & 0 & 0 & 0 & 0 \end{bmatrix} = \begin{bmatrix} \sigma_1^2 & 0 & 0 & 0 & 0 \\ 0 & \sigma_2^2 & 0 & 0 & 0 \\ 0 & 0 & \sigma_3^2 & 0 & 0 \\ 0 & 0 & 0 & 0 & 0 \\ 0 & 0 & 0 & 0 & 0 \end{bmatrix}$$

Note: The above example does not mean that the SVD should be computed from the eigenvalues and eigenvectors of AA^T, and $A^T A$, there are more computationally attractive methods based upon Householder reductions.

keymone slides 27

The Singular Value Decomposition: Numerical Examples, continued

Consider

$$A = \begin{bmatrix} 1 & 2 & 3 & 0 & 0 \\ 0 & 0 & 4 & 5 & 6 \\ 0 & 0 & 0 & 0 & 7 \\ 1 & 2 & 7 & 5 & 6 \end{bmatrix}, \; x = \begin{Bmatrix} 1 \\ 1 \\ 1 \\ 1 \\ 1 \end{Bmatrix}, \; y = \begin{Bmatrix} 6 \\ 15 \\ 7 \\ 21 \end{Bmatrix}.$$

Since A is rank deficient (rank(A) = 3), we cannot directly compute with the minimum norm or least square solutions, since neither $(A^T A)^{-1}$ nor $(AA^T)^{-1}$ exists. However, we can compute the Moore-Penrose generalized inverse

$$A^\dagger = V\Sigma^\dagger U^T, \quad \Sigma^\dagger = \begin{bmatrix} \dfrac{1}{\sigma_1} & 0 & 0 & 0 & 0 \\ 0 & \dfrac{1}{\sigma_2} & 0 & 0 & 0 \\ 0 & 0 & \dfrac{1}{\sigma_3} & 0 & 0 \\ 0 & 0 & 0 & 0 & 0 \end{bmatrix}, \; \textit{where } \sigma_i\textit{'s and U,V are on previous chart.}$$

Then $\quad \hat{x} = A^\dagger y = \begin{Bmatrix} 0.3209 \\ 0.6419 \\ 1.4651 \\ 0.6279 \\ 1.0000 \end{Bmatrix}$. *Note that* $|x| = 2.236$, $|\hat{x}| = 2.014$, $\hat{x}$ *is the minimum norm solution.*

keymone slides 26

569

Recall

$$A = \begin{bmatrix} 1 & 2 & 3 & 0 & 0 \\ 0 & 0 & 4 & 5 & 6 \\ 0 & 0 & 0 & 0 & 7 \\ 1 & 2 & 7 & 5 & 6 \end{bmatrix} = U\Sigma V^T = \begin{bmatrix} .1303 & -.4318 & .6806 & .5774 \\ .5908 & .0686 & -.5594 & .5574 \\ .3375 & .8227 & .4574 & .0000 \\ .7711 & -.3632 & .1212 & -.5774 \end{bmatrix} \begin{bmatrix} \sigma_1 & 0 & 0 & 0 & 0 \\ 0 & \sigma_2 & 0 & 0 & 0 \\ 0 & 0 & \sigma_3 & 0 & 0 \\ 0 & 0 & 0 & 0 & 0 \end{bmatrix} \begin{bmatrix} .0584 & -.1364 & .2708 & -.9506 & .0322 \\ .1169 & -.2729 & .5416 & .1746 & -.7669 \\ .5355 & -.6115 & .2205 & .2004 & .5005 \\ .4502 & -.2528 & -.7398 & .1604 & -.4004 \\ .72024 & .6849 & .1935 & .0000 & .0000 \end{bmatrix}^T$$

This 4x5 matrix is of rank 3, therefore there should be 3 lin. indep. rows and columns. The first 3 columns of **V** (associated with non-zero singular values), are an orthogonal set of vectors spanning the columns of *A*. The last 2 columns of **V** are orthogonal vectors spanning the (column) *null space* of *A*. All vectors *x* lying in the null space of *A* satisfy *Ax=0*. Note that:

$$A v_i = 0, \; i=3,4 \implies \begin{bmatrix} 1 & 2 & 3 & 0 & 0 \\ 0 & 0 & 4 & 5 & 6 \\ 0 & 0 & 0 & 0 & 7 \\ 1 & 2 & 7 & 5 & 6 \end{bmatrix} \begin{Bmatrix} -.9506 \\ .1746 \\ .2004 \\ .1604 \\ .0000 \end{Bmatrix} = \begin{Bmatrix} .0000 \\ .0000 \\ .0000 \\ .0000 \\ .0000 \end{Bmatrix}, \quad \begin{bmatrix} 1 & 2 & 3 & 0 & 0 \\ 0 & 0 & 4 & 5 & 6 \\ 0 & 0 & 0 & 0 & 7 \\ 1 & 2 & 7 & 5 & 6 \end{bmatrix} \begin{Bmatrix} .0322 \\ -.7669 \\ .5005 \\ -.4004 \\ .0000 \end{Bmatrix} = \begin{Bmatrix} .0000 \\ .0000 \\ .0000 \\ .0000 \\ .0000 \end{Bmatrix}, \text{ as advertised.}$$

In a similar fashion, we can show that the fourth column of *U* spans the (column) null space of A^T, or the (row) null space of *A*.

In terms of the partitioned matrices, $y = Ax$

can be written as

$$\begin{Bmatrix} y_1 \\ y_2 \end{Bmatrix} = \begin{pmatrix} A_{11} & A_{12} \\ A_{21} & A_{22} \end{pmatrix} \begin{Bmatrix} x_1 \\ x_2 \end{Bmatrix} \tag{1}$$

or as the two equations

$$y_1 = A_{11}x_1 + A_{12}x_2 \tag{2}$$

$$y_2 = A_{21}x_1 + A_{22}x_2$$

For the special case that $m=n$, $r=k$ and the square diagonal submatrices (A_{11}, A_{22}) are presumed nonsingular, we can solve Eqs. (1), (2) two different ways. In the process, we will establish some important matrix identities which express A^{-1} directly as a function of the A_{ij} submatrices:

FIRST FORM:

1) Solve Eq. (1) for $\quad x_1 = A_{11}^{-1} y_1 - A_{11}^{-1} A_{12} x_2$ $\hfill$ (3)

2) Subs. Eq. (3) into Eq. (2) and solve for

$$x_2 = -(A_{22} - A_{21} A_{11}^{-1} A_{12})^{-1} A_{21} A_{11}^{-1} y_1 + (A_{22} - A_{21} A_{11}^{-1} A_{12})^{-1} y_2 \qquad (4)$$

3) Substitute Eq. (4) into Eq. (3) to obtain

$$x_1 = [A_{11}^{-1} + A_{11}^{-1} A_{12} (A_{22} - A_{21} A_{11}^{-1} A_{12})^{-1} A_{21} A_{11}^{-1}] y_1 - A_{11}^{-1} A_{12} (A_{22} - A_{21} A_{11}^{-1} A_{12})^{-1} y_2 \quad (5)$$

4) It is evident from Eq. (1) that $\begin{Bmatrix} x_1 \\ x_2 \end{Bmatrix} = \begin{bmatrix} A_{11} & A_{12} \\ A_{21} & A_{22} \end{bmatrix}^{-1} \begin{Bmatrix} y_1 \\ y_2 \end{Bmatrix}$, so $\hfill$ (6)

$$\begin{bmatrix} A_{11} & A_{12} \\ A_{21} & A_{22} \end{bmatrix}^{-1} = \begin{bmatrix} A_{11}^{-1} + A_{11}^{-1} A_{12} (A_{22} - A_{21} A_{11}^{-1} A_{12})^{-1} A_{21} A_{11}^{-1} & -A_{11}^{-1} A_{12} (A_{22} - A_{21} A_{11}^{-1} A_{12})^{-1} \\ -(A_{22} - A_{21} A_{11}^{-1} A_{12})^{-1} A_{21} A_{11}^{-1} & (A_{22} - A_{21} A_{11}^{-1} A_{12})^{-1} \end{bmatrix} \quad (7)$$

SECOND FORM:

1) Solve (1) for $\quad x_2 = A_{22}^{-1} y_2 - A_{22}^{-1} A_{21} x_1$ $\hfill$ (8)

2) Subs Eq. (8) into Eq. (2) and solve for

$$x_1 = -(A_{11} - A_{12} A_{22}^{-1} A_{21})^{-1} A_{12} A_{21}^{-1} y_2 + (A_{11} - A_{12} A_{22}^{-1} A_{21})^{-1} y_1 \qquad (9)$$

3) Substitute Eq. (9) into Eq. (8) to obtain

$$x_2 = [A_{22}^{-1} + A_{22}^{-1} A_{21} (A_{11} - A_{12} A_{22}^{-1} A_{21})^{-1} A_{12} A_{22}^{-1}] y_2 - A_{22}^{-1} A_{21} (A_{11} - A_{12} A_{22}^{-1} A_{21})^{-1} y_1 \quad (10)$$

4) It is evident from Eq. (1) that $\begin{Bmatrix} x_1 \\ x_2 \end{Bmatrix} = \begin{bmatrix} A_{11} & A_{12} \\ A_{21} & A_{22} \end{bmatrix}^{-1} \begin{Bmatrix} y_1 \\ y_2 \end{Bmatrix}$, so

$$\begin{bmatrix} A_{11} & A_{12} \\ A_{21} & A_{22} \end{bmatrix}^{-1} = \begin{bmatrix} (A_{11} - A_{12} A_{22}^{-1} A_{21})^{-1} & -(A_{11} - A_{12} A_{22}^{-1} A_{21})^{-1} A_{12} A_{22}^{-1} \\ -A_{22}^{-1} A_{21} (A_{11} - A_{12} A_{22}^{-1} A_{21})^{-1} & A_{22}^{-1} + A_{22}^{-1} A_{21} (A_{11} - A_{12} A_{22}^{-1} A_{21})^{-1} A_{12} A_{22}^{-1} \end{bmatrix} \quad (11)$$

Matrix Inversion Lemma & Related Identities

Equating the four submatrices of (7) & (11) immediately establishes the following four identities:

$$(A_{11} - A_{12} A_{22}^{-1} A_{21})^{-1} = A_{11}^{-1} + A_{11}^{-1} A_{12} (A_{22} - A_{21} A_{11}^{-1} A_{12})^{-1} A_{21} A_{11}^{-1} \tag{12a}$$

$$(A_{11} - A_{12} A_{22}^{-1} A_{21})^{-1} A_{12} A_{22}^{-1} = A_{11}^{-1} A_{12} (A_{22} - A_{21} A_{11}^{-1} A_{12})^{-1} \tag{12b}$$

$$(A_{22} - A_{21} A_{11}^{-1} A_{12})^{-1} A_{21} A_{11}^{-1} = A_{22}^{-1} A_{21} (A_{11} - A_{12} A_{22}^{-1} A_{21})^{-1} \tag{12c}$$

$$(A_{22} - A_{21} A_{11}^{-1} A_{12})^{-1} = A_{22}^{-1} + A_{22}^{-1} A_{21} (A_{11} - A_{12} A_{22}^{-1} A_{21})^{-1} A_{12} A_{22}^{-1} \tag{12d}$$

If we note that the numbering of dependent and independent variables is arbitrary, then we conclude that *there are actually only two identities*. Note that Eq. (12d) follows from Eq. (12a) by interchanging indices 1 and 2. Similarly, Eq. (12c) can be obtained from Eq. (12b). The above identities can be viewed in a more general context if we adopt the view that the submatrices A_{ij} are themselves <u>general</u> matrices, subject only to having compatible dimensions and the existence of the inverses; we conclude Eqs. (12) provides the possibility of <u>updating</u> inverses of large matrices by computing smaller ones. Suppose we have a square (m, m) matrix B *and have previously* computed its inverse B^{-1}. We now consider the inverse $(B+E)^{-1}$, where E is a perturbation of B. Consider, for example $E = C^T D^{-1} C$, $C \in R^{nxm}$, $\in R^{nxn}$, and D is nonsingular. It follows from Eq. (12a) that the inverse $(B+E)^{-1}$ can be updated as

$$(B + C^T D^{-1} C)^{-1} = B^{-1} - B^{-1} C^T (D + C B^{-1} C^T)^{-1} C B^{-1} \tag{13}$$

This is the *matrix inversion lemma* used, for example, to update the covariance matrix in sequential estimation.

Generalized Eigenvalue Problem $\Longleftrightarrow$ Spectral Decomposition

Consider Two General Matricies $A \in R^{nxn}$, $B \in R^{nxn}$, and the Two Eigenvalue Problems:

("Right")	$Au = \lambda B u$	$\Longleftrightarrow$	$[A - \lambda B]u = 0$	(1)
("Left")	$A^T v = \lambda B^T v$	$\Longleftrightarrow$	$[A^T - \lambda B^T]v = 0$	(2)

For a non-trival solution, λ's must be roots of the characteristic equation

$$\det(A - \lambda B) = 0 \quad \Rightarrow \quad \{\lambda_1, \ldots, \lambda_n\} \tag{3}$$

Thus (1) and (2) written for each λ are

$$A u_i = \lambda_i B u_i \quad , \quad i = 1, 2, \ldots, n \tag{4}$$

$$A^T v_j = \lambda_j B^T v_j \quad , \quad j = 1, 2, \ldots, n \tag{5}$$

Pre-multiply (4) by v_j^T and post-multiply the transpose of (5) by u_i, to obtain

$$\left.\begin{cases} v_j^T A u_i = \lambda_i v_j^T B u_i \\ v_j^T A u_i = \lambda_j v_j^T B u_i \end{cases}\right\} \text{ Subtraction yields } \Rightarrow (\lambda_i - \lambda_j) v_j^T B u_i = 0, \textit{ for all } i, j \tag{6}$$

For $i \neq j$, and $\lambda_j \neq \lambda_i$, eqs. (6) provide the orthogonality conditions

$$v_j^T B u_i = 0 , \quad v_j^T A u_i = 0 \qquad \textit{for all } i \neq j \tag{7}$$

We elect to normalize the eigenvectors so that

$$v_i^T B u_i = 1 , \quad u_i^T A u_i = \lambda_i , \quad i = 1, 2, \ldots, n \tag{8}$$

For $A\,u_i = \lambda_i B\,u_i$ and $A^T v_j = \lambda_j B^T u_i$, we established

$$det[A - \lambda B] = 0 \quad \Rightarrow \quad \lambda_1, \,...,\, \lambda_n$$

$$v_i^T B\,u_i = \delta_{ij}\,, \quad v_j^T A\,u_i = \lambda_i \delta_{ij} \tag{9}$$

Defining the modal matrices of right and left eigenvectors

$$U = [u_1,\,...,\,u_n]\ \text{ and }\ V = [v_1,\,...,\,v_n] \tag{10}$$

Then Eq. (9) can be written as the diagonalizing transformations

$$V^T B U = I\,, \quad V^T A U = \Lambda = diag(\lambda_1,\,...,\,\lambda_n) \tag{11}$$

Eqns. (11) can be inverted to obtain the spectral decompositions

$$A = V^{-T} \Lambda U^{-1}\,, \quad B = V^{-T} U^{-1} \tag{12}$$

Note the special cases:

I. $B = I$	II. $B = I, A = A^T$	III. $B = B^T, A = A^T$
$\Rightarrow\ U^{-1} \equiv V^T$	$U \equiv V,\ so$	$U \equiv V,\ so$
	$U^T U = I,\ U^T A U = \Lambda$	$U^T B U = I,\ U^T A U = \Lambda$
	$A = U \Lambda U^T$	$A = U^{-T} \Lambda U^{-1}$
		$B = U^{-T} U^{-1}$

The Real Schur Decomposition

... A "Close Cousin" of the Spectral Decomposition...

An arbitrary square matrix $A \in R^{n\times n}$ can be factored as

$$A = U S\, U^T \qquad \Longleftarrow\Longrightarrow \qquad S = U^T A U$$

where $\quad U \in R^{n\times n}$ is orthogonal $(U^T U = U U^T = I)$ and

$S \in R^{n\times n}$ is a block "upper triangular matrix" of the typical form:

$$S = \begin{bmatrix} \lambda_1 & x & x & x & ... & x \\ & \lambda_2 & x & x & ... & x \\ & & \lambda_{3r} & \lambda_{3i} & ... & x \\ & & -\lambda_{3i} & \lambda_{3r} & ... & x \\ & \bigcirc & & & \ddots & x \\ & & & & & \lambda_n \end{bmatrix}$$

The real λ's appear on the diagonal of S whereas the conjugate pairs' real and imaginary parts appear in 2 by 2 sub-matrices. More generally, the conjugate pairs are the eigenvalues of a corresponding 2 by 2 diagonal block submatrix of S.

Example Schur Decomposition

$$A = \begin{bmatrix} -1.7588 & -0.4023 & 0.9528 & -0.2926 & -0.6437 & -0.4701 \\ 0.1175 & -2.9821 & -0.9294 & 0.0149 & 0.2429 & 0.2811 \\ 0.0339 & -0.0492 & -1.2245 & -0.2901 & -0.8919 & -0.2720 \\ 0.0411 & -0.0193 & -0.0493 & -1.6486 & 0.2672 & -0.5539 \\ 0.0234 & 0.0149 & 0.0380 & -0.2528 & -2.5594 & -1.0639 \\ -0.1491 & -0.0002 & 0.0150 & -2.0736 & 2.3685 & 1.9326 \end{bmatrix} = U S U^T$$

$$\lambda(A) = \{-2.9634, -1.1613, -1.8932, 1.7636, 1.7636, -1.9932 \pm 0.724i\}$$

Matlab generates from $> [U,S] = schur(A)$ the following results:

$$S = \begin{bmatrix} 1.7636 & -0.1493 & -0.1439 & -0.0154 & 0.1568 & -3.544 \\ 0 & -2.9634 & -0.1865 & -0.7249 & -0.0118 & -0.0302 \\ 0 & 0 & -1.1613 & -1.0873 & -0.8725 & -0.7061 \\ 0 & 0 & 0 & -1.8932 & 0.2659 & 0.4855 \\ 0 & 0 & 0 & 0 & -1.9932 & 0.7254 \\ 0 & 0 & 0 & 0 & -0.7254 & -1.9932 \end{bmatrix},$$

$$U = \begin{bmatrix} 0.0780 & 0.3091 & -0.8291 & -0.4586 & -0.0188 & -0.0135 \\ 0.0434 & 0.9504 & 0.2803 & 0.1257 & 0.0206 & -0.0018 \\ -0.0044 & 0.0240 & -0.4771 & 0.8763 & 0.0460 & 0.0411 \\ -0.1737 & 0.0184 & 0.0163 & 0.0409 & -0.9282 & 0.3255 \\ -0.2251 & 0.0036 & 0.0095 & 0.0624 & -0.2813 & -0.9307 \\ 0.9546 & -0.0137 & -0.0775 & -0.0170 & -0.2375 & -0.1610 \end{bmatrix}$$

An Important Application of the Schur Decomposition

● **Solution of algebraic Riccati Equations**

Find $P \in R^{n \times n}$ satisfying: $Q + A^T P + PA - PB R^{-1} B^T P = 0$

Solution for P can be constructed as follows: Compute the Schur Decom. of $Z = \begin{pmatrix} A & B R^{-1} B^T \\ -Q & -A^T \end{pmatrix} = U S U^T$

where S and U are partitioned into n by n sub matrices: $U = \begin{bmatrix} U_{11} & U_{12} \\ U_{21} & U_{22} \end{bmatrix}, \qquad S = \begin{bmatrix} S_{11} & S_{12} \\ 0 & S_{22} \end{bmatrix}$

with U ordered so that all real parts of S_{11} are negative and all real parts of S_{22} are positive.

The solution for P is then obtained by solving the linear system: $U_{21} P = U_{11}$

---

40 A.J. Laub, "A Schur Method for Solving Algebraic Riccati Equations," *IEEE Trans on Automatic Control*, Vol AC-24, No. 6, pp 913-921, Dec 1979.

Generalized Schur Decomposition ("The Q-Z Decomposition")

For general square matrices $A \in C^{n \times n}$, $B \in C^{n \times n}$, there exists unitary matrices Q, Z such that

$$QAZ = T \quad \text{and} \quad QBZ = S$$

where both S and T are upper triangular.

This decomposition also possesses the important property that the eigenvalues satisfying $|A - \lambda B| = 0$ are

$$\lambda(A,B) = \{ t_{ii} \, / \, s_{ii} \, , \quad s_{ii} \neq 0 \} \quad i = 1, 2, ..., n$$

Main Applications of the Q - Z Decomposition:

- Generalized Eigenvalue Problems

- Finding solutions of Riccatii and Lyapunov Equations

MATLAB Example Computation Using the Q - Z Algorithm

MATLAB Syntax: The operation $> \; [a,b,Q,Z,V] = qz(A,B)$

returns factors Q,Z,a,b in

$$QAZ = a = \text{upper triangular}$$
$$QBZ = b = \text{upper triangular}$$

and right the eigenvector matrix V satisfying $\; AV \, diag(a) = BV \, diag(b)$

Consider the example:
$$K = \begin{bmatrix} 4 & 6 & 14 \\ 6 & 14 & 36 \\ 14 & 36 & 98 \end{bmatrix}, \quad M = \begin{bmatrix} 3 & 3 & 5 \\ 3 & 5 & 9 \\ 5 & 9 & 17 \end{bmatrix}$$

Then $[k, m, Q, Z, V] = qz(K, M)$ generates the following numerical results:

$$k = \begin{bmatrix} 18.44 & -7.04 & 111.57 \\ 0.00 & -1.79 & -4.94 \\ 0.00 & 0.00 & -2.42 \end{bmatrix}, \quad m = \begin{bmatrix} 0.922 & 1.729 & 22.708 \\ 0.000 & -1.790 & -4.942 \\ 0.000 & 0.000 & -2.423 \end{bmatrix}, \quad V = \begin{bmatrix} -0.1111 & 1.0000 & -0.5841 \\ 1.0000 & -0.4310 & 1.0000 \\ -0.5556 & 0.0326 & -0.2684 \end{bmatrix}$$

$$Q = \begin{bmatrix} 0.1048 & 0.3145 & 0.9435 \\ -0.9437 & -0.2677 & 0.1941 \\ 0.3136 & -0.9107 & 0.2687 \end{bmatrix}, \quad Z = \begin{bmatrix} 0.0967 & 0.9780 & 0.1848 \\ -0.8701 & -0.0071 & 0.4929 \\ 0.4834 & -0.2084 & 0.8503 \end{bmatrix}$$

Note that $[K - \omega_i^2 M] v_i \;\Rightarrow\; \omega_i^2 = k(i, i) / m(i, i) = \{20, 1, 1\}$

Matrix Decompositions: A Summary

Name	Matrix must satisfy these conditions:					Form	Comments > (matlab operator syntax)
	General Matrix	Square Matrix	Sym. Pos. Definite	Non-Singular	Distinct λ's		
LU		x		x		$A = LU$	Gaussian Elimination > $[L,U] = lu(A)$
Cholesky		x	x	x		$A = LL^T$	Matrix Square Root > $[L] = chol(A)$
Spectral		x			x	$A = U \Lambda U^{-1}$	$[U,\Lambda] = eig(A)$
Schur		x				$A = US U^T$	U orthogonal, S block upper triangular > $[U,S] = schur(A)$
QR	x					$A = Q\begin{bmatrix} R \\ 0 \end{bmatrix}$	$Q^H Q = I$, R upper triangular > $[Q,R] = qr(A)$
SVD	x					$A = US V^H$	$V^H U = I,\ V^H V = I$ $S = \begin{bmatrix} \sigma\, 0 \\ 0\, 0 \end{bmatrix}$, $\sigma = diag(\sigma_1, ..., \sigma_r)$ $\sigma_i \geq 0, real$ > $[U,S,V] = svd(A)$

Overview of Matrix Decompositions' Most common Applications

Name	Most Common Applications
LU	Solution of linear equations
Cholesky	Solution of linear equations with pos. def. coeff. matrix, parameterization of pos. def. matrices
Spectral	Diagonalization of matrices, solution of eigenvalue problems
Schur	Triangularization of matrices, orthogonal projections, Solutions of eigenvalue problems, Solution of Riccati and Lyapunov equations
QR	Solution of least square problems (see note), determination of orthogonal vectors spanning the range space and null spaces of matrices
SVD	Solution of least square and minimum norm problems (see note), determination of orthogonal sub spaces, determination of rank and condition numbers, determination of the Moore-Penrose psuedo-inverse

Note: Of the decompositions listed, only the QR and SVD can accomodate easily over and under-determined systems of equations, including the rank-difficient and poorly-conditioned cases. The QR and SVD decompositions have been implemented and can often be carried through accurately for fully populated systems with dimensions ≈ 100.

Kronecker and Vec Notations & Operations

The Kronecker Product of $A \in R^{p \times q}$ and $B \in R^{m \times n}$ is an $pm \times qn$ matrix given by

$$
A \otimes B \equiv
\begin{bmatrix}
a_{11}B & a_{12}B & \dots & a_{1q}B \\
a_{21}B & a_{22}B & \dots & a_{2q}B \\
\vdots & \vdots & & \vdots \\
a_{p1}B & a_{p1}B & \dots & a_{pq}B
\end{bmatrix}
$$

Let $A_1, A_2, \dots, A_q$ denote the columns of $A = [A_1, A_2, \dots, A_q]$.

The $nq \times 1$ vector $vec(A)$ is defined as

$$
vec(A) =
\left\{
\begin{matrix}
A_1 \\
A_2 \\
\vdots \\
A_q
\end{matrix}
\right\}
= \text{the columns of } A \text{ stacked into a vector}
$$

Kronecker and Vec Operational Identities

	Identity	Reference
1.	$(A \otimes B)^T = A^T \otimes B^T$	1
2.	$(A \otimes B)^{-1} = A^{-1} \otimes B^{-1}$; A, B *square and non-singular*	1
3.	$(A \otimes B)^\dagger = A^\dagger \otimes B^\dagger$	4
4.	$(A \otimes B) \otimes C = A \otimes (B \otimes C)$	1
5.	$(A + C) \otimes (B + D) = A \otimes B + A \otimes D + C \otimes B + C \otimes D$	1
6.	$vec(ADB) = (B^T \otimes A)vec(D)$	5
7.	$(A \otimes B)(C \otimes D) = AC \otimes BD$	1
For Square Matrices:		
8.	$det(A \otimes B) = det(A)^m det(B)^n$, $A \in R^{n \times n}$, $B \in R^{m \times m}$	1
9.	$tr(A \otimes B) = tr(A)tr(B))$	1
10.	$exp(A \otimes B) = [exp(A)] \otimes [exp(B)]$	6
11.	if $\{\lambda_i, x_i\}$ are an {eigenvalue, eigenvector} pair of A $\quad\quad \{\mu_k, y_k\}$ are an {eigenvalue, eigenvector} pair of B then $\quad\quad \{\lambda_i \mu_k, y_k \otimes x_i\}$ are an {eigenvalue, eigenvector} pair of $B \otimes A$ and $\quad\quad \{\lambda_i + \mu_k, y_k \otimes x_i\}$ are an {eigenvalue, eigenvector} pair of $B \otimes A$	3

Kronecker Factorization

Consider the three curve fitting problems:

I. $\quad p(x_i) = \alpha_0 + \alpha_1 x_i + ... + \alpha_n x_i^n + ...,\ i = 1, 2, ...$

Least square solution: Where:

$$\alpha = (A_1^T A_1)^{-1} A_1^T p \qquad \alpha^T = [\alpha_0\ \alpha_1\ \alpha_2 .....] \qquad A_1 = \begin{bmatrix} 1 & x_1 & x_1^2 & ... \\ 1 & x_2 & x_2^2 & ... \\ : & : & : & \end{bmatrix}$$

$$= A_1^\dagger p \qquad p^T = [p(x_1)\ p(x_2) ....]$$

II. $\quad q(y_j) = \beta_0 + \beta_1 y_j + ... + \beta_m y_j^m + ...,\ j = 1, 2, ...$

Least square solution: Where:

$$\beta = (A_2^T A_2)^{-1} A_2^T q \qquad \beta^T = [\beta_0\ \beta_1\ \beta_2 .....] \qquad A_2 = \begin{bmatrix} 1 & y_1 & y_1^2 & ... \\ 1 & y_2 & y_2^2 & ... \\ : & : & : & \end{bmatrix}$$

$$= A_2^\dagger q \qquad q^T = [q(y_1)\ q(y_2) ....]$$

Kronecker Factorization Continued

III. $\quad r(x_i, y_j) = \gamma_{00} + \gamma_{01} x_i + \gamma_{20} x_i^2 + ... + \gamma_{01} y_j + \gamma_{02} y_j^2 + ... + \gamma_{nm} x_i^n y_j^m + ...,\quad i = 1, 2, ...\ j = 1, 2, ...$

Least square solution: Where:

$$\gamma = (A^T A)^{-1} A^T r \qquad \gamma^T = [\gamma_{00}\ \gamma_{10}\ ...\ \gamma_{nm} .....]$$

$$= A^\dagger r \qquad r^T = [r(x_1, y_1), r(x_2, y_1) ....]$$

$$A = \begin{bmatrix} 1 & x_1 & x_1^2 & ... & y_1 & y_1^2 & ... & x_1 y_1 & x_1^2 y_1 & ... & x_1^n y_1^m & ... \\ : & : & : & & : & : & & : & : & & : & \\ 1 & x_n & x_n^2 & ... & y_m & y_m^2 & ... & x_n y_m & x_n^2 y_m & ... & x_n^n y_m^m & ... \\ : & : & : & & : & : & & : & : & & : & \end{bmatrix}$$

It can be verified by substitution that $\qquad A \equiv A_2 \otimes A_1$

From the results in the previous table, it can be verified that

$$(A^T A)^{-1} A^T \equiv [(A_2^T A_2)^{-1} A_2^T] \otimes [(A_1^T A_1)^{-1} A_1^T], \text{ and more generally: } A^\dagger \equiv A_2^\dagger \otimes A_1^\dagger$$

Kronecker Factorization: A Numerical Example

$$A_1 = \begin{bmatrix} 1 & 0 & 0 \\ 1 & 1 & 1 \\ 1 & 2 & 4 \\ 1 & 3 & 9 \end{bmatrix}$$

$$A_2 = \begin{bmatrix} 1 & 0 & 0 & 0 \\ 1 & 1 & 1 & 1 \\ 1 & 2 & 4 & 8 \\ 1 & 3 & 9 & 27 \\ 1 & 4 & 16 & 64 \end{bmatrix}$$

$$A = A_2 \otimes A_1 = \begin{bmatrix}
1 & 0 & 0 & 0 & 0 & 0 & 0 & 0 & 0 & 0 & 0 & 0 \\
1 & 1 & 1 & 1 & 0 & 0 & 0 & 0 & 0 & 0 & 0 & 0 \\
1 & 2 & 4 & 8 & 0 & 0 & 0 & 0 & 0 & 0 & 0 & 0 \\
1 & 3 & 9 & 27 & 0 & 0 & 0 & 0 & 0 & 0 & 0 & 0 \\
1 & 4 & 16 & 64 & 0 & 0 & 0 & 0 & 0 & 0 & 0 & 0 \\
1 & 0 & 0 & 0 & 1 & 0 & 0 & 0 & 1 & 0 & 0 & 0 \\
1 & 1 & 1 & 1 & 1 & 1 & 1 & 1 & 1 & 1 & 1 & 1 \\
1 & 2 & 4 & 8 & 1 & 2 & 4 & 8 & 1 & 2 & 4 & 8 \\
1 & 3 & 9 & 27 & 1 & 3 & 9 & 27 & 1 & 3 & 9 & 27 \\
1 & 4 & 16 & 64 & 1 & 4 & 16 & 64 & 1 & 4 & 16 & 64 \\
1 & 0 & 0 & 0 & 2 & 0 & 0 & 0 & 4 & 0 & 0 & 0 \\
1 & 1 & 1 & 1 & 2 & 2 & 2 & 2 & 4 & 4 & 4 & 4 \\
1 & 2 & 4 & 8 & 2 & 4 & 8 & 16 & 4 & 8 & 16 & 32 \\
1 & 3 & 9 & 27 & 2 & 6 & 18 & 54 & 4 & 12 & 36 & 108 \\
1 & 4 & 16 & 64 & 2 & 8 & 32 & 128 & 4 & 16 & 64 & 256 \\
1 & 0 & 0 & 0 & 3 & 0 & 0 & 0 & 9 & 0 & 0 & 0 \\
1 & 1 & 1 & 1 & 3 & 3 & 3 & 3 & 9 & 9 & 9 & 9 \\
1 & 2 & 4 & 8 & 3 & 6 & 12 & 24 & 9 & 18 & 36 & 72 \\
1 & 3 & 9 & 27 & 3 & 9 & 27 & 81 & 9 & 27 & 81 & 243 \\
1 & 4 & 16 & 64 & 3 & 12 & 48 & 192 & 9 & 36 & 144 & 576
\end{bmatrix}$$

Using Matlab:

$$A^\dagger = pinv(A) \equiv (A^T A)^{-1} A^T$$

accurate to $\approx 2 \times 10^{-13}$

required $55,650$ flops

$$A_1^\dagger = pinv(A_1) \equiv (A_1^T A_1)^{-1} A_1^T$$

$$A_2^\dagger = pinv(A_2) \equiv (A_2^T A_2)^{-1} A_2^T$$

$$A^\dagger = A_2^\dagger \otimes A_1^\dagger, \quad \text{accurate to} \approx 9 \times 10^{-14}$$

required $3,167$ flops

$\Rightarrow$ Indeed, Whether or not a matrix is Kronecker factorable has enormous conseqences!

Sensitivity and Conditioning Issues

Condition No. of a square matrix $A \in C^{n \times n}$

$$k(A) = \sigma_1(A) / \sigma_n(A)$$

where

$$A = U \begin{bmatrix} \sigma_1 & 0 \\ 0 & \sigma_n \end{bmatrix} V^H, \quad \sigma_1 \geq \sigma_2 \geq \ldots \geq \sigma_n \geq 0$$

Notes:

o $\sigma_i(A) = 1$, $i = 1, 2 \ldots n$, if A is orthogonal $(A \in R^{n \times n}, A^T A = A A^T = I)$

   or if A is unitary $(A \in C^{n \times n}, A^H A = A A^H = I)$

o $\sigma_n(A) = 0$ if A is singular

o $1 \leq k(A) < \infty$

o A key result vis-a-vis eigenvalue conditioning (Bauer-Fike Theorem)

   $|\Delta\lambda|_{max} \leq k(X) \|\Delta A\|$, X is the modal matrix whose columns are the of eigenvectors of A

$\Rightarrow$ This is usually conservative, but it is useful as a robustness measure.

Partial Derivatives of Eigenvalues and Eigenvectors

For matrices with distinct eigenvalues, the following results can be derived:*

$$A\,x_i = \lambda_i\,B\,x_i \qquad and \qquad A^T y_i = \lambda_i\,B\,y_i, \ where \ A = A(p), \ B = B(p)$$

The bi-orthogonality conditions are $\qquad y_j^T A\,x_i = \lambda_i\,\delta_{ij}, \ \ y_j^T B\,x_i = \delta_{ij} \ , \ \ x_j^T B\,x_i = 1$

The eigenvalue first derivatives are* $\qquad \dfrac{\partial \lambda_i}{\partial p} = y_i^T \left(\dfrac{\partial A}{\partial p} - \lambda_i \dfrac{\partial B}{\partial p} \right) x_i \ , \ \ i = 1, 2, \ldots, n$

and the eigenvector partial derivatives are $\qquad \dfrac{\partial x_i}{\partial p} = \sum\limits_{k=1}^{n} \alpha_{ik}\,x_k \ , \ \ \dfrac{\partial y_i}{\partial p} = \sum\limits_{k=1}^{n} \gamma_{ik}\,y_k$

where the projection coefficients are

$$\alpha_{ik} = \begin{cases} \dfrac{y_j^T \left(\dfrac{\partial A}{\partial p} - \dfrac{\partial \lambda_i}{\partial p}B - \lambda_i \dfrac{\partial B}{\partial p} \right) x_i}{(\lambda_i - \lambda_k)}, & i \neq k \\[4ex] -\dfrac{1}{2}\left[x_i^T \dfrac{\partial B}{\partial p} x_i + \sum\limits_{j=1, j \neq k}^{n} \alpha_{ij}\,x_i^T (B + B^T) x_j \right], & i = k \end{cases}$$

$$\gamma_{ik} = \begin{cases} \dfrac{y_j^T \left(\dfrac{\partial A}{\partial p} - \dfrac{\partial \lambda_i}{\partial p}B - \lambda_i \dfrac{\partial B}{\partial p} \right) x_i}{(\lambda_i - \lambda_k)}, & i \neq k \\[4ex] -\alpha_{ii} - y_i^T \dfrac{\partial B}{\partial p} x_i \ , & i = k \end{cases}$$

*Lim, Junkins, and Wang, *"Re-examination of eigenvector derivatives"*, **AIAA J. Guid., Ctrl., & Dynamics**, Vol. 10, No. 6, Nov-Dec, 1987, pp581-587.

keystone.slides 51

EIGENVALUE SENSITIVITY: A Numerical Example

$$A = \begin{bmatrix} 0 & 0 & 0 & 1 & 0 & 0 \\ 0 & 0 & 0 & 0 & 1 & 0 \\ 0 & 0 & 0 & 0 & 0 & 1 \\ -3 & -3 & -5 & -.1 & 0 & 0 \\ -3 & -5 & -9 & 0 & -.1 & 0 \\ -5 & -9 & -17 & 0 & 0 & -.1 \end{bmatrix}, \qquad A_1 = A + E, \qquad E = \begin{bmatrix} 0 & 0 & 0 & 0 & 0 & 0 \\ 0 & 0 & 0 & 0 & 0 & 0 \\ 0 & 0 & 0 & 0 & 0 & 0 \\ 0 & 0 & 0 & \Delta p & 0 & 0 \\ 0 & 0 & 0 & 0 & 0 & 0 \\ 0 & 0 & 0 & 0 & 0 & 0 \end{bmatrix}, \ \Delta p = 0.1$$

The eigenvalues of A are: $\qquad \lambda_{1,2} = -0.0500 \pm 0.3446i \quad \lambda_{3,4} = -0.0500 \pm 1.1843i \quad \lambda_{5,6} = -0.0500 \pm 4.8447i$

The eigenvalues of A_1 are: $\qquad \lambda_{1,2} = -0.0478 \pm 0.3450i \quad \lambda_{3,4} = -0.0060 \pm 1.1852i \quad \lambda_{5,6} = -0.0463 \pm 4.8447i$

The actual eigenvalues variations are: $\Delta\lambda_{1,2} = 0.0022 \pm 0.0004i \quad \Delta\lambda_{3,4} = 0.0440 \pm 0.0009i \quad \Delta\lambda_{5,6} = 0.0037 \pm 0.0000i$

The linearly (analytical partials)
predicted eigenvalues variations are: $\Delta\lambda_{1,2} = 0.0023 \pm 0.0003i \quad \Delta\lambda_{3,4} = 0.0440 \pm 0.0019i \quad \Delta\lambda_{5,6} = 0.0037 \pm 0.0000i$

The bound on $\Delta\lambda$ based upon the condition number of A's modal matrix is:

$$X \text{ is the modal matrix of } A, \ \ k(X) = \sigma_{max}(X)/\sigma_{min}(X) = 4.84, \ \Longrightarrow \ |\Delta\lambda|_{max} \leq k(X)\|E\| = 0.484$$

Notes: (i) The linear prediciton is accurate in this case, but nonlinear behavior occurs often!
(ii) The condition number bound is conservative by an order of magnitude; this is typical.
However, the condition number remains a very useful robustness measure.

References

1. Bellman, R. **Introduction to Matrix Analysis**, New York, McGraw-Hill, 1960, Ch. 20.

2. Barnett, S., *Matrix Differential Equations and Kronecker Products*, **SIAM J. Appl. Math.**, Vol. 24, No. 1, Jan 1973.

3. Brewer, J., *Kronecker Products and Matrix Calculus in System Theory*, IEEE Trans. on Cir. and Sys. V. CAS-25, No. 9, Sept. 1978.

4. Rao, C. and Mitra, S., **Generalized Inverse of Matrices and It's Applications**, New York, John Wiley, 1971, Ch. 1.

5. Neudecker, H., *Some Theorems on Matrix Differentiation with Special Reference to Kronecker Products*, **J. Am. Stat. Assoc.**, V. 64, pp 953-963, 1969.

6. Neudecker, H., *A Note on Kronecker Matrix Products*, **SIAM J. Appl. Math.**, V. 17, No. 3, May 1969.

7. Lim, K., Junkins, J., and Wang, B., *Re-examination of Eigenvector Derivatives* **AIAA J. Guidance Control,& Dynamics**, v. 10, No. 6, pp 581-587, 1987.

8. Laub A., "A Schur Method for Solving Algebraic Riccati Equations," *IEEE Trans on Automatic Control,* Vol AC-24, No. 6, pp 913-921, Dec 1979.

9. Stewart, G. **Introduction to Matrix Computations**, Academic Press, New York, 1973.

10. Junkins, J., and Kim, Y. *First and Second Order Sensitivity of the Singular Value Decomposition*, submitted to **AIAA J. Guidance, Control, and Dynamics**, January, 1989.

11. Wilkerson, J. **The Algebraic Eigenvalue Problem**, Oxford Univ. Press, London, 1965.

Matrix Manipulation Software Packages

SOFTWARE	COMMENTS*
1. **MATRIX$_x$** by Integrated Systems, Inc.	Most convenient for controller design owing to "System Build," etc., block diagram user interfaces. Most expensive.
2. **PCMATLAB & PROMATLAB** by Mathworks, Inc.	Easiest to extend, owing to user-defined "m.files", etc. Intermediate price.
3. **MATRIX.EZ** by Analytical Dynamics Associates	Has most convenient "spreadsheet" style matrix editing & related I/O features. Least expensive.

All three packages are comparable vis-a-vis ease of use and reliability of performing the "30 most common" matrix algebra operations (solution of linear equations, eigenvalues, eigenvectors, QR, QZ, SVD, Cholesky, Schur, . . ., etc.); all use well-tested algorithms descending from LINPACK, EISPACK, etc., all are extendable, and all stem from the same "ancestor" (the MATLAB software developed by Cleve Moler, et al circa 1980±). All three are interpretive languages which consequently run comparably slower than equivalent C or Fortran compiled code, but all three are usually two orders of magnitude more efficient than conventional programming vis-a-vis ease of use and elapsed calendar time!

---

* Opinions expressed are qualitative, based upon less than infinite experience with each software package, a rigorous "shootout" contest on problems A-Z has not been attempted!

Attitude Dynamics: Basic Formulations

OUTLINE

Spacecraft Attitude Kinematics and Dynamics

Choice of coordinates

Euler's Principal Rotation Theorem

MATLAB.JLJ coordinate transformations

Eulerian dynamics

A nonlinear oscillator analog of rigid body dynamics

Dual Spin Momentum Transfer Maneuvers

Energy/momentum surfaces of Barba & Aubrun

Sensitivity of Nonlinear Systems

Linearized departure motion

The symplectic property

Choice of Coordinates: This Fundamental Issue is Often Overlooked

Question: Which of the following "physically equivalent" systems of differential equations
(for the general rotational motion of a rigid body) do you prefer to solve?

SET 1. 3-1-3 Euler Angles	SET 2. Euler Parameters and Orthogonal Components of Angular Velocity

$$\ddot{\phi} = \frac{s\psi}{s\theta}[r_1\dot{\phi}^2 s\theta c\theta c\psi - (1+r_1)\dot{\phi}\dot{\theta}c\theta s\psi - (1-r_1)\dot{\phi}\dot{\psi}s\theta c\psi + (1-r_1)\dot{\theta}\dot{\psi}s\psi + \tfrac{u_1}{I_1}]$$

$$+ \frac{c\psi}{s\theta}[r_2\dot{\phi}^2 s\theta c\theta s\psi - (1-r_2)\dot{\phi}\dot{\theta}c\theta c\psi + (1+r_2)\dot{\phi}\dot{\psi}s\theta s\psi + (1+r_2)\dot{\theta}\dot{\psi}c\psi + \tfrac{u_2}{I_2}]$$

$$\dot{\beta}_0 = \tfrac{1}{2}[-\omega_1\beta_1 - \omega_2\beta_2 - \omega_3\beta_3]$$

$$\dot{\beta}_1 = \tfrac{1}{2}[-\omega_1\beta_0 - \omega_2\beta_3 + \omega_3\beta_2]$$

$$\dot{\beta}_2 = \tfrac{1}{2}[+\omega_1\beta_3 + \omega_2\beta_0 - \omega_3\beta_1]$$

$$\ddot{\theta} = c\psi[r_1\dot{\phi}^2 s\theta c\theta c\psi - (1+r_1)\dot{\phi}\dot{\theta}c\theta s\psi - (1-r_1)\dot{\phi}\dot{\psi}s\theta c\psi + (1-r_1)\dot{\theta}\dot{\psi}s\psi + \tfrac{u_1}{I_1}]$$

$$- s\psi[r_2\dot{\phi}^2 s\theta c\theta s\psi - (1-r_2)\dot{\phi}\dot{\theta}c\theta c\psi + (1+r_2)\dot{\phi}\dot{\psi}s\theta s\psi + (1+r_2)\dot{\theta}\dot{\psi}c\psi + \tfrac{u_2}{I_2}]$$

$$\dot{\beta}_3 = \tfrac{1}{2}[-\omega_1\beta_2 + \omega_2\beta_1 - \omega_3\beta_0]$$

$$\dot{\omega}_1 = r_1\omega_2\omega_3 + \tfrac{u_1}{I_1}, \qquad r_1 \equiv \frac{I_2-I_3}{I_1}$$

$$\dot{\omega}_2 = r_2\omega_3\omega_1 + \tfrac{u_2}{I_2}, \qquad r_2 \equiv \frac{I_3-I_1}{I_2}$$

$$\ddot{\psi} = -\frac{s\psi c\theta}{s\theta}[r_1\dot{\phi}^2 s\theta c\theta c\psi - (1+r_1)\dot{\phi}\dot{\theta}c\theta s\psi - (1-r_1)\dot{\phi}\dot{\psi}s\theta c\psi + (1-r_1)\dot{\theta}\dot{\psi}s\psi + \tfrac{u_1}{I_1}]$$

$$- \frac{c\psi c\theta}{s\theta}[r_2\dot{\phi}^2 s\theta c\theta s\psi - (1-r_2)\dot{\phi}\dot{\theta}c\theta c\psi + (1+r_2)\dot{\phi}\dot{\psi}s\theta s\psi + (1+r_2)\dot{\theta}\dot{\psi}c\psi + \tfrac{u_2}{I_2}]$$

$$\dot{\omega}_3 = r_3\omega_1\omega_2 + \tfrac{u_3}{I_3}, \qquad r_3 \equiv \frac{I_1-I_2}{I_3}$$

$$+ s\theta[r_3\dot{\phi}^2 s^2\theta c\psi s\psi - r_3\dot{\theta}^2 s\psi c\psi + (1-r_3)\dot{\phi}\dot{\theta}s\theta + \tfrac{u_3}{I_3}], \quad s\equiv sin, c\equiv cos$$

$\Longrightarrow$ If you selected the one on the left, you probably spell "FORTRAN with a "PH" ! *(as in phortran !!!)*.

582

Euler's Principal Rotation Theorem

A rigid body (or a reference frame) can be brought from an arbitrary initial orientation to an arbitrary final orientation by rotating the body about a judicious line (the *principal line e*) through a judicious angle (the *principal angle* ϕ). The principal line is fixed in the body and in space.

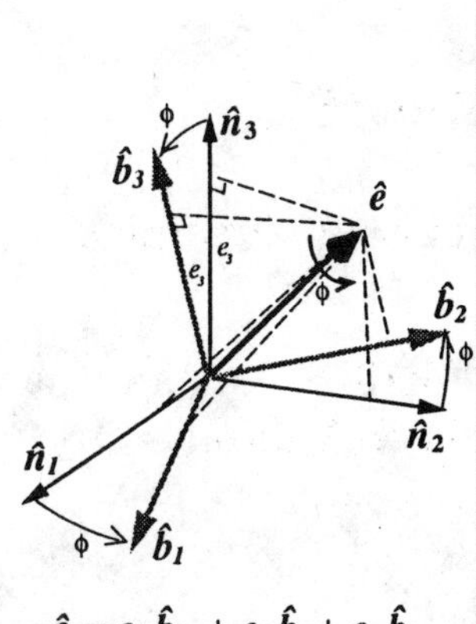

$$\hat{e} = e_1\,\hat{b}_1 + e_2\,\hat{b}_2 + e_3\,\hat{b}_3$$

$$\hat{e} = e_1\,\hat{n}_1 + e_2\,\hat{n}_2 + e_3\,\hat{n}_3$$

Projection of unit vectors: $\{\hat{b}\} = [C]\{\hat{n}\}$

Components of a vector e: $\hat{e} = e_{b1}\,\hat{b}_1 + e_{b2}\,\hat{b}_2 + e_{b3}\,\hat{b}_3$
project as $\hat{e} = e_{n1}\,\hat{n}_1 + e_{n2}\,\hat{n}_2 + e_{n3}\,\hat{n}_3$

$$\begin{Bmatrix} e_{b1} \\ e_{b2} \\ e_{b3} \end{Bmatrix} = [C] \begin{Bmatrix} e_{n1} \\ e_{n2} \\ e_{n3} \end{Bmatrix} \textit{ but according to Euler's Prin. Rot. Thm.}$$

$$e_{bi} \equiv e_{ni}$$

so we conclude that

$$\begin{Bmatrix} e_1 \\ e_2 \\ e_3 \end{Bmatrix} = [C] \begin{Bmatrix} e_1 \\ e_2 \\ e_3 \end{Bmatrix} \implies e = [C]e \implies e \textit{ must be an eigenvector of } [C]$$

> *The existence of the principal rotation vector requires that the direction cosine matrix have an eigenvalue of +1. ... Goldstein proves that all proper [C] matrices (corresponding to projections of right-handed triads of rigidly displaced unit vectors) do in fact have only one real eigenvalue and it is +1.*

Orientation as a Function of Principal Rotation Parameters

We have established that*

$$[C] = \begin{bmatrix} e_1^2(1-\cos\phi)+\cos\phi & e_1 e_2(1-\cos\phi)+e_3\sin\phi & e_1 e_3(1-\cos\phi)-e_2\sin\phi \\ e_2 e_1(1-\cos\phi)-e_3\sin\phi & e_2^2(1-\cos\phi)+\cos\phi & e_2 e_3(1-\cos\phi)+e_1\sin\phi \\ e_3 e_1(1-\cos\phi)+e_2\sin\phi & e_3 e_2(1-\cos\phi)-e_1\sin\phi & e_3^2(1-\cos\phi)+\cos\phi \end{bmatrix}$$

Since $e^T e = 1 = e_1^2 + e_2^2 + e_3^2$, we have only 3 dof, as expected.

The inverse transformation (given [C], determine e, ϕ) is straightforward.
The prin. rot. vector e is determined directly from $e=[C]e$, then careful
inspection of [C] reveals that

$$\cos\phi = (tr(C) -1)/2 = (c_{11} + c_{22} + c_{33} - 1)/2$$
$$\sin\phi = [e_3(c_{12}-c_{21}) + e_2(c_{31}-c_{13}) + e_1(c_{23}-c_{32})]/2 \implies \phi$$

Junkins & Turner, **Optimal Spacecraft Rotational Maneuvers, Elsevier, 1986.*

Euler (Quaternion) Parameterization of Attitude

Introducing Euler's (1776) definitions: $\quad \beta_0 = \cos\frac{\phi}{2}, \quad \beta_i = e_i \cos\frac{\phi}{2}, \quad i=1,2,3$

It is obvious that $\quad \sum_{i=0}^{3} \beta_i^2 = \beta_0^2 + \beta_1^2 + \beta_2^2 + \beta_3^2 = \beta^T \beta = 1.$

Making use of the half angle identities

$$\sin\phi = 2\sin\frac{\phi}{2}\cos\frac{\phi}{2} \quad and \quad \cos\phi = 2\cos s^2\frac{\phi}{2} - 1$$

$[C(\phi, e)]$ *becomes*

$$[C(\beta)] = \begin{bmatrix} \beta_0^2 + \beta_1^2 - \beta_2^2 - \beta_3^2 & 2(\beta_1\beta_2 + \beta_0\beta_3) & 2(\beta_1\beta_3 - \beta_0\beta_2) \\ 2(\beta_1\beta_2 - \beta_0\beta_3) & \beta_0^2 - \beta_1^2 + \beta_2^2 - \beta_3^2 & 2(\beta_2\beta_3 + \beta_0\beta_1) \\ 2(\beta_1\beta_3 + \beta_0\beta_2) & 2(\beta_2\beta_3 - \beta_0\beta_1) & \beta_0^2 - \beta_1^2 - \beta_2^2 + \beta_3^2 \end{bmatrix}$$

Also, we have the inverse relationships

$$\beta_i^2 = \frac{1}{4}(1 + 2c_{ii} - c_{oo}), \quad i=0,1,2,3, \quad c_{oo} \equiv tr[C]$$

$$\beta_0\beta_1 = (c_{23} - c_{23})/4, \qquad \beta_2\beta_3 = (c_{23} + c_{23})/4$$

$$\beta_0\beta_2 = (c_{31} - c_{13})/4, \qquad \beta_1\beta_3 = (c_{31} + c_{13})/4$$

$$\beta_0\beta_3 = (c_{12} - c_{21})/4, \qquad \beta_1\beta_2 = (c_{12} + c_{21})/4$$

Properties of Euler Parameters

Kinematics:

$$\begin{Bmatrix} \dot\beta_0 \\ \dot\beta_1 \\ \dot\beta_2 \\ \dot\beta_3 \end{Bmatrix} = \frac{1}{2}\begin{bmatrix} \beta_0 & -\beta_1 & -\beta_2 & -\beta_3 \\ \beta_1 & \beta_0 & -\beta_3 & \beta_2 \\ \beta_2 & \beta_3 & \beta_0 & -\beta_1 \\ \beta_3 & -\beta_1 & \beta_1 & \beta_0 \end{bmatrix}\begin{Bmatrix} 0 \\ \omega_1 \\ \omega_2 \\ \omega_3 \end{Bmatrix} \quad \overset{\text{\textit{"tranmsmutation"}}}{\Longrightarrow} \quad \begin{Bmatrix} \dot\beta_0 \\ \dot\beta_1 \\ \dot\beta_2 \\ \dot\beta_3 \end{Bmatrix} = \frac{1}{2}\begin{bmatrix} 0 & -\omega_1 & -\omega_2 & -\omega_3 \\ \omega_1 & 0 & \omega_3 & -\omega_2 \\ \omega_2 & -\omega_3 & 0 & \omega_1 \\ \omega_3 & \omega_2 & -\omega_1 & 0 \end{bmatrix}\begin{Bmatrix} \beta_0 \\ \beta_1 \\ \beta_2 \\ \beta_3 \end{Bmatrix}$$

$$or \quad \dot\beta = \Omega\beta$$

Successive Rotations:

$$\{b'\} = [C(\beta')]\{b\}, \quad \{b''\} = [C(\beta'')]\{b'\}, \quad \{b''\} = [C(\beta)]\{b\} \quad \Longrightarrow \quad [C(\beta)] = [C(\beta'')][C(\beta')]$$

$$\begin{Bmatrix} \beta_0 \\ \beta_1 \\ \beta_2 \\ \beta_3 \end{Bmatrix} = \begin{bmatrix} \beta_0'' & -\beta_1'' & -\beta_2'' & -\beta_3'' \\ \beta_1'' & \beta_0'' & \beta_3'' & -\beta_2'' \\ \beta_2'' & -\beta_3'' & \beta_0'' & \beta_1'' \\ \beta_3'' & \beta_1'' & -\beta_1'' & \beta_0'' \end{bmatrix}\begin{Bmatrix} \beta_0' \\ \beta_1' \\ \beta_2' \\ \beta_3' \end{Bmatrix} \quad \overset{\text{\textit{tranmsmutation}}}{\Longleftrightarrow\Longrightarrow} \quad \begin{Bmatrix} \beta_0 \\ \beta_1 \\ \beta_2 \\ \beta_3 \end{Bmatrix} = \begin{bmatrix} \beta_0' & -\beta_1' & -\beta_2' & -\beta_3 \\ \beta_1' & \beta_0' & -\beta_3 & \beta_2' \\ \beta_2' & \beta_3 & \beta_0' & -\beta_1' \\ \beta_3 & -\beta_2' & \beta_1' & \beta_0' \end{bmatrix}\begin{Bmatrix} \beta_0'' \\ \beta_1'' \\ \beta_2'' \\ \beta_3'' \end{Bmatrix}$$

Transformation from Euler Angles (e. g., 3-1-3 sequence):

$$\beta_0 = \cos\frac{\theta}{2}\cos\left(\frac{\phi+\psi}{2}\right) \qquad \beta_2 = \sin\frac{\theta}{2}\sin\left(\frac{\phi-\psi}{2}\right)$$

$$\beta_1 = \sin\frac{\theta}{2}\cos\left(\frac{\phi-\psi}{2}\right) \qquad \beta_3 = \cos\frac{\theta}{2}\sin\left(\frac{\phi+\psi}{2}\right)$$

For a general $\{i,j,k\}$ Euler angle sequence, see the general transformation in Junkins & Turner's 1986 text and implemented in MATLAB operator "betas(i,j,k,a1,a2,a3)"

Rodriguez Orientation Parameters

The Rodriguez Parameters are defined as $\quad q_i = e_i \tan \dfrac{\phi}{2} \equiv \beta_i / \beta_o \quad i=1,2,3$

Using the above definition and the results established already for the Euler parameters we can verify the following: The direction cosine matrix has the parameterization:

$$[C(q)] = \frac{1}{1+q_1^2+q_2^2+q_3^2} \begin{bmatrix} 1+q_1^2-q_2^2-q_3^2 & 2(q_1 q_2 + q_3) & 2(q_1 q_3 - q_2) \\ 2(q_1 q_2 - q_3) & 1-q_1^2+q_2^2-q_3^2 & 2(q_2 q_3 + q_1) \\ 2(q_1 q_3 + q_2) & 2(q_2 q_3 - q_1) & 1-q_1^2-q_2^2+q_3^2 \end{bmatrix}$$

Also, we have the kinematic relationships

$$\dot{q}_1 = \tfrac{1}{2}[\quad (1+q_1^2)\omega_1 \ + \ (q_1 q_2 - q_3)\omega_2 + (q_1 q_3 + q_2)\omega_3]$$

$$\dot{q}_2 = \tfrac{1}{2}[(q_1 q_2 + q_3)\omega_1 + \quad (1+q_2^2)\omega_2 + (q_2 q_3 - q_1)\omega_3]$$

$$\dot{q}_3 = \tfrac{1}{2}[(q_1 q_3 - q_2)\omega_1 + \ (q_2 q_3 + q_1)\omega_2 + \quad (1+q_3^2)\omega_3]$$

These orientation parameters are attractive because they involve only polynomial nonlinearities and they linearize well for small q's and ω's. Note the singularity at $\phi = n\pi$ where $q_i \Rightarrow \infty$. Vis–a–vis singularities, these parameters are less attrac-tive than Euler parameters (quaternions), but more attractive than any Euler angles.

MATLAB.JLJ Operators: Geometric & Kinematic Transformations

1. {i,j,k} seq. of euler angles (a₁, a₂, a₃) $\Longrightarrow$ direction cosine matrix C

> C = dircos(i, j, k, a₁, a₂, a₃)

2. {i,j,k} seq. of euler angles (a₁, a₂, a₃) $\Longrightarrow$ euler par. $\{\beta_0, \beta_1, \beta_2, \beta_3\}$

> beta = betas(i, j, k, a₁, a₂, a₃)

3. Euler parameters $\{\beta_0, \beta_1, \beta_2, \beta_3\}$ $\Longrightarrow$ direction cosine matrix C

> C = betatoc(beta)

4. Direction cosine matrix C $\Longrightarrow$ euler par. $\{\beta_0, \beta_1, \beta_2, \beta_3\}$

> beta = ctobeta(C)

```
function c=betatoc(b)
% writtem by j. l. junkins  12/30/88
% calculates elementary transformation
% from euler (quaternion) parameters to
% the direction cosine matrix.
c=zeros(3,3);
b0=b(1);
b1=b(2);
b2=b(3);
b3=b(4);
b02=b0*b0;
b12=b1*b1;
b22=b2*b2;
b32=b3*b3;
b1b2=b1*b2;
b1b3=b1*b3;
b2b3=b2*b3;
b0b1=b0*b1;
b0b2=b0*b2;
b0b3=b0*b3;
%
c(1,1)=b02+b12-b22-b32;
c(1,2)=2*(b1b2+b0b3);
c(1,3)=2*(b1b3-b0b2);
c(2,1)=2*(b1b2-b0b3);
c(2,2)=b02-b12+b22-b32;
c(2,3)=2*(b2b3+b0b1);
c(3,1)=2*(b1b3+b0b2);
c(3,2)=2*(b2b3-b0b1);
c(3,3)=b02-b12-b22+b32;
end


function b=betas(i,j,k,a1,a2,a3)
% writtem by j. l. junkins  12/30/88
% calculates elementary transformation from {i,j,k} euler angles
% {a1,a2,a3} to the euler (quaternion) parameters.
rfouri=rfour(i);
rfourj=rfour(j);
rfourk=rfour(k);
i4=eye(4,4);
rmat1 = i4*cos(a3/2) + rfourk*sin(a3/2);
rmat2 = i4*cos(a2/2) + rfourj*sin(a2/2);
rmat3 = i4*cos(a1/2) + rfouri*sin(a1/2);
rmat=rmat1*rmat2*rmat3;
b=rmat(:,1);
end


function rfour=rfour(i)
% writtem by j. l. junkins  12/30/88
% calculates elementary four x four matrices needed in the transformatio
%  from euler angles to euler (quaternion) parameters.
if i==1
  rfour = [0,-1,0,0;1,0,0,0;0,0,0,1;0,0,-1,0];
end
if i==2
  rfour = [0,0,-1,0;0,0,0,-1;1,0,0,0;0,1,0,0];
end
if i==3
  rfour = [0,0,0,-1;0,0,1,0;0,-1,0,0;1,0,0,0];
end
if i<1 error('rfour called with bad argument'); end
if i>3 error('rfour called with bad argument'); end
end
```

```
function b=ctobeta(c)
% writtem by j. l. junkins  12/30/88
% calculates elementary transformation
% from the direction cosine matrix c
% to the euler (quaternion) parameters.
cc=zeros(4);
cc(1)=c(1,1)+c(2,2)+c(3,3);
cc(2)=c(1,1);
cc(3)=c(2,2);
cc(3)=c(3,3);
b2=zeros(4);
for i=1:4
   b2(i)=(1+2*cc(i) - cc(1))/4;
end
[d,index]=sort(b2);
if index(4)==1
   b(1)=b2(1)^0.5;
   b(2)=(c(2,3)-c(3,2))/(4*b(1));
   b(3)=(c(3,1)-c(1,3))/(4*b(1));
   b(4)=(c(1,2)-c(2,1))/(4*b(1));
end
if index(4)==2
   b(2)=b2(2)^0.5;
   b(1)=(c(2,3)-c(3,2))/(4*b(2));
   b(3)=(c(1,2)-c(2,1))/(4*b(2));
   b(4)=(c(3,1)+c(1,3))/(4*b(2));
end
if index(4)==3
   b(3)=b2(3)^0.5;
   b(1)=(c(3,1)-c(1,3))/(4*b(3));
   b(2)=(c(1,2)+c(2,1))/(4*b(3));
   b(4)=(c(2,3)+c(3,2))/(4*b(3));
end
if index(4)==4
   b(4)=b2(4)^0.5;
   b(1)=(c(1,2)-c(2,1))/(4*b(4));
   b(2)=(c(3,1)+c(1,3))/(4*b(4));
   b(3)=(c(2,3)+c(3,2))/(4*b(4));
end
end
```

```
function c=dircos(i,j,k,a1,a2,a3)
% evaluates classical direction cosine matrix c(a1,a2,a3) for
% rotation sequence i-j-k
 r1=euler(i,a1);
 r2=euler(j,a2);
 r3=euler(k,a3);
c=r3*r2*r1;
end


function m=euler(i,a)
% calculates elementary rotation matrix corresponding to
% rotation thru angle a about axis i
if i==1
m= [1 0 0;0 cos(a) sin(a);0 -sin(a) cos(a)];
end
if i==2
m= [cos(a) 0 -sin(a);0 1 0;sin(a) 0 cos(a)];
end
if i==3
m= [cos(a) sin(a) 0;-sin(a) cos(a) 0;0 0 1];
end
if i>3 , 'error in euler, i must be 1,2, or 3', end
if i< 1, 'error in euler, i must be 1,2, or 3', end
end
```

The Eulerian Approach to Rotational Dynamics

Upon selecting your favorite (or most appropriate!) set of attitude coordinates, and doing some kinematics, you arrive at equations of the form (e. g., as shown here for the Euler *(quaternion)* parameters):

$$\dot{\beta}_0 = \tfrac{1}{2}[-\omega_1\beta_1 - \omega_2\beta_2 - \omega_3\beta_3]$$

$$\dot{\beta}_1 = \tfrac{1}{2}[-\omega_1\beta_0 - \omega_2\beta_3 + \omega_3\beta_2]$$

$$\dot{\beta}_2 = \tfrac{1}{2}[+\omega_1\beta_3 + \omega_2\beta_0 - \omega_3\beta_1]$$

$$\dot{\beta}_3 = \tfrac{1}{2}[-\omega_1\beta_2 + \omega_2\beta_1 - \omega_3\beta_0]$$

Upon applying the Eulerian Principle

$$(\text{external torque}) = \frac{d}{dt}(\text{angular momentum})_N$$

$$L = \frac{d}{dt}(H)_B + \omega_{B/N} \times H$$

and choosing body fixed principal axes to componentiate all vectors, you arrive at the following fundamental differential equations:

$$\dot{\omega}_1 = \left(\frac{I_2 - I_3}{I_1}\right)\omega_2\,\omega_3 + \frac{L_1}{I_1}$$

$$\dot{\omega}_2 = \left(\frac{I_3 - I_1}{I_2}\right)\omega_3\,\omega_1 + \frac{L_2}{I_2}$$

$$\dot{\omega}_3 = \left(\frac{I_1 - I_2}{I_3}\right)\omega_1\,\omega_2 + \frac{L_3}{I_3}$$

Of course, the results shown above are for the simplest case of a rigid body; these must be generalized, as appropriate in a particular application, to account for additional internal degrees of freedom and flexibility effects. The generalized coordinate formulations (e.g. Lagrangian formulations) are useful in these generalizations.

Energy and Momentum Integrals for Torque Free Motion

For the case of zero external torque, Euler's Eqs are: $\implies$

$$\dot{\omega}_1 = \left(\frac{I_2 - I_3}{I_1}\right)\omega_2\,\omega_3$$

$$\dot{\omega}_2 = \left(\frac{I_3 - I_1}{I_2}\right)\omega_3\,\omega_1$$

$$\dot{\omega}_3 = \left(\frac{I_1 - I_2}{I_3}\right)\omega_1\,\omega_2$$

These differential equations are nonlinear, but they can be solved (*a la Jacobi*) in terms of Jacobian elliptic functions. They have two exact integrals of obvious physical and practical significance:

> *Conservation of momentum:* $H^2 = I_1^2\,\omega_1^2 + I_2^2\,\omega_2^2 + I_3^2\,\omega_3^2$ (*momentum ellipsoid*)
>
> *Conservation of energy:* $2T = I_1\,\omega_1^2 + I_2\,\omega_2^2 + I_3\,\omega_3^2$ (*energy ellipsoid*)

Thus, the most general angular velocity history of a torque-free motion of a general rigid body generates the intersection curve of the momentum and energy ellipsoids. While this is "nice", it ain't necessarily easy to visualize! Matters can be simplified if one of the ellipsoids is mapped into a sphere. This is easily accomplished by introducing $H_i = I_i\,\omega_i$, the above equations then become:

$$H^2 = H_1^2 + H_2^2 + H_3^2 \qquad (\textit{momentum sphere})$$

$$1 = \frac{H_1^2}{2I_1 T} + \frac{H_2^2}{2I_2 T} + \frac{H_3^2}{2I_3 T} \qquad (\textit{energy ellipsoid})$$

It is constructive to consider angular momentum constant and address the behavior of intersection curves which arise as energy is swept over all physically admissible values. Toward this end, let us determine the maxima and minima of kinetic energy T subject to constant angular momentum H.

keystone-slide2 10

Energy Extrema for Constant Angular Momentum

We found the integrals $\qquad H^2 = H_1^2 + H_2^2 + H_3^2 \qquad$ *(momentum sphere)*

$$1 = \frac{H_1^2}{2I_1 T} + \frac{H_2^2}{2I_2 T} + \frac{H_3^2}{2I_3 T} \qquad \text{(energy ellipsoid)}$$

Eliminating H_1, from the momentum integral via $\qquad H_1^2 = H^2 - (H_2^2 + H_3^2)$

The energy along the intersection curve is given by $\quad T = \dfrac{H^2}{2I_1} + \dfrac{1}{2}\left(\dfrac{I_1-I_2}{I_1 I_2}\right)H_2^2 + \dfrac{1}{2}\left(\dfrac{I_1-I_3}{I_1 I_3}\right)H_3^2$

It is obvious that the energy is an extremum at the "pure spin" state point $\{H_1, H_2, H_3\}=\{\pm H, 0, 0\}$ and this point will be a maximum, mimimum, or saddle point depending upon the inertias. Specifically:

$\qquad$ *if* $\{ I_1 > I_2$ *and* $I_1 > I_3 \}$ *then* T *has a* <u>*minimum*</u> *value* $(T=\frac{H^2}{2I_1})$ $\qquad$ *(spin about* <u>*largest*</u> *inertia axis)*

$\qquad$ *if* $\{ I_1 < I_2$ *and* $I_1 < I_3 \}$ *then* T *has a* <u>*maximum*</u> *value* $(T=\frac{H^2}{2I_1})$ $\qquad$ *(spin about* <u>*least*</u> *inertia axis)*

$$if \left\{ \begin{array}{c} \{ I_1 > I_2 \text{ and } I_1 < I_3 \} \\ or \\ \{ I_1 < I_2 \text{ and } I_1 > I_3 \} \end{array} \right\} \text{ then } T \text{ has a } \underline{saddle} \text{ point } (T=\tfrac{H^2}{2I_1}) \text{ (spin about } \underline{intermediate} \text{ inertia axis)}$$

Intersection of the Energy Ellipsoid with The Momentum Sphere

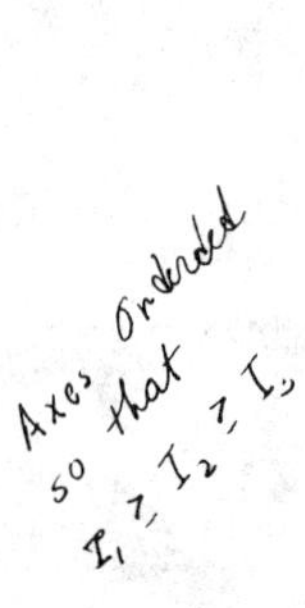

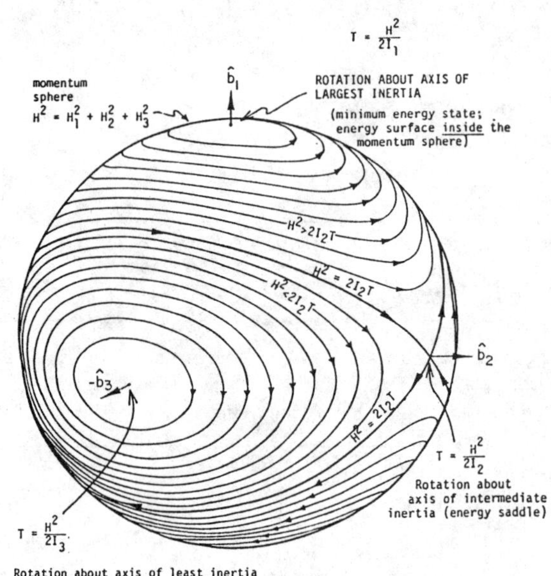

Roadmap to Establish the "Nonlinear Oscillator Analog" of Rigid Body Dynamics

$$\dot{\omega}_1 = \left(\tfrac{I_2-I_3}{I_1}\right)\omega_2\,\omega_3$$
$$\dot{\omega}_2 = \left(\tfrac{I_3-I_1}{I_2}\right)\omega_3\,\omega_1$$
$$\dot{\omega}_3 = \left(\tfrac{I_1-I_2}{I_3}\right)\omega_1\,\omega_2$$

energy & momentum integrals

$$H^2 = I_1^2\,\omega_1^2 + I_2^2\,\omega_2^2 + I_3^2\,\omega_3^2$$
$$2T = I_1\,\omega_1^2 + I_2\,\omega_2^2 + I_3\,\omega_3^2$$

$\Downarrow$ differentiate Euler's equations

$$\ddot{\omega}_1 = \left(\tfrac{I_2-I_3}{I_1}\right)\left[\dot{\omega}_2\,\omega_3 + \omega_2\,\dot{\omega}_3\right]$$
$$\ddot{\omega}_2 = \left(\tfrac{I_3-I_1}{I_2}\right)\left[\dot{\omega}_3\,\omega_1 + \omega_3\,\dot{\omega}_1\right]$$
$$\ddot{\omega}_3 = \left(\tfrac{I_1-I_2}{I_3}\right)\left[\dot{\omega}_1\,\omega_2 + \omega_1\,\dot{\omega}_2\right]$$

$\Downarrow$ substitute Euler's equations

$$\ddot{\omega}_1 = \left(\tfrac{I_2-I_3}{I_1}\right)\left[\left(\tfrac{I_1-I_2}{I_3}\right)\omega_1\,\omega_2^2 + \left(\tfrac{I_3-I_1}{I_2}\right)\omega_1\,\omega_3^2\right]$$
$$\ddot{\omega}_2 = \left(\tfrac{I_3-I_1}{I_2}\right)\left[\left(\tfrac{I_1-I_2}{I_3}\right)\omega_1^2\,\omega_2 + \left(\tfrac{I_2-I_3}{I_1}\right)\omega_2\,\omega_3^2\right]$$
$$\ddot{\omega}_3 = \left(\tfrac{I_1-I_2}{I_3}\right)\left[\left(\tfrac{I_3-I_1}{I_2}\right)\omega_1^2\,\omega_3 + \left(\tfrac{I_2-I_3}{I_1}\right)\omega_2^2\,\omega_3\right]$$

$\Downarrow$ even Jacobi would like this result!!!

$$\boxed{\ddot{\omega}_i = -A_i\,\omega_i - B_i\,\omega_i^3}\ ,\ i=1,2,3\ \ldots\ \textit{three uncoupled Duffing oscillators, without approximation!}$$

solve for each pair of ω's

$$\omega_2^2 = \left(\frac{2I_3T-H^2}{I_2I_3-I_2^2}\right) - \left(\frac{I_1I_3-I_2^2}{I_2I_3-I_2^2}\right)\omega_1^2$$
$$\omega_3^2 = \left(\frac{2I_2T-H^2}{I_2I_3-I_3^2}\right) - \left(\frac{I_1I_2-I_1^2}{I_2I_3-I_3^2}\right)\omega_1^2$$

$$\omega_1^2 = \left(\frac{2I_3T-H^2}{I_1I_3-I_1^2}\right) - \left(\frac{I_3I_2-I_2^2}{I_1I_3-I_1^2}\right)\omega_2^2$$
$$\omega_3^2 = \left(\frac{H^2-2I_1T}{I_3^2-I_1I_3}\right) - \left(\frac{I_2^2-I_1I_2}{I_3^2-I_1I_3}\right)\omega_2^2$$

$$\omega_1^2 = \left(\frac{2I_2T-H^2}{I_1I_2-I_1^2}\right) - \left(\frac{I_3I_2-I_3^2}{I_1I_2-I_1^2}\right)\omega_3^2$$
$$\omega_2^2 = \left(\frac{H^2-2I_1T}{I_2^2-I_1I_2}\right) - \left(\frac{I_3^2-I_1I_3}{I_2^2-I_1I_2}\right)\omega_3^2$$

Spring Constants for the Nonlinear Oscillators

Recall that we established the uncoupled oscillators

$$\dot{\omega}_i + A_i\,\omega_i + B_i\,\omega_i^3 = 0\ ,\ i=1,2,3$$

Each oscillator has an 'energy'–like integral:

$$\dot{\omega}_i^2 + A_i\,\omega_i^2 + \tfrac{1}{2}B_i\,\omega_i^4 = K_i\ ,\ i=1,2,3$$

Where the 'spring' and 'energy' constants have been determined as follows

i	A_i	B_i	K_i
1	$\dfrac{(I_1-I_2)\,(2I_3T-H^2) + (I_3-I_1)\,(H^2-2I_2T)}{I_1I_2I_3}$	$\dfrac{2(I_1-I_2)\,(I_1-I_3)}{I_2I_3}$	$\dfrac{(2I_2T-H^2)\,(H^2-2I_3T)}{I_1^{\,2}I_2I_3}$
2	$\dfrac{(I_2-I_3)\,(2I_1T-H^2) + (I_1-I_2)\,(H^2-2I_3T)}{I_1I_2I_3}$	$-\dfrac{2(I_1-I_2)\,(I_2-I_3)}{I_1I_3}$	$\dfrac{(2I_3T-H^2)\,(H^2-2I_1T)}{I_1I_2^{\,2}I_3}$
3	$\dfrac{(I_3-I_1)\,(2I_2T-H^2) + (I_2-I_3)\,(H^2-2I_1T)}{I_1I_2I_3}$	$\dfrac{2(I_1-I_3)\,(I_2-I_3)}{I_1I_2}$	$\dfrac{(2I_1T-H^2)\,(H^2-2I_2T)}{I_1I_2I_3^{\,3}}$

Note the oscillators are *uncoupled*, but they are not *independent!*

Using the axis ordering $I_1 > I_2 > I_3$, we have established that the spring constants satisfy

$$\Rightarrow\quad \boxed{A_1 \gtrless 0,\quad B_1 > 0;\qquad A_2 > 0,\quad B_2 < 0;\qquad A_3 \gtrless 0,\quad B_3 > 0}$$

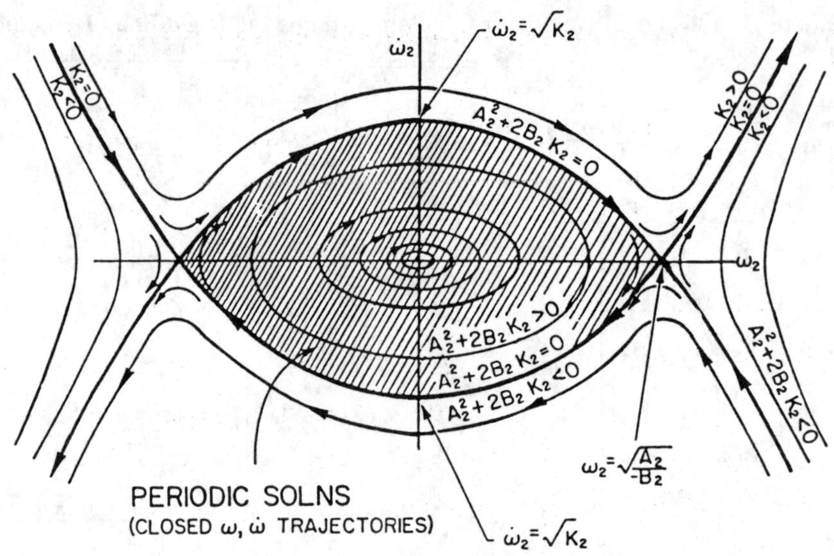

The $\{\omega_2, \dot{\omega}_2\}$ Phase Portrait

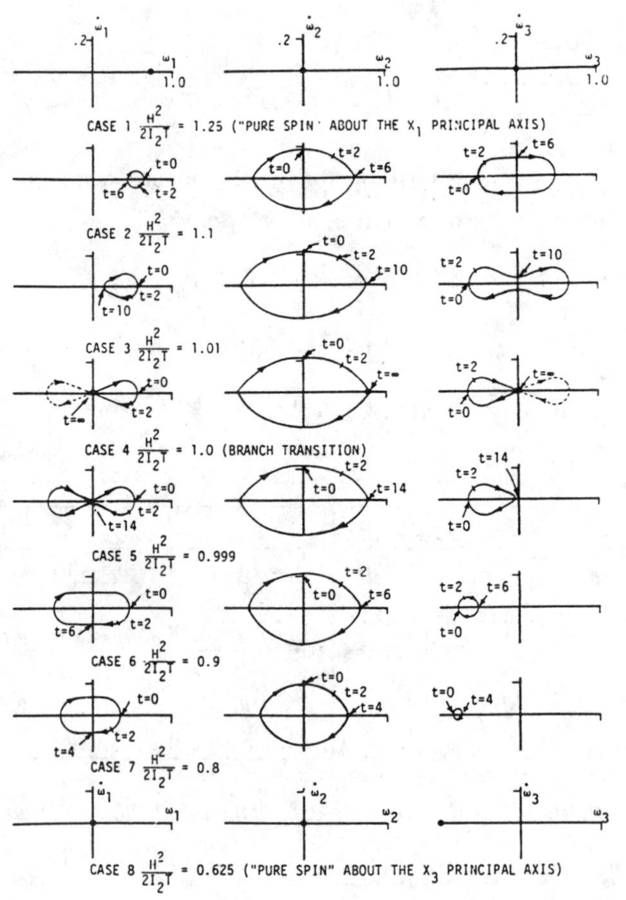

Figure 4.4 A Parametric Study of Rigid Body Dynamics in ω, $\dot{\omega}$ Space

Dynamics of Dual Spin Configurations

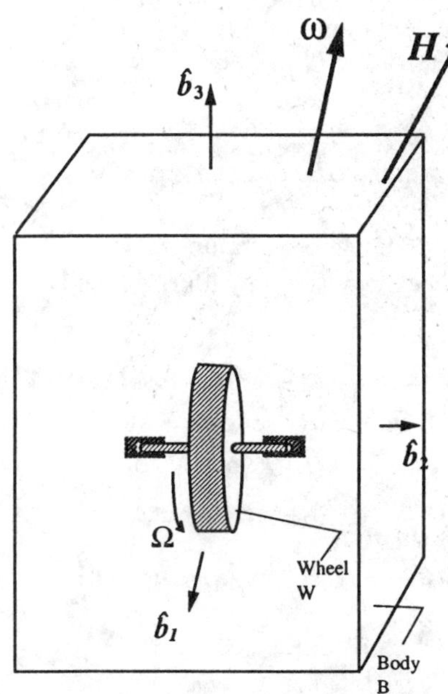

Angular Momentum: $\;H = H_{body} + H_{wheel}$

Euler's Equations:

$$L = \frac{d}{dt}(H)_N = \frac{d}{dt}(H)_B + \omega \times H$$

This leads to:

$$Roll:\quad L_1 = I_1\dot\omega_1 + (I_3 - I_2)\omega_2\omega_3 - \omega_3 h_2,\quad h_2 \equiv J_2\Omega,$$

$$Pitch:\quad L_2 = I_2\dot\omega_2 + (I_1 - I_3)\omega_3\omega_1 + h_2,\quad I_i = I_{Bi} + J_i$$

$$Yaw:\quad L_3 = I_3\dot\omega_3 + (I_2 - I_1)\omega_1\omega_2 - \omega_1 h_2$$

$$Wheel\ dynamics:\quad u = J_2(\dot\omega_2 + \dot\Omega) = J_2\dot\omega_2 + \dot h_2$$

Modified Pitch & Wheel dynamics:

$$L_2 = I_{B2}\dot\omega_2 + (I_1 - I_3)\omega_3\omega_1 + u$$

$$\dot h_2 = \left(\frac{J_2}{I_{B2}}\right)(I_1 - I_3)\omega_3\omega_1 - \left(\frac{J_2}{I_{B2}}\right)L_2 + \left(\frac{I_2}{I_{B2}}\right)u$$

Work/Energy Equation for the Dual Spin Configuration

The total energy of the system is given by

$$T = \tfrac{1}{2}[I_{B1}\omega_1^2 + I_{B2}\omega_2^2 + I_{B3}\omega_3^2] + \tfrac{1}{2}[J_1\omega_1^2 + J_2(\omega_2 + \Omega)^2 + J_3\omega_3^2]$$

or

$$T = \tfrac{1}{2}[I_1\omega_1^2 + I_2\omega_2^2 + I_3\omega_3^2] + \tfrac{1}{2}\frac{h_2^2}{J_2} + \omega_2 h_2$$

It can be verified that the work/energy eqn for this system is

$$\frac{dT}{dt} = \sum_{i=1}^{3} L_i\omega_i + \frac{I_{B2}}{J_2}h_2 u$$

Note the energy is constant if
    external torques are zero
    motor torque is zero *or* the wheel is locked.

This motivates an important geometrical device due to Barba and Aubrun for analyzing momentum transfer maneuvers of dual spin spacecraft.

"Flat Spin Recovery" Momentum Transfer Maneuvers of Dual Spin Configurations

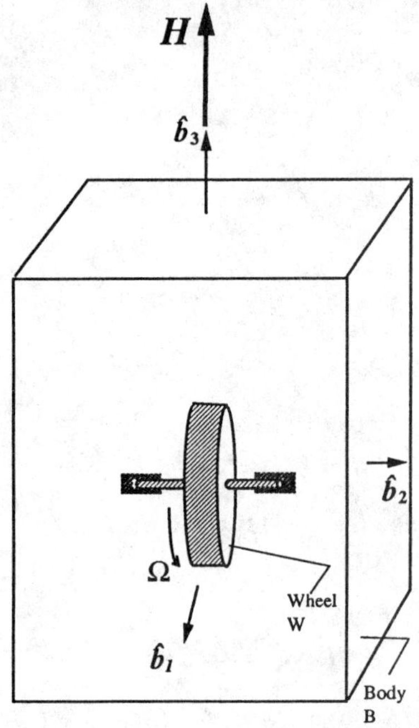

The given initial state:
The wheel is locked in B and B is spinning about it's largest axis of inertia.

The desired final state:
By judicious torquing of the wheel, transfer all angular momentum to the wheel. The result *must* be that a 90 deg re-orientation is achieved to align the wheel spin axis up with *H* which is inertially fixed.

Numerical integration of the modified Euler's Equations:

$$Roll:\quad I_1\dot{\omega}_1 = -(I_3-I_2)\omega_2\omega_3 + \omega_3 h_2$$

$$Pitch:\quad I_{B2}\dot{\omega}_2 = -(I_1-I_3)\omega_3\omega_1 - u$$

$$Yaw:\quad I_3\dot{\omega}_3 = -(I_2-I_1)\omega_1\omega_2 + \omega_1 h_2$$

$$Wheel\ dynamics:\quad h_2 = \left(\frac{J_2}{I_{B2}}\right)(I_1-I_3)\omega_3\omega_1 + \left(\frac{I_2}{I_{B2}}\right)u$$

yields the results on the following charts.

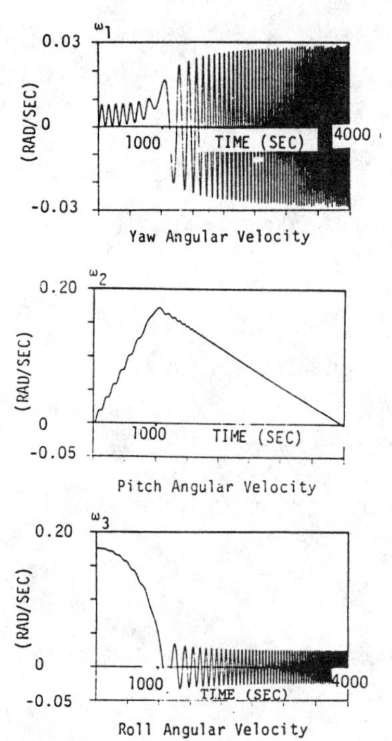

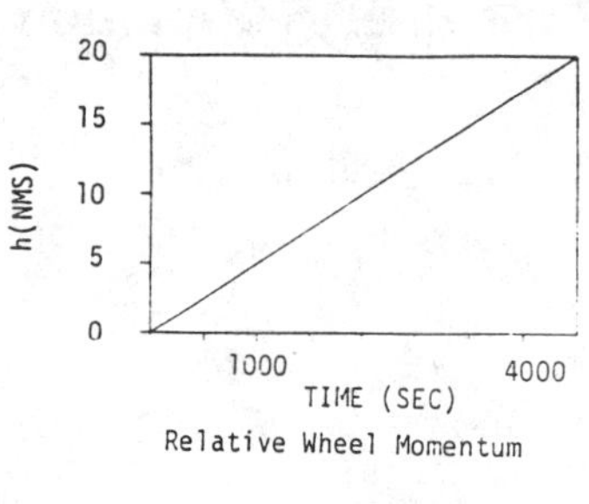

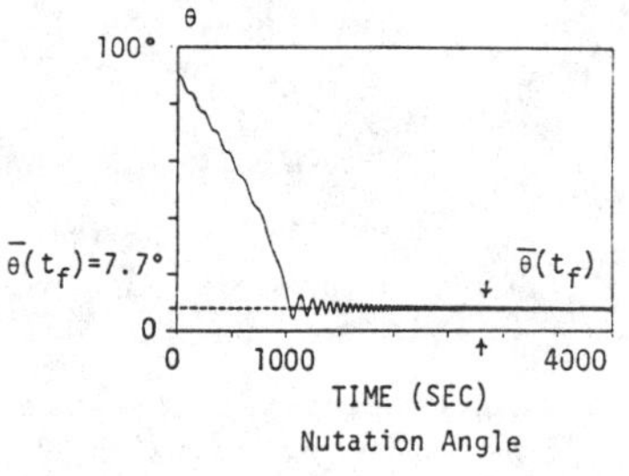

Figure 4.8 Flat Spin Recovery of a Dual Spin Spacecraft: The Maneuver of Barba and Aubrun

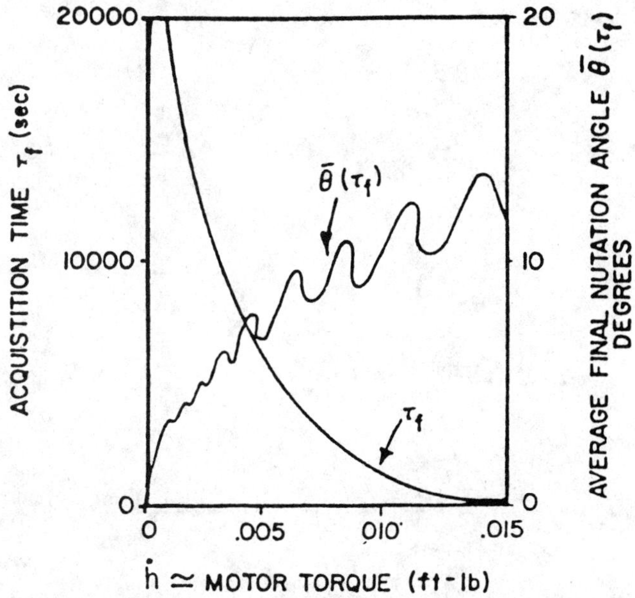

Energy and Momentum Surfaces for Momentum Transfer Maneuvers

Introducing T_s, the total energy excluding the rotor spin contribution

$$T_s = \frac{1}{2}[I_1\omega_1^2 + I_2\omega_2^2 + I_3\omega_3^2]$$

Barba and Aubrun (1975) showed that $\dfrac{dT_s}{dt} = -c\,\omega_2(t)$, $\qquad c \equiv h = J\Omega$

Thus if c is small, T_s changes slowly, modulated by $\omega_2(t)$. recall that

$$H^2 = H_1^2 + H_2^2 + H_3^2 = constant$$

Introducing
$$I_1\omega_1 = H_1$$
$$I_2\omega_2 = H_2 - h$$
$$I_3\omega_3 = H_3$$

Then the energy expression can be re-written as

$$1 = \frac{H_1^2}{2I_1 T_s} + \frac{(H_2 - h)^2}{2I_2 T_s} + \frac{H_3^2}{2I_3 T_s}$$

*Thus, at any instant, we can conceive of **H** as locating a point on the intersection of the constant momentum sphere and the instantaneous energy ellipsoid ... this provides insight!*

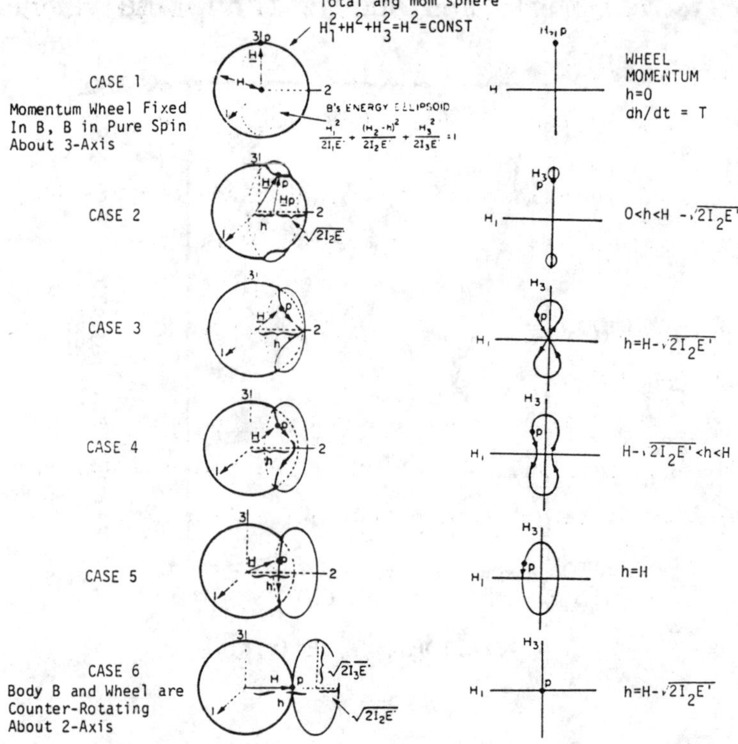

Figure 4.10 Intersection of the E^*- H Surfaces for a Dual Spin Spacecraft

Nonlinear Dynamical Systems: Linearized Departure Motion

Consider the n simultaneous nonlinear odes:

$$\dot{x}_i = f_i(x_1, x_2, \ldots, x_n, u_1, u_2, \ldots, u_m, p_1, p_2, \ldots, p_r, t) \quad i=1,2,\ldots,n \tag{1}$$

or

$$\dot{x} = f(x, u, p, t)$$

Using the nominal initial state $\{x_n(t_o)\}$, control history $\{u_n(t)\}$, and model parameter vector $\{p_n\}$, we can integrate the nominal trajectory by solving Eq. 1 -- indicated as

$$x_n(t) = x_n(t_o) + \int_{t_o}^{t} f(x_n(\tau), u_n(\tau), p_n, \tau)\, d\tau \tag{2}$$

Let a neighboring varied trajectory be described by

$$x(t) = x_n(t) + \delta x(t) \tag{3}$$

and where the varied control history is $u(t) = u_n(t) + \delta u(t)$

the varied model parameter vector is $p = p_n + \delta p$

Nonlinear Dynamical Systems: Linearized Departure Motion, cont.

The varied trajectory $x(t)$ is the solution of Eq. (1)

$$x(t) = x(t_o) + \int_{t_o}^{t} f(x_n(\tau) + \delta x(\tau), u_n(\tau) + \delta u(\tau), p_n + \delta p, \tau)\, d\tau$$

Subtracting (2) from (3) and expanding (3) in a first order Taylor series, we obtain

$$\delta x(t) = \delta x(t_o) + \int_{to}^{t} [A(*)\, \delta x(\tau), + B(*)\delta u(\tau) + D(*)\delta p]\, d\tau + \ldots \tag{4}$$

where $\{*\} \equiv \{x_n(\tau), u_n(\tau), p_n, \tau\}$, and the Jacobians: $A \equiv \left[\dfrac{\partial f}{\partial x}\right]$, $B \equiv \left[\dfrac{\partial f}{\partial u}\right]$, $D \equiv \left[\dfrac{\partial f}{\partial p}\right]$ evaluated along the nominal trajectory.

Differentiating Eq. (4) with respect to time, we obtain $\dfrac{d}{dt}(\delta x) = A\,\delta x + B\,\delta u + D\,\delta p$

This has essentially the same form as $\dot{x} = Ax + Bu$, so we can write the solution as

$$\delta x(t) = \Phi(t, t_o)\, \delta x(t_o) + \int_{t_o}^{t} \Phi(\tau, t)B(*)\, \delta u(\tau)d\tau + \Psi(t, t_o)\delta p$$

where $\quad \Phi(t, t_o) \equiv \left[\dfrac{\partial x(t)}{\partial x(t_o)}\right] \quad$ and $\quad \Psi(t, t_o) \equiv \left[\dfrac{\partial x(t)}{\partial p}\right]$

satisfy $\quad \dot{\Phi} = A\Phi, \quad \Phi(t_o, t_o) = I; \qquad\qquad \Psi = A\Psi + D, \quad \Psi(t_o, t_o) = 0 \tag{5}$

Sensitivity of Nonlinear Systems

Consider the system of n *nonlinear odes*

$$\dot{x}_i = f_i(x_1, x_2, \ldots, x_n, u_1, u_2, \ldots, u_m, p_1, p_2, \ldots, p_r, t) \quad i = 1,2,\ldots, n \quad or$$

$$\dot{x} = f(x, u, p, t) \qquad x(t_o) = x_o$$

model parameter vector
control vector
state vector

We're interested the sensitivities

$$\left[\dfrac{\partial x(t)}{\partial x(t_o)}\right] \equiv \begin{bmatrix} \dfrac{\partial x_1(t)}{\partial x_1(t_o)} & ---- & \dfrac{\partial x_1(t)}{\partial x_n(t_o)} \\ & ---- & \\ \dfrac{\partial x_n(t)}{\partial x_1(t_o)} & ---- & \dfrac{\partial x_n(t)}{\partial x_n(t_o)} \end{bmatrix} \equiv \text{the } n\text{x}n \text{ *state transition matrix*}$$

$$\left[\dfrac{\partial x(t)}{\partial p}\right] \equiv \begin{bmatrix} \dfrac{\partial x_1(t)}{\partial p_1} & ---- & \dfrac{\partial x_1(t)}{\partial p_r} \\ & ---- & \\ \dfrac{\partial x_n(t)}{\partial p_1} & ---- & \dfrac{\partial x_n(t)}{\partial p_r} \end{bmatrix} \equiv \text{the } n\text{x}x \text{ *model parameter sensitivity matrix*}$$

Sensitivity of Nonlinear Systems, cont.

Write $\dot{x} = f(x,u,p,t)$ in integral equation form

$$x(t) = x(t_o) + \int_{t_o}^{t} f(x(\tau),\ u(\tau),p,\ \tau)\ d\tau$$

$$\Phi(t,t_o) = \left[\frac{\partial x(t)}{\partial x(t_o)}\right] = I + \int_{t_o}^{t}\left[\left[\frac{\partial f(*)}{\partial x(\tau)}\right]\left[\frac{\partial x(\tau)}{\partial x(t_o)}\right]\right] d\tau$$

$$\Psi(t,t_o) = \left[\frac{\partial x(t)}{\partial p}\right] = O + \int_{t_o}^{t}\left[\left[\frac{\partial f(*)}{\partial x(\tau)}\right]\left[\frac{\partial x(\tau)}{\partial p}\right] + \left[\frac{\partial f(*)}{\partial p}\right]\right] d\tau$$

Differentiation w.r.t. time yields, with $\frac{d}{dt}(\Phi(t,t_o)) = \dot{\Phi}(t,t_o)$, $\frac{d}{dt}(\Psi(t,t_o)) = \dot{\Psi}(t,t_o)$

$$\dot{\Phi}(t,t_o) = A\,\Phi(t,t_o) \quad , \qquad \Phi(t_o,t_o) = I$$

$$\dot{\Psi}(t,t_o) = A\,\Psi(t,t_o) + D \ , \quad \Psi(t_o,t_o) = O$$

where $\quad A \equiv \left[\dfrac{\partial f(x,u,p,t)}{\partial x(t)}\right], \quad D \equiv \left[\dfrac{\partial f(x,u,p,t)}{\partial p}\right]$

are evaluated along some nominal non-linear trajectory of $\dot{x} = f(x,u,p,t)$.

The Symplectic ("Hamiltonian") Property

A matrix $A \in R^{2n \times 2n}$ is symplectic if it satisfies

$$\boxed{A^T J A = J}$$

Where the matrix J is defined as

$$J = \begin{bmatrix} O_n & I_n \\ -I_n & O_n \end{bmatrix}, \qquad
\begin{aligned} I_n &= n \times n \text{ identity matrix} \\ O_n &= n \times n \text{ zero matrix} \end{aligned}$$

and has the property $\qquad\qquad\qquad J^2 = -\,I_{2n}$

What is the consequence of a matrix satisfying the symplectic property? $\Rightarrow$ The inverse can be obtained without doing any arithmetic! Notice the definition provides immediately

$$A^{-1} = -J A^T J$$

The Symplectic Property, Continued

Denote the $n{\times}n$ submatrices of A as $\quad A = \begin{bmatrix} A_{11} & A_{12} \\ A_{21} & A_{22} \end{bmatrix}$

then
$$A^{-1} = -JA^{T}J = -\begin{bmatrix} 0 & I \\ -I & 0 \end{bmatrix}\begin{bmatrix} A_{11}^{T} & A_{21}^{T} \\ A_{12}^{T} & A_{22}^{T} \end{bmatrix}\begin{bmatrix} 0 & I \\ -I & 0 \end{bmatrix}$$

so
$$A^{-1} = \begin{bmatrix} A_{22}^{T} & -A_{12}^{T} \\ -A_{21}^{T} & A_{11}^{T} \end{bmatrix}$$

$\Rightarrow$ So a symplectic matrix is "the next best thing to an orthogonal matrix"

$\Rightarrow$ The practical importance of this property stems (in part) from the truth that *all* state transition matrices of conservative, "natural" coordinate mechanical systems have this property.

The Symplectic Property, Illustrated for the Orbit Problem

Acceleration coordinates:
$$\ddot{x} = -\frac{\partial V}{\partial x}$$
$$\ddot{y} = -\frac{\partial V}{\partial y}$$
$$\ddot{z} = -\frac{\partial V}{\partial z}$$

V = potential of force/unit mass, e.g. for 2-body problem,
$$V = \frac{-GM}{r}, \quad r^2 = x^2 + y^2 + z^2$$

State space form:

$$\begin{aligned} x_1 &\equiv x \\ x_2 &\equiv y \\ x_3 &\equiv z \\ x_4 &\equiv \dot{x} \\ x_5 &\equiv \dot{y} \\ x_6 &\equiv \dot{z} \end{aligned} \Rightarrow \begin{aligned} \dot{x}_1 &= x_4 &\equiv f_1 \\ \dot{x}_2 &= x_5 &\equiv f_2 \\ \dot{x}_3 &= x_6 &\equiv f_3 \\ \dot{x}_4 &= -\frac{\partial V}{\partial x_1} &\equiv f_4 \\ \dot{x}_5 &= -\frac{\partial V}{\partial x_2} &\equiv f_5 \\ \dot{x}_6 &= -\frac{\partial V}{\partial x_3} &\equiv f_6 \end{aligned}$$

$\Longleftrightarrow$ campactly: $\dot{x} = f(x)$

Linearized departure motion about a nominal trajectory $\{\dot{x}_n(t) = f(x_n), \quad x_n(t_0)\}$

$$\delta\dot{x} = A\,\delta x, \quad \delta x(t) = \Phi(t,t_0)\,\delta x(t_0)$$
$$\dot{\Phi} = A\Phi, \quad \text{with} \quad \Phi(t_0,t_0) = I$$

$$A = \begin{bmatrix} 0 & I \\ G & 0 \end{bmatrix}, \quad G \equiv -\begin{bmatrix} \frac{\partial^2 V}{\partial x_1 \partial x_1} & \frac{\partial^2 V}{\partial x_1 \partial x_2} & \frac{\partial^2 V}{\partial x_1 \partial x_3} \\ \frac{\partial^2 V}{\partial x_2 \partial x_1} & \frac{\partial^2 V}{\partial x_2 \partial x_2} & \frac{\partial^2 V}{\partial x_2 \partial x_3} \\ \frac{\partial^2 V}{\partial x_3 \partial x_1} & \frac{\partial^2 V}{\partial x_3 \partial x_2} & \frac{\partial^2 V}{\partial x_3 \partial x_3} \end{bmatrix}_n$$

$\Rightarrow$ Question: Is $\Phi(t,t_0)$ *symplectic?* ($\Phi^T J \Phi = J$?)

Is $\Phi(t, t_o)$ Symplectic ($\Phi^T J \Phi = J$) ?

We just established: $\quad \dot\Phi = A\Phi, \quad \Phi(t_0, t_0) = I, \quad A = \begin{bmatrix} 0 & I \\ G & 0 \end{bmatrix}, \quad G \equiv -\begin{bmatrix} \frac{\partial^2 v}{\partial a_1 \partial a_1} & \frac{\partial^2 v}{\partial a_1 \partial a_2} & \frac{\partial^2 v}{\partial a_1 \partial a_3} \\ \frac{\partial^2 v}{\partial a_2 \partial a_1} & \frac{\partial^2 v}{\partial a_2 \partial a_2} & \frac{\partial^2 v}{\partial a_2 \partial a_3} \\ \frac{\partial^2 v}{\partial a_3 \partial a_1} & \frac{\partial^2 v}{\partial a_3 \partial a_2} & \frac{\partial^2 v}{\partial a_3 \partial a_3} \end{bmatrix}$

Ask the question in two parts:

Part A: Does $\Phi(t_0, t_0)$ satisfy $\Phi^T(t_0, t_0) J \Phi(t_0, t_0) \overset{?}{=} J \implies I^T J I \overset{?}{=} J \implies J = J$ yes!

Part B: Is $\Phi^T J \Phi \overset{?}{=}$ constant? Investigate: $\frac{d}{dt}(\Phi^T J \Phi) = \dot\Phi^T J \Phi + \Phi^T J \dot\Phi$

$\qquad$ Substitute $\dot\Phi = A\Phi$ to obtain

$$\frac{d}{dt}(\Phi^T J \Phi) = \Phi^T[A^T J + J A]\Phi, \text{ then substitute } A = \begin{bmatrix} 0 & I \\ G & 0 \end{bmatrix}, \quad J = \begin{bmatrix} 0 & I \\ -I & 0 \end{bmatrix}, \text{ so}$$

$$\frac{d}{dt}(\Phi^T J \Phi) = \Phi^T\left[\begin{bmatrix} -G^T & 0 \\ 0 & I \end{bmatrix} + \begin{bmatrix} G & 0 \\ 0 & -I \end{bmatrix}\right]\Phi = \Phi^T\begin{bmatrix} G - G^T & 0 \\ 0 & 0 \end{bmatrix}\Phi = \Phi^T[0]\Phi = 0 \;!$$

Obviously, the key is the structure of A and symmetry of G(true for all conservative systems).

References

1. Junkins, J. L. and Turner, J. D., *Optimal Spacecraft Rotational Maneuvers*, Elsevier, Amsterdam, 1986.

2. Junkins, J.L., Jacobson, I.D., and Blanton, J.N., *A Nonlinear Oscillator Analog of Rigid Body Dynamics*, **Celestial Mechanics**, Vol. 7, No. 4, (April 1973), pp. 398-407.

3. Morton, H. S., Junkins, J. L., and Blanton, J. N., *Analytical Solutions for Euler Parameters*, **Cel. Mech.**, 10 (Sep. 1974), pp. 287-301.

4. Junkins, J. L., *Some Recent Developments in Attitude Dynamics/Maneuver Strategies for Multiple Momentum Wheel Satellites*, **Flight Mechanics/Estimation Theory**, NASA CP-2003 (October 1975), pp. 154-72.

5. Kraige, L. G., and Junkins, J. L., *Perturbation Formulations for Satellite Attitude Dynamics*, **Cel. Mech.**, 13, (June 1976), pp. 39-64.

6. Junkins, J. L. and Turner, J. D., *On the Analogy Between Orbital Dynamics and Rigid Body Dynamics*, **Journal of the Astronautical Sciences**, Vol. 27, No. 4, pp. 345-358, (Dec. 1979, appeared January 1980).

7. Junkins, J. L. and Turner, J. D., *Optimal Continuous Torque Attitude Maneuvers*, **AIAA Journal of Guidance and Control**, Vol. 3, No. 3, (May 1980), pp. 210-217.

8. Junkins, J. L., Rajaram, S. and Baracat, W. A., *Precision autonomous satellite attitude control using momentum transfer and magnetic torquing*, **J. of the Astronautical Sciences**, Vol. xxx, No. 1, pp. 31-48, also paper AAS-007, AAS Rocky Mountain Guidance and Control Conf. Jan. 1981, Keystone, CO.

9. Barba, P., and Aubrun, J., *Satellite Attitude Acquisition by Momentum Transfer*, Paper # AAS-75-053, Presented at the AAS/AIAA Astrodynamics Conference, Nassau, Bahamas, July, 1975.

10. Gebman, G., and Mingori, T., *Perturbation Solution for the Flat Spin Recovery of a Dual Spin Spacecraft*, **AIAA Journal**, Vol. 14, pp859-867, July, 1976.

Example Applications:
Optimal Open Loop Attitude Maneuvers

NOVA Minimum-time Magnetic Attitude Maneuvers

Optimal Momentum Transfer Maneuvers

Optimal Control of Dynamical Systems: Pontryagin's Principle

The objective is to find the optimal *admissible* control $u^*(t)$, and the corresponding optimal trajectory $x^*(t)$ which minimize the *performance functional*:

$$J = \int_{t_o}^{t_f} \left(\tau, x(\tau), u(\tau) \right) d\tau$$

subject to the physical requirement that the trajectory satisfy the mathematical model of the system: $\dot{x} = f(t, x, u)$ with the boundary conditions: $x(t_o) = x_o$, $x(t_f) = x_f$

Using the *Lagrange multiplier method*, we adjoin the constraint $f(t, x, u) - \dot{x} = 0$ to J to form the *augmented functional*

$$\bar{J} = \int_{t_o}^{t_f} [\, F + \lambda^T (f - \dot{x})\,] \, d\tau = \int_{t_o}^{t_f} [\, H - \lambda^T \dot{x}\,] \, d\tau \; ; \; H \equiv F + \lambda^T f = \text{the Hamiltioian}$$

Requiring the first variation $\delta \bar{J} = \bar{J}(x(t), \lambda(t), u(t)) - \bar{J}(x^*(t), \lambda^*(t), u^*(t))$ to vanish leads to the necessary conditions satisfied by the optimal control and the optimal trajectory:

state equation: $\dot{x} = \dfrac{\partial H}{\partial \lambda} = f(t, x, u)$ with the b. c.: $x(t_o) = x_o$, $x(t_f) = x_f$

adjoint equation: $\dot{\lambda} = -\dfrac{\partial H}{\partial x} = -\left[\dfrac{\partial f}{\partial x}\right]^T \lambda - \left\{\dfrac{\partial F}{\partial x}\right\}$, b. c. = ??

optimal control: *minimize H over <u>admissible</u> u, . . . for smooth unbounded admissible u:*

$$\text{solve } \dfrac{\partial H}{\partial u} = 0 \;\Rightarrow\; u = g(x(t), \lambda(t)).$$

Summary of Necessary Conditions for Fixed End Point Problems

$$J = {}_{t_o}\int^{t_f} F(x, \dot{x}, u)\, dt$$

<table>
<tr><td valign="top">

Hamiltonian:

$$H = F + \lambda^T f$$

state equation:

$$\dot{x} = \frac{\partial H}{\partial \lambda} = f(t, x, u), \quad \text{b. c.:} \quad x(t_o) = x_o, \; x(t_f) = x_f$$

adjoint equation:

$$\dot{\lambda} = -\frac{\partial H}{\partial x} = -\left[\frac{\partial f}{\partial x}\right]^T \lambda - \left\{\frac{\partial F}{\partial x}\right\}, \; \text{b. c.} = ??$$

optimal control:

find $u(t)$ to minimize the Hamiltonian. For a smooth

unbounded u: solve $\dfrac{\partial H}{\partial u} = 0 \;\Rightarrow\; u = g(x(t), \lambda(t))$

</td><td valign="top">

Two-Point Boundary-Value Problem:

set $x(t_o) = x_o$, $\lambda(t_o) = \lambda_o$, solve

$$\dot{x} = F(t, x, \lambda)$$

$$\dot{\lambda} = G(t, x, \lambda)$$

iterate λ_o until $x(t_f) = x_f$

</td></tr>
</table>

- *Prayers for convergence aren't as reliably answered as we'd like!!*
- *Difficulties may arise because of solution sensitivity, poor starting guess, multiple extrema, and sometimes, no solution exists.*
- *The differential equations are often stiff & difficult to solve numerically.*
- *Convergence enhancement via: homotopy algorithms & interactive methods.*

Classical Bang-Bang Control: Minimum Time Maneuvers of a Linear System

Consider minimizing the maneuver time $\displaystyle J = \int_{t_o}^{t_f} F(x(\tau), \dot{x}(\tau), u(\tau))\, d\tau = \int_{t_o}^{t_f} (1)\, d\tau = t_f - t_o$

subject to satisfying the scalar equation

$$\ddot{\theta} = u(t) = \frac{torque}{inertia}, \quad \text{where } |u(t)| \le u_{max}, \; \{\theta(t_o), \dot{\theta}(t_o)\} = \{\theta_o, \dot{\theta}_o\}, \; \{\theta(t_f), \dot{\theta}(t_f)\} = \{0, 0\}$$

Converting to state space notation, introduce the state variables $x_1 = \theta$ & $x_2 = \dot{\theta}$, *then*

$$\dot{x}_1 = f_1 = x_2$$

$$\dot{x}_2 = f_2 = u, \; |u| \le u_{max}$$

The Hamiltonian of this system is: $H = \lambda_1 f_1 + \lambda_2 f_2 = \lambda_1 x_2 + \lambda_2 u$ and the necessary conditions for the optimal controls & optimal trajectories are

State: Co-State:

$$\dot{x}_1 = \frac{\partial H}{\partial \lambda_1} = x_2 \quad ; x_1(t_o)=\theta_o, \, x_1(t_f)=0 \qquad \dot{\lambda}_1 = -\frac{\partial H}{\partial x_1} = 0 \;\Rightarrow\; \lambda_1 = a \text{ constant} = c_1$$

$$\dot{x}_2 = \frac{\partial H}{\partial \lambda_2} = u, \; |u| \le u_{max}; \, x_2(t_o)=\dot{\theta}_o, \, x_2(t_f)=0 \qquad \dot{\lambda}_2 = -\frac{\partial H}{\partial x_2} = -\lambda_1 \;\Rightarrow\; \lambda_2 = c_2 - c_1 t$$

<u>Optimal Control:</u> Find $u(t)$ with $|u| \le u_{max}$ which minimizes $H = \lambda_1 f_1 + \lambda_2 f_2 = \lambda_1 x_2 + \lambda_2 u$

... *Thus the optimal control is given by the linear "switch function:"*

$$u(t) = -u_{max}\, sign(\lambda_2) = -u_{max}\, sign(c_2 - c_1 t)$$

Notice, at most one control switch occurs; the control is either max positive or negative.

Construction of the "Bang-Bang" Switch Phase Portrait

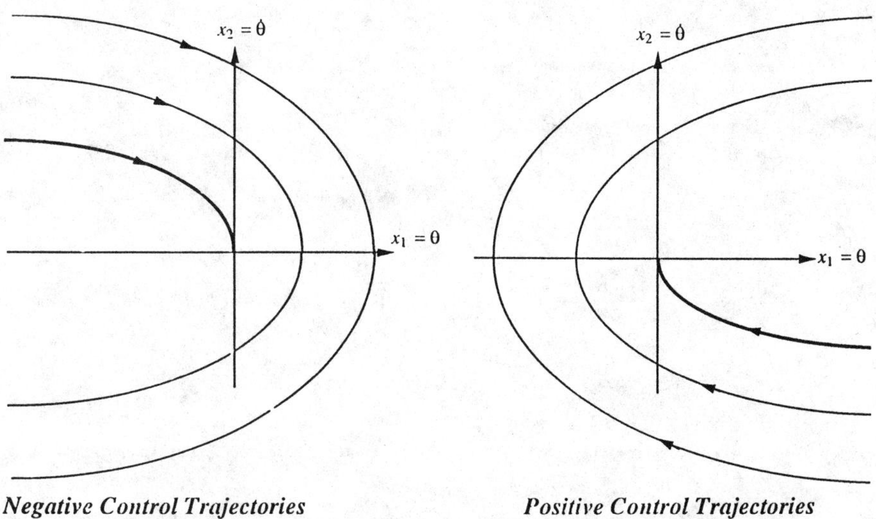

Negative Control Trajectories *Positive Control Trajectories*

>>>>>notice in each case . . . there is only one trajectory to the origin!<<<<<

Bang-Bang Control

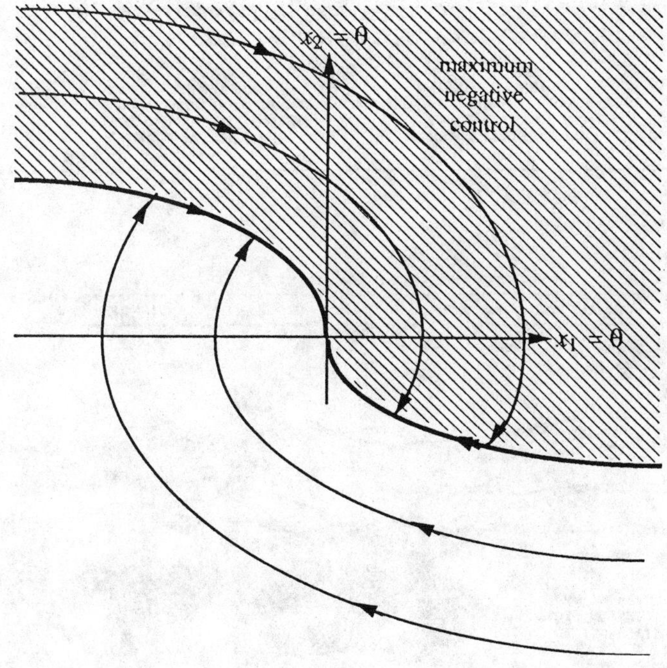

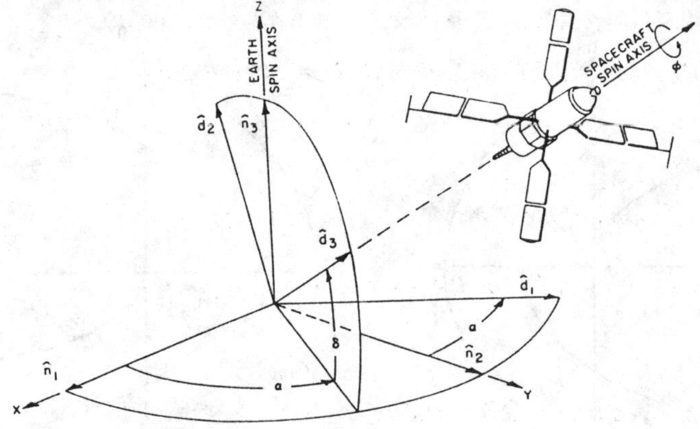

Figure 8.1 Geometry for Magnetic Maneuvers of a Spin Stabilized Symmetric Spacecraft

AAS 81-010

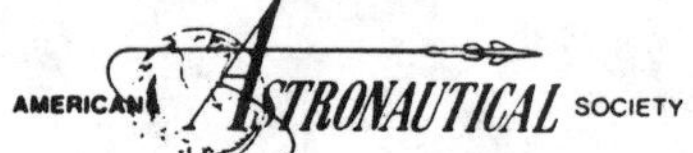

TIME OPTIMAL ATTITUDE MANEUVERS FOR THE NAVY NAVIGATION SATELLITES

DR. JOHN L. JUNKINS
VIRGINIA POLYTECHNIC INSTITUTE AND STATE UNIVERSITY,
BLACKSBURG, VIRGINIA

CHARLES E. WILLIAMS
THE JOHNS HOPKINS UNIVERSITY
APPLIED PHYSICS LABORATORY,
LAUREL, MARYLAND

ANNUAL ROCKY MOUNTAIN GUIDANCE AND CONTROL CONFERENCE

January 31 through February 4, 1981
Keystone, Colorado

Sponsored by
ROCKY MOUNTAIN SECTION
AMERICAN ASTRONAUTICAL SOCIETY

Magnetic Torque Model

$M = S/C \ dipole = p(t)M\,\hat{d}_3$, $p(t) = \pm 1$, *the polarity of the magnet*

$B = Earth's\ magnetic\ field\ vector = B_1\hat{d}_1 + B_2\hat{d}_2 + B_3\hat{d}_3$

$L = Torque\ acting\ on\ the\ S/C = M \times B = L_1\hat{d}_1 + L_2\hat{d}_2 + L_3\hat{d}_3$

*The Earth's field vector **B** is messy, because*

    (i)   It rotates with the earth and varies with orbital position.

    *(ii) We need to project **B** onto the instantaneous $\{\hat{d}_i\}$ axes as*

$$\begin{Bmatrix} B_1 \\ B_2 \\ B_3 \end{Bmatrix} = \begin{bmatrix} function\ (\alpha,\delta,\phi'(t),\theta(t)) \end{bmatrix} \begin{Bmatrix} B_{south} \\ B_{east} \\ B_{radial} \end{Bmatrix}$$

    *(iii) The model for **B** involves a messy function:*

$$\begin{Bmatrix} B_{south} \\ B_{east} \\ B_{radial} \end{Bmatrix} = \{function(r(t), \phi'(t), \lambda(t))\} = \text{4th order spherical harmonic expansion}$$

Neglecting Nutation & its Consequences for NOVA Magnetic Attitude Maneuvers

- Detailed simulations of NOVA attitude dynamics indicate that nutation angles ($< 5^{o}$) decays about one order of magnitude in 5 minutes (due to nutation damper effects).

- A typical large angle magnetic maneuver requires > 5 hours; simulations indicate that maximum nutation transients (at magnet switch times) seldom exceed 1°.

- As a consequence of the smallness of the nutation angle history (usually $<< 0.5^{o}$), it turns out to be an excellent approximation to ignore nutation and simply re-orient the angular momentum vector to the desired direction.

- Based upon this approximation, we achieve an order reduction from 6 to 2 and it is possible to establish a nice minimum-time control solution which we show below.

- As borne out in our preflight simulations, and as demonstrated by the successful mission, this approach is elegant, sufficiently accurate, and it works!

Equations of Motion: Precession of Angular Momentum Vector

Euler's Equation $\{\ L = \frac{d}{dt}(H)_N\ \}$, together with neglecting the nutation angle lead to

$$L_1(\alpha,\delta,t) = H\,\dot\alpha\,\cos\delta$$

$$L_2(\alpha,\delta,t) = H\,\dot\delta$$

$$L_3 = 0 = \dot H \;\Rightarrow\; H = \text{constant}$$

Therefore, the precession rates of the angular momentum vector is given by equations of the following functional form (significant algebra is required to establish $f_i(\alpha,\delta,t)$):

$$\dot\alpha = p(t)\,f_1(\alpha,\delta,t)$$

$$\dot\delta = p(t)\,f_2(\alpha,\delta,t)$$

Notice that the optimal control $p(t)$ appears in product with the two functions $f_i(\alpha,\delta,t)$, this has significant theoretical and practical importance, as we shall see below.

Minimum Time NOVA Attitude Maneuvers: Optimal Control Formulation

Find $p(t)$ to minimize: $\underset{|p(t)|\le 1}{}\ \ J = \int_{t_o}^{t_f}(1)dt = t_f - t_o = \textit{maneuver time}$

subject to: $\qquad \dot\alpha = p(t)\,f_1(\alpha,\delta,t), \qquad \alpha(t_o) = \alpha_o,\ \alpha(t_f) = \alpha_f$

$$\dot\delta = p(t)\,f_2(\alpha,\delta,t), \qquad \delta(t_o) = \delta_o,\ \delta(t_f) = \delta_f$$

The Hamiltonian for the above constrained minimization is

$$H \equiv 1 + p[\lambda_1 f_1 + \lambda_2 f_2], \;\Rightarrow\; p(t) = -sign[\lambda_1 f_1 + \lambda_2 f_2]$$

The co-state differential equations are:

$$\dot\lambda_1 = -\frac{\partial H}{\partial\alpha} = -p\left[\lambda_1\frac{\partial f_1}{\partial\alpha} + \lambda_2\frac{\partial f_2}{\partial\alpha}\right], \qquad \lambda_1(t_o) = ?$$

$$\dot\lambda_2 = -\frac{\partial H}{\partial\delta} = -p\left[\lambda_1\frac{\partial f_1}{\partial\delta} + \lambda_2\frac{\partial f_2}{\partial\delta}\right], \qquad \lambda_2(t_o) = ?$$

Summary of Optimal Maneuver Necessary Conditions

State equations (for right ascension & declination of H):

$$\dot{\alpha} = p(t)\, f_1(\alpha,\, \delta,\, t), \qquad \alpha(t_o) = \alpha_o,\ \alpha(t_f) = \alpha_f$$

$$\dot{\delta} = p(t)\, f_2(\alpha,\, \delta,\, t), \qquad \delta(t_o) = \delta_o,\ \delta(t_f) = \delta_f$$

Co-state equations:

$$\dot{\lambda}_1 = -\frac{\partial H}{\partial \alpha} = -p\left[\lambda_1 \frac{\partial f_1}{\partial \alpha} + \lambda_2 \frac{\partial f_2}{\partial \alpha}\right], \qquad \lambda_1(t_o) = ?$$

$$\dot{\lambda}_2 = -\frac{\partial H}{\partial \delta} = -p\left[\lambda_1 \frac{\partial f_1}{\partial \delta} + \lambda_2 \frac{\partial f_2}{\partial \delta}\right], \qquad \lambda_2(t_o) = ?$$

Optimal magnet polarity switch function: $p(t) = -sign[\lambda_1 f_1 + \lambda_2 f_2]$

Important observation: Notice that a scale factor c applied to λ_1 and λ_2 does not affect $p(t)$.

Also notice that if $\{\lambda_1$ and $\lambda_2\}$ satisfy the co-state equations then so does $\{c\lambda_1$ and $c\lambda_2\}$.

$\Rightarrow$ **Conclude λ_i do not have a unique magnitude, so we are free to impose a constraint ... we choose**

$$\lambda_1^2(t_o) + \lambda_2^2(t_o) = 1 \ \Rightarrow\ \lambda_1(t_o) = cos\gamma_o,\ \ \lambda_2(t_o) = sin\gamma_o\ \ \text{with}\ \ 0 \le \gamma_o \le 2\pi$$

Initial - Target State Inertial Coordinate System

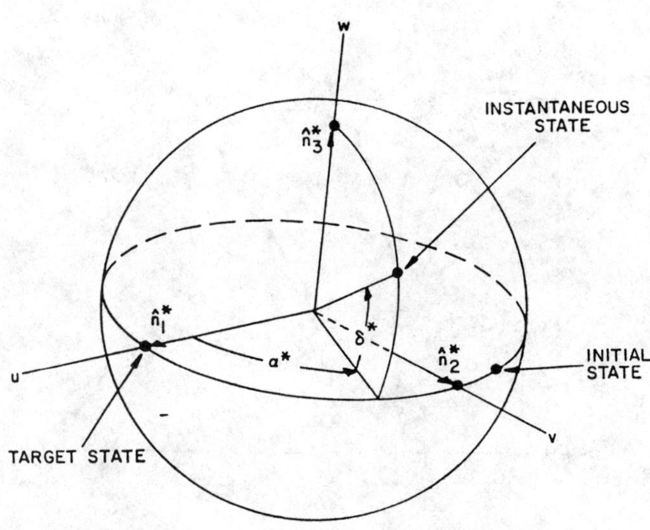

NOVA Configuration Parameters

TYPICAL NOVA PARAMETERS

t_0 = year 1980, day 320, hour 12, min 0 (GMT)

M = 69,600 pole-cm = 69.600 amp-m^2

$\dot{\phi}$ = 5 rpm

I_3 = 34 kg-m^2

a = orbit semimajor axis = 1.102818 Earth radii

e = eccentricity = 0.029192

i = inclination = 89.290 deg

ω = arg. of perigee = 24.870 deg

Ω = arg. of ascending node = 366.602 deg

di/dt = 0

$d\omega/dt$ = -3.536 deg/day

$d\Omega/dt$ = -0.0876 deg/day

Interactive Solution of the Two-Point Boundary Value Problem

extremal field map of minimum time maneuvers .... which one goes to the origin?

"first screen"

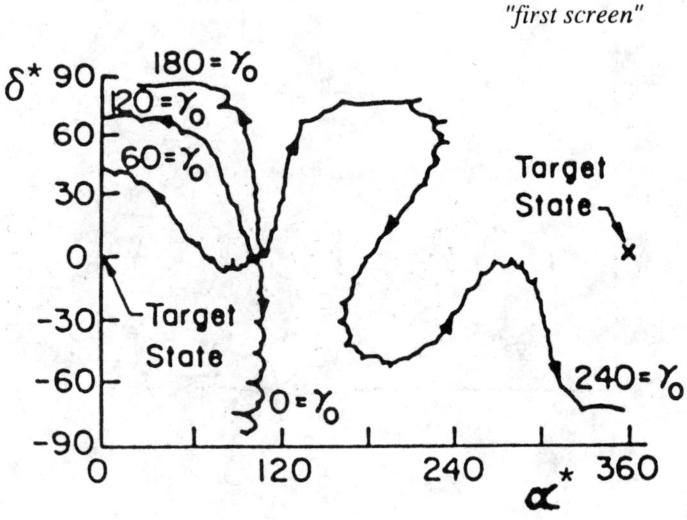

606

Interactive Solution of the Two-Point Boundary Value Problem

extremal field map of minimum time maneuvers .... which one goes to the origin?

"second screen"

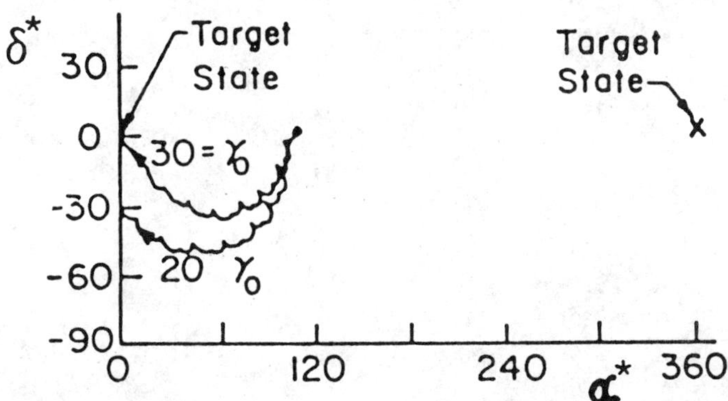

Non-unique Minimum-time Solutions near 180⁰ (~12 hour) Maneuvers

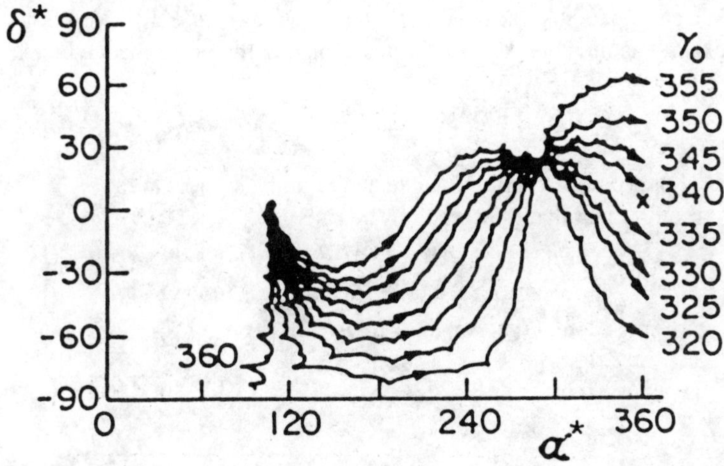

Command no.	Day	Hour	Minute	$p(t)$	$\alpha(t)$	$\delta(t)$
1	320	12	0	-1	45.2	35.1
2	320	12	17	1	39.7	29.7
3	320	12	45	-1	44.2	22.4
4	320	13	9	1	40.8	17.9
5	320	13	32	-1	42.9	10.5
6	320	13	57	1	36.5	4.1
7	320	14	23	-1	37.6	-2.8
8	320	14	49	1	31.0	-6.1
9	320	15	11	-1	30.8	-12.4
10	320	15	37	1	21.8	-17.2
11	320	16	1	-1	20.1	-22.6
12	320	16	28	1	9.8	-23.7
13	320	16	50	-1	7.0	-27.9
14	320	17	17	1	-4.7	-29.9
15	320	17	39	-1	-9.1	-32.7
16	320	18	8	1	-21.5	-31.0
17	320	18	30	-1	-26.0	-32.8
18	320	18	57	1	-38.3	-32.0
19	320	19	12	0	-42.7	-31.4

"Flat Spin Recovery" Momentum Transfer Maneuvers: Revisited

Question: What if we apply optimal control theory to the Barba/Aubrun maneuver ... "how much better" will the maneuver be accomplished?   Well, let's find out.

We anticipate that the final desired state (B at rest, $H = J\,\Omega$) is not exactly achievable , this motivates choosing the performance index:

$$J = \frac{w}{2}[\omega_1^2(t_f) + \omega_2^2(t_f) + \omega_3^2(t_f)] + \frac{1}{2}\int_{t_o}^{t_f} u^2(t)\,dt$$

Thus we seek to minimize the above index subject to the requirement that the equations be satisfied:

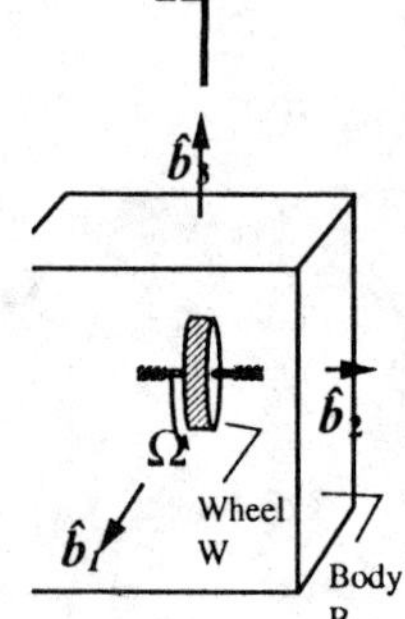

$$\dot{\omega}_1 = -r_1\omega_2\omega_3 + \omega_3 h/I_1 \qquad \equiv f_1(\omega_1,\omega_2,\omega_3,h,u)$$

$$\dot{\omega}_2 = -r_2\omega_3\omega_1 - u/I_{B2} \qquad \equiv f_2(\omega_1,\omega_2,\omega_3,h,u)$$

$$\dot{\omega}_3 = -r_3\omega_1\omega_2 + \omega_1 h/I_2 \qquad \equiv f_3(\omega_1,\omega_2,\omega_3,h,u)$$

$$\dot{h} = -r_4\omega_3\omega_1 + \left(\frac{I_2}{I_{B2}}\right)u \qquad \equiv f_4(\omega_1,\omega_2,\omega_3,h,u)$$

where

$$r_1 = \frac{I_3-I_2}{I_1}, \quad r_2 = \frac{I_1-I_3}{I_{B2}}, \quad r_3 = \frac{I_2-I_1}{I_3}, \quad r_4 = -r_2 J_2, \, , \quad h = J_2\Omega$$

Application of Pontryagin's Principle leads to the necessary conditions (following chart):

608

Momentum Transfer Maneuver: Necessary Conditions for Optimality

The Hamiltonian function is

$$H = \tfrac{1}{2}u^2 + \sum_{i=1}^{4}\lambda_i f_i, \quad \Rightarrow \quad \frac{\partial H}{\partial u} = 0 \quad \Rightarrow \quad optimal\ control:\ u = -\sum_{i=1}^{4}\lambda_i f_i$$

The state and co-state necessary conditions are the eight differential equations:

$$\dot{\omega}_1 = f_1(\omega_1, \omega_2, \omega_3, h, u),\ \textit{see previous chart,} \qquad \dot{\lambda}_1 = -\frac{\partial H}{\partial \omega_1} = r_2\,\omega_3\lambda_2 + r_3\,\omega_2\lambda_3 + r_4\,\omega_3\lambda_4$$

$$\dot{\omega}_2 = f_2(\omega_1, \omega_2, \omega_3, h, u) \qquad \dot{\lambda}_2 = -\frac{\partial H}{\partial \omega_2} = r_1\,\omega_3\lambda_1 + r_3\,\omega_1\lambda_3$$

$$\dot{\omega}_3 = f_3(\omega_1, \omega_2, \omega_3, h, u) \qquad \dot{\lambda}_3 = -\frac{\partial H}{\partial \omega_3} = r_1\,\omega_2\lambda_1 + r_2\,\omega_1\lambda_3 + r_4\,\omega_1\lambda_4$$

$$\dot{h} = f_4(\omega_1, \omega_2, \omega_3, h, u) \qquad \dot{\lambda}_4 = -\frac{\partial H}{\partial h} = -(\omega_3/I_1)\lambda_1 + (\omega_2/I_3)\lambda_3$$

Initial boundary conditions:

$$\omega_1(t_o)=0,\ \omega_2(t_o)=0,\ \omega_3(t_o)=\omega_{30},\ h(t_o)=0$$

Final boundary conditions:

$$\lambda_i(t_o)= \frac{\partial J}{\partial \omega_i(t_f)} = w\,\omega_i(t_f),\ h(t_f) = H = J_2\omega_{30}$$

Solve the TPBVP using the "method of particular solutions."

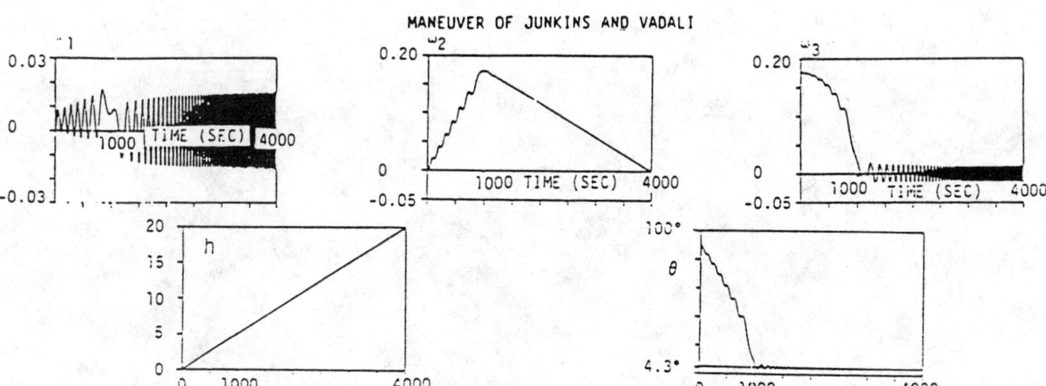

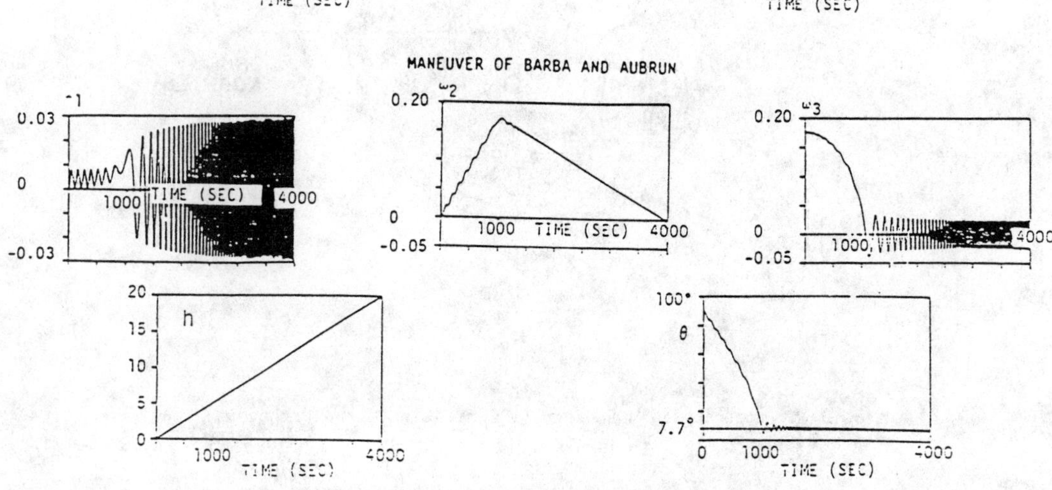

Figure 8.8 A Comparison of Two Flat Spin Recovery Maneuvers

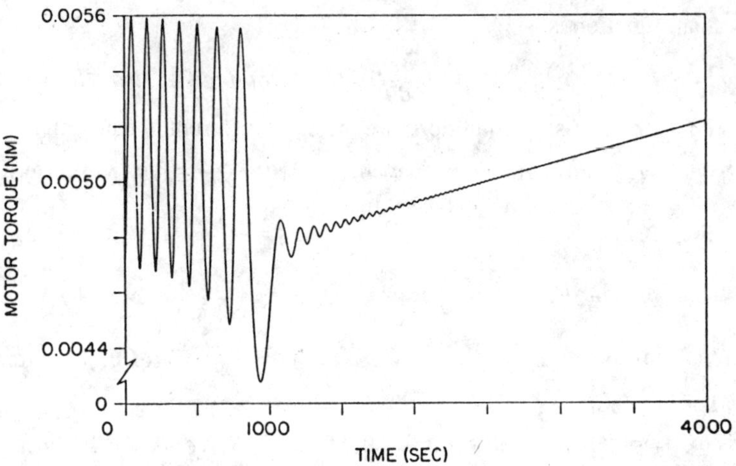

Figure 8.9 Optimal (Junkins and Vadali) Motor Torque History for a Flat Spin Recovery

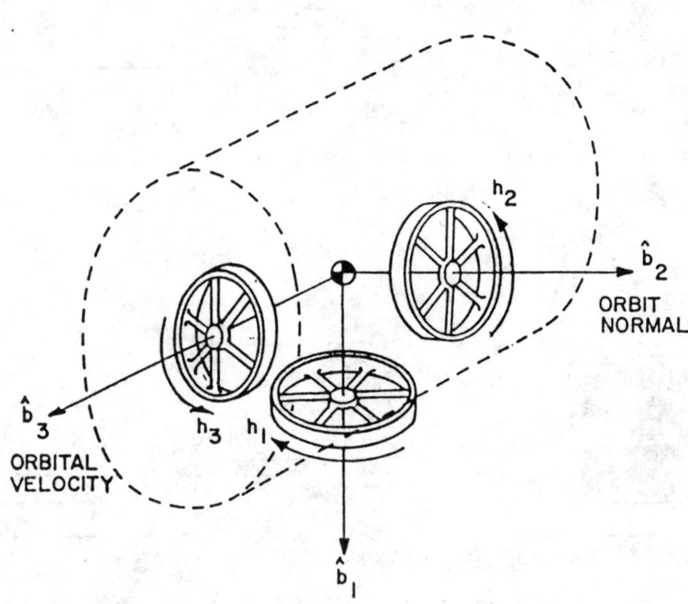

Figure 8.10 Asymmetric Rigid Spacecraft Containing Three Orthogonal Reaction Wheels

3-WHEEL CONFIGURATION : EQS. OF MOTION

$$\underline{L} = \frac{d}{dt}(\underline{H})_N \qquad \underline{H} = \underline{H}_B + \sum_{i=1}^{3}\underline{H}_{W_i}$$

$$(I_1 - J_a)\dot{\omega}_1 = (I_2 - I_3)\omega_2\omega_3 - h_3\omega_2 + h_2\omega_3 - u_1$$

$$(I_2 - J_a)\dot{\omega}_2 = (I_3 - I_1)\omega_3\omega_1 - h_1\omega_3 + h_3\omega_1 - u_2$$

$$(I_3 - J_a)\dot{\omega}_3 = (I_1 - I_2)\omega_1\omega_2 - h_2\omega_1 + h_1\omega_2 - u_3$$

$$\dot{h}_i = -J_a\dot{\omega}_i + u_i \quad , \quad i = 1,2,3$$

$$\begin{Bmatrix} \dot{\beta}_0 \\ \dot{\beta}_1 \\ \dot{\beta}_2 \\ \dot{\beta}_3 \end{Bmatrix} = \frac{1}{2} \begin{bmatrix} \beta_0 & -\beta_1 & -\beta_2 & -\beta_3 \\ \beta_1 & \beta_0 & -\beta_3 & \beta_2 \\ \beta_2 & \beta_3 & \beta_0 & -\beta_1 \\ \beta_3 & -\beta_2 & \beta_1 & \beta_0 \end{bmatrix} \begin{Bmatrix} 0 \\ \omega_1 \\ \omega_2 \\ \omega_3 \end{Bmatrix}$$

or

$$\dot{x} = f(x, u)$$

with

$$x = [\omega_1 \ \omega_2 \ \omega_3 \ h_1 \ h_2 \ h_3 \ \beta_0 \ \beta_1 \ \beta_2 \ \beta_3]^T.$$

Optimal Control Formulation

The first issue is the selection of a performance index. We consider three indices based upon quadratic penalties on the control magnitude and derivatives thereof:

$$J_1 = \frac{1}{2} \int_0^T (\sum_{i=1}^{3} u_i^2(t) dt = \int_0^T f_1 \ dt$$

$$J_2 = \frac{1}{2} \int_0^T (\sum_{i=1}^{3} \dot{u}_i^2(t)) dt = \int_0^T f_2 \ dt \quad , \quad \dot{u}_i = \frac{d}{dt} u_i$$

$$J_3 = \frac{1}{2} \int_0^T (\sum_{i=1}^{3} \ddot{u}_i^2(t)) dt = \int_0^T f_3 \ dt \quad , \quad \ddot{u}_i = \frac{d^2}{dt^2} u_i$$

In addition, for each optimal maneuver, we determine a positive measure of electrical energy expenditure, by computing the following integral (E):

$$E = \int_0^T \sum_{i=1}^{3} |u_i(t)\Omega_i(t)| dt$$

	Initial Conditions $(t_0 = 0)$	Final Conditions $(t_f = T = 100 \text{ sec.})$
β_0	.64278761	1
β_1	.44227597	0
β_2	.44227597	0
β_3	.44227597	0
ω_1	.01 r/s	0
ω_2	.005 r/s	0
ω_3	.001 r/s	0
Ω_1	0	H_{n1}/J_a
Ω_2	0	H_{n2}/J_a
Ω_3	0	H_{n3}/J_a

Performance Index	No. of Iterations	Performance Index J_i
J_1	6	$.248039 (N^2 m^2 s)$
J_2	5	$.001477 (N^2 m^2 / s^2)$
J_3	5	$.000021 \ (N^2 m^2 / s^3)$

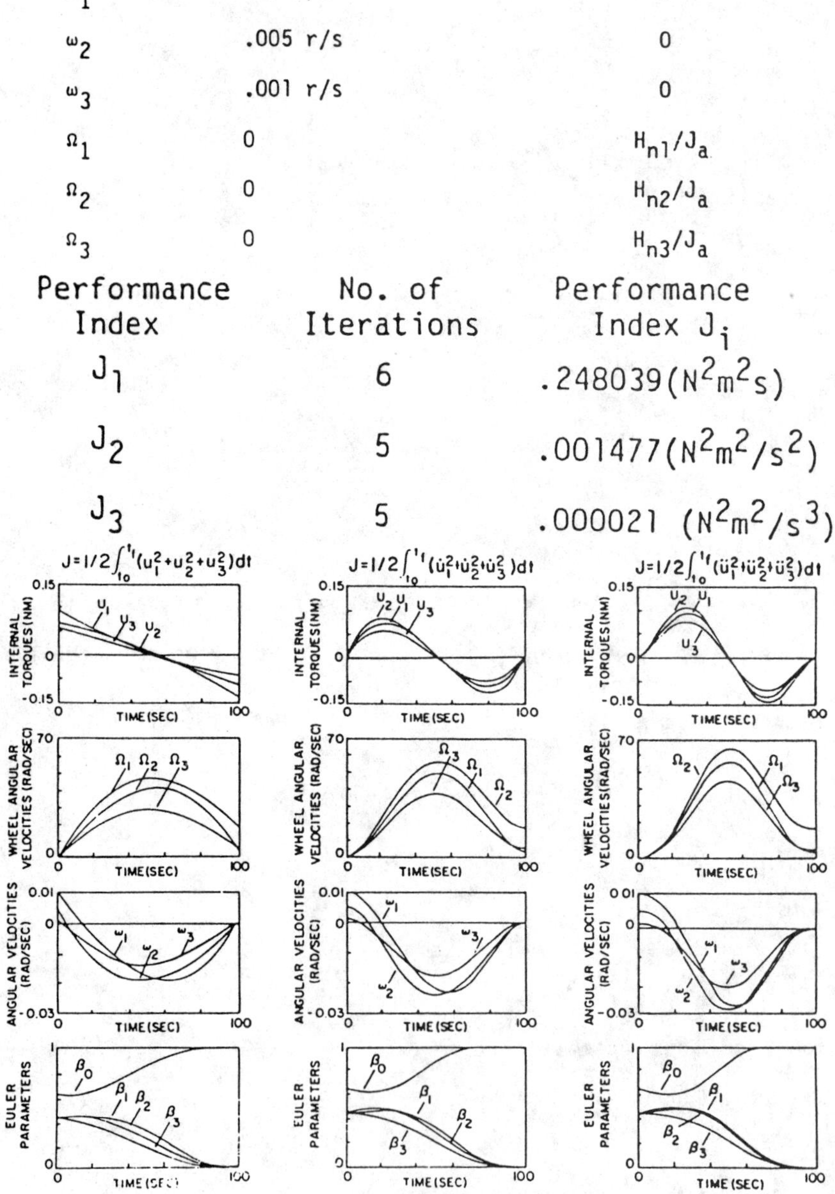

Figure 8.11   Three Optimal Momentum Transfer Maneuvers: Effect of Control Derivative Penalties in the Performance Index

References

1. Junkins, J. L. and Turner, J. D., *Optimal Spacecraft Rotational Maneuvers*, Elsevier, Amsterdam, 1986.

2. Junkins, J.L., Williams, C. E., *Time Optimal Attitude Maneuvers fir the Navt Navigation Satellites*, **AAS** Paper AAS81-010, presented at the Annual Rocky Mountain Guidance and Control Conference, Jan. 31-Feb. 4, 1981, Keystone, CO.

3. Junkins,J.L., Carrington,C.K. and Williams,C.E., *"Time Optimal Magnetic Attitude Maneuvers,"* **AIAA J. of Guidance and Control**, Vol. 4, No. 4, pp. 363-368, July-Aug. 1981.

4. Junkins, J.L., *"Optimal Feedback Slewing of Flexible Spacecraft,"* Tech. Comment, **AIAA Journal of Guidance, Control, and Dynamics**, Vol. 5, No. 3, p. 318, May-June 1982.

5. Vadali, S.R. and Junkins, J.L., *"Spacecraft Large Angle Rotational Maneuvers with Optimal Momentum Transfer,"* **Journal of the Astronautical Sciences**, Vol. XXXI, No. 2, pp. 217-235, April-June 1983.

6. Vadali, S.R. and Junkins, J.L., *"Optimal Open Loop and Stable Feedback Control of Rigid Spacecraft Attitude Maneuvers,"* AAS Paper #83-373, presented at the AAS/AIAA Astrodyn. Specialists Conference, Lake Placid, New York, August 22-25, 1983.

7. Vadali, S.R., Kraige, L.G. and Junkins, J.L., *"New Results on the Optimal Spacecraft Attitude Maneuver Problem,"* AIAA **Journal of Guidance, Control, and Dynamics**, Vol. 7, No. 3, pp. 378-380, May-June 1984.

8. Vadali, S. R., Junkins, J. L., *"Optimal Open Loop and Stable Feedback Control of Spacecraft Attitude Maneuvers,"* **J. of the Astronautical Sciences**, Vol.32, No.2, 105-122, Apr.-June 1984.

Feedback Control Design

Stability Robustness Measures

Eigenstructure Assignment Methods

Multiple Criteria Optimization

Feedback Control for Multi-Input Multi-Output Systems

Consider the System

$$\dot{x} = Ax + Bu + Dw, \quad u = - Gx \tag{1}$$

The closed loop system is then

$$\dot{x} = (A - BG)x + Dw \tag{2}$$

The central question is "how do we select the control gain G?" What we desire:

- Stability (first and formost!) $\quad\Rightarrow\quad$ all λ's in LHP.

- Robustness ("Stability in spite of ignorance!") $\quad\Rightarrow\quad$ <u>beyond</u> pole placement!

- "Small" average and peak control effort $\quad\Rightarrow\quad$ to be <u>physically realizable.</u>

- "Small" average and peak performance errors $\quad\Rightarrow\quad$ to be <u>practical!</u>

- "Rapid" settling from a <u>distribution</u> of initial cond. $\Rightarrow$ constraints on λ's.

- Finite bandwidth of sensors and actuators, $\quad\Rightarrow\quad$ transfer function constraints
  spectral properties of w . . . . .

Linear Dynamical Systems Subject to Arbitrary Disturbances

Consider the class of systems described by the system of differential equations:

$$\dot{x} = Ax + f(t, x(t)), \quad \textit{(A is a constant matrix)} \tag{27}$$

A central question which we'd like to answer is the following:

For a given A matrix, is it possible to guarantee that x(t) remains stable, <u>without any explicit information on f(t,x(t))</u>....so long as f(t,x(t)) is less than a bound determined by the matrix A?

Thanks to recent research, the answer to this question is yes.

Liapunov Stability Theorem

Given the class of dynamical systems described by the differential equation:

$$\dot{x}(t) = Ax(t) \,, A= \text{constant,}$$

the state $x = 0$ is asymptotically stable if and only if symmetric, positive-definite matrices P and Q exist which satisfy the matrix Liapunov Equation:

$$A^T P + PA = -Q$$

Given A and $Q=Q^T > 0$ the solution for P will be positive definite, if A is stable. The converse is not true (given stable A & P>0, the Q computed from the Liapunov equation may or may not be positive definite).

The above theorem has a geometrical interpretation in terms of the Liapunov function (which is a positive and decreasing measure of distance from $x = 0$):

$$V(x) = x^T P x \quad \text{.... note that } V(x) \geq 0 \text{ & also } \dot{V}(x) \leq 0 \text{ since}$$
$$\dot{V}(x) = \dot{x}^T P x + x^T P \dot{x} = x^T (A^T P + PA) x = - x^T Q x \text{ .....}$$

Perturbation Bound of Patel and Toda [1980]

Consider a system described by

$$\dot{x}(t) = Ax(t) + f(x(t), t), \tag{28}$$

where

    A is a constant matrix whose eigenvalues are in left half plane.

    $f(x(t), t)$ is an arbitrary nonlinear function, except we require $f(0,t) = 0$.

Question: Can we obtain bounds on f such that $x(t)$ remains stable?

Choose as a trial Liapunov function

$$V(x) = x^T P x, \tag{29}$$

where $P = P^T$ is a positive definite solution of the Liapunov Equation:

$$A^T P + PA = -2Q \tag{30}$$

Observe that existence of a $P = P^T > 0$ for $Q = Q^T > 0$ satisfying Eq. (30) is guaranteed as a consequence of the assumed stability for $f(x(t), t) = 0$.

Perturbation Bound of Patel and Toda [1980], continued...

From differentiation of $V(x) = x^T P x$, we get

$$\dot{V}(x) = \dot{x}^T P x + x^T P \dot{x}, \qquad\qquad \text{substituting } \dot{x}(t) = A x(t) + f(x(t), t),$$

$$= x^T(A^T P + PA)x + 2 f^T P x, \quad \text{substituting } A^T P + PA = -2Q, \text{ we get}$$

$$\dot{V}(x) = -2x^T Q x + 2 f^T P x \tag{31}$$

Thus $V(x) \le 0$ is guaranteed if

$$f^T P x \le x^T Q x \tag{32}$$

Patel & Toda showed that this condition (32) is satisfied if

$$\frac{|f|}{|x|} \le \frac{\min \lambda(Q)}{\max \lambda(P)} \equiv \mu_{PT} \tag{33}$$

Observations:
- *Conservatism is introduced by inequality analysis*
- $\mu_{PT} = \mu_{PT}(Q)$ *... what is the "best" Q ...* $\rightarrow Q = I$

$$\text{so } \mu_{PT} = \frac{1}{\max \lambda(P)} = \frac{1}{\bar{\sigma}(P)} \quad \textit{(Patel \& Toda, 1980)}$$

Special case:

$$\dot{x}(t) = \bar{A}x(t) + Ex = (\bar{A} + E)x$$

e.g.

$$\bar{A} = A + BG$$

remains stable if E is bounded by (Bauer - Fike theorem)

$$\| E \| \le \frac{-\max[\operatorname{Re} \lambda(\bar{A})]}{k(\Psi)}, \qquad \left\{ \text{compare with } \Delta E = \frac{\Delta \lambda}{\frac{\partial \lambda}{\partial E}} \right\}$$

where

$$k(\Psi) \equiv \| \Psi \| \; \| \Psi^{-1} \| = \text{the condition number of } \Psi$$

↳ closed loop modal matrix

Notice the relationships:

Optimum Stability Robustness ⟷ *Optimum Conditioning*

Stability Margin ⟷ *Robustness Limit*

Feedback Control for Multi-Input Multi-Output Systems

<u>*Quadratic Regulator Approach to find G in u = Gx:*</u>

minimize a functional such as

$$J = \int_o^\infty [x^T Q x + u^T R u]\, dt \; ; \quad Q=?, R=?$$

Note $G = G(Q,R)$, solution for the "optimal" gains requires solution of a matrix Riccati equation; thus the weight matrices provide a parametric family of stable controllers.

<u>*Eigenstructure Assignment Approach:*</u>

prescribe closed loop eigenvalues & target **eigenvectors (but how?!)**

Computation of G requires only *linear* operations; the closed loop eigenvalues are assigned exactly (to within arithemtic errors, for controllable systems), whereas the target eigenvectors are achieved to some degree of approximation.

For both of these approaches, we often require unrealistic insight to select the (often hundreds of) free parameters to satisfy all our objectives without iteration. For high dimensions, which approach is most suited to efficiently obtaining feasible/optimal solutions?

Generalized Quadratic Regulators

The cornerstones of *modern* control theory are linear, constant gain feedback controls determined by minimizing

$$J = \int_o^\infty (x^TQx + u^TRu)dt, \quad \text{with weight matrices } Q = Q^T \geq 0, \; R = R^T > 0 \tag{12}$$

subject to satisfying $\dot{x} = Ax + Bu$. This is the linear-quadratic regulator (LQR) problem. The solution is determined from Pontryagin's Principle[13]; it requires the positive-definite P matrix which satisfies the algebraic Riccati equation

$$PA + A^TP - PBR^{-1}B^TP + Q = 0 \tag{13}$$

The associated optimal control gain matrix is then

$$G = - R^{-1}B^TP \tag{14}$$

We now consider two recently developed generalizations of the quadratic regulator.

<u>Generalized LQR - Form 1</u>:

Recently[16], the following generalized quadratic index was introduced

$$J = \int_o^\infty \begin{Bmatrix} x \\ u \end{Bmatrix}^T W \begin{Bmatrix} x \\ u \end{Bmatrix} dt, \text{ with } W = W^T = \begin{bmatrix} Q+N^TRN & N^TR \\ RN & R \end{bmatrix} > 0, \quad Q = Q^T \geq 0, \; R = R^T > 0 \tag{15}$$

The Pontryagin necessary conditions(minimizing Eq. (15), subject to $\dot{x} = Ax + Bu$) lead to the generalized Riccati equation

$$P\bar{A} + \bar{A}^TP - PBR^{-1}B^TP + Q = 0, \text{ where } \bar{A} = A - BN \tag{16}$$

and the optimal gain matrix

$$G = - R^{-1}B^TP - N \tag{17}$$

Reference [16] shows that an arbitrary real $m{\times}n$ cross coupling weight matrix (N) can be chosen without de-stabilizing the closed loop system. Notice that N is essentially a pre-feedback gain, so it is evident that some redundancy must exist. In particular, [16] investigated $N = [UD^T\Phi \;\; VD^T\Phi]$, with U, V positive semi-definite $m{\times}m$ matrices.

617

Generalized Quadratic Regulators: Continued

<u>Generalized LQR - Form 2</u>:

The second modification of the LQR problem, introduced in [16], is obtained by minimizing the quadratic index

$$J = \int_0^\infty \begin{Bmatrix} x \\ u \end{Bmatrix}^T W \begin{Bmatrix} x \\ u \end{Bmatrix} dt, \; with \; W = W^T = \begin{bmatrix} Q+N^TRN+PBR^{-1}B^TP & N^TR \\ RN & R \end{bmatrix} > 0, \tag{18}$$

subject to $\dot{x} = Ax + Bu$.

The motivation for the final term $(PBR^{-1}B^TP)$ in the $1,1$ partition of W is that it results in an important simplification of the necessary conditions. In fact the the P matrix satisfies the simpler *linear* algebraic Lyapunov equation

$$P\bar{A} + \bar{A}^TP + Q = 0, \; where, \; as \; before \; \bar{A} = A - BN, \tag{19}$$

instead of the *nonlinear* algebraic Riccati Eq. (16). The optimal gain matrix is still given by Eq. (17). Notice that we choose the weights Q, R, N, and solve for P from Eq. (19), only then can we fully specify W in Eq. (18), since it depends upon P.

The weight matrices $Q, R,$ and N should be parameterized in such a way as to guarantee their definiteness properties and the controllability of A, this insures the existence of a solution for a positive-definite P satisfying Eq. (19). This problem is solved[16] by the Choleski factorizations: $Q = L_1L_1^T, \quad R^{-1} = L_2L_2^T, \quad U = L_3L_3^T, \quad V = L_4L_4^T, \quad N = [UD^T\Phi \;\; VD^T\Phi]$ (20)

<u>Generalized LQR - Form 2</u>, Continued:

Choleski factorizations:

$$Q = L_1L_1^T, \quad R^{-1} = L_2L_2^T, \quad U = L_3L_3^T, \quad V = L_4L_4^T, \quad N = [UD^T\Phi \;\; VD^T\Phi]$$

with

$$L_1 = \begin{bmatrix} q_{11}^2 & 0 & .. & 0 \\ q_{21} & q_{22}^2 & .. & 0 \\ : & : & : & : \\ q_{NN} & q_{NN} & .. & q_{NN}^2 \end{bmatrix}, \quad L_2 = \begin{bmatrix} r_{11}^2 & 0 & .. & 0 \\ r_{21} & r_{22}^2 & .. & 0 \\ : & : & : & : \\ r_{m1} & r_{m2} & .. & r_{mm}^2 \end{bmatrix},$$

$$L_3 = \begin{bmatrix} u_{11}^2 & 0 & .. & 0 \\ u_{21} & u_{22}^2 & .. & 0 \\ : & : & : & : \\ u_{m1} & u_{m2} & .. & u_{mm}^2 \end{bmatrix}, \quad L_4 = \begin{bmatrix} v_{11}^2 & 0 & .. & 0 \\ v_{21} & v_{22}^2 & .. & 0 \\ : & : & : & : \\ v_{m1} & v_{m2} & .. & v_{mm}^2 \end{bmatrix} \tag{21}$$

The high dimensionality of the parameter space consisting of the above, or any complete, weight matrix parameterization has proven a significant obstacle to numerical implementations with fully populated weight matrices. In [16], we introduced a *minimum modification homotopy* strategy for tuning the above matrices and demonstrated successful implementations. This method is employed where appropriate in the numerical studies reported below.

Constant Gain Feedback: Eigenstructure Assignment

Output feedback:

    linear feedback control: $u = Gy$,   *measured output:* $y = Sx$ (8)

    where $u \in R^{mx1}$, $y \in R^{rx1}$, $G \in R^{mxr}$, $S \in R^{rxn}$

Closed Loop System: *Substituting the feedback law of Eqs. (8) into* $\dot{x} = Ax + Bu$ *to obtain*

    $\dot{x} = [A + BGS]x$ (9)

The Closed Loop Eigenvalue Problem is

    $det\big[[A+BGS] - \lambda I\big] = 0$ $\Rightarrow$ *eigenvalues:* $\{\lambda_1, \lambda_2, \ldots \lambda_{2n}\}$ (10)

    $\big[[A+BGS] - \lambda_j I\big]\psi_j = 0$ $\Rightarrow$ *right eigenvectors:* $\{\psi_1, \psi_2, ., \psi_{2n}\}$ (11)

Assignment of Eigenvalues and Eigenvectors:

    Let the structural design parameter vector be denoted p, for a *specified family* of structures, system matrices have a known functional dependence upon p : $A = A(p)$, $B = B(p)$, $S = S(p)$; it is a modest extension conceptually to let the p vector include the control gains, so $G = G(p)$. As a consequence, the eigensolution depends upon p. There are two distinct classes of problems: 1. For a *fixed structure* (specified A, B, S), and a given family of gains $G(p)$, or,

        2. For a given *family of structures and controllers;* $A(p), B(p), S(p), G(p)$,
find a judicious design for p to impose constraints & optimality defined in the space of closed loop eigenvalues and eigenvectors.

Assignment of Eigenvalues & Eigenvectors: fixed A, B, S; $G = G(p)$

    There is a recently developed theoretical foundation which underlies existence, uniqueness and computation of G. The principle existence issues are captured in a theorem due to Srinathkumar[4,47]:

Eigenstructure Assignment Theorem.   *Given a controllable and observable dynamical system described by* $\dot{x} = Ax + Bu$, $y = Sx$, $u = Gy$, *{having N states, r sensors, and m actuators, with matrices B and S of full rank}, then through choosing values for the elements of G, max(m,r) closed-loop eigenvalues (of A + BGS) can be assigned, max(m,r) eigenvectors can be <u>partially</u> assigned, and min(m,r) entries in each eigenvector can be arbitrarily assigned.*

Remarks:

    Prior to results published by Bhattacharyya, Juang, Junkins, Kautsky, Laub, Lim, and Rew, et al during the 1980's, eigenvalue and eigenvector assignment was impractical, except for low order systems, even for the special case of *fixed* structures.

    Several recently developed algorithms have been successfully demonstrated on systems >30; all of these approaches either explicitly or implicitly focus upon achieving *small condition numbers for the matrix of closed loop eigenvectors.* We have found this to be a key requirement not only robust control, but also for <u>numerical robustness of the algorithms</u>.

Robust Eigenstructure Assignment Algorithm (REAL)

The closed loop eigenvalues and eigenvectors satisfy

$$(A - BG - \lambda_i I)\phi_i = 0, \quad i = 1, 2, \dots n \tag{3}$$

Consider the *inverse* eigenvalue problem (given A, B, λ's, & ϕ's, find G), we will find this problem easier, in one sense than the usual *foreward* problem, since we will find it involves only linear operations. Notice that Eq. (3) can be re-written in *Sylvester form* as

$$(A - \lambda_i I)\phi_i = B h_i, \quad \text{where} \ \ h_i = G\phi_i, \ \ i = 1, 2, \dots n \tag{4}, (5)$$

Introducing the following matrix definitions:

$$H = [h_1 \ .. \ h_n], \quad \Phi = [\phi_1 \ .. \ \phi_n], \quad \Lambda = \mathrm{diag}(\lambda_1 \ .. \ \lambda_n) \tag{6}$$

Then Eqs. (4) and (5) become

$$A\Phi - \Phi\Lambda = BH, \quad \text{where} \ \ H = G\Phi \tag{7}, (8)$$

Referring to Eqs. (7), (8), consider the following algorithm:

(1) Choose the closed loop eigenvalue matrix Λ, and the parameter matrix H.

(2) Solve Eq. (7) for Φ.

(3) Solve Eq. (8) for G.

As a practical matter the above algorithm is oversimplified, because there are an infinity of possible choices for the matrix H, and most of them lead to poor control laws!

In The Real World . . .

In order to circumvent the difficulty of guessing the matrix H, suppose we instead seek a *target* set of eigenvectors $(\hat{\phi}_i, i = 1,2, \dots n)$. We can find the h_i vectors which most nearly consistent with these target eigenvectors by least square inversions of Eq. (4) as

$$(A - \lambda_i I)\hat{\phi}_i = B h_i, \ \Rightarrow \ \hat{h}_i = (B^T B)^{-1} B^T (A - \lambda_i I)\hat{\phi}_i, \ i=1,2,\dots,n \tag{9}$$

We can now execute the three steps of the algorithm on the previous page, to the extent that the target eigenvectors are reasonable, we will obtain a reasonable design. It is apparent that we still have the responsibility of generating the h_i vectors indirectly by selecting judicious set of *target* eigenvectors $(\hat{\phi}_i, i = 1,2, \dots n)$. We have given significant analytical and computational attention to three choices for the target eigenvectors:

(i)   The modal matrix constructed[1, 2] from open loop eigenvectors.

(ii) The unitary matrix nearest[2] the modal matrix of open loop eigenvectors.

(iii) The unitary matrix (left singular vectors) resulting[1] from singular value decomposition of the matrix: $[U_1 U_2 \dots U_n]$, where U_i is a basis matrix spanning $(A - \lambda_i I)^{-1} B$.

$$\text{THREE CHOICES OF TARGET EIGEN-VECTORS}$$

1. OPEN LOOP EIGENVECTORS

$$[A - \lambda_{oi} I]\, \phi_{oi} \qquad i = 1, 2, \cdots n$$

$$\rightarrow \hat{\phi}_i = \phi_{oi} \,, \quad \hat{\Phi} = \Phi_o = [\, \phi_{o_1} \cdots \phi_{on} \,]$$

2. UNITARY MATRIX "NEAREST" THE OPEN-LOOP EIGENVECTORS

$$\Phi_o = U_o\, \Sigma_o\, V_o^H$$

$$\text{TARGET VECTORS:} \quad \hat{\Phi} = U_o\, V_o^H$$

THIS CHOICE MINIMIZES

$$\| \Phi_o - \hat{\Phi} \|$$

SUBJECT TO

$$\hat{\Phi}^H \hat{\Phi} = I$$

3. REW'S CHOICE : $\quad \hat{\Phi} = U_s$

$$\text{WHERE} \quad S = [\, U_1\ U_2 \cdots U_n \,] = U_s\, \Sigma_s\, V_s^H$$

$$U_i \sim \text{UNITARY MATRIX SPANNING}$$

$$(A - \lambda_i I)^{-1} B$$

$$\left. \begin{array}{l} \text{FEASIBLE} \\ \text{EIGEN-} \\ \text{VECTORS} \end{array} \right\} \Rightarrow \quad \phi_i = (A - \lambda_i I)^{-1} B\, \ell_i \,, \quad \ell_i = ?$$

Some Remarks ....

The explicit algorithms and justifications for determining H corresponding to the above three choices of target eigenvectors are given in the references. It is also evident that an initial choice can be subjected to further optimization or to impose other constraints. It is of significance to record here the following observations:

(a)   For cases in which the closed loop eigenvalues are to be only slightly perturbed from their open loop values, then choosing the open loop target eigenvectors(i) has been found to lead to the smallest gains, but not necessarily the most robust design.

(b)   For cases in which some of the closed loop eigenvalues are moved deep into the left half plane, then the two unitary vectors choices, (ii) and (iii), lead to consistently smaller condition numbers than (i) for the closed loop modal matrix, and usually smaller gain norms as well. To date a consistent pattern has not emerged as to which is best, both lead to robust designs with small gain norms.

We have found all three choices of target vectors typically lead to "good " designs for feedback control which places eigenvalues exactly and shapes eigenvectors approximately. Iteration will be often be required to finalize a design. This approach provides a *linear* parameterization of gains. Design iteration is thus relatively inexpensive.

TAMU Grid Structure: Alternate Grid Sensor and Actuator Locations

Preliminaries

Consider

$$M\ddot{z} + C\dot{z} + Kz = Du \tag{1}$$

The *structural eigenvalue problem* associated with $z = \phi e^{\lambda t}$ solutions for undamped free vibration special case of Eq. (1)

$$M\ddot{z} + Kz = 0 \tag{2}$$

is

$$\det[M\lambda^2 + K] = 0, \quad \Rightarrow \quad eigenvalues: \{\lambda_i^2 = -\omega_i^2, i = 1,2,\ldots n\} \tag{3}$$

$$[M\lambda_i^2 + K]\phi_i = 0, \quad \Rightarrow \quad eigenvectors: \{\phi_i, \quad i = 1,2,\ldots n\} \tag{4}$$

The eigenvectors are orthogonal with respect to M and K; we use the usual normalizations

$$[\Phi]^T M[\Phi] = [I] \quad \text{and} \quad [\Phi]^T K[\Phi] = [\omega^2] = diag\{\omega_1^2, \ldots, \omega_n^2\} \tag{5}$$

It is convenient to introduce a transformation of Eq. (1) into the modal state space associated with Eq. (2) as follows:

$$z = [\Phi]\eta, \quad \text{and} \quad \eta = [\Phi]^T Mz, \text{ this maps Eq. (1) into the the modal space equation of motion}$$

$$\ddot{\eta} + [\Phi]^T C[\Phi]\dot{\eta} + [\omega^2]\eta = [\Phi]^T Du, \qquad x = \begin{Bmatrix} \eta \\ \dot{\eta} \end{Bmatrix}, \textit{ the modal state vector} \tag{6}$$

Or

$$\dot{x} = Ax + Bu, \qquad \text{with} \qquad A = \begin{bmatrix} 0 & I \\ -[\omega^2] & -[\Phi]^T C[\Phi] \end{bmatrix}, \quad B = \begin{bmatrix} 0 \\ [\Phi]^T D \end{bmatrix} \tag{7}$$

MATLAB CODE FOR Robust Eigenstructure Assignment

```
%
function G = assign(A,B,Lambda,T)
%
%   eigenstructure assignment for full state feedback a la Junkins 5/4/88
%
%   x dot = Ax + BGx, Lambda = col(assigned eigenvalues), T = target
%                                                         eigenvectors
%   Notice, if complex eigenvalues are assigned, then these must be
%   assigned (in L) in complex conjugate pairs; the corresponding columns
%   of T must be chosen as corresponding complex conjugate pairs, otherwise
%   you will probably get back a gain matrix with large imaginary parts!!
[n,n1]=size(A);   L=eye(n);
%
%   Store eigenvlues into a diagonal matrix
for ii=1:n, L(ii,ii)=Lambda(ii); end;
%
% compute least square pseudo-inverse of B
%
Binv=pinv(B);
%
%   Determine parameter matrix (H = G*psi) which is consistent (in the
%   least square sense) with the assigned eigenvalues (L) & target
%   eigenvectors (T).
%
% First compute the least square solution for H
%
H=Binv*(T*L - A*T);
%
% Now, determine the achieved eigenvectors (psi) by inverting the matrix
% Lyapunov form of the eigenvalue problem ("Sylvesters' Equation"):
%   A*psi - psi*L = -B*H
%
psi = lyap(A,-L,B*H);
%
% Finally, we can compute gain matrix by inverting H = G*psi
%
G=H/psi;
%
%   The following statements were put in during debug, they may be found
%   useful someday!  We can compute G with only real arithmetic via
%      G=[real(H) imag(H)]*(pinv([real(psi) imag(psi)]'))'
%   In order to asess conditioning,it is useful to normalize psi via
%   for ii=1:n, PHI(:,ii)=psi(:,ii)/max(abs(psi(:,ii))); end
%      c=cond(PHI);
end
```

This is a "Matlab" SESSION using "assign" to demononstrate design
of a full state feedback law for a reduced order model, then
verifying it with an intermediate order model.

The physical system is the TAMU flexible grid experiment
with three reaction wheel actuators and six sensors.
The full order model is order 120. Two reduced order models
are used; both are modal space models (we used the modal matrix
of the undamped structure to generate the transformation).   The
first reduced order model considers the 1st 3 modes only and is of
order 6. The 2nd reduced order model considers 10 modes (order 20).

THE A MATRIX FOR THE 1ST MODEL IS
/ a

a =

             0   0   0   1.0000   0   0
             0   0   0   0   1.0000   0
             0   0   0   0   0   1.0000
      -31.9112   0   0   0   0   0
             0 -215.5905   0   0   0   0
             0   0 -927.7507   0   0   0

THE CONTROL INFLUENCE MATRIX IS
/ b

b =

             0   0   0
             0   0   0
             0   0   0
       -0.2230   -0.1912   -0.1912
             0   -0.1484   0.1484
        0.6040   0.0256   0.0256

THE TARGET (ASSIGNED) CLOSED LOOP EIGENVALUES ARE
/ lama

lama =

   -0.2640 - 5.6490i
   -0.2640 + 5.6490i
   -0.2000 -14.6830i
   -0.2000 +14.6830i
   -0.7750 -30.4590i
   -0.7750 +30.4590i

COMPUTE THE OPEN LOOP EIGENVECTORS AND EIGENVALUES
/ [va,lamda]=eig(a);
SORT THE EIGENVALUES IN INCREASING FREQUENCY MAGNITUDE
/ [lam,index]=sort(diag(lamda));
/ lam

lam =

        0 - 5.6490i
        0 + 5.6490i
        0 -14.6830i
        0 +14.6830i
        0 -30.4590i
        0 +30.4590i

SORT THE EIGENVECTORS ... USE THE OPEN LOOP EIGENVECTORS AS TARGETS
/ va=va(:,index);

CALL ASSIGN TO DESIGN GAIN MATRIX TO ASSIGN EIGENVALUES EXACTLY
AND EIGENVECTORS IN THE LEAST SQUARE SENSE
/ gain=assign(a,b,lama,va)

gain =

  Columns 1 through 4

  -0.0163 - 0.0000i   0.0000 - 0.0000i   -1.0462 - 0.0000i   -0.1232 - 0.0000i
   0.1917 + 0.0000i   0.1348 + 0.0000i   0.6100 + 0.0000i   1.4523 + 0.0000i
   0.1917 + 0.0000i   -0.1348 - 0.0000i   0.6100 + 0.0000i   1.4523 + 0.0000i

  Columns 5 through 6

  -0.0000 + 0.0000i   -2.6998 + 0.0000i
   1.3477 + 0.0000i   1.5741 - 0.0000i
  -1.3477 - 0.0000i   1.5741 - 0.0000i

DELETE THE NEGLIGIBLE IMAGINARY PARTS OF THE GAIN MATRIX
/ gain=real(gain);

CONFIRM THAT THE CLOSED LOOP EIGENVALUES WERE INDEED CORRECTLY
ASSIGNED:
/ eig(a+b*gain)

ans =

  -0.2640 + 5.6490i
  -0.2640 - 5.6490i
  -0.7750 +30.4590i
  -0.7750 -30.4590i
  -0.2000 +14.6830i
  -0.2000 -14.6830i

YES THEY ARE AS THEY SHOULD BE ...

  NOW THE CLOSED LOOP EIGENVECTORS AND EIGENVALUES ARE COMPUTED
  (AND SORTED) FOR THE VERIFICATION (20TH) ORDER DESIGN:
  / [VA,LAMDA]=eig(A+B*GAIN);
  / [LAM,index]=sort(diag(LAMDA));
  / LAM

  LAM =

    1.0e+002 *

    -0.0026 - 0.0565i
    -0.0026 + 0.0565i
    -0.0020 - 0.1468i
    -0.0020 + 0.1468i
    -0.0077 - 0.3046i
    -0.0077 + 0.3046i
     0.0000 - 0.3778i
     0.0000 + 0.3778i
     0.0000 - 0.4887i
     0.0000 + 0.4887i
     0.0000 - 0.8062i
     0.0000 + 0.8062i
     0.0000 - 0.8348i
     0.0000 + 0.8348i
    -0.0000 - 0.8365i
    -0.0000 + 0.8365i
     0.0000 - 1.0138i
     0.0000 + 1.0138i
     0.0000 - 1.2534i
     0.0000 + 1.2534i

  IT IS OBVIOUS THAT THE FIRST 6 ARE AS THEY WERE AND THE HIGHER
  EIGENVALUES ARE NEGLIGIBLY PERTURBED IN THIS CASE. THE VERIFICATION
  WITH THE FULL (120TH ORDER) SYSTEM YIELDED THE SAME CONCLUSION.

NOW CHECK THE NORM OF THE GAIN MATRIX AND THE CONDITION OF THE
INTERMEDIATE ORDER CLOSED LOOP MODAL MATRIX

```
/ norm(GAIN,'fro')

ans =

    4.6967

/ cond(VA)

ans =

   125.4001

/ diary off
```

Performance of Eigenstructure Assignment Algorithms for Test Set 1 (Mass-Spring Systems)

MODEL* NO.	MASS MATRIX	DESIRED EIGENVALUE** ω	ζ
1	$M = \mathrm{diag}[1,1,1]$	$\omega_1 = 1.44$ $\omega_2 = 4.47$ $\omega_3 = 6.92$	$\zeta_1 = \zeta_2 = \zeta_3 = 0.5$
2	$M = \mathrm{diag}[10^2,1,1]$	$\omega_1 = 0.89$ $\omega_2 = 2.33$ $\omega_3 = 6.76$	$\zeta_1 = \zeta_2 = \zeta_3 = 0.5$
3	$M = \mathrm{diag}[10^3,10^2,1]$	$\omega_1 = 0.09$ $\omega_2 = 0.95$ $\omega_3 = 4.70$	$\zeta_1 = \zeta_2 = \zeta_3 = 0.5$
4	$M = \mathrm{diag}[10^4,10^2,1]$	$\omega_1 = 0.03$ $\omega_2 = 0.31$ $\omega_3 = 4.49$	$\zeta_1 = \zeta_2 = \zeta_3 = 0.5$

* All the models have the same stiffness matrix and control influence matrix

MODEL NO.	$k(\Psi^o)$**	Algorithm I* $k(\Psi^C)$**	$\|G\|_f$	Algorithm II* $k(\Psi^C)$	$\|G\|_f$	Algorithm III* $k(\Psi^C)$	$\|G\|_f$
1	6.92	9.26	50.91	9.26	19.83	9.26	19.83
2	7.10	9.04	68.39	10.39	45.08	10.39	45.08
3	20.79	10.08	121.79	22.31	128.96	22.31	128.96
4	72.94	31.79	367.04	73.46	402.12	73.46	402.12

* Algorithm I: Eigenstructure assignment (REAL) using Rew's unitary target vectors.
Algorithm II: Eigenstructure assignment (REAL) using the open loop eigenvectors as targets.
Algorithm III: IMSC a la Meirovitch.

$$** \quad \Phi^T M \Phi = I, \ \Phi^T K \Phi = \mathrm{diag}(\omega^2), \ \Psi = \begin{bmatrix} \Phi & \dot{\Phi} \\ \Lambda\Phi & \dot{\Lambda\Phi} \end{bmatrix}$$

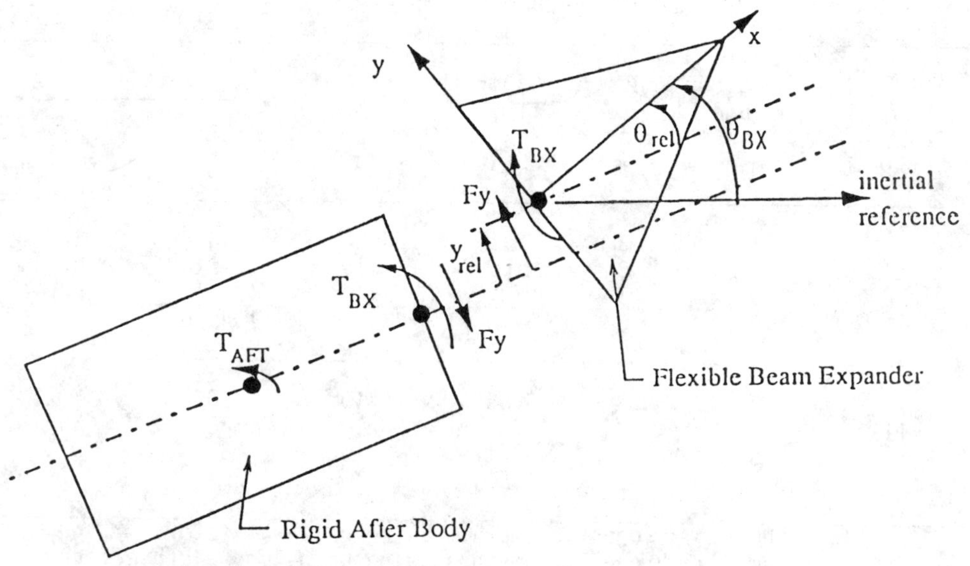

Figure 1   Configuration of R2P2 Model

Table 3   Test Example Set II (R2P2 Models)

| Open-Loop Eigenvalues | | Desired Eigenvalues | | | |
| | | Design I* | | Design II* | |
Freq (Hz)	Damping	Freq (Hz)	Damping	Freq (Hz)	Damping
.0	.0	.19	.93	.08	.87
.0	.0	.40	.81	.13	.80
.0	.0	.56	.75	.19	.70
3.58	.001	3.58	.49	3.58	.06
6.74	.001	6.74	.28	6.74	.03
11.90	.001	11.90	.17	11.90	.02
50.80	.001	–	–	–	–
3.59	.001	–	–	–	–

* Settling time for Design I   ≈ 0.5 second
  Settling time for Design II ≈ 2.0 second

Table 4 Performance of Eigenstructure Assignment Algorithms
(R2P2 Models)

Desired Eigenvalue Set (Table 3)	No. of Modes and No. of Controllers		Algorithm I*		Algorithm II*	
	n	m	$K(\phi)$**	$\|G\|_f$**	$K(\phi)$**	$\|G\|_f$**
I	3	3	9.31	$.24 \times 10^9$	9.42	$.24 \times 10^9$
	4	3	6×10^3	$.37 \times 10^9$	119×10^3	82×10^9
	5	3	5×10^3	9×10^9	137×10^3	90×10^9
	6	3	15×10^3	27×10^9	154×10^3	91×10^9
II	3	3	3.86	25×10^6	3.86	31×10^6
	4	3	474	34×10^6	582	55×10^6
	5	3	677	1×10^9	637	57×10^6
	6	3	2,264	11×10^9	657	57×10^6

*   Algorithm I: Projection method using unitary vectors
   Algorithm II: Projection method using open-loop eigenvectors
** $K(\phi)$: Condition number of the matrix of eigenvectors
   $\|G\|_f$: Frobenius norm of the gain matrices

Multiple Objective Optimization

State Error Energy:

$$J_s(H) = \int_o^\infty x^T Q_s x \, dt = trace(P_s X_o)$$

Control Effort Energy:

$$J_u(H) = \int_o^\infty u^T Q_u u \, dt = trace(P_u X_o)$$

Stability Robustness Measure:

$$J_e(H) = k(\Phi)$$

where

$$P_s^{ij} = -\underline{\phi}_i^H Q_s \underline{\phi}_j / (\lambda_i + \lambda_j), \quad X_o = E\{x_o x_o^T\}$$

$$P_u^{ij} = -h_i^H Q_u h_j / (\lambda_i + \lambda_j)$$

$$k(\Phi) = |\Phi| |\Phi^{-1}| = \overline{\sigma}(\Phi)/\underline{\sigma}(\Phi)$$

= condition number: the ratio of the maximum and minimum singular
values of the closed loop modal matrix Φ.

First and Second Order Sensitivity of the Singular Value Decomposition

John L. Junkins

Youdan Kim

Department of Aerospace Engineering
Texas A&M University
College Station, Texas 77843

submitted to
AIAA Journal of Guidance, Control, and Dynamics
January, 1988

TOWARD DERIVATIVES OF THE SINGULAR VALUE DECOMPOSITION

We restrict the initial discussion to the partial derivatives of a square $(2n) \times (2n)$ complex-valued matrix Φ, the *singular value decomposition* is the matrix factorization

$$\Phi = U\Sigma V^H \tag{1}$$

where U and V are $(2n) \times (2n)$ unitary matrices normalized so that

$$U^H U = I , \quad V^H V = I \tag{2, 3}$$

and $\Sigma = \text{diag}(\sigma_1 , \sigma_2 , \dots , \sigma_{2n})$ contains the real *singular values* $\sigma_1 \geq \sigma_2 , \geq \dots \geq \sigma_{2n} \geq 0$.

Since U and V are unitary, we can use Eqs. (2), (3) to re-arrange Eq. (1) as $\Sigma = U^H \Phi V$ (4)

The above four equations can be expressed as conditions upon the $(2n) \times 1$ complex column vectors of $U = [U_1 \ U_2 \ \dots \ U_{2n}]$ and $V = [V_1 \ V_2 \ \dots \ V_{2n}]$ as

$$\Phi V_i = U_i \sigma_i , \quad i = 1, 2, \dots, 2n , \quad \text{and} \quad \Phi^H U_i = V_i \sigma_i , \; i = 1, 2, \dots, 2n \tag{5, 6}$$

$$U_i^H U_j = \delta_{ij} , \qquad V_i^H V_j = \delta_{ij} , \quad \text{and} \quad \sigma_i \delta_{ij} = U_i^H \Phi V_j \tag{7, 8, 9}$$

keystone slide3   9

PARTIAL DERIVATIVES OF THE SINGULAR VALUES

We define the parameter vector $\mathbf{p} = [p_1, p_2, \dots, p_N]^T$, and assume $\Phi = \Phi(\mathbf{p})$ (11)

Differentiating Eqs.(5) and (6) with respect to a typical element p_k of $\mathbf{p}$, upon premultiplying the resulting two equations by U_i^H and V_i^H , gives

$$U_i^H \frac{\partial \Phi}{\partial p_k} V_i + U_i^H \Phi \frac{\partial V_i}{\partial p_k} = U_i^H \frac{\partial U_i}{\partial p_k} \sigma_i + U_i^H U_i \frac{\partial \sigma_i}{\partial p_k} \tag{12}$$

$$V_i^H \frac{\partial \Phi^H}{\partial p_k} U_i + V_i^H \Phi^H \frac{\partial U_i}{\partial p_k} = V_i^H \frac{\partial V_i}{\partial p_k} \sigma_i + V_i^H V_i \frac{\partial \sigma_i}{\partial p_k} \tag{13}$$

Adding together Eqs.(12), (13) gives

$$U_i^H \frac{\partial \Phi}{\partial p_k} V_i + V_i^H \frac{\partial \Phi^H}{\partial p_k} U_i + (U_i^H \Phi - V_i^H \sigma_i)\frac{\partial V_i}{\partial p_k} + (V_i^H \Phi^H - U_i^H \sigma_i)\frac{\partial U_i}{\partial p_k} = \frac{\partial \sigma_i}{\partial p_k}(U_i^H U_i + V_i^H V_i) \tag{14}$$

Using Eqs. (5) to (9) in Eq. (14), we obtain the following result (identical to [2]) for the gradient of the singular values

$$\frac{\partial \sigma_i}{\partial p_k} = \frac{1}{2}\left(U_i^H \frac{\partial \Phi}{\partial p_k} V_i + V_i^H \frac{\partial \Phi^H}{\partial p_k} U_i\right) = \text{Re}\left(U_i^H \frac{\partial \Phi}{\partial p_k} V_i\right) \tag{15}$$

keystone slide3   10

PARTIAL DERIVATIVES OF THE SINGULAR VALUES , continued

Upon differentiating Eq. (15) with respect to p_l , we obtain the following expression for the second partial derivatives of the singular values

$$\frac{\partial^2 \sigma_i}{\partial p_k \partial p_l} = \mathrm{Re}\left(\frac{\partial U_i^H}{\partial p_l} \frac{\partial \Phi}{\partial p_k} V_i + U_i^H \frac{\partial^2 \Phi}{\partial p_k \partial p_l} V_i + U_i^H \frac{\partial \Phi}{\partial p_k} \frac{\partial V_i}{\partial p_l} \right) \tag{16}$$

Since Eq. (16) involves the singular vector gradients $\frac{\partial U_i}{\partial p_l}$ and $\frac{\partial V_i}{\partial p_l}$, we must either evaluate them or eliminate them, we choose the latter approach. We choose to project $\frac{\partial U_i}{\partial p_l}$ and $\frac{\partial V_i}{\partial p_l}$ onto the corresponding sets of unitary vectors as

$$\frac{\partial U_i}{\partial p_l} = \sum_{j=1}^{2n} a_{ji}^l U_j \quad , \quad \text{and} \quad \frac{\partial V_i}{\partial p_l} = \sum_{j=1}^{2n} b_{ji}^l V_j \tag{17}$$

where a_{ji}^l and b_{ji}^l are scalar complex constant "components" of the singular vector gradients. These can be eliminated (see the for paper for details), so that Eq. (16) leads to (considerable algebra!)

$$\frac{\partial^2 \sigma_i}{\partial p_k \partial p_l} = \mathrm{Re}\left[\sum_{\substack{j=1 \\ j\neq i}}^{2n} \frac{1}{\sigma_i^2 - \sigma_j^2} \{ \sigma_i [(U_j^H \frac{\partial \Phi}{\partial p_l} V_i)^H (U_j^H \frac{\partial \Phi}{\partial p_k} V_i) + (U_i^H \frac{\partial \Phi}{\partial p_l} V_j)^H (U_i^H \frac{\partial \Phi}{\partial p_k} V_j)] + \sigma_j [(U_i^H \frac{\partial \Phi}{\partial p_l} V_j)(U_j^H \frac{\partial \Phi}{\partial p_k} V_i) + (U_j^H \frac{\partial \Phi}{\partial p_l} V_i)(U_i^H \frac{\partial \Phi}{\partial p_k} V_j)] \} + U_i^H \frac{\partial^2 \Phi}{\partial p_k \partial p_l} V_i \right] \cdot \frac{i}{\sigma_i} \mathrm{Im}(U_i^H \frac{\partial \Phi}{\partial p_l} V_i) \mathrm{Im}(U_i^H \frac{\partial \Phi}{\partial p_k} V_i) \tag{27}$$

HOMOTOPIC NONLINEAR PROGRAMMING METHOD FOR GAIN DESIGN

 An embedding method is used to define a continuous, one parameter family of problems which contain two important members: (i) a trivial or starting problem (one for which the solution is available) and (ii) the problem of interest.

For a fixed setting on the homotopy parameter, a generalized Newton process for solving the system of underdetermined nonlinear constraint equations: we locally minimizes the norm (sum square) of the correction vector required to satisfy locally linearized constraint functions

Homotopic continuation is used to define "portable objective values" for the constraint functions which are swept from the starting design to the final desired design.
    The converged intermediate solutions provide a sequence of ever more desirable designs and good starting iteratives for the next homotopy step.

Violated bounds of (active) inequality constraints are treated locally as equality constraints; locally inactive inequality constraints are ignored.

Multiple, competing objective functions (perf. measures) are be considered by treating them as constraints with objective values set at values which can be swept subject to satisfying all remaining design constraints, to create "multi-criterion tradeoff surfaces".

HOMOTOPIC NONLINEAR PROGRAMMING METHOD FOR GAIN DESIGN

The continuous one-parameter family (*homotopic map*) of problems is constructed in such a way that an imbedding parameter ($0 \leq \gamma \leq 1$) may be set to define any member of the family with $\gamma = 0$ generating the trivial problem and $\gamma = 1$ generating the problem of interest. By sweeping γ and controlling the γ increments, we control how closely spaced the neighboring solutions lie (in the space of the constraint functions), and the γ increments can be assigned adaptively based upon convergence progress.

Thus we can remain as near to neighboring converged solutions as is necessary to maintain sufficiently good starting iteratives. Therefore, if a local convergence cannot be achieved (for some γ value), we have eliminated the most common problem (of having a poor starting iterative), and we can focus on other, more interesting convergence issues, such as the non-existence of feasible solutions, local linear dependence of the linearized constraint equations, turning points, and other issues. For example, comparing the active constraint sets and their Jacobian between the apparently un-reachable set (the one for which convergence cannot be achieved) and the previous converged solution will usually reveal which constraints are in competition and will quite likely suggest avenues for revision of the problem or at least provide insights as to why the failure to converge occurred.

HOMOTOPIC NONLINEAR PROGRAMMING METHOD FOR GAIN DESIGN

We seek to minimize the function

$$J = J(\mathbf{p}) \tag{A1}$$

subject to the constraints

$$
\begin{aligned}
f_i(\mathbf{p}) &= f_{oi} , & i &= 1, 2, ..., m \\
g_j(\mathbf{p}) &\geq T_j , & j &= 1, 2, ...
\end{aligned}
\tag{A2}
$$

where f_{oi} is the "objective" value of the ith equality constraint, and T_j represents the upper bounds on the feasible region. Converging from an approximate starting solution $\mathbf{p}_{start}$ to a feasible solution satisfying the constraints of Eqs. (A2) may be approached by locally considering all of the equality constraints and the active (locally violated) subset of inequality constraints. The locally violated subset of inequality constraints are simply considered as additional equality constraints of the form

$$f_j(\mathbf{p}) = T_j , j = m+1, m+2, ..., r \tag{A3}$$

REMARKS ON THE TREATMENT OF INEQUALITY CONSTRAINTS

The subset of inequality constraints $g_j(p) \geq T_j$ included in Eqs. (A3), as local equality constraints, obviously depends upon which constraints are locally violated and thus will typically change during the differential correction process. So long as r (the number of locally violated constraints plus the m equality constraints) does not exceed N (the number of elements in $\mathbf{p}$) , then we can proceed with the discussion below, otherwise, only the N - m "most important" inequality constraints will be locally imposed, assuming the original ordering represents the "importance heirachy" of the inequality constraints. Adopting this approach permits the specification of greater than N - m inequality constraints, although not more than N - m can be active on any iteration, especially the last one.

Note that it is often the case that final convergence to a feasible solution satisfying (A2) is achieved even when local iterations encounter more than N - m locally active constraints (i. e., it is typical that only a small subset of the specified inequality constraints are active when convergence to a feasible solution is achieved, and only under special circumstances will this finally active subset of constraints be invariant during the local iterations). These ideas must also be viewed in the context of the homotopy procedure which we discuss below.

keystone slide3 29

MINIMUM NORM DIFFERENTIAL CORRECTION HOMOTOPY ALGORITHM

Our re-stated optimization problem is to minimize J of Eqn. (A1) subject to

$$\mathbf{f}(\mathbf{p}) = \mathbf{f}_0 \tag{A4}$$

where $\mathbf{f}(\mathbf{p})$ is the r x 1 vector of m equality constraints and the r - m active inequality constraints.

We initially address the problem of achieving a feasible solution near our starting solution, we introduce a "portable objective" vector $\mathbf{f}_p(\gamma)$ as

$$\mathbf{f}_p(\gamma) = \gamma\,\mathbf{f}_0 + (1-\gamma)\,\mathbf{f}(\mathbf{p}_{start}), \quad 0 \leq \gamma \leq 1 \tag{A5}$$

The linear homotopy map $\mathbf{H}(\mathbf{p}(\gamma), \gamma)$ is generated by replacing $\mathbf{f}_0$ by $\mathbf{f}_p$ from Eq. (A5) to obtain

$$\mathbf{H}(\mathbf{p}(\gamma), \gamma) \equiv \gamma\,\mathbf{f}_0 + (1-\gamma)\,\mathbf{f}(\mathbf{p}_{start}) - \mathbf{f}(\mathbf{p}(\gamma)) = \mathbf{0} \tag{A6}$$

Notice the γ boundary conditions satisfied by $\mathbf{H}(\mathbf{p}(\gamma), \gamma)$ in Eq. (A6):

at $\gamma = 0$, trivial problem: $\quad \mathbf{H}(\mathbf{p}(0), 0) \equiv \mathbf{f}(\mathbf{p}_{start}) - \mathbf{f}(\mathbf{p}(0)) = \mathbf{0}$

at $\gamma = 1$, problem of interest: $\mathbf{H}(\mathbf{p}(1), 1) \equiv \mathbf{f}_0 - \mathbf{f}(\mathbf{p}(1)) = \mathbf{0}$

keystone slide3 30

MINIMUM NORM DIFFERENTIAL CORRECTION HOMOTOPY ALGORITHM
Reaching the Feasible Region

An arbitrary guess $\mathbf{p}_{start}$ satisfies the homotopy map of Eq. (A6) for $\gamma = 0$, and if we can obtain a solution for $\gamma = 1$, then we have achieved a feasible solution. It is apparent that sweeping γ at a suitable increment generates a sequence of neighboring problems. Obviously, judicious γ increments make each "new" problem arbitrarily close to a neighboring converged problem.

So we find a local correction vector $\Delta\mathbf{p}$ to minimize $\Delta\mathbf{p}^T W \Delta\mathbf{p}$, subject to $\mathbf{H}(\mathbf{p}(\gamma),\gamma) + A\Delta\mathbf{p} = \mathbf{0}$, where $A \equiv \frac{\partial \mathbf{H}}{\partial \mathbf{p}} = -\frac{\partial f(\mathbf{p}(\gamma))}{\partial \mathbf{p}}$, and W is a suitable positive definite weight matrix. The solution provides the *minimum norm differential correction* [15]

$$\Delta\mathbf{p} = - W^{-1} A^T (A W^{-1} A^T)^{-1} \mathbf{H}(\mathbf{p}(\gamma)) \tag{A8}$$

and we use the recursion $\mathbf{p}_{new} = \mathbf{p}_{old} + \Delta\mathbf{p}$ iterate until Eq. (A6) is satisfied for each local γ value. Upon achieving a local convergence, γ is incremented by a prescribed amount. If local convergence is not achieved, the γ increment is reduced. The process is halted if (i) convergence to a feasible solution is achieved (at $\gamma = 1$), or (ii) local convergence cannot be achieved when we increment γ by some small tolerance $\Delta\gamma \leq \epsilon$.

MINIMUM NORM DIFFERENTIAL CORRECTION HOMOTOPY ALGORITHM
Optimizing within the Feasible Region

We now consider minimization of the performance index J subject to the local equality constraint of Eq. (A6). Assuming there are r < N active constraints, this can readily be accomplished [15] by simply introducing an objective value for the performance index J_o and treating it as the $(r + 1)^{th}$ equality constraint. In lieu of Eq. (A8), er make corrections using

$$\Delta\mathbf{p} = - W^{-1} \tilde{A}^T (\tilde{A} W^{-1} \tilde{A}^T)^{-1} \tilde{\mathbf{H}}(\mathbf{p}(\gamma)) \tag{A9}$$

$$\text{where} \quad \tilde{A} \equiv \begin{bmatrix} A \\ \cdots\cdots\cdots\cdots\cdots \\ \dfrac{\partial J}{\partial p_1} \ \dfrac{\partial J}{\partial p_2} \ \cdots \ \dfrac{\partial J}{\partial p_N} \end{bmatrix}, \ \tilde{\mathbf{H}}(\mathbf{p}(\gamma)) \equiv \begin{Bmatrix} \mathbf{H}(\mathbf{p}(\gamma)) \\ \cdots\cdots\cdots \\ J(\mathbf{p}(\gamma)) - J_o \end{Bmatrix} \tag{A10}$$

Upon achieving convergence to the feasible region, we reset $\gamma = 0$ and subsequently use Eq. (A9) to compute the differential corrections. For the iterations subsequent to the first feasible solution, a new homotopy process is established to drive the objective function to its minimum value, or alternatively, drive it to a "design goal" value J_{goal}:

$$J_o(\gamma) = \gamma J_{goal} + (1 - \gamma) J(\mathbf{p}_{first\ feasible\ solution}) \tag{A11}$$

MINIMUM NORM DIFFERENTIAL CORRECTION HOMOTOPY ALGORITHM
Goal Programming & Multi-Criterion optimization

The value for J_{goal} may be an actual goal, but is more typically interpreted as "the best one could possibly hope for", and may be taken as zero for a minimization problem. We increment γ from zero toward unity, with the size of the increments dictated by convergence progress of the differential corrections for each γ value. When convergence can not be achieved for a small tolerance increase in γ, we adopt the solution corresponding to the largest γ for which convergence was achieved as the constrained minimum. We have rigorously proven [15] that this process is mathematically equivalent to a gradient projection with appropriate strategies for correction step size control. However this algorithm is superior to the gradient projection method because the correction formulas are more easily programmed and the one parameter homotopic continuation process has been found much more attractive than step size control, in that it is inherently self-starting.

One can apply the above process to several objective functions by treating all goals but one as equality constraint values. The goal values assigned as equality constraint values can be swept to create a family of designs which *display the tradeoff between competing performance measures*. These surface summarizes performance tradeoffs for a *family of optimal designs*, and should be a useful device in practical applications.

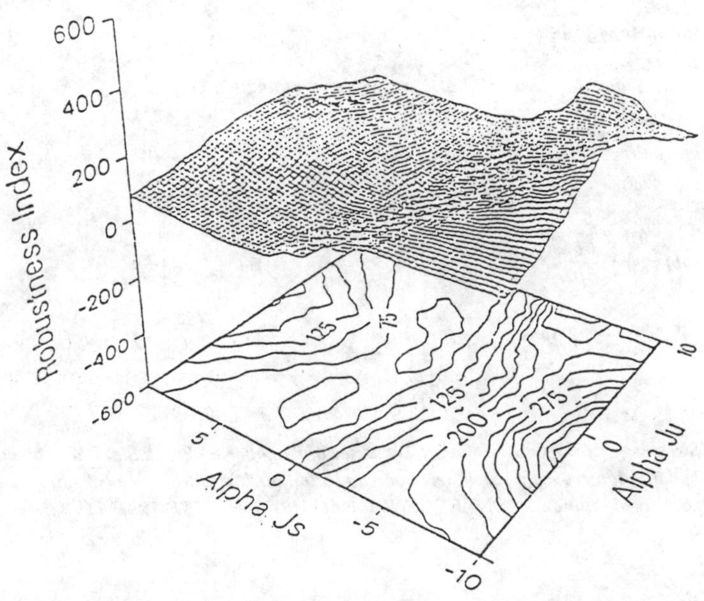

Configuration for Structure/Controller Design Optimization

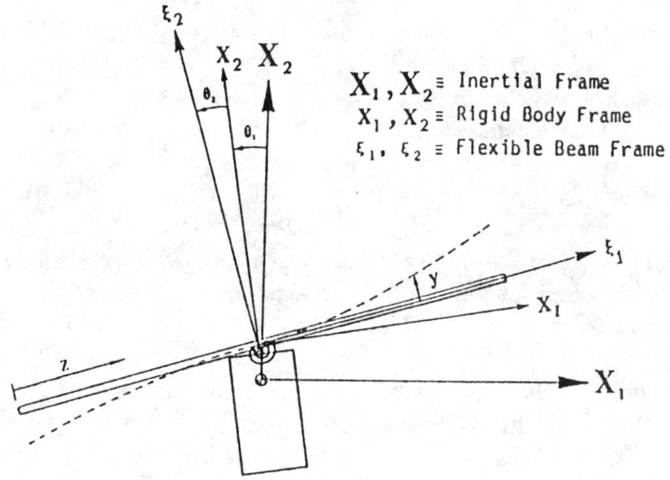

OPTIMIZATION VIA SEQUENTIAL LINEAR PROGRAMMING

Consider the general nonlinear programming problem:

maximize J(p)

subject to $f(p)$ {$\leq, =, \geq$} f^0

Solve via Sequential Linear Programming and continuation method

$$\text{maximize} \quad \sum_{j=1}^{np} \left[\left.\frac{\partial J}{\partial p_j}\right|_{p^{i-1}} \right] \Delta p_j$$

$$\text{subject to} \left[\left.\frac{\partial f}{\partial p}\right|_{p^{i-1}} \right] \Delta p \; \{\leq, =, \geq\} \; (1-\gamma_i)f(p^s) + \gamma_i f^o - f(p^{i-1})$$

$$-\varepsilon \leq \Delta p \leq \varepsilon$$

$$p^i = p^{i-1} + \Delta p; \quad p^o \equiv p^s$$

Sweeping γ from 0 to 1 defines a sequence of neighboring "portable" constrained optimization problems, each iteration can be initiated "arbitrarily near" a neignboring converged solution....CONVERGENCE FAILURES ARE INFORMATIVE!!

Table 6. Nominal Design Variables

DESIGN VARIABLE	SYMBOL	VALUE
actuator 2 location	a_1	5m
actuator 3 location	a_2	10m
actuator 4 location	a_3	15m
stiffness of torsional spring	k	500 n-m/rad
thickness of flexible beam	t_F	.1m
Young's modulus of beam	E	$.1482 \times 10^9$ N/m^2
mass density of rigid body	ρ_R	300 kg/m^3
output gain elements:		
G(1,1), G(2,1), G(3,1), G(4,1)		-1
all other G(i,j) elements		0

Table 7. Fixed structural parameters

PARAMETER	SYMBOL	VALUE
width of rigid body	w_R	1m
thickness of rigid body	t_R	3m
depth of rigid body	d_R	2m
width of flexible beam	w_F	20m
depth of flexible beam	d_F	1m
mass density of flexible beam	ρ_F	1799 kg/m^3
sensor 1 location	s_1	3m
sensor 2 location	s_2	7m
sensor 3 location	s_3	13m
sensor 4 location	s_4	17m

Table 8. Open loop and desired closed loop damped frequencies and damping factors

MODE	OPEN LOOP		DESIRED CLOSED LOOP	
#	ω_d (rad/s)	ζ	ω_d^* (rad/s)	ζ^*
1	.0056	.4819E-10	.1	.7
2	.2803	.1402E-5	.3	.1
3	.3443	.1718E-5	.45	.1
4	1.241	.6204E-5	1.0	.05
5	1.768	.8839E-5	1.5	.05
6	3.981	.1990E-4	4.0	.05
7	5.004	.2502E-4	$> \omega_{d_6}^* + .1$	.02
8	8.295	.4147E-4	unconstrained	.02
9	9.902	.4951E-4	unconstrained	.02

Table 9. Lower, upper and local step size bounds on design parameters

PARAMETER	SYMBOL	VALUE
lower bounds on actuator location	$a_1^l,\ a_2^l,\ a_3^l$	0m
upper bounds on actuator location	$a_1^u,\ a_2^u,\ a_3^u$	20m
lower bounds on spring stiffness	k^l	5 N-m/rad
lower bound on beam thickness	t_F^l	.01m
upper bound on beam thickness	t_F^u	2m
lower bound on beam stiffness	E^l	$.1480 \times 10^9$ N/m^2
upper bound on beam stiffness	E^u	$.1496 \times 10^9$ N/m^2
lower bound on rigid body density	ρ_R^l	50 kg/m^3
upper bound on rigid body density	ρ_R^u	1000 kg/m^3
local step size bounds:		
actuator location	$\Delta a_1,\ \Delta a_2,\ \Delta a_3$	.1m
spring stiffness	Δk	30 N-m/rad
beam thickness	Δt_F	.01m
beam stiffness	ΔE	$.1 \times 10^5$ N/m^2
rigid body density	$\Delta \rho_R$	40 kg/m^3
gain elements i=1,...,4; j=1,...,12	$\Delta G(i,j)$	10
frequency separation between mode 7 and 6	$\Delta \omega_{76}$	.1 rad/s

Locii of First Ten Closed Loop Eigenvalues during Optimization

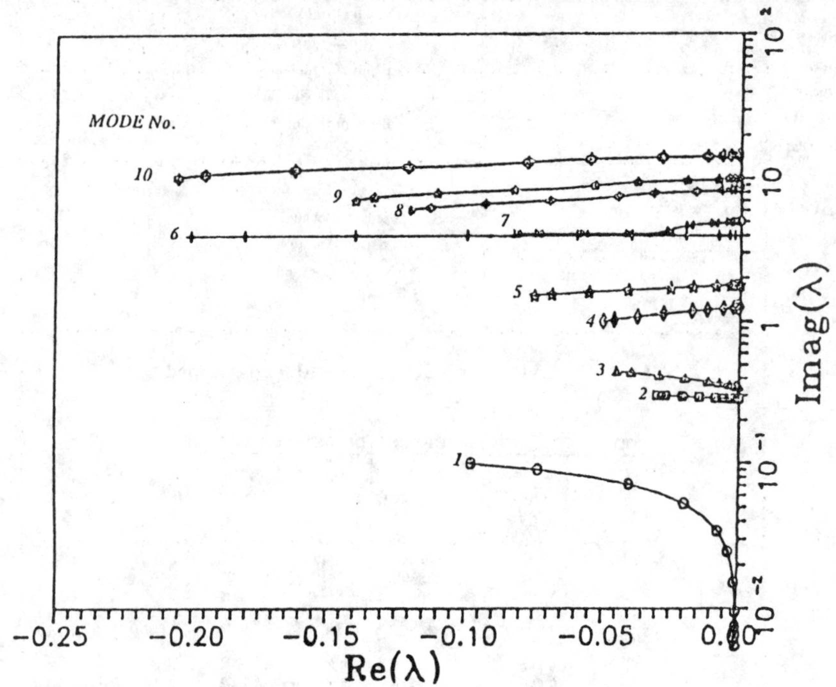

Performance Criteria Convergence Summary

		MINIMUM MASS DESIGN	MINIMUM SENSITIVITY DESIGN	MAXIMUM ROBUSTNESS DESIGN
TOTAL MASS	INITIAL DESIGN	5398	5398	5398
	CONVERGED DESIGN	2853	3336	4952
EIGENVALUE SENSITIVITY INDEX	INITIAL DESIGN	20.1	20.1	20.1
	CONVERGED DESIGN	130.6	6.6	20.7
STABILITY ROBUSTNESS INDEX	INITIAL DESIGN	0.251×10^{-7}	0.251×10^{-7}	0.251×10^{-7}
	CONVERGED DESIGN	0.138×10^{-3}	0.260×10^{-3}	0.275×10^{-2}
CONDITION NUMBER	INITIAL DESIGN	419	419	419
	CONVERGED DESIGN	329	453	96

Table 12. Actual Robustness of Three Designs

PARAMETER		PERCENT VARIATION CAUSING INSTABILITY		
		MASS DESIGN	*SENSITIVITY DESIGN*	*ROBUSTNESS DESIGN*
ACTUATOR LOCATIONS	a_1	6%	12%	16%
	a_2	1.5	9	8
	a_3	2.4	4	4
STRUCTURAL PARAMETERS	k	4	6	98
	t_F	3	5	10
	E	8	14	31
	ρ_R	6	24	28
SELECTED CONTROL GAINS	$G(1,1)$	9	9	67
	$G(1,3)$	12	50	28
	$G(2,8)$	43	31	91
	$G(3,9)$	38	51	25
	$G(4,11)$	24	8	31
55 PARAMETER AVERAGE		13%	18%	37%

Concluding Remarks

*New Methods for Simultaneous Optimization
of Dynamical Systems & Controllers*

*Multi-Criterion Optimization Viewpoint &
Associated Algorithms Developed*

*Robustness Indices Developed & Encouraging
Numerical Results Obtained*

*The Conservativeness of the Robustness Indices
does not (apparently) Invalidate Optimization!*

References

[1] Meirovitch, L., _Computational Method in Structural Dynamics_, Sijthoff and Noordhoff, 1980, Rockville, Maryland, U.S.A.

[2] Plaut, R.H. and Huseyin, K., "Derivatives of Eigenvalues and Eigenvectors in Non-Self-Adjoint Systems," _AIAA Journal_, Vol. 11, No. 2, pp. 250-251, Feb, 1973.

[3] Bodden, D.S. and Junkins, J.L., "Eigenvalue Optimization Algorithms for Structure/Controller Design Iterations," _AIAA Journal of Guidance, Control, and Dynamics_.

[4] Srinathkumar, S., "Eigenvalue/Eigenvector Assignment Using Output Feedback," _IEEE Transactions on Automatic Control_, Vol. AC-23, No. 1, 1978, pp. 79-81.

[5] Brogan, W.L., _Modern Control Theory_, Quantum Publishers, Inc., New York, New York, 1974, pp. 311-315.

[6] Bhattacharyya, S.P. and deSouza,E., "Pole Assignment via Sylvester's Equation," _Systems and Control Letters_, Vol. 1, No. 4, Jan. 1982, pp. 261-263.

[7] Cavin III, R.K. and Bhattacharyya, S.P., "Robust and Well-Conditioned Eigenstructure Assignment via Sylvester's Equation," _Journal of Optimal Control Applications and Methods_, Vol. 4, 1983, pp. 205-212.

[8] Potter, B. and D'Azzo, J.J., "Algorithm for Closed-Loop Eigenstructure Assignment by State Feedback in Multivariable Linear Systems," _Int. Journal of Control_, Vol. 27, No.6, 1978, pp. 943-947.

[9] Moore, B.C., "On the Flexibility Offered by State Feedback in Multivariable Systems Beyond Closed-Loop Eigenvalue Assignments," _IEEE Transaction on Automatic Control_, vol. AC-2:, 1976 pp. 689-692.

[10] Kautsky, J., Nichols, N.K. and Van Doorend, P., "Robust Pole Assignment in Linear State Feedback," _int. Journal of Control_, Vol. 41, No. 5, 1985, pp. 1129-1155.

[11] Wonham, W.M., "On Pole Assignment in Multiinput, Controllable Linear Systems," _IEEE Transactions on Automatic Control_, Vol. AC-12, 1976, pp. 660-665.

[12] Juang, J-N, Lim, K.B., and Junkins, J.L., "Robust Eigensystem Assignment," preprint of a paper submitted to 1987 AIAA Guidance and Control Conference, Dec. 1986.

[13] Pontryagin, L.S., Boltyanskii, V.G., Gamkrelidze, R.V., and Mishchenko, E.F., _The Mathematical Theory of Optimal Processes_, Interscience Publishers, Inc., New York, 1962.

[14] Potter, J.E., "Matrix Quadratic Solution," _SIAM J. of Applied Mathematics_, Vol. 14, No. 3, May 1966, pp. 496-501.

[15] Laub, A.J., "A Schur Method for Solving Algebraic Riccati Equations," _IEEE Transactions on Automatic Control_, Vol. AC-24, No. 6, Dec. 1976, pp. 913-921.

[16] Rew, D.W., and Junkins, J.L., "Multi-Criterion Approaches to Optimization of Linear Regulators," AIAA Paper No., 86-2198-CP, J. of the Astronautical Sciences, to appear, Aug. 1986.

[17] Wilkinson, J.H., _The Algebraic Eigenvalue Problem_, Oxford University Press, Oxford, 1965.

[18] Dongarra, J.J., Moler, C.B., Bunch, J.R., and Stewart, G.W., _LINPACK Users' Guide_, SIAM Publication, Philadelphia, 1979.

[19] Rew, D.W. and Junkins, J.L., "Robust Eigenstructure Assignment by a Projection Method: Application to Multi--Criterion Optimization," preprint of a paper submitted to 1987 AAS/AIAA Astrodynamics Conference, Feb. 1987.

[20] Oz, H. and Meirovitch, L., "Optimal Modal-Space Control of Flexible Gyroscopic Systems," _Journal of Guidance and Control_, Vol. 3, Nov.-Dec. 1980, pp. 220-229.

[21] Junkins, J.L. and Dunyak,J.P., "Continuation Methods for Enhancement of Optimization Algorithms," Presented to 19th Annual Meeting, Society of Engineering Science, University of Missouri, Rolla, Oct. 1982.

[22] Freudenberg, J.S., Looze, D.P. and Cruz, J.B., "Robustness Analysis Using Singular Sensitivities," _Int. Journal of Control_, Vol. 35, No. 1, 1982, pp. 95-116.

[23] Lim, K.B., and Junkins, J.L., "Robustness Optimization of Structural and Controller Parameters," paper No. AIAA-87-0791-CP, presented at AIAA 28th SDM Conference, Monterey, CA, April 6-8, 1987.

[24] Balas, M.J., "Trends in LSS Control Theory: Fondest Hopes, Wildest Dreams," _IEEE Transactions on Automatic Control_, Vol. Ac-27, No. 3, June 1982, pp. 522-535.

[25] Bekey, I. and Naugle, J.E., "Just Over the Horizon in Space," Astronautics and Aeronautics, May 1980, pp. 64-76.

[26] Khot, N.S., et al., "Optimal Structural Modifications to Enhance the Optimal Active Vibration Control of Large Flexible Structures," 26th Structures, Structural Dyn., and Materials Conf. Orlando, FL, April 15-17,1985.

[27] Junkins, J.L., Bodden D.S. and Turner, J.D., "A Unified Approach to Structure and Control System Design Iterations," Fourth International Conference on Applied Numerical Modeling, Tainan, Taiwan, Dec. 27-29, 1984.

[28] Haftka, R.T., et al., "Sensitivity of Optimized Control Systems to Minor Structural Modifications," 26th Structures, Structural Dynamics and Materials Conference, Orlando, FL, April 15-17, 1985.

[29] Junkins, J.L. and Rew, D.W., "A Simultaneous Structure/Control Design Iteration Method," American Controls Conference, Boston, MA, June 1985.

[30] Dantzig, G.B., _Linear Programming and Extensions_, Princeton University Press, 1963.

[31] Hadley, G., _Linear Programming_, Addison-Wesley Publishing Co., Inc., Reading, MA, 1962.

[32] Horta, L.G., Juang, J-N and Junkins, J.L., "A Sequential Linear Optimization Approach for Controller Design," AIAA Paper 85-1971-CP, 1985.

[33] Lim, K.B. and Junkins, J.L., "Minimum Sensitivity Eigenvalue Placement via Sequential Linear Programming," Proceedings of the Mountain Lake Dynamics and Control Institute, ed. by J.L. Junkins, Mountain Lake, VA, June 9-11, 1985.

[34] Newsom, J.R. and Mukhopadhyay, "A Multiloop Robust Controller Design Study Using Singular Value Gradients," _Journal of Guidance, Control, and Dynamics_, Vol. 8, No. 4, July-Aug., 1985, pp. 514-519.

[35] Howze, J.W. and Cavin, R.K., "Regulator Design with Modal Insensivity," _IEEE Transaction of Automatic Control_, Vol. Ac-24, No. 3, June 1979, pp. 466-9.

[36] Raman, K.V., "Modal Insensitivity with Optimality," Ph.D. Dissertation, Drexel University, Philadelphia, PA, 1984.

[37] Raman, K.V. and Calise, A.J., "Design of an Optimal Output Feedback Control System with Modal Insensitivity," AIAA Paper 84-1940, AIAA Guidance and Control Conference, Seattle, WA, Aug. 20-22, 1984.

[38] Lim, K.B. and Junkins, J.L., "Optimal Redesign of Dynamic Structures via Sequential Linear Programming," Fourth International Modal Analysis Conference, Los Angeles, CA, Feb. 3-6, 1986.

[39] Palacios-Gomez, R., Lasdon, L. and Engquist, M., "Nonlinear Optimization by Successive Linear Programming," Management Science, Vol. 28, No. 8, October, 1982, pp. 1106-1120.

[40] Palacios-Gomez, R., "The Solution of Nonlinear Optimization Problems Using Successive Linear Programming," Ph.D. Dissertation, The University of Texas, Austin, TX, 1980.

[41] Patel, R.B. and Toda, M., "Quantitative Measures of Robustness for Multivariable Systems," Proceedings of JACC, San Francisco, TP8-A, 1980.

[42] Noble, B. and Daniel, J.W., _Applied Linear Algebra_, Englewood Cliffs, NJ: Prentice-Hall, 1977.

[43] IMSL Reference Manual, International Mathematical and Statistical Library, Inc., 1982

[44] Chen, C.T., _Introduction to Linear Systems Theory_, Holt, Rinehart & Winston, Inc., 1970.

[45] Fleming, P., "Computer Aided Design of Regulators Using Multiobjective Optimization," preprint, University College of North Wales, November, 1986.

[46] Lim, K.B., "A Unified Approach to Structure and Controller Design Optimizations," Ph.D. Dissertation, Virginia Polytechnic Institute and State University, Blacksburg, VA, 1986.

[47] Sobel, K.M. and Shapiro, E.Y., "Application of Eigenstructure Assignment to Flight Control Design: Some Extension," _AIAA J. of Guidance, Control, and Dynamics_, Vol. 10, No. 1, Jan-Feb 1987, pp. 73-89.

Structural Identification and Minimum Sensitivity Vibration Suppression

Structural Identification
    Approach
    Stereo-Triangulation Deflection Measuring System
    Experimental Results

Minimum Sensitivity Output Feedback Design
    Approach and Optimization Algorithm
    Examples

Structural Model Identification
Parameterization Schemes

- **Physical / Geometrical Parameters**

  $\{EI\text{'}s, GJ\text{'}s,, \rho\text{'}s, etc.\}$

- **Equivalent Continuum Parameters**

  $\{\overline{EI}\text{'}s, \overline{GJ}\text{'}s,, \overline{\rho}\text{'}s, etc.\}$

- **Submatrix Scale Factors**

  $$M = M_o + \Sigma_i\, \mu_i M_i\ , \quad K = K_o + \Sigma_i\, \kappa_i K_i\ , \quad C = C_o + \Sigma_i\, \chi_i C_i$$

  where $\{M_i, K_i, C_i\}$ are prescribed and the scalars $\{\mu_i, \kappa_i, \chi_i\}$ are estimated.

- **Matrix Element Modifications**

  $$M_{ij} = M_{oij} + m_{ij}\ , \quad K_{ij} = K_{oij} + k_{ij}\ , \quad C_{ij} = C_{oij} + c_{ij}$$

Basic System Realization Concepts

Linear Autonomous System
$$\dot{x} = Ax + Bu$$
$$y = Cx$$

Time Response
$$x(t) = e^{At}\, x(t_0)t + \int_{t_0}^{t} e^{A(t-\tau)} Bu(\tau)d\tau$$
$$y(t) = C\, e^{At}\, x(t_0)t + \int_{t_0}^{t} Ce^{A(t-\tau)} Bu(\tau)d\tau$$

Laplace Transform
$$x(s) = (sI - A)^{-1} Bu(s)$$
$$y(s) = Cx(s) = C(sI - A)^{-1} Bu(s)$$
$$y(s) = G(s)\, u(s)$$

Transfer Function
$$G(s) = C(sI - A)^{-1} B$$

Freq. Response Function
$$G(j\omega) = C(j\omega I - A)^{-1} B$$

Notes

The triple (A,B,C) is not unique.

But $G(s)$ is unique.

(A_1, B_1, C_1) and (A_2, B_2, C_2) are said to be **equivalent** if any of the following statements are true:

- The transfer functions are equal:
  $G_1(s) = G_2(s)$, for all s
- The weighting patterns are the same:
  $C_1 e^{A_1 t} B_1 = C_2 e^{A_2 t} B_2$
- $C_1 A_1^k B_1 = C_2 A_2^k B_2$, for all k.

$\Rightarrow$ **Any** (A,B,C) which produces the correct transfer function $G(s) = C(sI - A)^{-1} B$ is said to be a "realization" of the system.

STRUCTURAL IDENTIFICATION

Consider the class of linear elastic structures suitably modeled by a finite element or similar discretization approach, leading to

$$M\ddot{x} + C\dot{x} + Kx = Bu \tag{1}$$

where x is an $n \times 1$ configuration vector, u is an $m \times 1$ excitation (or control force) vector.

Let the measured functions of the deflection be related to the instantaneous configuration by $\quad y = Lx \tag{2}$

Note that $M = M(p), K = K(p), C = C(p), B = B(p), L = L(p), p$ is a vector of uncertain model parameters.

The transfer function relating the input $U(s)$ to the output $Y(s)$ in the sense

$$Y(s) = G(s)U(s) \qquad \text{is} \qquad G(s) = L[s^2 M + sC + K]^{-1} B \tag{3, 4}$$

Since we are considering stable systems, the frequency response function is obtained from the transfer function by setting $s = j\omega$. For structural damping, the frequency response matrix is given by (*via spectral decomposition of* M, C, K):

$$G(i\omega) = L[-\omega^2 M + j\omega C + K]^{-1} B = \sum_{r=1}^{n} L\left[\frac{H_r}{\Omega_r^2 + 2j\omega\zeta_r \Omega_r - \omega^2}\right]B, \quad H_r(l,m) = \phi_{lr}\phi_{mr} \tag{4}$$

where $\{ -\zeta_r \Omega_r \pm j\Omega_r \sqrt{1-\zeta_r^2}, \phi_r\}$ are eigenvalues and eigenvectors satisfying $[K - \Omega^2 M]\phi = 0$ and

$$\Phi^T M \Phi = I, \quad \Phi^T K\Phi = diag(\Omega_1^2 \dots \Omega_n^2), \quad \Phi^T C\Phi = diag(2\zeta_1\Omega_1 \dots 2\zeta_n\Omega_n), \quad \Phi = [\phi_1 \dots \phi_n] \tag{5}$$

STRUCTURAL IDENTIFICATION: A FRONTAL ASSAULT

Consider the case that a force or moment is applied at a point, but the response at many measurement stations are available. For this case B and therefore $G(j\omega)$ are vectors. Suppose that the frequency response function (vector) is measured over a frequency range $\{\omega_{min} < \omega < \omega_{max}\}$ at the frequencies $\{\omega_1, \omega_2, \ldots, \omega_m\}$. Use the notation:

$$\tilde{G} = \begin{Bmatrix} \tilde{G}(\omega_1) \\ \vdots \\ \tilde{G}(\omega_m) \end{Bmatrix} = measured \text{ FRF}, \quad G(p) = \begin{Bmatrix} G(\omega_1, p) \\ \vdots \\ G(\omega_m, p) \end{Bmatrix} = computed \text{ FRF using model vector } p \tag{6}$$

and

$$\tilde{\Omega} = \{\tilde{\Omega}_1 \ \tilde{\Omega}_2 \ \ldots \}^T = measured \text{ free vibration } \Omega\text{'s}, \quad \Omega(p) = \{\Omega_1 \ \Omega_2 \ \ldots \}^T = computed \ \Omega\text{'s using } p. \tag{7}$$

We seek the optimal estimate of the model parameter vector p which minimizes

$$J = \frac{1}{2} \int_{\omega_{min}}^{\omega_{max}} \Delta G^T(p, \omega) W_G(\omega) \Delta G(p, \omega) \, d\omega + \frac{1}{2} \Delta \Omega^T W_\Omega \Delta \Omega, \quad \Delta G(p, \omega) \equiv \tilde{G}(\omega) - G(\omega, p), \ \Delta \Omega \equiv \tilde{\Omega} - \Omega(p) \tag{8}$$

For FRF measurements available at discrete frequencies, the integral can be replaced by a discrete summation; we seek to minimize a weighted sum square of the residuals between all measured and modeled FRF's and Ω's:

$$\Delta G \equiv \begin{Bmatrix} \tilde{G}(\omega_1) - G(\omega_1, p) \\ \vdots \\ \tilde{G}(\omega_m) - G(\omega_m, p) \end{Bmatrix} = \text{FRF residuals}, \quad \Delta \Omega \equiv \begin{Bmatrix} \tilde{\Omega}_1 - \Omega_1(p) \\ \tilde{\Omega}_2 - \Omega_2(p) \\ \vdots \end{Bmatrix} = \text{free vib. frequency residuals} \tag{9}$$

and we are led to the least squares differential correction algorithm:

$$\begin{Bmatrix} \Delta G \\ \Delta \Omega \end{Bmatrix} = A \, \Delta p + \ldots \implies \Delta p = A^\dagger \begin{Bmatrix} \Delta G \\ \Delta \Omega \end{Bmatrix}, \text{ where } A \equiv \begin{bmatrix} \frac{\partial G}{\partial p} \\ \frac{\partial \Omega}{\partial p} \end{bmatrix}, \text{ and } p_{new} = p_{old} + \Delta p \tag{10}$$

Potential trouble in toyland!! This approach works great, but only if: (i) the model is "good" & (ii) p_{start} is "close" to p.

keynote format 3

Creamer's Three Step Identification Process

Step 1: Identify a subset of the measured frequencies and mode shapes which correlate well with the corresponding modeled frequencies and mode shapes, do a least square correction correction of the stiffness parameterization to improve correlation if necessary.

Step 2: Find the normalization of the eigenvectors (mode shapes) which results in the best least square fit for the FRF, i. e., find $\{a_0, a_1, a_2, \ldots a_{m+1}\}$ to minimize

$$\int_{\omega_{min}}^{\omega_{max}} \left(\tilde{H}_{pq}(j\omega) - H_{pq_{model}}(j\omega) \right)^2 d\omega, \quad or \quad \sum_{\omega_{min}}^{\omega_{max}} \left(\tilde{H}_{pq}(j\omega_k) - H_{pq_{model}}(j\omega_k) \right)^2$$

where

$$H_{pq_{model}}(j\omega) = \frac{a_0}{\omega^2} + \sum_{r=1}^{m} \frac{\tilde{\phi}_{pr} \tilde{\phi}_{qr}}{\tilde{\omega}_r^2 - \omega^2} a_r + a_{m+1}$$

Step 3: Estimate values for the linear mass and stiffness parameterization to satisfy the orthonormality conditions in a least square sense; this leads to the following pair of linear equations for μ_r, *and* κ_r *in* $M = M_o + \Sigma_i \ \mu_i M_i$, $K = K_o + \Sigma_i \ \kappa_i K_i$:

643

PARAMETERIZATION OF THE STRUCTURE FOR IDENTIFICATION

Based upon experimental determination of a subset of the eigenvalues, eigenvectors and one or more elements of the frequency response function matrix $G(j\omega)$, Creamer and Junkins[3, 4] have developed an identification process whereby *linearly parameterized* mass and stiffness properties can be estimated to bring the computed eigenvalues, eigenvectors, and frequency response functions into least square agreement with the corresponding measurements, over a prescribed range of sample frequencies. Given measured motion of a sufficient number of points on the structure, the eigenstructure realization method of [5] can be used to find the minimum rank linear discrete-time model which represents the measured motion, from this linear model, eigenvalues(natural frequencies and damping) and eigenvectors (mode shapes) can be determined.

The mass and stiffness parameterization adopted for the present discussion is of the form

$$M = M_o + \sum_{r=1}^{p} \mu_r M_r \ , \quad K = K_o + \sum_{r=1}^{q} \kappa_r K_r \tag{6}$$

where μ_r, κ_r are scalars determined to bring the calculated and measured eigenvalues, eigenvectors, and frequency response functions into least square agreement. For the present discussion, ignore damping, and the possible uncertainty of B and L. Notice that the structure has been subjectively divided into substructures whose contributions to the nominal mass and stiffness matrices are scaled in unison; M_r, K_r are the nominal contributions of the prescribed substructures to the global assembly of $M_o = \Sigma M_r$ nd $K_o = \Sigma K_r$. These do not necessarily conform to physical substructures, but can be based upon collections of nominally identical members, and/or members made of the same material, etc. In the limit, of course, the sub-structures could be the finite elements themselves, but we usually find a much coarser parameterization to be highly satisfactory. The subjectivity involved in selecting substructures should not be viewed as a weakness of this approach, it is in fact a strength. The parameterization of Eq. (6) is essentially a "parameter linking" method [5] in which the engineer is permitted convenient latitude in modifying the dimensionality of the parameter estimation process to achieve accuracy and computational efficiency. Based upon the distribution of the energy for each modeled mode and the smallness of the associated residuals, it is possible to iteratively revise the substructuring decisions and introduce more detailed parameterizations for the substructures which dominate the most poorly modeled modes[3-6].

keystone format 4

$$\Phi^T M \Phi = I \ \Rightarrow$$

$$
\begin{bmatrix}
\phi_1^T M_1 \phi_1 & \phi_1^T M_2 \phi_1 & \cdots & \phi_1^T M_P \phi_1 \\
\vdots & \vdots & \cdots & \vdots \\
\phi_m^T M_1 \phi_m & \phi_m^T M_2 \phi_m & \cdots & \phi_m^T M_P \phi_m \\
\phi_1^T M_1 \phi_2 & \phi_1^T M_2 \phi_2 & \cdots & \phi_1^T M_P \phi_2 \\
\vdots & \vdots & \cdots & \vdots \\
\phi_{m-1}^T M_1 \phi_m & \phi_{m-1}^T M_2 \phi_m & \cdots & \phi_{m-1}^T M_P \phi_m
\end{bmatrix}
\begin{Bmatrix} \mu_1 \\ \mu_2 \\ \vdots \\ \mu_P \end{Bmatrix}
=
\begin{Bmatrix}
1/a_1 - \phi_1^T M_0 \phi_1 \\
\vdots \\
1/a_m - \phi_m^T M_0 \phi_m \\
- \phi_1^T M_0 \phi_2 \\
\vdots \\
- \phi_{m-1}^T M_0 \phi_m
\end{Bmatrix}
$$

and

$$\Phi^T K \Phi = \lceil \omega^2 \rfloor \ \Rightarrow$$

$$
\begin{bmatrix}
\phi_1^T K_1 \phi_1 & \phi_1^T K_2 \phi_1 & \cdots & \phi_1^T K_Q \phi_1 \\
\vdots & \vdots & \cdots & \vdots \\
\phi_m^T K_1 \phi_m & \phi_m^T K_2 \phi_m & \cdots & \phi_m^T K_Q \phi_m \\
\phi_1^T K_1 \phi_2 & \phi_1^T K_2 \phi_2 & \cdots & \phi_1^T K_Q \phi_2 \\
\vdots & \vdots & \cdots & \vdots \\
\phi_{m-1}^T K_1 \phi_m & \phi_{m-1}^T K_2 \phi_m & \cdots & \phi_{m-1}^T K_Q \phi_m
\end{bmatrix}
\begin{Bmatrix} \kappa_1 \\ \kappa_2 \\ \vdots \\ \kappa_Q \end{Bmatrix}
=
\begin{Bmatrix}
\omega_1^2/a_1 - \phi_1^T K_0 \phi_1 \\
\vdots \\
\omega_m^2/a_m - \phi_m^T K_0 \phi_m \\
- \phi_1^T K_0 \phi_2 \\
\vdots \\
- \phi_{m-1}^T K_0 \phi_m
\end{Bmatrix}
$$

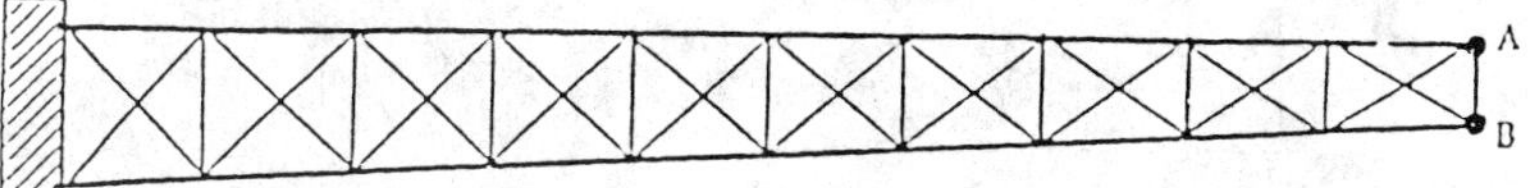

IDENTIFICATION OF A DAMPED TRUSS STRUCTURE

MEASUREMENTS: 8 open-loop eigenvalues
8 closed-loop eigenvalues
Longitudinal FRF between A and B
Transverse FRF between A and B

EIGENVALUES

MODE	measured $\bar{\lambda}$	apriori λ_0	identified model λ_f
1	-0.0872 + 6.73 i	6.11 i	-0.0872 + 6.93 i
2	-0.0890 + 34.59 i	31.45 i	-0.0888 + 35.46 i
3	-0.0891 + 87.58 i	78.89 i	-0.0898 + 88.29 i
4	-0.1042 + 117.11 i	106.67 i	-0.1039 + 119.39 i
5	-0.0907 + 156.49 i	142.12 i	-0.0906 + 157.86 i
6	-0.0914 + 240.33 i	217.54 i	-0.0915 + 240.05 i
7	-0.0929 + 332.09 i	302.00 i	-0.0925 + 331.39 i
8	-0.1039 + 359.29 i	325.25 i	-0.1042 + 363.28 i

Damped Truss Structure Identification

Frequency Response Function: Measured vs Initial Model vs Identified Model

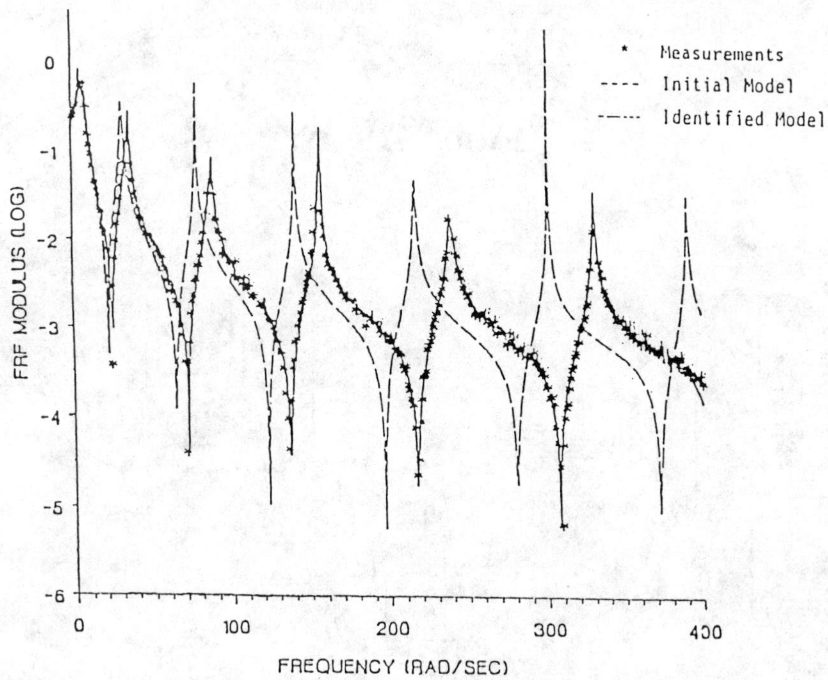

A Novel Approach to Structural Identification

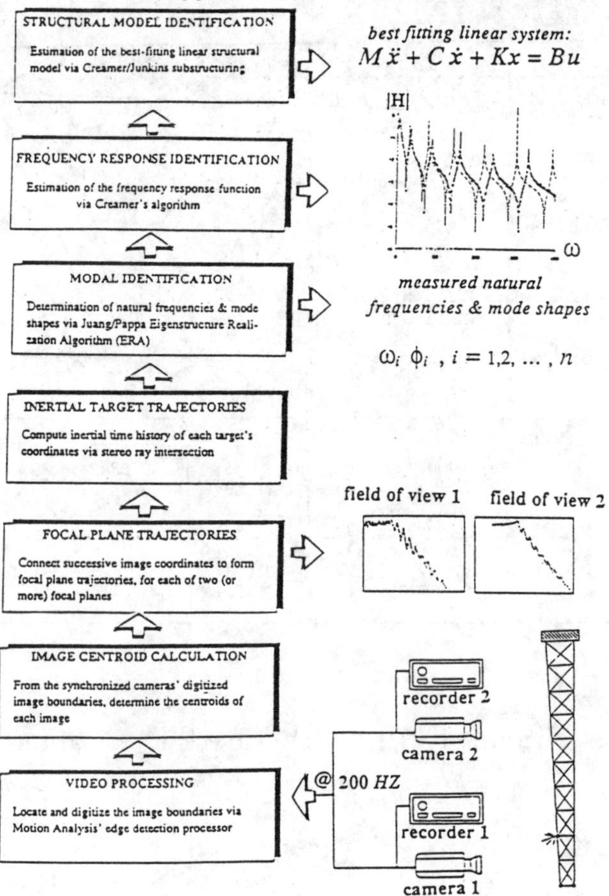

Motion Analysis System

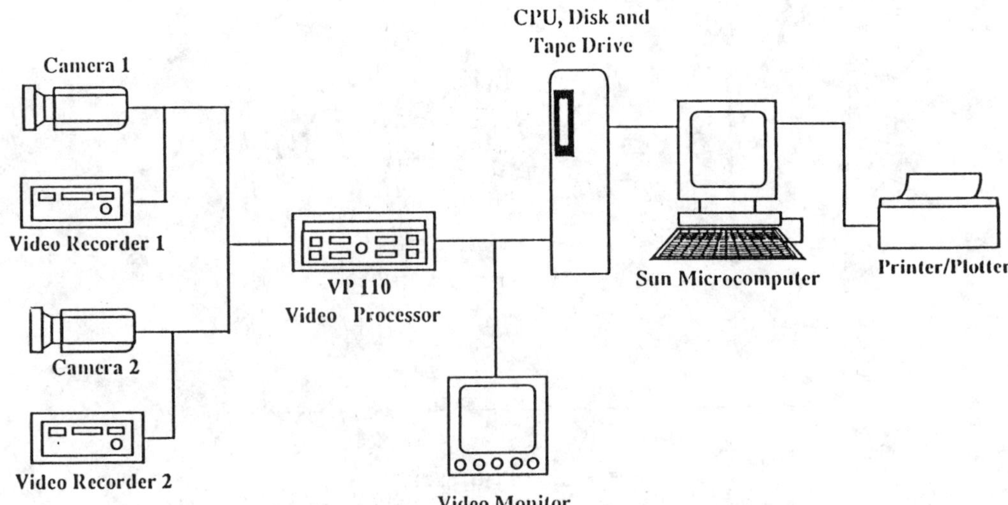

MAJOR HARDWARE ITEMS

ITEM	DESCRIPTION
Flexible Structure	- Monolithic 3003 H14 aluminum grid (5'x5') cantilevered in the vertical plane (clamped-free conditions)
markers (targets)	- 3M Scotchlite Reflective Sheeting #3290
Video Cameras (2)	- NAC model V-14B, 200/60 HZ, 2/3" MOS imaging CCD array with 320x244 pixels
Video Recorder (2)	- NAC model VTR V-32, 200HZ, configured for VHS cassettes
Video Processor	- Motion Analysis model VP-310 for threshold-based edge detection, hardware editing and filtering, digitizing image boundaries and data transfer
Computer	- SUN 2/120 with 42 megabyte hard disk and UNIX operating system

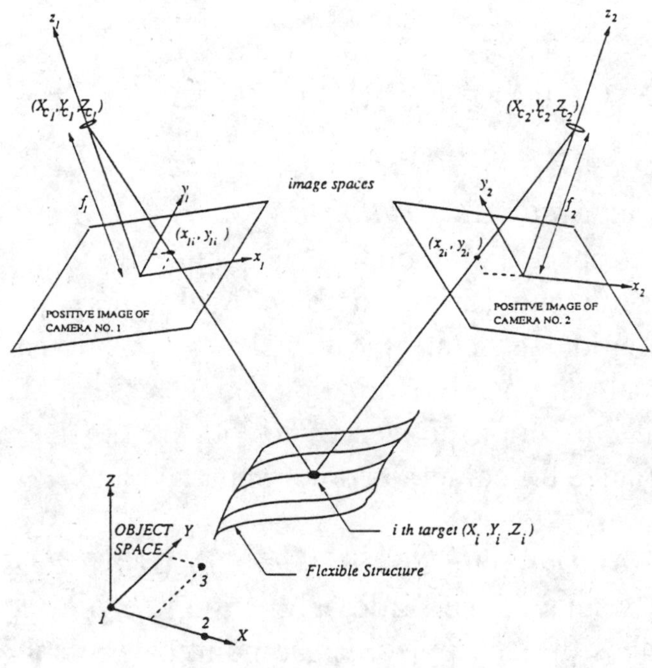

Figure 1. Stereo Triangulation Geometry

OPTICAL MEASUREMENT GEOMETRY AND STEREO TRIANGULATION

The transformation from laboratory rectangular coordinates (X,Y,Z) to focal plane image coordinates (x, y) is given by the *colinearity equations*

$$x = x_o - f\left[\frac{C_{11}(X-X_c) + C_{12}(Y-Y_c) + C_{13}(Z-Z_c)}{C_{31}(X-X_c) + C_{32}(Y-Y_c) + C_{33}(Z-Z_c)}\right] \equiv F(X,Y,Z;X_c,Y_c,Z_c;\phi,0,\psi;x_o,y_o,f)$$

$$y = y_o - f\left[\frac{C_{21}(X-X_c) + C_{22}(Y-Y_c) + C_{23}(Z-Z_c)}{C_{31}(X-X_c) + C_{32}(Y-Y_c) + C_{33}(Z-Z_c)}\right] \equiv G(X,Y,Z;X_c,Y_c,Z_c;\phi,0,\psi;x_o,y_o,f)$$

$$(1)$$

where the direction cosine matrix $[C]$ is parameterized in terms of 3-2-1 Euler angles as

$$[C] = \begin{bmatrix} C_{11} & C_{12} & C_{13} \\ C_{21} & C_{22} & C_{23} \\ C_{31} & C_{32} & C_{33} \end{bmatrix} = \begin{bmatrix} 1 & 0 & 0 \\ 0 & \cos\psi & \sin\psi \\ 0 & -\sin\psi & \cos\psi \end{bmatrix} \begin{bmatrix} \cos0 & 0 & -\sin0 \\ 0 & 1 & 0 \\ \sin0 & 0 & \cos0 \end{bmatrix} \begin{bmatrix} \cos\phi & \sin\phi & 0 \\ -\sin\phi & \cos\phi & 0 \\ 0 & 0 & 1 \end{bmatrix}$$

We adopt a double subscript notation for Eqs. (1), to denote the image coordinates of the i^t point measured in the jth camera's image space as

$$x_{ij} = F(X_i, Y_i, Z_i; X_{c_j}, Y_{c_j}, Z_{c_j}; \phi_j, 0_j, \psi_j; x_{o_j}, y_{o_j}, f_j)$$
$$y_{ij} = G(X_i, Y_i, Z_i; X_{c_j}, Y_{c_j}, Z_{c_j}; \phi_j, 0_j, \psi_j; x_{o_j}, y_{o_j}, f_j), \quad j = 1,2; \; i = 1,2,\ldots,N \qquad (2)$$

TWO PHOTOGRAMMETRIC ESTIMATION PROBLEMS

Problem One: *Static Calibration*

Each object space point imaged generates four measurements, these measurements can be combined with other a priori calibration information to determine the camera positions, orientation angles, focal lengths, and principle point offsets. Least square differential correction converges to yield estimates and associated covariance for camera geometric parameters.

Problem Two: *Dynamic Triangulation to Measure Deflections*

Given the results of the static calibration, the four equations for stereo images of each object space point can be solved for estimates of the object space coordinates; this can be done for each of several images, for each instant in time (e. g. at 200 HZ)

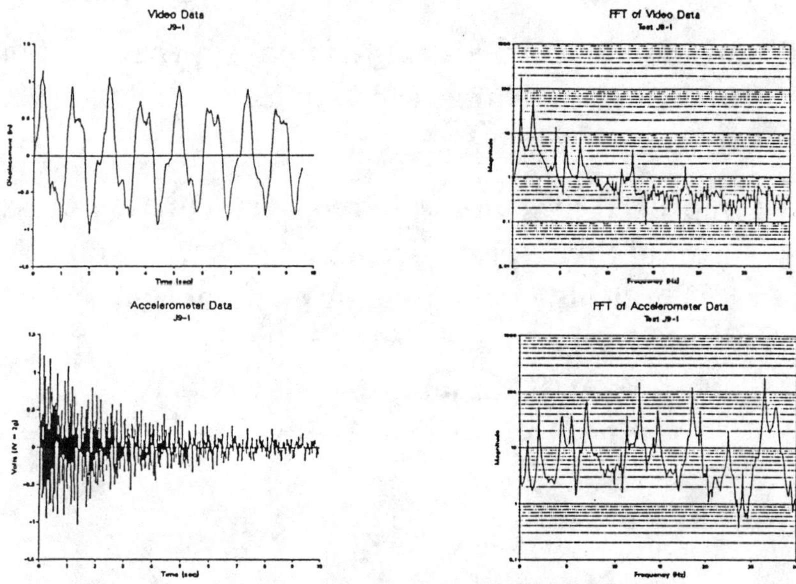

Figure 5. Comparision of Video-Derived Position Measurements with Accelerometer Measurements

The graphs shown above provide some insight into the advantages and disadvantages of the camera system versus an accelerometer. The video data is very useful at the low frequency large amplitude end of the spectrum. The accelerometer response is most reliable at the high frequency end of the spectrum. This data was taken on the AFAL structural identification test article which is similar to the TAMU structure. The video data was taken with a TAMU RCA TC2811 60 Hz video camera. An AFAL Endevco model 7751-500 accelerometer was also used.

Figure 4. Modeled, Measured, and Identified Natural Frequencies of the Frame Structure

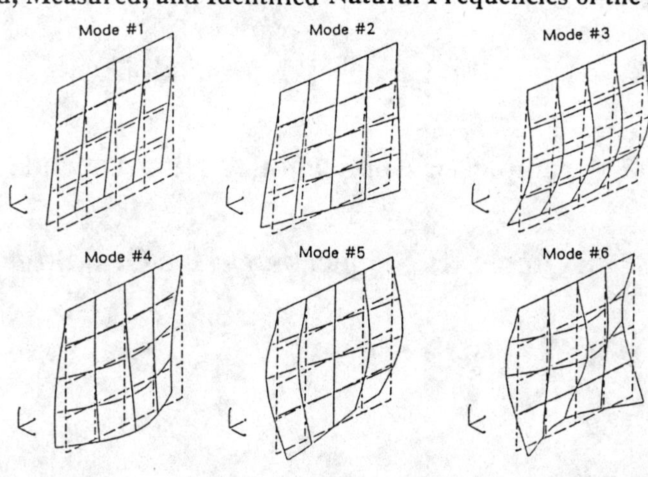

Mode No.	Modeled Value	Measured Value	Identified Value
1	.90 Hz	.92 Hz	.91 Hz
2	2.34	2.32	2.32
3	4.85	4.93	4.93
4	6.05	6.38	6.38
5	7.78	7.27	7.26

Concluding Remarks

- **Promising Structural Identification Approach Established**
    - **Analytical/Numerical Results**
    - **Experimental Verification**
- **An Improved System for Stereo Triangulation Measurement of Structural Vibration has been Established**
    - **Applicable to Measuring General Motion**
        - **Large Maneuvers & Deployment (e. g. RMS multi-body problems)**
        - **Small Vibratory Motions**

Minimum Sensitivity Controller Design using Symmetric Output Feedback Feedback Control

- Symmetric Feedback Formulation

- Example 1: Application to Hub/Appendage Structure

- Example II: Application to the Grid Structure

- Conclusions

keystone format 18

Minimum Sensitivity Symmetric Output Feedback Gain Design

Suppose that a mechanical system is defined by the equation of motion

$$M\ddot{x} + C\dot{x} + Kx = Du \tag{28}$$

where $x \in R^n$ and $u \in R^m$ are the configuration and control vectors, respectively. We introduce the symmetric or "structural" output feedback form of the control law

$$u = -(G_1 D^T x + G_2 D^T \dot{x}) \tag{29}$$

where G_1 and G_2 are $m \times m$ positive definite symmetric gain matrices. For all members of this special class of output feedback controllers, we can verify that asymptotic stability is guaranteed. The most familiar member of the family is the so-called "direct feedback" case of diagonal G_1 and G_2. More generally, fully populated gains allow every actuator to linearly operate on all sensor measurements and thereby provides more controller design freedom.

We admit the most general family of positive definite gain matrices by defining

$$G_1 = L_1 L_1^T \quad \text{and} \quad G_2 = L_2 L_2^T \quad \text{with} \quad
L_1 = \begin{bmatrix} q_{11}^2 & 0 & 0 & \cdots & 0 \\ q_{21} & q_{22}^2 & 0 & \cdots & 0 \\ q_{31} & q_{32} & q_{33}^2 & \cdots & 0 \\ \vdots & \vdots & \vdots & \vdots & \vdots \\ q_{m1} & q_{m2} & q_{m3} & \cdots & q_{mm}^2 \end{bmatrix}
\quad \text{and} \quad
L_2 = \begin{bmatrix} r_{11}^2 & 0 & 0 & \cdots & 0 \\ r_{21} & r_{22}^2 & 0 & \cdots & 0 \\ r_{31} & r_{32} & r_{33}^2 & \cdots & 0 \\ \vdots & \vdots & \vdots & \vdots & \vdots \\ r_{m1} & r_{m2} & r_{m3} & \cdots & r_{mm}^2 \end{bmatrix} \tag{31}$$

The corresponding global gain parameter vector then becomes

$$p = [q_{11}\; q_{21}\; \cdots\; q_{m1}\; q_{22}\; \cdots\; q_{mm}\; r_{11}\; r_{21}\; \cdots\; r_{m1}\; r_{22}\; \cdots\; r_{mm}]^T \tag{32}$$

We consider G_1 and G_2 to be functions of p, through Eqs. (31), it is clear that

$$G_1 = G_1(p) \quad \text{and} \quad G_2 = G_2(p) \tag{33}$$

The partial derivatives of the gain matrices w.r.t. the elements of p can be constructed easily.

Substitution of Eq. (29) into Eq. (28) gives the closed-loop system

$$M\ddot{x} + \hat{C}\dot{x} + \hat{K}x = 0 \tag{34}$$

where the closed-loop system matrices are

$$\hat{C} = C + DG_2 D^T \quad \text{and} \quad \hat{K} = K + DG_1 D^T \tag{35}$$

Notice that these closed loop matrices remain positive semi-definite, due to the symmetric structure of the feedback .

Continued

Considering a first-order state-space form of the system differential equations which is equivalent to the second-order closed-loop system of Eq. (34):

$$\mathbf{B}\dot{\mathbf{z}} = \mathbf{A}\mathbf{z} \tag{37}$$

where
$$\mathbf{z} \equiv \begin{Bmatrix} \mathbf{x} \\ \dot{\mathbf{x}} \end{Bmatrix}, \quad \mathbf{A} = \begin{bmatrix} [0] & M \\ -\hat{K} & -\hat{C} \end{bmatrix}, \quad \mathbf{B} = \begin{bmatrix} M & [0] \\ [0] & M \end{bmatrix} \quad \text{and} \quad \mathbf{A} = \mathbf{A}(\mathbf{p}) \tag{39}$$

The right and left eigenvalue problems associated with $\mathbf{z} = \phi\, e^{\lambda t}$ solutions of Eq. (37) are

$$\text{right:} \quad \lambda_i \mathbf{B}\phi_i = \mathbf{A}\phi_i$$
$$i = 1, 2, \ldots, 2n \tag{40}$$
$$\text{left:} \quad \lambda_i \mathbf{B}^T \psi_i = \mathbf{A}^T \psi_i$$

where we adopt the usual normalizations of the biorthogonality conditions

$$\phi_i^T \mathbf{B}\phi_i = 1, \quad i = 1, 2, \ldots 2n$$
$$\psi_j^T \mathbf{B}\phi_i = \delta_{ij}, \quad i,j = 1, 2, \ldots 2n \tag{41}$$

In order to apply gradient-based nonlinear programming algorithms, it is useful to compute partial derivatives of the eigenvalues and eigenvectors. Differentiating Eqs. (40), we find [7]

$$\frac{\partial \lambda_i}{\partial p_k} = \psi_i^T \left(\frac{\partial A}{\partial p_k} - \lambda_i \frac{\partial B}{\partial p_k} \right)\phi_i \,, \quad \frac{\partial \phi_i}{\partial p_k} = \sum_{j=1}^{2n} c_{ji}^k \phi_i = \Phi C^k, \quad \frac{\partial \psi_i}{\partial p_k} = \sum_{j=1}^{2n} d_{ji}^k \psi_i = \Psi D^k \tag{42}$$

with $\Phi = [\phi_1\,\phi_2\,...\,\phi_{2n}]$, $\Psi = [\psi_1\,\psi_2\,...\,\psi_{2n}]$, $C^k = [c_{ij}^k]$, $D^k = [d_{ij}^k]$, where

$$c_{ji}^k = \frac{1}{\lambda_i - \lambda_j} \psi_j^T \left(\frac{\partial A}{\partial p_k} - \lambda_i \frac{\partial B}{\partial p_k} \right)\phi_i \quad , j \neq i \qquad\qquad d_{ji}^k = \frac{1}{\lambda_i - \lambda_j} \psi_i^T \left(\frac{\partial A}{\partial p_k} - \lambda_i \frac{\partial B}{\partial p_k} \right)\phi_j \;, j \neq i$$
$$= -\frac{1}{2}\left[\phi_i^T \frac{\partial B}{\partial p_k} \phi_i + \sum_{\substack{j=1 \\ j \neq i}}^{2n} c_{ji}^k \phi_j^T (B + B^T)\phi_i \right], j = i \qquad\qquad = -\psi_i^T \frac{\partial B}{\partial p_k} \phi_i - c_{ii}^k \quad\quad , j = i \tag{43}$$

Second partial derivatives of the eigenvectors are obtained by differentiating Eq. (42) with

$$\frac{\partial^2 \Phi}{\partial p_k \partial p_l} = \frac{\partial \Phi}{\partial p_l} C^k + \Phi \frac{\partial C^k}{\partial p_l} \quad \text{where } C^k = [c_{ij}^k] \text{ and } \frac{\partial C^k}{\partial p_l} = \left[\frac{\partial c_{ij}^k}{\partial p_l} \right].$$

We can obtain expressions for $\frac{\partial c_{ij}^k}{\partial p_l}$ by differentiating as is developed in [7].

652

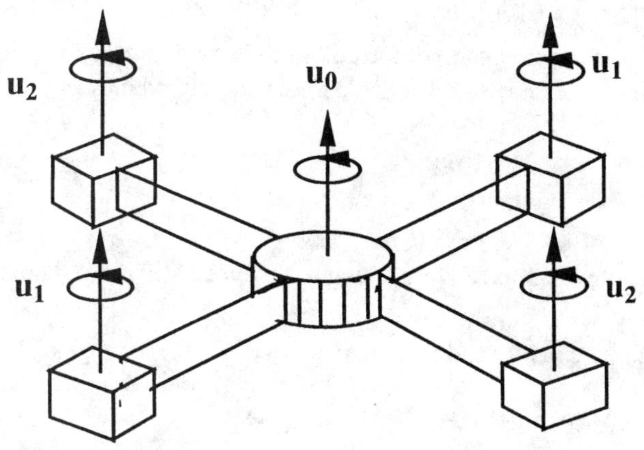

Fig. 1. DRAPER/RPL MODEL

Table 3. Draper/RPL Configuration Parameters

PARAMETER	SYMBOL	VALUE
Hub radius	r	1 ft
Rotary inertia of hub	J_h	8 slug-ft^2
Mass density of beams	ρ	5.22 slug/ft^3
Elastic modulus of the arms	E	1.1E+7 lb/in^2
Arm thickness	t	0.125 in
Arm height	h	6.0 in
Arm length	L	4.0 ft
Tip mass	m	0.156941 slug
Rotary inertia of tip masses	J_t	0.0018 slug-ft^2

Control Gain Optimization

We adopt the condition number $\aleph(\Phi^c)$ of the closed loop modal matrix Φ^c as our performance measure. This performance measure is related to the singular values as

$$J(\mathbf{p}) = \aleph(\Phi^c(\mathbf{p})) = \frac{\sigma_1(\Phi^c(\mathbf{p}))}{\sigma_{2n}(\Phi^c(\mathbf{p}))} \tag{46}$$

where σ_1 and σ_{2n} denote the max. and min. singular values of the right modal matrix Φ^c.

We also adopt the three equality and two inequality constraints on eigenvalue placement

$$\begin{aligned} f_i(\mathbf{p}) &= \omega_{io} - \omega_i(\mathbf{p}) = 0 \\ g_i(\mathbf{p}) &= \omega_i(\mathbf{p})\zeta_i(\mathbf{p}) \geq T_i \end{aligned} \quad , \quad i = 1, 2, 3 \tag{47}$$

where
$$\begin{aligned} \omega_i(\mathbf{p}) &= \mathrm{Im}(\lambda_i) \\ \zeta_i(\mathbf{p}) &= -\frac{\mathrm{Re}(\lambda_i)}{|\lambda_i|} \end{aligned} \tag{48}$$

The time constant and frequency constraints are imposed using a *homotopy* algorithm.

Draper/RPL Configuration: Equations of Motion

As an illustrative example, consider the planar rotational/vibrational dynamics of a rigid hub with four flexible appendages (Fig. 1). The eqs. of motion follow from Lagrange's equations:

$$\frac{d}{dt}\left(\frac{\partial T}{\partial \dot{x}_i}\right) - \frac{\partial T}{\partial x_i} + \frac{\partial V}{\partial x_i} = F_i \quad i=1,...n \tag{50}$$

where T is the system kinetic energy, V is the system potential energy, F_i is the i-th generalized force, and x_i is the i-th element of the configuration vector, $\mathbf{x}$.

$$\mathbf{x} = \{\, \theta \quad q_{11} \quad q_{21} \quad .... \quad q_{N1} \quad q_{21} \quad q_{22} \quad .... \quad q_{N2} \,\}^T \tag{51}$$

The q_{ij} are generalized beam coordinates which result from using a Ritz approach to approximate the local beam deflections, y_i as the truncated ("assumed modes") series

$$y_i(t,z) = \sum_{j=1}^{N} q_{ji}(t)\,\phi_j(z), \quad i=1,2, \ 0 \leq z \leq L \tag{52}$$

The comparison functions $\phi_j(z)$ in Eq.(52) are chosen as [16] which satisfy the geometric and physical boundary conditions of a clamped-free beam (clamped to the hub). We apply a torque u_1 to the hub, a torque u_2 to the end of appendages 1 and 3, and a torque u_3 to the end of the appendages 2 and 4.

System Discretization: Equations of Motion

Lagrange's eqns. lead to the following matrix form for the system equations of motion

$$\begin{bmatrix} J & M_{\theta q_1}^T & M_{\theta q_2}^T \\ M_{\theta q_1} & M_{q_1 q_1} & 0 \\ M_{\theta q_2} & 0 & M_{q_2 q_2} \end{bmatrix} \ddot{x} + \begin{bmatrix} 0 & 0 & 0 \\ 0 & K_{q_1 q_1} & 0 \\ 0 & 0 & K_{q_2 q_2} \end{bmatrix} x = \begin{bmatrix} 1 & 2 & 2 \\ 0 & 2\phi'(L) & 0 \\ 0 & 0 & 2\phi'(L) \end{bmatrix} u \tag{54}$$

or simply $\qquad M\ddot{x} + Kx = Du$ $\qquad\qquad\qquad\qquad\qquad\qquad$ (55)

where the elements of the matrices are given in terms of the shape functions as

$$J = J_h + 4\int_0^L \rho(x+r)^2\, dx + 4[J_t + m(r+L)^2]$$

$$[M_{\theta q_1}]_i = [M_{\theta q_2}]_i = 2\left[\int_0^L \rho(x+r)\phi_i(x)dx + m(r+L)\phi_i(L) + J_t\phi_i'(L)\right]$$

$$[M_{q_1 q_1}]_{ij} = [M_{q_2 q_2}]_{ij} = 2\left[\int_0^L \rho\phi_i(x)\phi_j(x)dx + m\phi_i(L)\phi_j(L) + J_t\phi_i'(L)\phi_j'(L)\right]$$

$$[K_{q_1 q_1}]_{ij} = [K_{q_2 q_2}]_{ij} = 2\int_0^L EI\phi_i''(x)\phi_j''(x)dx$$

Minimum Sensitivity Symmetric Output Feedback Gain Design

The symmetric feedback form of control guarantees stability, even though we may not have chosen a good reduced-order model.

Mechanical Second Order System

$$M\ddot{x} + Kx = Du$$

where $\quad M,\ K : (n \times n)$ symmetric matrices, $\quad D : (n \times m)$ matrix

It is desirable to develop reduced-order model to save on computational time when designing control laws for high order systems such as flexible structures.

Modal Coordinate Transformation

$$x(t) = \Phi\eta(t) = [\Phi_1 \ \ \Phi_2]\begin{pmatrix} \eta_1(t) \\ \eta_2(t) \end{pmatrix}$$

where $\quad \Phi$: Modal Matrix of Eigenvectors

$\qquad \eta_1$: n_r Modal Coordinates which are retained to design control law.

$\qquad \eta_2$: Residual Modal Coordinates.

Reduced-order Equation of Motion

$$\tilde{M}\ddot{\eta}_1 \;+\; \tilde{K}\eta_1 \;=\; \tilde{D}u$$

where $\quad \tilde{M} = \Phi_1^T M \Phi_1 = I, \quad \tilde{K} = \Phi_1^T K \Phi_1 = diag(\omega_1^2,\ldots,\omega_{n_r}^2), \quad \tilde{D} = \Phi_1^T D$

Symmetric or "Structural" Feedback Control

$$\mathbf{u} \;=\; -(\;\tilde{G}_1 \tilde{D}^T \eta_1 \;+\; \tilde{G}_2 \tilde{D}^T \dot{\eta}_1\;)$$

Introduce the Cholesky Decomposition Gain Parameterization

$$\tilde{G}_1 \;=\; L_1 L_1^T \qquad\qquad \tilde{G}_2 \;=\; L_2 L_2^T$$

where

$$
L_1 = \begin{bmatrix}
q_{11}^2 & 0 & 0 & \cdots & 0 \\
q_{21} & q_{22}^2 & 0 & \cdots & 0 \\
q_{31} & q_{32} & q_{33}^2 & \cdots & 0 \\
\vdots & \vdots & \vdots & \ddots & \vdots \\
q_{m1} & q_{m2} & q_{m3} & \cdots & q_{mm}^2
\end{bmatrix}
\qquad
L_2 = \begin{bmatrix}
r_{11}^2 & 0 & 0 & \cdots & 0 \\
r_{21} & r_{22}^2 & 0 & \cdots & 0 \\
r_{31} & r_{32} & r_{33}^2 & \cdots & 0 \\
\vdots & \vdots & \vdots & \ddots & \vdots \\
r_{m1} & r_{m2} & r_{m3} & \cdots & r_{mm}^2
\end{bmatrix}
$$

Let the gain parameter vector be

$$\mathbf{p} \;=\; \begin{bmatrix} q_{11} & q_{21} & \cdots & q_{m1} & q_{22} & \cdots & q_{mm} & r_{11} & r_{21} & \cdots & r_{mm} \end{bmatrix}^T$$

Reduced-order System $(n_r = 10)$

$$\tilde{M} = \Phi_1^T M \Phi_1 = I$$

$$\tilde{K} = \Phi_1^T K \Phi_1 = diag(\omega_1^2,\ldots,\omega_{n_r}^2)$$

where

$$\omega_1 = 5.649, \omega_2 = 14.683, \omega_3 = 30.459, \omega_4 = 37.785, \omega_5 = 48.852$$

$$\omega_6 = 80.470, \omega_7 = 82.958, \omega_8 = 83.658, \omega_9 = 102.454, \omega_{10} = 125.740$$

$$
\tilde{D} = \Phi_1^T D = \begin{bmatrix}
-0.2230 & -0.1912 & -0.1912 \\
0.0000 & -0.1484 & 0.1484 \\
0.6039 & 0.0256 & 0.0256 \\
0.0782 & 0.0316 & 0.0316 \\
0.0000 & -0.1264 & 0.1264 \\
-0.5645 & -0.3207 & -0.3207 \\
0.3505 & -0.8570 & -0.8570 \\
0.0000 & -0.0947 & 0.0947 \\
0.0000 & 0.8310 & -0.8310 \\
0.0000 & -0.3557 & 0.3557
\end{bmatrix}
$$

Closed-loop System

$$\ddot{\eta}_1 \;+\; \hat{C}\dot{\eta}_1 \;+\; \hat{K}\eta_1 \;=\; 0$$

where $\quad \hat{C} = \tilde{D}\tilde{G}_2\tilde{D}^{T}, \quad \hat{K} = \tilde{K} + \tilde{D}\tilde{G}_1\tilde{D}^{T} = diag(\omega_1^2,\ldots,\omega_{n_r}^2) + \tilde{D}\tilde{G}_1\tilde{D}^{T}$

First-order State-space Form of the Closed-loop System

$$B\dot{z} \;=\; A(p)z$$

where

$$z = \begin{pmatrix} \eta_1 \\ \dot{\eta}_1 \end{pmatrix}, \quad A = \begin{bmatrix} 0 & I \\ -\hat{K} & -\hat{C} \end{bmatrix}, \quad B = \begin{bmatrix} I & 0 \\ 0 & I \end{bmatrix}$$

We can compute closed-loop eigenvector/eigenvalue and the partial derivative of them with respect to the gain parameter vector.

Two-Stage Homotopic Nonlinear Programming Algorithm

To solve the above nonlinear programming problem, utilize the minimum norm differential correction algorithm and the homotopy technique.

First Stage : Seek the Nearest Feasible Solution (without working about the performance measure)

$$f_i(\mathbf{p}) \;=\; f_i^o \qquad i = 1,\ldots,m$$

$$g_i(\mathbf{p}) \;\geq\; T_i \qquad i = 1,\ldots,m$$

locally violated ("active") subset of the inequality constraints are simply considered as additional constraints of the form

$$f_j(\mathbf{p}) \;=\; T_j \qquad j = m+1,\ldots,r$$

Re-stated problem (only equality constraints)

$$\mathbf{f}(\mathbf{p}) \;=\; \mathbf{f_o}$$

Inactive inequality constraints are locally ignored.

Introduce a "portable objective" vector of constraint objectives

$$\mathbf{f_p}(\gamma) = \gamma \mathbf{f_o} + (1 - \gamma)\mathbf{f}(\mathbf{p_{start}})$$

where γ : homotopy parameter between 0 and 1

Define **Homotopy Map**

$$H(\mathbf{p}(\gamma), \gamma) \equiv \gamma \mathbf{f_o} + (1 - \gamma)\mathbf{f}(\mathbf{p_{start}}) - \mathbf{f}(\mathbf{p}(\gamma)) = 0$$

at $\gamma = 0$: trivial problem $(\mathbf{p}(0) \equiv \mathbf{p_{start}})$

at $\gamma = 1$: problem of interest $(\mathbf{p}(1) = $ a solution satisfying $\mathbf{f_o} - \mathbf{f}(\mathbf{p}(1)) = 0)$

Minimum Norm Differential Correction vector

$$\Delta\mathbf{p} = -W^{-1}A^{T}(AW^{-1}A^{T})^{-1}H(\mathbf{p}(\gamma))$$

where $A \equiv \frac{\partial H}{\partial \mathbf{p}},$ W : Positive definite weighting matrix

Recursion Formula

$$\mathbf{p_{new}} \quad = \quad \mathbf{p_{old}} \quad + \quad \Delta\mathbf{p}$$

Iterate with γ fixed until local convergence achieved, increment γ only after local convergence. We get the feasible solution when we get the converged solution at $\gamma = 1$.

Second Stage :   Seek the constrained minimum solution.

Treat objective funciton(s) as additional constraint(s).

Add new equality constraints to the original constrained problem (first stage problem).

$$J_o(\gamma) = \gamma J_{goal} + (1 - \gamma)J(\mathbf{p}_{first \quad feasible \quad sol.})$$

where J_{goal} is an actual goal, or the best one could possibly hope for.

We get the constrained minimum solution when we get the converged solution corresponding to the possibly largest γ.

We adopted N=20 and n_r=9. We design the control gain matrices $\tilde{G}_1$ and $\tilde{G}_2$ by using the 18th reduced order model and use the (42nd) order system to verify the controller design.

We adopt the following time constants for inequality constraints on the first three closed loop eigenvalues: τ_1=0.2, τ_2=0.2, and τ_3=0.25; the objective value for the closed loop frequency of the rigid body mode frequency is taken as 0.3.

The homptopic optimization process was initiated by simply adopting identity matrices for the G_i starting iteratives. The characteristics of the open loop and closed loop systems are summarized in Table 4. The final converged position and velocity gain matrices are

$$G_1 = \begin{bmatrix} 1.8342 & 0.2592 & 0.2601 \\ 0.2592 & 0.0409 & 0.0405 \\ 0.2601 & 0.0405 & 0.0403 \end{bmatrix} \qquad \|G_1\|_f = 1.9080$$

$$G_2 = \begin{bmatrix} 7.4284 & 0.4579 & 0.8335 \\ 0.4579 & 0.2411 & 0.0211 \\ 0.8335 & 0.0211 & 0.3509 \end{bmatrix} \qquad \|G_2\|_f = 7.5612$$

Table 4. The Characteristics of the Open-loop and Closed-loop System. (First nine modes of the full order system)

OPEN LOOP		CLOSED LOOP	
ω_i	$\zeta_i \omega_i$	ω_i	$\zeta_i \omega_i$
0	0	0.3000	0.2000
4.3724	0	4.3726	0.2061
7.9070	0	7.9071	0.2539
51.4510	0	52.0041	5.3354
52.8058	0	53.0653	8.0875
157.5566	0	158.0285	14.0467
158.3659	0	158.5273	21.4808
313.7099	0	308.8039	34.8815
314.3402	0	311.7689	22.6602
Condition No.	∞	Condition No.	37210.8

Concluding Remarks: Symmetric Output Feedback Control

- **Generalizes "direct velocity feedback" (DVFB)**
  ... more "knobs"(gains) >> use to impose additional control constraints
  and/or to minimize sensitivity
  ... shares fundamental robustness properties with DVFB:
  stable irregardless of model error & no state estimator req'd
- **Algorithm established to seek out the symmetric position and velocity feedback gains which minimize condition # of closed loop modal matrix**
- **Closed loop system still "looks like a structure"**
  Idea: We could use co-colocated sensor actuator pairs first to
  derive a symmetric control law with substantial stability margins ...
  then use ?? (your choice!) approach to further optimize control to
  include feedback from remaining (non-co-located) sensors
- **Convergence to feasible or optimal design is not guaranteed, even tho stability is! >>> Future work here.**

keystone.format 18

6.0 REFERENCES

1. <u>Manual of Photogrammetry</u> (4th Edition), published by the American Society of Photogrammetry, Falls Church, VA(1980).

2. Junkins, J. L. <u>Optimal Estimation of Dynamical Systems</u>, Sijthoff-Noordhoff, Leyden, the Netherlands (1978).

3. Creamer, N. G., <u>Identification of Flexible Structures.</u>, Ph. D. Dissertation, Engineering Mechanics, Virginia Tech, 1987.

4. Creamer, N. G., and J. L. Junkins, "An Identification Method for Flexible Structures", to appear, <u>AIAA J. of Guidance , Control, and Dynamics</u>, also presented at the AIAA/ ASME/ASCE/AHS 28th Structures, Structural Dynamics, and Materials (SDM) Conference, Monterey, CA, April 6-8, 1987.

5. Deneman, E., Hasselman, T., Sun, C., Juang, J., Junkins, J., Udwadia, F., Venkayya, V., Kamat, M., "Identification of Large Space Structures on Orbit", Final report under Air Force Contract F04611-85-C-0092 with the American Society of Civil Engineering, Air Force Rocket Propulsion Laboratory Report No. AFRPL TR-86-054, 342 pp, August 1986.

6. White, C. W. and Maytum, B. D., "Eigensolution Sensitivity to Parametric Model Perturbations," <u>Shock and Vibration Bulletin</u>, Bulletin 46, Part 5, August 1976, pp123-133.

7. James, G. H., "An Optical Sensor System for Measuring Structural Dynamics with applications to System Identification," final report under 1987 USAF-UES Summer Faculty/Graduate Student Research Program Contract No. F49620-85-C-0013, Texas A&M University, Department of Aerospace Engineering, August 28, 1987.

8. Das, A., Srange, W. T., Schlaegel, and Ward, J. M., "Experiment in Modeling and Parameter Estimation of Flexible Structures", Second NASA/DOD Control-Structure Interation Conference, Colorado Springs, Co., November 17-19, 1987.

9. Paz, M., <u>Structural Dynamics Theory and Computation</u>, 2nd Ed., Van Nostrand Reinhold Co., New York, (1985).

Feedback Control of Nonlinear and Distributed Parameter Systems

- **Symbolic Manipulation via Macsyma**

- **A Quasi-Analytical Perturbation Method for Nonlinear Feedback Control**

- **A Liapunov Approach to Control of Distributed Parameter Systems**

- **TAMU Flexible Structure Slewing Experiment**

MACSYMA

"... a computer-automated graduate student!!"

Expanding and factoring a multivariate polynomial:

```
(C1)  (Y^2+X)^2*(Y-X);
                                        2      2
(D1)                      (Y - X) (Y  + X)
(C2)  EXPAND(D1);
            5      4         3       2  2     2        3
(D2)    Y  - X Y  + 2 X Y  - 2 X  Y  + X  Y - X
(C3)  FACTOR(D2);
                                    2      2
(D3)                      (Y - X)  (Y  + X)
```

Solving simultaneous algebraic equations:

```
(C1)  [X*Y = 1-A,Y+X = 2];
(D1)                  [X Y = 1 - A,  Y + X = 2]

(C2)  SOLVE(D1,[X,Y]);
(D2)  [[X = 1 - SQRT(A),  Y = SQRT(A) + 1],

              [X = SQRT(A) + 1,  Y = 1 - SQRT(A)]]
```

Trigonometric simplification in two stages:

```
(C1)  SIN(X+2*Y);
(D1)                      SIN(2 Y + X)

(C2)  TRIGEXPAND(D1);
(D2)          COS(X) SIN(2 Y) + SIN(X) COS(2 Y)

(C3)  TRIGEXPAND(D2);
                    2           2
(D3) SIN(X) (COS (Y) - SIN (Y))

            + 2 COS(X) COS(Y) SIN(Y)
```

Differentiation:

```
(C1)  'DIFF(X^X^X,X)=DIFF(X^X^X,X);
            X       X
      d    X       X    X                              X - 1
(D1)  -- (X  ) = X   (X   LOG(X) (LOG(X) + 1) + X       )
      dX
```

Integrating a non-tabulated function:

```
(C1)  F(X):=(LOG(X)-1)/(LOG(X)^2-X^2);
                              LOG(X) - 1
(D1)                  F(X) := ------------
                              2        2
                           LOG (X) - X

(C2)  INTEGRATE(F(X),X);
            LOG(LOG(X) + X)    LOG(LOG(X) - X)
(D2)        --------------- - ---------------
                  2                  2
```

Computing a Laplace transform:

```
(C1)  F(T):=%E^(-B*T)*SIN(W*T)*T;

                             - B T
(D1)              F(T) := %E        SIN(W T)  T
(C2)  LAPLACE(F(T),T,S);

                       (2 S + 2 B) W
(D2)          -----------------------------------
                  2     2             2 2
              (W   + S   + 2 B S + B )
```

Solving an ordinary differential equation:

```
(C1)  EQ:'DIFF(Y,T,2)=-(Y+E*Y^3)-SIN(2*T);
              2
            d Y           3
(D1)        --- = - E Y  - Y - SIN(2 T)
              2
            dT
(C2)  IC:['AT(Y,T=0)=A,'AT('DIFF(Y,T),T=0)=0];

                               !
              !            dY!
(D10)       [Y!       = A, --!        = 0]
             !T = 0        dT!
                            !T = 0
```

Solve a simpler case exactly (e = 0):

```
(C3)  ODE(EQ,Y,T),E=0;
                SIN(2 T)
(D3)        Y = -------- + %K1 SIN(T) + %K2 COS(T)
                   3


(C4)  IC2(D3,T=0,Y=A,'DIFF(Y,T)=0);
                SIN(2 T)    2 SIN(T)
(D4)        Y = -------- - -------- + A COS(T)
                   3          3
```

Arbitrary-precision floating point arithmetic and exact algebraic arithmetic avoid numerical problems:

```
(C1)  (A:10^30*SQRT(6),B:10^25*SQRT(2),

                  C:10^5*SQRT(3))$

(C2)  FLOAT(A)-FLOAT(B)*FLOAT(C);
(D2)               1.5111573e23

(C3)  (FPPREC:35,BFLOAT(A)-BFLOAT(B)*BFLOAT(C));
(D3)               0.0B0

(C4)  RADCAN(A-B*C);
(D4)                     0
```

Numerical integration by the Romberg method:

```
(C1)  F(X):=1/(X^5+X+1)$
(C2)  ROMBERG(F(X),X,0,1.5);
(D2)                        0.75293845
```

Interpolate roots of equations:

```
(C1)  EQN:COS(X)^3 = X  $
(C2)  ROOT_BY_BISECTION(EQN,X,0,1);
(D2)                        0.5824401
```

Generate 'optimized' Fortran code:

```
(C1)  EXPR:45.9*X^4-13.6*X^3+76.2*X^2-3.4*SIN(X)$
(C2)  FORTRAN(Y=HORNER(EXPR,X)),KEEPFLOAT:TRUE$
        Y = X**2*(X*(45.9*X-13.6)+76.2)-3.4*SIN(X)
```

An Asymptotic Expansion Approach to Nonlinear Feedback Control

Problem I

Find an optimal feedback control law to compute $u(x)$ for the system described by

$$\dot{x} = -x + \alpha x^2 + u, \quad \alpha > 0 \tag{1}$$

which minimizes the performance measure

$$J = \frac{1}{2} \int_0^{t_f} \left(x^2 + u^2 \right) dt \tag{2}$$

Problem II

Find an optimal feedback control law to compute $u_1(x_1, x_2)$, $u_2(x_1, x_2)$ for the system

$$\dot{x}_1 = -x_1 + x_1 x_2 + x_2^2 + u_1$$

$$\dot{x}_2 = -x_2 + x_1 x_2 + x_1^2 + u_2$$

(3)

which minimizes the performance measure

$$J = \frac{1}{2} \int_0^{t_f} \sum_{i=1}^{2} \left(x_i^2 + u_i^2 \right) dt$$

(4)

Note that Eqs. (3) can be written in an alternate matrix format as

$$\dot{x} = A_1 x_1 + A_2 x_2 + B u \,,$$

(5)

where

$$x_1^T = x^T = [x_1 \ x_2], \quad x_2^T = [x_1^2 \ x_1 x_2 \ x_2^2], \quad u^T = [u_1 \ u_2], \quad A_1 = \begin{bmatrix} 1 & 0 \\ 0 & 1 \end{bmatrix}, \quad A_2 = \begin{bmatrix} 1 & 1 & 0 \\ 0 & 1 & 1 \end{bmatrix}, \quad B = \begin{bmatrix} 1 & 0 \\ 0 & 1 \end{bmatrix}$$

As we show below, the above notation generalizes and ultimately provides a convenient way basis for an algebraic method which solves a large family of nonlinear control problems. The notation should not obscure the fact that Eq. (5) is nonlinear! ... look at elements of x_2 ...

Problem III (Spacecraft Attitude feedback control for large angular maneuvers)

Find a feedback law for the torques $u_i(q_1, q_2, q_3, \omega_1, \omega_2, \omega_3)$, for $i = 1,2,3$ to control the system

$$\dot{q}_1 = \frac{1}{2}[(1+q_1^2)\omega_1 + (q_1 q_2 - q_3)\omega_2 + (q_1 q_3 + q_2)\omega_3]$$

$$\dot{q}_2 = \frac{1}{2}[(q_1 q_2 + q_3)\omega_1 + (1+q_2^2)\omega_2 + (q_2 q_3 - q_1)\omega_3]$$

$$\dot{q}_3 = \frac{1}{2}[(q_1 q_3 - q_2)\omega_1 + (q_2 q_3 + q_1)\omega_2 + (1+q_3^2)\omega_3]$$

$$\dot{\omega}_1 = \left(\frac{I_2 - I_3}{I_1}\right)\omega_2 \omega_3 + \left(\frac{1}{I_1}\right)u_1$$

$$\dot{\omega}_2 = \left(\frac{I_3 - I_1}{I_2}\right)\omega_3 \omega_1 + \left(\frac{1}{I_2}\right)u_2$$

$$\dot{\omega}_3 = \left(\frac{I_1 - I_2}{I_3}\right)\omega_1 \omega_2 + \left(\frac{1}{I_3}\right)u_3$$

$$\Longleftrightarrow \quad \dot{x} = \sum_{i=1}^{3} A_i x_i + B u \qquad (6)$$

which minimizes $J = \frac{1}{2} \int_o^{\infty} \left(x^T Q x + u^T R u \right) dt$, with $Q, R > 0$ (weight matrices), (7)

The state and control vectors are: $x^T = [q_1 \ q_2 \ q_3 \ \omega_1 \ \omega_2 \ \omega_3]$, $u^T = [u_1 \ u_2 \ u_3]$

and the x_i are column vectors and the A_i are sparse matrices, as given below:

$$x_1 = x = [x_1 \ x_2 \ x_3 \ x_4 \ x_5 \ x_6]^T \quad (6\times1) \qquad \left\{ \ \dot{x} = \sum_{i=1}^{3} A_i x_i + B u \right.$$

$$(21\times1)$$
$$x_2 = [x_1^2 \ x_1 x_2 \ x_1 x_3 \ x_1 x_4 \ x_1 x_5 \ x_1 x_6 \ x_2^2 \ x_2 x_3 \ x_2 x_4 \ x_2 x_5 \ x_2 x_6 \ x_3^2 \ x_3 x_4 \ x_3 x_5 \ x_3 x_6 \ x_4^2 \ x_4 x_5 \ x_4 x_6 \ x_5^2 \ x_5 x_6 \ x_6^2]^T$$

$$(56\times1)$$
$$x_3 = [\,x_1^3 \ x_1^2 x_2 \ x_1^2 x_3 \ x_1^2 x_4 \ x_1^2 x_5 \ x_1^2 x_6 \ x_1 x_2^2 \ x_1 x_2 x_3 \ x_1 x_2 x_4 \ x_1 x_2 x_5 \ x_1 x_2 x_6 \ x_1 x_3^2 \ x_1 x_3 x_4 \ x_1 x_3 x_5 \ x_1 x_3 x_6 \ x_1 x_4^2 \ x_1 x_4 x_5 \ x_1 x_4 x_6 \ x_1 x_5^2 \ x_1 x_5 x_6 \ x_1 x_6^2 \ x_2^3 \ x_2^2 x_3 \ x_2^2 x_4 \ x_2^2 x_5 \ x_2^2 x_6$$
$$x_2 x_3^2 \ x_2 x_3 x_4 \ x_2 x_3 x_5 \ x_2 x_3 x_6 \ x_2 x_4^2 \ x_2 x_4 x_5 \ x_2 x_4 x_6 \ x_2 x_5^2 \ x_2 x_5 x_6 \ x_2 x_6^2 \ x_3^3 \ x_3^2 x_4 \ x_3^2 x_5 \ x_3^2 x_6 \ x_3 x_4^2 \ x_3 x_4 x_5 \ x_3 x_5^2 \ x_3 x_5 x_6 \ x_3 x_6^2 \ x_4^3 \ x_4^2 x_5 \ x_4^2 x_6 \ x_4 x_5^2 \ x_4 x_5 x_6 \ x_4 x_6^2 \ x_5^3 \ x_5^2 x_6 \ x_5 x_6^2 \ x_6^3\,]^T$$

The non−zero elements of the A_i and B are:

A_i very sparse!

$$\overset{6\times6}{A_1}(1,4) = \overset{6\times6}{A_1}(2,5) = \overset{6\times6}{A_1}(3,6) = \tfrac{1}{2}$$

$$\overset{6\times21}{A_2}(1,11) = \overset{6\times21}{A_2}(2,13) = \overset{6\times21}{A_2}(3,5) = \tfrac{1}{2}$$

$$\overset{6\times21}{A_2}(3,14) = \overset{6\times21}{A_2}(2,6) = \overset{6\times21}{A_2}(3,9) = -\tfrac{1}{2}$$

$$\overset{6\times21}{A_2}(4,20) = \left(\frac{I_2 - I_3}{I_1}\right)$$

$$\overset{6\times21}{A_2}(5,18) = \left(\frac{I_3 - I_1}{I_2}\right)$$

$$\overset{6\times21}{A_2}(6,17) = \left(\frac{I_1 - I_2}{I_3}\right)$$

$$\overset{6\times56}{A_3}(1,4) = \overset{6\times56}{A_3}(1,10) = \overset{6\times56}{A_3}(1,15) = \tfrac{1}{2}$$

$$\overset{6\times56}{A_3}(2,9) = \overset{6\times56}{A_3}(2,25) = \overset{6\times56}{A_3}(2,30) = \tfrac{1}{2}$$

$$\overset{6\times56}{A_3}(3,13) = \overset{6\times56}{A_3}(3,29) = \overset{6\times56}{A_3}(3,37) = \tfrac{1}{2}$$

$$B = \begin{bmatrix} \frac{1}{I_1} & 0 & 0 \\ 0 & \frac{1}{I_2} & 0 \\ 0 & 0 & \frac{1}{I_3} \end{bmatrix}$$

A Solution Process Based Upon Computer Symbol Manipulation

We have developed a general approach which applies the optimal control necessary conditions (Pontryagin's Principle and Pontryagin's necessary conditions), and generates *symbolically* the differential and algebraic equations governing the optimal nonlinear feedback controls *for the class of nonlinear dynamical systems* described by differential equations of the form

$$\dot{x} = \sum_{i=1}^{M} A_i x_i + B u \tag{10}$$

where x is an $n\times1$ state vector, u is an $m\times1$ control vector, M is the highest degree of polynomial nonlinearity, x_i is a column vector containing all distinct polynomial combinations (of degree i) of the elements of the state vector x. Our approach leads directly to symbolic equations satisfied by the optimal feedback gain matrices G_i in the polynomial expansion

$$u = \sum_{i=1}^{N} G_i x_i, \text{ where } \ G_i = -R^{-1} B^T K_i, \ \text{ and } \ \lambda = \sum_{i=1}^{N} K_i x_i \tag{11}$$

The $n\times1$ λ is a vector of Lagrange multipliers. It is significant to note that we can find, via symbol manipulation, a set of sequentially solvable, *general algebraic and/or differential equations* satisfied by the gains K_i where the matrices A_i, B, Q, R , or subsets thereof, can appear as *algebraic parameters*.

Solution of Problem I

We seek an optimal feedback control law to compute $u(x)$ for

$$\dot{x} = -x + \alpha x^2 + u, \quad \alpha > 0 \tag{12}$$

which minimizes the index $\quad J = \frac{1}{2} \int_0^{t_f} \left(x^2 + u^2 \right) dt \tag{13}$

$$H = \tfrac{1}{2}(x^2 + u^2) + \lambda(-x + \alpha x^2 + u)$$

$$\frac{\partial H}{\partial u} = 0 \quad \Rightarrow \quad u = -\lambda$$

The Pontryagin Necessary conditions are $\tag{15}$

$$\dot{\lambda} = -\frac{\partial H}{\partial x} = -x + \lambda - 2\alpha x \lambda$$

$$\dot{x} = \frac{\partial H}{\partial \lambda} = -x - \lambda + \alpha x^2$$

We seek a feedback form of the control law, specifically, we seek the polynomial gains K_i in

$$-u = \lambda = \sum_{i=1}^{N} K_i x^i \tag{16}$$

Substituting Eq. (16) into Eqs. (15), we are led to the homogeneous condition:

Substituting Eq. (16) into Eqs. (15), we are led to the homogeneous condition:

$[\dot{K}_1 - 2K_1 - K_1^2 + 1]x + [\dot{K}_2 - 3(1+K_1)K_2 + 3\alpha K_1]x^2 +$

$[\dot{K}_3 - 4(1+K_1)K_3 + 4\alpha K_1 - 2K_2^2]x^3 + \ldots + [\dot{K}_N - (N+1)(1+K_1)K_N - F_N(\alpha, K_1, \ldots, K_{N-1})]x^N = 0$

Since the above equation must hold at every point in the state space (i. e. , for all x), we conclude that all [ ]'ed coefficients must vanish independently, this provides

$$
\begin{array}{lll}
\dot{K}_1 - 2K_1 - K_1^2 + 1 = 0 & \Rightarrow & K_1 \\[4pt]
\dot{K}_2 - 3(1+K_1)K_2 = -3\alpha K_1 & \Rightarrow & K_2 \\[4pt]
\dot{K}_3 - 4(1+K_1)K_3 = -4\alpha K_1 + 2K_2^2 & \Rightarrow & K_3 \\[4pt]
\qquad\qquad \vdots & & \\[4pt]
\dot{K}_N - (N+1)(1+K_1)K_N = F_N(\alpha, K_1, K_2, \ldots, K_{N-1}) & \Rightarrow & K_N
\end{array} \tag{17}
$$

Notice the following structure and properties of Eqs. (17) :

● These equations can be solved *sequentially* for the K_i, all but the first are linear equations.

● The eqns can solved to arbitrary order, we have developed explicit recursions for F_N.

● The first equation for K_1, is a scalar Riccati equation (no surprise here!).

● If we impose $\alpha = 0$, we can verify that the optimal control reduces to $u = -K_1 x$.

● If $t_f \to \infty$, we can show that all K_i approach constants.

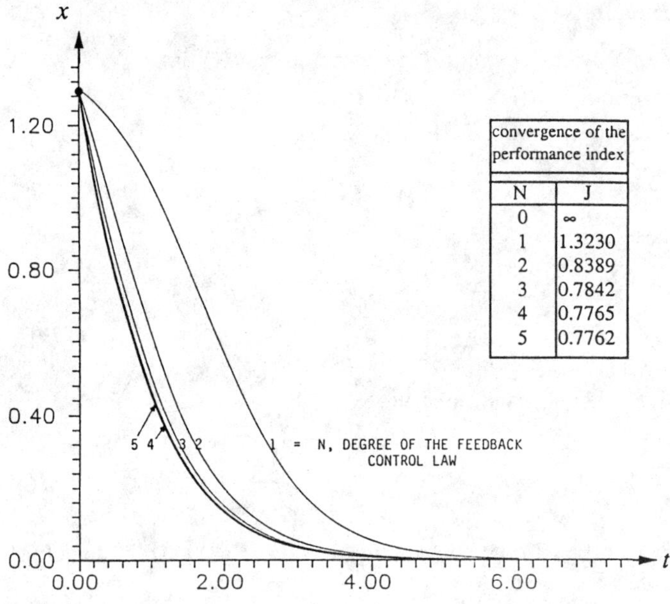

N	J
0	∞
1	1.3230
2	0.8389
3	0.7842
4	0.7765
5	0.7762

Figure 1. State Trajectory Versus Time: Effects of
Increasing the Degree of Nonlinear Feedback
($x(0) = 1.3$)

Solution of Problem II

We seek an optimal feedback control law to compute $u_1(x_1, x_2)$, $u_2(x_1, x_2)$ for the system:

$$\dot{x}_1 = -x_1 + x_1 x_2 + x_2^2 + u_1$$
$$\dot{x}_2 = -x_2 + x_1 x_2 + x_1^2 + u_2$$

which minimizes $J = \frac{1}{2} \int_0^\infty \sum_{i=1}^2 (x_i^2 + u_i^2)\, dt = \frac{1}{2}\int_0^\infty (x^T Q x + u^T R u)\, dt$ (18)

Eqs. (18) can be written as $\dot{x} = A_1 x_1 + A_2 x_2 + B u$, $Q = R = I$ (20)

where $x_1^T = x^T = [x_1 \ x_2]$, $x_2^T = [x_1^2 \ x_1 x_2 \ x_2^2]$, $u^T = [u_1 \ u_2]$, $A_1 = \begin{bmatrix} -1 & 0 \\ 0 & -1 \end{bmatrix}$, $A_2 = \begin{bmatrix} 1 & 1 & 0 \\ 0 & 1 & 1 \end{bmatrix}$, $B = \begin{bmatrix} 1 & 0 \\ 0 & 1 \end{bmatrix}$.

The Hamiltonian function is $H = \frac{1}{2}\left(x^T x + u^T u\right) + \lambda^T (A_1 x_1 + A_2 x_2 + B u)$ (21)

The Pontryagin nec. cond. are
$$\begin{cases} \dfrac{\partial H}{\partial u} = 0 \implies u = -B^T \lambda \\[2mm] \dot{\lambda} = -\dfrac{\partial H}{\partial x} = -x - A_1^T \lambda - \begin{bmatrix} 2x_1 & x_2 & 0 \\ 0 & x_1 & 2x_2 \end{bmatrix} A_2^T \lambda \\[2mm] \dot{x} = \dfrac{\partial H}{\partial \lambda} = A_1 x_1 + A_2 x_2 - B B^T \lambda \end{cases}$$ (22)

We seek to determine the feedback gains K_i in the control law expansion

$$u = -B^T\lambda = \sum_{i=1}^{N} G_i x_i, \text{ where } \lambda = \sum_{i=1}^{N} K_i x_i \text{ , } G_i = -B^T K_i \tag{23}$$

Substitution of the expansion of Eqs. (23) into the necessary conditions of Eq. (24) and collecting terms leads directly to a system of two homogeneous conditions of the following form

$$F_{1i}(K_1)x_1 + F_{2i}(K_1)x_2 + F_{11i}(K_1, K_2)x_1^2 + F_{12i}(K_1, K_2)x_1 x_2 + F_{22i}(K_1, K_2)x_2^2 + \ldots = 0 \text{ , } i = 1, 2$$

Requiring these conditions to hold at every point in the state space, we can set the coefficients of all powers and products of the elements of x to zero. Upon carrying through the algebra, we find the linear terms yield four algebraic equations $F_{11}(K_1) = F_{12}(K_1) = F_{21}(K_1) = F_{22}(K_1) = 0$, which are precisely the four elements of the Riccati equation [4,5]

$$K_1 A_1 + A_1^T K_1 - K_1 B R^{-1} B^T K_1 + Q = 0 \tag{24}$$

The Riccati equation can be solved for the symmetric linear gain matrix K_1. Setting the six quadratic term's coefficients to zero yields $F_{11i}(K_1, K_2) = F_{12i}(K_1, K_2) = F_{22i}(K_1, K_2) = 0$, $i = 1, 2$; we find that these six algebraic equations are linear in the six distinct elements of K_2 , and can be brought to the form of the linear system

$$[L_2(K_1)] \text{vec}\{K_2\} = R_2(K_1) \tag{25}$$

where $[L_2(K_1)]$ is a 6x6 matrix whose elements are functions of K_1, $R_2(K_1)$ is a 6x1 vector whose elements depend upon K_1, and $\text{vec}\{K_2\}$ is a 6x1 vector whose elements *are the six distinct elements of K_2*. Obviously, if $[L_2(K_1)]$ is of full rank, Eq. (25) can be inverted for $\text{vec}\{K_2\}$. To conserve space, we do not write out the algebraic equations for the elements of these matrices. We find that Eq. (25) generalizes; the higher order gains are determined by a *sequence* of linear equations of the form

$$[L_k(K_1, K_2, \ldots, K_{k-1})] \text{vec}\{K_k\} = R_k(K_1, K_2, \ldots, K_{k-1}), \quad k = 2, 3, \ldots N \tag{28}$$

For the case of particular A_i, B, of Eqs. (20), and $Q = R = I$, we have carried through the above developments and find the following numerical values for the first five control gains ($G_i = -R^T K_i$):

$$G_1 = -\begin{bmatrix} 0.41421 & 0 \\ 0 & 0.41421 \end{bmatrix}, \quad G_2 = -\begin{bmatrix} 0 & 0.39052 & 0.19526 \\ 0.19526 & 0.39052 & 0 \end{bmatrix}, \quad G_3 = -\begin{bmatrix} 0.12459 & 0.27022 & 0.22222 & 0.09007 \\ 0.09007 & 0.22222 & 0.27022 & 0.12459 \end{bmatrix}$$

$$G_4 = -\begin{bmatrix} 0.05125 & 0.17517 & 0.26213 & 0.17475 & 0.04379 \\ 0.04379 & 0.17475 & 0.26213 & 0.17517 & 0.05125 \end{bmatrix}, \quad G_5 = -\begin{bmatrix} 0.01656 & 0.08537 & 0.16605 & 0.16225 & 0.08302 & 0.01707 \\ 0.01707 & 0.08302 & 0.16225 & 0.16605 & 0.08537 & 0.01656 \end{bmatrix}$$

Since this system is highly nonlinear, simply ignoring the nonlinear terms and deriving an approximate linear optimal control law will be valid only near the origin. In Figure 2 we show graphs of the stable region (shaded) vs N and typical trajectories of the state and control variables for typical initial conditions, for controllers based upon linear ($N=1$) through quintic ($N=5$)

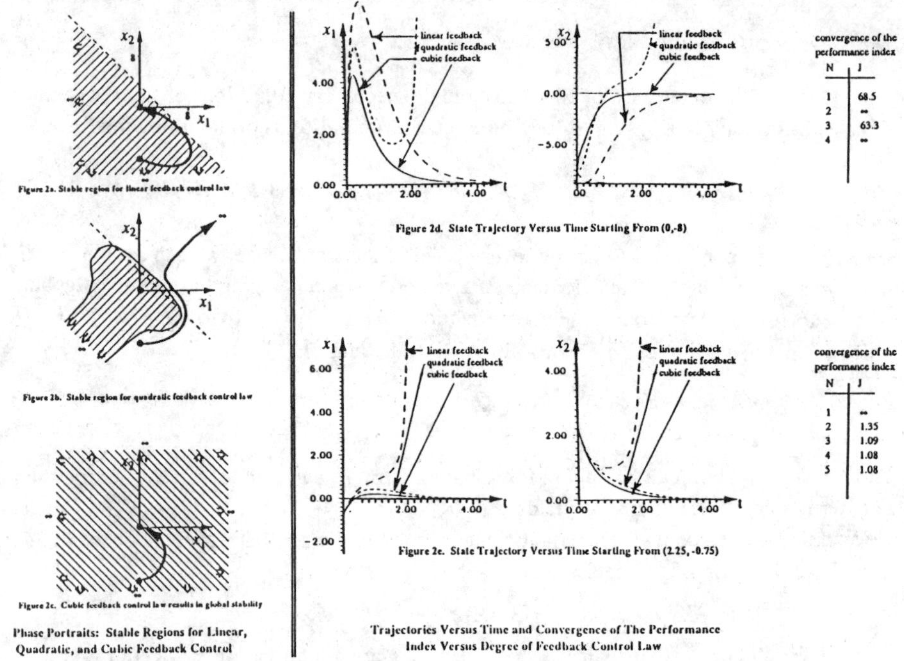

Figure 2. Stability Maps and Typical Trajectories for Problem II: Effects of Increasing The Degree of Nonlinear Feedback

$\mathcal{SOLUTION}$ OF $\mathbf{PROBLEM\ III}$ (NONLINEAR FEEDBACK LAW FOR LARGE & ATTITUDE MANEUVERS)

We seek to determine the feedback gains K_i in the control law expansion

$$\underline{u} = -R^{-1}B^T\underline{\lambda} \quad , \quad \text{for} \quad R = I$$

$$u = -B^T\lambda, \text{ where } \lambda = \sum_{i=1}^{N} K_i x_i \tag{29}$$

Upon substituting Eq. (29) into Eqs. (22), we are led to a matrix Riccati equation for the linear gain (K_1) matrix, and all subsequent matrices are given by *linear* equations in the form:

$$\dot{K}_i + [H_i(K_1)]K_i = F_i(K_1, K_2, \ldots, K_{i-1}), i = 2, 3,, \ldots, N$$

which can be solved sequentially for the gains.

For the first two gain matrices ($G_i = - R^{-1} B^T K_i$), where the K_i are from the expansion of Eq. (11), we find the following numerical values $(Q = R = I)$

$$G_1 = - \begin{bmatrix} 1.000 & 0 & 0 & 1.414 & 0 & 0 \\ 0 & 1.000 & 0 & 0 & 1.353 & 0 \\ 0 & 0 & 1.000 & 0 & 0 & 1.386 \end{bmatrix},$$

$$G_2 = -10^{-5} \begin{bmatrix} 0 & 0 & 0 & 0 & 0 & 0 & 0 & -3184 & 0 & 0 & -978 & 0 & 0 & 1893 & 0 & 0 & 0 & 0 & 0 & -.2078 & 0 \\ 0 & 0 & -2764 & 0 & 0 & -1118 & 0 & 0 & 0 & 0 & 0 & 2280 & 0 & 0 & 0 & 0 & -.2504 & 0 & 0 & 0 \\ 0 & 5948 & 0 & 0 & -1009 & 0 & 0 & 0 & -1063 & 0 & 0 & 0 & 0 & 0 & 0 & 0 & -.2259 & 0 & 0 & 0 & 0 \end{bmatrix}$$

Due to space limitations, we show only a few digits and do not display the third order gain matrix which was also computed. As is apparent, the particular system dynamics, inertias, and identity weight matrices resulted in sparse gain matrices, this pattern carries over to the cubic gains. This sparsity pattern is unaffected by variations of the elements in the diagonal inertia, Q, and R matrices, although, obviously, the numerical values of the gains are affected. The sparsity pattern of the optimal gain matrices allows us to write down the form of the feedback control law explicitly, as follows:

$$u_1 = \underbrace{G_1(1,1)q_1 + G_1(1,4)\omega_1}_{} + \underbrace{G_2(1,8)q_2 q_3 + G_2(1,11)q_2 \omega_3 + G_2(1,14)q_3 \omega_2 + G_2(1,20)\omega_2 \omega_3}_{} + \cdots$$
$$u_2 = \underbrace{G_1(2,2)q_2 + G_1(2,5)\omega_2}_{} + \underbrace{G_2(2,3)q_3 q_1 + G_2(2,13)q_3 \omega_1 + G_2(2,6)q_1 \omega_3 + G_2(2,18)\omega_3 \omega_1}_{} + \cdots$$
$$u_3 = \underbrace{G_1(3,3)q_3 + G_1(3,6)\omega_3}_{\text{linear feedback terms}} + \underbrace{G_2(3,2)q_1 q_2 + G_2(3,5)q_1 \omega_2 + G_2(3,9)q_2 \omega_1 + G_2(3,17)\omega_1 \omega_2}_{\text{quadratic (note "gyroscopic" structure) feedback terms}} + \underbrace{\cdots}_{\text{3rd \& HOT}}$$

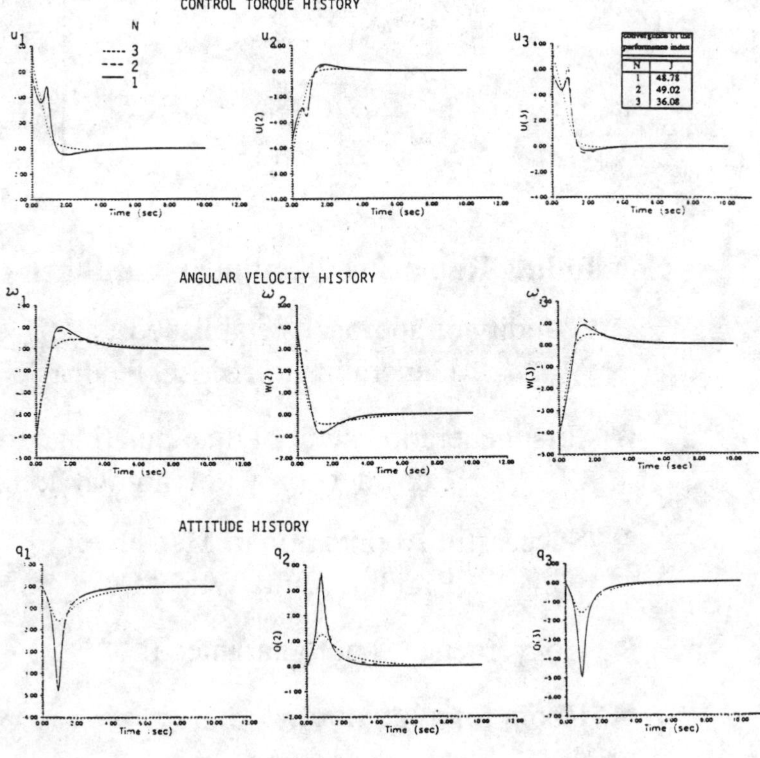

FIGURE 3   Perturbation Feedback Controlled Spacecraft Attitude Maneuver

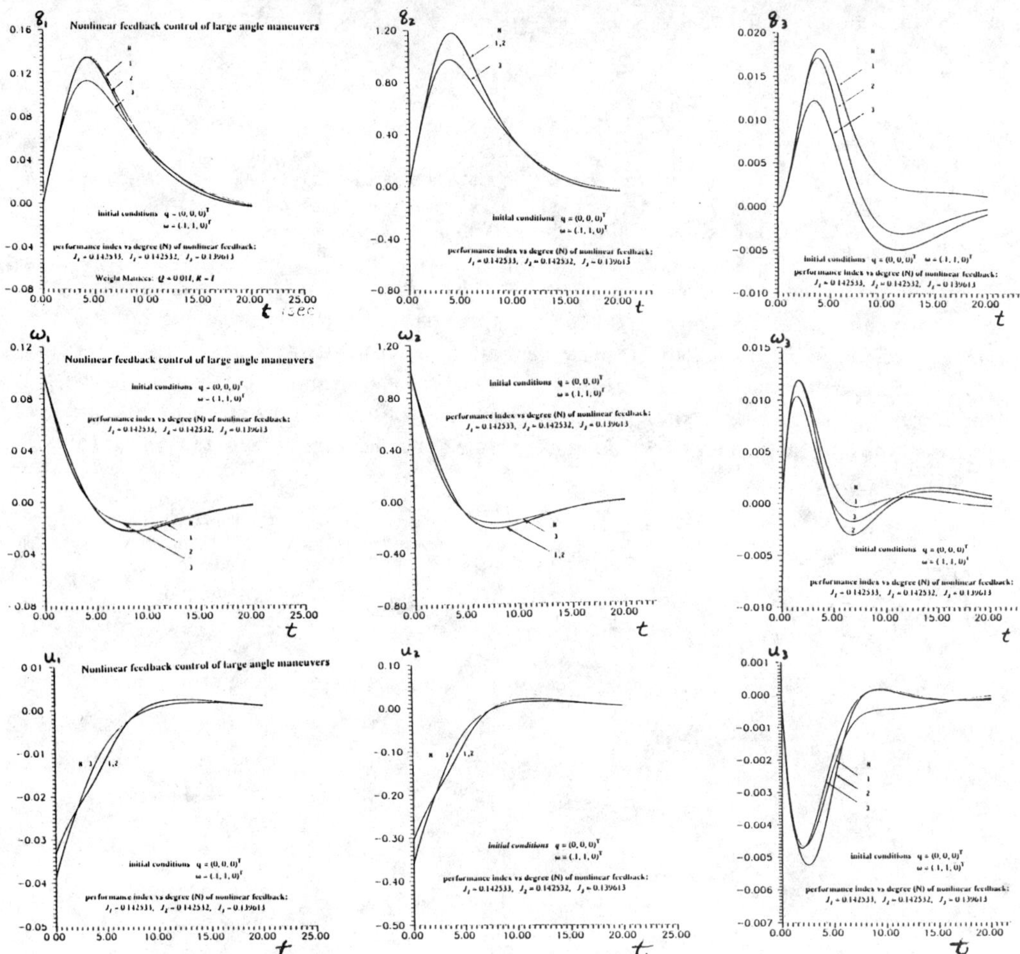

Concluding Remarks: Nonlinear Feedback Control

- Validity of approach established:
  Generalization of LQ feedback

- First generation symbol manipulation code:
  Considers only polynomial nonlinearities

- Successful Application to Maneuver:
  Rigid & Multi-Wheel S/C

- Convergence is not guaranteed!

- Room for a lot more work.

672

A Liapunov Stable Feedback Control Law for Large Angle Maneuvers

In Search of a Judicious Liapunov Function & a Stabilizing Control Law ...

For a rigid spacecraft, the equations of motion are

$$I_1 \dot{\omega}_1 = (I_2 - I_3)\omega_2\omega_3 + u_1 \qquad 2\dot{q}_1 = \omega_1 - \omega_2 q_3 + \omega_3 q_2 + q_1(q_1\omega_1 + q_2\omega_2 + q_3\omega_3)$$

$$I_2 \dot{\omega}_2 = (I_3 - I_1)\omega_3\omega_1 + u_2 \qquad 2\dot{q}_2 = \omega_2 - \omega_3 q_1 + \omega_1 q_3 + q_2(q_1\omega_1 + q_2\omega_2 + q_3\omega_3) \quad (1)$$

$$I_3 \dot{\omega}_3 = (I_1 - I_2)\omega_1\omega_2 + u_3 \qquad 2\dot{q}_3 = \omega_3 - \omega_1 q_2 + \omega_2 q_1 + q_3(q_1\omega_1 + q_2\omega_2 + q_3\omega_3)$$

Motivated by the total system energy, we investigate the Liapunov function

$$U = \frac{1}{2}(I_1\omega_1^2 + I_2\omega_2^2 + I_3\omega_3^2) + A(q_1^2 + q_2^2 + q_3^2) \equiv \text{kinetic energy} + A\tan^2\frac{\phi}{2} \qquad (2)$$

It is obvious that U is positive definite except at the desired state $q_i = \omega_i = 0$. Differentiation of (2) and substitution of Eqs. (1) leads directly to the following (beautiful!) expression for $\dot{U}$

$$\dot{U} = \sum_{i=1}^{3} \omega_i[u_i + Aq_i(1 + q_1^2 + q_2^2 + q_3^2)] \qquad (3)$$

From which we find a globally stabilizing feedback control for large angle attitude control:

$$u_i = -[k_i\omega_i + Aq_i(1 + q_1^2 + q_2^2 + q_3^2)], \ i=1,2,3 \quad \Rightarrow \quad \dot{U} = -(k_1\omega_1^2 + k_2\omega_2^2 + k_2\omega_2^2) \quad (4)$$

"Liapunov Law" Feedback Controlled Large Angle Maneuver

$$q(0) = (5,5,5)^T, \ \omega(0) = (0,0,0)^T, \ k_1 = k_2 = k_3 = A = 1$$

Rodriguez Attitude Parameters   **Angular Velocity**   **Control Torques**

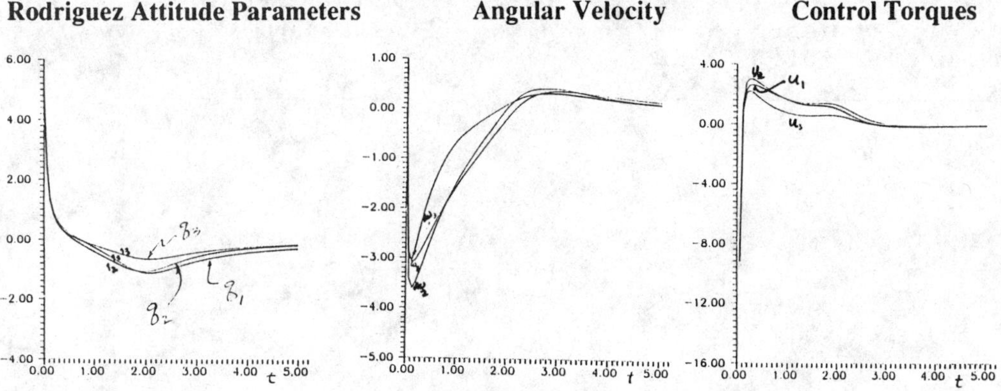

673

A Liapunov Approach to Control of Distributed Parameter Systems

Application to a Hub/Appendage Structure

TAMU Hub/Appendage Formulation of Equations of Motion

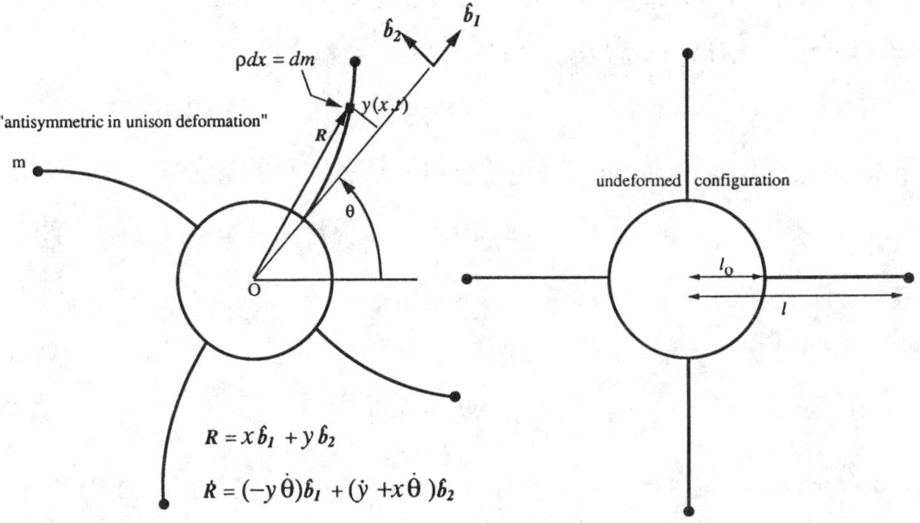

$$R = x\,\hat{b}_1 + y\,\hat{b}_2$$

$$\dot{R} = (-y\,\dot{\theta})\hat{b}_1 + (\dot{y} + x\,\dot{\theta})\hat{b}_2$$

$$L = T - V = T_{hub} + T_{beam} + T_{tip} - V_{beam} = \tfrac{1}{2}I_{hub}\dot{\theta}^2 + 4\left[\tfrac{1}{2}\int_{l_o}^{l}\dot{R}\cdot\dot{R}\rho\,dx + \tfrac{1}{2}m\dot{R}_{tip}\dot{R}_{tip} - \tfrac{1}{2}\int_{l_o}^{l}EI\left(\frac{\partial^2 y}{\partial x^2}\right)^2\rho\,dx\right]$$

Apply the Extended Hamilton's Principle: $\int_{t_1}^{t_2}(\delta L + \delta W)\,dt + BCs \implies PDE\ equations\ of\ motion \implies$

TAMU Hub/Appendage Equations of Motion

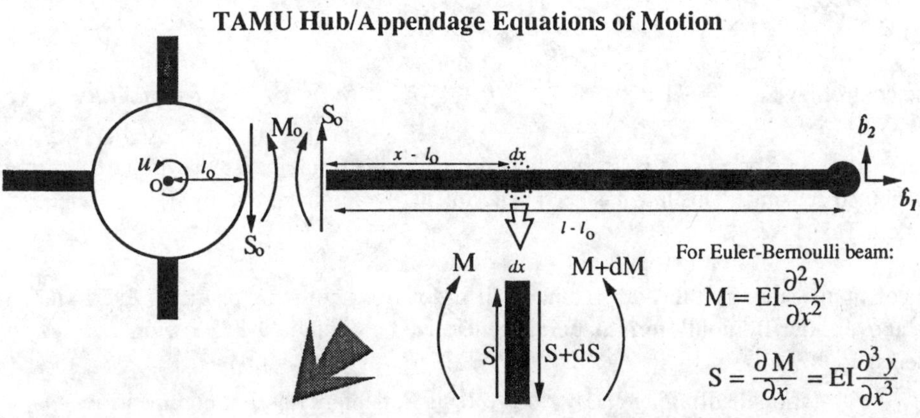

Equations of motion:

$$I_{hub}\frac{d^2\theta}{dt^2} = u + 4(M_o - S_o l_o) \quad + HOT$$

$$-(M_o - S_o l_o) = \int_{l_o}^{l}\rho x\left(\frac{\partial^2 y}{\partial t^2} + x\frac{d^2\theta}{dt^2}\right)dx + ml\left(l\frac{d^2\theta}{dt^2} + \frac{d^2 y}{dt^2}\Big|_l\right) + HOT$$

$$\rho\left(\frac{\partial^2 y}{\partial t^2} + x\frac{d^2\theta}{dt^2}\right) + EI\frac{\partial^4 y}{\partial x^4} = 0 \quad + HOT$$

Boundary conditions:

$$at\ x = l_o:\ y = \frac{\partial y}{\partial x} = 0$$

$$at\ x = l:$$

$$(moment)\ \frac{\partial^2 y}{\partial x^2} = 0$$

$$(shear)\ \frac{\partial^3 y}{\partial x^3} = \frac{m}{EI}\left(l\frac{d^2\theta}{dt^2} + \frac{d^2 y}{dt^2}\Big|_l\right)$$

Motivated by the total system energy

$$2E = 2(T+V) = I_{hub}\left(\frac{d\theta}{dt}\right)^2 + 4\left[\int_{l_o}^{l}\rho\left(\frac{\partial y}{\partial t} + x\frac{d\theta}{dt}\right)^2 dx + \int_{l_o}^{l}EI\left(\frac{\partial^2 y}{\partial x^2}\right)^2 dx + m\left(l\frac{d\theta}{dt} + \frac{dy}{dt}|_l\right)^2\right]$$

and the recent work of Fujii, we "investigate the wisdom" of the candidate Liapunov fct.:

$$2U = a_1 I_{hub}\dot{\theta}^2 + a_2(\theta - \theta_o)^2 + 4a_3\left[\int_{l_o}^{l}\rho\left(\frac{\partial y}{\partial t} + x\dot{\theta}\right)^2 dx + \int_{l_o}^{l}EI\left(\frac{\partial^2 y}{\partial x^2}\right)^2 dx + m(l\dot{\theta} + \frac{dy}{dt}|_l)^2\right] \quad (4)$$

It is obvious by inspection that choosing $a_i > 0$ guarantees that U>0, and that U=0 is the desired state. Differentiation, substitution of the Eqs of motion & some calculus lead to

$$\dot{U} = \frac{dU}{dt} = \dot{\theta}\left[a_1 u + a_2(\theta - \theta_o) + 4(a_3 - a_1)(l_o S_o - M_o)\right] \quad (5)$$

Since we require that $\dot{U}<0$ to guarantee stability, we choose the control torque u as

$$u = -\frac{1}{a_1}\left[a_2(\theta - \theta_o) + a_4\dot{\theta} + 4(a_3 - a_1)(l_o S_o - M_o)\right]$$

or, we see that the following *linear, discrete* feedback law globally stabilizes this system:

$$u = -\left[g_1(\theta - \theta_o) + g_2\dot{\theta} + g_3(l_o S_o - M_o)\right], \quad g_i \geq 0 \text{ for stability} \quad (6)$$

Discussion

The control law: $u = -\left[g_1(\theta - \theta_o) + g_2\dot{\theta} + g_3(l_o S_o - M_o)\right], \quad g_i \geq 0$ *for stability*   (6)

is very attractive because it is linear and controls the distributed system without requiring distributed sensing. This law has been experimentally demonstrated by Fujii and in our work at Texas A&M.

It is of significance that this law maintains it's globally stabilizing character even when the above Euler/Bernoulli formulation is generalized to include the following *nonlinear* effects:

        rotational stiffening and coriolis effects (kinematic/inertial nonlinearities)
        aerodynamic damping/drag

as well as a more general accounting for the following linear effects:

        modifications of the formulation to include shear deformation and rotary inertia
        finite inertia of the tip mass.

In short, closed-loop stability using this law is very forgiving of modeling assumptions!

Concluding Remarks: Liapunov-based Control for DPS

- A significant example has been considered

- Linear output feedback law found for DPS w/o discretizing the structure ... no spillover!

- Successful experimental demonstration Maneuver

- Has been applied successfully to nonlinear DPS

- Room for more work: Stability vs Optimality

PUBLICATIONS OF THE AMERICAN ASTRONAUTICAL SOCIETY

Following are the principal publications of the American Astronautical
Society:

JOURNAL OF THE ASTRONAUTICAL SCIENCES (1954-)

Published quarterly and distributed by AAS Business Office, 6212-B Old
Keene Mill Court, Springfield, VA 22152. Back issues available from
Univelt, Inc., P.O. Box 28130, San Diego, CA 92128.

SPACE TIMES (1986-)

Published bi-monthly and distributed by AAS Business Office, 6212-B Old
Keene Mill Court, Springfield, VA 22152., Virginia 22152

AAS NEWSLETTER (1962-1985)

Incorporated in *Space Times*. Back issues available from AAS Business
Office, 6212-B Old Keene Mill Court, Springfield, VA 22152.

ASTRONAUTICAL SCIENCES REVIEW (1959-1962)

Incorporated in *Space Times*. Back issues still available from Univelt,
Inc., P.O. Box 28130, San Diego, CA 92128.

ADVANCES IN THE ASTRONAUTICAL SCIENCES (1957-)

Proceedings of major AAS technical meetings. Published and distributed
for the American Astronautical Society by Univelt, Inc., P.O. Box 28130,
San Diego, CA 92128.

SCIENCE AND TECHNOLOGY SERIES (1964-)

Supplement to *Advances in the Astronautical Sciences*. Proceedings and
monographs, most of them based on AAS technical meetings. Published and
distributed for the American Astronautical Society by Univelt, Inc., P.O.
Box 28130, San Diego, CA 92128

AAS HISTORY SERIES (1977-)

Supplement to *Advances in the Astronautical Sciences*. Selected works in
the field of aerospace history under the editorship of R. Cargill Hall.
Published and distributed for the American Astronautical Society by
Univelt, Inc., P.O. Box 28130, San Diego, CA 92128.

AAS MICROFICHE SERIES (1968-)

Supplement to *Advances in the Astronautical Sciences*. Consists principally of technical papers not included in the hard-copy volume. Published and distributed for the American Astronautical Society by Univelt,
Inc., P.O. Box 28130, San Diego, CA 92128.

Subscriptions to the *Journal* and the *Space Times* should be ordered from
the AAS Business Office. Back issues of the *Journal* and all books and
microfiche should be ordered from Univelt, Inc.

ADVANCES IN THE ASTRONAUTICAL SCIENCES SERIES (1957-   )

ISSN 0065-3438, LIBRARY OF CONGRESS CARD NO. 57-43769

Proceedings of Major AAS Technical Meetings

Vol. 1 Third Annual AAS Meeting, Dec. 6-7, 1956, New York, NY, 1957, 184p., ed. Norman V. Petersen, Microfiche only, $20 *(ISBN 0-87703-002-2)*

Vol. 2 Fourth Annual AAS Meeting, Jan. 29-31, 1958, New York, NY, 1958, 440p., eds. Norman V. Petersen, Horace Jacobs, Microfiche only, $20 *(ISBN 0-87703-003-0)*

Vol. 3 First Western National AAS Meeting, Aug. 18-19, 1958 530p., eds. Norman V. Petersen, Horace Jacobs, Microfiche only, $20 *(ISBN 0-87703-004-9)*

Vol. 4 Fifth Annual AAS Meeting, Dec. 27-31, 1958, Washington, D.C., 1959, 462p., ed. Horace Jacobs, Microfiche only, $20 *(ISBN 0-87703-005-7)*

Vol. 5 Second Western National AAS Meeting, Aug. 4-5, 1959, Los Angeles, CA, 1960, 364p., ed. Horace Jacobs, Microfiche only, $20 *(ISBN 0-87703-006-3)*

Vol. 6 Sixth Annual AAS Meeting, Jan. 18-21, 1960, New York, NY, 1961, 968p., eds. Horace Jacobs and Eric Burgess, Hard Cover $45 *(ISBN 0-87703-007-3)*

Vol. 7 Third Western National AAS Meeting, Aug. 4-5, 1960, Seattle, WA, 1961, 464p., eds. Horace Jacobs and Eric Burgess, Microfiche only, $20 *(ISBN 0-87703-008-1)*

Vol. 8 Seventh Annual AAS Meeting, Jan. 16-18, 1961, Dallas, TX, 1963, 602p., ed. Horace Jacobs, Microfiche only, $20 *(ISBN 0-87703-009-X)*

Vol. 9 Fourth Western Regional AAS Meeting, Aug. 1-3, 1961, San Francisco, CA, 1963, 910p., ed. Eric Burgess, Hard Cover $45 *(ISBN 0-87703-010-3)*

Vol. 10 Manned Lunar Flight (AAS/AAAS Symposium) Dec. 19, 1961, Denver, CO, 1963, 310p., eds. George W. Morgenthaler and Horace Jacobs, Hard Cover $35 *(ISBN 0-87703-011-1)*

Vol. 11 Eighth Annual AAS Meeting, Jan. 16-18, 1962, Washington, D.C., 1963, 808p., ed. Horace Jacobs, Hard Cover $45 *(ISBN 0-87703-012-X)*

Vol. 12 Scientific Satellites - Mission and Design (AAS/AAAS Symposium), Dec. 27, 1962, Philadelphia, PA, 1963, 262p., ed. Irving E. Jeter, Hard Cover $25 *(ISBN 0-87703-013-8)*

Vol. 13 Interplantetary Missions, 9th Annual AAS Meeting, Jan. 15-17, 1963, Los Angeles, CA, 1963, 690p., ed. Eric Burgess, Hard Cover $45 *(ISBN 0-87703-014-6)*

Vol. 14 Second AAS Symposium on Physical and Biological Phenomena under Zero G Conditions, Jan. 18, 1963, Los Angeles, CA, 1963, 382p., eds. Elliot T. Benedikt and Robert W. Halliburton, Hard Cover $30 *(ISBN 0-87703-015-4)*

Vol. 15 Exploration of Mars Symposium, Jun. 6-7, 1963, Denver, CO, 1963, 634p., ed. George W. Morgenthaler, Hard Cover $45 *(ISBN 0-87703-016-2)*

Vol. 16 Space Rendezvous, Rescue, and Recovery Symposium, Sept. 10-12, 1963, Edwards, CA 1963, 1408p., ed. Norman V. Petersen, Hard Cover, **Part 1**, 1028p., $45 *(ISBN 0-87703-017-0)*; **Part 2**, 380p., $30 *(ISBN 0-87703-018-9)*

Vol. 17 Bioastronautics - Fundamental and Practical Problems (AAS/AAAS Symposium), Dec. 30, 1963, Cleveland, OH, 1964, 128p., ed. William C. Kaufman, Microfiche only, $10 *(ISBN 0-87703-019-7)*

Vol. 18 Lunar Flight Programs, 10th Annual AAS Meeting, May 4-7, 1964, New York, NY, 1964, 630p., ed. Ross Fleisig, Hard Cover $45 *(ISBN 0-87703-020-0)*

Vol. 19 Unmanned Exploration of the Solar System Symposium, Feb. 8-10, 1965, Denver, CO, 1965, 1000p., eds. George W. Morgenthaler, Robert G. Morra, Hard Cover $45 *(ISBN 0-87703-021-9)*

Vol. 20 Post Apollo Exploration, 11th Annual AAS Meeting, May 3-6, 1965, Chicago, IL, 1966, 1220p., ed. Francis Narin, Microfiche only, **Part l**, 572p., $30 *(ISBN 0-87703-022-7)*; **Part 2**, 648p., $35 *(ISBN 0-87703-023-5)*

Vol. 21 Practical Space Applications Symposium, Feb. 21-23, 1966, San Diego, CA, 1967, 508p., ed. Lawrence L. Kavanau, Hard Cover $40 *(ISBN 0-87703-024-3)*

Vol. 22 The Search for Extraterrestrial Life, 12th Annual AAS Meeting, May 23-25, 1966, Anaheim, CA, 1967, 388p., ed. James S. Hanrahan, Microfiche only $30 *(ISBN 0-87703-025-1)*; Microfiche Suppl. (Vol. 1 AAS Microfiche Series) $12 *(ISBN 0-87703-132-0)*

Vol. 23 Commercial Utilization of Space, 13th Annual AAS Meeting, May 1-3, 1967, Dallas, TX, 1968, 512p., eds. J. Ray Gilmer, Alfred M. Mayo, Ross C. Peavey, Hard Cover *(ISBN 0-87703-026-X)*; plus Microfiche Suppl. (Vol. 3 AAS Microfiche Series) $60 *(ISBN 0-87703-216-5)*

Vol. 24 Exploitation of Space for Experimental Research, 14th Annual AAS Meeting, May 13-15, 1968, Dedham, MA, 1968, 363p., ed. Harry Zuckerberg, Hard Cover $30 *(ISBN 0-87703-027-8)*

Vol. 25 Advanced Space Experiments, Sept. 16-18, 1968, Ann Arbor, MI, 1969, 530p., eds. O. Lyle Tiffany and Eugene M. Zaitzeff, Hard Cover $40 *(ISBN 0-87703-028-6)*

Vol. 26 Planning Challenges of the 70's in Space, 15th Annual AAS Meeting, Jun. 17-20, 1969, Denver, CO, 1970, 470p., eds. George W. Morgenthaler and Robert G. Morra, Hard Cover $35 *(ISBN 0-87703-053-7)*; Microfiche Suppl. (Vol. 14 AAS Microfiche Series) $20 *(ISBN 0-87703-130-4)*

Vol. 27/28 Space Stations (**v27**) and Space Shuttles and Interplanetary Missions (**v28**), 16th Annual AAS Meeting, Jun. 8-10, 1970, Anaheim, CA, 1970, **Vol. 27**, eds. Lewis Larmore and Robert L. Gervais, 6060p., Hard Cover $45 *(ISBN 0-87703-054-5)*; **Vol. 28**, eds. Lewis Larmore and Robert L. Gervais, 488p., Hard Cover $35 *(ISBN 0-87703-055-3)*

Vol. 29 The Outer Solar System, 17th Annual AAS Meeting, Jun. 28-30, 1971, Seattle, WA, 1971, 1358p., $85; ed. Juris Vagners, Hard Cover, **Part 1**, 618p., $40 *(ISBN 0-87703-059-6)*; **Part 2**, 740p., $45 *(ISBN 0-87703-060-X)*

Vol. 30 International Congress of Space Benefits, 19th Annual AAS Meeting, Jun. 19-21, 1973, Dallas, TX, 1974, 528p., ed. Francis S. Johnson, Hard Cover $40 *(ISBN 0-87703-065-0)*

Vol. 31 The Skylab Results, 20th Annual AAS Meeting, Aug. 20-22, 1974, Los Angeles, CA, 1975, 1174p., eds. William C. Schneider and Thomas E. Hanes, Microfiche only (ISBN 0-87703-072-3); Plus Microfiche Suppl. (Vol. 22 AAS Microfiche Series) $60 *(ISBN 0-87703-043-6)*

Vol. 32 Space Shuttle Missions of the 80's, 21st Annual AAS Meeting, Aug. 26-28, 1975, Denver, CO, 1977, 1364p., eds. William J. Bursnall, George W. Morgenthaler, Gerald E. Simonson, Hard Cover, **Part 1**, 598p., $40 *(ISBN 0-87703-078-2)*; Hard Cover, **Part 2**, 766p., $55 *(ISBN 0-87703-087-1)*; Microfiche Suppl. (Vol. 25 AAS Microfiche Series $65 *(ISBN 0-87703-133-9)*

Vol. 33 AAS/AIAA Astrodynamics Conference, July 28-30, 1975, Nassau, Bahamas, 1976, 390p., eds. William F. Powers, Herbert E. Rauch, Byron D. Tapley, Carmelo E. Velez, Hard Cover $35 *(ISBN 0-87703-079-0)*; Microfiche Suppl. (Vol. 26 AAS Microfiche Series) $40 *(ISBN 0-87703-142-8)*

Vol. 34 Apollo Soyuz Mission Report, 1977, 336p., ed. Chester M. Lee, Hard Cover $35 *(ISBN 0-87703-089-8)*

Vol. 35 The Bicentennial Space Symposium - New Themes for Space: Mankind's Future Needs and Aspirations, 22nd AAS Meeting, Oct. 6-8, 1976, Washington, D.C., 1977, 242p., ed. William C. Schneider, Hard Cover $25 *(ISBN 0-87703-090-1)*

Vol. 36 The Industrialization of Space, 23rd Annual AAS Meeting, Oct. 18-20, 1977, San Francisco, CA, 1978, 1160p., eds. Richard A. Van Patten, Paul Siegler, Edward V.B. Stearns, Hard Cover, **Part 1**, 610p., $55 *(ISBN 0-87703-094-4)*; Hard Cover, **Part 2**, 550p., $45 *(ISBN 0-87703-095-2)*; Microfiche Suppl. (Vol. 28 AAS Microfiche Series) $15 *(ISBN 0-87703-121-5)*

Vol. 37 Space Shuttle and Spacelab Utilization, What are the Near-Term and Long-Term Benefits for Mankind?, 16th Goddard Memorial Symposium, 24th Annual AAS Meeting, March 8-10, 1978, Washington, D.C., 1978, 865p., eds. George W. Morgenthaler and Manfred Hollstein, Hard Cover, **Part 1**, 400p., $40 *(ISBN 0-87703-096-0)*; Hard Cover, **Part 2**, 465p., $45 *(ISBN 0-87703-097-9)*

Vol. 38 The Future U.S. Space Program, 25th Anniversary Conference, Oct. 20 - Nov. 2, 1978, Houston, TX, 1979, 880p., eds. Richard S. Johnston, Albert Naumann, Jr., Clay W. G. Fulcher, Hard Cover, **Part 1**, 444p., $45 *(ISBN 0-87703-098-7)*; Hard Cover, **Part 2**, 436p., $40 *(ISBN 0-87703-099-5)*; Microfiche Suppl. (Vol. 30 AAS Microfiche Series) $15 *(ISBN 0-87703-129-0)*

Vol. 39 Guidance and Control 1979, Feb. 24-28, 1979, Keystone, CO, 1979, 492p., ed. Robert D. Culp, Hard Cover $45 *(ISBN 0-87703-100-2)*; Microfiche Suppl. (Vol. 31 AAS Microfiche Series) $10 *(ISBN 0-87703-128-2)*

Vol. 40 AAS/AIAA Astrodynamics Conference, Jun. 25-27, 1979, Provincetown, MA, 1980, 996p., eds. Paul A. Penzo, Bernard Kaufman, Louis Friedman, Richard Battin, Hard Cover, **Part 1**, 494p., $45 *(ISBN 0-87703-107-X)*; Soft Cover $35 *(ISBN 0-87703-108-8)*; Hard Cover, **Part 2**, 502p., $45 *(ISBN 0-87703-109-6)*; Soft Cover $35 *(ISBN 0-87703-110-X)*; Microfiche Suppl. (Vol. 32 AAS Microfiche Series) $20 *(ISBN 0-87703-139-8)*

Vol. 41 Space Shuttle: Dawn of an Era, 26th Annual AAS Meeting, Oct. 29-Nov. 1, 1979, Los Angeles, CA, 1980, 980p., eds. William F. Rector, III and Paul A. Penzo, Hard Cover, **Part 1**, 452p., $45 *(ISBN 0-87703-111-8)*; Soft Cover $35 *(ISBN 0-87703-112-6)*; Hard Cover, **Part 2**, 528p., $55 *(ISBN 0-87703-113-4)*; Soft Cover $40 *(ISBN 0-87703-114-2)*; Microfiche Suppl. (Vol. 33 AAS Microfiche Series) $10 *(ISBN 0-87703-136-3)*

Vol. 42 Guidance and Control 1980, Feb. 17-21, 1980, Keystone, CO, 1980, 738p., ed. Louis A. Morine, Hard Cover $60 *(ISBN 0-87703-137-1)*; Soft Cover $45 *(ISBN 0-87703-138-X)*

Vol. 43 Shuttle/Spacelab - The New Transportation System and its Utilization, (3rd DGLR/AAS Symposium), Apr. 28-30, 1980, Hannover, Germany, 1981, 342p., eds. Dietrich E. Koelle and George V. Butler, Hard Cover $45 *(ISBN 0-87703-144-4)*; Soft Cover $35 *(ISBN 0-87703-146-0)*

Vol. 44 Space--Enhancing Technological Leadership, 27th Annual AAS Meeting, Oct. 20-23, 1980, Boston, MA, 1981, 580p., ed. Lawrence P. Greene, Hard Cover $65 *(ISBN 0-87703-147-9)*; Soft Cover $50 *(ISBN 0-87703-148-7)*; Microfiche Suppl. (Vol. 35 AAS Microfiche Series) $10 *(ISBN 0-87703-164-9)*

Vol. 45 Guidance and Control 1981, Jan. 31- Feb. 4, 1981, Keystone, CO, 1981, 506p., ed. Edward J. Bauman, Hard Cover $60 *(ISBN 0-87703-150-9)*; Soft Cover $50 *(ISBN 0-87703-151-7)*; Microfiche Suppl. (Vol. 36 AAS Microfiche Series) $15 *(ISBN 0-87703-156-8)*

Vol. 46 AAS/AIAA Astrodynamics Conference, Aug. 3-5, 1981, North Lake Tahoe, NV, 1982, 1124p., eds. Alan L. Friedlander, Paul J. Cefola, Bernard Kaufman, Walt Williamson, G.T. Tseng, Hard Cover, **Part 1**, 552p., $55 *(ISBN 0-87703-159-2)*; Soft Cover $45 *(ISBN 0-87703-160-6)*; Hard Cover, **Part 2**, 572p., $55 *(ISBN 0-87703-161-4)*; Soft Cover $45 *(ISBN 0-87703-162-2)*; Microfiche Suppl. (Vol. 37 AAS Microfiche Series) $40 *(ISBN 0-87703-163-0)*

Vol. 47 Leadership in Space - For Benefits on Earth, 28th Annual AAS Meeting, Oct. 26-29, 1981, San Diego, CA, 1982, 310p., ed. William F. Rector, III, Hard Cover $45 *(ISBN 0-87703-168-1)*; Soft Cover $35 *(ISBN 0-87703-169-X)*

Vol. 48 Guidance And Control 1982, Jan. 30 - Feb. 3, 1982, Keystone, CO, 1982, 558p., eds. Robert D. Culp, Edward J. Bauman, W. E. Dorroh, Jr., Hard Cover $65 *(ISBN 0-87703-170-3)*; Soft Cover $50 *(ISBN 0-87703-171-1)*; Microfiche Suppl. (Vol. 38 AAS Michrofiche Series) $10 *(ISBN 0-87703-180-0)*

Vol. 49 Spacelab, Space Platforms, and the Future, Fourth AAS/DGLR Symposium and 20th Goddard Memorial Symposium, Mar. 17-19, 1982, Greenbelt, MD, 1982, 502p., eds. Peter M. Bainum, Dietrich E. Koelle, Hard Cover $55 *(ISBN 0-87703-174-6)*; Soft Cover $45 *(ISBN 0-87703-175-4)*; Microfiche Suppl. (Vol. 42 AAS Microfiche Series) $15 *(ISBN 0-87703-181-9)*

Vol. 50 Proceedings on an International Symposium on Engineering Sciences and Mechanics, Dec. 29-31, Tainan, Taiwan, 1983, **two parts**, 1570p., eds. Han-Min Hsia, Richard W. Longman, You-Li Chou, Hard Cover $120 *(ISBN 0-87703-176-2)*; Microfiche Suppl. (Vol. 43 AAS Microfiche Series) $10 *(ISBN 0-87703-215-7)*

Vol. 51 Guidance and Control 1983, Feb. 5-9, 1983, Keystone, CO, 1983, 494p., eds. Edward J. Bauman, Zubin W. Emsley, Hard Cover $60 *(ISBN 0-87703-182-7)*; Soft Cover $50 *(ISBN 0-87703-183-5)*; Microfiche Suppl. (Vol. 44 AAS Microfiche Series) $10 *(ISBN 0-87703-214-9)*

Vol. 52 Developing the Space Frontier, 29th Annual AAS Meeting, Oct. 25-27, 1982, Houston, TX, 1983, 436p., eds. Albert Naumann, Grover Alexander, Hard Cover $55 *(ISBN 0-87703-189-4)*

Vol. 53 Space Manufacturing 1983, May 9-12, 1983, Princeton, NJ, 1983, 496p., eds. James D. Burke, April S. Whitt, Hard Cover $60 *(ISBN 0-87703-188-6)*; Soft Cover $50 *(0-87703-189-4)*

Vol. 54 AAS/AIAA Astrodynamics Conference, Aug. 22-25, 1983, Lake Placid, NY, 1984, **two parts**, 1370p., eds. G.T. Tseng, Paul J. Cefola, Peter M. Bainum, David A. Levinson, Hard Cover $120 *(ISBN 0-87703-190-8)*; Soft Cover $90 *(ISBN 0-87703-191-6)*; Microfiche Suppl. (Vol. 45 AAS Microfiche Series) $40 *(ISBN 0-87703-192-4)*

Vol. 55 Guidance and Control 1984, Feb. 4-8, 1984, Keystone, CO, 1984, 500p., eds. Robert D. Culp, Parker S. Stafford, Hard Cover $60 *(ISBN 0-87703-199-1)*; Soft Cover $50 *(ISBN 0-87703-200-9)*; Microfiche Suppl. (Vol. 48 AAS Microfiche Series $15 *(ISBN 0-87703-201-7)*

Vol. 56 From Spacelab to Space Station, Fifth DGLR/AAS Symposium, Oct. 3-5, 1984, Hamburg, Germany, 1985, 270p., eds. H. Stoewer, Peter M. Bainum, Microfiche Only $30 *(ISBN 0-87703-209-2)*

Vol. 57 Guidance and Control 1985, Feb. 2-6, 1985, Keystone, CO, 1985, 618p., eds. Robert D. Culp, Edward J. Bauman, Charles A. Cullian, Hard Cover $65 *(ISBN 0-87703-211-4)*; Soft Cover $50 *(ISBN 0-87703-212-2)*; Microfiche Suppl. (Vol. 50 AAS Microfiche Series) $15 *(ISBN 0-87703-213-0)*

Vol. 58 AAS/AIAA Astrodynamics Conference, Aug. 12-15, 1985, Vail, CO, 1986, **two parts**, 1556p., eds. Bernard Kaufman, Joseph J.F. Liu, Robert A. Calico, Felix R. Hoots, Hard Cover $140 *(ISBN 0-87703-245-9)*; Soft Cover $110 *(ISBN 0-87703-246-7)*; Microfiche Suppl. (Vol. 51 AAS Microfiche Series); $60 *(ISBN 0-87703-247-5)*

Vol. 59 Space Station Beyond IOC, 32nd Annual AAS Meeting, Nov 6-7, 1985, Los Angeles, CA, 1986, 188p., ed. M. Jack Friedenthal, Hard Cover $40 *(ISBN 0-87703-252-1)*; Soft Cover $30 *(ISBN 0-87703-253-X)*

Vol. 60 Space Exploitation and Utilization, First AAS/JRS Symposium, Dec. 15-19, 1985, Honolulu, HI, 1986, 740p., eds. Gayle L. May, Peter M. Bainum, Kenji Ikeda, Tamiya Nomura, Tatsuo Yamanaka, Ryojiro Akiba, Hard Cover $70 *(ISBN 0-87703-254-8)*; Soft Cover $55 *(ISBN 0-87703-255-6)*; Microfiche Suppl. (Vol. 52 AAS Microfiche Series) $10 *(ISBN 0-87703-256-4)*

Vol. 61 Guidance and Control 1986, Feb. 1-5, 1986, Keystone, CO, 1986, 460p., eds. Robert D. Culp, John C. Durrett, Hard Cover $60 *(ISBN 0-87703-257-2)*; Soft Cover $50 *(ISBN 0-87703-258-0)*; Microfiche Suppl. (Vol. 53 AAS Microfiche Series) $10 *(ISBN 0-87703-259-9)*

Vol. 62 Tethers in Space, Proceedings of First International Conference on Tethers in Space (NASA & PSN Sponsors; AIAA, AAS, & AIDAA Co-Sponsors), Sept. 17-19, 1986, Arlington, VA, 1987, 784p., eds. Peter M. Bainum, Ivan Bekey, Luciano Guerriero, Paul A. Penzo, Hard Cover $80 *(ISBN 0-87703-264-5)*; Soft Cover $70 *(ISBN 0-87703-265-3)*

Vol. 63 Guidance and Control 1987, Jan. 31 - Feb. 4, 1987, Keystone, CO, 1987, 638p., eds. Robert D. Culp, Terry J. Kelly, Hard Cover $75 *(ISBN 0-87703-268-8)*; Soft Cover $60 *(ISBN 0-87703-269-6)*

Vol. 64 Aerospace Century XXI, 33rd AAS Annual Meeting, Oct. 26-29, 1986, Boulder, CO, 1987, **all three parts**, Hard Cover $225 *(ISBN 0-87703-276-9)*; Soft Cover $180 *(ISBN 0-87703-277-7)*; **Part I**, Space Missions and Policy, 686p., eds. George W. Morgenthaler, Gayle L. May, Hard Cover $75 *(ISBN 0-87703-279-3)*; Soft Cover $60 *(ISBN 0-87703-282-3)*; **Part II**, Space Flight Technologies, 608p., eds. George W. Morgenthaler, W. Kent Tobiska, Hard Cover $75 *(ISBN 0-87703-280-7)*; Soft Cover $60 *(ISBN 0-87703-283-1)*; **Part III**, Space Sciences, Applications, and Commercial Developments, 724p., eds. George W. Morgenthaler, Jean N. Koster, Hard Cover $75 *(ISBN 0-87703-281-5)*; Soft Cover $60 *(ISBN 0-87703-284-X)*; Microfiche Suppl. (Vol. 54 AAS Microfiche Series) $25 *(ISBN 0-87703-278-5)*

Vol. 65 AAS/AIAA Astrodynamics Conference, Aug. 10-13, 1987, Kalispell, MT, 1988, **two parts**, 1774p., eds. John K. Soldner, Arun K. Misra, Robert E. Lindberg, Walton Williamson, Hard Cover $180 *(ISBN 0-87703-285-8)*; Soft Cover $150 *(ISBN 0-87703-286-6)*; Microfiche Suppl. (Vol. 55 AAS Microfiche Series); $70 *(ISBN 0-87703-287-4)*

Vol. 66 Guidance and Control 1988, Jan. 30 - Feb. 3, 1988, Keystone, CO, 1988, 576p., eds. Robert D. Culp, Paul L. Shattuck, Hard Cover $75 *(ISBN 0-87703-288-2)*; Soft Cover $60 *(ISBN 0-87703-289-0)*; Microfiche Suppl. (Vol. 56 AAS Microfiche Series) $10 *(ISBN 0-87703-290-4)*

Vol. 67 Space - A New Community of Opportunity, 34th AAS Annual Meeting, Nov. 3-5, 1987, Houston, TX, 1989, 472p., eds. William G. Straight, Henry N. Bowes, Hard Cover $70 *(ISBN 0-87703-297-1)*; Soft Cover $55 *(ISBN 0-87703-298-X)*

Vol. 68 Guidance and Control 1989, Feb. 4-8, 1989, Keystone, CO, 1989, 708p., eds. Robert D. Culp, Robert A. Lewis, Hard Cover $85 *(ISBN 0-87703-299-8)*; Soft Cover $70 *(ISBN 0-87703-300-5)*

Order from Univelt, Inc., P.O. Box 28130, San Diego, California 92128

INDEX

NUMERICAL INDEX

AAS 89-021   A Comparison of Hardware and Software Simulation of an Agile Pointing
                   System, W. Scherer, P. Van Atta, J. Miller, G. Steiner, R. Briggs

AAS 89-022 to -023 Not Available

AAS 89-024   Zero-Lock^{TM} Laser Gyro, M. Fernandez, B. Ebner, N. Dahlen

AAS 89-025   Expert System Application to Improve Satellite Autonomy, J. M. Barry,
                   L. Rasiavicius, T. Gathmann

AAS 89-026 to -029 Not Assigned

AAS 89-030   Mirror Line-of-Sight on a Moving Base, J. J. Rodden

AAS 89-031   Not Available

AAS 89-032   Space-Stabilized Beam Pointing for a Bifocal Satellite Experiment,
                   D. C. Redding, T. T. Chien, E. H. Kopf, E. C. Wong

AAS 89-033   The Circumstellar Imaging Telescope Image Motion Compensation System:
                   Ultra-Precise Control on the Space Station Platform, G. E. Sevaston, M. M. Socha,
                   A. Eisenman

AAS 89-034   Not Available

AAS 89-035   Zenith Star: A Controls Challenge, L. A. Morine

AAS 89-036   Precision Pointing and Inertial Line-of-Sight Stabilization Using
                   Fine-Steering Mirror, and Strap-Down Inertial Sensors, A. A. Gupta, L. M. Germann

AAS 89-037   Not Available

AAS 98-038 to -039 Not Assigned

AAS 89-040   The National Aero-Space Plane, The Guidance and Control Engineer's
                   Dream or Nightmare?, F. Sanchez

AAS 89-041   Air-Launched Orbital Booster GN&C Requirements: The Pegasus Design
                   Experience, D. Rovner, A. Elias

AAS 89-042   Not Available

AAS 89-043   Demonstrating System for Servicing Satellites, L. M. Jenkins

AAS 89-044   Hypervelocity Orbital Intercept Guidance Using Certainty Control, S. Alfano,
                   C. E. Fosha, Jr.

AAS 89-045   Line-of-Sight Stabilization for the Dynamic Magellan Spacecraft,
                   C. O. Swanson, J. Kass, A. Greiner

AAS 89-046   Pinpoint Landing Concepts for the Mars Rover Sample Return Mission,
                   A. R. Klumpp

AAS 89-047   A Comparison of Intercept Algorithms, S. Alfano, C. E. Fosha, Jr.

AAS 89-048 to -049 Not Assigned

AAS 89-050   Delta 181 Pointing and Tracking Experience, T. S. Englar, J. C. Ray

AAS 89-051   Launch Delay Impact on the Galileo Attitude Control System, J. L. Chodas,
                    D. M. Weisenberg

AAS 89-052   A Fast Attitude Recovery System for Communications Satellites:
                    Experience of EUTELSAT 5, L. Van Holtz, M. Burton, R. C. Rogers

AAS 89-053   Not Available

AAS 89-054   Instability of Gravity Gradient Spacecraft in Full Sun Orbit: Flight
                    Experience from the Polar Bear Mission, A. Lewis, A. Streland

AAS 89-055   Space Missile Guidance and Control Simulation and Flight Testing,
                    D. C. Ductor, E. B. Wallace, C. H. Dillon

AAS 89-056 to -059 Not Assigned

AAS 89-060   Applications of Modern State Space Analysis in Spacecraft Dynamics,
                    Estimation and Control - Tutorial, J. L. Junkins

AAS 89-061 to -099 Not Assigned

AUTHOR INDEX[*]

---

* For each author the paper number is given. The page numbers refer to Volume 68, <u>Advances in the Astronautical Sciences</u>.